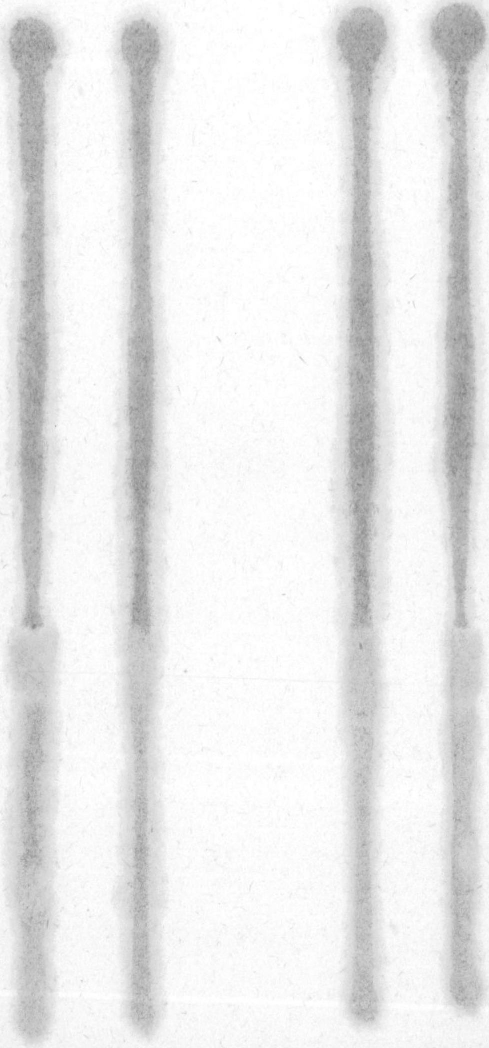

A Specialist Periodical Report

Molecular Spectroscopy

Volume 2

A Review of the Literature published during 1972 and early 1973

Senior Reporters

R. F. Barrow, *Physical Chemistry Laboratory, University of Oxford*
D. A. Long, *School of Chemistry, University of Bradford*
D. J. Millen, *Department of Chemistry, University College, London*

Reporters

R. T. Bailey, *University of Strathclyde*
J. Behringer, *University of Munich, West Germany*
M. S. Child, *University of Oxford*
F. R. Cruickshank, *University of Strathclyde*
R. E. Hester, *University of York*
J. W. Johns, *National Research Council of Canada, Ottawa, Ontario, Canada*
A. C. Legon, *University College, London*
E. A. Mallia, *University of Oxford*
W. B. Person, *University of Florida, U.S.A.*
D. Steele, *Royal Holloway College, University of London*

The Chemical Society
Burlington House, London, W1V 0BN

ISBN: 0 85186 516 X
Library of Congress Catalog Card No. 72–92545

Printed in Great Britain by
Billing & Sons Limited, Guildford and London

Foreword

This is the second of a series of annual volumes dealing with molecular spectroscopy: that is microwave spectroscopy, electronic spectroscopy, and infrared and Raman spectroscopy. As we explained in the Foreword to the first volume, the sheer volume of the literature relating to this extensive subject, makes it quite impracticable to attempt, annually, a complete coverage of all aspects of molecular spectroscopy. Our policy, therefore, has been to select those areas of molecular spectroscopy whose review we consider would be particularly timely and interesting. The topics chosen can then be given a detailed and critical analysis so that the reviews present not only a condensate of the literature but also set the subject in perspective. With these objectives in mind the literature for a number of the topics dealt with in this volume has been reviewed over a period of several years.

We are pleased to record our thanks again to our Reporters and the Editorial Staff of The Chemical Society for their assistance and collaboration in the preparation of this volume.

R.F.B.
D.A.L.
D.J.M.

March 1974

Contents

1

Microwave Spectroscopy

BY A. C. LEGON AND D. J. MILLEN

This Report preserves continuity with the Report in Volume 1, by covering papers included in *Chemical Titles* for 1972. In an attempt to reduce the interval between publication of papers and their inclusion in the Report, the present Report also covers papers included in *Chemical Titles* up to March 1973, that is up to and including No. 6 for 1973. The Report is structured in the same way as previously. After reviews of diatomic and triatomic molecules there are sections on inorganic and organic molecules, and then sections on a number of special topics.

1 Diatomic Molecules

A number of studies have been made on diatomic molecules leading to Dunham coefficients and rather precise values of r_e. For a number of molecules, hyperfine structures have been measured for vibrationally excited states and the dependence of nuclear quadrupole coupling on vibrational state has been explored. Some further dipole moments have been measured and for some molecules the sign of the dipole moment has been determined, leading to a result of general interest in at least one case.

Microwave spectra are reported for the first time for AlBr and AlI, and improved measurements have been obtained for AlCl.[1] Dunham coefficients have been evaluated and r_e/Å values obtained as follows:

AlCl	AlBr	AlI
2.130 11(3)	2.294 80(3)	2.537 09(3)

The nuclear quadrupole coupling coefficient is also reported for the iodine nucleus in AlI. Measurements have been made of the $J = 0 \rightarrow 1$ transitions of four isotopic species of GaCl, including the hyperfine structure.[2] Transitions for vibrational states up to $v = 3$ have been measured and Dunham coefficients as well as rotational constants have been obtained. A value of 2.201 681(3) Å has been obtained for r_e. The dependence of nuclear quadrupole coupling coefficients on vibrational state is given in Table 1. Hyperfine structure for InF has been measured, with a high-resolution-beam electric resonance spectrometer, for the $J = 1$ and $J = 2$ rotational states

[1] F. C. Wyse and W. Gordy, *J. Chem. Phys.*, 1972, **56**, 2130.
[2] E. Tiemann, M. Grasshoff, and J. Hoeft, *Z. Naturforsch.*, 1972, **27a**, 753.

Table 1 *Vibrational dependence of nuclear quadrupole coupling coefficients for* GaCl

$$^{35}\text{Cl}: eq_vQ = [-13.20 - 0.20(v + \tfrac{1}{2}) \pm 0.15]\ \text{MHz}$$
$$^{69}\text{Ga}: eq_vQ = [-92.40 + 0.68(v + \tfrac{1}{2}) \pm 0.14]\ \text{MHz}$$

of several vibrational levels, and *eqQ* values are reported.[3] Thallium halides have been the subject of a number of studies. The Stark effect for TlF has been studied with a molecular-beam electric resonance spectrometer [4] and the dipole moment has been obtained, among other parameters. Deviations from the Stark effect for a rigid-rotor model are discussed and additional terms arising from anharmonicity, centrifugal distortion, vibration–rotation interaction, and electronic polarizability are examined. The dipole moments/D of TlBr and TlI have also been determined [5] giving:

TlF	TlBr	TlI
4.2282(8)	4.493 ± 0.050	4.607 ± 0.070

Molecular-beam studies on TlCl include an examination of the Stark effect of different isotopic species in various vibrational states.[6]

The microwave spectrum of CS has been used *via* the Zeeman effect to determine the sign of the electric dipole moment as —CS+, in the same direction as found for CO.[7] The microwave spectrum of CSe has been reported for the first time and molecular *g*-values and the electric quadrupole moment have been evaluated from the observed Zeeman effect.[7] For silicon monosulphide the sign of the dipole moment has also been shown to be +SiS—.[8] The equilibrium internuclear separation is found to be $r_e =$ 1.929 254(3) Å. A further study has been made of GeTe and hyperfine structures have been measured for the $J = 2 \rightarrow 3$ transition, leading to eqQ = 153.1 MHz.[9] Two independent studies have been made for phosphorus mononitride, PN. The $J = 0 \rightarrow 1$ transition has been measured for a number of vibrational states and Dunham coefficients have been evaluated, giving r_e = 1.490 85 Å.[10] In another study, rotational transitions from $J = 1 \rightarrow 2$ to $J = 7 \rightarrow 8$ have been observed in the millimetre and submillimetre regions.[11] The analysis of the results leads to r_e = 1.490 80(2) Å, in close agreement with value from the study of the $J = 0 \rightarrow 1$ transitions. A new analysis has been made of all frequency data at present available for the oxygen molecule, from microwave, submillimetre, and infrared spectroscopy, including some recent measurements. A least-squares analysis was

[3] R. H. Hammerle, R. van Ausdal, and J. C. Zorn, *J. Chem. Phys.*, 1972, **57**, 4068.
[4] H. Dijkerman, W. Flegel, G. Graeff, and B. Moenter, *Z. Naturforsch.*, 1972, **27a**, 100.
[5] E. Tiemann, *Z. Naturforsch.*, 1971, **26a**, 1809.
[6] R. Ley and W. Schauer, *Z. Naturforsch.*, 1972, **27a**, 77.
[7] J. McGurk, H. Tigelaar, S. L. Rock, C. L. Norris, and W. H. Flygare, *J. Chem. Phys.*, 1973, **58**, 1420.
[8] E. Tiemann, E. Renwanz, J. Hoeft, and T. Toerring, *Z. Naturforsch.*, 1972, **27a**, 1566.
[9] E. Tiemann, J. Hoeft, and T. Toerring, *Z. Naturforsch.*, 1971, **26a**, 1928.
[10] J. Hoeft, E. Tiemann, and T. Toerring, *Z. Naturforsch.*, 1972, **27a**, 703.
[11] F. C. Wyse, E. L. Manson, and W. Gordy, *J. Chem. Phys.*, 1972, **57**, 1106.

used to fit 25 microwave frequencies and three submillimetre and i.r. frequencies and a new set of parameters for the oxygen molecule has been obtained in this way.[12] A gas-phase electron paramagnetic resonance study of oxygen is mentioned in a later section of the report.[13]

Chlorine and bromine monofluorides have been further examined.[14] Molecular *g*-values and molecular quadrupole moments have been evaluated from the observed Zeeman effect. The isotopic dependence of the *g*-values gives the sign of the electric dipole moment for chlorine monofluoride as —ClF+, which is of interest in view of simple arguments based on the idea of electronegativity and also for comparison with conclusions drawn from the chlorine nuclear quadrupole coupling coefficient. The dipole moment itself has been remeasured [$\mu = 0.888\,12(2)$ D] and the hyperfine structure examined using a molecular-beam electric resonance spectrometer.[15] The chlorine nuclear quadrupole coupling coefficient [$eqQ = -145.782\,11(9)$ MHz] and spin–spin interaction constants have been reported.

Hyperfine structures have been observed for $^{133}Cs^{35}Cl$,[16] $^{133}Cs^{79}Br$,[17] and $^{133}Cs^{127}I$.[18] The dependence on vibrational state is summarized in Table 2.

Table 2 *Vibrational dependence of halogen nuclear quadrupole coupling coefficients on vibrational states for caesium halides*

$$eq_vQ_{Cl} = +\ 1.830 - 0.118(v + \tfrac{1}{2}) \pm 0.150 \text{ MHz}$$
$$eq_vQ_{Br} = -\ 6.79 + 0.73(v + \tfrac{1}{2}) \pm 0.15 \text{ MHz}$$
$$eq_vQ_{I} = -14.28 - 2.10(v + \tfrac{1}{2}) \pm 0.35 \text{ MHz}$$

CsF has been studied using a molecular-beam electric resonance spectrometer.[19] The analysis of the results includes the following for the nuclear quadrupole coupling for caesium:

$$eq_vQ_{Cs} = 1245.2 - 16.2(v + \tfrac{1}{2}) + 0.3(v + \tfrac{1}{2})^2 \text{ kHz}$$

Other alkali-metal halides investigated by molecular-beam spectrometry include LiCl,[20] KF,[4] and RbF.[21]

2 Triatomic Molecules

A new molecule has been added during the year to the list of linear triatomic molecules investigated by microwave spectroscopy. Thioborine, HBS, an unstable molecule, which is isoelectronic with methylidenephosphine, HCP,

[12] W. M. Welch and M. Mizushima, *Phys. Rev.* (*A*), 1972, **5**, 2692.
[13] T. J. Cook, B. R. Zegarski, W. B. Breckenridge, and T. A. Miller, *J. Chem. Phys.*, 1973, **58**, 1548.
[14] J. J. Ewing, H. L. Tigelaar, and W. H. Flygare, *J. Chem. Phys.*, 1972, **56**, 1957.
[15] R. E. Davis and J. S. Muenter, *J. Chem. Phys.*, 1972, **57**, 2836.
[16] J. Hoeft, E. Tiemann, and T. Toerring, *Z. Naturforsch.*, 1972, **27a**, 1515.
[17] E. Tiemann, T. Toerring, and J. Hoeft, *Z. Naturforsch.*, 1972, **27a**, 702.
[18] J. Hoeft, E. Tiemann, and T. Toerring, *Z. Naturforsch.*, 1972, **27a**, 1017.
[19] H. G. Bennewitz, R. Haerten, O. Klais, and G. Muller, *Z. Phys.*, 1971, **249**, 168.
[20] T. F. Gallagher, R. C. Hilborn, and N. F. Ramsey, *J. Chem. Phys.*, 1972, **56**, 5972.
[21] J. H. Gitbaum and R. Schoenwasser, *Z. Naturforsch.*, 1972, **27a**, 92.

shows a rotational spectrum typical of a linear rotor.[22] Rotational and centrifugal distortion constants have been reported for eight isotopic species and *l*-doubling constants for four species. r_s Structures have been calculated taking each isotopic species in turn as the parent molecule and all these calculations give structures falling within a range very similar to the range found for HCN. The mean r_s values are r(B—H) = 1.1692 and r(S—B) = 1.5994 Å. Both HCN and HCP have been the subject of molecular Zeeman studies and molecular quadrupole moments have been evaluated.[23]

A similar study has been made for some of the halogen cyanides.[14] The molecular quadrupole moments (in units of 10^{-26} e.s.u. cm^2) for the various cyanides and for methylidenephosphine are as follows:

^{35}ClN	^{79}BrCN	^{127}ICN	HCN	HCP
−3.87 ± 1.0	−6.02 ± 1.1	−7.33 ± 1.1	3.1 ± 0.6	4.4 ± 1.2

The isotopic dependence of the molecular *g*-values has also been used to show that for chlorine cyanide the sign of the dipole moment is +ClCN—. For iodine cyanide a detailed re-examination has been made of the microwave spectrum.[24] Rotational constants for $^{127}I^{12}C^{14}N$ and $^{127}I^{12}C^{15}N$ have been obtained for nine vibrational states and *l*-doubling constants for three states. This amount of information has made possible the calculation of a number of rotational parameters, and at the same time the change of the iodine nuclear quadrupole coupling coefficient with vibrational state has been examined.

Among bent triatomic molecules, H_2O and H_2S have continued to be further studied. Microwave spectra of the radioactive species HTO and T_2O have been observed for the first time.[25] Twenty-six lines have been measured for HTO and five for T_2O, and the rotational constants given in Table 3

Table 3 *Rotational constants*/MHz *for* HTO *and* T_2O

Species	*A*	*B*	*C*
HTO	677 860.5	198 198.8	150 465.3
T_2O	338 808	145 670	100 262

evaluated. Revised rotational and distortion constants have been evaluated for $H_2{}^{16}O$ as a result of a submillimetre microwave study.[26] New transitions have also been reported in the submillimetre region for $H_2{}^{18}O$,[27] and the distortion constants obtained for this species have been compared with those calculated on the basis of $H_2{}^{16}O$ data. Excellent agreement is obtained, which is gratifying since the evaluations made for $H_2{}^{16}O$ and $H_2{}^{18}O$ include very different types of i.r. data as well as different higher-order terms in the Hamiltonian.

[22] E. F. Pearson and R. V. McCormick, *J. Chem. Phys.*, 1973, **58**, 1619.
[23] S. L. Hartford, W. C. Allen, C. L. Norris, E. F. Pearson, and W. H. Flygare, *Chem. Phys. Letters*, 1973, **18**, 153.
[24] J. B. Simpson, J. G. Smith, and D. H. Whiffen, *J. Mol. Spectroscopy*, 1972, **44**, 558.
[25] J. Bellet, G. Steenbeckeliers, and P. Stouffs, *Compt. rend.*, 1972, **275**, *B*, 501.
[26] F. C. de Lucia, P. Helminger, R. L. Cook, and W. Gordy, *Phys. Rev.* (*A*), 1972, **6**, 1324.
[27] F. C. de Lucia, P. Helminger, R. L. Cook, and W. Gordy, *Phys. Rev.* (*A*), 1972, **5**, 487.

A very detailed study of centrifugal distortion for $D_2{}^{16}O$ has been made by Steenbeckeliers and Bellet.[28] Twenty-two constants have been determined for the ground state and nine for the ν_2 state. The discussion includes an examination of the contribution of the more important distortion terms to the energies of a number of rotational states. Dipole moment measurements have been made for D_2O and HDO.[29] The dipole moment of D_2O has been determined from the Stark effect on microwave transitions and confirmed by high-resolution Stark spectroscopy using the HCN 337 μm laser. The dipole moment of HDO has also been determined from microwave measurements. The work indicates that the rotationless dipole of water increases slightly on deuteriation, and this is accounted for in terms of anharmonicity of the bending mode, analogously with the interpretation of a similar effect found for ammonia. Stark coefficients of millimetre-wave transitions for H_2O, D_2O, and HDO have also been reported.[30] Anharmonic force constants for water have been evaluated using data on H_2O, D_2O, and HDO.[31] Centrifugal distortion has also been investigated for H_2S and D_2S. For H_2S the distortion terms of order P^4 have been obtained and a partial set of order P^6 and P^{10} was found to be necessary in the analysis.[32] For D_2S a millimetre and submillimetre study has led to the measurement of sixty-six transitions, and these have been analysed with a twenty-two parameter distortion treatment.[33] However, it is found that only a partial set of terms of order P^6 and P^{10} is necessary. Hyperfine structure for deuteriated hydrogen selenide has been investigated by beam maser spectroscopy and quadrupole constants and spin–rotation and spin–spin coupling constants involving deuterium nuclei have been evaluated.[34]

The microwave spectrum has been reported for the first time of two isotopic species of hypofluorous acid, HOF and DOF.[35] The r_s structural parameters are found to be:

r(O—H)	r(O—F)	$\widehat{\text{HOF}}$
0.964 ± 0.01 Å	1.442 ± 0.001 Å	97.2 ± 0.6°

Interestingly the HOF angle is smaller than the angles in either H_2O or F_2O, which parallels the finding for HOCl, by comparison with H_2O and Cl_2O. In a further study, more than forty lines each for both HOF and DOF have been measured and a centrifugal distortion treatment has been applied, leading to the quartic distortion constants.[36]

[28] G. Steenbeckeliers and J. Bellet, *J. Mol. Spectroscopy*, 1973, **45**, 10.
[29] A. H. Brittain, A. P. Cox, G. Duxbury, T. G. Hersey, and R. G. Jones, *Mol. Phys.*, 1972, **14**, 843.
[30] Y. Beers and G. P. Klein, *J. Res. Nat. Bur. Stand. Sect. A*, 1972, **76**, 521.
[31] D. P. Smith and J. Overend, *Spectrochim. Acta*, 1972, **28A**, 471.
[32] P. Helminger, R. L. Cook, and F. C. de Lucia, *J. Chem. Phys.*, 1972, **56**, 4581.
[33] R. L. Cook, F. C. de Lucia, and P. Helminger, *J. Mol. Spectroscopy*, 1972, **41**, 123.
[34] S. Chandra and A. Dymanus, *Chem. Phys. Letters*, 1972, **13**, 105.
[35] H. Kim, E. F. Pearson, and E. H. Appleman, *J. Chem. Phys.*, 1972, **56**, 1.
[36] E. F. Pearson and H. Kim, *J. Chem. Phys.*, 1972, **57**, 4230.

The microwave spectrum of the DNO molecule, which has been previously studied by electronic spectroscopy, has been obtained by using a flow technique.[37] The rotational constants resulting from the analysis are in good agreement with those found for the ground state from the electronic spectrum. The nitrogen nuclear quadrupole coupling coefficients are found to be similar to those for nitrosyl fluoride and nitrous acid, and the dipole moment of deuteriated nitroxyl is found to have a value of 1.70 ± 0.05 D. Following the work on DNO the spectrum of HNO has also been studied (see ref. 383). Two independent measurements have been made of the dipole moment of NSCl. A value of $1.83(6) \pm 0.036$ D has been obtained for $^{15}N^{32}S^{35}Cl$ [38] and 1.87 ± 0.02 D for $^{14}N^{32}S^{35}Cl$.[39] Centrifugal distortion coefficients have been obtained from an analysis of twenty-eight transitions for $^{14}N^{32}S^{35}Cl$.[38] By combining these with vibrational frequencies and inertial defects force constants have been obtained:

$$F_{11} = 10.032 \text{ mdyn Å}^{-1} \qquad F_{22} = 1.44(4) \text{ mdyn Å}^{-1}$$
$$F_{33} = 0.74(9) \text{ mdyn Å} \quad \text{and} \quad F_{23} = -0.08(6) \text{ mdyn}$$

The microwave spectrum of GeF_2 has been further investigated.[40] Rotational constants have been obtained for vibrationally excited states for $^{70}GeF_2$, $^{72}GeF_2$, and $^{74}GeF_2$ and a revised r_e structure has been obtained:

$$r(\text{Ge—F}) = 1.7320(9) \text{ Å} \qquad \widehat{\text{FGeF}} = 97.148°$$

A centrifugal distortion analysis has been made and the question of the best use of centrifugal distortion constants and i.r. data in the calculation of the harmonic force field is discussed. Nuclear quadrupole coupling coefficients have been obtained for $^{73}GeF_2$. A revised calculation [41] has been made of the cubic potential constants for SeO_2 using vibration–rotation interaction constants and new values of vibrational frequencies reported for matrix-isolated SeO_2. The dipole moment has also been measured: $\mu = 2.62 \pm 0.05$ D. The millimetre-wave spectra of KOH and KOD have been examined,[42] four ground-state frequencies being measured for the former and three for the latter. These are consistent with a set of rotational constants given in Table 4, but are not consistent with the frequency reported earlier for the

Table 4 *Rotational constants*/MHz *for* KOH *and* KOD

	B_0	D_0
KOH	8208.679 ± 0.010	0.01219 ± 0.00006
KOD	7494.827 ± 0.010	0.00946 ± 0.00006
NaOH	12567.054 ± 0.010	0.02872 ± 0.00005

[37] K. Takagi and S. Saito, *J. Mol. Spectroscopy*, 1972, **44**, 81.
[38] S. Mizumoto, J. Izumi, T. Beppu, and E. Hirota, *Bull. Chem. Soc. Japan*, 1972, **45**, 786.
[39] A. Guarnieri, *Z. Naturforsch.*, 1971, **26a**, 1246.
[40] H. Takeo and R. F. Curl, *J. Mol. Spectroscopy*, 1972, **43**, 21.
[41] H. Takeo, E. Hirota, and A. Mueller, *J. Mol. Spectroscopy*, 1972, **41**, 420.
[42] E. F. Pearson and M. B. Trueblood, *J. Chem. Phys.*, 1973, **58**, 826.

$J = 1 \rightarrow 2$ transition. The pattern of satellites is qualitatively similar to those found previously for CsOH, RbOH and LiOH and is interpreted in the same way. A similar interpretation for NaOH leads to values included in Table 4 (see ref. 385). Other papers on linear triatomic molecules deal with vibration–rotation interaction [43, 44] and with centrifugal distortion of *l*-doublets.[45]

3 Inorganic Molecules

There has been a growth in the number of papers on inorganic compounds, which reflects an increasing interest in applications in this area. A considerable number of phosphorus compounds have been studied, several papers deal with silicon compounds, and a number deal with germanium derivatives. As well as papers on compounds of the more familiar elements, nitrogen, sulphur, and the halogens, papers have also appeared on compounds of vanadium and tellurium.

Some further boron compounds have been studied, including the microwave identification [46] of the molecule BF_2OH. The analysis of the spectrum shows the molecule has a planar structure with the parameters given in Table 5.

Table 5 *Structural parameters for* BF_2OH

	Bond lengths/Å		*Angles*/°
r(B—F)	1.32(3)	$\widehat{FBF}$	118.0
r(B—O)	1.34(4)	$\widehat{FBO}$	122.8 and 119.2
r(O—H)	0.941	$\widehat{BOH}$	114.1

The dipole moment $\mu = 1.86 \pm 0.02$ D. As shown in Figure 1, the BF_2 group is tilted away from the OH group in the same way as the NO_2 group is in the nitric acid molecule. For BF_3 itself, *l*-type resonance has been reported and some cubic potential constants have been evaluated.[47] Two further borane adducts have been investigated. Trimethylphosphine-borane, $(CH_3)_3P,BH_3$, shows a symmetric-rotor spectrum. By using deuterium substitution and ^{10}B and ^{11}B isotopic species, the molecule is found to have the parameters collected in Table 6.[48]

Table 6 *Structural parameters for* $(CH_3)_3P,BH_3$

	Bond lengths/Å		*Angles*/°
r(P—B)	1.901 ± 0.007	$\widehat{CPC}$	105.0 ± 0.4
r(P—C)	1.819 ± 0.010	$\widehat{HCH}$	109.3 ± 1.0
r(C—H)	1.08 ± 0.02	$\widehat{HBH}$	113.5 ± 0.5
r(B—H)	1.212 ± 0.010		

43 A. Y. Dzyublik, *Optika i Spectroskopiya*, 1973, **34**, 237.
44 U. I. Polyakov, *Optika i Spectroskopiya*, 1972, **33**, 457.
45 M. R. Aliev, *Optika i Spectroskopiya*, 1972, **33**, 795.
46 H. Takeo and R. F. Curl, *J. Chem. Phys.*, 1972, **56**, 4314.
47 F. M. Massi, *J. Mol. Spectroscopy*, 1972, **43**, 168.
48 P. J. Bryan and R. L. Kuczkowski, *Inorg. Chem.*, 1972, **11**, 553.

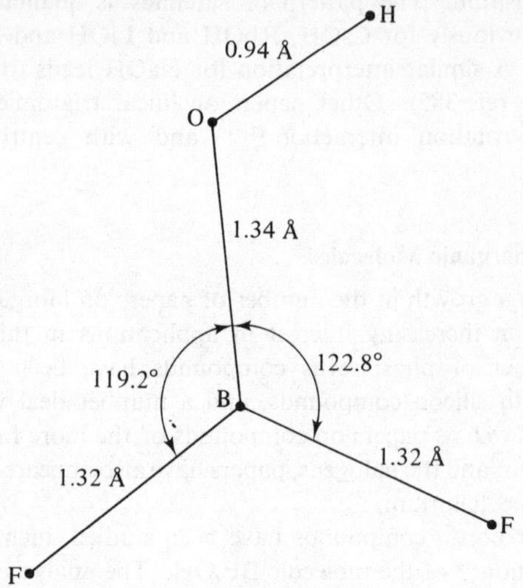

Figure 1

The structure of the closely related methylphosphine-borane, CH_3PH_3,BH_3, has also been determined.[48] Substitution co-ordinates were obtained for all the atoms except those of phosphorus which were calculated from first-moment conditions. The resulting structure, which has a plane of symmetry, is shown in Figure 2. The methyl group is tilted towards the borane group and the origin of this tilt is considered in terms of electrostatic interactions. The dipole moments are found to be 4.99 ± 0.2 D for $(CH_3)_3P,BH_3$ and 4.66 ± 0.05 D for CH_3PH_2,BH_3.

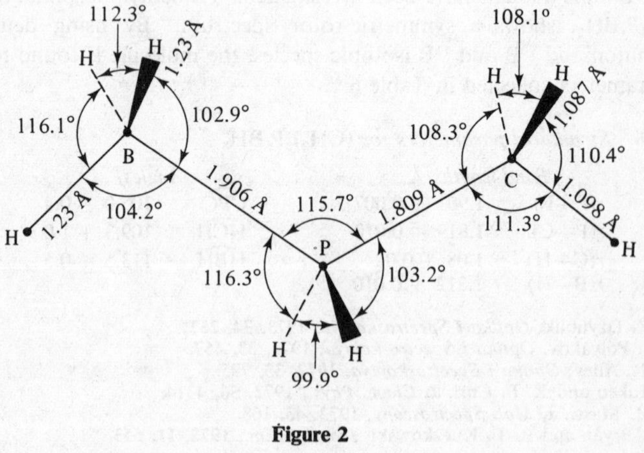

Figure 2

A number of papers have appeared on isocyanates, and a comparison of structures of isocyanates has been made.[49] Chlorine isocyanate, ClNCO, has been shown to have a planar structure with the parameters listed in Table 7, having Cl and O *trans* to one another.[49] An interesting feature of the structure is the bend in the NCO chain of about 8° away from the chlorine atom. A preliminary report [50] has also appeared on the spectrum of the recently prepared cyanogen isocyanate, NCNCO.

Table 7 *Structural parameters for* ClCNO

	Bond lengths/Å		*Angles*
r(Cl—N)	1.703 ± 0.11	$\widehat{NCO}$	171°24′ ± 1°30′
r(C—O)	1.165 ± 0.008	$\widehat{ClNC}$	119°22′ ± 1°
r(N—C)	1.218 ± 0.012		

Trimethylsilyl isocyanate, Me_3SiNCO, is found to have a rotational spectrum characteristic of a symmetric rotor, indicating that the equilibrium configuration of the Si—N=C=O chain is either linear or very close to linear.[51] There is a very low-frequency bending mode which has been estimated at 64 ± 15 cm^{-1} from the intensities of satellites, and the distinction between linear and non-linear equilibrium configuration is not easily made. The problem has been discussed in connection with an electron-diffraction study of the large-amplitude vibrations in both H_3SiNCO and H_3SiNCS.[52] For the potential function which is proposed for the thiocyanate the energy required to change the Si—N=C angle from 180 to 140° is less than 1 kcal mol^{-1}. The parent molecule of this series of compounds, isocyanic acid, HNCO, has been the subject of a high-resolution molecular-beam maser study.[53] Nitrogen nuclear quadrupole coupling constants were determined and deuteron quadrupole couplings obtained for DNCO. Further lines have been measured in the microwave spectrum of HNCO in connection with radio-astronomical interests.[54] The isomeric linear molecule fulminic acid, HCNO, has been the subject of a detailed study of rotational spectra in vibrationally excited states.[55] The analysis has given unperturbed rotational constants B_v and centrifugal distortion constants D_v for several vibrational states, and vibrational anharmonicity constants have also been obtained. A potential function containing a hump which disturbs the isotropic two-dimensional oscillator potential is proposed for the low-lying degenerate bending mode. Cyanogen azide, NCN_3, has a spectrum consistent with a planar molecule. Independent studies lead to a dipole moment μ = 2.96 ±

[49] W. H. Hocking and M. C. L. Gerry, *J. Mol. Spectroscopy*, 1972, **42**, 547.
[50] W. H. Hocking and M. C. L. Gerry, *J.C.S. Chem. Comm.*, 1973, 47.
[51] A. J. Careless, M. C. Green, and H. W. Kroto, *Chem. Phys. Letters*, 1972, **16**, 414.
[52] C. Glidwell, A. G. Robiette. and G. M. Sheldrick, *Chem. Phys. Letters*, 1972, **15**, 526.
[53] S. G. Kukolich, A. C. Nelson, and B. S. Yamanashi, *J. Amer. Chem. Soc.*, 1971, **93**, 6769.
[54] W. H. Hocking, M. C. L. Gerry, and G. Winnewisser, *Astrophys. J.*, 1972, **174**, L93.
[55] M. Winnewisser and B. P. Winnewisser, *J. Mol. Spectroscopy*, 1972, **41**, 143.

0.07 D [56] and 2.99 ± 0.03 D.[57] Structural parameters [56] based on certain assumptions are given in Table 8. The $r(C\equiv N)$ distance is an r_s value, and the remaining parameters were derived on the assumption that the N_3 chain is linear and that the terminal N—N bond length is 1.133 Å, as in hydrazoic acid. Interestingly the N—C≡N chain appears to be bent, which parallels earlier findings for $CH_2(CN)_2$ and $S(CN)_2$.

Table 8 *Structural parameters for cyanogen azide*

Bond lengths/Å			*Angles*
$r_s(C\equiv N)$	1.164 ± 0.005	$\widehat{CNC}$	120°13′ ± 1°
$r_0(N=N)$	1.252 ± 0.01	$\widehat{NCN}$	176° 2′ ± 2°

A number of papers have appeared on silicon compounds, leading to further knowledge about silane derivatives in particular. Ground-state rotational constants B_0, D_0^J, and D_0^{JK} have been obtained for $^{28}SiH_3D$ from a high-resolution i.r. study.[58] A Zeeman study has been made of methylsilane ($MeSiH_3$ and $MeSiD_3$).[59] It proves not to be possible to decide unambiguously from the Zeeman evidence on the correct quadrupole moment or on the sign of the electric dipole moment. However, several arguments favour the sign $+CH_3SiH_3-$, contrary to the relative electronegativities of carbon and silicon, and this conclusion has been supported by non-empirical calculations on the electronic structure.[60] The finding may be understood in terms of the C—H and Si—H bond moments, the octupole moment of methane and silane indicating polarities +H—C— and +Si—H— respectively.[59] The study of alkylsilanes has been extended from the previously studied methyl-, ethyl-, and dimethyl-silane to include methylethylsilane.[61] This is believed to exist in two rotational isomeric forms, but assignments have so far been made in the microwave spectrum only for the *trans*-isomer. The rotational constants can be fitted with structural parameters transferred from Me_2SiH_2 and Et_2SiH_3 with small adjustments in the $\widehat{CCSi}$, $\widehat{CCH}$, and $\widehat{SiCH}$ angles. The dipole moment is found to be 0.758 ± 0.005 D, which is almost identical with that for dimethylsilane.

The spectra of the following halogeno-derivatives of silane have been investigated: CH_3SiF_3, $(CH_3)_3SiCl$, CD_3SiH_2Cl, and $CH_2{=}CHSiF_3$. A re-investigation [62] of CH_3SiF_3 includes the study of ^{13}C isotopic species and a consequent re-determination of the structural parameters leads to the values listed in Table 9. The dipole moment is found to be 2.33 ± 0.10 D.

56 C. C. Costain and H. W. Kroto, *Canad. J. Phys.*, 1972, **50**, 1453.
57 K. Bolton, R. D. Brown, and F. R. Burden, *Chem. Phys. Letters*, 1972, **15**, 79.
58 R. W. Lovejoy and W. B. Olson, *J. Chem. Phys.*, 1972, **57**, 2224.
59 R. L. Shoemaker and W. H. Flygare, *J. Amer. Chem. Soc.*, 1972, **94**, 684.
60 D. H. Liskow and H. F. Schaefer, *J. Amer. Chem. Soc.*, 1972, **94**, 6641.
61 M. Hayashi and C. Matsumura, *Bull. Chem. Soc. Japan*, 1972, **45**, 732; M. Hayashi, K. Ohno, and H. Murata, *ibid.*, 1973, **46**, 684.
62 J. R. Durig, Y. S. Li, and C. C. Tong, *J. Mol. Structure*, 1972, **14**, 255.

Table 9 *Structural parameters for* CH_3SiF_3

Bond lengths/Å		*Angles*	
r(C—H)	1.081 ± 0.004	$\widehat{HCSi}$	111° 1′ ± 30′
r(C—Si)	1.812 ± 0.014	$\widehat{FSiC}$	112°20′ ± 1°6′
r(Si—F)	1.574 ± 0.007		

Four isotopic species of Me_3SiCl have been examined, *viz*. the ^{35}Cl and ^{37}Cl species for the normal and the fully deuteriated molecules.[63] The four rotational constants were fitted by a least-squares programme which, with assumed C—H bond distances and $\widehat{HCSi}$ angles as for Me_3SiH, gave the following parameters:

$$r(\text{Si—Cl}) = 2.022 \pm 0.05\ \text{Å} \qquad \widehat{\text{ClSiC}} = 110.5 \pm 1°$$
$$r(\text{Si—C}) = 1.857 \pm 0.005\ \text{Å}$$

The chlorine nuclear quadrupole coupling coefficient for ^{35}Cl has a value of −46.9 ± 1.5 MHz. The study of the microwave spectra of four isotopic species of CD_3SiH_2Cl ($^{29}Si^{35}Cl$, $^{30}Si^{35}Cl$, $^{28}Si^{35}Cl$, and $^{28}Si^{37}Cl$) has allowed the substitution parameters for Si and Cl to be calculated in two ways, leading to r(Si—Cl) = 2.049 ± 0.013 and 2.052 ± 0.005 Å.[64] Centrifugal distortion and chlorine nuclear quadrupole coupling constants have also been reported. The microwave spectrum of vinyltrifluorosilane, $CH_2{=}CHSiF_3$, has been analysed to give rotational constants for the ground state and five vibrationally excited states.[65] The dipole moment is found to have a value of 2.33 D.

Some further derivatives of germane have also been investigated. For fluorogermane, GeH_3F, the parameters in Table 10 have been reported.[66]

Table 10 *Parameters for* GeH_3F

r(Ge—F) = 1.74 ± 0.01 Å	μ = 2.33 ± 0.06 D
r(Ge—H) = 1.47 ± 0.01 Å	$eQq(^{73}Ge)$ = −93.0(3) ± 0.10 MHz

The structure is calculated on the assumption of regular tetrahedral geometry. The microwave spectrum of iodogermane, GeH_3I, has been reported and bromogermane, GeH_3Br, has been re-investigated.[67] Isotopic species for both Ge and Br allow the germanium–bromine bond length to be obtained as r(Ge—Br) = 2.2970 ± 0.0002 Å. Because only a single iodine isotope occurs naturally, the same method cannot be used for GeH_3I. The assumption is made that the Ge—H projection on the symmetry axis is the same as in GeH_3Br and this gives r(Ge—I) = 2.5075 ± 0.0005 Å. Nuclear quadrupole coupling constants are also obtained, and the spectrum of GeH_3Br in

63 J. R. Durig, R. D. Carter, and Y. S. Li, *J. Mol. Spectroscopy*, 1972, **44**, 18.
64 W. Zeil, R. Gegenheimer, P. Ferrer, and M. Dakkouri, *Z. Naturforsch.*, 1972, **27a**, 1150.
65 H. Jones and R. F. Curl, *J. Mol. Spectroscopy*, 1972, **41**, 226.
66 L. C. Krisher, J. A. Morrison, and W. A. Watson, *J. Chem. Phys.*, 1972, **57**, 1357.
67 S. N. Wolf and L. C. Krisher, *J. Chem. Phys.*, 1972, **56**, 1040.

excited vibrational states has been reported.[68] The microwave spectrum of trimethylgermane has been analysed [69] and rotational constants have been obtained for Me_3GeH, and with certain assumptions the heavy-atom dispositions have also been determined, giving the parameters in Table 11.

Table 11 *Parameters for* Me_3GeH *and* Me_3GeBr

Me_3GeX	r(Ge—C)/Å	r(Ge—X)/Å	$\widehat{CGeX}$/°
Me_3GeH	1.947 ± 0.006	1.532 ± 0.001	109.3 ± 0.1
Me_3GeBr	1.936 ± 0.006	2.323 ± 0.001	106.3 ± 0.1

The structure of trimethylbromogermane has also been obtained [70] from a study of its microwave spectrum and the parameters are included in Table 11.

Nitrogen compounds on which papers have appeared include hydroxylamine, methylhydrazine, monochloramine, nitrous acid, and dinitrogen trioxide. The microwave spectrum of hydroxylamine reveals only the *trans*-isomer.[71] Only a single rotational spectrum is observed for NHDOH (and for NHDOD) and so it follows that the molecule has a plane of symmetry which bisects the $\widehat{HNH}$ angle and includes the $\widehat{NOH}$ group. The six structural parameters given in Table 12 have been obtained from a least-squares method

Table 12 *Structural parameters for* NH_2OH

	Bond lengths/Å		*Angles*
r(N—H)	1.016 ± 0.008	$\widehat{HNH}$	107° 6′ ± 30′
r(N—O)	1.453 ± 0.002	$\widehat{HNO}$	103°15′ ± 30′
r(O—H)	0.962 ± 0.005	$\widehat{NOH}$	101°22′ ± 30′

using thirteen observed moments of inertia. The dipole moment is found to be $\mu = 0.59 \pm 0.05$ D and nitrogen nuclear quadrupole coupling coefficients have also been obtained. The observed transitions show no splittings due to inversion, indicating a higher barrier than in ammonia. It appears that electronegative substituents lead to an increased barrier for it has been found previously that no splitting is observed for NHF_2 nor has any been reported for NH_2Cl. A further investigation [72] of NH_2Cl leads to the r_s structure in Table 13. Nitrogen and chlorine nuclear quadrupole coupling

Table 13 *Structural parameters for* NH_2Cl

	Bond lengths/Å		*Angles*
r(N—H)	1.017 ± 0.005	$\widehat{HNCl}$	103°41′ ± 22′
r(N—Cl)	1.7480 ± 0.0001	$\widehat{HNH}$	107° ± 2°

constants are also reported. A number of theoretical papers on the calculation of the barrier height in ammonia have appeared, including a comparison

[68] L. C. Krisher and S. N. Wolf, *J. Chem. Phys.*, 1973, **58**, 396.
[69] J. R. Durig, M. M. Chen, and J. B. Turner, *J. Phys. Chem.*, 1973, **77**, 227.
[70] Y. S. Li and J. R. Durig, *Inorg. Chem.*, 1973, **12**, 306.
[71] S. Tsunekawa, *J. Phys. Soc. Japan*, 1972, **33**, 167.
[72] G. Cazzoli, D. G. Lister, and P. G. Favero, *J. Mol. Spectroscopy*, 1972, **42**, 286.

of NH_3 and CH_3^-.[73] Further work on the high-resolution spectrum of ammonia obtained by using a beam maser spectrometer shows splittings which are attributed to a difference in molecular *g*-values for the upper and lower inversion states with $J = 3$ and $K = 2$.[74] A reinterpretation of observed hyperfine patterns has also been given.[75]

The potential function hindering the N—N torsional motion in methylhydrazine, $MeNH{\cdot}NH_2$, has been determined through the use of microwave and far-i.r. spectroscopic data.[76] The microwave work provides the torsion angles of the *cis*- and *trans*-rotamers and the energy difference of the two potential minima and the i.r. spectrum provides torsional frequencies. Calculations lead to a *trans* barrier height of 3.58 ± 0.07 kcal mol^{-1} and a *cis* barrier of 8.66 ± 0.86 kcal mol^{-1}. Some re-assignment of the far-i.r. spectrum has been made on the basis of the potential function.

A detailed study has been made of centrifugal distortion and its use in the calculation of potential functions for both *cis*- and *trans*-forms of nitrous acid, both the HONO and the DONO species being examined in each case.[77] Distortion constants have been obtained for *cis*- and *trans*-isomers of each species and have been used together with vibrational frequencies to evaluate the quadratic potential functions. Coriolis coupling constants have been calculated and these, taken with the vibrational frequencies, allow the calculation of average moments of inertia, from which the structures of the average configurations have been determined. A further examination [78] has also been made of the anhydride of nitrous acid, dinitrogen trioxide, N_2O_3. Nuclear quadrupole coupling coefficients have been evaluated for both nitrogen nuclei and discussed in terms of the bonding in the molecule. Another paper deals with the use of vibration–rotation interaction constants and inertial defects for calculating force constants for nitryl fluoride.[79]

An increasing number of papers are appearing on phosphorus compounds. Phosphorous oxytrichloride has been re-investigated, and a number of compounds derived from PF_3 have been studied. The spectrum of PF_2Cl has been analysed to give rotational constants for both of the chlorine isotopic species and these lead to the structural parameters in Table 14.[80] The dipole moment was found to be $\mu = 0.89 \pm 0.01$ D and the nuclear quadrupole

Table 14 *Structural parameters for* PF_2Cl

Bond lengths/Å		*Angles*/°	
r(P—F)	1.571 ± 0.003	$\widehat{FPF}$	97.3 ± 0.2
r(P—Cl)	2.030 ± 0.006	$\widehat{PFCl}$	99.2 ± 0.2

73 R. E. Kari and I. G. Csizmadi, *J. Chem. Phys.*, 1972, **56**, 4337.
74 S. G. Kukolich and K. H. Castleton, *Chem. Phys. Letters*, 1973, **18**, 408.
75 J. T. Hougen, *J. Chem. Phys.*, 1972, **57**, 4207.
76 R. P. Lattimer and M. D. Harmony, *J. Amer. Chem. Soc.*, 1972, **94**, 351.
77 D. J. Finnigan, A. P. Cox, A. H. Brittain, and J. G. Smith, *J. C. S. Faraday II*, 1972, **68**, 548.
78 A. P. Cox and D. J. Finnigan, *J. C. S. Faraday II*, 1973, **69**, 49.
79 V. F. Somova and L. M. Sverdlov, *Optika i Spektroskopiya*, 1972, **33**, 642.
80 A. H. Brittain, J. E. Smith, and R. H. Schwendeman, *Inorg. Chem.*, 1971, **11**, 39.

coupling coefficients have been interpreted to indicate a minimum of 1.5% π-bond character for the P—Cl bond. The spectrum of the recently prepared aminodifluorophosphine, PF_2NH_2, has also been observed and its analysis provides strong evidence in favour of a planar PNH_2 configuration.[81] Figure 3

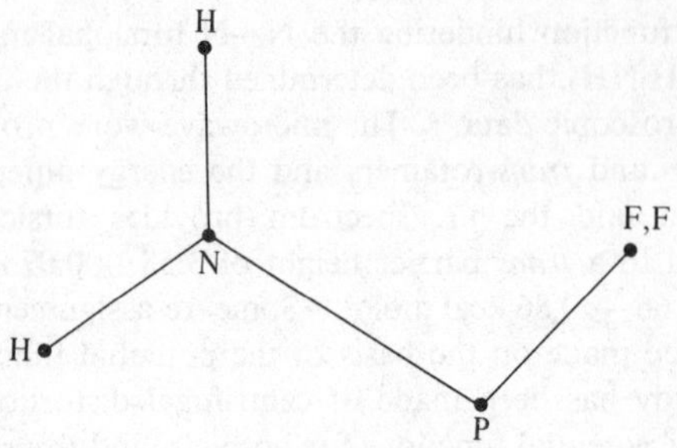

Figure 3

shows a projection of the molecule in the plane of symmetry containing the NH_2 group. Structural parameters in Table 15 have been obtained by making use of ^{15}N- and ^{2}H-substituted species. This appears to be the first case of a

Table 15 *Structural parameters for* PF_2NH_2

Bond lengths/Å		*Angles*/°	
r(P—F)	1.587 ± 0.004	$\widehat{FPF}$	94.6 ± 0.2
r(P—N)	1.650 ± 0.004	$\widehat{PFN}$	100.6 ± 0.2
r(N—H) (*cis*)	1.002 ± 0.005	$\widehat{PNH}$ (*cis*)	123.1 ± 0.2
r(N—H) (*trans*)	0.981 ± 0.005	$\widehat{PNH}$ (*trans*)	119.7 ± 0.4
		$\widehat{HNH}$	117.2 ± 0.4

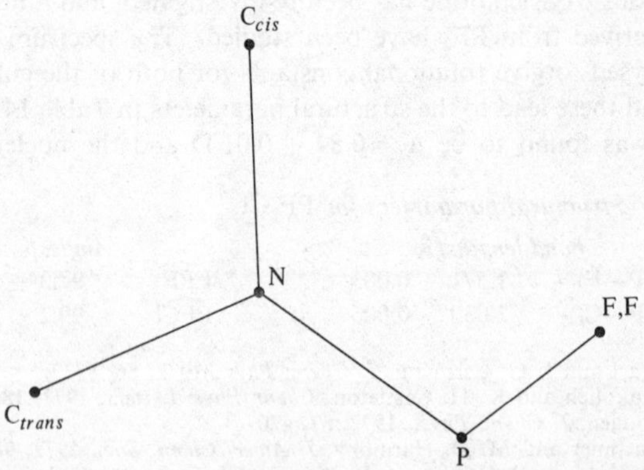

Figure 4

[81] A. H. Brittain, J. E. Smith, P. L. Lee, K. Cohn, and R. H. Schwendeman, *J. Amer. Chem. Soc.*, 1971, **93**, 6772.

free molecule in which a planar NH_2 configuration has been found; it is attributed to (*d–p*) π-bonding. A microwave study has also been made of the dimethyl derivative, dimethylaminodifluorophosphine, PF_2NMe_2, and a structure similar to that for the parent amino-compound is consistent with the experimental results.[82] A projection of the molecule in its plane of symmetry is shown in Figure 4 and the structural parameters obtained on the basis of assumed parameters for the methyl group and the N—C bond lengths are given in Table 16. An electron-diffraction study for both of these molecules, PF_2NH_2 and PF_2NMe_2, led to a non-planar *gauche* structure in each

Table 16 *Structural parameters for* PF_2NMe_2

Bond lengths/Å		*Angles*/°	
r(P—F)	1.57	$\widehat{FPF}$	95.3
r(P—N)	1.66	$\widehat{NPF}$	99.8
		$\widehat{PNC}$ (*cis*)	124.5
		$\widehat{PNC}$ (*trans*)	121.3

case.[83] On the other hand, the attempt to get a reasonable fit of the microwave data for F_2PNMe_2 to a *gauche* conformation did not prove successful.[82]

The microwave spectrum of the cyano-derivative PF_2CN has also been analysed and structural parameters in Table 17 were obtained by making use of ^{13}C and ^{15}N species.[84] This structure has also been investigated by an electron-diffraction study.[85] The parameters obtained by the two methods

Table 17 *Structural parameters for* PF_2CN

Bond lengths/Å		*Angles*/°	
r(P—F)	1.566 ± 0.007	$\widehat{FPF}$	99.2 ± 0.2
r(P—C)	1.815 ± 0.005	$\widehat{FPC}$	96.9 ± 0.2
r(C—N)	1.157 ± 0.003	$\widehat{PCN}$	171.2 ± 0.8

are in reasonable agreement though not all are within the sum of experimental uncertainties. The assumption of a linear PCN arrangement in the electron-diffraction work probably contributes to this discrepancy. Striking features of the structure as determined spectroscopically are firstly, as illustrated in Figure 5, that the PCN chain is non-linear and secondly that the P—C bond length is unusually short. The situation is reminiscent of $S(CN)_2$, where quite analogous features are found.[86] An interpretation of these features for the PCN chain has been made in terms of π-bonding; interestingly the nitrogen nuclear quadrupole coupling indicates an asymmetric perturbation to the bonding.[84]

[82] P. Forti, D. Damiani, and P. G. Favero, *J. Amer. Chem. Soc.*, 1973, **95**, 756.
[83] G. C. Holywell, D. W. H. Rankin, B. Beagley, and J. M. Freeman, *J. Chem. Soc. (A)*, 1971, 785.
[84] P. L. Lee, K. Cohn, and R. H. Schwendeman, *Inorg. Chem.*, 1972, **11**, 1917.
[85] G. C. Holywell and D. W. H. Rankin, *J. Mol. Structure*, 1971, **9**, 11.
[86] L. Pierce, R. Nelson, and C. Thomas, *J. Chem. Phys.*, 1965, **43**, 3423.

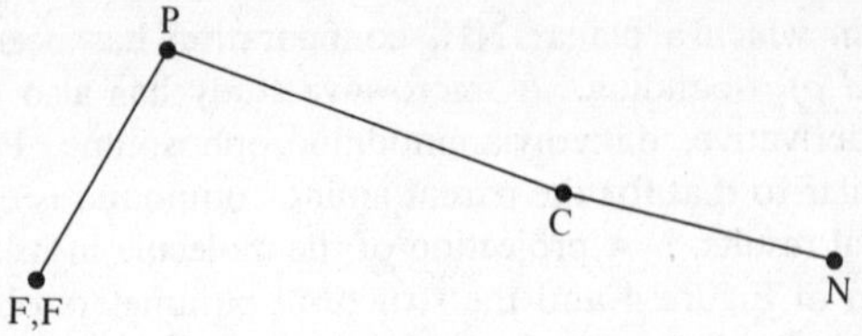

Figure 5

The structure of another related molecule, hydrothiophosphonyl difluoride, $SPHF_2$, has been obtained from the microwave spectra of the isotopic species $^{32}SPHF_2$, $^{34}SPHF_2$, and $^{32}SPDF_2$. The S, P, and H atoms define a molecular plane of symmetry as shown in Figure 6, and the parameters listed in Table 18 were obtained.[87]

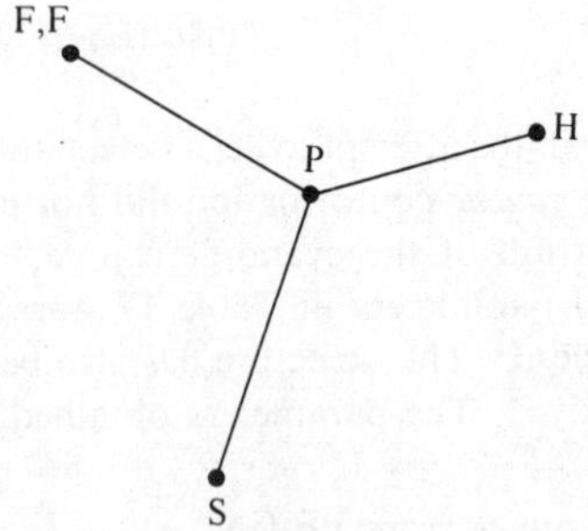

Figure 6

Table 18 *Structural parameters for* $SPHF_2$

Bond lengths/Å		*Angles*/°	
r(P—H)	1.392 ± 0.005	$\widehat{SPF}$	117.4 ± 0.2
r(P—F)	1.551 ± 0.005	$\widehat{SPH}$	119.2 ± 0.2
r(S—P)	1.867 ± 0.005	$\widehat{FPF}$	98.6 ± 0.2

The spectrum of $POCl_3$ has been re-examined,[88] and the structural parameters are listed in Table 19 with those of vanadyl chloride, $VOCl_3$, which has now also been examined by microwave spectroscopy.[89] The parameters for the latter molecule were obtained from the spectra of the two symmetric rotor species and the asymmetric rotor $OV^{35}Cl_2{}^{37}Cl$. The vanadium valence angle $\widehat{ClVCl}$ is about 8° larger than the phosphorus valence angle $\widehat{ClPCl}$.

Among sulphur compounds some detailed studies have been reported on hydrogen persulphide and thionyl halides. An interesting feature to emerge

[87] C. R. Nave and J. Sheridan, *J. Mol. Structure*, 1973, **15**, 391.
[88] Y. S. Li, M. M. Chen, and J. R. Durig, *J. Mol. Structure*, 1972, **14**, 261.
[89] K. Karakida, K. Kuchitsu, and C. Matsumura, *Chem. Letters*, 1972, 293.

Table 19 *Structural parameters for* $POCl_3$ *and* $VOCl_3$

Bond lengths/Å		*Angles*/°	
r(P—O)	1.455 ± 0.005	$\widehat{ClPCl}$	103.7 ± 0.2
r(P—Cl)	1.989 ± 0.002		
r(V—O)	1.595 ± 0.005	$\widehat{ClVCl}$	111.8 ± 0.2
r(V—Cl)	2.131 ± 0.001		

from the analyses of the spectra of HSSH [90] and DSSH [91] is the finding that K levels are split by an amount which is several orders of magnitude larger than expected from the asymmetry parameter. The origin of the effect is a centrifugal distortion splitting of the K levels. Normally asymmetry splitting is much larger than that arising from centrifugal distortion but in these cases not only is splitting of the $K = 2$ levels large but the sequence of levels is reversed as compared with the order expected from asymmetry splitting.[92] Some 350 lines have been measured and it is found that centrifugal distortion treatment in terms of six quartic constants is sufficient to account for the observations.[91,93] The spectrum of HSSD has also been observed and analysed in the same way.[94] There is insufficient isotopic information to allow the determination of a complete r_s structure but the parameters in Table 20 have been obtained. The barrier to internal rotation in hydrogen peroxide has been discussed from a theoretical point of view.[95]

Table 20 *Structural parameters for* H_2S_2 *and* D_2S_2

Bond lengths/Å		*Angles*	
r(S—S)	2.055 ± 0.001	$\widehat{SSH}$	91°20′ ± 3′
r(S—H)	1.327 ± 0.003	η(HSSH)	90°36′ ± 3′
r(S—D)	1.324 ± 0.003	η(DSSD)	90°14′ ± 2′
		η(HSSD)	90°25′ ± 5′

Thionyl halides have been examined and centrifugal distortion coefficients evaluated. In one study [96] more than 100 transitions observed in the region 40—70 GHz have been used in an attempt to determine quartic and sextic distortion parameters for thionyl fluoride, SOF_2. Another study uses some 250 transitions for thionyl fluoride but advises caution in assessing the significance of the sextic coefficients.[97] It is found that if the sextic coefficients are omitted then the difference between observed and calculated frequencies rises from a few hundredths of a megahertz to several megahertz. Although inclusion of about four sextics appears to be necessary to fit all the lines

[90] G. Winnewisser and P. Helminger, *J. Chem. Phys.*, 1972, **56**, 2967.
[91] G. Winnewisser and P. Helminger, *J. Chem. Phys.*, 1972, **56**, 2954.
[92] G. Winnewisser, *J. Chem. Phys.*, 1972, **57**, 1803.
[93] G. Winnewisser, *J. Chem. Phys.*, 1972, **56**, 2944.
[94] G. Winnewisser, *J. Mol. Spectroscopy*, 1972, **41**, 534.
[95] G. Guidotti, U. Lamanna, M. Maestro, and R. Moccia, *Theor. Chim. Acta*, 1972, **27**, 55; J. P. Rank and H. Johansen, *ibid.*, 1972, **24**, 334.
[96] A. Dubrelle and J. L. Destombes, *J. Mol. Structure*, 1972, **14**, 461.
[97] N. J. D. Lucas and J. G. Smith, *J. Mol. Spectroscopy*, 1972, **43**, 327.

measured, various combinations of four sextics are about equally good; evidently there is a strong correlation among the individual sextic coefficients. However, it seems that this does not upset a reliable determination of the quartic coefficients. These quartic coefficients have been used with i.r. data in an attempt to determine the harmonic force fields, and, with some constraints because of lack of data, this has been achieved. The structural parameters (r_0) have also been refined and are given in Table 21. The quoted errors

Table 21 *Structural parameters for* SOF_2

	Bond lengths/Å		*Angles/°*
r(S—O)	1.4127 ± 0.0003	$\widehat{FSF}$	92.83 ± 0.02
r(S—F)	1.5854 ± 0.0002	$\widehat{FSO}$	106.82 ± 0.03

here represent the limits in which the r_0 concept is valid; they are not observational errors but relate to the internal consistency of the I_0 values used in their calculation. The microwave spectra of thionyl fluoride in two excited vibrational states has been the subject of another investigation.[98] An observed perturbation is attributed to a Coriolis reasonance between the two levels and an estimate of the coupling parameter has been made. A centrifugal distortion treatment has also been made for thionyl chloride, $SOCl_2$, some eighty transitions being fitted by a set of quartic and sextic coefficients.[99] The spectrum of a further pentafluoride chloride of a Group VI element has been observed and analysed. Altogether rotational constants have been obtained for twelve isotopic species of TeF_5Cl.[100] These lead to the first Te—Cl distance for a Te^{VI} molecule. The bond length is in fact well determined, the mean of thirty r_s values giving r(Te—Cl) = 2.250 ± 0.002 Å. A unique determination of the remaining three parameters is not possible but if the axial and equatorial Te—F bonds are assumed to be of equal length then the structure shown in Figure 7 is obtained, the parameters

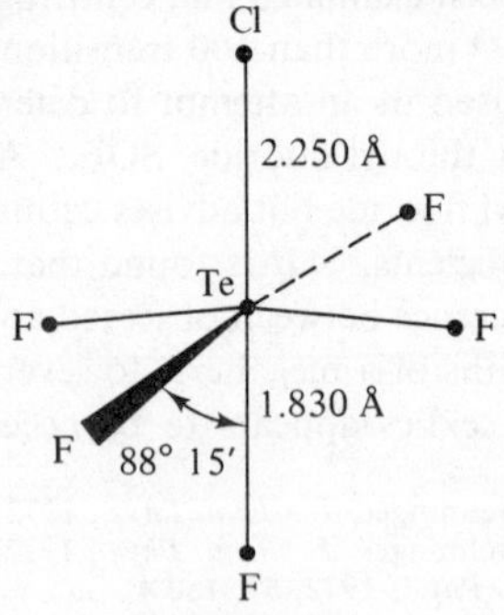

Figure 7

[98] D. F. Rimmer, J. G. Smith, and D. H. Whiffen, *J. Mol. Spectroscopy*, 1973, **45**, 114.
[99] A. Dubrelle and D. Boucher, *Compt. rend.*, 1972, **274**, *B*, 1426.
[100] A. C. Legon, *J. C. S. Faraday II*, 1973, **69**, 29.

being given by r(Te—F) = 1.8305 ± 0.0012 Å and $\widehat{F_{eq}TeF_{ax}}$ = 88°15′ ± 7′. There are good reasons for supposing that the angle probably lies between these same limits since implausible differences in the two Te—F bond lengths would be required to take the value outside these limits. Thus the Te—F equatorial bonds are bent away from the chlorine atom, in an analogous way to that proposed in structures for SF_5Cl and WF_5Cl.

A further investigation has been made of BrF_5, which has previously been shown to have C_{4v} symmetry, and a new feature has now been observed.[101] It is predicted for molecules of C_{4v} symmetry that there should be a splitting of the $K = 2$ lines, and this effect has now been observed for BrF_5. The analysis of the spectrum leads to the following values of the centrifugal distortion constants: $D_J = 0.57 \pm 0.01$ kHz; $D_{JK} = -0.50 \pm 0.02$ kHz; and $R_6 = 0.07 \pm 0.005$ kHz. Electron-diffraction studies [102] have been made of BrF_5 and IF_5 and the results combined with microwave data to give the structural parameters in Table 22.

Table 22 *Structural parameters for* BrF_5 *and* IF_5

XF_5	$r(X—F)_{eq}$/Å	$r(X—F)_{ax}$/Å	$\widehat{F_{ax}XF_{eq}}$/°
BrF_5	1.774 ± 0.003	1.689 ± 0.008	84.8 ± 0.1
IF_5	1.869 ± 0.005	1.844 ± 0.005	81.9 ± 0.1

The microwave spectrum of chloryl fluoride has been analysed for two isotopic species.[103] The molecule has a pyramidal configuration with chlorine at the apex. The structural parameters which have been obtained are given in Table 23.

Table 23 *Structural parameters for* $FClO_2$

Bond lengths/Å		*Angles*/°	
r(Cl—F)	1.664 ± 0.030	$\widehat{OClO}$	113.5 ± 2.0
r(Cl—O)	1.434 ± 0.015	$\widehat{FClO}$	103.2 ± 1.5

The hydrogen-bonded dimer formed by hydrogen fluoride has been investigated by the molecular-beam electric resonance method.[104] Parameters obtained for the three isotopic species $(HF)_2$, $(DF)_2$, and HFDF are given in Table 24. The spectrum of DFHF was not observed. These results

Table 24 *Parameters for* $(HF)_2$, $(DF)_2$, *and* HFDF

Parameter	$(HF)_2$	$(DF)_2$	HFDF
$\frac{1}{2}(B + C)$/MHz	6504.8 ± 2.0	6252.194 ± 0.002	6500.1 ± 0.1
μ_a/D	2.987 ± 0.003	2.9919 ± 0.0006	3.029 ± 0.003
$[eqQ_D]_a$/kHz		110 ± 8	270 ± 30

[101] R. H. Bradley, P. N. Brier, and M. J. Whittle, *J. Mol. Spectroscopy*, 1972, **44**, 536.
[102] A. G. Robiette, R. H. Bradley, and P. N. Brier, *Chem. Comm.*, 1971, 1567.
[103] C. R. Parent and M. C. L. Gerry, *J. C. S. Chem. Comm.*, 1972, 285.
[104] T. R. Dyke, B. J. Howard, and W. Klemperer, *J. Chem. Phys.*, 1972, **56**, 2442.

are interpreted in terms of a semi-rigid model with large-amplitude motion of the hydrogen and deuterium atoms. Tunnelling motion causes a splitting of the rotational levels for $(HF)_2$ and $(DF)_2$ but not for HFDF. By assuming the H—F bond distance to be the same as in the free monomer it is found that the F—F distance is 2.79 ± 0.05 Å and that the end HF unit is bent 60—70° from the F—F axis. The microwave spectrum of hydrogen fluoride has also been examined and lines have been assigned to a hexamer and to a heptamer.[105]

4 Organic Molecules

Acyclic Molecules without Internal Rotation.—Following Duncan's paper [106] of 1970 in which the equilibrium geometries of the methyl halides were evaluated and critically discussed, Matsuura and Overend [107] present an equilibrium structure for methyl iodide. These latter authors have now observed sufficient i.r. bands of methyl iodide under high resolution to ensure that the vibration–rotation constants α_i^A and α_i^B for all of the fundamental modes *i* are available for sufficient isotopic species from direct spectroscopic observation (microwave and i.r.). Thus, recourse to assumptions such as the Coriolis sum rule for A_0 values (as employed by Duncan) is unnecessary. The results of Matsuura and Overend are r_e(C—H) = 1.084 ± 0.001 Å, r_e(C—I) = 2.132 ± 0.001 Å, and $(\widehat{HCH})_e$ = 111°12′ ± 9′, identical within experimental error with those of Duncan.

The α_i^B values and the iodine nuclear quadrupole coupling constants for the vibrationally excited states ν_2, ν_3, ν_5, ν_6, $2\nu_3$, and $\nu_3 + \nu_6$ of the species CD_3I are reported from a thorough analysis of the $J = 3 \leftarrow 2$ transitions by Kuczkowski.[108] The degenerate states ν_5, ν_6, and $\nu_3 + \nu_6$ are treated by the method of Grenier-Besson and Amat to give the *l*-doubling constant, among other parameters. The α_i^B values for the fundamentals are in good agreement with those used by Matsuura and Overend [107] in the r_e structure determination.

Kukolich and Nelson [109, 110] have carried out high-resolution measurements of the hyperfine structure in the $J = 1 \rightarrow 0$ transitions of $Me^{35}Cl$ and $Me^{37}Cl$, using a molecular-beam maser spectrometer to give very accurate values of the chlorine nuclear quadrupole coupling constants and values for the nuclear spin to molecular rotation magnetic coupling constants. A comparison of the $eq_{zz}Q(^{35}Cl)$ values along the C—Cl bond direction (*z*) in the series $CH_3{}^{35}Cl$, $CDH_2{}^{35}Cl$, and $CD_3{}^{35}Cl$ shows a linear decrease; the difference between the first and last (180 kHz) is about fifty times the experimental error. Such a change represents a decrease in electric field gradient along *z*

[105] U. V. Reichert and H. Hartmann, *Z. Naturforsch.*, 1972, **27a**, 983.
[106] J. L. Duncan, *J. Mol. Structure*, 1970, **6**, 447.
[107] H. Matsuura and J. Overend, *J. Chem. Phys.*, 1972, **56**, 5725.
[108] R. L. Kuczkowski, *J. Mol. Spectroscopy*, 1973, **45**, 261.
[109] S. G. Kukolich and A. C. Nelson, *J. Amer. Chem. Soc.*, 1973, **95**, 680.
[110] S. G. Kukolich and A. C. Nelson, *J. Chem. Phys.*, 1972, **57**, 4052.

at the ^{35}Cl nucleus of 0.2%. The authors attribute this reduction to an indirect effect involving changes in the chemical properties of the methyl group on deuterium substitution. Apparently, the shortening of the C—D distances on deuteriation reduces the electron-withdrawing power of the carbon–hydrogen bonds and thereby decreases the electronegativity of the methyl group, leading to increased ionic character for the C—Cl bond. Calculation using Townes–Dailey theory indicates that an increased ionic character of 0.17% for the C—Cl bond in $CD_3{}^{35}Cl$ over that in $CH_3{}^{35}Cl$ is sufficient to account for the decreased nuclear quadrupole coupling in the former. The pure rotational spectra of the [^{79}Br]- and [^{81}Br]-species of methyl bromide [111] have now been resolved in the far-i.r. region between 17 and 43 cm^{-1}.

Kukolich and his collaborators have published several other studies involving a molecular-beam maser spectrometer. The inherently high resolution of the technique has been used in methyl isocyanide ($J = 1 \rightarrow 0$ transitions) [112, 113] to obtain the small nitrogen nuclear quadrupole coupling constant ($eq_{zz}Q$), the nitrogen nuclear spin–rotation magnetic coupling constant (C_N), the corresponding quantity (C_H) for the hydrogen nuclei, and the hydrogen spin–spin interaction constant (D_{HH}) with high precision. From a conventional millimetre-wave investigation of transitions of methyl isocyanide up to $J = 7 \leftarrow 6$ in the $\nu_8 = 1$ and $\nu_8 = 2$ vibrationally excited states, on the other hand, the quantities $B^* = 10\,092.3169$, $D_J = 0.00492$, $D_{JK} = 0.2271$, $\rho^* = 0.05349$, and $|q^8{}_{eff}| = 6.937$ MHz have been determined by Bauer, Bogey, and Maes [114] using a Grenier-Besson–Amat type of analysis.

The previously reported high-resolution work on the hyperfine structure in fluoroform by Kukolich, Nelson, and Ruben [115] has recently been criticized by Reynders, Ellenbroek, and Dymanus.[116] These latter authors, also using a beam-maser spectrometer, report a sharp difference in the C_F and C_H spin–rotation constants obtained by each group when both use the $J = 1 \rightarrow 0$ transition. Kukolich and Ruben [117] have subsequently re-investigated the hyperfine structure and report results in close agreement with the Netherlands group. In this context, the molecular-beam electric resonance spectrum of methyl fluoride [118] measured in several rotational states is worth reporting. Spin–rotation constants C_F and C_H for the fluorine and hydrogen nuclei respectively are derived together with an accurate value $\mu = 1.8585(5)$ D for the dipole moment, a result in excellent agreement with the previous value.[119]

[111] J. Presenti and M. Sergent-Rozey, *Compt. rend.*, 1972, **275**,*B*, 145.
[112] S. G. Kukolich, *Chem. Phys. Letters*, 1971, **10**, 52.
[113] S. G. Kukolich, *J. Chem. Phys.*, 1972, **57**, 869.
[114] A. Bauer, M. Bogey, and S. Maes, *J. Phys. (Paris)*, 1971, **32**, 763.
[115] S. G. Kukolich, A. C. Nelson, and D. J. Ruben, *J. Mol. Spectroscopy*, 1971, **40**, 33.
[116] J. M. H. Reynders, A. W. Ellenbroek, and A. Dymanus, *Chem. Phys. Letters*, 1972, **17**, 351.
[117] S. G. Kukolich and D. J. Ruben, *J. Mol. Spectroscopy*, 1972, **44**, 607.
[118] S. C. Wofsy, J. S. Muenter, and W. Klemperer, *J. Chem. Phys.*, 1971, **55**, 2014.
[119] P. Steiner and W. Gordy, *J. Mol. Spectroscopy*, 1966, **21**, 291.

Another halogenated methane which comes under the scrutiny of Kukolich's molecular-beam maser spectrometer is methylene fluoride.[120] This time the molecular Zeeman effect of the $3_{03} \rightarrow 2_{12}$ rotational transition is observed in emission by placing the spectrometer cavity in a 100 kG magnetic field. The unusually high resolution possible provides more precise molecular susceptibility anisotropies, $2\chi_{aa} - \chi_{bb} - \chi_{cc} = -4.16 \pm 0.02$ and $2\chi_{bb} - \chi_{cc} - \chi_{aa} = 0.58 \pm 0.01 \times 10^{-7}$ J G^{-2} mol^{-1}, than previously available[121] and when these are combined with the previously reported molecular *g*-values [121] the improved molecular quadrupole tensor components $Q_{aa} = -3.95 \pm 0.14$, $Q_{bb} = 2.05 \pm 0.16$, and $Q_{cc} = 1.90 \pm 0.18$ ($\times 10^{-26}$ e.s.u. cm^2) are obtained.

Some further substituted methanes have also been investigated through more conventional microwave spectroscopic techniques. An essentially complete r_s structure for nitromethane (1) is presented by Cox and Waring [122]

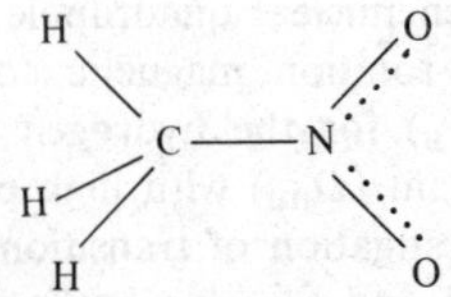

(1)

from a microwave study of the parent molecule and five isotopically substituted species. They use the $m = 0$ internal rotation state to obtain accurate rigid-rotor rotational constants by means of a Watson-type centrifugal distortion analysis. This state behaves just like the usual effective rigid rotor except that, because of the almost free internal rotation of the methyl group, I_a derived from the $m = 0$ spectrum corresponds only to the moment of inertia of the NO_2 framework about the molecular symmetry axis. Consequently, the quantity $\Delta = I_c - I_b - I_a(NO_2)$ is analogous to that of a planar asymmetric rotor in that the contribution of the hydrogen atoms to I_c and I_b is the same, and indeed the value 0.2015 a.m.u Å^2 for the parent species and similar values for the heavy-atom isotopically substituted species strongly indicate coplanarity of the heavy atoms. More conclusive evidence of planarity is derived from nuclear-spin statistical-weight effects in transitions of the CH_3NO_2 and CD_3NO_2 species. The quantitative structure which results from isotopic substitution is:

$$r(\text{N—O}) = 1.224 \pm 0.005 \text{ Å} \qquad \widehat{\text{ONO}} = 125.3 \pm 0.3^\circ$$
$$r(\text{C—H}) = 1.088 \text{ Å} \qquad \widehat{\text{HCH}} = 107.2^\circ$$
$$r(\text{C—N}) = 1.489 \pm 0.005 \text{ Å}$$

[120] S. G. Kukolich and A. C. Nelson, *J. Chem. Phys.*, 1972, **56**, 4446.
[121] R. P. Blickensderfer, J. H. S. Wang, and W. H. Flygare, *J. Chem. Phys.*, 1969, **51**, 3196.
[122] A. P. Cox and S. Waring, *J. C. S. Faraday II*, 1972, **68**, 1060.

The authors also point out by means of a comparison of several nitro-group structures, as in Table 25, an interesting correlation between the electronegativity of X in XNO_2 and the NO_2 group structural parameters. In FNO_2 the fluorine atom is highly electronegative and had previously been assumed [123] to allow important contributions from the ionic valence-bond

Table 25 *Structural parameters of nitro-groups*

Species	N—O *bond length*/Å	$\widehat{ONO}$/°	O · · · O *distance*/Å	*Electronegativity of attached group*
NO_2^-	1.240	114.9	2.090	0.8
$C_6H_5NO_2$	1.227	124.4	2.171	2.5
CH_3NO_2	1.224	125.3	2.175	2.5
$ClNO_2$	1.202	130.1	2.181	3.0
NH_2NO_2	(1.206)	130.1	2.185	3.0
N_2O_3	1.210	129.8	2.190	3.0
CH_3ONO_2	1.207	129.5	2.182	3.5
HNO_3	1.205	130.3	2.184	3.5
NO_2	1.193	134·1	2.198	—
FNO_2	1.180	136.0	2.188	4.0
NO_2^+	1.09	180.0	2.180	>4.0

structure (2) in order to account for the relatively large angle ONO (136°) and the relatively short NO bond (1.180 Å). On the other hand, for nitromethane the less electronegative CH_3 group is expected to allow important contributions from 'nitrite-ion type' valence-bond structures (3) in addition to the usual structures (4). This is used to rationalize the smaller angle ONO and longer NO bond in nitromethane. The intermediate electronegativity of chlorine fits conveniently with the nitro-group structure, which is seen in $ClNO_2$ to be intermediate between those in FNO_2 and CH_3NO_2.

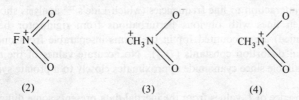

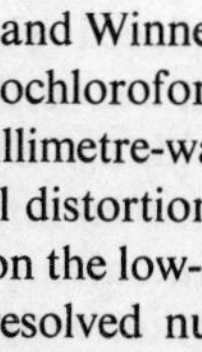
(2)

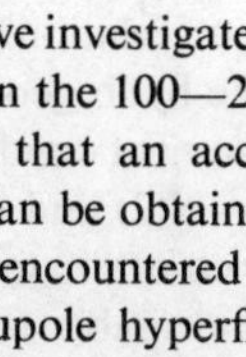
(3)

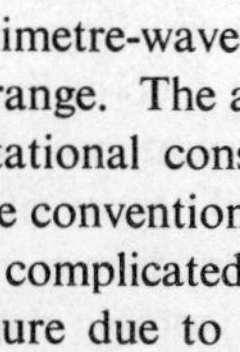
(4)

McLay and Winnewisser [124] have investigated the millimetre-wave spectrum of deuteriochloroform ($CDCl_3$) in the 100—200 GHz range. The advantage of the millimetre-wave study is that an accurate rotational constant and centrifugal distortion constants can be obtained. In the conventional microwave region the low-J transitions encountered possess a complicated and only partially resolved nuclear quadrupole hyperfine structure due to the three chlorine nuclei, whereas the K-splitting due to centrifugal distortion is small. Fortunately, for transitions with $J > 15$ in the millimetre region the hyperfine

[123] A. C. Legon and D. J. Millen, *J. Chem. Soc.* (*A*), 1968, 1736.
[124] D. B. McLay and G. Winnewisser, *J. Mol. Spectroscopy*, 1972, **44**, 32.

structure is sufficiently condensed that the individual K components of the transition are well resolved, with no apparent hyperfine broadening. Then the B_0 value and centrifugal distortion constants can be accurately determined from the K-splitting, which is just like that of a molecule without coupling nuclei. The reported constants are $B_0 = 3250.292 \pm 0.0015$ MHz, $D_J = 1.414 \pm 0.005$ KHz, and $D_{JK} = -2.314 \pm 0.006$ KHz.

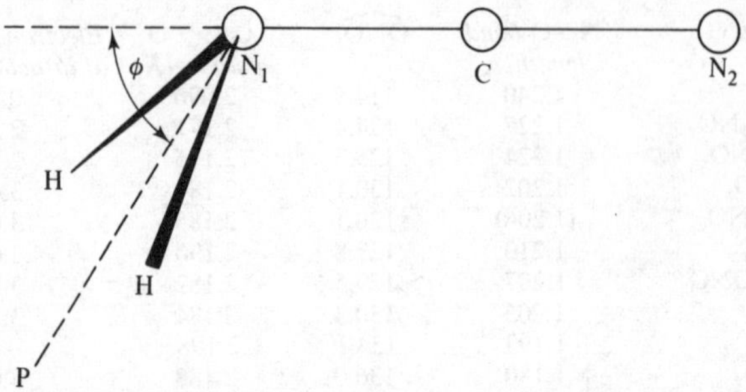

Figure 8 PN_1 = *bisector of* HN_1H

The simple but important molecule cyanamide (Figure 8) continues to excite considerable interest. A detailed analysis of the conclusions that can be derived from the μ_a spectrum of cyanamide is given by Tyler, Sheridan, and Costain.[125] Rotational constants B and C have been obtained for the isotopic varieties NH_2CN, NHDCN, ND_2CN, $^{15}NH_2CN$, $NH_2C^{15}N$, $^{15}ND_2CN$, and $ND_2C^{15}N$ for the vibrational ground state and, for the first three species, in the first excited inversion level. The method of analysis of the spectral data in each case is such that the contributions from inversion–rotation interaction to line frequencies (which Lide's [126] analysis shows exist even when lines with obvious perturbations from rigid-rotor behaviour are excluded) are accounted for in the terms inseparable from one of the centrifugal distortion constants (D_{JK}). No accurate values of the constant A are available since cyanamide approximates closely to a prolate symmetric top.

The absence of I_a values from the inertial data presents some difficulties in the structure determination, but relationships are produced which allow essentially r_s co-ordinates for the hydrogen atoms and r_s, a co-ordinates for the nitrogen atoms. The first-moment condition then yields a_C. By assumption of collinearity of N(1)—C—N(2) and use of first-moment and product of inertia conditions, the small c co-ordinates of the heavy atoms are obtained. Sufficient isotopic data is also available for the calculation with reference to ND_2CN as the parent species to be carried out. The two structures, the mean of which is shown in Table 26, are in close agreement.

125 J. K. Tyler, J. Sheridan, and C. C. Costain, *J. Mol. Spectroscopy*, 1972, **43**, 248.
126 D. R. Lide, jun., *J. Mol. Spectroscopy*, 1962, **8**, 142.

Table 26 *Structure of cyanamide*

r[N(1)—H]	1.001 ± 0.015 Å
r[N(2)—N(1)]	2.506 ± 0.002 Å
r[N(1)—C]	1.346 ± 0.005 Å
r[C—N(2)]	1.160 ± 0.005 Å
$\widehat{\text{HN(1)H}}$	113°31′ ± 2°
ϕ (see Figure 8)	37°58′ ± 1°

Some discussion of the potential function governing the NH_2 inversion was also possible. Reference to Lide's theory [126] indicates which $v = 0$ state transitions can interact with the $v = 1$ state transitions observed to suffer appreciable perturbations from rigid-rotor behaviour. Knowing the rotational constants and the pairs of transitions which interact, approximate values of 50, 33, and 15 cm^{-1} have been estimated for the $v = 1 \leftarrow v = 0$ vibrational separations of NH_2CN, NHDCN, and ND_2CN respectively. Given a simple model of the inversion motion at the nitrogen atom, the reduced mass is calculated and a Manning potential used to obtain a best fit to the NH_2CN, NHDCN, and ND_2CN $1 \leftarrow 0$ vibrational separations. The barrier height to inversion obtained is $V_0 = 710$ cm^{-1}. Although this value is considerable disagreement with $V_0 = 467 \pm 30$ cm^{-1} from the i.r. study,[127] the authors point to the extreme sensitivity of V_0 from the microwave data to the value of the out-of-plane angle corresponding to the minima in the potential function, which is only approximated by ϕ. In consequence of the extra vibrational data used in the derivation of the potential function from the i.r. study, this identification with ϕ is unnecessary and the lower barrier therefrom is to be preferred.

Attanasio, Bauder, and Günthard [128] have used the exact solution of the rotation–inversion problem by infinite matrix diagonalization as a procedure to attack the cyanamide problem. They employ experimental microwave data consisting of pure rotational transitions in the $v = 0$ and $v = 1$ inversion levels and rotation–inversion transitions between rotational levels in the $v = 0$ and $v = 1$ states of ND_2CN (data collected mainly by Tyler). However, all of the data could not be fitted by their procedure and the authors conclude that no set of dynamical constants within the semi-rigid model used could reproduce the observed transitions satisfactorily. Hence, a unique value of V_0 is not possible, but if either $v = 0$ or $v = 1$ rotational transitions only are used with the intersystem transitions the values of V_0 obtained are 735.5 and 700.6 cm^{-1} respectively.

The group at the Swiss Federal Institute has also carried out *ab initio* LCAO–MO–SCF calculations on cyanamide in order to determine the molecular electric dipole moment [129] as a function of ϕ and $\widehat{\text{HN(1)H}}$ (see Figure 8). The value obtained for $\widehat{\text{HN(1)H}} = 115°$ and $\phi = 39°$ (*cf.* Table 26 above)

[127] T. R. Jones and N. Sheppard, *Chem. Comm.*, 1970, 715.
[128] A. Attanasio, A. Bauder, and Hs. H. Günthard, *Mol. Phys.*, 1972, **24**, 889.
[129] A. Attanasio, Hs. H. Günthard, and Tae-Kyu Ha, *Mol. Phys.*, 1972, **24**, 215.

is 4.625 D, in good agreement with $\mu_a = 4.32 \pm 0.08$ as reported by Tyler, Sheridan, and Costain.[125] The protons determine the positive end of the dipole.

A close relative of cyanamide, difluorocyanamide (NF_2CN), and its phosphorus analogue (PF_2CN) are the subjects of microwave structure determination by Schwendeman and his co-workers. For difluorocyanamide [130] the spectra of the isotopic species NF_2CN, $NF_2{}^{13}CN$, $NF_2C^{15}N$, and $^{15}NF_2CN$ have been analysed and two types of atomic co-ordinate derived from the observed moments of inertia. First, assuming a plane of symmetry for the molecule, the nine non-zero co-ordinates are fitted to the moments of inertia (excluding the imprecisely determined I_a values for the isotopically substituted species), the two centre of mass relations, and the one product of inertia relation to give an essentially r_0 structure. Secondly, r_s co-ordinates (*a* and *c*) are derived for the carbon and two nitrogen atoms while the fluorine atom co-ordinates are fixed by auxiliary moment conditions. The two structures agree closely (Table 27). The dipole moment values measured are $\mu_a = 1.03 \pm 0.02$, $\mu_c = 0.39 \pm 0.10$, and $\mu = 1.10 \pm 0.02$ D. The similar investigation of PF_2CN,[84] with less isotopic substitution possible, yields the

Table 27 *Molecular structures*[a] *of* XF_2CN (X = N *or* P)

	X = N		X = P
	r_0	r_s	
r(C≡N)	1.151 ± 0.004	1.158 ± 0.004	1.157 ± 0.003
r(X—C)	1.392 ± 0.009	1.386 ± 0.009	1.815 ± 0.005
r(X—F)	1.398 ± 0.005	1.399 ± 0.008	1.566 ± 0.007
$\widehat{XCN}$	169.7 ± 2.1	173.9 ± 2.2	171.2 ± 0.8
$\widehat{CXF}$	104.7 ± 0.6	105.4 ± 0.7	96.9 ± 0.2
$\widehat{FXF}$	102.9 ± 0.4	102.8 ± 0.5	99.2 ± 0.2

[a] Bond distances in Å, angles in degrees.

structure shown in Table 27. There are two noteworthy features of these structures. In both molecules the X—C—N atoms are not collinear but the C≡N group is tilted by about 10° away from the fluorine atoms. This can be explained on the basis of either a bonding interaction between the N or P atom electron lone pair and the CN group π-system or an electrostatic repulsion between the fluorine atoms and the CN group π-system. Similar behaviour has been reported in $S(CN)_2$.[86] The other significant structural feature is the much more pronounced pyramidal nature of the bonds attached to the nitrogen atom in NF_2CN than in NH_2CN, a change also observed [131] if HNF_2 is compared with ammonia.

The observed rotational microwave spectra of aminoacetonitrile for the isotopic species NH_2CH_2CN, $NHDCH_2CN$, and ND_2CH_2CN, reported by

[130] P. L. Lee, K. Cohn, and R. H. Schwendeman, *Inorg. Chem.*, 1972, **11**, 1920.
[131] D. R. Lide, jun., *J. Chem. Phys.*, 1963, **38**, 456.

Macdonald and Tyler,[132] are interpreted in terms of a molecular conformation in which the amino and methylene groups have a *trans* orientation with respect to each other as in (5). Negative values of χ_{aa} observed for both the nitrogen nuclei and a quadrupole hyperfine structure for the $1_{01} \leftarrow 0_{00}$ transition as in hydroxyacetonitrile,[133] which also has the *trans* structure, are further evidence. The measured dipole moment components are $\mu_a = 2.55 \pm 0.02$, $\mu_b = 0.30 \pm 0.02$, and $\mu_{total} = 2.57 \pm 0.03$ D.

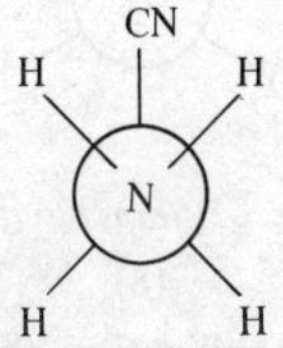

(5)

The conformations of some ethane derivatives existing in a dominant rotameric form have been established recently. The rotational spectrum of but a single isomer of ethyl thiocyanate could be identified by Bjørseth and Marstokk,[134] corresponding to the *gauche*-rotamer (6). A rather thorough analysis of the observed μ_a spectrum with a Watson five-constant centrifugal distortion treatment gave rise to rotational constants which could be reproduced by assuming structural parameters for the relevant parts of the molecule from n-propyl cyanide and methyl thiocyanate if the dihedral angle (measured from the *anti*-form) for ethyl thiocyanate was 122° (*i.e.* the spectrum corresponds to the *gauche*-form). The failure to find a spectrum consistent with the *anti*-form is interpreted to mean that the *gauche* is the more stable of the two forms by at least 0.8 kcal mol^{-1}. Vibrational satellites in the spectrum assigned to the methyl-group torsion exhibit no splitting for $v \leqslant 2$, which places a lower limit of $V_3 = 3.8$ kcal mol^{-1} on the barrier to internal rotation of the CH_3 group, a value consistent with 4.1 kcal mol^{-1} from the i.r. frequency of the methyl torsion. The dipole moment components $\mu_a = 3.80 \pm 0.10$, $\mu_b = 1.13 \pm 0.05$, $\mu_c = 0.61 \pm 0.04$, and $\mu = 4.01 \pm 0.12$ D have been derived from Stark-effect measurements.

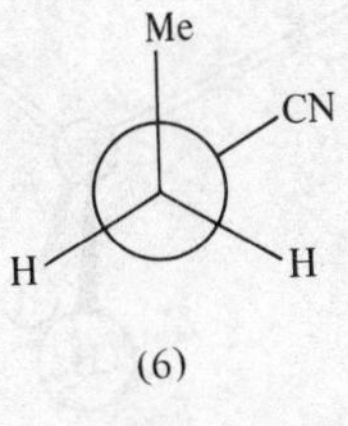

(6)

[132] J. N. Macdonald and J. K. Tyler, *J. C. S. Chem. Comm.*, 1972, 995.
[133] J. K. Tyler and D. G. Lister, *Chem. Comm.*, 1971, 1350.
[134] A. Bjørseth and K. M. Marstokk, *J. Mol. Structure*, 1972, **11**, 15.

For trifluoroethylamine, ($CF_3CH_2NH_2$), if only staggered configurations with respect to rotation about the C—N bond are considered, two equivalent *gauche*-forms, (7) and (9), and one *trans*-form (8) are possible.

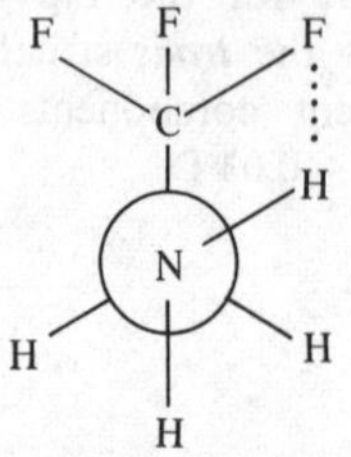

gauche
(7)

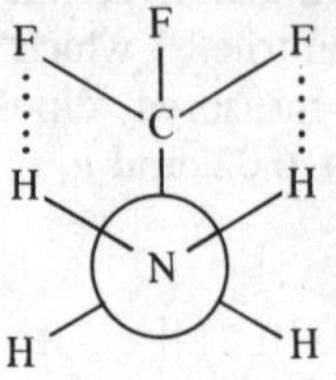

trans
(8)

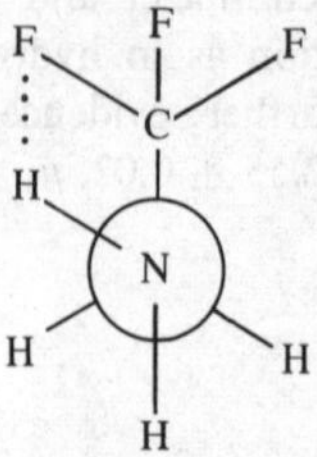

gauche
(9)

Clearly, in the *trans*-form stabilization through two hydrogen bonds is likely whereas only one such interaction need be considered in the *gauche*-forms. Warren and Wilson [135] observe and assign near-prolate μ_a-type spectra belonging to a single rotamer for the $CF_3CH_2NH_2$, CF_3CH_2NHD, and $CF_3CH_2ND_2$ species. No other rotameric form of these molecules could be detected. Although the isotopic substitution allows the amino-proton co-ordinates to be determined, in order to determine the configuration of the observed rotamer the CF_3 group parameters and the C—C bond distance are assumed from CF_3CH_3 and the CH_2-group, NH_2-group, and C—N-bond parameters are assumed from CH_3NH_2. The angle CCN is varied until the observed rotational constants are reproduced. Since the amino-proton co-ordinates do not strongly affect the calculated rotational constants, the NH_2-group parameters are subsequently varied to fit the amino-proton co-ordinates so that the plausible structure found for the molecule is the *trans*-form shown in Table 28 and Figure 9. The measured

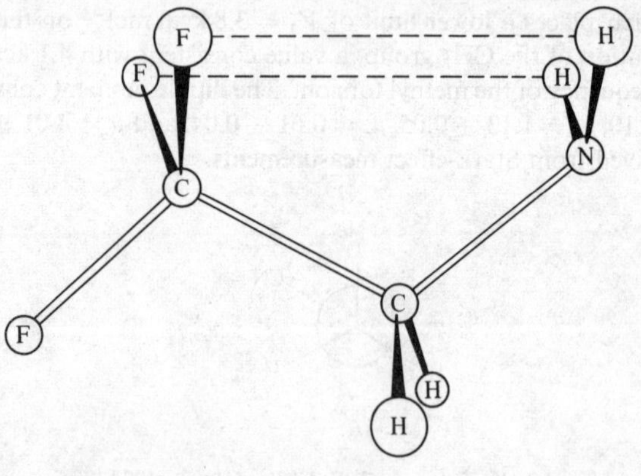

Figure 9 trans-*Form of trifluoroethylamine*

[135] I. D. Warren and E. B. Wilson, jun., *J. Chem. Phys.*, 1972, **56**, 2137.

dipole moment components are $\mu_a = 1.13 \pm 0.03$, $\mu_b = 0.28 \pm 0.03$, and $\mu_c = 0.0 \pm 0.05$ D. The authors point out that, because of relative dipole moment components, if the *gauche*-form were as populated as in ethylamine the lines of the *gauche*-form of trifluoroethylamine would be stronger than those of the *trans*-form. A stabilization of the *trans*-form through interaction of the amino-protons and fluorine atoms is therefore indicated. The contribution of a covalent interaction to this stabilization is ruled out on the basis of the N—H bond length essentially unchanged from that in methylamine and an F · · · H distance of 2.65 Å, which is 0.10 Å in excess of the sum of the hydrogen and fluorine van der Waals radii. A dipole–dipole interaction is suggested by the almost parallel N—H and C—F bonds.

Table 28 *Plausible structure of trifluoroethylamine*

Bond lengths/Å		*Angles*	
r(C—F)	1.335[a]	$\widehat{FCF}$	108°30′[a]
r(C—C)	1.530[a]	$\widehat{CCF}$	111° 2′[a]
r(C—H)	1.093[a]	$\widehat{CCH}$	109°28′[a]
r(C—N)	1.474[a]	$\widehat{HCH}$	109°28′[a]
r(N—H)	1.01 ± 0.02	$\widehat{CCN}$	113°50′ ± 1°
r(H · · · F)	2.65 ± 0.02	$\widehat{CNH}$	110° ± 2°
		$\widehat{HNH}$	110°30′ ± 1°
		$\widehat{NHF}$	95° ± 5°
		$\widehat{NH{\cdot}CF}$	5° ± 5°

[a] Assumed (see text).

Buckley and Brochu[136] have identified a single dominant conformer of 2-methoxyethanol ($CH_3OCH_2CH_2OH$) from the spectra of the parent and the hydroxy-group-deuteriated species. Six stable conformations of this compound arising from internal rotations about the C—C and C—O(H) bonds are possible, even ignoring those resulting from rotation of the methyl group as a whole about the C—O (ether) bond. Conformations (10), (11), and (12), referred to rotation about the C—C bond, are described by capital letters whereas (13), (14), and (15), referred to rotation about the C—O(H) bond, are given by lower-case letters.

From a consideration of (i) the agreement between the observed rotational constants and those calculated on the assumption of structural parameters for the ether part of the molecule from the dimethyl ether and for the alcohol part from 2-fluoroethanol, (ii) the hydroxyl proton Kraitchman co-ordinates, and (iii) the observed dipole moment components $\mu_a = 2.03 \pm 0.02$, $\mu_b = 1.15 \pm 0.02$, and $\mu_c \approx 0.25$ D, Buckley and Brochu assign the observed spectrum to a *GAUCHE*(*G*)–*gauche*(*g*)-form in which the methyl group has been fixed in its *trans* position so as to make the COCC chain coplanar. A better fit to the rotational constants, hydroxyl proton co-ordinates, and dipole

[136] P. Buckley and M. Brochu, *Canad. J. Chem.*, 1972, **50**, 1149.

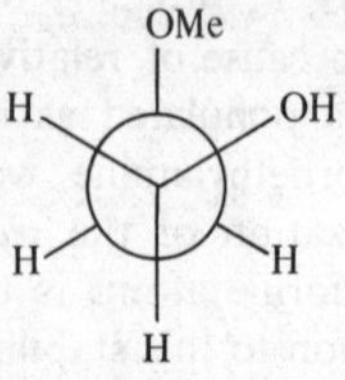

GAUCHE (*G*)

(10)

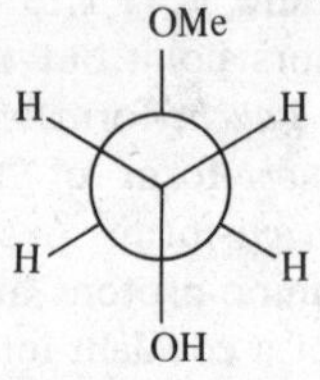

TRANS (*T*)

(11)

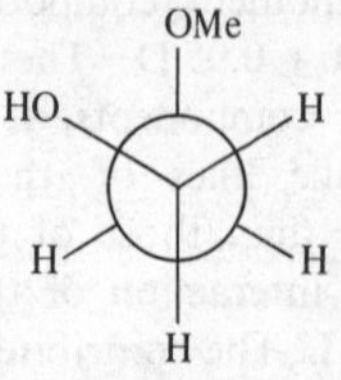

GAUCHE (*G*)

(12)

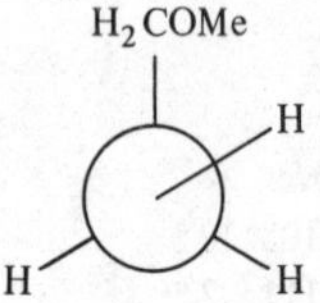

gauche (*g*)

(13)

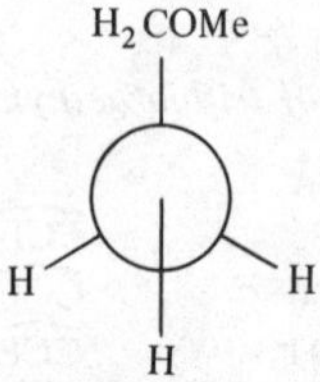

trans (*t*)

(14)

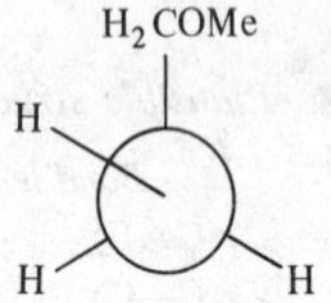

gauche (*g′*)

(15)

moment components is obtained if the dihedral angles θ = C(H_3)OC/CCO, ϕ = OCC/CCO, and γ = CCO/COH (see Figure 10) are varied slightly from their standard values $\theta = 0°$, $\phi = 60°$, and $\gamma = 60°$ respectively for the all-*gauche*-form to give $\theta = 8 \pm 3°$, $\phi = 57 \pm 3°$, and $\gamma = 45 \pm 5°$. In the assigned conformation, the preferred orientation of the methyl group about its own axis is assumed to be staggered with respect to the nearest methylene group. It should be noted that in the established all-*gauche* configuration (Figure 10), the hydroxyl hydrogen is approximately aligned with the nearest tetrahedrally disposed sp^3 lone pair of electrons on the ether oxygen atom.

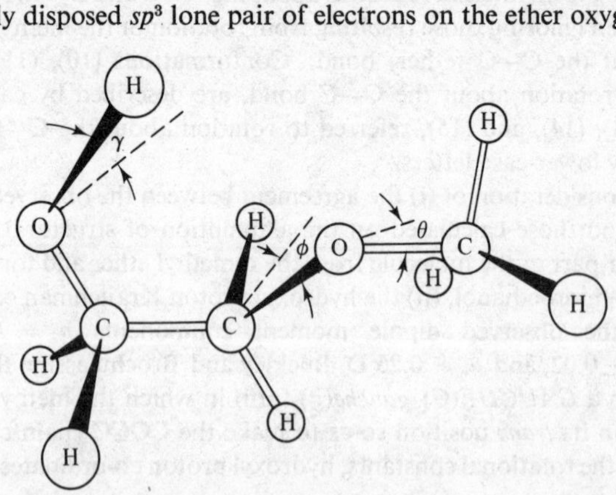

Figure 10 *All*-gauche *conformation of 2-methoxyethanol*

Some further work on the microwave spectrum of 2-fluoroethanol [137] has been carried out and a structure similar to that used by Buckton and Azrak [138] in their earlier publication has been found to be consistent with the observed spectrum.

Azodicarbonitrile has the possibility of *cis*- and *trans*-isomers, (16) and (17). The i.r. and Raman spectra recorded [139] under a variety of conditions strongly indicate C_{2h} symmetry for the compound. No evidence for the *cis*-form was found. In the microwave spectrum [139] only seven transitions appeared and these could not be assigned to the *cis*-isomer, which is thereby restricted to exist in < 1% abundance.

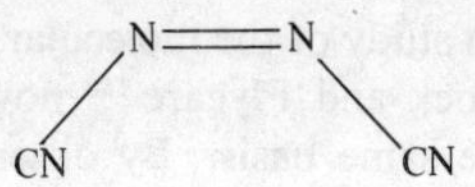

cis

(16)

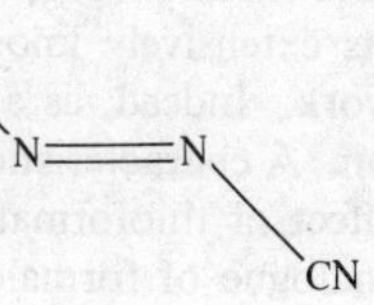

trans

(17)

A preliminary publication [140] concerned with pivalonitrile (18) is adorned with an impressive reproduction of well-resolved K structure in the $J = 11 \leftarrow 10$ transition of this symmetric rotor. A strong satellite series of a low-lying bending mode also exhibits such structure in each transition. Of course, the symmetric-top nature of the spectrum establishes collinearity of the C—C≡N→O atoms and by suitable structural assumptions for the —C≡N→O group and the t-butyl group the author derives r(C—C≡) $= 1.458 \pm 0.01$ Å from $B_0 = 1459.227$ MHz. This parameter is longer than in CH_3—CNO (1.442 Å), a change consistent with the family of C—X bond lengthenings in going from CH_3X to $(CH_3)_3C$—X compounds.

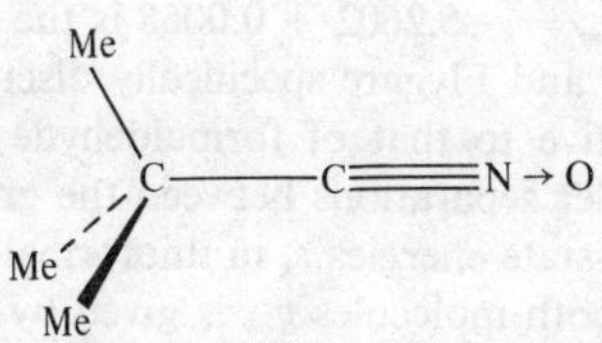

(18)

Investigations of small molecules containing the C=O group (or its sulphur

[137] C. D. Kadzhar, G. A. Abdullaev, and L. M. Imanov, *Izvest. Akad. Nauk Azerb. S.S.R., Ser. fiz.-mat. tekh. Nauk*, 1971, 114.
[138] K. S. Buckton and R. G. Azrak, *J. Chem. Phys.*, 1970, **52**, 5652.
[139] B. Bak and P. Jansen, *J. Mol. Structure*, 1972, **11**, 25.
[140] M. Winnewisser, *Chem. Phys. Letters*, 1971, **11**, 515.

analogue) are again in evidence, particularly Zeeman studies. The simplest aldehyde, formadehyde, is the subject of a Zeeman study [141] in which the splittings in the presence of the magnetic field of the radio-frequency K-doublet transitions 2_{20}—2_{21}, 3_{21}—3_{22}, 7_{34}—7_{36}, and 8_{35}—8_{36} at 71.143, 355.569, 136.925, and 300.869 MHz respectively are measured and the diagonal elements of the molecular g-tensor obtained. Of the experimental results, $g_{aa} = -2.898 \pm 0.005$ is in excellent agreement with a previous measurement [142] but $g_{bb} = 0.956$ and $g_{cc} = -0.322 \pm 0.003$ appear not to be so, presumably because these elements cannot be determined independently from the transitions in the present study.

The above-mentioned study of formaldehyde provides confirmation of some of its extensively known magnetic properties derived from previous Zeeman work. Indeed, as a result, much is known of its electronic charge distribution. A characteristically thorough study of the molecular rotational Zeeman effect of thioformaldehyde by Rock and Flygare [143] now puts the sulphur analogue of formaldehyde on the same basis. By observing first- and second-order magnetic field effects for rotational transitions, molecular g-values ($g_{aa} = -5.2602 \pm 0.0068$, $g_{bb} = -0.1337 \pm 0.0004$, and $g_{cc} = -0.0239 \pm 0.0004$, where a is the symmetry axis and c the out-of-plane axis) and the molecular magnetic susceptibility anisotropies $2\chi_{aa} - \chi_{bb} - \chi_{cc} = 52.3 \pm 1.1$ and $2\chi_{bb} - \chi_{aa} - \chi_{cc} = -5.1 \pm 0.7$ ($\times 10^{-13}$ J G^{-2} mol^{-1}) are derived. These five magnetic parameters are then combined with the known molecular structure to give the molecular quadrupole moments $Q_{aa} = 3.0 \pm 0.7$, $Q_{bb} = -2.4 \pm 0.5$, and $Q_{cc} = -0.6 \pm 1.1$ ($\times 10^{-26}$ e.s.u. cm^2). The authors are then able to calculate numerous other molecular properties including anisotropies of the second moments of charge distribution, the second moments of the electronic charge distribution themselves (if one of the values, $\langle c^2 \rangle$, is estimated by summing individual free-atom diamagnetic susceptibilities), and the diagonal elements of the diamagnetic and paramagnetic susceptibility tensors.

It turns out that $g_{aa} = -5.2602 \pm 0.0068$ is the largest molecular g-value yet measured. Rock and Flygare specifically discussed the reasons for this enhanced value relative to that of formaldehyde (which is already large) on the basis of smaller separations between the ground-state energy ε_0 and electronically excited-state energies ε_k in thioformaldehyde than in formaldehyde itself, since in both molecules g_{aa} is given by

$$g_{aa} = 1.00 + (2M_p/mI_{aa}) \sum_{k>0} [\,|\langle 0 | L_a | k \rangle|^2/\varepsilon_0 - \varepsilon_k]$$

where M_p is the proton mass, m the electron mass, I_{aa} the moment of inertia about the a-axis, and L_a the electronic angular momentum operator projected on the a-axis, and the sum over k includes all electronically excited states. The authors also use this equation to establish rather neatly the absolute signs of the g-values since the second term is always negative and therefore

[141] J.-C. Chardon, S. Luneau, and G. Mandret, *Compt. rend.*, 1972, **275**, *B*, 541.
[142] W. Huttner, Mei-Kuo-Lo, and W. H. Flygare, *J. Chem. Phys.*, 1968, **48**, 1206.
[143] S. L. Rock and W. H. Flygare, *J. Chem. Phys.*, 1972, **56**, 4723.

$g_{aa} \not> 1$. Hence, a value of magnitude 5.2602 can only be negative. Another interesting point to emerge from the discussion is that the second moments of charge distribution $\langle a^2 \rangle$ are 21.1 and 10.4 ($\times 10^{-16}$ cm^2) in thioformaldehyde and formaldehyde respectively, indicating an elongation of the electron cloud in the former, which is a trend also apparent when CO and CS are compared.

The recent detection of thioformaldehyde in interstellar space has spurred workers at N.B.S. to attempt a catalogue of all transitions of the molecule within the range of radiotelescopes. As part of this effort, Beers, Klein, Kirchoff, and Johnson [144] have looked at $H_2{}^{12}C^{32}S$ in the millimetre-wave range 100—250 GHz. Twenty-five newly assigned *R*-branch transitions complement all previously measured transitions in a Watson-type centrifugal distortion treatment. A high correlation of the τ_{aaaa} value with the rotational constant *A* dictates that the well-determined value of the former (from a high-resolution i.r. study [145]) should be used as a datum point in the least squares. It is concluded that the rotational and centrifugal distortion constants so obtained should predict all transitions (except for one *Q*-branch series) with $J \leqslant 30$ and $\nu < 300$ GHz with a standard deviation of < 0.5 MHz.

Methyleneimine ($CH_2{=}NH$), a molecule isoelectronic with formaldehyde, has been identified as a transient species in the gas phase (half-life ~ 0.1 s at 30 μmHg pressure) through its microwave spectrum.[146] Initially, production of this species was *via* the mixing of fluorine atoms derived from a microwave discharge of CF_4 or SF_6 with methylamine downstream of the discharge. A hydrogen atom is abstracted from the methylamine by a fluorine atom. Later, the pyrolysis of methylamine proved to be a useful source of methyleneimine. Twenty-seven transitions, including the $1_{01} \leftarrow 0_{00}$, have been identified and give rotational constants consistent with a reasonable model structure as well as quadrupole coupling constants due to a single nitrogen nucleus. The quantity $\Delta = I_c{}^0 - I_b{}^0 - I_a{}^0 = 0.05592$ a.m.u. Å^2 is strong evidence of molecular planarity.

Evidence for the existence of the *cis*-form of glyoxal (19) was recently obtained from an electronic band of glyoxal at 4875 Å by Currie and Ramsay,[147]

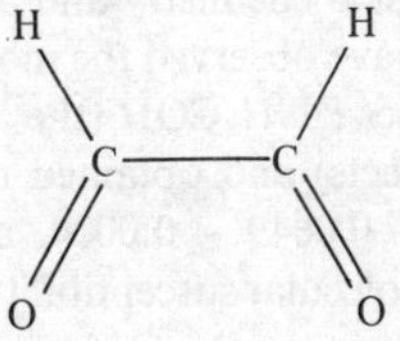

(19)

[144] Y. Beers, G. P. Klein, W. H. Kirchoff, and D. R. Johnson, *J. Mol. Spectroscopy*, 1972, **44**, 553.
[145] J. W. C. Johns and W. B. Olson, *J. Mol. Spectroscopy*, 1971, **39**, 479.
[146] D. R. Johnson and F. J. Lovas, *Chem. Phys. Letters*, 1972, **15**, 65.
[147] G. N. Currie and D. A. Ramsay, *Canad. J. Phys.*, 1971, **49**, 317.

who also published some rotational constants. With these constants as a starting-point, Durig, Tong, and Li [148] have successfully searched for the microwave rotational spectrum of *cis*-glyoxal. A small value of the inertia defect $\Delta = -0.193$ a.m.u. Å^2 points to the *cis* rather than the other dipolar form (*gauche*). Intensity measurements of satellites associated with the first excited torsional level relative to the ground state give 114 ± 8 cm^{-1} for the $1 \leftarrow 0$ torsion transition, which is to be compared with the value 128 cm^{-1} for the corresponding transition of *trans*-glyoxal. The molecular dipole moment is found to be $\mu_b = 4.8 \pm 0.2$ D with the assumption $\mu_a = \mu_c = 0$ on symmetry grounds.

The very slightly asymmetric rotor keten exhibits an *a*-type microwave spectrum from which the rotational constant A and the distortion constant D_K are ill determined. Johns, Stone, and Winnewisser [149] remedy this by a high-resolution study of the perpendicular bands ν_7 and $(\nu_2 + \nu_8)$ in the i.r. spectrum and examination of the millimetre-wave spectrum up to 200 GHz. Ground-state combination differences from the former and new millimetre transitions from the latter in conjunction with the previously known microwave spectrum lead to a precise set of ground-state molecular constants (with A now particularly well determined). As a result the authors confidently predict several microwave and millimetre-wave rotational transitions, as yet unmeasured, which are possibly of use in astrophysical investigations.

An investigation of the rotational Zeeman effect in the microwave spectrum of dimethylketen [150] by the group at Kiel has given the molecular *g*-values ($g_{aa} = \pm 0.020$, $g_{bb} = \pm 0.0165$, and $g_{cc} = \pm 0.0126$) and the molecular magnetic susceptibility anisotropies ($2\chi_{aa} - \chi_{bb} - \chi_{cc} = -1.0 \pm 1.0$ and $2\chi_{bb} - \chi_{cc} - \chi_{aa} = -0.8 \pm 1.2[\times 10^{-13}$ J G^{-2} mol^{-1}]). Also reported are the diagonal elements of the molecular quadrupole and paramagnetic susceptibility tensors. In a very similar study the same group [151] take perdeuteriodimethyl sulphide as the subject. Earlier measurements of Benson and Flygare [152] on Me_2S are also improved by additional measurements. From the change of *g*-values on deuteriation the negative end of the molecular electric dipole is assigned to the sulphur atom.

Two papers concerned with the microwave spectrum of formamide in fact have their origins in biological chemistry and astrophysics respectively.

Flygare and Tigelaar [153] have observed the molecular Zeeman effect in transitions of the isotopic species $^{15}NH_2COH$ (the ^{15}N isotope avoids complicating nuclear quadrupole effects) and obtained the molecular *g*-values ($g_{aa} = -0.2843 \pm 0.0011$, $g_{bb} = -0.0649 \pm 0.0004$, and $g_{cc} = -0.0117 \pm 0.0004$) and the two independent molecular susceptibility anisotropies $2\chi_{aa} - \chi_{bb} - \chi_{cc}$

148 J. R. Durig, C. C. Tong, and Y. S. Li, *J. Chem. Phys.*, 1972, **57**, 4425.
149 J. W. C. Johns, J. M. R. Stone, and G. Winnewisser, *J. Mol. Spectroscopy*, 1972, **42**, 523.
150 D. H. Sutter, L. Charpentier, and H. Dreizler, *Z. Naturforsch.*, 1972, **27a**, 597.
151 E. Hamer, D. H. Sutter, and H. Dreizler, *Z. Naturforsch.*, 1972, **27a**, 1159.
152 R. C. Benson and W. H. Flygare, *J. Chem. Phys.*, 1970, **52**, 5291.
153 H. L. Tigelaar and W. H. Flygare, *J. Amer. Chem. Soc.*, 1972, **94**, 343.

$= 2.2 \pm 0.7$ and $2\chi_{bb} - \chi_{aa} - \chi_{cc} = 8.0 \pm 0.5$ $(\times 10^{-13}\ \mathrm{J\ G^{-2}\ mol^{-1}})$. They then take formamide as a model for the amide linkage in poly-(L-alanine). This polypeptide (and others) shows two n.m.r. peaks for the α-proton in helix-breaking solvents; the upfield peak, it is agreed, is assigned to α-protons in the helical form whereas controversy surrounds the downfield peak. Some argue that the downfield peak arises from α-protons in the random coil, the shift being caused by the consequent re-orientation of amide planes adjacent to the proton. Others attribute the shift to solvation of the amide planes. Using the molecular susceptibility anisotropy of formamide as transferable to the amide groups, Flygare and Tigelaar calculate using McConnell's formula [154] that the expected shift caused by the helix to random coil transition in poly-(L-alanine) is slightly *upfield* rather than the observed downfield shift, thus supporting the solvation theory.

Rotational transitions of formamide have also been observed in emission in space recently.[155] Kirchoff and Johnson [156] have therefore carried out a very accurate centrifugal distortion analysis of transitions in the microwave spectrum of the molecule in order to be able to predict, with low uncertainty, frequencies of transitions of further astrophysical interest. They critically assess the existing microwave data and measure 22 new transitions in the ground state of the $^{14}NH_2{}^{12}CH^{16}O$ species. The centrifugal distortion constants of the closely related molecule carbamyl fluoride (NH_2CFO) have also been evaluated recently.[157]

Conformational problems in some carboxylic acids are receiving attention. Van Eijck, van der Plaats, and van Roon [158] report the molecular structure of two conformers of monofluoroacetic acid in a long and distinguished paper. The *trans*-conformer (20) gives rise to a strong microwave spectrum whereas that arising from the *cis*-conformer (21) is much weaker. Invariance of the

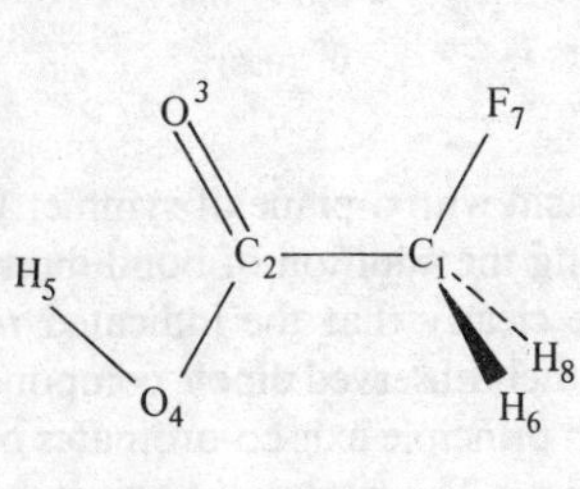

trans

(20)

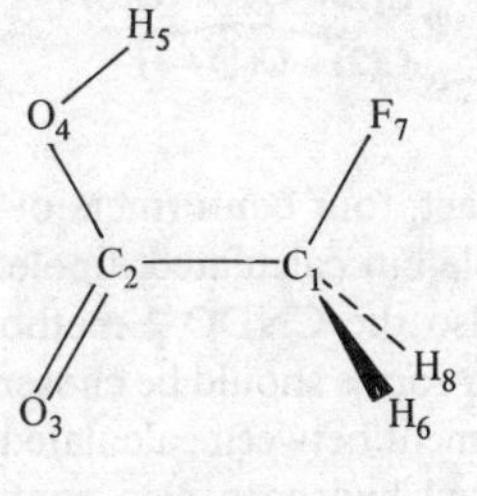

cis

(21)

[154] H. M. McConnell, *J. Chem. Phys.*, 1957, **27**, 226.
[155] R. H. Rubin, G. W. Swenson, jun., R. C. Benson, H. L. Tigelaar, and W. H. Flygare, *Astrophys. J.*, 1971, **169**, L39.
[156] W. H. Kirchoff and D. R. Johnson, *J. Mol. Spectroscopy*, 1973, **45**, 159.
[157] S. P. Srivastava and S. L. Srivastava, *Indian J. Pure Appl. Phys.*, 1971, **9**, 849.
[158] B. P. van Eijck, G. van der Plaats, and P. H. van Roon, *J. Mol. Structure*, 1972, **11**, 67.

quantity $I_a + I_b - I_c$ to isotopic substitution of the hydroxyl hydrogen and the oxygen atoms clearly establishes that both have a plane of symmetry. For both isomers, satellites associated with the torsional vibration and an in-plane skeletal mode where identified. The smooth variation of the rotational constants of the *trans*-form with the torsional quantum number indicates that the potential function is essentially harmonic in the lower part and relative intensity measurements indicate the equal spacing to be 104 cm^{-1} for the *trans*-form and 112 cm^{-1} for the *cis*-form. Stark-effect measurements in the *cis*- and *trans*-forms lead to the dipole moments $\mu_a = 1.582 \pm 0.004$, $\mu_b = 2.434 \pm 0.007$, and $\mu = 2.903 \pm 0.007$ D for *trans* and 2.739 ± 0.007, 0.426 ± 0.010, and 2.772 ± 0.008 D for the corresponding quantities of the *cis*-isomer.

Table 29 *Molecular parameters for fluoroacetic acid* [H′ *is projection of* H(6) *and* H(8) *on the molecular plane*]

Bond lengths/Å	trans	cis-trans
C(2)=C(3)	1.202 ± 0.004	0 (ass)
C(2)—O(4)	1.344 ± 0.010	0 (ass)
C(1)—C(2)	1.534 ± 0.008	0 (ass)
C(1)—F(7)	1.387 ± 0.010	0 (ass)
O(4)—H(5)	0.971 ± 0.004	−0.043 ± 0.008
C(1)—H(6, 8)	1.073	0 (ass)
Angles/°		
$\widehat{C(1)—C(2)=O(3)}$	126.1 ± 0.7	−6.3 ± 0.7
$\widehat{C(1)—C(2)—O(4)}$	108.8 ± 0.4	+6.3 ± 0.7
$\widehat{C(2)—C(1)—C(7)}$	109.2 ± 1.0	+2.5 ± 0.4
$\widehat{C(2)—O(4)—H(5)}$	105.9 ± 0.5	+3.0 ± 0.8
$\widehat{C(2)—C(1)—H'}$	125° (ass)	0° (ass)

In fact, four conformers of fluoroacetic acid with a plane of symmetry are possible but calculated dipole moments using the addition of bond moments and also the CNDO/2 method show quite clearly that the indicated *trans*- and *cis*-forms should be chosen by virtue of their observed dipole components. Agreement between calculated and observed principle axis co-ordinates of the hydroxyl hydrogen also confirms this choice. The molecular structures of both isomers are determined by a combination of microwave spectral data with electron-diffraction data. Each bond length, except *r*(O—H), is taken to be the same in both conformations and some assumptions about the CH_2-group parameters are made. The resulting structures are recorded in Table 29. The energy difference between the *cis*- and *trans*-forms from relative intensities in the microwave spectra is estimated as 604 ± 28 cal mol^{-1}, corresponding to $70 \pm 2\%$ of the *trans*-form at 385 K. A potential energy function based on a simple model for *cis*–*trans* conversion and fitted to the estimated torsional frequencies of each form and to the *cis–trans* energy difference gives

$$V_1 = 1155 \pm 180 \text{ cal mol}^{-1},\ V_2 = 5300 \pm 350 \text{ cal mol}^{-1}, \text{ and}$$
$$V_3 = -580 \pm 180 \text{ cal mol}^{-1}$$

for the first three terms in the Fourier expansion. The energy barrier separating the two forms based on the three-term function is 5.7 kcal mol^{-1}.

Glyoxylic acid has three likely conformations in which the heavy atoms are coplanar, *viz. trans*-1 (22), *trans*-2 (23), and *cis* (24). The microwave spectrum of one conformer has been detected by Marstokk and Møllendal.[159]

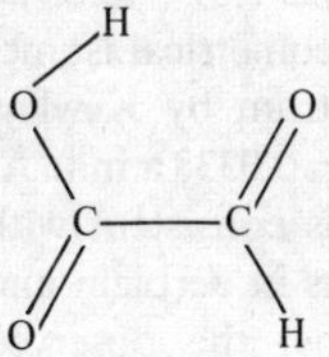

trans - 1
(22)

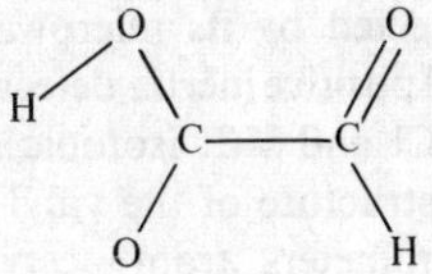

trans - 2
(23)

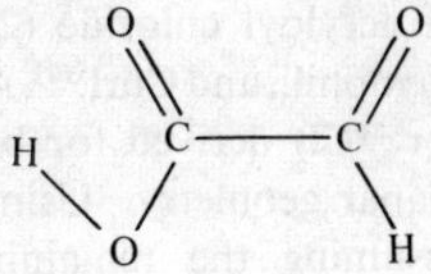

cis
(24)

The observed rotational constants agree closely with those of the *trans*-forms calculated from structural parameters transferred from oxalic acid and glyoxal. The observed inertia defect $\Delta = 0.075\,78$ a.m.u. Å^2 is suggestive of a planar arrangement of the atoms, and the smooth variation of rotational constants derived from satellites assigned to successive states of the C—C torsional mode confirms a planar equilibrium configuration. Relative intensity measurements give the $1 \leftarrow 0$ separation in the torsional mode as 167 ± 12 cm^{-1}, a value to be compared with 122 cm^{-1} from an i.r. combination band. The dipole moment components of the observed conformer are $\mu_a = 1.85 \pm 0.03$, $\mu_b = 0.20 \pm 0.10$, and $\mu = 1.86 \pm 0.04$ D. The isotopic species OHC—CO_2D was studied in order to differentiate between the two *trans*-forms. The principal axis co-ordinates of the carboxyl hydrogen atom point unequivocally to the *trans*-1 form, presumably stabilized by an intramolecular hydrogen bond [see (22)] as well as conjugation.

Propiolic acid (25) is a planar molecule as indicated by the inertia defects $\Delta = 0.2430, 0.2369$, and 0.2496 a.m.u. Å^2 of the isotopic species HC≡CCO_2H, HC≡CCO_2D, and DC≡CCO_2H derived from a study of their microwave

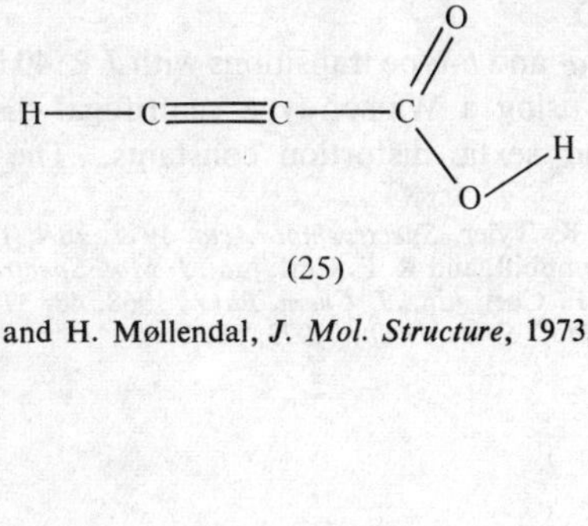

(25)

[159] K.-M. Marstokk and H. Møllendal, *J. Mol. Structure*, 1973, **15**, 137.

spectra by Lister and Tyler.[160] The principal axis co-ordinates of the hydrogen atoms derived from the isotopic moments of inertia are in excellent agreement with those calculated from an assumed structure based on the accurately determined microwave structures of propynal and formic acid and show without doubt that the hydroxyl proton adopts the *cis* configuration with respect to the carbonyl group, as in formic acid. The dipole moment components are measured from the $1_{11} \leftarrow 0_{00}$ and $3_{21} \leftarrow 3_{12}$ transitions of $HC{\equiv}CCO_2H$ and the $2_{12} \leftarrow 1_{01}$ and $3_{21} \leftarrow 3_{12}$ transitions of $HC{\equiv}CCO_2D$. The values are $\mu_a = 0.80 \pm 0.02$ and $\mu_b = 1.38 \pm 0.02$ D for the former species and $\mu_a = 0.76 \pm 0.02$ and $\mu_b = 1.40 \pm 0.02$ D for the latter; μ_{total} is 1.59 ± 0.03 D.

A planar *s-trans*-form has been established for the one geometrical isomer of acryloyl chloride (26) detected by its microwave spectrum by Kewley, Hemphill, and Curl.[161] A small positive inertia defect (*e.g.* $\Delta = 0.0333$ a.m.u. Å^2 for ^{35}Cl) derived for both ^{35}Cl and ^{37}Cl isotopic species is consistent with planar geometry. Using the structure of the vinyl group as in acrolein and assuming the remaining parameters from acetyl fluoride, the observed rotational constants demonstrably correspond with those calculated for the *s-trans*-form, Unlike acryloyl fluoride,[162] the *s-cis*-form is as yet unidentified.

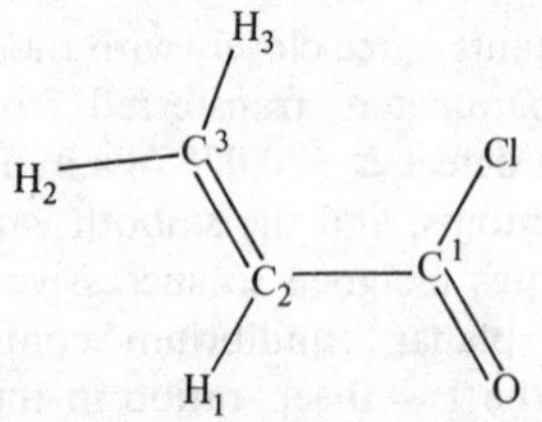

(26)

A partial structure is determined by allowing four of the assumed parameters to vary until a best fit to the rotational constants of the ^{35}Cl and ^{37}Cl species is achieved. The final parameters, r[C(2)—C(1)] = 1.476 Å, r[C(1)—Cl] = 1.816 Å, $\widehat{C(3){=}C(2){-}C(1)} = 122.6°$, and $\widehat{C(2){-}C(1){-}Cl} = 116.3°$, do not differ greatly from those assumed initially. Relative intensity measurements lead to 95 cm^{-1} for the $1 \leftarrow 0$ separation associated with the C—C torsion, a value in excellent agreement with that estimated from Δ (94 cm^{-1}) but, as expected, lower than that of acryloyl fluoride[162] (115 cm^{-1}). Dipole moment components are $\mu_a = 2.76 \pm 0.02$, $\mu_b = 1.45 \pm 0.03$, and $\mu_{total} = 3.12 \pm 0.04$ D.

A large number of *a*- and *b*-type transitions with $J \leqslant 40$ in vinyl fluoride[163] have been analysed using a Watson-type centrifugal distortion treatment including quartic and sextic distortion constants. The observed quartic

[160] D. G. Lister and J. K. Tyler, *Spectrochim. Acta*, 1972, **28A,** 1423.
[161] R. Kewley, D. C. Hemphill, and R. F. Curl, jun., *J. Mol. Spectroscopy*, 1972, **44,** 443.
[162] J. J. Keirns and R. F. Curl, jun., *J. Chem. Phys.*, 1968, **48,** 3773.
[163] M. C. L. Gerry, *J. Mol. Spectroscopy*, 1973, **45,** 71.

coefficients agree quite well with those calculated from a valence force field of vinyl fluoride published previously while the observed inertia defect ($\Delta = 0.0894$ a.m.u. Å^2) is in moderate agreement with the value $\Delta = 0.0592$ a.m.u. Å^2 calculated on the basis of a Urey–Bradley force field.[164] A preliminary report of the microwave spectrum of the four-atom cumulene chlorobutatriene [165] shows the rotational constants to be entirely consistent with a planar model in which the bond lengths are transferred from an electron-diffraction study of butatriene and a microwave study of vinyl chloride. Rotational constants for 1,1,2-trifluoropropylene are also reported.[166]

Three halogenated acetylenes have come under investigation. Allen and Flygare [167] examined the high-field molecular Zeeman effect of the $J = 1 \leftarrow 0$ transitions of the ^{35}Cl and ^{37}Cl species of chloroacetylene to give the molecular g-values, $^{35}g_\perp = -0.00630 \pm 0.00014$ and $^{37}g_\perp = -0.00601 \pm 0.00008$, the magnetic susceptibility anisotropy $(\chi_\parallel - \chi_\perp) = (9.3 \pm 0.5) \times 10^{-13}$ J G^{-2} mol^{-1}, and the molecular quadrupole moment $Q_\parallel = (8.8 \pm 0.4) \times 10^{-26}$ e.s.u. cm^2. The $g_\perp$-values of the ^{35}Cl and ^{37}Cl isotopes and the shift of centre of mass between these two isotopic species are combined to give a magnitude 0.3 ± 0.2 D for the electric dipole moment (*cf.* 0.44 D from the Stark effect [168]) and an unambiguous direction $-ClC_2H+$. A similar study of bromoacetylene has been made.[23] As a result of their work on 1-chloro-3,3,3-trifluoropropyne (27), Bjørseth and Marstokk [169]

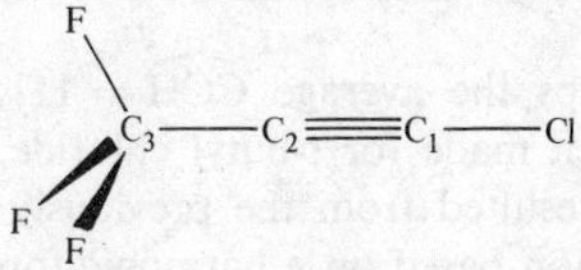

(27)

present a profusion of parameters for this molecule. For the most abundant species, $^{35}ClC{\equiv}C{-}CF_3$, the chlorine nuclear quadrupole coupling constant is $eqQ = -79.4 \pm 3$ MHz and the α-values for three vibrational satellite series are $\alpha_1 = -1.740 \pm 0.007$, $\alpha_2 = -1.05 \pm 0.02$, and $\alpha_3 = -0.71 \pm 0.02$ MHz. For the mode ν_1 five quanta are observed and from the l-doubling in the $\nu_1 = 1$ state, the l-type doubling constant is $q_1 = 0.586 \pm 0.004$ MHz, which leads to $\nu_1 = 97 \pm 6$ cm^{-1} from the familiar expression $q = B_e^2 a'/\omega$. On the other hand, $\nu_2 = 300 \pm 50$ cm^{-1} and $\nu_3 = 380 \pm 60$ cm^{-1}, according to relative intensity measurements. In addition, rotational constants for the isotopic species

$$CF_3{-}C{\equiv}C{-}^{37}Cl,\quad ^{13}CF_3{-}C{\equiv}C{-}^{37}Cl,\ \text{and}\ CF_3{-}C{\equiv}^{13}C{-}^{37}Cl$$

164 S. Jeyapandian and G. A. Savari Raj, *J. Mol. Structure*, 1972, **14**, 27.
165 F. Karlsson, R. Vestin, and A. Borg, *Acta Chem. Scand.*, 1972, **26**, 3390.
166 I. A. Mukhtarov and V. A. Kuliev, *Izvest. Akad. Nauk Azerb. S.S.R. Ser. fiz.-mat. tekh. Nauk*, 1971, 27.
167 W. C. Allen and W. H. Flygare, *Chem. Phys. Letters*, 1972, **15**, 461.
168 J. K. Tyler and J. Sheridan, *Trans. Faraday Soc.*, 1963, **59**, 2661.
169 A. Bjørseth and K.-M. Marstokk, *J. Mol. Structure*, 1972, **13**, 191.

allow the r_s parameters r_s[C(1)—Cl] = 1.629 ± 0.006 Å and r_s[C(1)···C(3)] = 2.647 ± 0.004 Å to be derived. The latter value is short compared with those in Me_3C—C≡C—Cl and Me—C≡C—Cl and indicates that either C(3)—C(2) or C(2)—C(1) (or both) is short. Alternatively, given the assumption of $\widehat{FCF}$ = 107.0 ± 1.0° (as in CF_3—C≡C—Me), the following r_0 structure can be determined by fitting all the rotational constants:

$$r_0[\mathrm{C(1)—Cl}] = 1.627 \pm 0.009,\ r_0[\mathrm{C(1)}{\equiv}\mathrm{C(2)}] = 1.199 \pm 0.005,$$
$$r_0[\mathrm{C(2)—C(3)}] = 1.453 \pm 0.002,\text{ and } r_0[\mathrm{C(3)—F}] = 1.336 \pm 0.006\ \text{Å}.$$

Stark-effect measurements lead to μ = 1.73 ± 0.20 D, which is reproduced within experimental error by taking the difference of the moments of CF_3—C≡C—H (2.36 D) [170] and H—C≡C—Cl (0.44 D).[168]

Acyclic Molecules with Internal Rotation.—*Hydrocarbons and Halogeno-derivatives*. A zero-point average structure for isobutane, $(CH_3)_3CH$, has been obtained by combining new electron-diffraction data with microwave spectroscopic rotational constants obtained previously.[171] The results show that the tertiary C—H distance (r_g = 1.122 ± 0.006 Å) is substantially longer than the average methyl C—H distance (r_g = 1.113 ± 0.002 Å). Other structural parameters are:

$$r_g(\mathrm{C—C}) = 1.535 \pm 0.001\ \text{Å and } \widehat{\mathrm{CCC}} = 110.8 \pm 0.2°$$

and for the methyl groups the average $\widehat{CCH}$ = 111.4 ± 0.2°. A similar investigation has also been made for t-butyl chloride, $(CH_3)_3CCl$. In this case an r_z structure has resulted from the previously obtained microwave constants using a calculation based on a harmonic force field.[172] The parameters calculated in this way are in excellent agreement with those obtained directly from electron-diffraction work.

A detailed study [173] has been made of rotational isomerism in pent-1-yne, $CH_3CH_2CH_2C$≡CH. The spectrum can be attributed to two rotational isomers, one having the methyl group *trans* with respect to the acetylene group, as shown in Figure 11, and the other the *gauche* conformation. The latter, which has a dihedral angle of 115° ± 3° from the *trans* position, is slightly more stable than the *trans*-form, its ground state falling 27 ± 36 cm^{-1} lower. The potential barrier between the two wells is estimated to be 3.0 ± 1.3 kcal mol^{-1} from a potential function derived from the torsional frequencies found for both *trans*- and *gauche*-forms. Some plausible structural parameters are put forward and the dipole moments are found to be μ_{trans} = 0.842 ± 0.010 and μ_{gauche} = 0.769 ± 0.028 D. The microwave

[170] J. N. Shoolery, R. G. Shulman, W. F. Sheehan, jun., V. Schomaker, and D. M. Yost, *J. Chem. Phys.*, 1951, **19**, 1364.
[171] R. L. Hilderbrandt and J. D. Wieser, *J. Mol. Structure*, 1972, **15**, 27.
[172] R. L. Hilderbrandt and J. D. Wieser, *J. Chem. Phys.*, 1972, **56**, 1143.
[173] F. J. Wodarczyk and E. B. Wilson, *J. Chem. Phys.*, 1972, **56**, 166.

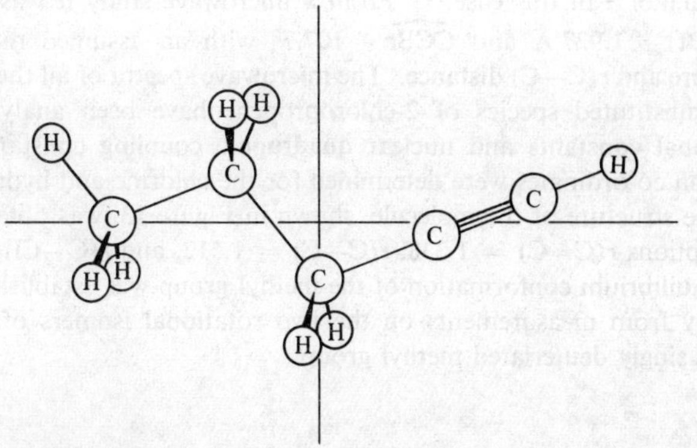

Figure 11 trans-*Pent*-1-*yne*

spectrum of the related molecule *trans*-pent-3-en-1-yne shown in Figure 12 has been analysed.[174] The dipole moment is found to be 1.06 $\pm$ 0.05 D and the barrier to internal rotation of the methyl group is 1.903 $\pm$ 0.050 kcal mol^{-1}.

New information has been obtained about both 1,1,1-trichloroethane, CH_3CCl_3,[175] and 1,1,1-tribromoethane, CH_3CBr_3.[176] From the relative intensities of satellite lines, torsional-mode frequencies of 276 and 199 cm^{-1} were found for the CH_3CCl_3 and CD_3CCl_3 molecules respectively. From these frequencies the barrier to internal rotation for the chloride was calculated to be 5.1 kcal mol^{-1}. For CH_3CBr_3 a torsional frequency of 304 cm^{-1} has been obtained from an i.r. and Raman study of the solid, leading to a barrier of

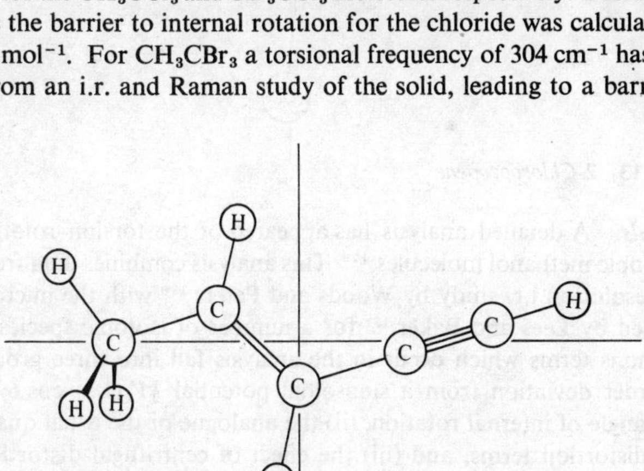

Figure 12 trans-*Pent*-3-*en*-1-*yne*

[174] R. G. Ford and L. B. Szalanski, *J. Mol. Spectroscopy*, 1972, **42**, 344.
[175] J. R. Durig, M. M. Chen, and Y. S. Li, *J. Mol. Structure*, 1972, **15**, 37.
[176] Y. S. Li, K. L. Kizer, and J. R. Durig, *J. Mol. Spectroscopy*, 1972, **42**, 430.

6.0 kcal mol^{-1} in this case.[177] From a microwave study it was found that r(C—Br) = 1.927 Å and $\widehat{CCBr}$ = 107.7° with an assumed methyl group structure and r(C—C) distance. The microwave spectra of all the deuterium-monosubstituted species of 2-chloropropene have been analysed to give rotational constants and nuclear quadrupole coupling constants.[178] Substitution co-ordinates were determined for the chlorine and hydrogen atoms and the structure of the molecule, shown in Figure 13, was obtained on the assumptions r(C=C) = 1.338, r(C—C) = 1.512, and r(C—Cl) = 1.735 Å. The equilibrium conformation of the methyl group was established unambiguously from measurements on the two rotational isomers of the species with a singly deuteriated methyl group.

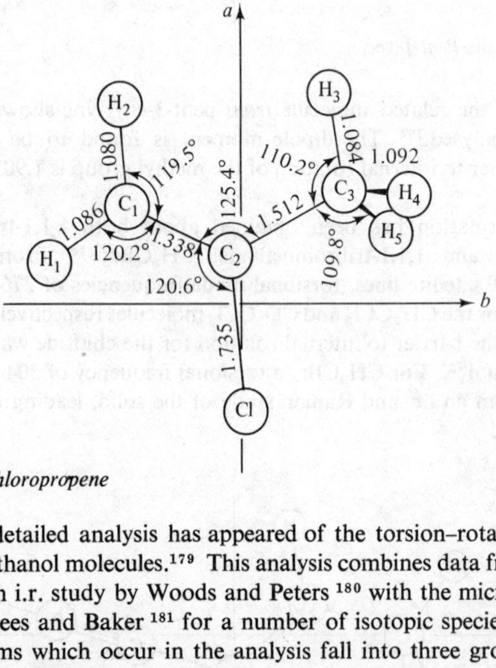

Figure 13 2-*Chloropropene*

Alcohols. A detailed analysis has appeared of the torsion–rotation spectra of isotopic methanol molecules.[179] This analysis combines data from a recent high-resolution i.r. study by Woods and Peters [180] with the microwave data obtained by Lees and Baker [181] for a number of isotopic species. The perturbations terms which occur in the analysis fall into three groups: (i) the first-order deviation from a sinusoidal potential $\frac{1}{2}V_6(1 - \cos 6\gamma)$, where γ is the angle of internal rotation, (ii) the analogue of the usual quartic centrifugal distortion terms, and (iii) the effect of centrifugal distortions on the

177 J. R. Durig, S. M. Craven, C. W. Hawley, and J. Bragen, J. *Chem. Phys.*, 1972, **57**, 131.
178 W. Good, R. J. Conan, A. Bauder, and Hs. H. Günthard, *J. Mol. Spectroscopy*, 1972, **41**, 381.
179 Y. Y. Kwan and D. M. Dennison, *J. Mol. Spectroscopy*, 1972, **43**, 291.
180 D. R. Woods and C. W. Peters, quoted by Y. Y. Kwan and D. M. Dennison in ref. 179.
181 R. M. Lees and J. G. Baker, *J. Chem. Phys.*, 1968, **48**, 5299.

effective height of the potential barrier. The evaluation of the perturbation constants is dependent on information relating to excited states of the torsional oscillation, and for the species CD_3OD Lees has recently reported frequencies for the $J = 0 \rightarrow 1$ transition up to the fourth excited torsional state.[182] The result of the analysis by Kwan and Dennison [179] is a set of ten constants for each of four isotopic species. The interrelation of these constants has been examined and a set of twenty-one constants has been derived which provide a good fit for the data on the four isotopic species. An interesting feature to emerge from the analysis is that the torsional barrier is highly sinusoidal in form, with $V_6/V_3 = -0.0002 \pm 0.0006$. Lees has also recently determined the barrier height for two excited vibrational states of methanol. For the first excited state of the C—O stretching mode the barrier height is found to be 397.9 ± 3 cm^{-1}.[183] For an excited state at 933 ± 150 cm^{-1}, which is most probably the CH_3 in-plane rocking mode, $V_3 = 557 \pm 35$ cm^{-1}, an increase of 50% above the ground-state value of 376 cm^{-1}.[184] Centrifugal distortion constants have also been reported for four isotopic species of methanol,[185] and spin–rotation and spin–spin coupling constants have been obtained by beam maser spectroscopy of three transitions for CH_3OH.[186]

A further investigation [187] has been made of the spectrum of the *gauche*-rotamer of ethanol. In addition to *a*-type lines, some *c*-type lines have also been observed. The latter lead to a direct determination of the tunnelling frequency between the equivalent *gauche* configurations, and the barrier hindering internal rotation has been calculated on the assumption that the threefold term is dominant, giving $V_3 = 420.8$ cm^{-1}. Rotational isomerism in propanol has been further discussed [188] and frequencies and relative intensities of lines have been reported for t-butyl alcohol.[189]

Other Molecules. The determination of zero-point average structures of molecules which contain a symmetric internal rotor has been examined, and it is shown how microwave and electron-diffraction data may be combined for this purpose.[190] Application of the method has been made to acetaldehyde and the zero-point average structure has been determined. Similar treatments have been applied to acetone,[191] acetyl chloride,[192] and acetyl bromide.[192] The r_z parameters for the three molecules are collated in Table 30.

182 R. M. Lees, *J. Chem. Phys.*, 1972, **56**, 5887.
183 R. M. Lees, *J. Chem. Phys.*, 1972, **57**, 2249.
184 R. M. Lees, *J. Chem. Phys.*, 1972, **57**, 824.
185 I. Y. Zemlyanukhina and L. M. Sverdlov, *Optics and Spectroscopy*, 1972, **32**, 433.
186 J. E. M. Heuval and A. Dymanus, *J. Mol. Spectroscopy*, 1973, **45**, 282.
187 R. K. Kakar and P. J. Seibt, *J. Chem. Phys.*, 1972, **57**, 4060.
188 A. A. Abdurakhmanov, R. A. Ragimova, and L. M. Imanov, *Izvest. Akad. Nauk Azerb. S.S.R., fiz.-mat. tekh. Nauk*, 1971, 9; *ibid.*, 1972, 3.
189 A. A. Abdurakhmanov, E. I. Veliyalin and L. M. Imanov, *Zhur. strukt. Khim.*, 1972, **13**, 25; C. O. Kadzhar, G. A. Abdullaev, and L. M. Imanov, *Izvest. Akad. Nauk Azerb. S.S.R., fiz.-mat. tekh. Nauk*, 1971, 46.
190 T. Iijima and S. Tsuchiya, *J. Mol. Spectroscopy*, 1972, **44**, 88.
191 T. Iijima, *Bull. Chem. Soc. Japan*, 1972, **45**, 3526.
192 S. Tsuchiya and T. Iijima, *J. Mol. Structure*, 1972, **13**, 327.

Table 30 *Structural parameters for* CH_3CXO

CH_3CXO	$r(C—C)$/Å	$r(C{=}O)$/Å	$r(C—X)$/Å	$\widehat{OCX}$/°	$\widehat{CCX}$/°	$\widehat{CHC}$/°
CH_3CHO	1.512 ± 0.004	1.207 ± 0.004	1.114 ± 0.009	120.3 ± 0.5	115.3 ± 0.3	109.6 ± 0.9
CH_3COCl	1.505 ± 0.003	1.185 ± 0.003	1.796 ± 0.002	121.2 ± 0.6	111.6 ± 0.6	108.8 ± 0.8
CH_3COBr	1.516 ± 0.003	1.181 ± 0.003	1.974 ± 0.003	122.3 ± 1.5	110.1 ± 1.5	109.9 ± 0.1
CH_3COCH_3	1.517 ± 0.003	1.210 ± 0.003	1.517 ± 0.003	122.0 ± 0.3	116.0 ± 0.3	108.5 ± 0.5

The methyl group is tilted towards the oxygen atom by 1.3 ± 1.0° for the chloride, by 1.9 ± 1.0° for the bromide, and by 2.0 ± 0.5° in acetone. The barriers V_3 to internal rotation have been revised to 1.260 and 1.256 kcal mol^{-1} for the chloride and bromide respectively. The structure and dipole moment of dimethyl sulphone, $(CH_3)_2SO_2$, have been determined.[193] With the assumption that the methyl groups have the same structure as in dimethyl sulphide the parameters given in Table 31 have been obtained. The value of the dipole moment is found to be 4.432 ± 0.041 D.

Table 31 *Structural parameters for* $(CH_3)_2SO_2$

	Bond lengths/Å		*Angles*
r(C—S)	1.777 ± 0.006	$\widehat{CSC}$	103°17′ ± 10′
r(S—O)	1.431 ± 0.004	$\widehat{OSO}$	121° 1′ ± 15′

The microwave spectrum of a molecule having two non-equivalent methyl groups, namely *trans-N*-methylethylideneimine, $CH_3CH{=}NCH_3$, has been analysed in detail. The theory for overall and internal rotation has been extended to cover molecules with two non-equivalent symmetric internal rotors and internal rotation splittings have been calculated using the internal-axis method (IAM).[194] The following values are reported for the methyl-group barriers: C—CH_3, 1.642 ± 0.019, and N—CH_3, 2.109 ± 0.084 kcal mol^{-1}. The molecular dipole moment is found to be 1.499 ± 0.007 D and nitrogen nuclear quadrupole coupling coefficients are also reported. The barrier to internal rotation in dimethyl ether has been obtained as 2.50 kcal mol^{-1} by treating CH_3OCD_3 as single-top problem.[195] The inertial defect of dimethylnitrosamine shows the nuclear framework to be planar, in agreement with earlier electron-diffraction results, and the dipole moment is found to be 4.22 ± 0.02 D.[196]

The microwave spectrum of trifluoro(trifluoromethyl)silane, CF_3SiF_3, has been analysed and the barrier to internal rotation found to be 489 ± 50 cm^{-1}, corresponding to a torsional frequency of 37.0 ± 2.0 cm^{-1}.[197] The theory of coupling between internal rotation and a degenerate vibration in a symmetric rotor has been re-examined and revised expressions have been obtained for *l*-type doubling.[198] The treatment has been used to refine the analysis of the spectrum of methylsilane, CH_3SiH_3.

Experimental Evaluation and Theoretical Calculation of Barriers. A number of papers have appeared which deal with the analysis of spectra of molecules for which there are interactions between internal and overall rotation. The

[193] S. Saito and F. Makino, *Bull. Chem. Soc. Japan*, 1972, **45**, 92.
[194] J. Meier, A. Bauder, and Hs. H. Günthard, *J. Chem. Phys.*, 1972, **57**, 1219.
[195] J. R. Durig and Y. S. Li, *J. Mol. Structure*, 1972, **13**, 459.
[196] F. Scappini, A. Guarnieri, H. Dreizler, and R. Rademacher, *Z. Naturforsch.*, 1972, **27a**, 1329.
[197] D. R. Lide, D. R. Johnson, K. G. Sharp, and T. D. Coyle, *J. Chem. Phys.*, 1972, **57**, 3699.
[198] E. Hirota, *J. Mol. Spectroscopy*, 1972, **43**, 36.

evaluation of average structures for such molecules,[190, 191] the treatment of molecules involving two non-equivalent tops,[194] and the coupling of internal rotation and a degenerate vibration have already been mentioned.[198] Internal rotation–quadrupole interaction in asymmetric rotors has also been treated and typical effects have been discussed for the low- and high-barrier cases.[199]

An improved method has been developed for determining barriers to internal rotation from relative intensity measurements as a function of temperature.[200] The technique, which is discussed in more detail in the section on instrumentation, uses a special Stark absorption cell and employs the antimodulation method of measuring intensities. The method has been applied to a range of molecules where barriers are well known and an accuracy of 3—5% has been achieved.

Among the theoretical calculations of barrier heights there is an application of molecular orbital theory to a wide range of molecules of spectroscopic interest.[201] Another paper examines the adequacy of molecular orbital theory for predicting barriers.[202] Particular molecules which have been examined theoretically include ethane,[203] ethyl fluoride,[204] formamide,[205] methyl vinyl ether,[206] 3-fluoropropene,[206] acetaldehyde, and some of its chloro- and fluoro-derivatives.[207] An atom–atom approach has been made for calculating torsional oscillations in ethane derivatives,[208] and a review has appeared of theoretical approaches to the calculation of the barrier in ethane.[209]

5 Cyclic Molecules

Three-membered Rings.—Schwendeman has been investigating several carbonyl derivatives of cyclopropane. In papers concerned with cyclopropanecarbaldehyde[210] and cyclopropanecarbonyl fluoride[211] (reviewed last year) the torsional potential functions governing the twofold rotation of the planar carbonyl grouping with respect to the ring were derived and interpreted. Schwendeman and Lee[212] have now extended these studies to cyclopropyl methyl ketone (28) in an attempt to increase understanding of such potentials. Assignment difficulties in the very rich spectrum were overcome by the identification of characteristic *E,A* doublets arising from methyl-group internal rotation in the $J_{0,J} - J_{1,J-1}$ series and by double-resonance experi-

199 M. Ribeaud, A. Bauder, and Hs. H. Günthard, *J. Mol. Spectroscopy*, 1972, **42**, 441.
200 E. Ruitenberg, *J. Mol. Spectroscopy*, 1972, **42**, 161.
201 L. Radom, W. J. Hehre, and J. A. Pople, *J. Amer. Chem. Soc.*, 1972, **94**, 2371.
202 L. C. Allen and J. Arents, *J. Chem. Phys.*, 1972, **57**, 1818.
203 U. Wahlgreen and K. H. Johnson, *J. Chem. Phys.*, 1972, **56**, 3715.
204 W. E. Palke, *Chem. Phys. Letters*, 1972, **15**, 244.
205 P. N. Skancke and I. Aanesland, *Acta Chem. Scand.*, 1972, **26**, 2614.
206 B. Cadioli and U. Pincelli, *J. C. S. Faraday II*, 1972, **68**, 991.
207 H. Møllendal, *Acta Chem. Scand.*, 1972, **26**, 3804.
208 M. I. Dakhis, V. G. Dashevskii, *Doklady Akad. Nauk S.S.S.R.* 1972, **203**, 369.
209 J. P. Lowe, *Science*, 1973, **179**, 527.
210 H. N. Volltrauer and R. H. Schwendeman, *J. Chem. Phys.*, 1971, **54**, 260.
211 H. N. Volltrauer and R. H. Schwendeman, *J. Chem. Phys.*, 1971, **54**, 268.
212 P. L. Lee and R. H. Schwendeman, *J. Mol. Spectroscopy*, 1972, **41**, 84.

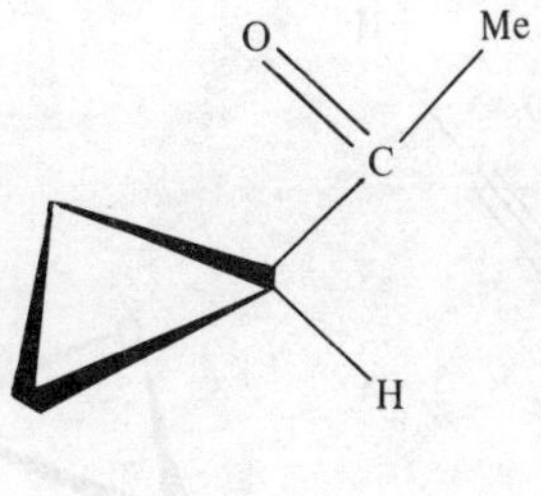

(28)

ments. Effective rigid-rotor constants for the *A*-species transitions are interpreted as belonging to the *cis*-form (28) rather than the *trans*-form (rotation of the $COCH_3$ group by 180° from *cis*) or an *anti-gauche*-form (rotation of the $COCH_3$ group by about 150° from *cis*) according to a reasonable model structure calculation. Indeed, the last form (suggested as a minor-component rotamer in an electron-diffraction study) is ruled out by establishing from $P_{cc} = (I_b + I_a - I_c)/2$ that the molecule has a symmetry plane. After subtracting out contributions from the two out of-plane hydrogens of the methyl group, P_{cc} becomes identical with those of the *cis*- and *trans*-forms of both cyclopropanecarbaldehyde and cyclopropanecarbonyl fluoride (P_{cc} = 19.75 — 19.80 a.m.u. Å²). The internal-rotation *E–A* splittings in the vibrational ground-state transitions lead to a barrier to internal rotation of the methyl group of $V_3 = 1180 \pm 20$ cal mol^{-1} when analysed according to the Herschbach formulation (see Vol. 1, p. 20 of these Reports for a discussion of this formulation) and the methyl-group moment of inertia is assumed as 3.12 a.m.u. Å². Such a barrier is characteristic of several CH_3COX compounds.

Unfortunately no other rotamer of cyclopropyl methyl ketone could be detected and hence the barrier hindering internal rotation to a *trans*- or an *anti-gauche*-form was not determined. However, rotational constants for the first three excited states of the CH_3CO-group torsional motion are a smoothly varying function of the torsional quantum number and hence indicate a fairly high barrier to the motion. The dipole moment components reported are $\mu_a = 0.47 \pm 0.17$, $\mu_b = 2.58 \pm 0.26$, and $\mu_{total} = 2.62 \pm 0.26$ D, with the vector associated with the latter value making an angle of ~1° with the CO axis.

Structure determinations in the isoelectronic compounds cyclopropylacetylene (29) and epoxybutyne (30) pose interesting questions about conjugation of the cyclopropyl ring and attached groups. According to the Walsh model of cyclopropane each carbon atom is sp^2-hybridized and the pure *p*-orbital on each carbon atom lies in the plane of the ring, able to overlap with similar *p*-orbitals on adjacent carbon atoms. Two of the three sp^2-hybridized orbitals on each carbon form bonds with hydrogen atoms above and below the plane of the ring whereas the third is directed toward the ring

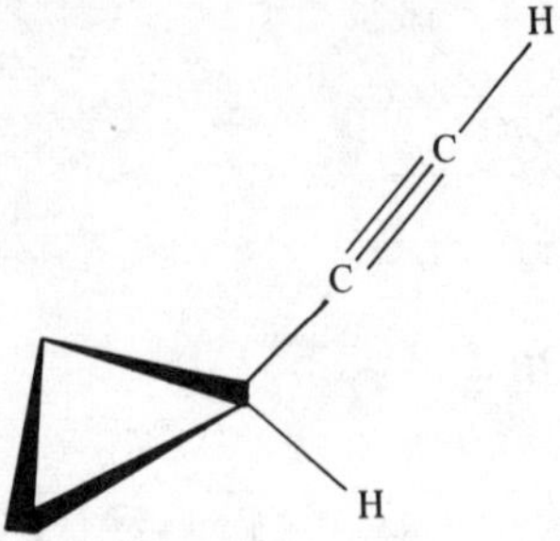

(29)

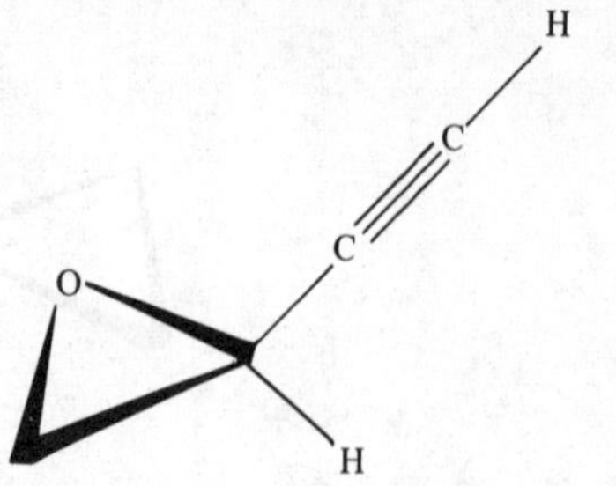

(30)

centre to form an intra-annular multicentre bond. Two types of conjugation of substituent groups with the ring are then postulated. First, a pure *p*-orbital in the ring plane can conjugate with a properly oriented *p*-orbital in a substituent atom; hence the rationalization of the bisected forms (*i.e. cis*, *trans*, or both) in cyclopropyl methyl ketone, cyclopropanecarbaldehyde, cyclopropanecarbonyl fluoride and others. Secondly, conjugation between excess electron density on a substituent and the intra-annular multicentre bond is postulated. The observed conformations and the shortenings of C—X bonds in ▷—X, where X = Cl, Br, N(H_2), or P(H_2), relative to CH_3X compounds have been so rationalized.[213]

With the study of the microwave spectra of (29) ($C_3H_5C{\equiv}CH$ and $C_3H_5C{\equiv}CD$ species) by Collins, Britt, and Boggs[214] and of (30) ($C_2H_3OC{\equiv}CH$ and $C_2H_3OC{\equiv}CD$ species) by Collins and Boggs[215] there is evidence that the above rationalizations are not without exception. In each compound the moments of inertia for both species have been combined with some reasonable structural assumptions to calculate the molecular parameters of most interest. The main conclusion is that $r(\mathrm{C{-}C}{\equiv})$ $= 1.466 \pm 0.018$ and 1.484 ± 0.02 Å for cyclopropylacetylene and epoxybutyne respectively are *longer* than in the corresponding methyl analogue ($CH_3C{\equiv}CH$, 1.458 Å), in contrast to the series ▷—X, where X = Cl, Br, N(H_2), or P(H_2), in which r(C—X) is 0.041, 0.04, 0.046,[216] and 0.029 Å[213] *shorter* than in the corresponding CH_3X. It is interesting that cyclopropyl cyanide[217] [isoelectronic with both (29) and (30)] also has r(C—CN) lengthened by 0.01 Å relative to methyl cyanide. The dipole moment components are $\mu_a = 0.891 \pm 0.01$, $\mu_c = 0.048 \pm 0.01$, and $\mu_{\mathrm{total}} = 0.892 \pm 0.01$ D for cyclopropylacetylene and $\mu_a = 0.316$, $\mu_b = 1.656$, $\mu_c = 0.618$; and μ_{total} $= 1.795$ D for epoxybutyne.

[213] L. A. Dinsmore, C. O. Britt, and J. E. Boggs, *J. Chem. Phys.*, 1971, **54**, 915.
[214] M. J. Collins, C. O. Britt, and J. E. Boggs, *J. Chem. Phys.*, 1972, **56**, 4262.
[215] M. J. Collins and J. E. Boggs, *J. Chem. Phys.*, 1972, **57**, 3811.
[216] D. K. Hendriksen and M. D. Harmony, *J. Chem. Phys.*, 1969, **51**, 700.
[217] J. P. Friend and B. P. Dailey, *J. Chem. Phys.*, 1958, **29**, 577.

An r_0 structure has been determined [218] for ethylene sulphone (31) from the observed ground-state moments of inertia of the $\overline{CH_2CH_2{}^{32}SO_2}$ and $\overline{CH_2CH_2{}^{34}SO_2}$ isotopic species. With the assumptions $r(C—H) = 1.08$ Å, $\widehat{HCH} = 116°$, and $\widehat{H_2CC} = 152°$ (from the microwave structure of ethylene

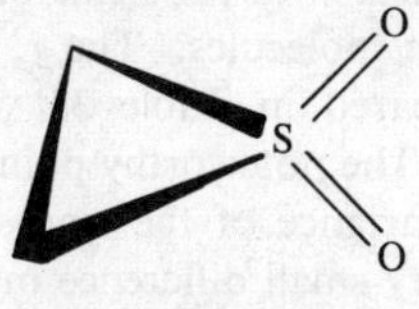

(31)

sulphide), the structure so derived is that displayed in Table 32, wherein it is compared with those of sulphur dioxide, ethylene sulphide, and ethylene sulphoxide. The errors quoted for the sulphone are generated from reasonable ranges of the assumed parameters. The striking feature of Table 32 is the progressive lengthening of the C—C bond from sulphide to sulphone. Normally C—C bonds in three-membered rings are shorter than in ethane, the only other exception being the C—C bond opposite the carbonyl in cyclopropanone [r(C—C) = 1.575 Å]. The dipole moment of ethylene sulphone from the Stark effect is 4.47 ± 0.02 D.

Table 32 *Structure of ethylene sulphone and related molecules*

Species	r(C—C)/Å	r(C—S)/Å	r(S—O)/Å	$\widehat{OSO}$°
SO_2			1.408	119
$(CH_2)_2S$	1.492	1.819		
$(CH_2)_2SO$	1.504	1.822	1.483	
$(CH_2)_2SO_2$	1.586 ± 0.01	1.76 ± 0.02	1.42 ± 0.02	124 ± 2

Much painstaking isotopic preparative and spectroscopic work has allowed Bak and Skaarup to derive very precise r_s structures for the closely related aziridine [32; X = H(1)] [219] and 1-chloroaziridine (32; X = Cl).[220] The

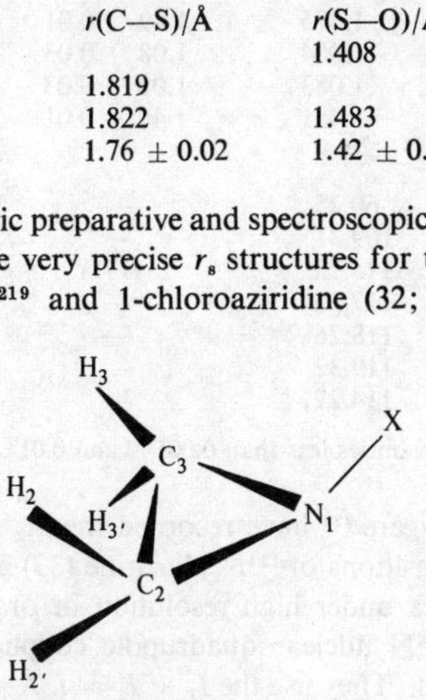

(32)

[218] H. Kim, *J. Chem. Phys.*, 1972, **57**, 1075.
[219] B. Bak and S. Skaarup, *J. Mol. Structure*, 1971, **10**, 385.
[220] B. Bak and S. Skaarup, *J. Mol. Structure*, 1972, **12**, 259.

spectra of the five different isotopically substituted species necessary for a complete r_s structure were observed in both molecules. In aziridine enrichment was carried out chemically in each case whereas the presence of chlorine made this necessary for only four isotopes in 1-chloroaziridine. Recourse to the pair of moment conditions $\Sigma m_i a_i = \Sigma m_i a_i c_i = 0$ (or their equivalent) was necessary in order to deal with problems of two small substitution co-ordinates occurring in both molecules. The r_s structures of aziridine and 1-chloroaziridine are compared in Table 33 with an electron-diffraction structure [221] of the former. The noteworthy points concerning the structures is the almost complete invariance of the ring structure to replacement of H(1) by Cl, except for a very small difference involving the C(2)—H(2) and C(2)—H(2′) distances of the 1-chloro-compound. A slight structural distortion is also noted and is almost identical for both compounds. The angle between the planes containing N(1)C(2)C(3) and H(2)C(2)H(2′) is near to 87° in both molecules, so that the H(2) and H(3) protons are tilted together and H(2′) and H(3′) are tilted apart. Moreover, the plane through H(2)C(2)H(2′) does not bisect the angle N(1)C(2)C(3) but divides it in the ratio 33.7° and 26.2° in aziridine and 37.6° and 22.5° in 1-chloroaziridine.

Table 33 *Structures of aziridine and 1-chloroaziridine*

	Aziridine [X=H(1)]		1-*chloroaziridine* (X=Cl)
	Microwave[a]	*Electron diffraction*	*Microwave*[a]
Distances/Å			
N(1)—X	1.016	1.05 ± 0.07	1.738_4
N(1)—C(2)	1.475	1.49 ± 0.01	1.488_7
C(2)—H(2)	1.084	1.08 ± 0.03	1.092_9
C(2)—H(2′)	1.083	1.08 ± 0.03	1.079_2
C(2)—C(3)	1.481	1.48 ± 0.01	1.483_9
Angles/°			
C(2)N(1)C(3)	60.25	—	59.78
XN(1)C(2)	109.31	—	111.70
H(2)C(2)H(2′)	115.72	117	117.11
H(2)C(2)C(3)	117.75	—	117.26
H(2)C(2)N(1)	118.26	—	115.18
H(2′)C(2)C(3)	119.32	—	120.38
H(2′)C(2)N(1)	114.27	—	113.66

[a] Experimental uncertainties less than 0.001 Å and 0.01°.

Pochan and Flygare [222] have recorded the $4_{22} \rightarrow 4_{23}$, $3_{22} \rightarrow 3_{21}$, $5_{33} \rightarrow 5_{32}$, and $7_{35} \rightarrow 7_{34}$ transitions of [$^{14}N_2$]diazirine (33) and the $0_{00} \rightarrow 1_{01}$ transition of the [^{13}C]-species under high resolution in order to obtain the diagonal elements of the ^{14}N nuclear quadrupole coupling tensor in the principal inertial axis system. They use the $I_1 + I_2 = I, I + J = F$ coupling scheme to analyse their results and obtain $\chi_{aa} = 0.196 \pm 0.150$, $\chi_{bb} = -3.010 \pm 0.150$,

[221] M. Igarashi, *Bull. Chem. Soc. Japan*, 1961, **34**, 369.
[222] J. M. Pochan and W. H. Flygare, *J. Phys. Chem.*, 1972, **76**, 2249.

and $\chi_{cc} = 2.814 \pm 0.300$ MHz. The $0_{00} \rightarrow 1_{01}$ transition of the [^{13}C]-species depends only on χ_{aa} and its unresolved character at very low pressure was used to rule out a set of coupling constants with large χ_{aa} which also initially fitted the data. The coupling constants are in fair agreement with the results of an *ab initio* calculation: $\chi_{aa} = -0.8$, $\chi_{bb} = -1.88$, and $\chi_{cc} = +2.69$ MHz.

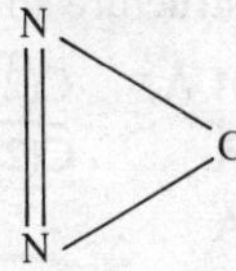

(33)

The authors attempt a simple localized description of the bonding in diazirine to compute the ^{14}N electric field gradients. Of the bonding schemes considered the best appears to be the Walsh picture using the *sp*-hybrid orbitals on the nitrogen atoms in the plane of the ring (one pointing away from the other N and containing a lone pair, the other entering into a C—N σ-bond) with p_z and p_y unhybridized (z is perpendicular to the ring plane).

Four-membered Rings.—The three experimental papers in this area published during the past year all have four-membered rings containing a carbonyl group and an oxygen atom as their subject.

Oxetan-3-one (34) is the latest in a series of four-membered rings studied by Harris and his co-workers.[223] Although Gibson and Harris have analysed the rotational spectrum of this molecule in the ground state and in the first five vibrationally excited states of the ring-bending mode to give rotational constants, they do not derive a potential function for the bending mode by the method of fitting the observed rotational constant variation by the familiar expansion in expectation values $\langle z^n \rangle_{vv}$ of the reduced vibrational co-ordinate z. Rather, they present arguments why the coefficient b of the quartic term in a potential function of the type $V(z) = h\nu_0(z^2 + bz^4)$ should not be well determined by such a procedure when, as in the present case, the dependence of the rotational constants on the ring-bending quantum number is only very slightly non-linear. Instead, they fit the constants and obtain the coefficients β_0, β_2, and β_4 in $B_v = \beta_0 + \beta_2\langle z^2 \rangle_{vv} + \beta_4\langle z^4 \rangle_{vv}$ by calculating the $\langle z^n \rangle_{vv}$ values using the potential function derived from the far-i.r. spectrum.

Observation of the [$^{13}C_1$] and [$^{13}C_2$] isotopically substituted species in natural

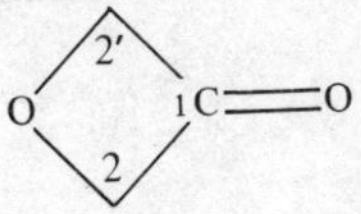

(34)

[223] J. S. Gibson and D. O. Harris, *J. Chem. Phys.*, 1972, **57**, 2318.

abundance allows the r_s co-ordinates of these atoms, and therefore the r_s parameters $r[\mathrm{C(1)—C(2)}] = 1.522 \pm 0.002$ Å and $\widehat{\mathrm{C(2)C(1)C(2')}} = 88.06° \pm 0.15°$, to be determined. The oxygen atom positions are calculated by means of the moment equations, after the data have been extrapolated to the hypothetical planar state, using the diagnostic least-squares procedure introduced by Curl.[224] The final structure so obtained is:

$$r[\mathrm{C(1)—C(2)}] = 1.524 \pm 0.001 \text{ Å}, \quad \widehat{\mathrm{C(2)C(1)C(2')}} = 88.23 \pm 0.11°$$
$$r[\mathrm{C(2)—O}] = 1.441 \pm 0.013 \text{ Å}, \quad \widehat{\mathrm{C(2)OC(2')}} = 94.82°$$
$$r[\mathrm{C(1)=O}] = 1.222 \pm 0.015 \text{ Å}$$

if the values $r[\mathrm{C(2)—H}] = 1.09$ Å and $\widehat{\mathrm{HC(2)H}} = 110°$ are assumed. The dipole moment measured is $\mu = 0.887 \pm 0.005$ D. An approximate model for the ring-bending vibration is discussed and the potential function for the vibration is interpreted in terms of contributions from the ring angle deformations and torsional interactions.

In a study[225] of the related β-propiolactone (35), Boggia, Favero, and Sorarrain measure rotational constants for the first four excited states of the ring-bending vibration. Using the relative intensity measuring facility of the Hewlett Packard 8400C spectrometer, they carefully derive the vibrational separations/cm^{-1} for the first four levels as:

$V =$	$1 \leftarrow 0$	$2 \leftarrow 1$	$3 \leftarrow 2$	$4 \leftarrow 3$
	158 ± 2	188 ± 5	167 ± 8	185 ± 13

They attribute the irregular level spacing to the existence of a double-minimum function for the vibration. A slight zig-zag behaviour of the rotational constants is also consistent with this conclusion. A function $V(z) = h\nu_0[z^2/2 + \lambda \exp(-\beta z^2)]$ with $\lambda = 0.39$, $\beta = 3.5$, and $h\nu_0 = 183$ cm^{-1} as final parameters is obtained by fitting the vibrational data and the rotational constants. The barrier height for this double-minimum function is 19 cm^{-1}, with the ground vibrational state 105 cm^{-1} above the maximum: hence the mild zig-zag in the rotational constants. Tigelaar, Gierke, and Flygare[226] deduce from the molecular Zeeman effect the g-values, molecular magnetic susceptibility anisotropies ($2\chi_{aa} - \chi_{bb} - \chi_{cc}$ and $2\chi_{bb} - \chi_{aa} - \chi_{cc}$), the diagonal elements of the molecular quadrupole tensor (Q_{aa}, Q_{bb}, Q_{cc}), and the diagonal elements of the diamagnetic susceptibility tensor ($\chi^{d}_{aa}, \chi^{d}_{bb}, \chi^{d}_{cc}$) for β-propiolactone (35) and diketen (36). A method whereby molecular electric

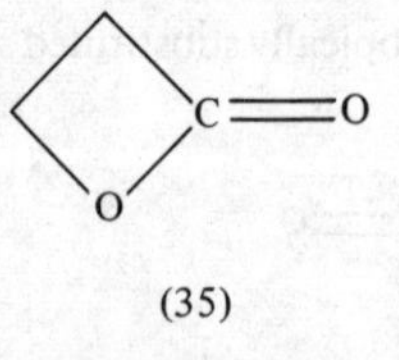

(35)

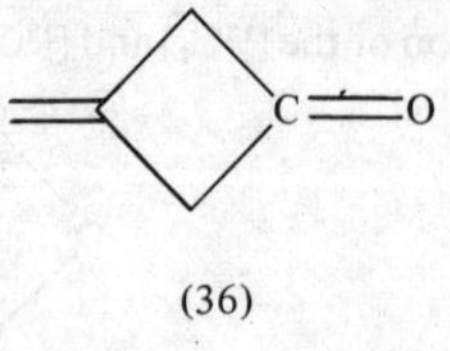

(36)

[224] R. F. Curl, jun., *J. Comput. Phys.*, 1970, **6**, 367.
[225] L. M. Boggia, P. G. Favero, and O. M. Sorarrain, *Chem. Phys. Letters*, 1971, **12**, 382.
[226] H. L. Tigelaar, T. D. Gierke, and W. H. Flygare, *J. Chem. Phys.*, 1972, **56**, 1966.

dipole and quadrupole moments and χ^{d}_{aa}, χ^{d}_{bb}, and χ^{d}_{cc} values can be calculated by summing empirically derived atom dipole moments has been developed by the same authors.[227] Using an established set of reduced atom dipoles, deduced originally on the basis of the successful prediction of the properties of many molecules, good agreement is achieved between observed and calculated values of the above molecular properties. Also of interest is a comparison of observed and calculated values of the out-of-plane minus average in-plane magnetic susceptibilities $\Delta\chi = \chi_{cc} - (\chi_{aa} + \chi_{bb})/2$.[226] Flygare and his co-workers have already shown that for unstrained, acyclic compounds it is possible to reproduce a large number $\Delta\chi_{obs}$ values simply by the summation of the local group $\Delta\chi$ values which compose the molecule. For β-propiolactone and diketen (as well as other four-membered rings) the ring strain apparently gives rise to a positive contribution to $\Delta\chi$ since $\Delta\chi_{obs}$ is larger (more positive) than $\Delta\chi_{calc}$ from local group values. This is interpreted in terms of an increased paramagnetic susceptibility anisotropy arising from the raising of the ground-state energy because of the ring strain.

Although not strictly related to microwave spectroscopy, there are some papers dealing with the theory of ring-puckering vibrations which should be mentioned. Günthard and his co-workers discuss a new set of puckering co-ordinates for four-membered-ring molecules[228] and also describe the exact solution of the rotation–ring-puckering problem for a cyclobutane-type four-membered ring by infinite matrix diagonalization,[229] while Malloy[230] proposes a semi-rigid model for the ring-puckering vibration in some pseudo-four-membered-ring molecules such as cyclopentene.

Five-membered Rings.—Fulvene (37), the non-alternant isomer of benzene, is of some theoretical interest and a detailed molecular geometry is therefore important as a basis for calculations. A complete and precise geometry is reported by Baron, Brown, Burden, Domaille, and Kent[231] from the micro-

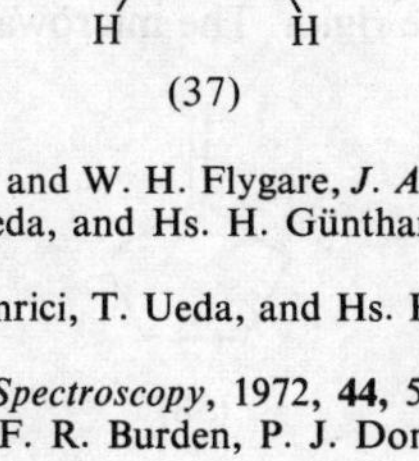

(37)

[227] T. D. Gierke, H. L. Tigelaar, and W. H. Flygare, *J. Amer. Chem. Soc.*, 1972, **94**, 330.
[228] F. Baltagi, A. Bauder, T. Ueda, and Hs. H. Günthard, *J. Mol. Spectroscopy*, 1972, **42**, 112.
[229] F. Baltagi, A. Bauder, P. Henrici, T. Ueda, and Hs. H. Günthard, *Mol. Phys.*, 1972, **24**, 945.
[230] T. B. Malloy, jun., *J. Mol. Spectroscopy*, 1972, **44**, 504.
[231] P. A. Baron, R. D. Brown, F. R. Burden, P. J. Domaille, and J. E. Kent, *J. Mol. Spectroscopy*, 1972, **43**, 401.

wave spectra of seven different isotopic species, *viz*. the normal species and the [1-^{13}C]-, [2-^{13}C]-, [3-^{13}C]-, [6-^{13}C]-, [2-^{2}H]-, and [3-^{2}H]-species. The [^{13}C]-species were detected in natural abundance whereas the deuterio-species were synthesized. Rotational constants were derived from a rigid-rotor analysis with some attention given to ensure satisfactory determination of A. As required, the [1-^{13}C]- and [6-^{13}C]-species leave A essentially invariant and all species exhibit a small positive inertia defect (in the range 0.005—0.01 a.m.u. Å^2). r_s Co-ordinates have been determined for all atoms with the exception of the small a co-ordinate of C(2) and C(5) and the co-ordinates of H(6), which are derived from simultaneous solution of the moment of inertia and first-moment conditions using all other r_s co-ordinates. The essentially r_s structure reported is in Table 34. It is interesting that the C—C bond lengths are almost identical with those of 6,6′-dimethylfulvene derived from an electron-diffraction study and also that the C(2)C(3)C(4)C(5) molecular fragment bond lengths are very similar to those in butadiene and cyclopentadiene. The latter observation draws from the authors the opinion that fulvene is best considered as composed of weakly coupled ethylenic entities rather than as a delocalized system. The very precisely measured dipole moment of fulvene is 0.4236 ± 0.0013 D.

Table 34 *Bond lengths/Å and bond angles/° of fulvene*

r[C(1)—C(6)]	1.3485 ± 0.0005
r[C(1)—C(2)] = r[C(1)—C(5)]	1.470 ± 0.001
r[C(2)—C(3)] = r[C(4)—C(5)]	1.355 ± 0.001
r[C(3)—C(4)]	1.476 ± 0.002
r[C(2)—H(2)] = r[C(5)—H(5)]	1.078 ± 0.001
r[C(3)—H(3)] = r[C(4)—H(4)]	1.080 ± 0.001
r[C(6)—H(6)]	1.13 ± 0.02
$\widehat{C(2)C(1)C(5)}$	106.62 ± 0.12
$\widehat{C(1)C(2)C(3)}$ = $\widehat{C(1)C(5)C(4)}$	107.71 ± 0.13
$\widehat{C(1)C(2)H(2)}$ = $\widehat{C(1)C(5)H(5)}$	124.66 ± 0.15
$\widehat{C(2)C(3)H(3)}$ = $\widehat{C(5)C(4)H(4)}$	126.39 ± 0.18
$\widehat{H(6)C(6)H(6)'}$	117 ± 5

According to Durig, Li, and Carreira,[232] methylenecyclopentane (38) is conformationally very similar to cyclopentanone and not, as reported in an i.r. study, significantly more rigid. The microwave spectrum of (38) is of the

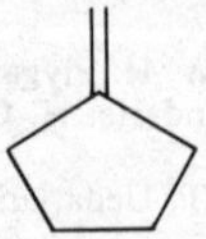

(38)

[232] J. R. Durig, Y. S. Li, and L. A. Carreira, *J. Chem. Phys.*, 1972, **57**, 1896.

effective rigid-rotor type exhibiting one strong satellite progression, assigned to successive excited states of the ring-bending mode (ν_B), and another weaker satellite, assigned to the first excited state of the ring-twisting mode (ν_T). The two modes appear to be quite independent, with an essentially harmonic ν_B indicated by almost linear dependence of the rotational constants of the associated satellite series on the corresponding quantum number. The previous i.r. study was unable to decide between the possibilities of 68 cm^{-1} and 158 cm^{-1} for the $1 \leftarrow 0$ energy separation in the bending mode, but microwave relative-intensity measurements give 63 ± 10 cm^{-1} and 270 cm^{-1} respectively for the bending and twisting modes, so that the first of the i.r. values must be chosen. The measured dipole moment components, $\mu_a = 0.60 \pm 0.01$ and $\mu_c = 0.0 \pm 0.06$ D, and the fact that the rotational constants of methylenecyclopentane are closely reproduced when the twisted ring atom structure from cyclopentanone and the methylene structure from 2-methylpropene are assumed point to a permanently twisted ring, as does the closely similar behaviour of ν_B and ν_T in methylenecyclopentane and cyclopentanone. There is no evidence that the former is significantly more rigid, rather the contrary if at all.

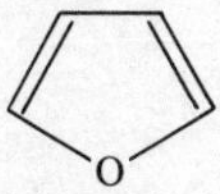

(39)

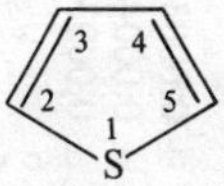

(40)

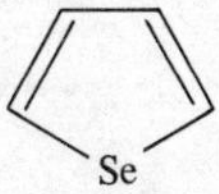

(41)

Some conclusions from microwave spectroscopy about the electronic properties of furan (39), thiophen (40), and selenophen (41) are of interest. Rotational Zeeman studies of [2,5-$^2H_2$]furan [233] and selenophen [234] give rise to molecular *g*-values, molecular magnetic susceptibility anisotropies ($2\chi_{aa} - \chi_{bb} - \chi_{cc}$ and $2\chi_{bb} - \chi_{cc} - \chi_{aa}$), diagonal elements of the electric quadrupole moment tensor (Q_{aa}, Q_{bb}, Q_{cc}), and several other valuable molecular properties. The first two of these are recorded in Table 35 for furan (previously studied by the Kiel group), [2,5-$^2H_2$]furan, and [^{80}Se]selenophen. The latter molecule was studied independently by the groups at Kiel and Urbana but the jointly published results exhibit satisfying agreement.

In principle, the change of *g*-values and rotational constants of furan on deuteriation can be used to determine the sign of the dipole moment and, although there is an unhappy near-compensation in the two changes, it is tentatively concluded that the negative end of the dipole points at the oxygen atom. The out-of-plane minus the average in-plane magnetic susceptibility, $\Delta\chi = \chi_{cc} - \frac{1}{2}(\chi_{aa} + \chi_{bb})$, as a criterion of aromaticity has been reviewed

[233] E. Hamer, D. Sutter, and H. Dreizler, *Z. Naturforsch.*, 1972, **27a**, 705.
[234] W. Czieslik, D. Sutter, H. Dreizler, C. L. Norris, S. L. Rock, and W. H. Flygare, *Z. Naturforsch.*, 1972, **27a**, 1691.

Table 35 *Zeeman parameters of furan and selenophen (susceptibility anisotropies in units of* 10^{-13} J G^{-2} mol^{-1})

Parameter	*Furan*	[2,5-$^{2}H_2$]*Furan**	[^{80}Se]*Selenophen* *Kiel*	[^{80}Se]*Selenophen* *Urbana*
g_{aa}	−0.0911 ± 0.0007	−0.08875 ± 0.0003	−0.0844 ± 0.0014	−0.0854 ± 0.0010
g_{bb}	−0.0913 ± 0.0002	−0.07793 ± 0.0004	−0.0432 ± 0.0012	−0.0423 ± 0.0010
g_{cc}	+0.0511 ± 0.0001	+0.04692 ± 0.0003	0.0365 ± 0.0010	0.0365 ± 0.0010
$2\chi_{aa} - \chi_{bb} - \chi_{cc}$	43.0 ± 0.2	33.5 ± 0.6	50.56 ± 1.0	49.74 ± 1.2
$2\chi_{bb} - \chi_{cc} - \chi_{aa}$	34.4 ± 0.2	43.4 ± 0.5	53.10 ± 1.6	50.56 ± 2.0

* Note that the *a*- and *b*-axes for the two species of furan change on deuteriation.

previously.[235] Exhaltation of the experimentally determined $\Delta\chi$ over the sums of assigned group or atom $\Delta\chi$ values is taken as evidence of a ring current. Since local atomic $\Delta\chi$ values for S, Se, and O are not very different, the change in observed $\Delta\chi$ from —38.7 through —50.1 to —51.0 for furan, thiophen, and selenophen points to aromaticities of the order furan < thiophen ~ selenophen.

3-Methylthiophen[236] exhibits a fairly low barrier to internal rotation of the methyl group (V_3 = 740 cal mol^{-1}), as calculated from appreciable *A–E* splittings of transitions in the vibrational ground state using Herschbach's theory and the methyl-group moment of inertia I_α = 3.19 a.m.u. Å^2. This barrier is to be compared with 1088 cal mol^{-1} in 3-methylfuran. There is possibly a connection between the increased aromaticity and the decrease of V_3 for the 3-methyl compound in passing from furan to thiophen. The same authors also report the dipole moments $\mu = 0.55 \pm 0.01$ D for thiophen and $\mu_a = 0.92 \pm 0.03$, $\mu_b = 0.21 \pm 0.12$, and $\mu_{total} = 0.95 \pm 0.06$ D for 3-methylthiophen, which are to be compared with $\mu_{total} = 0.55 \pm 0.04$ and 0.82 D respectively from dielectric constant measurements.

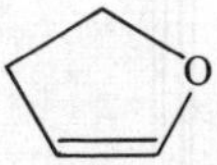

(42)

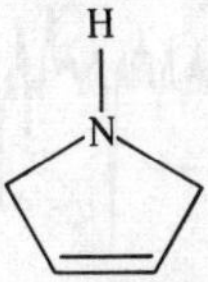

(43)

Conflicting interpretations of the ring-puckering mode of 2,3-dihydrofuran (42) have been neatly adjudicated by Durig, Li, and Tong[237] from the microwave spectrum. According to their interpretation of difference bands in the mid-i.r. spectrum, Ueda and Shimanouchi[238] report 16.58, 65.33, and 44.52 cm^{-1} for the fundamental and first two hot bands of the ring-puckering mode whereas Green[239] assigns the frequencies 19.1, 88.5, and 84.2 cm^{-1} to these bands in the far-i.r. spectrum, although both groups evolve the same barrier height for the ring inversion. Durig *et al.* observe a satellite series in the microwave spectrum which they attribute to the first five excited states of the ring-puckering mode. The spacing of the $v = 2$ state satellite (assigned *via* its relative intensity) is obviously irregular, as shown for one transition in Figure 14, indicating a differential perturbation of this state relative to the others even though all obey rigid-rotor behaviour. The effect is reflected in the distinct kink at the $v = 2$ state shown in the

[235] 'Molecular Spectroscopy', ed. R. F. Barrow, D. A. Long, and D. J. Millen (Specialist Periodical Reports), The Chemical Society, London, 1973, Vol. 1, p. 41.
[236] T. Ogata and K. Kozima, *J. Mol. Spectroscopy*, 1972, **42**, 38.
[237] J. R. Durig, Y. S. Li, and C. K. Tong, *J. Chem. Phys.*, 1972, **56**, 5692.
[238] T. Ueda and T. Shimanouchi, *J. Chem. Phys.*, 1967, **47**, 5018.
[239] W. H. Green, *J. Chem. Phys.*, 1969, **50**, 1619.

variation of the rotational constants with v, which indicates this state to be near the top of a barrier to ring inversion and $v = 3$, 4, and 5 to be above it. Relative-intensity measurements give 20, 85, and 86 cm^{-1} for the first three vibrational separations of the mode, amply confirming the far-i.r. assignment. The potential function $V/cm^{-1} = 28.3(x^4 - 3.44x^2)$ (x is a reduced co-ordinate) determined from the vibrational separations gives a barrier to the planar ring of 83 cm^{-1}. Evidently, the replacement of CH_2 by O in going from cyclopentene to 2,3-dihydrofuran reduces the unfavourable eclipsed methylene hydrogen interactions in the planar ring and reduces the barrier from 232 cm^{-1} in the former molecule.[240] The electric dipole moment varies little with V and all values are close to those of the ground state: $\mu_a = 1.30 \pm 0.02$, $\mu_b = 0.25 \pm 0.02$, $\mu_c = 0.00 \pm 0.01$, and $\mu_{total} = 1.32 \pm 0.02$ D.

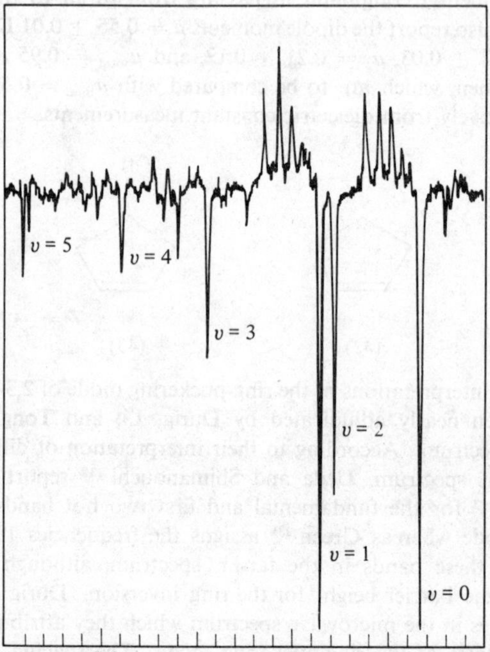

Figure 14 *The $7_{43} \leftarrow 7_{62}$ transition for $v = 0$—5 states of the ring-puckering mode in 2,3-dihydrofuran*

(Reproduced with permission from *J. Chem. Phys.*, 1972, **56**, 5692)

The microwave spectrum of another molecule isoelectronic with cyclopentene, 2,5-dihydropyrrole (43),[241] has also been assigned in the vibrational ground state and the first excited state of the ring-inversion vibration. A large value of $\Delta = I_c^0 - I_b^0 - I_a^0$ indicates the ring to be permanently bent.

[240] L. H. Scharpen, *J. Chem. Phys.*, 1968, **48**, 3552.
[241] C. R. Nave and K. P. Pullen, *Chem. Phys. Letters*, 1972, **12**, 499.

The additional assignments of the rotational spectra of the ^{15}N, ^{13}C, and ^{2}H singly substituted species (the first two in natural abundance) of 1,3,4-oxadiazole (44) now published[242] allow a detailed and precise molecular structure to be determined. An analysis under high resolution of quadrupole hyperfine structure arising from the two equivalent ^{14}N nuclei leads to $\chi_{aa} = -3.80 \pm 0.01$, $\chi_{bb} = 2.02 \pm 0.02$, and $\chi_{cc} = 1.78 \pm 0.02$ MHz and unperturbed frequencies for transitions. Moreover, ^{13}C and ^{2}H substitution allows a knowledge of the off-diagonal element, $\chi_{ab} = 2.65 \pm 0.05$ MHz,

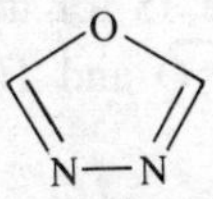

(44)

so that necessary corrections for unresolved hyperfine structure in the [^{15}N]-species can be made through coupling constants obtained by applying the appropriate rotation to the coupling tensor of the parent. Thus a centrifugal distortion analysis of unperturbed transition frequencies of all isotopic species is possible to give precise rotational constants leading to the following structure: r(C—O) = 1.348, r(C—N) = 1.297, r(N—N) = 1.399, and r(C—H) = 1.075 Å, $\widehat{COC}$ = 102.0, $\widehat{OCN}$ = 113.4, $\widehat{CNN}$ = 105.6, $\widehat{OCH}$ = 118.1, and $\widehat{NCH}$ = 128.5°. Uncertainties are less than 0.001 Å and 0.1°. Although the absence of substitution at O requires that its co-ordinate be determined using r_s co-ordinates of all other atoms in $\Sigma_i m_i a_i$, the reported structure is effectively of the r_s type. Through a comparison of ring bond lengths with those in similar positions in furan (I), 1,2,5-oxadiazole (II), 1,3,4-thiadiazole (III), thiophen (IV), and 1,2,5-thiadiazole (V), the authors suggest orders of electron delocalization in these molecules relative to cyclopentadiene and benzene as: cyclopentadiene < (I) < 1,3,4-oxadiazole ~ (II) for oxygen compounds and (III) < (IV) < (V) < benzene for sulphur rings.

A microwave study[243, 244] of ethylene ozonide (45) (1,2,4-trioxacyclopentane) has been undertaken to elucidate the mechanism of ozonolysis of ethylene from isotopic labelling. As part of the study, Gillies and Kuczkowski report a detailed structure from isotopic substitution. (The mechanistic aspect is dealt with in Section 13.) The μ_b spectrum of the parent molecule

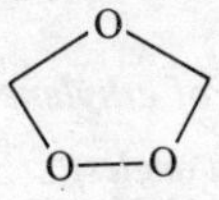

(45)

[242] L. Nygaard, R. L. Hansen, J. Tormod Nielsen, J. Rastrup-Anderson, G. O. Sorensen, and P. A. Steiner, *J. Mol. Structure*, 1972, **12**, 59.
[243] C. W. Gillies and R. L. Kuczkowski, *J. Amer. Chem. Soc.*, 1972, **94**, 7609.
[244] C. W. Gillies and R. L. Kuczkowski, *J. Amer. Chem. Soc.*, 1972, **94**, 6337.

exhibits statistical-weight effects consistent with the 5:3 ratio of intensities expected for transitions originating in symmetric states to those from antisymmetrical states (all other intensity factors having been accounted for) if the molecule has a C_2 axis. Moreover, the increase in $P_{cc} = \frac{1}{2}(I_a + I_b - I_c)$ on ^{13}C substitution shows that the carbon atoms do not lie in a symmetry plane and, taken together with the above-mentioned nuclear spin statistics, establishes a C_2 point group for the molecule. Such symmetry is consistent with the observed two conformers of the monodeuteriated species $HD_A\overline{COOCH_2O}$ and $HD_B\overline{COOCH_2O}$ (see Figure 15 for notation) and of the *trans* dideuteriated species ($HD_A\overline{COOCHD_AO}$ and $HD_B\overline{COOCHD_BO}$). The detailed

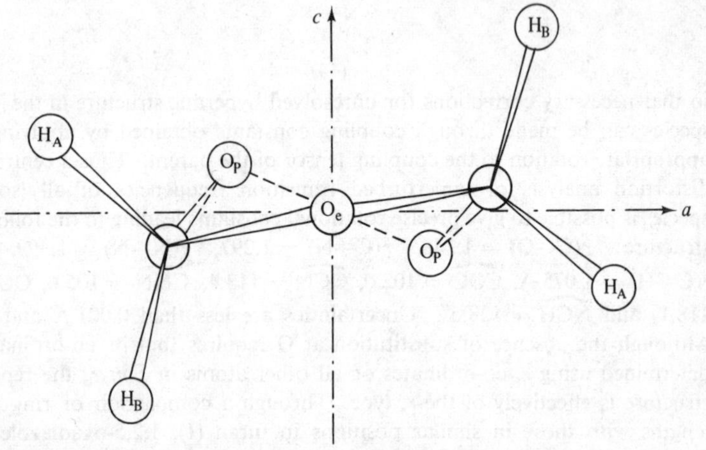

Figure 15 *Structure of ethylene ozonide: projection in the* ac *plane*

structure determined from seven isotopic species is recorded in Table 36 and shown in projection on the *ac* plane in Figure 15. The ring is bent in a half-chair configuration with the peroxy-oxygen atoms (O_p) twisted considerably from the CO_eC plane. The observation of three successive, well-behaved satellites associated with $v = 0$—3 of the ring-bending vibration (confirmed by the correct nuclear spin statistics) sets a lower limit of 500—600 cm^{-1} on the barrier to pseudorotation in this ring. A value of 1.09 ± 0.01 D is reported for the dipole moment ($\mu_b = \mu_{total}$).

Table 36 *Structural parameters of ethylene ozonide*

Bond lengths/Å		*Bond angles/°*		*Dihedral angles/°*	
C—H_A	1.094 ± 0.005	COC	102.83 ± 0.44	$C_1O_eC_2O_p$ =	−16.60 ± 0.40
C—H_B	1.094 ± 0.005	COO	99.23 ± 0.38	$C_1O_pO_pC_2$ =	−50.24 ± 1.26
C—O_p	1.395 ± 0.006	OCO	106.25 ± 0.64	$O_eCO_pO_p$ =	41.27 ± 0.96
C—O_e	1.436 ± 0.006	HCH	112.93 ± 0.48	$C_1O_eC_2H_A$ =	−132.09 ± 0.50
O_p—O_p	1.470 ± 0.015	O_eCH_A	109.43 ± 0.34	$C_1O_eC_2H_B$ =	103.90 ± 0.98
		O_eCH_B	109.08 ± 0.64		

Six-membered Rings.—Several papers have apperared dealing with mono- and di-substituted benzenes.

Mirri and Caminati [245] complete their investigation of the percentage double-bond character in the C—X bonds of the halogenobenzenes with the publication of accurate diagonal elements of the principal-axis quadrupole coupling tensor for the chlorine and bromine nuclei in chloro- and bromobenzene. In order to obtain very precise values of χ_{aa}, χ_{bb}, and χ_{cc} a small second-order correction is necessary in bromobenzene but such effects are negligible in chlorobenzene. The asymmetry parameter $\eta = (\chi_{bb} - \chi_{cc})/\chi_{aa}$, recorded with the χ's for the whole family X = Cl, Br, or I in Table 37, shows each C—X bond to possess cylindrical symmetry in the bond ($a = z$) direction while the difference in the number of electrons in the np_x and np_y orbitals ($x \perp$ plane of ring) given by $\frac{2}{3}(\chi_{cc} - \chi_{bb})/eq_{n10}Q$ indicates a small but decreasing percentage of double-bond character δ for np_x, assuming np_y in the ring plane cannot be involved in π-bonding.

In the course of an analysis of the dependence of the observed rotational constants on the vibrational quantum number associated with the NO_2 torsional mode (ν_{36}) of nitrobenzene using their rotation–internal-rotation theory, Ribeaud, Bauder, and Günthard [246] encounter an interesting difficulty. They find that the exact solution of the rotation–internal-rotation problem consisting of a C_{2v} frame (phenyl group and nitrogen atom) and a C_{2v} rotor (two oxygen atoms) cannot explain the observed variation of the rotational constants, particularly A, with the torsional state. However, when the rotation–internal-rotation approach is extended so that the angle ONO may vary synchronously with the torsional angle, a very satisfactory quantitative agreement for the variation of A (and B and C) is obtained. It is found that the angle ONO widens by 1.3° in going from the coplanar to the perpendicular configuration. Such a striking improvement in the analysis makes it tempting to consider the relaxation of the ONO angle as a significant phenomenon associated with large-amplitude internal motion.

The torsional motion of the CFO group against the benzene ring in benzoyl fluoride (46) is considered in a detailed study of that molecule by Kakar.[247] Seven torsionally excited states (vibrational quantum number τ = 1—7) in a vibrational satellite series as well as the ground-state spectrum are assigned.

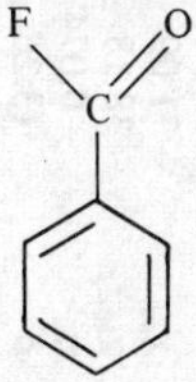

(46)

[245] W. Caminati and A. M. Mirri, *Chem. Phys., Letters*, 1971, **12**, 127.
[246] M. Ribeaud, A. Bauder, and Hs. H. Günthard, *Mol. Phys.*, 1972, **23**, 235.
[247] R. K. Kakar, *J. Chem. Phys.*, 1972, **56**, 1189.

Table 37 *Coupling constants*/MHz *and* C—X *double-bond character in halogenobenzenes*

	[^{35}Cl]*Cholorobenzene*	[^{79}Br]- *and* [^{81}Br]-*Bromobenzene*		*Iodobenzene*
χ_{aa}	-71.09 ± 10	558.9 ± 1.3	464.1 ± 1.8	-1892.1 ± 2.2
χ_{bb}	38.18 ± 0.49	-292.5 ± 0.5	-242.7 ± 0.7	976.2 ± 1.5
η	-0.074 ± 0.015	-0.046 ± 0.004	-0.046 ± 0.007	-0.031 ± 0.0025
δ	$(3.26 \pm 0.67)\%$	$(2.26 \pm 0.19)\%$	$(2.21 \pm 0.33)\%$	$(1.75 \pm 0.14)\%$

The vibrational satellites tentatively assigned to the out-of-plane wag (168 ± 10 cm^{-1}) and the in-plane bend (215 ± 10 cm^{-1}) of the CFO group are also observed. A planar molecular equilibrium structure is established from the inertia defect, $\Delta_0 = -0.325$ a.m.u. Å^2, which although negative becomes equal to $+0.286$ a.m.u. Å^2 on extrapolation to the hypothetical torsionless state, a behaviour pattern also exhibited in benzaldehyde and nitrobenzene. By use of the expression of Oka and Morino for Δ_τ, the change of inertia defect with torsional state is shown simply to be $\Delta_{\tau+1} - \Delta_\tau = -h/2\pi^2 c\omega_\tau$, which for $\tau = 1$ gives the torsional separation $\omega_\tau = 55.2$ cm^{-1}. Moreover, careful relative-intensity measurements among torsional satellites using a Hewlett Packard 8460 spectrometer give $\omega_\tau = 59.5 \pm 3$, 117 ± 7, 177 ± 10, 236 ± 15, 295 ± 15, 354 ± 17, and 411 ± 20 cm^{-1} for the separations of the $\tau = 1, 2, 3, 4, 5, 6$, and 7 states respectively from the ground state. This almost harmonic behaviour is consistent with the near-linear variation of the rotational constants with τ and a consideration of this latter variation leads, in combination with the $\tau = 1 \leftarrow 0$ separation of 59.5 cm^{-1}, to the twofold barrier $V_2 = 1560 \pm 160$ cm^{-1} for the torsional motion, which value should be compared with 1175 cm^{-1} in phenol, 1350 cm^{-1} in nitrosobenzene, and 1000 ± 500 cm^{-1} in nitrobenzene.

Partial structures of some interest are reported for two disubstituted benzenes. The microwave spectrum of four isotopic species of *m*-fluoroaniline (47),[248] containing the groups NH_2, *cis*-NHD, *trans*-NHD, and ND_2, are assigned by Cazzoli, Diamani, and Lister in order to characterize the amino-group structure. The r_s co-ordinates of the amino hydrogen atoms referred to the principal axis system of both the NH_2 and the ND_2 species are calculated. The amino-group geometry is determined by assuming r(N—H) = 1.00 Å as in aniline and by choosing the FC_6H_4N fragment geometry from aniline and fluorobenzene so that the moments of inertia of all isotopic species are reproduced to within 1%. The parameter of most interest is ϕ, the angle between the bisector of the HNH angle and the extension of the CN bond. The value 36.2° (based on the NH_2 species as the parent molecule) compares with 37.6° in aniline and 46.4° in *p*-fluoroaniline, which confirms a steeper R—NH_2 pyramid in the latter molecule and is consistent with the extra possibility of a mesomeric interaction of the nitrogen atom and the fluorine

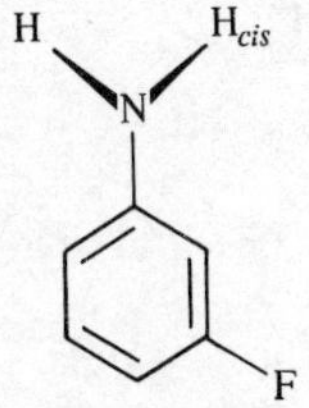

(47)

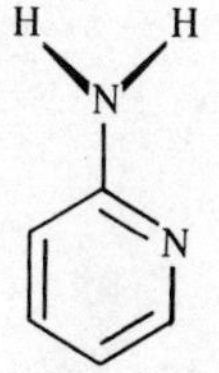

(48)

[248] G. Cazzoli, D. Damiani, and D. G. Lister, *J. C. S. Faraday II*, 1973, **69**, 119.

atom through the ring therein. Such an increase in ϕ in *p*-fluoroaniline has, in fact, been predicted on the basis of *ab initio* calculations by Hehre, Radom, and Pople.[249]

It is of interest to compare here a closely related investigation of 2-aminopyridine (48) by Kydd and Mills.[250] Again, four isotopic species are analysed, R—NH_2, the two R—NHD species, and the R—ND_2 species, to obtain r_s co-ordinates of the amino-group hydrogen atoms. The striking result is that one hydrogen atom is consistently 0.08 Å closer to the *ab* inertial plane (and therefore to the ring plane since the two planes nearly coincide) than the other, whether the NH_2 or ND_2 species is used as the parent molecule. Moreover, the same atom is closer to the *bc* plane and, although it is not known which of the two hydrogen atoms this is, it seems likely the NH_2 group is distorted so that the hydrogen nearer to the ring nitrogen lies closer to the plane. The angle ϕ (defined above) is estimated to be 31.6° on the basis of some structural assumptions and the observed inertial constants, and so 2-aminopyridine has a more planar NH_2 group than has aniline, as might be expected from increased conjugation of the ring and amino-group owing to the greater electronegativity of the ring nitrogen. The transitions in the rotational spectrum of (48) also exhibit a strong satellite, assigned to be the 0^- component of the inversion doublet (the ground state is assigned as 0^+). Relative intensity measurements give 135 ± 25 and $95 \pm 30\ cm^{-1}$ for the separation of the 0^+—0^- inversion levels in the NH_2 and ND_2 species respectively, compared with 40 cm^{-1} in (NH_2) aniline.

From an analysis of the microwave spectra of the normal, —OD, and —$CH^{18}O$ species of salicylaldehyde (49), Jones and Curl [251] first established coplanarity of all the atoms from the small *negative* inertia defects in the three species which, as discussed above, is a characteristic observation for a benzene ring in conjunction with a coplanar group capable of large torsional motions about the planar equilibrium structure. The data also allow the Kraitchman co-ordinates of the carbonyl oxygen and the phenolic proton to be located. The distance between these atoms, which follows immediately, is 1.76 Å and is less than the sum of their van der Waals' radii. Further calculations illustrate an interesting point. A molecular model based on an *X*-ray diffraction structure of salicylic acid and the formaldehyde structure

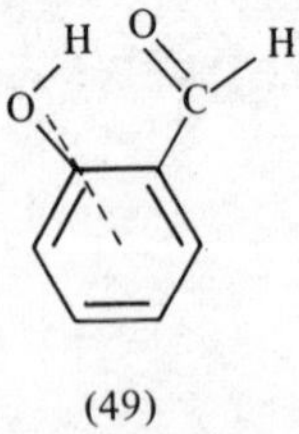

(49)

[249] W. J. Hehre, L. Radom, and J. A. Pople, *J. C. S. Chem. Comm.*, 1972, 669.
[250] R. A. Kydd and I. M. Mills, *J. Mol. Spectroscopy*, 1972, **42**, 320.
[251] H. Jones and R. F. Curl, jun., *J. Mol. Spectroscopy*, 1972, **42**, 65.

for the aldehyde group can only reproduce the observed moments of inertia and the Kraitchman co-ordinates of the two atoms simultaneously if the C—O bond is tilted away by 6° from the line bisecting the ring angle, as indicated in (49). Microwave spectra of *m*-chlorofluorobenzene [252] have also been reported.

The energy difference W_c between the vibrational ground states of the axial and equatorial conformers of cyclohexyl fluoride (50) as isolated molecules in the gas phase is now available.[253] Scharpen has measured very carefully the relative intensities of rotational transitions of the axial and equatorial forms using the partial power saturation technique described by Harrington.[254]

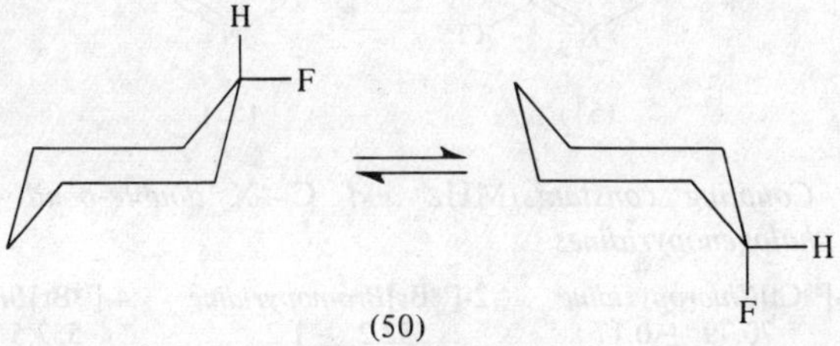

(50)

The precision of the measured intensity ratios was good and the error estimated from reproducibility was 1—2%. Instrumental uncertainties and reported errors in the measured dipole moment values contribute more heavily so that the result is quoted as $W_c = 259 \pm 28$ cal mol^{-1}. This is in excellent agreement with free-energy differences measured from ^{1}H and ^{19}F n.m.r. studies in solution as 276 ± 15 and 248 ± 12 cal mol^{-1} respectively. (The differences in rotational and vibrational partition functions between the conformers necessary to convert W_c to a free-energy difference are estimated to have little net effect on the numerical value.)

Four papers concerning halogenopyridines are published. Sharma and Doraiswamy [255] discuss the detailed determination of rotational and centrifugal distortion constants for 2-fluoropyridine. Scappini and Guarnieri [256] report rotational constants and χ_{aa}, χ_{bb}, and χ_{cc} values for the ^{35}Cl nucleus in 2-[^{35}Cl]chloropyridine (51). Some assignment difficulties are solved using the RFMDR technique in which radiofrequency transitions such as $2_{20} \leftarrow 2_{21}$ are strongly pumped and μ_a microwave transitions such as $3_{31} \leftarrow 2_{20}$ and $3_{30} \leftarrow 2_{21}$ are observed. A planar structure is dictated by the inertia defect $\Delta = 0.0385$ a.m.u. Å^2 and the values r(C—C) = 1.72 ± 0.01 Å and $\widehat{\text{NCCl}} = 116°25' \pm 10'$ are derived by fitting the rotational constants after the pyridine structure is assumed. The near coincidence of the *a*-axis and the C—Cl bond axis allows the χ_{aa}, χ_{bb}, and χ_{cc} values for the ^{35}Cl nucleus to be used in the usual way to determine the asymmetry parameter η and the degree

[252] K. K. Kirty and S. L. Srivastava, *Indian J. Pure Appl. Phys.*, 1972, **10**, 533.
[253] L. H. Scharpen, *J. Amer. Chem. Soc.*, 1972, **94**, 3737.
[254] H. W. Harrington, *J. Chem. Phys.*, 1967, **46**, 3698; 1968, **49**, 3023.
[255] S. D. Sharma and S. Doraiswamy, *Proc. Indian Acad. Sci., Sect. A.*, 1972, **76**, 221.
[256] F. Scappini and A. Guarnieri, *Z. Naturforsch.*, 1972, **27a**, 1011.

of double-bond character δ for the carbon–chlorine bond (in the manner and with the definitions of η and δ given in connection with 1-halogenobenzenes earlier in this section). The results are compared in Table 38 with those from a similar study of 2-bromopyridine and 4-bromopyridine (52) by Caminati and Forti.[257] The C—X bond characteristics appear to be very similar in this family of molecules.

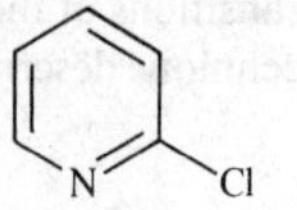

(51)

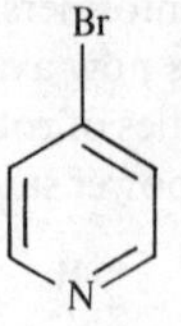

(52)

Table 38 *Coupling constants*/MHz *and* C—X *double-bond character in halogenopyridines*

	2-[^{35}Cl]*Chloropyridine*	2-[^{79}Br]*Bromopyridine*	4-[^{79}Br]*Bromopyridine*
χ_{aa}	-70.79 ± 0.17	552.2 ± 1.2	557.5 ± 1.3
χ_{bb}	39.01 ± 0.45	-304.4 ± 0.6	-300.5 ± 0.6
η	-0.10 ± 0.01	-0.102 ± 0.004	-0.078 ± 0.005
δ	4%	4.77%	3.58%

Phosphabenzene (53), the phosphorus analogue of pyridine, exhibits only a μ_a rotational spectrum.[258] The derived inertial constants lead to the inertia defect $\Delta = 0.0522$ a.m.u. Å^2, a value consistent with the expected molecular planarity. Moreover, a planar molecule with a C_2 axis about which rotation exchanges two pairs of equivalent protons is indicated by an observed 10:6 intensity alternation between contiguous transition pairs such as 6_{06}—7_{17} and 6_{16}—7_{07} in the vibrational ground state. Although a complete structure determination is not currently possible, the ranges for the parameters of interest [r(C—P) = 1.70—1.73 Å and $\widehat{C(5)PC(1)}$ = 101—104°] have been established from the inertial data with the following assumptions guided by by *X*-ray studies of substituted phosphabenzenes: r(C—H) = 1.08 Å, $\widehat{C(2)C(3)C(4)}$ = 122°, $\widehat{C(1)C(2)C(3)}$ = 123.6, the C—H bond bisects the corresponding ring angle, and the C—C bonds are restricted to the range 1.40—1.44 Å (*cf.* 1.397 Å in benzene).

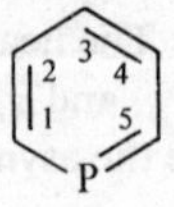

(53)

[257] W. Caminati and P. Forti, *Chem. Phys. Letters*, 1972, **15**, 343.
[258] R. L. Kuczkowski and A. J. Ashe, *J. Mol. Spectroscopy*, 1972, **42**, 457.

Thus, the P—C bond length is intermediate between the single-bond length (1.843 Å in trimethylphosphine) and the double-bond length (1.66 Å in $Ph_3P{=}CH_2$) as might be expected. The two most intense vibrational satellites in phosphabenzene correspond to vibration modes of 290 ± 40 and 325 ± 40 cm^{-1} according to relative intensity measurements and have been assigned to ν_{16b} (B_2) and ν_{16a} (A_2) respectively by analogy with the lowest two modes of pyridine, correct behaviour of the inertia defect, and correct nuclear statistical weight patterns. The dipole moment of phosphabenzene is 1.54 ± 0.02 D.

The microwave spectrum of a single conformer of 1,3-dioxan is assigned by Kewley.[259] The observation from Stark-effect measurements of a value of μ_a very close to zero ($\mu_a < 0.06 \pm 0.1$, $\mu_b = 1.61 \pm 0,02$, $\mu_c = 1.29 \pm 0.01$, and $\mu_{total} = 2.06 \pm 0.03$ D) rules out the boat form (55) or any twisted form since these require non-zero values of all three components. Further arguments based on reasonable models, the observed moments of inertia, and the dipole moment components militate in favour of the chair form (54) and against the other boat form (56) or a half-chair or a flattened chair form. Several vibrational satellites are identified and their intensities relative to the ground-state transitions measured to give values 257 ± 25, 290 ± 25, 432 ± 50, 439 ± 50, and 472 ± 50 cm^{-1} for the five lowest modes of vibration in the molecule.

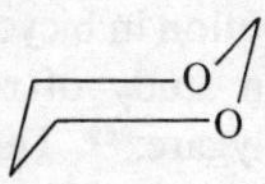

(54)

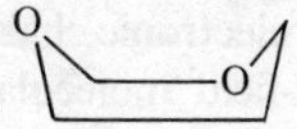

(55)

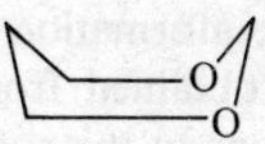

(56)

Fused Rings.—The rotational spectra of bicyclic and tricyclic systems have recently evoked an increased interest.

The small bicyclic hydrocarbon bicyclo[2,1,0]pentane (57) is investigated thoroughly by Suenram and Harmony.[260] A noteworthy feature of this work is that it has been possible, through a long and skilful synthesis, to produce an isotopic mixture of composition 10% [1-^{13}C]-, 10% [2-^{13}C]-, 5% [5-^{13}C]-, and 75%[^{12}C]-bicyclo[2,1,0]pentane starting only with barium carbonate of 60% ^{13}C enrichment. Another feature is the use of a computer-controlled spectrometer with signal averaging and storage as an aid in the search

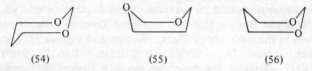

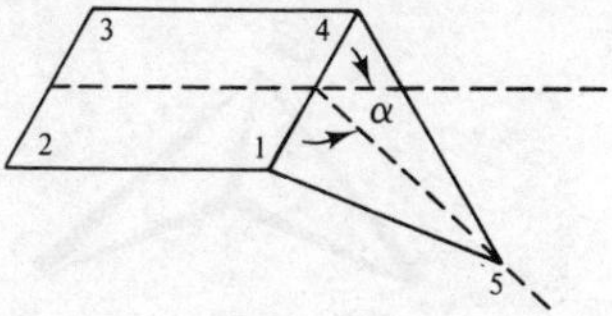

(57)

[259] R. Kewley, *Canad. J. Chem.*, 1972, **50**, 1690.
[260] R. D. Suenram and M. D. Harmony, *J. Chem. Phys.*, 1972, **56**, 3837.

and measurement of some of the weaker ^{13}C transitions. The two observations of significance are that only *c*-type transitions can be observed and that substitution in the 5-position leaves $I_a + I_c - I_b$ invariant. Thus the molecule has a plane of symmetry (*ac* plane) with the *a*-component of the dipole moment presumably too small to allow observable μ_a-transitions. The isotopic distribution in the synthesis product is in accord with a plane of symmetry. The r_s structure of the carbon skeleton reported is:

$$r[\text{C(1)—C(4)}] = 1.536 \pm 0.001 \text{ Å}, r[\text{C(2)—C(3)}] = 1.565 \pm 0.001 \text{ Å}$$
$$r[\text{C(1)—C(2)}] = r[\text{C(3)—C(4)}] = 1.528 \pm 0.002 \text{ Å}, r[\text{C(1)—C(5)}] =$$
$$r[\text{C(4)—C(5)}] = 1.507 \pm 0.002 \text{ Å, and } \alpha = 67.26 \pm 0.18^\circ$$

The planar cyclobutane ring entity thus has three C—C bonds shorter than 1.548 Å in cyclobutane and one longer. It is pointed out that Σr(C—C) for cyclobutane rings in several bicyclic compounds is sensibly constant so that in a given environment any necessary bond shortening (or lengthening) must apparently be accompanied by a corresponding lengthening (or shortening) elsewhere in the ring. The skeletal structure of bicyclo[2,1,0]pentane is in fact in serious disagreement with a recent electron-diffraction study[261] but an analysis of the radial distribution function by Suenram and Harmony leads them to propose that a false local minimum has been found in the original least-squares fit. The dipole moment components are $\mu_a = 0.00 \pm 0.01$ D, $\mu_b = 0$ (by symmetry), and $\mu_c = 0.26 \pm 0.01$ D.

Some information about the electronic charge distribution in bicyclobutane (58) is obtained from a high-field molecular Zeeman study of rotational transitions in this molecule by Gierke, Benson, and Flygare.[262] They derive molecular *g*-values ($g_{aa} = 0.0593 \pm 0.0002$, $g_{bb} = 0.0025 \pm 0.0002$, and $g_{cc} = 0.0412 \pm 0.0002$), the molecular magnetic susceptibility anisotropies ($2\chi_{aa} - \chi_{bb} - \chi_{cc} = -5.9 \pm 0.3$ and $2\chi_{bb} - \chi_{aa} - \chi_{cc} = 21.1 \pm 0.3$ in units of 10^{-13} J G^{-2} mol^{-1}) and molecular electric quadrupole moments ($Q_{aa} = 1.3 \pm 0.2$, $Q_{bb} = -2.6 \pm 0.3$, and $Q_{cc} = 1.3 \pm 0.4$ in units of 10^{-26} e.s.u. cm^2). Arguing from the various contributions to the *g*-values, these authors interpret the positive *g*-values as indicating the electrons in bicyclobutane to be nearer the centre of mass than the nuclei, *i.e.* there exists a concentration of elecronic charge beneath the bridgehead. Such a charge distribution is consistent with both the enhanced reactivity of the bridgehead bond and the observation that electrophilic attack occurs from beneath the bridgehead.

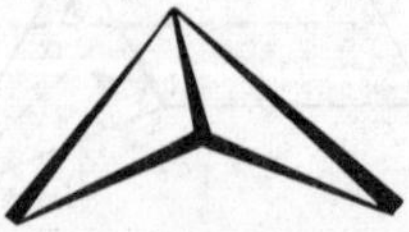

(58)

[261] R. K. Bohn and Y. H. Tai, *J. Amer. Chem. Soc.*, 1970, **92**, 6447.
[262] T. D. Gierke, R. C. Benson, and W. H. Flygare, *J. Amer. Chem. Soc.*, 1972, **94**, 339.

Following his work with bicyclo[2,1,1]hexan-2-one, Coffey [263] now reports the rotational constants and dipole moment of the isomeric bicyclo[2,1,1]-hexan-5-one (59). The interesting conclusion is that the dipole moment $\mu_{total} = 3.18 \pm 0.02$ D, which is estimated to make an angle of only $1.7 \pm 1.7°$ with the C=O bond direction, is anomalously high and similar in this respect to those of cyclopentanone and bicyclo[2,1,1]hexan-2-one. This confirms the author in his opinion that such molecules containing a cyclopentanone entity have anomalously high dipole moments compared with most carbonyl compounds.

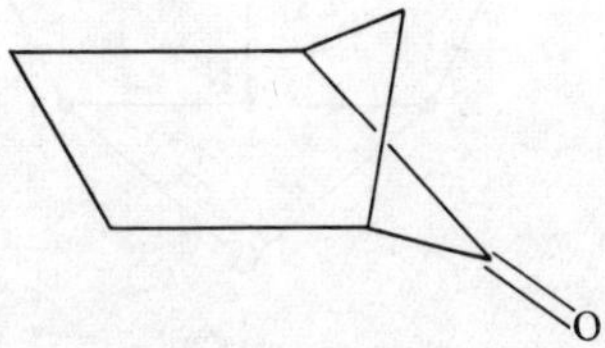

(59)

The large number of independent structural parameters of cyclohexene oxide (Figure 16) make a detailed determination from microwave spectroscopy a formidable task. However, Ikeda, Kewley, and Curl [264] are able to use the ground-state rotational constants of the common isotopic species to establish a twisted carbon atom skeleton for the molecule. Their assumptions, based on previous studies of cyclohexene oxide and related molecules, are (with reference to Figure 16) as follows:

(*a*) r[C(1)—C(2)] = r[C(2)—C(3)] = r[C(3)—C(4)] = 1.53 Å, r[C(1)—C(6)] = 1.48 Å, and r[C(1)—O] = 1.42 Å;
(*b*) atoms C(5)C(6)C(1)C(2) are coplanar;
(*c*) $\widehat{C(5)C(6)C(1)} = \widehat{C(6)C(1)C(2)}$, $\widehat{C(4)C(5)C(6)} = \widehat{C(1)C(2)C(3)}$, $\widehat{C(5)C(4)C(3)} = \widehat{C(2)C(3)C(4)}$;
(*d*) $\widehat{C(6)C(1)C(2)} = 122°$;
(*e*) standard hydrogen parameters, to which the reported conclusions are insensitive.

The remaining skeletal parameters are the angle $\beta = 180° - $ LMN, the twist angle τ between C(2)—C(5) and C(3)—C(4), and the dihedral angle θ between the epoxy-ring and the C(5)C(6)C(1)C(2) plane. In view of a probable C_2 symmetry for the carbon ring ($\beta = 0$), the rotational constants are fitted to τ and θ for several β values near to zero. The fit is best with β so near to zero that it is assumed to be identically so, in which case $\tau = 34°26'$ and $\theta = 68°23'$. The results of an electron-diffraction study are close [$\beta = 0$ (assumed), $\tau = 33°26'$, and $\theta = 75°34'$] while θ is similar to the values near to 75° reported for *cis*-2,3-epoxybutane and propene oxide. The twisted

[263] D. Coffey, *J. Mol. Spectroscopy*, 1972, **42**, 47.
[264] T. Ikeda, R. Kewley, and R. F. Curl, jun., *J. Mol. Spectroscopy*, 1972, **44**, 459.

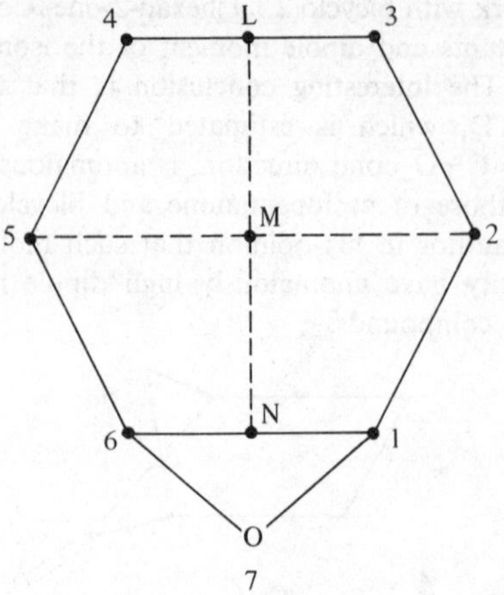

Figure 16

ring is consistent with indications of a high barrier to inversion of the ring by twisting, as inferred from the twisting-mode frequency of 280 cm^{-1} from satellite relative intensities and the liquid-phase i.r. spectrum. The dipole moment components are $\mu_a = 1.152 \pm 0.006$, $\mu_b = 0.18 \pm 0.08$, $\mu_c = 1.52 \pm 0.01$, and $\mu_{total} = 1.91 \pm 0.02$ D.

Christen, Bauder, and Günthard [265] have returned to the study of azulene (60). They analyse exhaustively and with high sensitivity the rotational spectrum in the ground state and six vibrationally excited states ($\nu_{23} = 1$, $\nu_{23} = 2$, $\nu_{48} = 1$, $\nu_{48} = 2$, $\nu_{23} + \nu_{48}$, and $\nu_{47} = 1$) with a centrifugal distortion treatment appropriate to a planar molecule. In addition, by a careful application of the Harrington [254] partial power saturation technique for relative-intensity measurements (with some minor modifications discussed), these authors determine the $1 \leftarrow 0$ vibrational separations for the three lowest vibrational modes of azulene, ν_{48} (B_2), ν_{23} (A_2), and ν_{47} (B_2), to be 180.3 ± 1.9, 188.0 ± 2.4, and 272.7 ± 5.0 cm^{-1} respectively. Their measurements exhibit an excellent internal consistency. Assignment of the satellites to a particular mode is aided by nuclear-spin statistical-weight considerations, for the weight

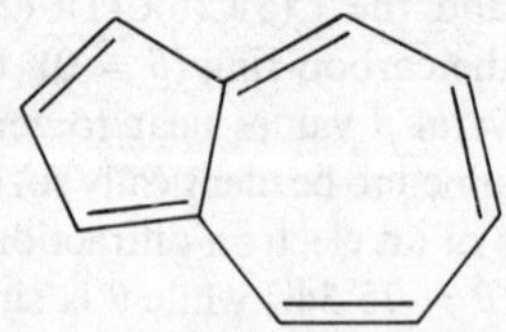

(60)

[265] P. Christen, A. Bauder, and Hs. H. Günthard, *J. Mol. Spectroscopy*, 1972, **43**, 1.

ratio between the ground and first excited states of B_1 or B_2 species modes for the planar molecule in which three pairs of equivalent hydrogen nuclei are exchanged by a C_2 rotation is 9/7 for rotational transitions originating in states with K_{-1} even and 7/9 for those with K_{-1} odd. On the other hand, no change is expected between the ground state and excited states of A_1 or A_2 modes. This work settles the assignment problem in the far-i.r. spectrum of a single crystal of the compound, wherein four absorption bands at 275 (vw), 189, 180, and 170 cm^{-1} respectively were observed but the assignment of the ν_{48}, ν_{23}, and ν_{47} modes could not be made.

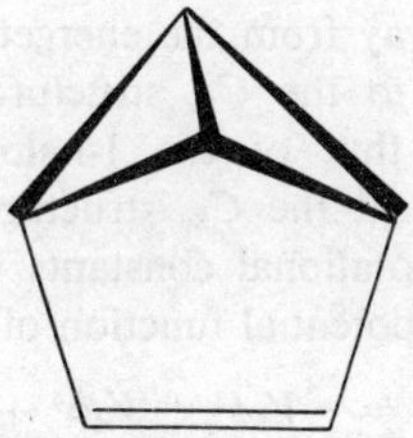

(61)

The investigation of one isomer of benzene (fulvene: see pp. 53—54) is already reported. A preliminary analysis of another isomer, benzvalene (61) (tricyclo[3,1,0,0^{2,6}]hex-3-ene), has also appeared.[266] Suenram and Harmony report rotational constants and the dipole moment $\mu = \mu_a = 0.883$ D, which are consistent with a molecular model formed by combination of bicyclobutane (58) and cyclobutene fragments. In particular, the sum of the dipole moments of bicyclobutane (0.68 D) and cyclobutene (0.139 D) gives a value in good agreement with that observed for (61). If the unsaturated end of the cyclobutene fragment defines the negative end of the dipole, the negative end of the bicyclobutane fragment lies beneath the bridgehead bonds, in agreement with conclusions from the Zeeman work on bicyclobutane already mentioned.[262]

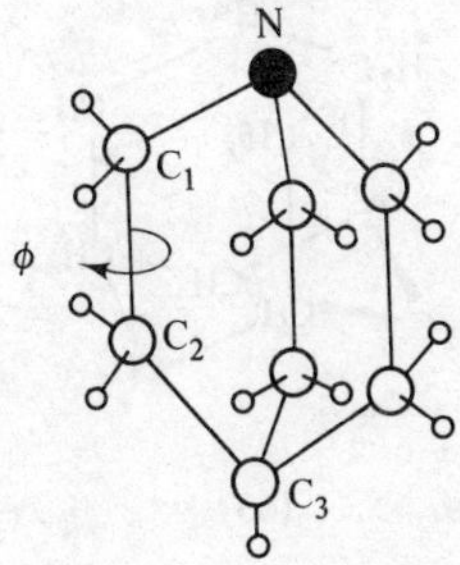

(62)

[266] R. D. Suenram and M. D. Harmony, *J. Amer. Chem. Soc.*, 1972, **94**, 5915.

Hirota and Suenaga[267] have extended their treatment of the 1-halogeno-bicyclo[2,2,2]octanes[268] to include 1-azabicyclo[2,2,2]octane (62) (quinuclidine). Each rotational transition of this symmetric-top molecule exhibits a strong satellite progression assigned to successive excited states of the torsional motion in which the molecule twists about the symmetry axis [the torsional angle ϕ is defined as the angle between the NC(1)C(2) and the C(1)C(2)C(3) planes]. Irregular spacing between the $v = 0, 1$, and 2 members of the progression is attributed to the existence of a shallow double-minimum potential function governing the torsional motion and therefore indicates an equilibrium structure of C_3 symmetry in which the C(1) and C(2) methylene groups are twisted away from the energetically unfavourable eclipsed configuration appropriate to the C_{3v} structure. The satellite irregularity is attenuated relative to that of the 1-halogenobicyclo[2,2,2]octanes,[268] suggesting a lower barrier at the C_{3v} structure in the present case. The quantitative variation of rotational constants with the torsional quantum number can be fitted by a potential function of the form

$$V(\phi) = -V_2\phi^2 + V_4\phi^4 + V_6\phi^6$$

to give the height of the barrier to the C_{3v} molecule $V_0 = 39 \pm 15$ cal mol^{-1} and the equilibrium torsional angle $\phi_0 = 10.7 \pm 1.7°$. The corresponding values for 1-fluorobicyclo[2,2,2]octane,[268] 191 ± 52 cal mol^{-1} and $16.4 \pm 1.6°$, indicate that the equilibrium structure is nearer to C_{3v} symmetry in 1-azabicyclo[2,2,2]octane.

Evidence of a differential distortion of the cage structure in 1-fluoroadamantane relative to other 1-substituted adamantanes (63; X=Cl, Br, I, or CN) is suggested in a recent publication.[269] From the symmetric-top rotational spectra of the [$^{12}C^{14}N$]-, [$^{13}C^{14}N$]-, and [$^{12}C^{15}N$]-isotopic species of 1-cyano-adamantane (63; X=CN) it is straightforward to derive r_s co-ordinates of the C and N atoms and therefore to subtract the contribution of these atoms

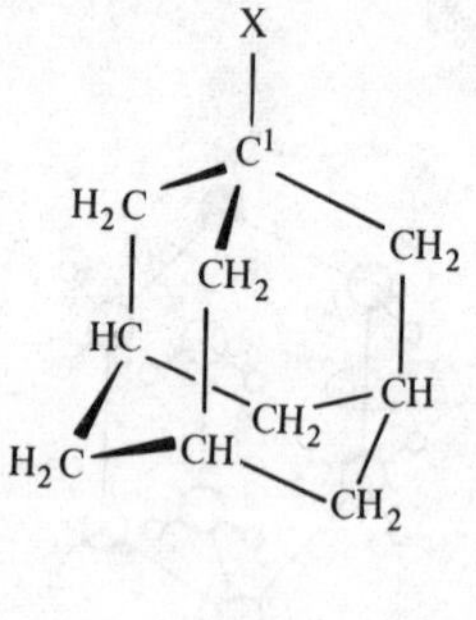

(63)

[267] E. Hirota and S. Suenaga, *J. Mol. Spectroscopy*, 1972, **42**, 127.
[268] E. Hirota, *J. Mol. Spectroscopy*, 1972, **38**, 367.
[269] D. Chadwick, A. C. Legon, and D. J. Millen, *J. C. S. Faraday II*, 1972, **68**, 2064.

from the parent-molecule moment of inertia, I_b, to give the moment of inertia of the adamantyl radical referred to the *molecular* principal axis system. Further, the application of the parallel axis theorem generates I^*_{rad}, the moment of inertia of the radical referred to its own principal axis system. Such a procedure when also applied in the cases X = Cl or Br leads to I^*_{rad} values invariant for X = Cl, Br, or CN, suggesting an adamantyl cage structure sensibly independent of the nature of the 1-substituent. Accordingly, to proceed further it seems reasonable to assume a cage with all C—C bond distances equal, all angles 109°28′ and all C—H distances at 1.09 Å, as in adamantane itself. Then, the experimental I^*_{rad} values lead to 1.541 ± 0.001, 1.542 ± 0.002, and 1.543 ± 0.002 Å for the C—C distance in the cases X = Cl, Br, and CN respectively. The C—X distances simultaneously generated are shown in Table 39 compared with those of the corresponding methyl, ethyl, and t-butyl compounds (from microwave spectroscopy).

Table 39 *Comparison of r(C—X)/Å in adamantyl, methyl, ethyl, and t-butyl compounds*

	Methyl	*Ethyl*	*t-Butyl*	*Adamantyl*
F	1.385 ± 0.001	1.398 ± 0.005	1.43 ± 0.02	1.370 ± 0.004
Cl	1.781 ± 0.002	1.788 ± 0.002	1.803 ± 0.002	1.790 ± 0.002
Br	1.939 ± 0.001	1.950 ± 0.004	1.98*	1.947 ± 0.006
I	2.139	2.139 ± 0.005	2.19*	2.160 ± 0.002
CN	1.458 ± 0.001	1.473 ± 0.002	1.495 ± 0.015	1.466 ± 0.003

* Values calculated from experimental moments of inertia assuming $\widehat{CCC}$ = 111.5°, which is close to that found in other t-butyl compounds.

Unfortunately, the absence of stable isotopes other than ^{19}F and ^{127}I precludes extension of the procedure to X = F or I. However, assuming the cage structure as in the 1-chloro-compound the *r*(C—F) and *r*(C—I) values shown in Table 39 are obtained. We note that, for all cases but X = F, the adamantyl *r*(C—X) values agree closely with the ethyl values. The C—F distance is anomalously short, suggesting that in the case of the electronegative fluorine atom the assumption of cage invariance is invalid. As a simple model for the distortion induced by the fluorine atom, the bridgehead CC(1)C angles are allowed to vary by moving the C(1) atom along the molecular symmetry axis while the positions of all other atoms are unchanged. A value of *r*(C—F) in 1-fluoroadamantane equal to that in ethyl fluoride can then be obtained if the bridgehead CC(1)C angles are increased by 1° relative to those in the chloride.

Bjørseth, Drew, Marstokk and Møllendal report [270] the structure of the cyclopentadienylberyllium chloride molecule (64) from an analysis of the isotopic species $C_5H_5Be^{35}Cl$, $C_5H_5Be^{37}Cl$, $C_4{}^{13}CH_5Be^{35}Cl$, and $C_5H_4DBe^{35}Cl$.

[270] A. Bjørseth, D. A. Drew, K. M. Marstokk, and H. Møllendal, *J. Mol. Structure*, 1972, **13**, 233.

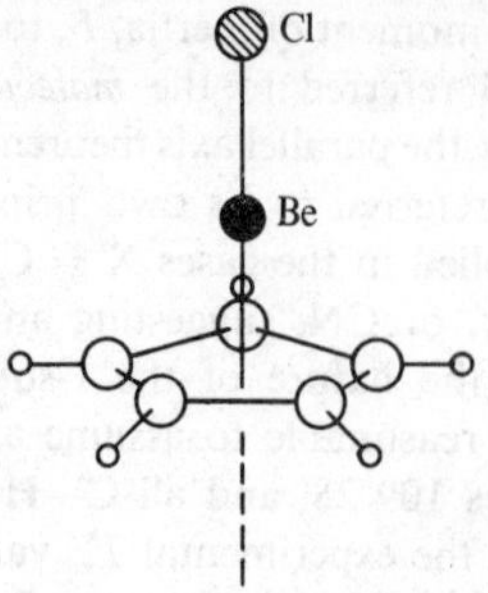

(64)

The symmetric-top nature of the rotational spectrum of the parent and ^{37}Cl species is consistent with a structure of C_{5v} symmetry. Moreover, the regular variation of a vibrational satellite series associated with successive quanta of the cyclopentadienyl-ring–beryllium–chlorine bend indicates that the vibration is quite harmonic with no potential barrier at the fivefold symmetry axis. The structural parameters recorded in Table 40 are of the r_0 variety and reproduce all observed rotational constants better than 0.1%. Also recorded in Table 40 is the result of an electron-diffraction investigation of the cyclopentadienylberyllium chloride structure.[271] The agreement between the two techniques is excellent. The Be—Cl bond is longer than in $BeCl_2$. Moreover, the dipole moment, 4.25 ± 0.16 D, is high and the ^{35}Cl nuclear quadrupole coupling constant, $eqQ = 22 \pm 2$ MHz, is low. All this is evidence for an ionic Be—Cl bond.

Table 40 *Structural parameters of cyclopentadienylberyllium chloride*/Å

	Microwave, r_0	*Electron diffraction*, r_g
C—H*	1.090	1.097
C—C	1.424	1.424
Be—Cl	1.839	1.837
Be-ring(h)	1.485	1.484
Cl···C	3.538	3.535

* Hydrogen atoms taken to be coplanar with ring carbon atoms.

6 Molecules in States other than $^1\Sigma$

The spectra of a number of diatomic molecules have been studied by one or more of various techniques, *viz.* microwave spectroscopy, gas-phase electron paramagnetic resonance, molecular-beam electric resonance spectroscopy, laser magnetic resonance, and the use of pure rotational lasers. The microwave spectrum of the BrO radical has been further examined[272] and transitions have been measured for the first excited vibrational state of the $^2\Pi_{\frac{3}{2}}$ ground electronic state. The results have been combined with results of a

[271] D. A. Drew and A. Haaland, *Chem. Comm.*, 1971, 1551.
[272] T. Amano, A. Yoshinaga, and E. Hirota, *J. Mol. Spectroscopy*, 1972, **44**, 594.

re-analysis of the ground vibrational state to give the equilibrium internuclear distance $r_e = 1.717(1) \pm 0.0013$ Å. Vibrationally excited states of the oxygen molecule have been investigated by gas-phase electron paramagnetic resonance spectroscopy [13] and an improved set of parameters for the oxygen molecule has been obtained. Molecular-beam electric resonance studies have been made of Λ-doubling for the $a^3\Pi$ state for CO and the $^2\Pi_{\frac{3}{2}}$ and $^2\Pi_{\frac{1}{2}}$ states of NO. Hyperfine structure has been measured for NO and it is shown how it may be accounted for theoretically.[273] For CO, Λ-doubling has been observed for a number of transitions with $v = 0$—5. The vibrational dependence of the Λ-doubling is accounted for and it is shown that the $v = 4$ and $v = 5$ states are perturbed by the $v = 0$ and $v = 1$ states of the $a'^3\Sigma^+$ electronic state. The analysis leads to a number of molecular parameters, including information about dipole moments.[274] The dipole moment of the $a^3\Pi$ state is obtained as function of v and J. For the $a'^3\Sigma^+$ state it is found that $\mu = 1.06 \pm 0.2$ D and that the sign is —CO+ whereas for the $^3\Pi$ state $\mu = 1.37$ D and the sign is +CO—. The NF species in the $^1\Delta$ state has been studied by electron paramagnetic resonance spectroscopy.[275] Hyperfine coupling constants of the ^{14}N and ^{19}F nuclei have been determined and the dipole moment is found to be 0.37 ± 0.06 D. A further study [276] of SH by the paramagnetic resonance technique has led to information about the second rotational state, $J = 5/2$, and molecular parameters have been obtained from the analysis of the spectrum. For OH and OD pure rotational laser oscillations have been observed.[277] A total of 44 lines have been measured for OH and 17 for OD in the region 500—800 cm^{-1}. Accurate values of rotational and centrifugal distortion constants have been obtained. A Stark-modulated spectrometer for the study of free radicals has been described.[278] The cell consists of a Pyrex tube containing plane-parallel Stark electrodes which are covered with Teflon to reduce recombination. Transitions with the spectrometer have been observed for NS, SO, and OH.

A further microwave spectroscopic study [279] of the NCO radical has identified transitions in the first excited bending mode of the Δ state and effective rotational constants for the $^2\Delta_{\frac{3}{2}}$ and $^2\Delta_{\frac{5}{2}}$ states have been obtained. Two Zeeman components of several rotational transitions of nitrogen dioxide have been examined by laser magnetic resonance spectroscopy.[280] Hyperfine components of four rotational transitions have been assigned and isotropic and anisotropic components of the electron-spin g-factor have been determined.

[273] W. L. Meerts and A. Dymanus, *J. Mol. Spectroscopy*, 1972, **44**, 320.
[274] B. G. Wicke, R. W. Field, and W. Klemperer, *J. Chem. Phys.*, 1972, **56**, 5758.
[275] A. H. Curran, R. G. Macdonald, A. J. Stone, and B. A. Thrush, *Proc. Roy. Soc.*, 1973, **A332**, 355.
[276] J. M. Brown and P. J. Thistlethwaite, *Mol. Phys.*, 1972, **23**, 635.
[277] T. W. Ducas, L. T. Geoffrion, R. M. Osgood, and A. Javan, *Appl. Phys. Letters*, 1972, **21**, 42.
[278] C. Marliere, J. Burie, and J. L. Destombes, *Compt. rend.*, 1972, **275**, *B*, 315.
[279] T. Amano and E. Hirota, *J. Chem. Phys.*, 1973, **57**, 5608.
[280] R. F. Curl, jun., K. M. Evenson, and J. S. Wells, *J. Chem. Phys.*, 1972, **56**, 5143.

The effective Hamiltonian for a linear molecule in an orbitally degenerate state has been re-examined.[281] It is established that the magnetic moment in a particular rotational state is determined by two parameters. A recalculation is made on this basis for the SO molecule in the $^1\Delta$ state. The theory of the Zeeman effect in doublet states has been developed for near-symmetric rotors [282] and spin–orbit coupling for small molecules in orbitally degenerate states has been examined theoretically.[283] Calculations have been made of hyperfine coupling constants for a number of free radicals,[284] including the formyl radical [285] and a number of inorganic radicals.[286]

7 Analysis of Spectra and the Evaluation of Molecular Parameters

The question of devising a systematic method for the analysis of microwave spectra has been considered and a method put forward which combines the use of the Stark effect and double resonance.[287] As a starting-point the procedure requires the recognition in the spectrum (as observed under the usual conditions for which $\Delta M = 0$) of a line with a single Stark component. This transition (ν_1) belongs to a set of six possible assignments, three being *R*-branch transitions, $J = 0 \rightarrow 1$, and three being *Q*-branches, $J = 1 \rightarrow 1$, for all of which only a single Stark component arises. Use is then made of the fact that whichever of these six assignments is correct, a transition occurs at $3\nu_1$. Of the six possible assignments for this line, three are *R*-branch transitions and three are *Q*-branch, and the Stark effect can be used to make the distinction. The next step is to use double resonance, pumping with frequency ν_1, to locate a particular $J = 1 \rightarrow 2$ transition depending on the particular $J = 1$ level already identified by the Stark effect. Use of the procedure would depend on identifying a transition with a single Stark component and could require working in a much wider range of the spectrum than is usual. Three papers dealing with the theory and interpretation of spectra of molecules with internal rotation have been mentioned earlier in the Report. One deals with molecules having two non-equivalent internal rotors,[194] another with internal-rotation–nuclear-quadrupole interaction in asymmetric rotor spectra,[199] and the third with the coupling between internal rotation and a degenerate vibration in a symmetric rotor molecule.[198] A comprehensive discussion has been given of the determination of nuclear quadrupole coupling constants from microwave spectra.[288] Matrix elements

281 J. M. Brown and H. Uehara, *Mol. Phys.*, 1972, **24**, 1169.
282 J. C. D. Brand, A. M. Garcia, and C. di Lauro, *J. Mol. Spectroscopy*, 1972, **43**, 199.
283 W. H. Moores and R. McWeeny, *Proc. Roy. Soc.*, 1973, **A332**, 365.
284 G. M. Zhudomirov and N. D. Chuvylkin, *Chem. Phys. Letters*, 1972, **14**, 52.
285 N. D. Chuvylkin, G. M. Zhudomirov, and P. V. Schastner, *Zhur. struct. Khim.*, 1972, **13**, 602.
286 D. C. McCain and W. S. Palke, *J. Chem. Phys.*, 1972, **56**, 4957.
287 G. Roussy, J. Demaison, and J. Barriol, *Spectrochim. Acta*, 1972, **28A**, 897.
288 W. Zeil, *Fortschr. Chem. Forsch.*, 1972, **30**, 103.

are obtained using Wigner–Racah formalism. Exact calculations and the use of perturbation theory are discussed both for a single quadrupolar nucleus and for two quadrupolar nuclei.

In the evaluation of molecular parameters it is often useful to be able to check that the correct number of parameters is being used, as for example the number of cubic or quartic potential constants for a particular molecule. A systematic group-theoretical method of evaluating these numbers has been given.[289] The number of structural parameters and harmonic, cubic, and quartic constants can readily be obtained using these results from the characters of the representation Γ_{vib} of the vibrational displacements of the molecule. Two papers deal with the problem of obtaining atomic Cartesian co-ordinates from internal co-ordinates of atoms in molecules. One is concerned with the derivation of formulae which give the co-ordinates in terms of the structural parameters and angles of rotation, including molecules with asymmetric tops and tops on tops.[290] The second paper is concerned with the situation where a structure (monomer or polymer) is defined by a set of internal co-ordinate values, possessing the full symmetry of the structure but not calculated with sufficient accuracy to be consistent when more co-ordinates are specified than there are internal degrees of freedom.[291] The required atomic Cartesian co-ordinates are obtained by adjusting a trial set so that the internal co-ordinates relating the atoms are the best least-squares fit to the specified values. An earlier method for calculating derivatives of Cartesian co-ordinates with respect to internal co-ordinates has been extended to provide a general machine algorithm for the calculation of derivatives of moments of inertia with respect to internal co-ordinates.[292] The familiar problem of obtaining the structure of a symmetric rotor by use of isotopic substitution has been re-examined in the context of disubstitution.[176] Kraitchman's method is applied to the situation where two of three equivalent atoms are substituted isotopically. The use of these equations by themselves, or in conjunction with single substitution, or substitution for all three atoms is examined. For methylbromoform the rotational constants which have been determined proved insufficiently accurate for application of the method, but its use for trichloroacetonitrile leads to a structure in reasonable agreement with that previously reported. Two papers have appeared on the determination of average structures of molecules in which internal rotation occurs, the first dealing with the case of a single internal symmetric rotor [190] and the second with molecules having two symmetric internal rotors.[191] The results of applications of the method are given in another section of the Report. The theory of higher-order Stark effects of symmetric rotors has also been discussed.[293]

[289] J. K. G. Watson, *J. Mol. Spectroscopy*, 1972, **41**, 229.
[290] U. V. Sorokin and P. G. Maslov, *Doklady Akad. Nauk S.S.S.R.*, 1972, **202**, 384.
[291] M. Gardner and D. E. Rogers, *J. C. S. Faraday II*, 1972, **68**, 1410.
[292] J. J. Jacob and H. P. Thompson, *J. Mol. Structure*, 1972, **12**, 298.
[293] L. Tomutza and M. Mizushima, *J. Quant. Spectroscopy Radiative Transfer*, 1972, **12**, 925.

8 Vibration–Rotation Interaction

The Born–Oppenheimer approximation for a diatomic molecule has been further examined. Bunker[294] has given a treatment in which it is shown, even allowing for the breakdown of the Born–Oppenheimer approximation, that for a diatomic molecule in a $^1\Sigma$ electronic ground state the rotation–vibration term values, $F(v,J)$, can still be expanded as a Dunham series:

$$F(v,J) = \sum_{ij} Y_{ij}(v + \tfrac{1}{2})^i[J(J+1)]^j$$

Expressions have been obtained for the Dunham coefficients which involve additional parameters of two types, adiabatic and non-adiabatic, these describing the extent to which the Born–Oppenheimer approximation breaks down. Explicit expressions for the Dunham coefficients in terms of these additional parameters as well as the usual parameters have been derived, and their application for the determination of precise values for bond lengths and force constants of diatomic molecules in $^1\Sigma$ ground states has been discussed. The resulting corrections to r_e values are of the order of B_e^2/ω_e^2 times r_e and are negligible in all but the most precise work. The technique is applied to the i.r. and microwave data available for the hydrogen halides. It is also shown how the theory can account for the discrepancy that exists between the theoretical (*ab initio*) and experimental vibration–rotation energies of H_2 and D_2 in their electronic ground states. The isotopic dependence of equilibrium rotational constants and bond lengths (for molecules in $^1\Sigma$ states) has also been examined by Watson.[295] Watson's treatment involves three internuclear distances: r_e, r_e^{ad}, and r_e^{BO}. The first is obtained from the Dunham coefficient Y_{01} through $Y_{01} = \hbar^2/2\mu r_e^2$. The second, r_e^{ad}, is the bond length at the minimum of a defined adiabatic potential. The third, r_e^{BO}, refers to the minimum of the Born–Oppenheimer potential and unlike the other two is isotopically invariant. It is shown that:

$$Y_{01} = \frac{\hbar^2}{2\mu(r_e^{ad})^2}\left\{1 + \frac{\Delta Y_{01}^{(D)}}{B_e} + \frac{m_e}{m_p} g_J\right\}$$

where $\Delta Y_{01}^{(D)}$ is the difference between Y_{01} and B_e. It is also shown that:

$$r_e^{ad} = r_e^{BO}\left\{1 + m_e\left(\frac{d_a^{ad}}{M_a} + \frac{d_b^{ad}}{M_b}\right)\right\}$$

for a diatomic molecule whose constituent atoms have masses M_a and M_b. The first equation allows the determination of r_e^{ad} for each isotopic species and the second allows the determination of the isotopically invariant r_e^{BO}. It is emphasized that this second equation contains two parameters, d_a^{ad} and d_b^{ad}, one depending on each nucleus, and that for a complete solution of this equation it is necessary to have isotopic substitution at both atoms.

[294] P. R. Bunker, *J. Mol. Spectroscopy*, 1972, **42**, 478.
[295] J. K. G. Watson, *J. Mol. Spectroscopy*, 1973, **45**, 99.

It is pointed out that in earlier treatments d_a^{ad} and d_b^{ad} have generally been assumed equal, leading to the result that the correction term in the equation would be proportional to μ^{-1}. Applications of the results of the treatment are made to CO, HCl, and LiCl. To test the conclusion that the adiabatic bond length depends linearly on M_a^{-1} and M_b^{-1} calls for highly accurate information relating to both isotopic substitutions, but positive experimental support for this form of the dependence is found for CO.

The effect of the breakdown of the Born–Oppenheimer approximation on expressions for the dependence of the dipole moment of a diatomic molecule on vibrational state has also been examined, and so too has the dependence on vibrational state of the nuclear quadrupole coupling coefficient for a nucleus in a diatomic molecule.[296] After allowing for the Born–Oppenheimer approximation the vibrational dependence of the dipole moment can be expressed as a power series:

$$\mu_v = [\mu_e + (B_e/\omega_e)^2\mu_e] + \mu_1[(B_e/\omega_e)(v + \tfrac{1}{2})] + \mu_2[(B_e/\omega_e)(v + \tfrac{1}{2})]^2$$

A similar expression gives the dependence of the nuclear quadrupole coupling constant on vibrational state:

$$eQq = [eQq_e + (B_e/\omega_e)^2 eQq_e] + eQq_1[(B_e/\omega_e)(v + \tfrac{1}{2})] + eQq_2[(B_e/\omega_e)(v + \tfrac{1}{2})]^2$$

In the dipole moment expression the constants μ_e, μ_1, and μ_2 can be expressed in terms of the anharmonic potential constants and the dipole moment derivatives; the corresponding constants in the nuclear quadrupole coupling can be expressed in a similar way.

The work of Hougen, Bunker, and Johns [297] on a vibration–rotation Hamiltonian for a triatomic molecule, which allows explicitly for a large-amplitude bending vibration, has been extended.[298] Instead of the usual separation of the Hamiltonian in zeroth-order into a rotation part and a bending–stretching part, the separation is made into a rotation–bending part and a stretching part. To do this the elements of the usual μ-matrix, related to the instantaneous moment of inertia matrix, are expressed explicitly in terms of the bending co-ordinate and expanded as a power series in the two stretching co-ordinates, instead of the usual expansion in terms of all three normal co-ordinates. Similarly the potential function is expanded as a power series in the two stretching co-ordinates (with coefficients that are functions of the bending co-ordinate). In this way a zeroth-order Hamiltonian is obtained which describes a triatomic molecule as a rigid rotor that is simultaneously bending with each bond length held fixed. This Hamiltonian has advantages over the usual rigid-rotor harmonic oscillator approach, proving to be a closer description of the physics of the motion, since interaction terms are found to be much less important. The present paper con-

296 P. R. Bunker, *J. Mol. Spectroscopy*, 1973, **45**, 151.
297 J. T. Hougen, P. R. Bunker, and J. W. C. Johns, *J. Mol. Spectroscopy*, 1970, **34**, 136.
298 P. R. Bunker and J. M. R. Stone, *J. Mol. Spectroscopy*, 1972, **41**, 310.

siders the rotation–bending energy levels, $E(v_2,J)$, for this model, and a computer program has been written to evaluate these. The calculation involves, at most, five parameters, *viz.* the two bond lengths, the inter-bond angle, the quadratic bending force constant, and the potential energy barrier opposing the straightening of the molecule. No vibration–rotation interaction constants are explicitly included, the most important of these being implicit in the model. The rotation–bending energy levels of HCN and DCN in their bent $\tilde{A}$ electronic states have been calculated and so too have the levels for H_2O, D_2O, and HDO in their ground electronic states. Very satisfactory agreement between theory and experiment is obtained. A comparison has also been made of the use of curvilinear and rectlinear models in the determination of barriers to linearity in triatomic molecules.[299] Bend–stretch interactions are ignored *explicitly* in each model but are built into the rectilinear model *implicitly* by the requirement of a constant bond projection. Bending-mode transitions calculated for the two models are compared with experimental values for H_2O and D_2O in their ground electronic states and for the $^1B_1$ state. Similar comparisons are made for SO_2 in three excited states. In all cases the curvilinear model demonstrates a potential more nearly invariant to isotopic substitution than does the rectilinear model.

The problem of dealing with large-amplitude vibrations has been examined from the point of view of choice of rotating axes.[300] An analysis is made of the effect of change of definition of the rotating axes in terms of convergence of properties of perturbation expressions. After a discussion on finding the best axes, applications are made to an internal rotor problem, *gauche*-propionaldehyde, and to three small ring molecules, cyclopentene, trimethylene sulphide, and methylenecyclobutane. Another paper dealing with large-amplitude vibrations extends the technique introduced for triatomic molecules [297, 298] to obtain a Hamiltonian of a vibrating–rotating inverting XYZ_2 structure and applies the results to an excited electronic state of formaldehyde.[301]

A study has been made of vibration–rotation interaction for the bent XYZ molecule to second-order approximation using the Darling–Dennison Hamiltonian in the Amat–Nielsen–Goldsmith expansion.[302] The quantities $a_s^{\alpha\beta}$ and $(A_{ss'}^{\alpha\beta})'$, which enter the expansions for the instantaneous moments and products of inertia in the normal co-ordinates, have been examined. Compact expressions are obtained for $a_s^{\alpha\beta}$ and attention is drawn to sum rules in relation to the $(A_{ss'}^{\alpha\beta})'$, which simplify their calculation. A discussion of the fourth-order approximation is also given. Using the same kind of Hamiltonian for a planar asymmetric rotor, Chan and Parker [303] have obtained explicit expressions for the vibration–rotation constants, α, in terms

299 F. B. Brown, *J. Chem. Phys.*, 1973, **58**, 827.
300 H. M. Pickett, *J. Chem. Phys.*, 1972, **56**, 1715.
301 D. C. Moule and V. S. R. Rao, *J. Mol. Spectroscopy*, 1973, **45**, 120.
302 M. Y. Chan and P. M. Parker, *J. Mol. Spectroscopy*, 1972, **42**, 53.
303 M. Y. Chan and P. M. Parker, *J. Mol. Spectroscopy*, 1972, **42**, 449.

of the $a_s^{\alpha\alpha}$ and $(A_{ss}^{\alpha\alpha})'$. They go on to show that certain sums, involving the a values and moments of inertia, namely

$$I_x^2 a_n^x + I_y^2 a_n^y - I_z^2 a_n^z$$

are independent of the cubic constants. General expressions for these sums are obtained in terms of the vibrational frequencies and Coriolis coupling constants, and particular forms are given for XYZ and planar XY_2Z molecules. Expressions for the grand sum of these sums over all normal vibrations are also given.

The use of the method of contact transformation to obtain a reduced Hamiltonian for a near-symmetric rotor has been discussed.[304] It is required that terms appear in a reduced Hamiltonian should be linearly independent of each other. Matrix elements are classified into three types: (i) those diagonal with respect to the zeroth-order Hamiltonian, (ii) those with both diagonal and off-diagonal elements, and (iii) those which are purely off-diagonal. It is shown that in the jth order ($j \geqslant 2$), $j + 1$ of the first type, $2j - 1$ of the second type, and none of the third type remain in a reduced Hamiltonian. The non-degenerate case was considered by Watson who found that $2j + 3$ terms remain in the reduced Hamiltonian of a non-planar asymmetric rotor.[305]

A number of papers related to Coriolis interactions have appeared. Expressions have been obtained for Coriolis coupling coefficients for XY_2 and X_3 molecules of C_{2v} symmetry in parametric form.[306] Transformation properties of Coriolis coefficients for polyatomic molecules have been examined [307] and expressions have been obtained for the anomalous values of effective rotational constants found for states perturbed by first-order Coriolis resonance.[308]

The interesting prediction, made independently by Watson [309] and by Fox,[310] that a very small dipole moment is allowed to some types of formally non-polar molecules in their vibronic ground states through the mechanism of distortion of the molecule by centrifugal forces was reported in Volume 1 of these Reports. Although Watson was also concerned with molecules belonging to point groups D_{3h}, D_n, D_{2d}, and C_{3v}, both authors paid special attention to T_d molecules and, choosing methane as an example, made predictions of the magnitude of the so-called centrifugal distortion dipole moment, the frequencies of the $\Delta J = +1$ pure rotational transitions of this spherical top allowed by the effect and, most importantly, their intensities. The predictions have been amply fulfilled. The calculated values of the ground-state dipole moment of methane were soon confirmed by the experi-

[304] J. J. Choi and F. W. Birss, *J. Chem. Phys.*, 1972, **56,** 1937.
[305] J. K. G. Watson, *J. Chem. Phys.*, 1967, **46,** 1935.
[306] A. Alix and L. Bernard, *Z. Naturforsch.*, 1972, **27a,** 593.
[307] A. P. Aleksandrov, *Optika i Spektroskopiya*, 1973, **34,** 60.
[308] K. Sarka, *J. Mol. Spectroscopy*, 1972, **41,** 233.
[309] J. K. G. Watson, *J. Mol. Spectroscopy*, 1971, **40,** 536.
[310] K. Fox, *Phys. Rev. Letters*, 1971, **27,** 233.

mental measurement of Ozier[311] (as already reported). Now Rosenberg, Ozier, and Kudian[312] have observed a series of $J+1 \leftarrow J$ pure rotational transitions of methane in the far-i.r. region. The frequencies observed for the range $9 \leqslant J \leqslant 12$ agree within 0.1 cm^{-1} with predictions from the expression $2B_0(J+1) - D(J+1)^3$ using the best values of $B_0 = 5.240\,03 \pm 0.000\,06$ cm^{-1} and $D = (1.002 \pm 0.004) \times 10^{-4}$ cm^{-1} currently available while the excess of 0.35 cm^{-1} of calculated over observed values in the range $13 \leqslant J \leqslant 16$ is readily explained by the neglect of higher terms in the equation. Moreover, the absolute and relative intensities of lines predicted by Watson and Fox are substantiated by intensity measurements in the observed spectrum.

Recently, Watson and Dorney[313] and Fox[314] have, independently and using different theoretical approaches, considered another aspect of forbidden spectra in the vibronic ground state of T_d molecules. Their attention is now concentrated on the $\Delta J = 0$ transitions which are also allowed by the small centrifugal distortion dipole moment. The origin of such transitions in a spherical-top molecule lies in high-order terms in the rotational Hamiltonian. The term values for the energy of the vibronic ground state of a tetrahedral molecule are

$$E = B_0J(J+1) - DJ^2(J+1)^2 + D_t f(J,K)$$

where the symbols take their usual meanings, D_t is a constant, and $f(J,K)$ (to use Watson's notation) is an eigenvalue of the fine-structure Hamiltonian, the effect of which is to split the ground-state rotational levels into several components classified according to the T_d group representations (A_1, A_2, E, F_1, and F_2). It is transitions between the fine-structure components of a given J allowed by the centrifugal distortion dipole moment that the authors consider. Dorney and Watson use $D_t = 4.0 \times 10^{-6}$ cm^{-1} while Fox uses 4.403×10^{-6} cm^{-1} to calculate transitions from the above expression which are both allowed by the selection rules ($\Delta k = 2$ and $E \leftrightarrow E$, $A \leftrightarrow A$, $F \leftrightarrow F$) and which fall in the microwave region. Allowing for the different D_t values, their results are in agreement. Both authors compute dipole moment matrix elements and tabulate the intensities of the allowed transitions. The result of the calculations is a rich and weak spectral pattern, with absorption coefficients $\gamma \leqslant 10^{-10}$ cm^{-1} for most transitions, making it unlikely that they will be detected by conventional techniques. Dorney and Watson additionally consider the Stark effects of the $\Delta J = 0$ transitions, show that E species lines have linear Stark effects, and tabulate their Stark coefficients. Fox also includes in his paper a discussion of other effects, such as Mizushima–Venkateswarlu-type transitions and collision-induced absorptions, which might occur in the microwave region and obscure the transitions of interest.

[311] I. Ozier, *Phys. Rev. Letters*, 1971, **27**, 1329.
[312] A. Rosenberg, I. Ozier, and A. K. Kudian, *J. Chem. Phys.*, 1972, **57**, 568.
[313] A. J. Dorney and J. K. G. Watson, *J. Mol. Spectroscopy*, 1972, **42**, 135.
[314] K. Fox, *Phys. Rev. (A)*, 1972, **6**, 907.

A paper which also discusses the possibility of rotational spectra in the vibronic ground state for molecules of D_n, D_{2d}, D_{3h}, S_4, and T as well as T_d symmetry is published by Aliev.[315]

Several papers have appeared dealing with the centrifugal distortion analysis of microwave spectra and the use of centrifugal distortion constants in molecular force field calculations.

Kirchoff[316] has carried out an extremely thorough investigation of all aspects of centrifugal distortion analysis. In a very long paper the theoretical distortion treatment of Watson is reviewed in some detail and its relation to the Kivelson–Wilson theory discussed. In another section the calculation of force fields for bent symmetric XYX triatomic molecules from centrifugal distortion data is reviewed and the work of several authors brought together and rationalized. Also discussed are the application of the least-squares method to centrifugal distortion analyses, problems encountered such as rounding errors, and the use of the statistics associated with the least-squares method, for example as a means of detecting misassignments and mismeasurements. In later parts of the paper the various points discussed are illustrated by application to particular molecules. For example, the detection of misassignments and mismeasurements in the SiF_2, NSF, and OF_2 spectra is dealt with and the calculation of force fields in SO_2, F_2O, and SiF_2 discussed.

Two papers already mentioned in Section 3 but worthy of further review in the present context are the centrifugal distortion analyses of *cis*- and *trans*-nitrous acid and of thionyl fluoride. In the former case, Finnigan, Cox, Brittain, and Smith[77] analyse the microwave spectra of *cis*- and *trans*-nitrous acids and their deuteriated species using first-order distortion treatments due to Hill and Edwards and to Watson. The constants derived from both treatments are in agreement and are used with i.r. fundamentals and their isotopic shifts to determine eleven out of sixteen possible harmonic force constants for the molecule. This represents the first case of a planar tetra-atomic molecule of other than C_{2v} symmetry being so treated. Similarly, Smith and Lucas[97] determine centrifugal distortion constants from a Watson analysis of many lines in the microwave spectrum of the pyramidal tetra-atomic molecule thionyl fluoride. The distortion constants are employed with vibrational fundamental frequencies to determine ten of thirteen possible harmonic force constants (three are constrained to preset values). This is the first microwave-i.r. derivation of a harmonic force field for a non-planar asymmetric rotor.

Other publications in the area of centrifugal distortion deal with the upper and lower limits of the centrifugal distortion constants in symmetric-top-type molecules[317] and with the calculation of centrifugal expansion constants for XY_3 molecules of C_{3v} and D_{3h} symmetry.[318]

315 M. R. Aliev, *Pis'ma Zhur. eskp. i teor. Fiz.*, 1971, **14**, 600.
316 W. H. Kirchoff, *J. Mol. Spectroscopy*, 1972, **41**, 333.
317 M. R. Aliev, *Optika i Spektroskopiya*, 1972, **33**, 858.
318 V. S. Timoshinin and I. N. Godnev, *Izvest. V. U. Z. Fiz.*, 1972, **15**, 141.

9 Linewidths, Lineshapes, and Pressure Broadening

The measurement and interpretation of linewidths of rotational transitions has continued to attract attention. In principle such studies are sources of information about intermolecular forces and the transfer of rotational energy during collisions. They are also beginning to link up with double-resonance studies which can lead to rates of collision-induced energy transfer and selection rules governing such transfers. In practice, accurate measurements of linewidths have proved experimentally demanding. Theoretical interpretations have generally been based on an electric multipole expansion and through this has come the information about intermolecular forces. However, questions about the reproducibility of the experimental measurements and the adequacy of the model underlying theories of pressure broadening have both limited the extraction of molecular quadrupole moments or other parameters relating to intermolecular forces from pressure-broadening studies. During the year developments have been reported on the experimental measurement of lineshapes and linewidths, and at the same time tests of the adequacy of particular theories have been made for a number of systems.

Netterfield, Parsons, and Roberts [319] have described the design of a spectrometer for lineshape studies and give results of applications to the investigation of ammonia and carbonyl sulphide lines. The klystron source of the spectrometer is modulated by three signals on the reflector: one is a saw-tooth ramp with a repetition rate of one cycle every few seconds; the second is a frequency-stabilizing signal; and the third is a sinusoidal frequency of $25/n$ kHz where n is integral. The output signal of an IN 26 crystal detector is fed into an amplifier tuned to a frequency of 25 kHz and in this way outputs from the amplifier are obtained for the first, second, third, and nth harmonics. These amplitudes reflect the lineshape being studied. Comparison of experimental values is made with theoretically predicted values on the basis of Lorentz, Gaussian, and hard-sphere lineshapes. Measurements are reported for self-broadening of the $(J,K) = (6,6)$ inversion line of ammonia, broadening for the same line by helium, and for self-broadening of the $J = 1 \rightarrow 2$ transition of OCS. The results show that to within experimental error the shape of these lines can be adequately described as Lorentzian corrected for the Doppler effect. The majority of work on line broadening has in fact assumed a Lorentzian shape and made use of one or two measured points on the observed line to determine the half-width of the line. In practice there is nearly always a good deal of difficulty in determining the true base line for the spectrum because of the presence of Stark lobes, vibrational satellites, and other weak lines, and also for instrumental reasons. It has been pointed out [320] that if the line is truly Lorentzian all the points on the line contain information about the half-width and so it would be more reasonable to make use of all this

[319] R. P. Netterfield, R. W. Parsons, and J. A. Roberts, *J. Phys.* (*B*), 1972, **5**, 146.
[320] D. S. Olson, C. O. Britt, V. Prakash, and J. E. Boggs, *J. Phys.* (*B*), 1973, **6**, 206.

information. This has been done by recording the entire lineshape and fitting it numerically to the lineshape function using the centre frequency and half-width as parameters, and so in this way the need to determine the base line is avoided. The technique has been applied to the $J = 1 \rightarrow 2$ transition of OCS under conditions self-broadening and when broadening is by CH_3F. The $J = 0 \rightarrow 1$ transition of CH_3F has likewise been examined under conditions of self-broadening and when broadened by OCS. Calculated half-widths on the basis of the theory of Murphy and Boggs are in good agreement with the experimental results except that there is a small but apparently real discrepancy in the case of OCS self-broadening.

Pressure broadening has been investigated for a number of systems and measured half-widths of lines have been compared with values calculated theoretically.[321] Linewidths have been investigated for four rotational lines of ethylene oxide when broadened by three polar molecules of widely differing dipole moments, *viz.* C_2H_4O, OCS, and CH_3CN.[322] Comparison is made with values calculated theoretically, taking only dipole–dipole interaction into account, on the basis of the theories of Murphy and Boggs and of Anderson. The theoretical widths from the former are smaller than from the latter, but are still larger than the experimental values. A similar study [323] of ethylene oxide lines has been made for broadening by two quadrupolar molecules, N_2 and H_2. Molecular quadrupole moments were calculated on the basis of both theories, and the values are found to be in reasonable agreement with values found by other methods. A similar investigation has been made of the J,K 0,0 $\rightarrow$ 1,0 transition of methyl bromide in which again both dipolar and quadrupolar broadening are examined. For broadening by CH_3Br, OCS, and CH_3CN, theoretical calculations, on the basis of dipole–dipole interactions only, lead for both theories to rather larger linewidths than are observed experimentally, the values obtained from the Murphy and Boggs theory being somewhat nearer to the measured value.[324] On the other hand for broadening of the same methyl bromide line by quadrupolar molecules (N_2, H_2, CO_2, O_2, C_6H_6, and CS_2) both theories generally lead to smaller linewidth parameters than the measured values.[325] Molecular quadrupole moments are calculated and values of quadrupole moments calculated by other methods are tabulated for comparison. For broadening by the spherical rotor CCl_4, both theories lead, on the basis of dispersion forces, to much lower linewidths than are found experimentally. Comparison of experimental and theoretically calculated linewidth parameters for self-broadening have also been made for CH_3CN [326] and CH_2F_2.[327]

The theory of Murphy and Boggs has been used to provide a method for

[321] Krishnaji, S. L. Srivastava, and P. C. Pandey, *Chem. Phys. Letters*, 1972, **13**, 372.
[322] P. C. Pandey and S. L. Srivastava, *J. Chem. Phys.*, 1972, **57**, 3282.
[323] P. C. Pandey and S. L. Srivastava, *J. Chem. Phys.*, 1973, **58**, 1630.
[324] P. C. Pandey and S. L. Srivastava, *J. Phys. (B)*, 1972, **5**, 997.
[325] P. C. Pandey and S. L. Srivastava, *J. Phys. (B)*, 1972, **5**, 2074.
[326] G. P. Srivastava and H. O. Gautam, *Indian J. Pure Appl. Phys.*, 1972, **10**, 442.
[327] P. C. Pandey and S. L. Srivastava, *J. Phys. (B)*, 1972, **5**, 1427.

calculating rates of collision-induced transitions between molecular rotational levels.[328] Comparison of the theory with experiment is made through measured intensities in modulated double-resonance experiments. Values calculated for transitions appropriate to double-resonance experiments for H_2CO, HDCO, H_2CCO, HCN, and DCN are in satisfactory agreement with the experimental results. Another paper related to theory of energy redistribution in molecular collisions has derived a vibration–rotation Hamiltonian for bimolecular collisions, analogous to the Wilson–Howard vibration–rotation Hamiltonian for isolated molecules.[329] Group theory is applied to the symmetry operations for the Hamiltonian and selection rules are examined. The selection rule $\Delta K = \pm 3n$ for He–NH_3 collisions is readily obtained. Though other qualitative rules which have been determined experimentally cannot be obtained without some numerical calculation, the group-theory treatment is expected to reduce the extent of such numerical calculations. Other theoretical papers include a simplified version of Anderson's theory for a diatomic molecule perturbed by a linear molecule with a very small rotational constant.[330] Anderson's theory has also been examined for asymmetric rotors and attention is drawn to the inclusion of the asymmetry of the electric charge distribution.[331] The behaviour of the rotational absorption spectrum of a linear molecule in a non-polar buffer gas, as the frequency of collision increases, has been examined theoretically. The behaviour is traced from the region where sharp lines are obtained, through overlapping lines, to the region where the Debye theory is valid.[332] Another approach to the effect of collisions on rotation considers the problem in terms of memory functions.[333]

Two papers have appeared on pressure broadening of the O_2 microwave spectrum. In one a general expression for the calculation of pressure broadening is derived in the strong-collision model using a procedure which avoids perturbation expansions, impact-parameter cut-offs, and straight paths.[334] Excellent agreement is obtained with experiment. The other work on oxygen [335] has been an examination of absorption in the region 48—81 GHz over the pressure range 3.7—51 atm, either as pure oxygen or as mixtures with argon or nitrogen, by use of a Fabry–Perot interferometer. Good agreement was found with the theoretically calculated lineshape. A number of other papers also deal with problems related to atmospheric pressure broadening. Line broadening of rotational fine structure in i.r. bands of SO_2 and NO_2 by nitrogen and oxygen has been examined [336] in terms of the

[328] V. Prakash and J. E. Boggs, *J. Chem. Phys.*, 1972, **57**, 2599.
[329] J. T. Hougen, *J. Chem. Phys.*, 1972, **56**, 6245.
[330] M. Giraud, D. Robert, and L. Galatry, *J. Chem. Phys.*, 1972, **57**, 144.
[331] G. Yamamoto and T. Aoki, *J. Quant. Spectroscopy Radiative Transfer*, 1972, **12**, 227.
[332] E. P. Gross and J. Otieno-Malo, *J. Chem. Phys.*, 1972, **57**, 2229.
[333] F. Bliot and E. Constant, *Chem. Phys. Letters*, 1973, **18**, 253.
[334] U. Mingelgrin, R. G. Gordon, L. Frenkel, and T. E. Sullivan, *J. Chem. Phys.*, 1972, **57**, 2923.
[335] T. H. Dillon and J. T. Godfrey, *Phys, Rev.* (*A*), 1972, **5**, 599.
[336] G. D. T. Tejwani, *J. Chem. Phys.*, 1972, **57**, 4676.

Anderson theory as amplified by Tsao and Curnutte.[337] Broadening of CO_2 lines by N_2 has also been considered.[338, 339] Half-widths of HCl rotation lines have been measured in the far-i.r. and broadening by helium investigated.[340] Other papers [341, 342] report broadening studies in both the i.r. and microwave regions.

10 Double Resonance

There has been a considerable number of publications concerned with microwave double resonance since Volume 1 of these Reports. The whole range of combinations from microwave–radiofrequency fields, through microwave–microwave and microwave–i.r., to microwave–optical double-resonance spectroscopy has been covered.

The one paper dealing with microwave–radiofrequency double resonance describes an experiment by Schwarz and Dreizler [343] wherein the small Stark splittings between the $M = 1$ and $M = 2$ components of the 2_{11} level and the 2_{12} level in formaldehyde are pumped with a strong radiofrequency field. The Stark components of the $2_{11} \leftarrow 2_{12}$ transition constitute the signal transition and the fields are so arranged that the microwave radiation is polarized parallel to the applied Stark electric field while the radiofrequency field is polarized perpendicular to these. When the pump radiation is at resonance with the Stark $M = 1$ to $M = 2$ splitting in the levels, observation of the signal shows a characteristic double-resonance pattern for the $M = 1$ and $M = 2$ components of the $2_{11} \leftarrow 2_{12}$ transition.

The technique of modulated microwave double-resonance (MMDR) spectroscopy, introduced recently by Wilson and his co-workers,[344] uses source modulation of the pump. In such experiments the pump field is switched on and off resonance by, for example, a square wave applied to a klystron reflector. The signal field is then monitored at the modulation frequency so that a signal will only be detected if the population difference in the signal levels varies with time at the modulation frequency. The utility of this technique is that if the signal and pump transitions have no levels in common an MMDR signal is detected only if there is a collisionally induced population transfer between the pump and signal levels (four-level system). At low pressure of the sample gas, when few collisions occur between molecules during the pump-on period, no double-resonance signal can result. At slightly higher pressures when a few collisions occur during the pump-on period there will be a time lag while the effect of the pump field is transferred

337 C. J. Tao and B. Curnutte, *J. Quant. Spectroscopy Radiative Transfer*, 1962, **2**, 41.
338 C. Boulet, E. Arie, J. P. Bounich, and N. Lacome, *Canad. J. Phys.*, 1972, **50**, 2178.
339 A. Arych and A. Sorgen, *Phys. Rev.* (*A*), 1972, **5**, 1967.
340 J. Pourcin, *J. Quant. Spectroscopy Radiative Transfer*, 1972, **12**, 1617.
341 F. Bliot, C. Abbar, and E. Constant, *Mol. Phys.*, 1972, **24**, 241.
342 M. Cattani, *Anais Acad. brasil. Cienc.*, 1971, **43**, 51.
343 R. Schwarz and H. Dreizler, *Z. Naturforsch.*, 1972, **27a**, 708.
344 R. C. Woods, A. M. Ronn, and E. B. Wilson, *Rev. Sci. Instr.*, 1966, **37**, 927.

slowly to other rotational levels. As a result a phase shift of the double-resonance signal relative to the applied modulation will occur. At high pressures, many collisions occur during the modulation period and so the phase shift should approach 0 or 180° depending on whether the pump causes a decrease or increase in the absorption. Measurement of the intensity of the MMDR signal relative to the conventional Stark amplitude of the signal transition and the phase lags as a function of pressure gives much information about the selection rules and the mechanism of collisional energy transfer.

Cohen and Wilson have now published two papers applying the study of the above-described MMDR technique to rotational energy transfer. In the first of these [345] the double-resonance relative intensities and phase shifts have been determined as a function of pressure for the HCN *l*-doublet transitions up to $\Delta J = +3$ from the pump transition ($J = 10$) using both pure HCN gas and its mixture with one of several different noble gases. The intensity data show that the behaviour of helium as a collision partner differs distinctly from that of argon or xenon. For the last two the relative intensity of the MMDR signals for all ΔJ gradually decrease as the noble-gas partial pressure is increased whereas for helium the $\Delta J = +1$ intensity decays but increases are recorded for $\Delta J = +2$ and $\Delta J = +3$ signals. This is evidence that population is being transferred between states with definite preference, the strongest being for $\Delta J = 2$, $\pm \leftrightarrow \pm$, where $\pm$ indicates parities of the *l*-doublet components. Another conclusion of interest is that the curves of phase shift *versus* pressure for the mixtures approach the high-pressure limit more slowly than for pure HCN, indicating that HCN–HCN collisions are more effective in rotational energy transfer than HCN–noble-gas collisions. Moreover, the approach is more rapid for the heavier noble gas.

In their second paper, Cohen and Wilson [346] study rotational energy transfer in trimethylene oxide, β-propiolactone, cyclobutanone, and *cis*-difluoroethylene with the MMDR technique. The main conclusion from this study is that the predominant collisional transitions observed in these molecules are those expected from consideration of first-order dipole–dipole interactions. A particular conclusion from studying the MMDR relative intensities in vibrationally excited transitions is that no evidence can be found for a rotational relaxation mechanism *via* a resonant vibrational energy transfer process, indicating the cross-section for pure rotational relaxation to be much larger than that for vibrational relaxation processes. Also, an investigation of the *M*-dependence of rotational energy transfer by saturation of individual *M* components of the pump transition while observing the various *M* components of the signal shows that ΔM selection rules are determined by the ΔM changes that minimize angular momentum reorientation during collision.

[345] J. B. Cohen and E. B. Wilson, *J. Chem. Phys.*, 1973, **58**, 442.
[346] J. B. Cohen and E. B. Wilson, *J. Chem. Phys.*, 1973, **58**, 456.

Seibt[347] also uses the MMDR technique to look at rotational relaxation in ethanol. A MMDR spectrometer, similar to that of Woods, Ronn, and Wilson,[344] is described in some detail. The summarized conclusions of this work are that collision-induced transitions in both C_2H_6OH and CH_2DCH_2OH species obey normal dipole selection rules, that *a*-dipole and $\Delta J = 2$ transitions induced by collision in *trans*-ethanol are much less probable than *b*-dipole transitions, and that a comparison of linewidth and double-resonance data for ethanol indicates dipole–dipole interactions to be the principal mechanism for rotational transitions induced by collision.

Flynn[348] has sounded a cautionary note in the use of MMDR spectroscopy as a technique in the search for pairs of transitions sharing a common level. In this situation the pump source is tuned over a frequency range and those frequency pairs for the pump and signal sources at which double-resonance signals occur are noted. It is generally assumed that in such experiments a detectable double-resonance signal is observed only when the modulated pump source has a frequency within a few linewidths of the resonance frequency of a transition sharing a common level with the signal transition, at whose frequency the double-resonance signal occurs. Flynn, however, shows that for molecules with intense microwave absorption spectra, double-resonance signals can appear even if the pump field is as much as a few hundred MHz off resonance. He also derives an expression for the double-resonance lineshape when the pump is far from resonance. In a comment on Flynn's paper, Macke and Glorieux[349] suggest that the effect described is related to the high-frequency Stark effect arising from the non-resonant pump field. Such effects have been treated theoretically by Glorieux, Legrand, Macke, and Messelyn.[350]

In a conventional microwave–microwave double-resonance experiment, Redon and Fourrier[351] pump selected *M*-value ($\Delta M = 0$) components in the Stark effect of $J = 1 \leftarrow 0$, $2 \leftarrow 1$, and $3 \leftarrow 2$ transitions of OCS while observing Stark components of various *M* values ($\Delta M = 0$) in a different $J + 1 \leftarrow J$ transition chosen so as to ensure that the signal and pump transitions share no common level. The relative change in intensity of the signal transition in the presence of the pump, $\Delta I/I$, is observed and the signs of the various changes are explained on the basis of dipolar and quadrupolar collision-induced transitions from pump to signal levels. An expression for the absorption coefficient in two-photon resonance transitions is deduced under general conditions by Chiarini, Martinelli, Santucci, and Bucci.[352] No particular restrictions are imposed on the levels systems or the frequencies or intensities of the fields.

[347] P. J. Seibt, *J. Chem. Phys.*, 1972, **57**, 1343.
[348] G. W. Flynn, *J. Mol. Spectroscopy*, 1972, **43**, 353.
[349] B. Macke and P. Glorieux, *J. Mol. Spectroscopy*, 1973, **45**, 302.
[350] P. Glorieux, J. Legrand, B. Macke, and J. Messelyn, *J. Quant. Spectroscopy Radiative Transfer*, 1972, **12**, 731.
[351] M. Redon and M. Fourrier, *Chem. Phys. Letters*, 1972, **17**, 114.
[352] F. Chiarini, M. Martinelli, S. Santucci, and P. Bucci, *Phys. Rev.* (*A*), 1972, **6**, 1300.

Four papers reporting microwave–i.r. double-resonance experiments have appeared. We reported last year an elegant method, devised by Oka and Shimizu,[353] which allows an i.r. laser transition to be tuned by several hundred MHz into exact resonance with a molecular absorption. Briefly, the i.r. laser frequency ν_L indicated in Figure 17 is off resonance with levels 2 and 3 by $\Delta\nu$ (perhaps ~300 MHz). If, however, the level system is simultaneously irradiated with a sufficiently high-powered microwave radiation at $\nu_{12} + \Delta\nu$ and the i.r. laser, the two-photon transition ν_{13} can become strongly allowed. Thus, the non-linear process effectively 'tunes' the laser into resonance. When a gas capable of absorbing i.r. laser radiation is placed within the laser cavity, the Doppler-broadened absorption profile of the i.r. transition exhibits a very sharp 'blip' at its centre called the Lamb dip.

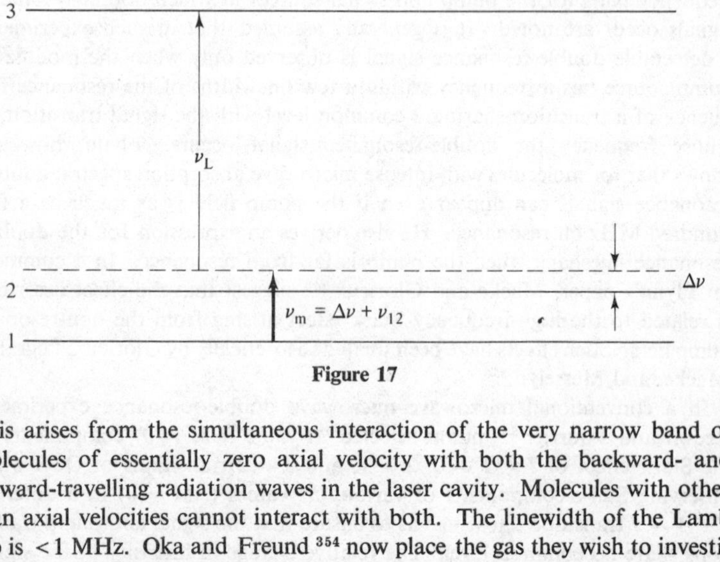

Figure 17

This arises from the simultaneous interaction of the very narrow band of molecules of essentially zero axial velocity with both the backward- and forward-travelling radiation waves in the laser cavity. Molecules with other than axial velocities cannot interact with both. The linewidth of the Lamb dip is <1 MHz. Oka and Freund [354] now place the gas they wish to investigate, for example $^{15}NH_3$, within the laser cavity and simultaneously apply to it the requisite saturating microwave radiation at $\Delta\nu$ so as to 'tune' the laser into resonance with the molecular transition and therefore cause the Lamb dip to appear. The appearance of the Lamb dip is restricted to a very narrow range of $\Delta\nu$ and hence $\Delta\nu$ can be very accurately measured. Given a stabilized laser of accurately known frequency ν_L, the molecular absorption frequency can therefore be measured very accurately. In the present experiment the $\nu_2[{}^qQ(4,4)]$ i.r. transition of $^{15}NH_3$, the $P(15)$ N_2O laser line, ν_m = 23 360 MHz, and $\Delta\nu = +314$ MHz are used in one example.

The problem of achieving exact coincidence between a monochromatic laser and an i.r. absorption in low-pressure gas is dealt with by other authors

[353] T. Oka and T. Shimizu, *Appl. Phys. Letters*, 1971, **19**, 88.
[354] S. M. Freund and T. Oka, *Appl. Phys. Letters*, 1972, **21**, 60.

using other techniques. Takami and Shimoda [355] use a 3.5 μm He–Xe gas laser which can be tuned by virtue of the Zeeman effect to obtain an exact coincidence with the $6_{06}(\nu_5 = 1) \leftarrow 5_{15}$(ground state) transition of formaldehyde. The microwave transition $5_{14} \leftarrow 5_{15}$ in the ground state is used as the pump and is saturated with microwave radiation at 72 409 MHz during the experiments. The i.r. laser absorption is used as the signal transition in which changes of intensity are sought as a result of the microwave saturation. An interesting result of this work is the pressure dependence of the double-resonance signal. When the pressure of formaldehyde is sufficiently low that the i.r. as well as the microwave transition is saturated, the double-resonance signal changes sign. The theory of this double saturation effect is discussed.

On the other hand, Fourrier and Redon [356] achieve exact coincidence between vibration–rotation transitions of the $\nu_2 = 1 \leftarrow 0$ band of ammonia and CO_2 or N_2O laser lines by applying high, very homogeneous electric fields to the NH_3 gas in a specially designed cell. The Stark effect shifts a particular vibration–rotation transition into resonance with the high-powered laser. The Stark components in the ($\nu_2 = 0$) ground-state inversion transitions of ammonia are then monitored as signal transitions.

Beautiful microwave–optical double resonance (MODR) experiments have been performed by Field, Bradford, Broida, and Harris. In the first experiment of its kind,[357] these authors use the 496.5 nm line of an argon-ion laser, which coincides with the $R(1)$ line of the $A^1\Sigma$–$X^1\Sigma$ (7,0) band of gaseous barium oxide, to pump barium oxide molecules from the $J'' = 1$ level to the $J' = 2$ level as shown in Figure 18. The result is an intense fluorescence from the $J' = 2$ $A^1\Sigma$ $v' = 7$ state, as indicated. Microwave radiation is introduced into the region of the fluorescing gaseous barium oxide and when tuned through the region near 37 403.9 MHz is found to enhance the fluorescence so giving rise to a MODR signal of half-width 3.5 MHz. The microwave radiation, it is firmly established in a later paper,[358] pumps the $J'' = 1 \leftarrow 2$ transition and therefore increases the population of the $J'' = 1$ state, leading to the enhanced A–X (7,1) band fluorescence indicated in Figure 18.

In a second experiment the same authors [358] report the first observations of microwave rotational transitions in a short-lived electronically excited state of a diatomic molecule. The argon-ion laser is again used to pump molecules as indicated in Figure 18. Now using microwave radiation centred at 44 891.4 and 29 927.6 MHz (in separate experiments), molecules of barium oxide are pumped from the $J' = 2$ to the $J' = 3$ and from the $J' = 2$ to the $J' = 1$ levels respectively of the $A^1\Sigma$ $v' = 7$ state (see Figure 18). This population transfer results in a transfer of fluorescence intensity from the

[355] M. Takami and K. Shimoda, *Jap. J. Appl. Phys.*, 1972, **11**, 1648.
[356] M. Fourrier and M. Redon, *Appl. Phys. Letters*, 1972, **21**, 463.
[357] R. W. Field, R. S. Bradford, D. O. Harris, and H. P. Broida, *J. Chem. Phys.*, 1972, **56**, 4712.
[358] R. W. Field, R. S. Bradford, H. P. Broida, and D. O. Harris, *J. Chem. Phys.*, 1972, **57**, 2209.

$R(1)$ and $P(3)$ lines into the $R(2)$ and $P(4)$ lines of the A–X (7,1) photoluminescence band when, for example, the $J' = 3 \leftarrow 2$ transition is pumped. Given the observed microwave frequencies, the rotational constant B_7 for the $A^1\Sigma$ state is estimated to be 7482.01 MHz.

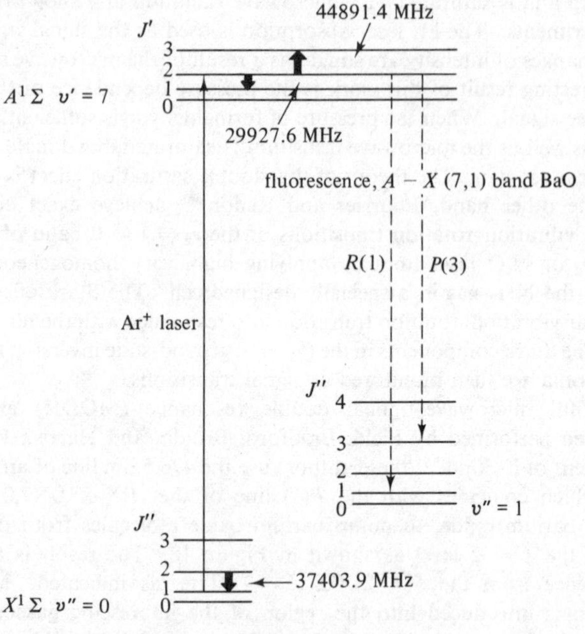

Figure 18

A phenomenon which is not strictly double resonance (but which may be treated here because the partial saturation of a transition is involved) is that concerned with transient effects in an absorption, such as can be seen during very short time periods after partially saturating microwave radiation is switched on or off resonance. A typical experiment described in an experimental paper on the subject by Macke and Glorieux [359] gives an idea of what is involved. Low-pressure OCS gas (17 mTorr) is subject to microwave radiation at a frequency ν_0 which is exactly resonant with the $J = 2 \leftarrow 1$ ground-state transition. A static electric field of very fast rise time and strength about 1000 V cm^{-1} is switched on and the behaviour of the absorption at ν_0 in the 10 μs after switch-on is observed by means of a Boxcar integrator. The absorption signal does not fall off smoothly but exhibits an exponentially damped oscillatory behaviour, *i.e.* relaxation wiggles. In another experiment by the same authors, irradiation is at ν_S, the frequency of a Stark component when the field is on, and the Stark field is suddenly

[359] B. Macke and P. Glorieux, *Chem. Phys. Letters*, 1972, **14**, 85.

switched off. The transient behaviour of absorption at ν_s after switch-off is then observed.

Macke and Glorieux [359] offer a simple qualitative picture of their observed transient phenomena. For example, in the second experiment described above, when the Stark field is switched off the molecules excited to the upper level of the Stark component previously by the microwave radiation field at ν_s now make a rapid adiabatic shift to the new state, *i.e.* the $J = 2$ zero-field level. Molecules thus in the $J = 2$ zero-field level can then emit radiation coherently at ν_0, the zero-field $J = 2 \rightarrow 1$ transition frequency, and this beats with the ν_s radiation during the decay process, giving rise to the exponentially damped oscillation described. A theoretical treatment of such transient phenomena has also been given by Macke and Glorieux.[360] Schwendeman and Pickett [361] give an alternative theoretical treatment of the behaviour of the absorption coefficient in microwave spectroscopy under conditions of partial saturation at times shortly after the onset of radiation. They too show that the absorption coefficient varies with time as an exponentially damped sine wave.

Transient phenomena are also observed in an i.r.–microwave 'double-resonance' experiment described by Levy, Wang, Kukolich, and Steinfeld.[362] A very short i.r. pulse from a laser is absorbed by low-pressure ammonia gas. The transient behaviour of a microwave absorption having a level in common with the i.r. absorption is observed in the 10 μs immediately after receipt of the pulse. In fact, the ${}^{Q}Q(8,7)$ line of the ν_2 band in ammonia is pumped by a pulse from the $P(13)$ line of an N_2O laser at 927.739 cm^{-1}. The $J = 8$, $K = 7$ inversion doublet in the ground state is monitored by low-powered microwave radiation. Again, the microwave absorption signal exhibits an exponentially damped sinusoidal oscillation after the pulse. The same paper treats the observed phenomena theoretically.

11 Astrophysics

The search for small molecules in interstellar space by means of their rotational transitions continues to be a field of frenetic activity. However, the number of molecules reported that have been undetected hitherto is small, as judged from papers in the only journal in this field abstracted by *Chemical Titles* (the *Astrophysical Journal*). The past year seems rather to have been one of consolidation, with known lines of known molecules being detected in new sources and new lines of known molecules being reported from known sources. Moreover, a considerable effort is being made by microwave spectroscopists to measure, with high precision, laboratory rest frequencies in molecules of current or possible future astronomical interest. It is interesting

360 B. Macke and P. Glorieux, *Chem. Phys. Letters*, 1973, **18**, 91.
361 R. H. Schwendeman and H. M. Pickett, *J. Chem. Phys.*, 1972, **57**, 3511.
362 J. M. Levy, J. H.-S. Wang, S. G. Kukolich, and J. I. Steinfeld, *Phys. Rev. Letters*, 1972, **29**, 395.

that of twenty papers abstracted in the *Astrophysical Journal* in the period covered, thirteen are concerned with interstellar observations while seven involve laboratory measurements.

Interstellar hydrogen sulphide ($H_2{}^{32}S$) has been detected for the first time [363] through the 1_{10}–1_{01} transition in emission at 168.76 GHz, bringing the number of sulphur molecules so detected to four (H_2S, OCS, CS, and H_2CS). From the detection of hydrogen sulphide in seven Galactic sources it is found to be about as abundant as formaldehyde. An additional line of silicon monoxide ($J = 2 \rightarrow 1$) at 86 847 MHz is detected [364] in Sgr.B2 while this molecule is detected in Ori.A for the first time through the same transition. Some interesting conclusions about isotopic ratios have been made. Observations of $J = 1 \rightarrow 0$ emission lines of $^{13}C^{16}O$ and $^{12}C^{18}O$ from nine sources [365] lead to ratios of the line intensities in each case within a factor of two of the terrestrial ratio. On the other hand, observations of the same transition in four interstellar dark clouds [366] gives a high apparent value ($\sim$0.3) for the $^{13}C/^{12}C$ ratio owing to large optical depth densities of carbon monoxide. The nitrogen nuclear quadrupole hyperfine components $F = 0 \rightarrow 1$, $2 \rightarrow 1$, $1 \rightarrow 1$ of the $J = 1 \rightarrow 0$ transition of $D^{12}C^{14}N$ in the Orion A cloud have been resolved [367] and the observed intensities of the three components are in agreement with theoretical line strengths. A comparison of the $D^{12}C^{14}N$ intensity with those from recent observations [368] of $H^{13}C^{14}N$ and $H^{12}C^{15}N$ leads to the interesting result that the ratio $D/H = 6 \times 10^{-3}$ in hydrocyanic acid is 40 times greater than the terrestrial value. An explanation of this high ratio based on chemical fractionation due to the different zero-point vibrational energies of HCN and DCN is advanced. The positive identification of the $J = 1 \rightarrow 0$ transition of $H^{12}C^{14}N$ and the $J = 2 \rightarrow 1$ transition of $^{12}C^{32}S$ in NGC 2264 has allowed the radial velocity of the cloud to be determined [369] and the rest frequency of two unidentified lines to be established more precisely.

The 'anomalous' 1_{11}–1_{10} line at 4830 MHz in formaldehyde, so called because it appears in absorption, continues to excite interest. Thaddeus [370] has investigated the quantum mechanics of the collisional-pumping process proposed by Townes and Cheung [371] as the cause of this 'anomalous' absorption in diffuse dark nebulae and concludes that quantum effects probably

[363] P. Thaddeus, M. L. Kutner, A. A. Penzias, R. W. Wilson, and K. B. Jefferts, *Astrophys. J.*, 1972, **176**, L73.
[364] D. F. Dickinson, *Astrophys. J.*, 1972, **175**, L43.
[365] A. A. Penzias, K. B. Jefferts, R. W. Wilson, H. S. Liszt, and P. M. Solomon, *Astrophys. J.*, 1972, **178**, L35.
[366] A. A. Penzias, P. M. Solomon, K. B. Jefferts, and R. W. Wilson, *Astrophys. J.*, 1972, **174**, L43.
[367] R. W. Wilson, A. A. Penzias, K. B. Jefferts, and P. M. Solomon, *Astrophys. J.*, 1973, **179**, L107.
[368] R. W. Wilson, A. A. Penzias, K. B. Jefferts, P. Thaddeus, and M. L. Kutner, *Astrophys. J.*, 1972, **176**, L77.
[369] B. Zuckerman, M. Morris, P. Palmer, and B. E. Turner, *Astrophys. J.*, 1972, **173**, L125.
[370] P. Thaddeus, *Astrophys. J.*, 1972, **173**, 317.
[371] C. H. Townes and A. C. Cheung, *Astropys. J.*, 1969, **157**, L103.

limit the proposed process to regions where the kinetic temperature exceeds 45 K. A new experimental fact about the 1_{11}–1_{10} formaldehyde absorption in at least one dust cloud is that there exists an anomalous intensity distribution among the hyperfine components of the transition.[372] Hitherto, equilibrium line intensities have been attributed to these components in attempts to explain the 'anomalous' absorption. From a survey of nine Galactic sources, isocyanic acid ($H^{14}N^{12}C^{16}O$) has been observed through the $4_{04} \rightarrow 3_{02}$ transition [373] only in the directions of Sgr.B2 and possibly W51. A further transition, $1_{01} \rightarrow 0_{00}$, in the former direction has now been identified.[374] Methanol remains the most detected of interstellar molecules with the identification of the $4_1 \rightarrow 3_0\,(E_2)$ [375] and $5_1 \rightarrow 4_0\,(E_2)$ [376] transitions in the Sgr.B2 direction while the $1_1 \rightarrow 0_0$ (*A* and *E* species) [377] lines are observed in both the Sgr.B2 and Sgr.A directions.

Complementary to these extensive interstellar observations are the aforementioned laboratory measurements of rest frequencies which can be discussed in two categories. In the first are measurements concerned with those molecules already detected in interstellar space in an effort to improve previously known frequencies or to establish, with high precision, transition frequencies of likely interest. In the second category are the measurement of transitions in small molecules considered as likely to exist in space.

Included in the first classification are measurements to a precision exceeding 1 part in 10^6 of millimetre transitions of formaldehyde [378] ($H_2{}^{12}CO$ and $H_2{}^{13}CO$) and measurements of *a*-type, *R*-branch transitions in isocyanic acid [379] ($H^{14}N^{12}C^{16}O$, $H^{15}N^{12}C^{16}O$, $H^{14}N^{13}C^{16}O$, and $H^{14}N^{12}C^{18}O$). In both cases centrifugal distortion analyses are used to predict further transitions of interest in radioastronomy. Also of interest in this context is a critical review of the microwave spectra of formaldehyde, formamide, and thioformaldehyde by Johnson, Lovas, and Kirchoff.[380] The authors tabulate all available data for the spectra of the observed isotopic species of these molecules and report a detailed centrifugal distortion analysis for each of the most abundant species, from which predictions of further transitions are made with high confidence limits. High-resolution laboratory measurements are reported for the 1_{11}–1_{10} methanol line [381] at 834.267 ± 0.002 MHz and the four transitions between the $F = 2{,}1$ hyperfine levels of the $^2\Pi_{\frac{3}{2}}$, $J = \frac{3}{2}$, doublet

[372] N. H. Dieter, *Astrophys. J.*, 1972, **178,** L133.
[373] L. E. Snyder and D. Buhl, *Astrophys. J.*, 1972, **177,** 619.
[374] D. Buhl, L. E. Snyder, and J. Edrich, *Astrophys., J.* 1972, **177,** 625.
[375] A. E. Turner, M. A. Gordon, and G. T. Wrixon, *Astrophys. J.*, 1972, **177,** 609.
[376] B. Zuckerman, B. E. Turner, D. R. Johnson, P. Palmer, and M. Morris, *Astrophys. J.*, 1972, **177,** 601.
[377] A. H. Barrett, R. N. Martin, P. C. Myers, and P. R. Schwartz, *Astrophys. J.*, 1972, **178,** L23.
[378] R. B. Nerf, jun., *Astrophys. J.*, 1972, **174,** 467.
[379] W. H. Hocking, M. C. L. Gerry, and G. Winnewisser, *Astrophys. J.*, 1972, **174,** L93.
[380] D. R. Johnson, F. J. Lovas, and W. H. Kirchoff, *J. Phys. and Chem. Ref. Data*, 1973, **1,** 1011.
[381] H. E. Radford, *Astrophys. J.*, 1972, **174,** 207.

state of OH,[382] the latter from a beam maser study with frequencies accurate to 100 Hz.

Molecular species included in the second category (*i.e.* not already observed in interstellar space) are HNO, NaOH, and the HCO and DCO radicals. The $1_{01} \leftarrow 0_{00}$ transition of HNO at 81 477.49 ± 0.1 MHz is presented as astronomically the most important for that molecule.[383] The $1_{01} \leftarrow 0_{00}$ transitions of the HCO and DCO radicals are reported for the first time,[384] with the $J = \frac{3}{2} \leftarrow \frac{1}{2}$, $F = 2 \leftarrow 1$ transition at 86 670.55 ± 0.20 MHz in HCO singled out for special mention. The rotational spectrum of sodium hydroxide[385] has been observed for the first time and the ground-state constants $B_0 = 12\,567.054 \pm 0.01$ MHz and $D_0 = 0.028\,72 \pm 0.000\,05$ MHz obtained by fitting transitions to the familiar linear-rotor expression.

12 Instrumentation

The relative intensities of microwave rotational transitions in a given compound are replete with information about vibrational separations, barriers to internal rotation of one part of the molecule against the other, and barriers to inversion. In determining, for example, a barrier to internal rotation from microwave spectroscopy one can measure the intensity I_0 of a given rotational transition in the ground state relative to the intensity I_v of the same transition in a torsionally excited state, v. If a Boltzmann population distribution exists, then the energy difference $E_v - E_0$ between the two states is given by

$$I_0/I_v = f \exp |(E_v - E_0)/kT|$$

where the proportionality factor f is a function of, amongst other things, the ratio $(l_{eff})_0/(l_{eff})_v$ of the effective cell length at the two transition frequencies, the two linewidth parameters $(\Delta\nu)_0/(\Delta\nu)_v$, and the squares of the transition dipole moment matrix elements $\langle i|\mu|j\rangle_0^2/\langle i|\mu|j\rangle_v^2$. Given that I_0/I_v can be accurately determined, f must be known with sufficient accuracy that $E_v - E_0$ may be well determined. In general, the above-mentioned ratios contributing to f are not known. In particular, the effective cell length will, because of reflections, vary in an unknown way with frequency and make the dominant contribution to the error in f.

Ruitenberg[200] has proposed a neat solution to the problem of determining f. He measures (I_0/I_v) as a function of temperature T so that, under the assumption of temperature-invariance of effective cell lengths, a plot of $\ln (I_0/I_v)$ *versus* $1/T$ yields a straight line of slope $(E_v - E_0)/k$ and eliminates the need to know f. Tests of this procedure are reported using a conventional Stark-modulated spectrometer and the anti-modulation method to measure

[382] J. J. ter Meulen and A. Dymanus, *Astrophys. J.*, 1972, **172**, L21.
[383] S. Saito and K. Takagi, *Astrophys. J.*, 1972, **175**, L47.
[384] S. Saito, *Astrophys. J.*, 1972, **178**, L95.
[385] E. F. Pearson and M. B. Trueblood, *Astrophys. J.*, 1973, **179**, L145.

relative intensities to within 1.5%. In order that the effective cell length be as independent of temperature and frequency as possible, a new type of Stark cell without a mid-electrode and in which one of the broad walls of the waveguide is separated from the rest of the cell by Teflon tape has been developed. Vibrational separations from the ring-bending vibrational satellites in trimethylene oxide and barriers determined from torsional satellite intensities in methylsilane, methoxyethyne, and thio[*Me*-$^2H_1$]methoxyacetylene ($HC{\equiv}CSCH_2D$) give good agreement with i.r. data in the first of these and barrier heights determined from the frequency-splitting method for the remainder. The barrier heights determined by the two methods are estimated to have comparable accuracies ($\sim$4%).

Relative-intensity measurements must be carried out under conditions of constant crystal-diode detector current, as originally pointed out by Esbitt and Wilson.[386] Seibt [387] now describes as an aid to such measurement a crystal-current leveller, for use with the Hewlett Packard 8460 A spectrometer in particular, but easily adaptable to other spectrometers. In essence, the leveller monitors the detector crystal current and controls a rotary wave attenuator.

A commercial microwave spectrometer designed with chemical analysis in mind is described by Cuthbert *et al.*[388] The instrument is of the Stark-modulation variety employing a B.W.O. source which achieves a high FM stability through very stable power supplies. The absorption cell is of the expanded-Lide type made from two channel sections of waveguide held together by spring loading against insulated spacing pieces. No central Stark electrode is therefore necessary. The result is a rigid, easily pumped cell which can be heated and cooled without loss of performance and which is readily dismantled for cleaning. A feature of the specially designed sample-handling system is an automatic gas-pressure regulator which enables reproducible doses of gaseous sample to be introduced into the cell, obviously important from the analytical viewpoint. Similarly, facilities for the injection of reproducible quantities of liquid are available. The important analytical feature of rotational transition intensity measurement under high-resolution conditions has been approached in a novel way. An automatic electronic integration system measures the line area between the peak maximum and a point which is a known fraction of the peak height above the baseline. From the known Lorentzian lineshape, the area between these points is normalized so that the extrapolated total peak area is printed out. The end-point of the integration can be varied until a decrease of the extrapolated peak area indicates that a point has been reached where overlap with adjacent features is becoming important. The area projected at the last point before the incidence of detectable overlap is thus to be chosen for intensity purposes.

[386] A. S. Esbitt and E. B. Wilson, jun., *Rev. Sci. Instr.*, 1963, **34**, 901.
[387] P. Seibt, *Rev. Sci. Instr.*, 1971, **42**, 1895.
[388] J. Cuthbert, E. J. Denney, C. Silk, R. Stratford, J. Farren, T. L. Jones, D. Pooley, R. K. Webster, and F. H. Wells, *J. Phys.* (*E*), 1972, **5**, 698.

As a result of an effort to find an absorption cell for microwave double-resonance experiments which is freer from reflection, insertion loss, and sample-adsorption problems than the conventional extended waveguide systems, Lee and White[389] describe a cell consisting of two Fabry–Perot interferometers mounted at right angles to each other and excited independently by separate microwave radiation sources so that the microwaves propagate at right angles but with their electric vectors parallel. The interferometer mirrors (of the semi-confocal variety) are gold plated and are contained in a vacuum-tight stainless steel chamber provided with requisite gas and radiation inlet and outlet ports. The pump frequency is derived from a klystron which is source-modulated by a 33 kHz square wave and fed into one cavity while the low-power signal frequency is fed into the other. Each interferometer cavity is tuned to the required resonant frequency. Only when a double-resonance effect occurs in the cell is the modulation of the pump cavity power transferred to the absorption which is occurring in the signal transition cavity. The variation of the absorption in the signal cavity is then detected by a phase-sensitive detector tuned to 33 kHz. Initial experiments with the $1_{11} \leftarrow 0_{00}$ and $2_{02} \leftarrow 1_{11}$ transitions of methylene chloride indicate a sensitivity comparable with that of a Stark-modulated spectrometer.

Other papers concerned with certain aspects of microwave instrumentation have appeared. Beers and Klein[30] describe a parallel-plate waveguide cell for the precise measurement of the Stark effect of water in the millimetre region (where the Stark effect is not generally used). A d.c. Stark voltage is applied across the plates and superimposed on this is a small sinusoidal a.c. field which allows phase-sensitive detection. Particular attention is paid to accurate calibration of this cell with the $J = 7 \leftarrow 6$ transition of OCS and the correction of measured frequencies to allow for field inhomogeneities and the effects of modulation. Lainé and Bardo[390] describe multiple population inversion in an ammonia beam maser as part of a general study of masers.

13 Analytical Applications

A recent review of microwave spectroscopy[391] emphasizes those aspects which bear on analysis and surveys the major types of applications that have been made. The appearance of a commercial microwave spectrometer with design features which facilitate analytical applications has been referred to earlier in the Report.[388]

Further applications of microwave spectroscopy have been made in investigations of hydrogen-exchange reactions. The distribution of deuterium in [$^2H_1$]propene obtained from the exchange reaction between deuteriomethanol (CH_3OD) and propene in the presence of homogeneous Pt, Rh, and Ni

389 M. C. Lee and W. F. White, *Rev. Sci. Instr.*, 1972, **43**, 638.
390 D. C. Lainé and W. S. Bardo *J. Phys.* (*B*), 1971, **4**, 1738.
391 L. H. Scharpen and V. W. Laurie, *Analyt. Chem.*, 1972, **44**, 378R.

catalysts has been determined.[392, 393] The five species of [$^2H_1$]propene, differing in the location of the deuterium atom, were distinguished by their microwave spectra, and the percentage composition was determined from intensity measurements. In making the measurements a microwave spectrometer was interfaced with a computer. In this way computer programs were used to automate the analytical procedure as far as possible, and at the same time increased accuracy was obtained through digital averaging of signals. The relative standard deviation of the analytical result was about 2% of the values for the deuterio-species, concentrations ranging from 1.5 to 17% of the total samples. The hydrogen-exchange reaction between propene and toluene-*p*-sulphonic acid has also been investigated by a microwave spectroscopic technique.[394] In this case the analysis was made of [$^2H_2$]propene species as well as [$^2H_1$]propene species. In another investigation[395] the direct exchange between propene and deuterium over alumina has been followed by microwave spectroscopy. In the initial stages the *cis*- and *trans*-[1-$^2H_1$]-species of propene were the only products of the exchange reaction, but as the reaction progressed the [3-$^2H_1$]-species appeared with increasing concentration. When *cis*-[$^2H_1$]propene was circulated over alumina, isotope scrambling and double-bond migration could be detected and followed. A comparative rate study showed that these processes occurred considerably more slowly for [3-$^2H_1$]propene. The catalytic effect of zinc oxide on reactions of propene has been studied by microwave spectroscopy.[396] It appears that a π-alkyl intermediate occurs in the exchange reaction between propene and deuterium under these conditions, whereas the formation of [$^2H_2$]propene proceeds by a simple addition mechanism. In another analytical application of microwave spectroscopy the method has been used to determine the amounts of the separate isomers of [$^2H_1$]propene and [$^2H_2$]propene in a mechanistic investigation of reactions of olefins with electron-donor complexes.[397]

Another mechanistic study which has used microwave spectroscopic methods is an application to the mechanism of ozonolysis of ethylene in some work which has been referred to earlier in the Report.[243, 244] An interesting qualitative application has been reported in the preparation of deuterium-labelled 2-chloropropenes.[178] The addition of DCl to $CH_3C{\equiv}CH$ led to a product different from that obtained from the addition of HCl to $CH_3C{\equiv}CD$. Both *cis*- and *trans*-isomers of $CHD{=}CCl{\cdot}CH_3$ were prepared in pure form in this work and so it was possible to investigate the stereospecificity of the addition of hydrogen chloride to a triple bond under ideal conditions where steric effects were completely excluded. It was shown that this reaction is a *trans* addition according to Markovnikoff's rule.

[392] L. H. Scharpen, R. F. Rauskolb, and C. A. Tolman, *Analyt. Chem.*, 1972, **44**, 2010.
[393] C. A. Tolman and L. H. Scharpen, *J. C. S. Dalton*, 1973, 584.
[394] T. Kondo, M. Ichikawa, S. Saito, and K. Tamaru, *J. Phys. Chem.*, 1973, **77**, 299.
[395] Y. Sakurai, T. Onishi, and T. Tamaru, *Bull. Chem. Soc. Japan*, 1972, **45**, 980.
[396] K. Tamaru, S. Naito, T. Kondo, and M. Ichikawa, *J. Phys. Chem.*, 1972, **76**, 2184.
[397] M. Ichikawa, S. Naito, S. Saito, and K. Tamaru, *J. C. S. Faraday I*, 1973, **69**, 685.

2
Theories of Resonance Raman Scattering*

BY J. BEHRINGER

1 Introduction

This Report is intended to give a survey of the progress made with theories of RR scattering† during recent years. Systematic reviews of this field of research have not yet been published and therefore reference to the literature over a long period has been necessary. The reader may regard this article as continuing or supplementing, as far as theory is concerned, the reviews given by this author [1, 2] and by others.[3, 4]

As a result of the availability of lasers, RR spectroscopy has in the past decade achieved unexpected success in investigating the structure and behaviour of widely different systems. New experimental techniques have considerably reduced, and in many cases virtually removed, the most annoying obstacle in observing RR spectra—the energy loss due to absorption with its equally disturbing after-effects of fluorescence emission and radiolysis. Recently, by excitation with optical frequencies, excellent RR spectra have been obtained from deeply coloured compounds which formerly were regarded as completely inaccessible to Raman spectroscopy. The experimental progress in this branch of spectroscopy has been so fast that it seemed to some that the theoretical advances were not keeping pace with it. Perhaps this Report will serve as a defence of the theoretician by demonstrating that the number of theoretical models already put forward for the explanation of resonance light-scattering effects is at this moment too large rather than too small, and that only with a very large amount of experimental data can the applicability of these models be tested. It is only now that the necessary data are accumulating in sufficient quantity to enable such tests to be made.

RR spectroscopy is interesting not only for chemical purposes but also

* To the 85th birthday (Aug. 1st, 1974) of Prof. Walther Gerlach, Munich.

† Abbreviations: RE = Raman Effect, RR = Resonance Raman, RRE = Resonance Raman Effect, RF = Resonance Fluorescence.

[1] J. Behringer and J. Brandmüller, *Z. Elektrochem.*, 1956, **60**, 643 (NCAR Library Translation, No. 141.)

[2] J. Behringer, 'Observed Resonance Raman Spectra', in 'Raman Spectroscopy Theory and Practice', ed. H. A. Szymanski, Plenum Press, New York, 1967, Vol. 1, pp. 168—223.

[3] R. E. Hester, *Analyt. Chem., Ann. Revs.*, 1972, **44**, 490R; P. P. Shorygin, 'Combination Scattering of Light and Conjugation', *Uspekhi Khim.*, 1971, **40**, 604—739.

[4] P. P. Shorygin, 'Combination Scattering of Light Near and Far From Resonance', *Uspekhi fiz. Nauk*, 1973, **109**, 293—332.

from a more fundamental physical point of view. Whereas the ordinary (non-resonance) RE can be easily distinguished from other types of interaction between light and matter, there arise peculiar difficulties in defining clearly the meaning of RR scattering, at least in the rigorous resonance case. It will be necessary therefore to review critically in the next section the terminology and concepts at present used for distinguishing the different kinds of reaction of systems upon irradiation with light.

Apart from a few necessary quotations this Report will consciously exclude a discussion of experimental work. It is hoped that experimental techniques and results in RR spectroscopy can be reviewed in a later Report.

2 The Relation of Resonance Raman Scattering to Resonance Fluorescence

It seems to be necessary to start the discussion of theoretical RR developments by a critical review of the terminology, which at this moment is far from being uniform in the literature. A fundamental problem—but in the opinion of this author at least partly only a pseudo-problem, arising from ambiguous definitions—is the relation of the RRE to resonance fluorescence (RF). This problem cannot be solved without careful consideration of the history of the concepts used for classifying the diverse types of 'secondary radiation' emitted by corpuscular matter after excitation by 'primary light'. The terms 'Raman scattering', RR scattering', and 'RF' are intended to describe particular modes of interaction between a system of particles and a radiation field. In the quantum mechanical view of the radiation field as a system of non-interacting quanta (photons) occupying different field modes, both these effects formally can be described as simple transformations of the field mode occupation numbers (or briefly as transformations of photons) accompanied by a change of state of the particle system interacting with the field. In the older classical nomenclature the emission of light caused by irradiation of light is termed 'photoluminescence' (*cf*. ref. 5, p. 1, ref. 6, p. 1) and it is clear that before the advent of quantum theory and the prediction (Smekal, 1923) and discovery (Raman, Landsberg, and Mandelstam, 1928) of the RE, any observed phenomenon akin to RR scattering would have been considered as 'photoluminescence'. Now if RF, which was already known and investigated around 1900, were indeed nothing else than RR scattering under special conditions, we should have the strange fact that the RRE had been discovered and investigated long before the ordinary RE was known and nobody was conscious of the real importance of this early work! Such an anticipation of the RRE by late 19th or early 20th century physics might seem impossible to those accustomed to the characteristics of the ordinary RE. It should be noted, however, first, that from an experimental

[5] P. Pringsheim, 'Fluoreszenz und Phosphoreszenz im Lichte der neueren Atomtheorie', Springer, Berlin, 1921.

[6] P. Pringsheim, 'Fluorescence and Phosphorescence', Interscience, New York, 1949.

point of view the intensities in the rigorous RR case are several (up to seven and more) orders of magnitude larger than in the ordinary RE, so that an observation of RR phenomena by much weaker instrumentation cannot be wholly excluded, and secondly that from the theoretical point of view, the pre-quantum theories developed for the explanation of fluorescence, and particularly of RF, must be considered as classical descriptions not of fluorescence in the sense accepted today but of light scattering. The confusion in terminology was still more increased by the attempts to distinguish verbally secondary radiation that is displaced from that which is not displaced in frequency relative to the primary light. It is *this* distinction which in the past stimulated the introduction of the concept of fluorescence as well as of Raman scattering.

In order to clarify the terminological confusion encountered in the recent literature let us first give a survey of the most important concepts connected with RR spectroscopy and their historical background.

The term 'fluorescence' is due to Stokes [7] (1852), who by this term obviously wanted to accentuate the fact discovered by him that in this phenomenon (which can be observed for example with certain impure species of 'fluorite' CaF_2) there is a bulk frequency difference between the re-emitted light and the exciting radiation ('Stokes' rule'). Evidently, however, this basic idea behind the introduction of the concept of fluorescence was forgotten a few decades later by physicists and the term 'fluorescence' became customary for the designation of any kind of 'diffusion' (*i.e.* angular spreading) of light whether accompanied by 'dispersion' (*i.e.* colour separation) or not. (In French 'diffusion' is equivalent to 'scattering'.) Lommel [8] developed a theory of 'fluorescence phenomena, the colour of which depends on the nature of the body as well as on the frequency of the exciting light', but this theory actually is a 'complete mathematical theory of light scattering' (Placzek, ref. 9, p. 206). Lommel called the unshifted part of the scattered light 'isochromatic fluorescence'—an expression completely contrary to Stokes' idea! Despite its shortcomings (the displaced frequencies were traced back to anharmonicities) Lommel's classical theory is particularly interesting for its inclusion of the special case of resonance excitation. Here we have indeed the first (though insufficient) theory of the RRE. Before Lommel, the unshifted part of light scattering usually was called 'Tyndall scattering'. Tyndall [10] himself had studied only the scattering of light by colloidal suspensions and aerosols. The idea that light can be scattered by atoms or molecules themselves is due to Maxwell. In a letter of August 28th, 1873, he encouraged Rayleigh to make his studies (which began about 1871) on light scattering by particles suspended in air so as to 'obtain data about the size of the

[7] G. G. Stokes, *Phil. Trans.*, 1852, **142,** II, 463.
[8] E. Lommel, *Wiedemanns Ann. Physik*, 1878, **3**, 251; *Poggendorffs Ann.*, 1871, **143,** 26.
[9] G. Placzek, 'Rayleigh-Streuung und Raman-Effekt', in 'Handbuch der Radiologie', ed. E. Marx, Akad. Verlagsges., Leipzig, 1934, Vol. VI, Part II, Ch. 3, pp. 205—374.
[10] J. Tyndall, *Proc. Roy. Soc.*, 1868, **17,** 92, 222, 317; *Phil. Mag.*, 1869, **37,** 384; **38,** 156.

molecules of air' (for literature referring to Rayleigh's work see Kerker [11]). It was only in 1899, however, that Rayleigh in his famous paper [12] explaining the blue colour of the sky confirmed the essential correctness of Maxwell's conjecture. It therefore must be taken as granted that the term 'Rayleigh scattering', honouring Rayleigh's pioneer work on unshifted molecular light scattering, did not come into use before the turn of the century. The molecular origin of 'Rayleigh scattering' was not properly appreciated in the first decades of this century. Even today, in many publications, 'Rayleigh scattering' simply and abstractly means light scattering from statistically distributed, approximately spherical, non-absorbing particles which are small in comparison to the wavelength. It goes without saying that any connection between Rayleigh scattering and the RE could not be appreciated before the latter's discovery. In Kohlrausch's book of 1931,[13] which represents the first comprehensive treatise on the RE, the terms 'Tyndall scattering' and 'Rayleigh scattering' are used synonymously for unshifted molecular light scattering (ref. 13, p. 1).

Classically, Rayleigh scattering is explained in terms of the forced vibrations of the virtual electronic oscillators within the molecule with the frequency of, and in definite phase correlation with, the exciting wave. This necessarily leads to coherence between the secondary (scattered) and primary (incoming) waves. (Quantum mechanically, Rayleigh scattering may contain an incoherent part, *cf.* ref. 9, p. 223.) This coherence of course is not destroyed in the case of resonance, *i.e.* if the exciting frequency equals the eigenfrequency of the electronic oscillator. In this case the phase lags of the electronic oscillator with respect to the incoming wave and of the scattered wave with respect to the electronic oscillator simply both take the value of $\pi/2$ so that the scattered wave is opposite in phase to the exciting wave. Another important property of this resonance case is the high scattering intensity due to the vibration of the electronic oscillator with its maximal amplitude (finite only by damping). It is this intensive resonance scattering, unshifted in frequency but coherent with the primary wave, which was called 'resonance radiation' or 'RF' (*cf.* ref. 6, p. 7 and ref. 14, p. 517). This amounted to saying that RF could be equated with resonance Rayleigh scattering. It is noteworthy that the term 'fluorescence' in this word combination again does not correspond to Stokes' intention. Predictions of this resonance radiation are contained in the theoretical papers of Lommel and Rayleigh, but it was not before Wood's extensive work on gases and vapours starting in 1905 (for a reference list see Pringsheim [6]) that the existence of this phenomenon was conclusively proved by experiment. (Lommel already had observed the photoluminescence of I_2 vapour, but he did not recognize its resonance properties.) However, Wood's experiments in most cases showed

[11] M. Kerker, 'The Scattering of Light', Academic Press, New York, 1969.
[12] Lord Rayleigh, *Phil. Mag.*, 1899, **47**, 375.
[13] K. W. F. Kohlrausch, 'Der Smekal–Raman-Effekt', Springer, Berlin, 1931.
[14] M. Born, 'Optik', Springer, Berlin, 1933.

more than had been expected on the basis of Rayleigh's theory: in addition to the 'resonance line' exactly equal in frequency to the monochromatic exciting line, other lines of different frequency appeared, forming a so-called 'resonance series'. This concept of a 'resonance series of fluorescence' represents an extension of the concept of 'RF' toward Stokes' concept of 'fluorescence', but on the other hand it clearly violates the original definition of RF claiming equality of exciting and scattered frequencies.

Another attack on the old idea of 'RF' was started by the advent of quantum theory. The original quantum theory conceived RF as the succession in time of two independent processes of absorption and emission. The molecule, by the absorption of a quantum of light, was though to be *actually* (*i.e.* under conservation of energy according to Bohr's frequency condition) raised into an excited state and after a period of time determined only in statistical average to fall back into a lower state emitting a quantum of light. If the final state is identical with the initial state before the absorption, the so-called 'resonance line' will be emitted, if not, another line of the 'resonance series'. Which line is emitted by an individual molecule is governed only by the laws of probability. However, the statistics of the lifetime of the excited state destroy any definite phase relation between the absorbed and emitted quantum. Consequently, RF even in the case of coincident initial and final states, cannot be coherent.[15] This contradiction to the earlier understanding of RF led to the suspicion that the equation of RF with resonance Rayleigh scattering could not be correct. In fact, it was occasionally even thought that the simultaneous and parallel existence of both phenomena (classically understood coherent resonance Rayleigh scattering and quantum mechanically interpreted incoherent RF) must be accepted (*cf.* ref. 6, p. 8, 239 ff. and ref. 16, pp. 189, 223).

Dirac's theory of radiation [17] provided a way out of this unsatisfactory situation which obviously was only a demonstration of the still partly obscure correspondence between classical and quantum theories. Weisskopf in his famous 'Göttinger Dissertation' of 1931 [18] arrived at the result that the idea must be rejected 'that RF is an independent succession of absorption and spontaneous emission'. An equivalent statement was made and proved earlier for ordinary light scattering by Placzek,[19] *cf.* ref. 9, p. 222. (In this paper [19] difficulties for the limiting case of RF are mentioned.) Weisskopf, by quantum mechanical arguments, was also able to restore the classical attribute of coherence to RF. Backed now by quantum theory, RF again appeared as the resonance case of Rayleigh scattering. This interpretation

[15] *Cf.* N. Bohr, H. A. Kramers, and J. C. Slater, *Phil. Mag.*, 1924, **47**, 822; *Z. Physik*. 1924, **24**, 69.
[16] P. Pringsheim, 'Anregung von Lichtemission durch Einstrahlung', in 'Handbuch der Physik', ed. H. Geiger and K. Scheel, Springer, Berlin, 1933, Vol. XXIII, Part 1, pp. 185—322.
[17] P. A. M. Dirac, *Proc. Roy. Soc.*, 1927, **A114**, 243, 710.
[18] V. Weisskopf, *Ann. Physik*, 1931, **9**, 23.
[19] G. Placzek, *Z. Physik*, 1929, **38**, 585.

of RF remained predominant up to the present. Heitler's frequently used book [20] in a somewhat modernized formulation repeats Weisskopf's arguments (p. 196 ff.) and states on p. 203 that 'RF represents a single coherent quantum process' and therefore must be regarded as identical with coherent Rayleigh scattering (*cf. ibid.*, p. 189).

However, despite this seemingly satisfactory result the riddle is not yet completely solved. This became evident after the appearance of the RE on the scene (1923 theoretically, 1928 experimentally). The term 'RE' is due to Pringsheim (ref. 21, p. 600). It seems that this designation for the new effect was already in use in 1931 in western countries. (Raman received the Nobel prize in 1930.) However, Kohlrausch's book,[13] published in 1931, in an attempt to give credit also to Smekal, received the title 'Der Smekal–Raman-Effekt'. In Russia the original designation 'combination scattering of light' has been maintained right up to the present. ('Combination' here means that the scattered frequency is the sum or difference of the exciting frequency and an eigenfrequency—electronic, vibrational, rotational, *etc.*—of the scattering system.) The term 'RRE' to this author's knowledge was coined by Shorygin or by one of his co-workers. In 1960 the expression 'RR spectra' appeared [22] probably for the first time in the literature and a 1962 paper of Shorygin [23] was entitled 'RRE'. It is clear that only after this very late appearance of the term 'RRE' could the question be formally asked (although the problem itself existed already before): what is the relation of RF to RR scattering? Today there are widely varying opinions as to the answer to this question. We shall only set out the two most typical contradictory opinions, both of which nevertheless seem to be possible.

One could on the one hand say that RF is (at least in the wider sense including the 'resonance series of fluorescence') identical with RR scattering. The reasoning is as follows: RF (in the restricted sense of equal exciting and scattered frequencies) classically and quantum mechanically is considered as identical with resonance Rayleigh scattering. Rayleigh scattering quite generally is the special case of Raman scattering corresponding to equal energies of the initial and final states of the scattering system (*cf.* the common theory for both, *e.g.* in ref. 20, pp. 189, 192). Therefore, if in the notion of RF there are included in addition to the 'resonance line' the frequency-shifted lines of the 'resonance series', no difference can be found between this 'RF' and RR scattering. This opinion obviously was shared by Placzek. In Sections 12 and 13 of ref. 9, discussing light scattering by atoms, he mentions several examples for non-shifted (Rayleigh) and shifted (Raman) resonance scattering, and exactly the same examples are presented by Pringsheim (ref. 6, p. 32 ff. and elsewhere) under the heading of 'Resonance Lines'

[20] W. Heitler, 'The Quantum Theory of Radiation', Clarendon Press, Oxford, 3rd edn. 1954.
[21] P. Pringsheim, *Naturwiss.*, 1928, **16**, 597.
[22] P. P. Shorygin and L. L. Krushinsky, *Doklady Akad. Nauk S.S.S.R.*, 1960, **133**, 337.
[23] P. P. Shorygin, *Pure Appl. Chem.*, 1962, **4**, 87.

(of fluorescence); *cf.* also in ref. 9 Sections 7 (the end) and 25 (p. 371 ff.) with the references to Pringsheim's publications. Placzek (p. 371) does not criticize at all Pringsheim's opinion [24] that there is a continuous 'transition from the Raman spectrum to the resonance spectrum'. Note that Pringsheim does not say: from the 'RR spectrum'—this term was unknown in 1929—but: from the (ordinary non-resonance) 'Raman spectrum'. The RR spectrum as the limiting case in the Raman spectrum is for him identical with the resonance spectra of fluorescence.

One could on the contrary assert that RF is a phenomenon completely different from RR scattering. Here the reasoning is that RF is a special case of fluorescence. As is well known, fluorescence has properties markedly different from those of scattering. The most remarkable of these distinguishing properties is the possibility of fluorescence quenching. Quenching has also been proved experimentally for the so-called resonance spectra (see *e.g.* ref. 6, pp. 89—123). Moreover, resonance spectra of gases on increase of gas pressure gradually change into normal fluorescence spectra, which means that the vibrational and rotational energy of the fluorescing molecule in the excited electronic state is redistributed before emission according to thermal equilibrium.[25, 26] Such influences on fluorescence spectra are only possible if there is a sufficient time interval between absorption and re-emission so that additional (radiative or radiationless) processes stealing energy may intervene. Thus 'the elementary process underlying the RE is clearly to be distinguished from the process of fluorescence' (Herzberg ref. 27, p. 85). In fluorescence (including RF) the double-photon process of excitation and re-emission disintegrates into a *real* absorption followed (after a sufficient period of time controlled only by laws of statistics) by a *real* emission. 'Real' means that the intermediate excited state can be identified experimentally by a measurement of (at least) its energy (*cf.* ref. 20, p. 202 ff.). In light scattering (Rayleigh, Raman, including RR) such a bisection of the elementary process which transforms the incident photon into the scattered one is impossible. It would destroy essential properties of the observed scattered radiation, *e.g.* its linewidth. It is generally assumed that the lifetime of the intermediate state in fluorescence is at least 10^{-9} s (ref. 26, p. 8, ref. 28, p. 1) whereas in light scattering after Smekal (*cf.* ref. 19, p. 585) it has only the order of magnitude of the vibrational period of the light wave, *i.e.* less than 10^{-14} s (*cf.* ref. 13, p. 5). [It must be emphasized, however, that the

[24] P. Pringsheim, 'Ramanspektren', in 'Handbuch der Physik', ed. H. Geiger and K. Scheel, Springer, Berlin, 1929, Vol. XXI, p. 632.

[25] *Cf.* T. Förster, 'Fluoreszenz organischer Verbindungen', Vandenhoeck und Ruprecht, Göttingen, 1951; J. D. Winefordner and J. M. Mansfield, 'Atomic Fluorescence Flame Spectrometry', in 'Fluorescence Theory, Instrumentation, and Practice', ed. G. G. Guilbault, Dekker, New York, 1967, pp. 565—625.

[26] *Cf.* E. J. Bowen, 'Luminescence in Chemistry', Van Nostrand, London, 1968.

[27] G. Herzberg, 'Molecular Spectra and Molecular Structure', Vol. I: 'Spectra of Diatomic Molecules', Van Nostrand, Princeton, 2nd edn., 1950.

[28] N. Riehl and H. Vogel, in 'Einführung in die Lumineszenz', Thiemig, Munich, 1971, pp. 1—34.

applicability of the concept of 'lifetime' (which presupposes an exponentially decaying probability function) to the intermediate state(s) is problematic; in non-resonance scattering it is not applicable at all.]

Disregarding views intermediate between these two opinions it must be acknowledged that there are rather convincing arguments for both sides. Which, though, is finally correct?

It seems to this author that the rapid scientific progress in both fluorescence and Raman spectroscopy which has led to the accumulation of many new results has overrun the formation in due time of a terminology adequate for the classification of the phenomena discovered. All that can be done now is to establish scientifically acceptable criteria (both theoretical and experimental) capable of distinguishing clearly the observed phenomena and then to try to arrive at an agreement about appropriate designations for the phenomena. Holzer, Murphy, and Bernstein [29] published a table (Table I, p. 400 of ref. 29) presenting characteristic 'differences in the observation of RF and RRE'. The criteria given there and adopted also by other authors are mainly external and are taken from and related to special experiments (perhaps excepting the behaviour with regard to quenching) and therefore they certainly are not generally applicable. This is in particular true for the band envelope, the intensities in the overtone series, and the depolarization ratios (for the latter *cf.* ref. 4, p. 325). Historically it is interesting that in the first papers of Raman [30] on the newly discovered effect the supposed differences in line (or band) intensities and depolarization ratios were taken as important criteria for distinguishing Raman scattering from fluorescence. However, these criteria were soon criticized by others as unreliable (*cf.* ref. 21, p. 600).

In the present author's opinion, the distinction between scattering (including the RRE) and fluorescence excited by radiation (including RF) ought to start from the fundamental theoretical question of whether the elementary process (which in every case involves the absorption of one photon and the emission of one photon) is an inseparable double-photon process (not allowing the experimental detection and identification of one distinct intermediate state) or the succession (interrupted by a statistically indeterminate time interval with eventual intervening transitions in the excited state) of two independent, single-photon processes. It is only in the second case that the absorption and the emission can be termed 'real'; in the first case it is usual to call them 'virtual'. In modern understanding fluorescence (like luminescence in general) is essentially a kind of light *emission* and so for this reason the separability of absorption and emission is indispensable for fluorescence. The preceding process of excitation by light absorption is completely unimportant for the properties of fluorescence emission. Let us quote Pringsheim (ref. 6, p. 9): 'The radiation emitted by an atom or a molecule depends only

[29] W. Holzer, W. F. Murphy, and H. J. Bernstein, *J. Chem. Phys.*, 1970, **52**, 399.
[30] C. V. Raman and H. P. Krishnan, *Nature*, 1928, **121**, 501; C. V. Raman, *Indian J. Phys.*, 1928, **2**, 387.

on its state of excitation and on the probabilities of transitions from this state to those of lower energy. It does not depend on the mode of excitation by which the system has been brought into the excited state. In this sense there is no difference between fluorescence and any other kind of light emission by the same atoms or molecules caused by collisions with electrons, by chemical processes, or by thermal agitation. The characteristic properties of as pectral line or a band . . . must be the same in every case.' This is quite different for scattering. Here absorption and emission must be considered as inseparably united. At least in the non-resonance case a definite 'meta-stationary' (Smekal) or 'virtual' intermediate level obeying the law of energy conservation (Bohr's frequency rule) does not exist. The scattered radiation cannot be described without the parameters characterizing the partial process of absorption. The scattering atom or molecule when (virtually) emitting 'remembers' which photon it had absorbed (*cf.* ref. 20, p. 201), nay even this word 'remembers' is misleading because there is no definite succession in time of the absorption and the emission. It appears to be natural to consider RF as a special case of fluorescence, although here the prefix 'resonance' constitutes a clear relation to the exciting absorption process. ['Resonance' here means primarily that the exciting wave (and not the scattered one) agrees in frequency with an eigenfrequency of the scatterer.] Therefore (despite the latter restriction) for RF also the double-photon process should, at least in principle, be actually resolvable in two constituent single-photon processes.

Summing up, we propose to accept the following basic criterion for distinguishing light scattering (including Rayleigh and Raman in the non-resonance and resonance cases) from fluorescence (including RF in the more restricted sense of the 'resonance line' and in the wider sense of the 'resonance spectra of fluorescence'): the transformation of light by interaction with a particle system (atom, molecule, crystal, *etc.*) shall be called 'fluorescence' if the elementary process separates into independent and individually measurable partial processes of absorption and emission, and 'scattering' if it does not so separate. [Although there are elementary scattering processes involving more than two photons (stimulated RE, hyper-RE) our attention in this context will be directed only towards double-photon processes.] It is clear that the above criterion ought to be translated into the language of experimentalists. However, this can be done only after a solid theoretical foundation has been established. It is also only then that a pertinent answer can be given to the often discussed question: is there a continuous transition from scattering to fluorescence or, more specifically, from the RRE to RF? In the terminology proposed above, this question can be formulated as follows: are there any conditions under which the double-photon process of scattering dissolves into two independent parts? It might seem that the answer to this question had already been anticipated (in the form of 'no') by Weisskopf [18] (see above). This is not true, however. A close and careful re-examination of the quantum mechanics of double-photon processes shows that there exist indeed conditions for a possible decay of a double-photon scattering process into two independent single processes of absorption and emission. The main necessary condition is rigorous resonance excitation. It should be noted that with this condition it is no longer possible to distinguish the RE from fluorescence by the fact that 'the RE can take place for any frequency of the incident light, whereas fluorescence can occur only for the absorption frequencies' (ref. 27, p. 86) or, in other words, that the whole Raman spectrum is displaced in the absolute frequency scale when the exciting frequency is displaced, whereas the fluorescence frequencies are fixed. This, incidentally, was historically the decisive criterion for considering Raman scattering as a phenomenon different from fluorescence; see ref. 21, p. 600. The condition of rigorous resonance excitation is not sufficient, however. It is also necessary that the intermediate level be non-degenerate and that it lie in the discrete part of the energy spectrum of the scattering system. Further conditions regarding line- and level widths must be specified by the exact theory. We shall return to these questions in Sections 3E and 4D.

3 Classical Theories of the Vibrational Resonance Raman Effect

Only quantum theory is truly adequate for atomic and molecular phenomena, and therefore a totally satisfactory description of the RRE (and particularly of RR intensities) cannot renounce the methods of quantum mechanics. Nevertheless classical considerations also yield useful information about vibrational and rotational Raman scattering. Although Smekal's prediction of the RE was based on quantum mechanical arguments, Cabannes and Rocard, shortly after Raman's discovery, showed that the essential features of the ordinary RE can be interpreted classically. Subsequently, Placzek, in his polarizability theory, developed this idea for vibrational Raman scattering (ref. 9, Section 14 *et seq.*; *cf.* also ref. 14, pp. 390—403), thus securing almost universal application of classical considerations in practical ordinary Raman spectroscopy. Although Placzek himself, starting from exact quantum mechanical arguments, explicitly stated the application limits of his polarizability theory, thus clearly excluding the resonance case (ref. 9, pp. 270, 366), it has proved to be possible and useful to look for modifications of the polarizability theory which will extend its applicability to the case of approaching resonance. Placzek himself in Sections 7 and 25 of ref. 9 gave much valuable advice for developing systematic theories of the RRE. As we shall see at the end of Section 3A, the justification for such endeavours lies in the fact that there is a remarkable correspondence between the classical and the quantum mechanical dispersion formulae. During recent years Shorygin and his co-workers have published a long series of papers on classical methods of treating the vibrational RRE. Since considerations of this type are probably not well known we shall present a selection of them in a self-contained and, in comparison with Shorygin's work, partly modified and generalized version. It will be seen that classical theory, despite its unquestionable deficiencies, can give at least qualitatively satisfactory interpretations for certain important properties of the vibrational RRE. The principle of correspondence always was a great stimulus for progress in quantum mechanics. The results of classical RR theory seem to contain certain features which still lack an adequate quantum mechanical interpretation. Clearly there is no classical theory for the electronic RE, either ordinary or resonance. Likewise the rotational RRE cannot be treated classically because the rotational energy quanta are too small and there is nothing comparable to the Franck–Condon principle in rotational molecular spectroscopy.

A. The Polarizability Theory and its Limitations.—It is only the electronic polarizability α which is responsible for the scattering of visible light by atoms and molecules. From a classical point of view the RE originates from the parametrical dependence of this polarizability on the instantaneous nuclear positions. The well-known classical dispersion theory of dielectric media (*cf. e.g.* ref. 14, p. 469 ff., ref. 31, p. 89 ff., and ref. 32, p. 269 ff.) gives for α the expression

$$\alpha = \frac{e^2}{m} \sum_r f_r \, [\omega_r^2 - \omega_0^2 + i\omega_0 \Gamma_r]^{-1} \tag{1}$$

where the index r refers to the different virtual electronic oscillators in the atom or molecule, with f_r, ω_r, and Γ_r representing their oscillator strengths, angular frequencies, and damping constants, respectively. e and m are the electronic charge and mass and ω_0 is the angular frequency of the incident wave. In the general case the polarizability is a tensor (or dyad)

$$\boldsymbol{\alpha} = (\alpha_{\rho\sigma})$$

with ρ and σ molecule- or space-fixed Cartesian co-ordinates. For the sake of simplicity we shall restrict ourselves here to the isotropic case $\alpha_{\rho\sigma} = \alpha\delta_{\rho\sigma}$. The extension of the following arguments to the general case is tiresome but not difficult (*cf.* ref. 14, p. 397).

For vibrating molecules f_r, ω_r, and Γ_r are changing their values periodically with the change of the configuration of the nuclear skeleton. Introducing a system of normal co-ordinates Q_l [33, 34] and denoting a definite nuclear configuration by the 'normal co-ordinate vector'

$$\boldsymbol{Q} \equiv \{Q_1, \ldots, Q_l, \ldots, Q_L\} \tag{2}$$

(with $L = 3N_j - 6$ for non-linear and $L = 3N_j - 5$ for linear molecules; N_j is the number of nuclei j in the molecule) we have to consider f_r, ω_r, Γ_r and consequently also α as functions of $\boldsymbol{Q}$. The basic idea of Placzek's

[31] M. Born and E. Wolf, 'Principles of Optics', Pergamon, London, 1959.
[32] M. Garbuny, 'Optical Physics', Academic Press, New York, 1965.
[33] H. Eyring, J. Walter, and G. E. Kimball, 'Quantum Chemistry', Wiley, New York, 1944, p. 16.
[34] L. A. Woodward, 'Introduction to the Theory of Molecular Vibrations and Vibrational Spectroscopy', Clarendon Press, Oxford, 1972, p. 13.

polarizability theory is to expand α into a power series about the equilibrium configuration $\boldsymbol{Q} = \boldsymbol{0}$. This can be formally done without difficulty by writing

$$\alpha(\boldsymbol{Q}) = \alpha(\boldsymbol{0}) + \boldsymbol{Q} \cdot \boldsymbol{\nabla}_0 \alpha + \tfrac{1}{2}(\boldsymbol{Q} \cdot \boldsymbol{\nabla}_0)^2 \alpha + \ldots \equiv e^{\boldsymbol{Q} \cdot \boldsymbol{\nabla}_0} \alpha \qquad (3)$$

Here

$$\boldsymbol{\nabla}_0 \equiv \left\{ \frac{\partial}{\partial Q_1}, \ldots, \frac{\partial}{\partial Q_L} \right\}_{\boldsymbol{Q}=\boldsymbol{0}}$$

and $e^{\boldsymbol{Q} \cdot \boldsymbol{\nabla}_0}$ is a symbolical differential operator.

The crucial question is whether equation (3) converges or not, and, in the positive case, how fast. The exceedingly successful application of the polarizability theory to the ordinary RE leads to the expectation that at least under non-resonance conditions the convergence of (3) must be extremely good, for already the quadratic terms can be neglected in most cases. (These terms correspond to the first overtones and simplest combination tones, which usually are observed with only very weak relative intensities in ordinary Raman spectra.) However, the conditions valid for the ordinary RE are more or less invalidated in the case of the RRE and it is therefore necessary to examine more carefully the convergence of (3). Obviously if the convergence becomes bad the appearance of more and stronger overtones and combination tones is to be expected. Let us first show this in detail by writing (3) in a more explicit form. [For the present argument we may still assume essentially non-resonance excitation, *i.e.* $\omega_r > \omega_0$ for all r. Then for $\Gamma_r \ll \omega_r$ the imaginary term $i\omega_0 \Gamma_r$ in (1) can be neglected and α in this approximation is still a real quantity.] Abbreviating

$$\alpha(\boldsymbol{0}) \equiv \alpha_0, \left(\frac{\partial \alpha}{\partial Q_1}\right)_0 \equiv \alpha_0^{(1)}, \left(\frac{\partial^2 \alpha}{\partial Q_1^2}\right)_0 \equiv \alpha_0^{(11)}, \left(\frac{\partial^3 \alpha}{\partial Q_1^2 \partial Q_2}\right)_0 \equiv \alpha_0^{(112)}, \textit{etc.} \qquad (4)$$

(3) becomes

$$\begin{aligned} \alpha = {} & \alpha_0 + \alpha_0^{(1)} Q_1 + \ldots + \tfrac{1}{2}\alpha_0^{(11)} Q_1^2 + \ldots + \alpha_0^{(12)} Q_1 Q_2 + \ldots + \\ & + \tfrac{1}{6}\alpha_0^{(111)} Q_1^3 + \ldots + \tfrac{1}{2}\alpha_0^{(112)} Q_1^2 Q_2 + \ldots + \alpha_0^{(123)} Q_1 Q_2 Q_3 + \ldots \end{aligned} \qquad (5)$$

Here the Q's in a first ('harmonic') approximation are sinusoidally varying functions of the time t

$$Q_l = Q_{l0} \cos(\omega_l t + \delta_l) = \tfrac{1}{2} Q_{l0}(e^{i\varphi_l} + e^{-i\varphi_l}), \text{ with } \varphi_l \equiv \omega_l t + \delta_l \qquad (6)$$

We emphasize that Q_{l0} is the maximum amplitude of the normal vibration l. The molecular dipole moment induced by the incident electric field strength

$$\boldsymbol{E} = \boldsymbol{E}_0 \cos \omega_0 t = \mathrm{Re}(\boldsymbol{E}_0 e^{i\omega_0 t}), \ \boldsymbol{E}_0 \text{ real} \qquad (7)$$

is given by

$$\boldsymbol{\mu} = \mathrm{Re}(\alpha \boldsymbol{E}) \qquad (8)$$

and its magnitude by

$$\mu \equiv |\boldsymbol{\mu}| = \mathrm{Re}(\alpha E_0 e^{i\omega_0 t}), \ E_0 \equiv |\boldsymbol{E}_0| \qquad (9)$$

The intensity scattered into the solid angle 4π (total energy emitted by the vibrating dipole moment per unit of time) according to classical radiation theory in Gaussian units is given by

$$I = \overline{2\ddot{\mu}^2}/3c^3 \tag{10}$$

where c is the velocity of light and the bar designates the time average.

Inserting first (6) into (5) and then (5) into (9) we find, after a somewhat tedious calculation,

$$\begin{aligned}
\mu = {} & [\alpha_0 + \tfrac{1}{4}\alpha_0^{(11)}Q_{10}^2 + \tfrac{1}{64}\alpha_0^{(1111)}Q_{10}^4 + \tfrac{1}{16}\alpha_0^{(1122)}Q_{10}^2Q_{20}^2 + \ldots]E_0 \cos \omega_0 t \\
& + [\tfrac{1}{2}\alpha_0^{(1)}Q_{10} + \tfrac{1}{16}\alpha_0^{(111)}Q_{10}^3 + \tfrac{1}{8}\alpha_0^{(122)}Q_{10}Q_{20}^2 + \ldots]E_0\{\cos[(\omega_0 - \omega_1)t - \delta_1] \\
& + \cos[(\omega_0 + \omega_1)t + \delta_1]\} + [\tfrac{1}{8}\alpha_0^{(11)}Q_{10}^2 + \tfrac{1}{96}\alpha_0^{(1111)}Q_{10}^4 + \tfrac{1}{32}\alpha_0^{(1122)}Q_{10}^2Q_{20}^2 \\
& + \ldots]E_0\{\cos[(\omega_0 - 2\omega_1)t - 2\delta_1] + \cos[(\omega_0 + 2\omega_1)t + 2\delta_1]\} \\
& + [\tfrac{1}{4}\alpha_0^{(12)}Q_{10}Q_{20} + \tfrac{1}{32}\alpha_0^{(1112)}Q_{10}^3Q_{20} \\
& \qquad + \ldots]E_0\{\cos[(\omega_0 - \omega_1 - \omega_2)t - \delta_1 - \delta_2] \\
& \qquad\qquad + \cos[(\omega_0 + \omega_1 - \omega_2)t + \delta_1 - \delta_2] \\
& \qquad\qquad + \cos[(\omega_0 - \omega_1 + \omega_2)t - \delta_1 + \delta_2] \\
& \qquad\qquad + \cos[(\omega_0 + \omega_1 + \omega_2)t + \delta_1 + \delta_2]\} + \ldots
\end{aligned} \tag{11}$$

If this series converges sufficiently well all but the zeroth and first powers of the Q_{l0} may be neglected and we get

$$\begin{aligned}
\mu = \alpha_0 E_0 \cos \omega_0 t + \sum_l \tfrac{1}{2}\left(\frac{\partial \alpha}{\partial Q_l}\right)_0 Q_{l0}E_0\{\cos[(\omega_0 - \omega_l)t - \delta_l] \\
+ \cos[(\omega_0 + \omega_l)t + \delta_l]\}
\end{aligned} \tag{12}$$

It is this linear approximation of (11) which was obtained by Cabannes and Rocard [35] in essentially the same way and which, as the basis of polarizability theory, is reproduced in most introductory books on Raman spectroscopy. It describes the undisplaced Rayleigh scattering (first term) and the Stokes and anti-Stokes Raman fundamentals (second term). According to this formula, Rayleigh scattering is determined by the *actual* polarizability tensor $\boldsymbol{\alpha}_0$, and the Raman fundamentals (corresponding to the normal vibrations l) by the *derived* tensor $\left(\dfrac{\partial \boldsymbol{\alpha}}{\partial Q_l}\right)_0$.

These results being well known we now turn to the contrary case of bad convergence of (11). Then the higher powers and products of the Q_{l0} no longer play a negligible role and it is immediately recognized from (11) that a variety of overtones and combination tones will appear.

[The general term of (11) is

$$\left[\sum_{(k_1,\ldots,k_L)} Z_{k_1,\ldots,k_L}\left(\frac{\partial^{k_1+\ldots+k_L}\alpha}{\partial Q_1^{k_1}\cdots\partial Q_L^{k_L}}\right)_0 Q_{10}^{k_1}\cdots Q_{L0}^{k_L}\right]$$

$$E_0\{\sum_{\pm} \cos[\omega_0 t \pm r_1(\omega_1 t + \delta_1) \pm \ldots \pm r_L(\omega_L t + \delta_L)]\} \tag{13}$$

Regard the set of integers $r_l \geqslant 0$ as given and fixed. The subscript $\pm$ on the

[35] Y. Rocard, *Compt. rend.*, 1928, **186**, 1107; J. Cabannes, *ibid.*, pp. 1201, 1714; J. Cabannes and Y. Rocard, *J. Phys. Radium*, 1929, **10**, 52.

last summation symbol means that the summation has to be extended over all sign combinations of the following terms. The numbers k_l in the first term are determined by

$$k_l = r_l + 2s_l \tag{14}$$

with s_l (for all l) to be successively chosen equal to 0, 1, 2, ... The numerical coefficients of the first sum are given by

$$Z_{k_1, \ldots, k_L} = \frac{1}{k_1! \ldots k_L!} \frac{1}{2^{k_1 + \ldots + k_L}} \binom{k_1}{s_1} \cdots \binom{k_L}{s_L} \qquad (15)]$$

There is no reason why the convergence of (11) should be equally bad for all fundamentals Q_l. Consequently in general not all vibrations will display overtones and combination tones. It would be desirable to find out for what fundamentals and under what circumstances the convergence of (11) becomes unsatisfactory. A preliminary answer can be found by a closer examination of the explicit expression (1) for $\alpha(Q)$. Introducing the angular frequency

$$\bar{\omega}_r \equiv (\omega_r^2 - \tfrac{1}{4}\Gamma_r^2)^{\frac{1}{2}} \tag{16}$$

of the damped virtual electronic oscillator, (1) by a simple algebraic procedure may be transformed into

$$\alpha = \frac{e^2}{m} \sum_r \frac{f_r}{2\omega_0 - \mathrm{i}\Gamma_r} \left[\frac{1}{\bar{\omega}_r - \omega_0 + \frac{\mathrm{i}}{2}\Gamma_r} + \frac{1}{-\bar{\omega}_r - \omega_0 + \frac{\mathrm{i}}{2}\Gamma_r} \right] \tag{17}$$

Here f_r, $\bar{\omega}_r$, Γ_r are functions of $\mathbf{Q}$. (For later use we keep here the imaginary terms, but throughout Section 3A they may be considered as vanishing.) If we make the plausible assumption that Γ_r does not depend on $\mathbf{Q}$ and that f_r and $\bar{\omega}_r$ are well-behaved (*i.e.* analytical) functions of $\mathbf{Q}$ near $\mathbf{Q} = \mathbf{0}$, we may roughly approximate

$$\Gamma_r \approx \text{const.} \tag{18}$$

$$f_r \approx f_r(\mathbf{0}) + \mathbf{Q} \cdot \nabla_0 f_r \equiv f_{r0} + \textstyle\sum_l f_{r0}^{(l)} Q_l \tag{19}$$

$$\bar{\omega}_r \approx \bar{\omega}_r(\mathbf{0}) + \mathbf{Q} \cdot \nabla_0 \bar{\omega}_r \equiv \bar{\omega}_{r0} + \textstyle\sum_l \bar{\omega}_{r0}^{(l)} Q_l \tag{20}$$

with the abbreviations

$$f_{r0} \equiv f_r(\mathbf{0}) \,, \; f_{r0}^{(l)} \equiv \left(\frac{\partial f_r}{\partial Q_l} \right)_0 \tag{21}$$

and similarly for $\bar{\omega}_r$.
Introducing these expressions in (17) the bracket takes the form

$$[\quad] = \frac{1}{\zeta_+ + z} + \frac{1}{\zeta_- - z} \tag{22}$$

with

$$\zeta_\pm \equiv \pm\bar{\omega}_{r0} - \omega_0 + \tfrac{\mathrm{i}}{2}\Gamma_r \tag{23}$$

and

$$z \equiv \mathbf{Q} \cdot \nabla_0 \bar{\omega}_r = \textstyle\sum_l \bar{\omega}_{r0}^{(l)} Q_l \tag{24}$$

Equation (22) can be expanded in a Maclaurin's series about $z = 0$ (*i.e.* $\boldsymbol{Q} = \boldsymbol{0}$):

$$[\quad] = \sum_{n=0}^{\infty} \frac{(-z)^n}{\zeta_+^{n+1}} + \sum_{n=0}^{\infty} \frac{z^n}{\zeta_-^{n+1}} \tag{25}$$

However, this series represents the original function only within (and eventually on the boundary of) the convergence region defined by

$$|z| < |\zeta_+| \text{ and } |z| < |\zeta_-| \tag{26}$$

Because $|\zeta_+| \ll |\zeta_-|$ (the ω's are always positive and Γ_r is relatively small) the second of these conditions is included in the first one. The convergence of (25) and indirectly of the power-series expansion of α in this approximation therefore may be judged from the sole criterion

$$|z| < |\zeta_+| \text{ or explicitly } |\boldsymbol{Q} \cdot \boldsymbol{\nabla}_0 \bar{\omega}_r| < {}_+[(\bar{\omega}_{r0} - \omega_0)^2 + \tfrac{1}{4}\Gamma_r^2]^{\frac{1}{2}} \tag{27}$$

which must be fulfilled for every r. $\boldsymbol{Q}$ here must be understood as $\boldsymbol{Q}_0 \equiv \{Q_{10}, \ldots, Q_{L0}\}$, for (27) must be valid for all times t [*cf.* (6)].

Equation (27) is both a sufficient and (apart from the physically unimportant boundary of the convergence region) necessary condition for convergence. There are several interesting inferences to be drawn from (27). Inspection of the right-hand side (RHS) shows that most critical are those r's whose $\bar{\omega}_{r0}$ lie closest to ω_0. If (under inclusion of the values of Γ_r) the value of the square root for one r or a few r's gets exceptionally small we may bestow on the corresponding virtual electronic oscillator(s) the attribute 'resonance'. The occurrence of resonance of course depends on the choice of ω_0. In the resonance case, which we shall discuss below, the second term in the bracket of (17) may be neglected and the sum over r reduced to the one or several r in resonance. The discussion of the LHS of (27) is more complicated, but also more informative. We first note that (27) could easily be made more restrictive (at the same time giving up the property of necessariness) by replacing $|\boldsymbol{Q} \cdot \boldsymbol{\nabla}_0 \bar{\omega}_r|$ by $|\boldsymbol{Q}_0||\boldsymbol{\nabla}_0 \bar{\omega}_r|$. Then an explicit convergence radius ρ can be formulated. The polarizability series converges with certainty if

$$|\boldsymbol{Q}_0| \equiv (\textstyle\sum_l Q_{l0}^2)^{\frac{1}{2}} < \rho = \left[(\bar{\omega}_{r0} - \omega_0)^2 + \tfrac{1}{4}\Gamma_r^2)/\textstyle\sum_l \left(\dfrac{\partial \bar{\omega}_r}{\partial Q_l}\right)_0^2\right]^{\frac{1}{2}} \tag{28}$$

For the special case of non-resonance excitation this formula includes the condition given by Placzek on p. 270 of ref. 9 for the applicability of the polarizability theory. ($|\boldsymbol{Q}_0||\boldsymbol{\nabla}_0 \bar{\omega}_r|$ has the order of magnitude of typical vibrational angular frequencies.) However, an important aspect of (27) is lost by the replacement we have just considered. $\boldsymbol{\nabla}_0 \bar{\omega}_r$ is the gradient of $\bar{\omega}_r(\boldsymbol{Q})$ at $\boldsymbol{Q} = \boldsymbol{0}$; its magnitude is largest in the direction of the steepest ascent of the function $\bar{\omega}_r(\boldsymbol{Q})$. Now consider two extreme cases: if $\boldsymbol{Q}$ is orthogonal to this gradient, (27) is always fulfilled, no matter how large $|\boldsymbol{Q}|$ is, because the scalar product vanishes. However, if $\boldsymbol{Q}$ is parallel to the gradient, $|\boldsymbol{Q}|$ may only take relatively small values without endangering the convergence of the polarizability series. In nature the direction of Q (*i.e.* the relative excitation amplitudes of the diverse nuclear normal vibrations) normally is not open to choice (unless selective excitation of individual vibrations can be realized), but we perceive that the convergence condition (27) will most easily be violated by those components Q_l of Q in whose direction $\bar{\omega}_r(Q)$ increases most. These Q_l most likely then will produce overtones and combination tones, particularly if (with the beginning of resonance) ω_0 approaches a distinct $\bar{\omega}_{r0}$ and the RHS of (27) therefore becomes small. For specific molecules a judgment of the behaviour of $\bar{\omega}_r(Q)$ can be made only with the assistance of quantum mechanical methods. The quantity $\hbar\bar{\omega}_r$ represents the energy of transition between the ground (=initial) state and an electronic (=intermediate) state. In order to test the above results the variation of the transition energy with the change of the nuclear configuration (*i.e.* the potential surfaces of the electronic states participating in the scattering process) must be known. It is qualitatively clear, however, that it is not necessarily the totally symmetric normal co-ordinate which is parallel to $\mathbf{V}_0\bar{\omega}_r$ and which therefore always must be favoured in the RRE.

If the expressions (18) *et seq.* explicitly are inserted in (17) (with the second term neglected) and comparison is made with (5) we find for the constant part of α and its first partial derivative with respect to the normal co-ordinate Q_l at the nuclear equilibrium configuration, in linear approximation, the expressions

$$\alpha_0 = \frac{e^2}{m}\sum_r \frac{f_{r0}}{(2\omega_0 - \mathrm{i}\Gamma_r)(\bar{\omega}_{r0} - \omega_0 + \frac{\mathrm{i}}{2}\Gamma_r)} \tag{29}$$

$$\left(\frac{\partial\alpha}{\partial Q_l}\right)_0 \equiv \alpha_0^{(l)} = \frac{e^2}{m}\sum_r \frac{1}{(2\omega_0 - \mathrm{i}\Gamma_r)(\bar{\omega}_{r0} - \omega_0 + \frac{\mathrm{i}}{2}\Gamma_r)}\left[f_{r0}^{(l)} - \frac{f_{r0}\bar{\omega}_{r0}^{(l)}}{\bar{\omega}_{r0} - \omega_0 + \frac{\mathrm{i}}{2}\Gamma_r}\right] \tag{30}$$

The last equation (with insignificant modifications) was first used by Shorygin [36] to describe the incipient (or pre-) RRE (*cf.* ref. 2, p. 187 ff. and refs. 22 and 37). (30) is composed of two terms. The importance of these terms depends on the magnitudes of $f_{r0}^{(l)}$ and $f_{r0}\bar{\omega}_{r0}^{(l)}$, respectively. $f_{r0}^{(l)}$ (or $\bar{\omega}_{r0}^{(l)}$) describes the rate of change of f_r (or $\bar{\omega}_r$, respectively) due to the distortion of the nuclear equilibrium configuration in the direction of Q_l. If $f_{r0}\bar{\omega}_{r0}^{(l)}$ does not vanish (*i.e.* in particular if the pure electronic transition is allowed), the increase of the second term with ω_0 approaching $\bar{\omega}_{r0}$ exceeds that of the first term and then the second term will be more important for generation of the pre-RRE. By numerical estimations Shorygin concluded that the first term in (30) may in most cases be neglected if ω_0 is close enough to resonance (*cf.* ref. 38, p. 202). In addition we see that with this presupposition and if for only one r

[36] P. P. Shorygin, *Zhur. fiz. Khim.*, 1947, **21**, 1125; 1951, **25**, 341; *Izvest. Akad. Nauk S.S.S.R., Ser. fiz.*, 1948, **12**, 576; *Doklady Akad. Nauk S.S.S.R.*, 1951, **78**, 469.
[37] P. Shorygin and L. Krushinsky, *Z. Physik*, 1958, **150**, 332.
[38] P. P. Shorygin and T. M. Ivanova, *Optika i Spektroskopiya*, 1968, **25**, 200.

in the sum the condition of resonance is satisfied, $\alpha_0^{(l)}$ becomes roughly proportional to $\bar{\omega}_{r0}^{(l)}$. This means that for comparable vibrational amplitudes Q_{l0}, those normal vibrations will show up most intensely in the pre-RRE which contribute most to the change of $\bar{\omega}_r$ near nuclear equilibrium. Expressions similar to (29) and (30) may also be found without difficulty for all the higher (including mixed) derivatives of α which refer to the overtones and combination tones.

The convergence of the series (11) (but with the restriction to only one normal co-ordinate Q) was first discussed in detail by Krushinsky.[39] Krushinsky for this special case of diatomic molecules also derived explicit values for the radius of convergence ρ valid under specific assumptions for the shapes of the two potential curves $U_g(Q), U_e(Q)$ of the electronic ground state g and the excited resonance state e. In the simplest case of congruent parabolae the denominator in (28) turns out to be

$$\left|\frac{\mathrm{d}\bar{\omega}_r}{\mathrm{d}Q}\right| = \frac{k}{\hbar c}|b| \tag{31}$$

where $k = \dfrac{\mathrm{d}^2 U_g}{\mathrm{d}Q^2}$ and b is the distance in direction Q of the two potential minima. Thus ρ for a given exciting frequency is proportional to $|b|^{-1}$ which shows that, under the special assumptions made, the RRE will occur sooner for large relative displacements of the potential minima. For $\omega_0 = \bar{\omega}_{r0}$, ρ in this case takes the minimum value

$$\rho_{\min} = \frac{\hbar c \Gamma_r}{2k|b|} \tag{32}$$

so that qualitatively a large damping constant Γ_r of the resonance state should reduce the chances for the RRE.

Krushinsky and Shorygin[40] have also discussed in detail the correspondence of the expansions of the polarizability for diatomic molecules with the quantum mechanical dispersion theory. Kondilenko, Pogorelov, Strizhevsky, and Shinkaryova,[41] starting from the Kramers–Heisenberg dispersion formula and using an expansion similar to (25) (but also taking account of only one normal co-ordinate), derived expressions also for the non-diagonal polarizability components (non-isotropic case) $\alpha_{\rho\sigma}$, $\rho \neq \sigma$, for the Raman fundamental and the first overtone.[42]

The correspondence of (1) and all conclusions drawn from it with the quantum mechanical polarizability tensor (and also the discrepancies) can

[39] L. L. Krushinsky, *Optika i Spektroskopiya*, 1963, **14**, 767.
[40] L. L. Krushinsky and P. P. Shorygin, *Optika i Spektroskopiya*, 1965, **19**, 562.
[41] I. I. Kondilenko, V. E. Pogorelov, V. L. Strizhevsky, and E. E. Shinkaryova, *Optika i Spektroskopiya*, 1969, **26**, 203.
[42] *Cf.* also I. I. Kondilenko, P. A. Korotkov, and V. L. Strizhevsky, *Optika i Spektroskopiya*, 1960, **9**, 26; I. I. Kondilenko and V. L. Strizhevsky, *ibid.*, 1961, **11**, 262; 1965, **18**, 938.

be perceived in the following way: equation (1) is already a generalization of the basic equation

$$\alpha_k = \frac{e_k^2}{m_k}\,[\omega_k^2 - \omega_0^2 + \mathrm{i}\omega_0\Gamma_k]^{-1} \tag{33}$$

for the polarizability of a charged particle (charge e_k, mass m_k) harmonically bound to a fixed site and performing a forced vibration (damping constant Γ_k) under the influence of an alternating electric field (7) ['scattering centre'; for a derivation of (33) *cf.* ref. 32, p. 269 ff.]. Imagining that an atom or molecule in general consists of many particles k (electrons i and nuclei j) and taking account of the geometrical additivity of dipole moments (and therefore also of polarizabilities), the atomic or molecular polarizability becomes simply the sum of (33) over all k. Now for nuclei the quotients e_k^2/m_k and the angular frequency squares ω_k^2 are about three or four orders of magnitude smaller than for electrons and therefore, if ω_0 falls into the visible part of the spectrum, the contribution of the nuclei to the polarizability can be neglected. Then it is essentially only the electronic polarizability

$$\alpha_{\mathrm{el}} = \frac{e^2}{m}\sum_{i=1}^{N_i}[\omega_i^2 - \omega_0^2 + \mathrm{i}\omega_0\Gamma_i]^{-1} \tag{34}$$

(N_i is number of electrons within the molecule) which accounts for light scattering. However, (34), in view of experimental results, proves to be very imperfect. A certain degree of improvement can be attained by sacrificing the independence of the motions of the single electrons and regarding the total electronic cloud as a compound oscillator performing a multitude of joint oscillations. Indexing these 'virtual oscillators' by r and supplying each one with an effectivity factor f_r (Ladenburg's [43] 'oscillator strength'), (34) can be replaced by (1). If the identification of (34) with (1) is justified, multiplication of both formulae by ω_0 and subsequent integration over ω_0 from 0 to ∞ must give the same result, *i.e.* we ought to have the sum rule for oscillator strengths

$$\sum_r f_r = N_i \tag{35}$$

The conjectures leading to (1) and (35) are backed by quantum mechanics. (1) turns out to be a special case of the quantum mechanical polarizability tensor (written in dyadic form)

$$\boldsymbol{\alpha}_{nm} = \frac{1}{\hbar}\sum_r\left(\frac{\boldsymbol{\mu}_{rn}\,\boldsymbol{\mu}_{mr}}{\omega_{rm} - \omega_0 + \frac{\mathrm{i}}{2}\Gamma_r} + \frac{\boldsymbol{\mu}_{mr}\,\boldsymbol{\mu}_{rn}}{\omega_{rn} + \omega_0 + \frac{\mathrm{i}}{2}\Gamma_r}\right) \tag{36}$$

Here m, r, and n are the initial, intermediate, and final molecular states involved in the scattering transition, and $\boldsymbol{\mu}_{rn}$ and $\boldsymbol{\mu}_{mr}$ are the matrix elements of the electric dipole moment operator. Putting $m = n$ (Rayleigh scattering),

[43] R. Ladenburg, *Z. Physik*, 1921, **4**, 451; K. Ladenburg and F. Reiche, *Naturwiss.*, 1923, **11**, 584.

assuming $\boldsymbol{a}_{nm}$ to be spherically symmetric (*i.e.* its space components $a_{\rho\sigma,mn} = a\delta_{\rho\sigma}$), and abbreviating

$$\omega_r^2 \equiv \omega_{rm}^2 - \tfrac{1}{4}\Gamma_r^2 \qquad (37)$$

(36) then gives

$$a = a_{\rho\rho,mm} = \frac{1}{\hbar}\sum_r \mu_{rm,\rho}^2 \frac{2\omega_{rm} + \mathrm{i}\Gamma_r}{\omega_r^2 - \omega_0^2 + \mathrm{i}\omega_{rm}\Gamma_r} \qquad (38)$$

If $\Gamma_r \ll \omega_{rm}$, $\mathrm{i}\Gamma_r$ in the numerator can be neglected and ω_{rm} in the denominator can be approximated first by ω_r and then, because of the sharp maximum of (38) at $\omega_0 = \omega_r$, by ω_0. Then (38) becomes identical with (1) under the condition

$$\frac{e^2}{m} f_r = \frac{2}{\hbar}\mu_{rm,\rho}^2\omega_{rm} = \frac{2}{3\hbar}\left|\mu_{rm}\right|^2 \omega_{rm} = \frac{1}{\pi} B_{rm}\hbar\omega_{rm} \qquad (39)$$

B_{rm} is the Einstein transition probability coefficient for induced absorption or emission occurring between the states m and r. It is proved in quantum theory that the magnitude f_r given by (39) indeed satisfies the sum rule (35) (see *e.g.* ref. 44, p. 336 ff.). Of course f_r and ω_r [as given by (37)] in this connection must be understood as referring to the *pair* of energy levels m,r whereas Γ_r and the summation index r in (1) and (37) refer only to the intermediate level r. In a further step the correspondence between the classical formula (1) and the quantum mechanical expression (36) could be improved by omitting the restriction of $\boldsymbol{a}_{nm}$ to its isotropic part and attributing to the oscillator strengths f_r also the property of a tensor: $f_r \rightarrow (f_{r,\rho\sigma})$. Then also (1) would become a tensor $\boldsymbol{a} = (a_{\rho\sigma})$ and (8) would have to be replaced by

$$\mu_\rho = \mathrm{Re}(\sum_\sigma a_{\rho\sigma}E_\sigma) \qquad (40)$$

In summary we can conclude that the classical equation (1), in spite of certain deficiencies, reflects the essential content of the quantum mechanical expression (36) for the polarizability in the dipole approximation. It is therefore justifiable to have confidence in the qualitative results drawn from a discussion of (1).

B. Amplitude and Phase Modulation of the Polarizability.—In Section 3A it was assumed [see the remark before (4)] that $\omega_r > \omega_0$ for all r and that the imaginary term $\mathrm{i}\omega_0\Gamma_r$ in (1) can be neglected. This is certainly no longer permitted if ω_0 approaches closely to ω_r for at least one r. Therefore near to resonance ($\omega_r - \omega_0 \lesssim \Gamma_r$) or in rigorous resonance ($\omega_0 = \omega_r$) a is a *complex* quantity. Consequently the derivatives of a with respect to the normal coordinates Q_l are also complex quantities and the expansion (11), where they were considered as real, is no longer correct. Now any complex number z

[44] H. A. Bethe and E. E. Salpeter, 'Quantum Mechanics of One- and Two-Electron Systems', in 'Encyclopedia of Physics' ed. S. Flügge, Springer, Berlin, 1957, Vol. 35, pp. 88—446.

can be written in the form $z = re^{i\delta}$ where $r = |z|$ is the 'modulus' and δ the 'argument' or 'phase' of z. The parameters f_r, ω_r, and Γ_r of α through the nuclear vibrations being more or less periodically dependent on the time t, now not only the modulus of α but also its phase becomes dependent on t. Thus the amplitude modulation of α prevailing in the non-resonance case is supplemented or perhaps even replaced by a phase modulation when going over to resonance. The recognition of this change is due to Shorygin and Krushinsky (ref. 4, p. 315 and refs. 22, 23, and 39).

Let us consider this idea a little more systematically. For simplicity we again take α as a scalar quantity (isotropic case). We shall first, analogously to the procedure in Section 3A, treat this problem in an abstract way using the unspecified function

$$\alpha = \alpha(Q) = |\alpha(Q)|e^{i\delta(Q)} \tag{41}$$

Later, by using the explicit form (1) or (17) for α, we shall show that in the rigorous RRE the phase modulation even becomes more important than the amplitude modulation. For reasons of simplicity we shall restrict this discussion to the case of a diatomic molecule with only one normal vibration

$$Q \equiv Q_l = Q_0 \cos \omega_l t \tag{42}$$

instead of $\boldsymbol{Q}$. (In order to avoid confusion we always retain the subscript l on the nuclear vibration angular frequency ω_l.) Then in linear approximation we have for the modulus and the phase of α

$$|\alpha(Q)| = \alpha_0 + \alpha_1 Q \quad \text{with real constants} \quad \alpha_0 \equiv |\alpha(0)|,\ \alpha_1 \equiv \left(\frac{\partial \alpha}{\partial Q}\right)_{Q=0} \tag{43}$$

$$\delta(Q) = \delta_0 + \delta_1 Q \quad \text{with real constants} \quad \delta_0 \equiv \delta(0),\ \delta_1 \equiv \left(\frac{\partial \delta}{\partial Q}\right)_{Q=0} \tag{44}$$

Insertion of $\alpha = \alpha(Q)$ in (9) gives

$$\begin{aligned}\mu &= \mathrm{Re}(\alpha_0 + \alpha_1 Q)E_0 e^{i(\omega_0 t + \delta_0 + \delta_1 Q)} = E_0(\alpha_0 + \alpha_1 Q)\cos(\omega_0 t + \delta_0 + \delta_1 Q)\\ &= E_0(\alpha_0 + \alpha_1 Q_0 \cos \omega_l t)\cos(\omega_0 t + \delta_0 + \delta_1 Q_0 \cos \omega_l t)\end{aligned} \tag{45}$$

This rather unpleasant function of t will be broken up by considering two special cases: first we consider pure amplitude modulation of the dipole moment μ, *i.e.* the case $\delta_1 = 0$, and then pure phase modulation, *i.e.* $\alpha_1 = 0$.

In the first case we find at once by simple trigonometry

$$\begin{aligned}\mu = E_0\alpha_0 \cos(\omega_0 t + \delta_0) + \tfrac{1}{2}E_0\alpha_1 Q_0\{&\cos[(\omega_0 + \omega_l)t + \delta_0]\\ &+ \cos[(\omega_0 - \omega_l)t + \delta_0]\}\end{aligned} \tag{46}$$

This is again the original formula of Cabannes and Rocard already mentioned in (12). The customary classical theory of the RE is thus founded upon the concept of amplitude modulation of the electric dipole moment by the nuclear vibrations.

The second case, $a_1 = 0$, is more interesting though more involved. In order to discuss the expression

$$\mu = E_0 a_0 \cos(\omega_0 t + \delta_0 + \delta_1 Q_0 \cos \omega_l t) \quad (47)$$

we suitably abbreviate

$$\zeta_1 \equiv \delta_1 Q_0, \theta = \omega_l t \quad (48)$$

and go back to the complex formulation of μ. Then

$$\mu = E_0 a_0 \, \mathrm{Re}(e^{i(\omega_0 t + \delta_0)} e^{i\zeta \cos\theta}) \quad (49)$$

The last factor appears in a well-known definition of the Bessel functions of the first kind [45]

$$J_n(\zeta) = \frac{1}{2\pi} e^{-in\pi/2} \int_0^{2\pi} e^{i\zeta \cos\theta} \cos n\theta \mathrm{d}\theta \quad (50)$$

For extraction of $e^{i\zeta \cos\theta}$ from this integral we recall the fact that any even function $f(\theta)$ which (apart from a finite number of points) is continuous within $0 < \theta < 2\pi$ can be expanded in a Fourier series of cosines

$$f(\theta) = \tfrac{1}{2} a_0 + \sum_{n=1}^{\infty} a_n \cos n\theta \quad (51)$$

where the coefficients are given by

$$a_n = \frac{1}{\pi} \int_0^{2\pi} f(\theta) \cos n\theta \mathrm{d}\theta, \; n = 0, 1, \ldots \quad (52)$$

Identifying $f(\theta)$ with $e^{i\zeta \cos\theta}$ we find

$$e^{i\zeta \cos\theta} = J_0(\zeta) + 2 \sum_{n=1}^{\infty} e^{in\pi/2} J_n(\zeta) \cos n\theta \quad (53)$$

Inserting this result in (49) we have

$$\mu = E_0 a_0 \{J_0(\zeta) \cos(\omega_0 t + \delta_0) + 2 \sum_{n=1}^{\infty} J_n(\zeta) \cos(\omega_0 t + \delta_0 + \tfrac{1}{2} n\pi) \cos n\theta\}$$

or, by use of (48) and the trigonometric addition theorems,

$$\begin{aligned} \mu = E_0 a_0 \{ & J_0(\delta_1 Q_0) \cos(\omega_0 t + \delta_0) \\ & + \sum_{n=1}^{\infty} J_n(\delta_1 Q_0) \cos[(\omega_0 + n\omega_l)t + \delta_0 + \tfrac{1}{2} n\pi] \\ & + \sum_{n=1}^{\infty} J_n(\delta_1 Q_0) \cos[(\omega_0 - n\omega_l)t + \delta_0 + \tfrac{1}{2} n\pi]\} \end{aligned} \quad (54)$$

Here the first term gives Rayleigh scattering, the second one anti-Stokes, and the third one Stokes Raman scattering. Now an infinite multitude of

[45] See, *e.g.* P. M. Morse and H. Feshbach, 'Methods of Theoretical Physics', McGraw-Hill, New York, 1953, Part I, p. 620.

overtones appears. The intensity of the $(n-1)$th overtone (anti-Stokes and Stokes equally) is determined by the square of $J_n(\delta_1^r Q_0)$. For big arguments the values of J_n vary strongly with n, so there is not even a qualitative general rule about the relative intensities of different overtones. For small ζ, however, we can approximate

$$J_n(\zeta) \approx \zeta^n/2^n n! \tag{55}$$

and the intensities of the overtones fall off exponentially with their order. This is similar to the distribution of overtone intensities for pure amplitude modulation [*cf.* (11)]. Incidentally we should note the peculiar phase shift augmented by $\frac{\pi}{2}$ from overtone to overtone in (54). In the general case μ will show both amplitude and phase modulation. It is evident how the above two extreme cases must be combined for this general case.

We now shall demonstrate that in the case of rigorous resonance excitation the phase modulation of α or μ prevails over the amplitude modulation. This constitutes a very remarkable difference between the ordinary RE and the (rigorous) RRE. It will be sufficient to show this by means of a simplified model of the rigorous RRE. Let us return to (54) for a diatomic molecule and take from (20)

$$\bar{\omega}_r = \bar{\omega}_r(0) + \left(\frac{\mathrm{d}\bar{\omega}_r}{\mathrm{d}Q}\right)_0 Q \equiv \bar{\omega}_{r0} + \bar{\omega}_{r0}'Q \tag{56}$$

Then for strict resonance excitation ($\omega_0 = \bar{\omega}_{r0}$), if there is only one resonance electronic oscillator r and if the damping constants are small enough to avoid mixing with neighbouring r's we can reduce the sum (17) to one term and write

$$\alpha = \frac{e^2}{m}\frac{f_r}{2\omega_0 - \mathrm{i}\Gamma_r}\frac{1}{\bar{\omega}_{r0}'Q + \frac{\mathrm{i}}{2}\Gamma_r} = \frac{e^2}{m}\frac{f_r}{2\omega_0 - \mathrm{i}\Gamma_r}\frac{2}{\Gamma_r}\frac{1}{\xi + \mathrm{i}} \tag{57}$$

with

$$\xi = \frac{2\bar{\omega}_{r0}'Q}{\Gamma_r} \tag{58}$$

The conformal mapping effected by the complex function

$$\eta \equiv |\eta|\mathrm{e}^{\mathrm{i}\varphi} = \frac{1}{\xi + \mathrm{i}} \tag{59}$$

is shown in Figure 1. The section $\xi = (-1, +1)$ of the real axis is mapped on to the semicircle with the radius $\frac{1}{2}$ about the centre $-\frac{\mathrm{i}}{2}$ and extending from $-\frac{1}{2}(\mathrm{i} + 1)$ over $-\mathrm{i}$ to $\frac{1}{2}(1 - \mathrm{i})$. It is immediately recognized that the point η oscillates about $-\mathrm{i}$ on this semicircle when ξ oscillates about 0 on the real axis. If the oscillation amplitude of ξ is sufficiently small, there will only be a change of the argument φ of η, whereas the modulus $|\eta|$ practically remains constant. This proves that phase modulation is characteristic for the RRE.

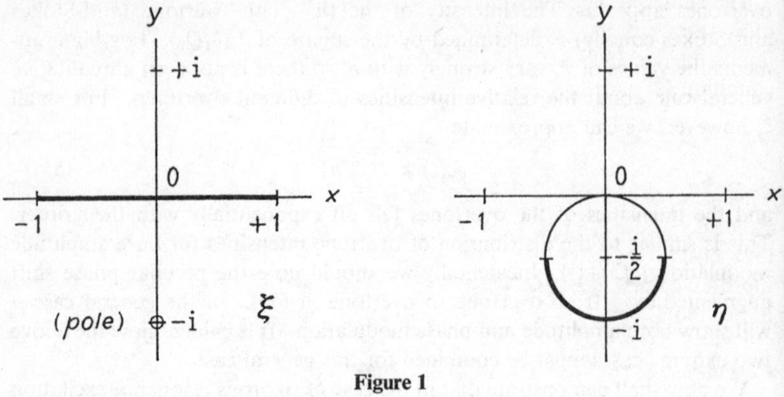

Figure 1

The Maclaurin's expansion of (59) is

$$\eta = \sum_{n=0}^{\infty} e^{i\frac{\pi}{2}(n-1)}\xi^{n} \tag{60}$$

and it converges within

$$|\xi| < 1 \tag{61}$$

After some manipulation this gives by (57) (the terms $-i\Gamma_r$ in the first denominator can be neglected and $f_r = f_{r0}$ is taken to be a constant > 0) the polarizability expansion

$$a = a(Q) = \sum_{n=0}^{\infty} \frac{1}{n!}\left(\frac{d^n a}{dQ^n}\right)_0 Q^n \equiv \sum_{n=0}^{\infty} \frac{1}{n!}\left|\left(\frac{d^n a}{dQ^n}\right)_0\right| e^{i\delta_n} Q^n \tag{62}$$

with

$$\left|\left(\frac{d^n a}{dQ^n}\right)_0\right| = n!\,\frac{e^2 f_{r0}}{m\omega_0 \Gamma_r}\left(\frac{2\bar{\omega}'_{r0}}{\Gamma_r}\right)^n$$

and

$$\delta_n = \tfrac{\pi}{2}(n-1) \tag{63}$$

Equation (63) shows that $\left(\frac{d^n a}{dQ^n}\right)_0$ is real for odd n and purely imaginary for even n. The Raman fundamentals and even overtones are determined by the odd polarizability derivatives, whereas Rayleigh scattering and the odd Raman overtones are determined by the even ones [*cf.* (11)]. Thus, in the rigorous RRE, the real and imaginary parts of a are alternatively responsible for successive Raman overtones [including Raman fundamentals as zeroth overtones and Rayleigh scattering as (—1)th]. δ_n is the phase shift of the induced dipole moment corresponding to the $(n-1)$th overtone (Stokes and anti-Stokes) with respect to the same induced dipole moment in non-resonance scattering.

The simple discussion above demonstrates that a power-series expansion of α, *i.e.* a modified application of the polarizability theory, is not entirely impossible for the RRE. However, attention must be paid to the convergence condition (61) or, explicitly,

$$Q_0 < \rho = \left|\frac{\Gamma_r}{2\bar{\omega}'_{r0}}\right| \tag{64}$$

which corresponds to the resonance case of (28). Equation (64) might be violated. Then in order to reveal the overtone structure of the RR spectra, (57) must be expanded in a series of a different kind, *e.g.* a Laurent series, or, more advantageously, a Fourier series. The latter possibility will be considered in Section 3C.

C. Fourier Expansion of the Polarizability.—The restrictions in the convergence of power series for the polarizability α can be escaped by expanding α in a Fourier series. This method was first used by Krushinsky.[46] In order to give the essential ideas of this method as briefly as possible, we, unlike Krushinsky, shall employ some function theory.

Let us again restrict ourselves to the discussion of the rigorous vibrational RRE occurring with a diatomic molecule. The knowledge from the preceding discussions that a multitude of Raman overtones with angular frequencies $\omega_0 + n\omega_l$, $n = 0, \pm 1, \ldots$ will turn up suggests looking for a formulation of the electric dipole moment which from the very beginning contains the factor $e^{i(\omega_0 + n\omega_l)t}$. This is rendered feasible by a Fourier expansion of the polarizability using the set of complex functions

$$g_n \equiv g_n(t) = e^{in\omega_l t} \tag{65}$$

The orthonormality relations are (* indicates conjugate complex)

$$\frac{1}{T_l}\int_0^{T_l} g_n^* g_{n'} \mathrm{d}t = \delta_{nn'} \quad \text{with} \quad T_l \equiv 2\pi/\omega_l \tag{66}$$

We obtain the expansion

$$\alpha \equiv \alpha[Q(t)] = \sum_{n=-\infty}^{+\infty} a_n g_n \tag{67}$$

with the Fourier coefficients

$$a_n \equiv |a_n| e^{i\theta_n} = \frac{1}{T_l}\int_0^{T_l} \alpha g_n^* \mathrm{d}t \tag{68}$$

By (9) the scattering dipole moment then is

$$\begin{aligned}\mu = \mathrm{Re}(\alpha E_0 e^{i\omega_0 t}) &= E_0 \mathrm{Re} \sum_{n=-\infty}^{+\infty} a_n e^{i(\omega_0 + n\omega_l)t} \\ &= E_0 \sum_{n=-\infty}^{+\infty} |a_n| \cos[(\omega_0 + n\omega_l)t + \theta_n]\end{aligned} \tag{69}$$

[46] L. L. Krushinsky, *Optika i Spektroskopiya*, 1963, **15**, 747.

Now let us turn to the rigorous resonance case using (57) in the unessentially simplified form

$$\alpha(Q) = \frac{e^2 f_{r0}}{2m\omega_0} \frac{1}{\bar{\omega}'_{r0} Q + \frac{1}{2}\mathrm{i}\Gamma_r} \quad \text{with} \quad Q = Q_0 \cos \omega_l t \tag{70}$$

Abbreviating

$$a \equiv \frac{e^2 f_{r0}}{2m\omega_0 \bar{\omega}'_{r0} Q_0}, \qquad b \equiv \frac{\Gamma_r}{2\bar{\omega}'_{r0} Q_0} \tag{71}$$

(real constants), we have

$$\alpha = \frac{a}{\cos \omega_l t + \mathrm{i}b} \tag{72}$$

Substituting

$$\mathrm{e}^{\mathrm{i}\omega_l t} \equiv z, \; \mathrm{i}\omega_l z \mathrm{d}t = \mathrm{d}z \tag{73}$$

we get

$$\alpha_n \equiv |\alpha_n| \mathrm{e}^{\mathrm{i}\theta_n} = \frac{1}{2\pi \mathrm{i}} \oint \frac{2az^n \mathrm{d}z}{z^2 + 2\mathrm{i}bz + 1} \tag{74}$$

Here the integral has to be taken in the positive sense around the closed unit circle in the complex z-plane. We evaluate (74) by means of the calculus of residues, presupposing $n \geqslant 0$. The denominator in the integral vanishes at

$$z = \mathrm{i}[-b \pm (b^2 + 1)^{\frac{1}{2}}] \tag{75}$$

One of the zeros, namely

$$z_\varepsilon = \mathrm{i}[-b + \varepsilon(b^2 + 1)^{\frac{1}{2}}], \; \varepsilon \equiv \frac{b}{|b|} \tag{76}$$

because of the triangle inequality for complex numbers always lies in the interior of the unit circle, whereas the other always lies outside. The residue of the integrand at z_ε is $2az_\varepsilon^n/(z_\varepsilon - z_{-\varepsilon})$ and so we have

$$\alpha_n = \mathrm{i}^{n-1} \frac{a[-b + \varepsilon(b^2 + 1)^{\frac{1}{2}}]^n}{\varepsilon(b^2 + 1)^{\frac{1}{2}}}, \; n \geqslant 0 \tag{77}$$

The fraction is real-valued and represents, apart from an eventual minus sign, $|\alpha_n|$. We note that $\alpha_{-n} = \alpha_{+n}$, which may be derived from the relation $(\alpha_{-n})^* = (\alpha^*)_n$ [following from (65) and (68)] and the observation that $\alpha \rightarrow \alpha^*$ by (72) and (76) is equivalent to a change of sign of b and ε. For $a, b \gtrless 0$ [in (71) only $\bar{\omega}'_{r0}$ can become negative] (77) yields the phase

$$\theta_n = \tfrac{\pi}{2}(\pm |n| - 1) \quad (+ \text{ or } - \text{ for the fraction being } >0 \text{ or } <0) \tag{78}$$

The meaning of θ_n is different from δ_n in (63), but it leads to the same qualitative conclusions. In resonance Rayleigh scattering ($n = 0$) the phase shift of the vibrating dipole moment μ with respect to the incoming wave is $-\frac{\pi}{2}$, as must be expected for a driven oscillator.

For physical discussion (71) must be inserted into (77). We omit quoting the complicated formal result, and instead consider two extreme special cases presupposing $\bar{\omega}'_{r0} > 0$ in (71), *i.e.* $b > 0$, $\varepsilon = 1$. For $b > 1$ or $Q_0 < \rho \equiv \frac{\Gamma_r}{2\bar{\omega}'_{r0}}$ [*cf.* (64)] $|a_n|$, $n \geqslant 0$ may be approximated by

$$|a_n| = \frac{a}{b}\left(\frac{1}{2b}\right)^n = \frac{e^2 f_{r0}}{m\omega_0 \Gamma_r}\left(\frac{\bar{\omega}'_{r0} Q_0}{\Gamma_r}\right)^n \tag{79}$$

and for $b < 1$ or $Q_0 > \rho$ by

$$|a_n| = a = \frac{e^2 f_{r0}}{2m\omega_0 \bar{\omega}'_{r0} Q_0} \tag{80}$$

Equations (79) and (80) show that by increase of Q_0 $|a_n|$ first increases proportionally to Q_0^n, but finally decreases as $1/Q_0$. This result supplements Sections 3A and 3B where we only arrived at the qualitative statement that for large vibration amplitudes strong overtones of any order should show up. As to the dependence of the overtone intensity on their order n, we obtain from (77) for $b > 0$

$$\left|\frac{a_{n+1}}{a_n}\right| = -b + (b^2 + 1)^{\frac{1}{2}} \tag{81}$$

By (69) the intensity ratio of two successive overtones is (apart from a minor frequency dependence) proportional to the square of this expression. (81) is always < 1, but for $b \to 0$ it approaches unity. This means that for $\frac{\Gamma_r}{\bar{\omega}'_{r0} Q_0} \to 0$ (*i.e.* for small damping constants Γ_r and large gradients of $\bar{\omega}_r$ near $Q = 0$) overtones of very high order may appear with an intensity comparable to that of the Raman fundamental or even the Rayleigh band.

D. The Method of Parametrically Excited Electronic Oscillations.—In order to allow better for the dynamic interaction between the electronic and the nuclear motions of the molecule, Krushinsky and Shorygin [47] have used the method of parametrically excited (or rheolinear) vibrations to treat the RRE on a classical basis (*cf.* also ref. 4, p. 316 ff.). We can only briefly outline this approach and refer the reader to the original literature for more details. Reducing the electronic system of the molecule to one elastically bound electron, we have the elementary equation of motion

$$\ddot{x} + \Gamma\dot{x} + \omega^2 x = \frac{e}{m}E(t) = \frac{e}{m}E_0 \cos \omega_0 t \tag{82}$$

x is the displacement of the electron from the equilibrium position under the influence, and in the direction, of the electric field strength. If Γ and ω were taken constant, the stationary solution of (82) (indexed by k) would

[47] L. L. Krushinsky and P. P. Shorygin, *Optika i Spektroskopiya*, 1964, **16**, 30.

result in the equation (33) basic to (1). In this context, however, we shall replace ω by a periodical function $\omega(t)$ of time varying with the rhythm of the nuclear vibrations:

$$\omega(t) = \omega + \omega' Q,\ Q = Q_0 \cos \omega_l t \tag{83}$$

In (83) Γ, ω (written without the argument t), and ω' are constants and Q_0 is assumed to be so small that terms with Q_0^n, $n > 1$ in the expansion of $\omega(t)$ can be neglected. Γ is taken $\ll \omega$. Inserting (83) into (82), substituting

$$x \equiv \mathrm{e}^{-\frac{1}{2}\Gamma t} y,\ t \equiv 2z/\omega_l \tag{84}$$

and abbreviating

$$\bar{\omega} \equiv (\omega^2 - \tfrac{1}{4}\Gamma^2)^{\frac{1}{2}} \approx \omega \tag{85}$$

we have for y, Mathieu's differential equation

$$\frac{\mathrm{d}^2 y}{\mathrm{d}z^2} + \left(\frac{2\bar{\omega}}{\omega_l}\right)^2 \left(1 + \frac{2\omega'}{\bar{\omega}} Q_0 \cos 2z\right) y = \frac{4e}{m\omega_l^2} E_0 \mathrm{e}^{\Gamma z/\omega_l} \cos \frac{2\omega_0}{\omega_l} z \equiv f(z) \tag{86}$$

First, suitably applying the method of Wentzel, Kramers, and Brillouin, a stable solution y_0 of the reduced (homogeneous) equation [corresponding to (86) with $f(z) = 0$] is obtained:

$$y_0(z) = \left[\frac{2\bar{\omega}}{\omega_l}\left(1 + \frac{\omega'}{\bar{\omega}} Q_0 \cos \omega_l t\right)\right]^{-\frac{1}{2}} \left[A \cos\left(\bar{\omega} t + \frac{\omega'}{\omega_l} Q_0 \sin \omega_l t\right) + B \sin\left(\bar{\omega} t + \frac{\omega'}{\omega_l} Q_0 \sin \omega_l t\right)\right] \tag{87}$$

with arbitrary integration constants A and B. The last factor, similar to (45), can be expanded in a Fourier series with Bessel function amplitudes which after insertion in (84) results in

$$x_0(t) = \mathrm{e}^{-\frac{1}{2}\Gamma t} \left[\frac{2\bar{\omega}}{\bar{\omega}_l}\left(1 + \frac{\omega'}{\omega} Q_0 \cos \omega_l t\right)\right]^{-\frac{1}{2}} \sum_{r=-\infty}^{+\infty} J_r\left(\frac{\omega'}{\omega_l} Q_0\right) [A \cos(\bar{\omega} + r\omega_l)t + B \sin(\bar{\omega} + r\omega_l)t] \tag{88}$$

The factor in the first bracket in (88) corresponds to amplitude modulation and the sum in (88) to phase modulation of the oscillating electronic dipole moment.

In order to find the stationary solution of the complete equation (86) the method of variation of constants can be used. The result is

$$x_1(t) = \frac{eE_0}{2m\bar{\omega}} \sum_{n=-\infty}^{+\infty} \sum_{r=-\infty}^{+\infty} J_r\left(\frac{\omega'}{\omega_l} Q_0\right) J_{r+n}\left(\frac{\omega'}{\omega_l} Q_0\right) \left[\frac{\cos[(\omega_0 - n\omega_l)t + \varphi_+]}{[(\bar{\omega} + r\omega_l + \omega_0)^2 + \frac{1}{4}\Gamma^2]^{\frac{1}{2}}} + \frac{\cos[(\omega_0 + n\omega_l)t - \varphi_-]}{[(\bar{\omega} + r\omega_l - \omega_0)^2 + \frac{1}{4}\Gamma^2]^{\frac{1}{2}}}\right] \tag{89}$$

with

$$\tan \varphi_\pm = \frac{\Gamma/2}{\bar{\omega} + r\omega_l \pm \omega_0} \tag{90}$$

From (89) there follows at once the scattering dipole moment $\mu = ex$ which may be compared with (69). The individual terms of the sum over n, apart from their cosine factors, ought to correspond to (77). It is immediately seen that (89) contains as an additional feature the frequency shifts $r\omega_l$ in the denominators. Physically these refer to the vibrational structure of the electronic absorption band. Thus the classical formula (89) displays a surprising similarity to the quantum mechanical dispersion formula of Kramers and Heisenberg [*cf.* (36)]. This similarity is particularly impressive in the case of non-resonance excitation when $\Gamma \approx 0$, $\varphi_\pm \approx 0$ can be inserted. Changing n into $-n$ in the last term of (89), omitting the argument $\omega' Q_0/\omega_l$ from J_n, and finally interchanging the two terms in the brackets we have

$$\mu = \frac{e^2 E_0}{2m\bar{\omega}} \sum_{n,r} \left[\frac{J_r J_{r-n}}{\bar{\omega} + r\omega_l - \omega_0} + \frac{J_r J_{r+n}}{\bar{\omega} + r\omega_l + \omega_0} \right] \cos(\omega_0 - n\omega_l)t \tag{91}$$

It was shown by Andreyev and Shorygin [48] that for small $\omega' Q_0/\omega_l$ the analogy with the Kramers–Heisenberg formula goes over into formal coincidence. In the case of resonance excitation the first term in the brackets of (89) can be neglected. For rigorous resonance ($\omega_0 = \bar{\omega}$) we get

$$\mu = \sum_{n=-\infty}^{+\infty} \mu_{n0} \cos[(\omega_0 + n\omega_l)t - \varphi_-] \tag{92}$$

with

$$\mu_{n0} = \frac{e^2 E_0}{2m\bar{\omega}} \sum_{r=-\infty}^{+\infty} \frac{J_r J_{r+n}}{[r^2\omega_l^2 + \frac{1}{4}\Gamma^2]^{\frac{1}{2}}} \tag{93}$$

The intensity of the $(n-1)$th overtone in the rigorous RR spectrum is, apart from the factor $(\omega_0 + n\omega_l)^4$, proportional to the square of μ_{n0}.

If $\frac{1}{2}\Gamma \gg \omega_l$, the denominator in the sum terms of (93) is only weakly dependent on r and therefore for $n \neq 0$ the sum over $J_r J_{r+n}$ is practically 'destroyed by interference'. This means that under this condition the Raman overtones are less intense than the Raman fundamental and the latter is less intense than the Rayleigh band. If, however, $\frac{1}{2}\Gamma \ll \omega_l$, all terms with $r \neq 0$ in (93) can be neglected and we have

$$\mu_{n0} = \frac{e^2 E_0}{m\bar{\omega}\Gamma} J_0 J_n \tag{94}$$

Depending on the argument $\omega' Q_0/\omega_l$ of the Bessel functions, μ_{n0} for $n \neq 0$ may now become comparable with, or even larger than, μ_{n0} with $n = 0$. There will be an intensity distribution among the Raman overtones (inclusive of the fundamental and Rayleigh band) similar to that known from the resonance spectra of fluorescence.

[48] N. S. Andreyev and P. P. Shorygin, *Optika i Spektroskopiya*, 1967, **22**, 714; P. P. Shorygin and N. S. Andreyev, *ibid.*, 1967, **23**, 687.

E. The Distinction of Scattering and Fluorescence by Classical Theory.—In an extension of the work mentioned in Section 3D Shorygin and Andreyev [48] have discussed the case where the electric field strength exciting the damped oscillator is itself a *damped* function of the time. In this case $E(t)$ in the RHS of (82) has the form

$$E(t) = E_0 e^{-\frac{1}{2}Gt} \cos(\omega_0 t + \varphi_0) \tag{95}$$

where G is the damping constant for the light wave and φ_0 an additional phase constant. The general solution of the corresponding equation of motion similar to (82) was obtained for Γ and ω^2 both being modulated by a superposition of arbitrary nuclear vibrations (several normal co-ordinates). The solution differs greatly for the following two cases

$$\text{Case I}: \Gamma \neq G \qquad \text{and Case II}: \Gamma = G \tag{96}$$

We omit a quotation of the rather complicated formulae which form the solution and give only short reports on the qualitative results for the two cases (96) separately. For simplicity we take Γ as time-independent and restrict ourselves to only one normal co-ordinate with angular frequency ω_l.

In Case I ($|\Gamma - G| > 0$) the solution is similar to that given in Section 3D (where we had $\Gamma > G = 0$), and it consists of two parts $x_0(t)$ and $x_1(t)$, $x_0(t)$ representing the general solution of the homogeneous equation of motion (*i.e.* physically the transient free vibration of the oscillator) and $x_1(t)$ representing a special solution of the inhomogeneous equation (*i.e.* physically the quasi-stationary or forced vibration). x_0 and x_1 are both infinite sums [*cf.* (88) and (89)]. Under the condition $\frac{\omega'}{\bar{\omega}} Q_0 \ll 1$ [see the denominator in (88)] the time dependence of these two functions is expressed by factors of the form $e^{-\frac{1}{2}\Gamma t} e^{-i(\bar{\omega} + r\omega_l)t}$ in the individual terms of x_0 and $e^{-\frac{1}{2}Gt} e^{-i(\omega_0 + n\omega_l)t}$ in those of x_1. Now the Fourier transform of a function of the type

$$\varphi(t) = \begin{cases} e^{-\alpha t} e^{-i\beta t}, & t \geqslant 0 \\ 0, & t < 0 \end{cases} \tag{97}$$

is

$$\tilde{\varphi}(\Omega) = \frac{1}{2\pi} \int_{-\infty}^{+\infty} \varphi(t) e^{i\Omega t} dt = \frac{-1/(2\pi)}{i(\Omega - \beta) - \alpha} \tag{98}$$

The square of its modulus,

$$|\tilde{\varphi}(\Omega)|^2 = \frac{1/(4\pi^2)}{(\Omega - \beta)^2 + \alpha^2} \tag{99}$$

represents a Lorentz distribution with the maximum at $\Omega = \beta$ and the half-width 2α. This shows that the spectral distribution $\tilde{x}_0(\Omega)$ of $x_0(t)$ is a sum of functions like (98) peaked at $\Omega = \bar{\omega} + r\omega_l$ (r is an integer) and all with identical half-widths Γ. The 'weight' of each term (indexed r) in this sum is given by amplitude factors like $J_r\left(\frac{\omega'}{\omega_l}Q_0\right)$ in (88). $|\tilde{x}_0(\Omega)|^2$ can be viewed as a classical description of the fluorescence spectrum which is emitted by the freely vibrating charge. The index r numerates the peaks of the vibrational structure of the fluorescence spectrum. $\bar{\omega}$ corresponds to the pure electronic transition. Γ is the half-width of the individual vibrational maxima. It is clear that the vibrational structure is only resolved when $\Gamma < \omega_l$. A similar consideration can be made for x_1. The peaks of $|\tilde{x}_1(\Omega)|^2$ are centred at $\Omega = \omega_0 + n\omega_l$. This is characteristic for Raman scattering because the vibrational bands now are displaced by the constant values $0, \pm\omega_l, \pm 2\omega_l, \ldots$ from the *exciting* angular frequency ω_0 and not from the spectrally fixed pure electronic transition angular frequency $\bar{\omega}$. Moreover, the half-width of the peaks of $|\tilde{x}_1(\Omega)|^2$ now is equal to the half-width G of the exciting line [see (95)]. The resolution of the vibrational structure now is only good when $G < \omega_l$. The intensity of the individual vibrational bands $\omega_0 + n\omega_l$, $n = 0, \pm1, \ldots$ (Rayleigh, Raman fundamental, Raman overtones) is determined by Bessel function products and denominators similar to those in (89). An interesting feature of the explicit formulae (not quoted here) is that their denominators contain $\Gamma - G$ instead of Γ [see (89)] and so resemble the quantum mechanical Breit–Wigner formula used in nuclear physics.[49] Also in other respects (except some intensity questions) the correspondence (by letting $\hbar \to 0$) of the classical formula for $x_1 = \mu/e$ with the quantum mechanical formula for $a = \mu/E_0$ can be shown.[48] In practice it will be desirable to separate fluorescence from scattering. It will be seen, however, that this is not always possible. The motion of the scattering particle in Case I is described by the sum of the instantaneous amplitudes $x_0(t) + x_1(t)$, and it is only under certain conditions that a separation of the two terms makes sense. First of all it should be noted that the measured intensity is proportional to the *square* of the second derivative with respect to time of this sum and not to the second derivative itself [*cf.* (10)]. In this square cross terms will appear, mixing fluorescence and scattering. A separation of the two kinds of secondary radiation in general is only possible if $G \gg \Gamma$ (excitation by a very broad line with respect to the damping constant Γ of the oscillator) or $G \ll \Gamma$ (excitation by a very narrow line), for then either the amplitude x_1 for scattering or the amplitude x_0 for fluorescence will decay relatively fast. (The case usually occurring in Raman spectroscopy is the second one.) However, even if this condition is satisfied, a *spectral* separation will only be possible if the two spectra corresponding to $|\tilde{x}_0(\Omega)|^2$ and $|\tilde{x}_1(\Omega)|^2$ do not substantially overlap, *i.e.* when ω_0 differs strongly from $\bar{\omega}$. More precisely, this non-resonance case occurs when $\omega_0 \approx \bar{\omega} + r'\omega_l$ with $|r'|$ so large that the corresponding vibrational peak in the absorption spectrum, because of the smallness of $J_{r'}\left(\frac{\omega'}{\omega_l}Q_0\right)$ [see (88) and (89)], has practically vanishing intensity.

[49] G. Breit, *Rev. Mod. Phys.*, 1933, **5**, 91; G. Breit and E. Wigner, *Phys. Rev.*, 1936, **49**, 519.

So we can state that the fluorescence and scattering spectra which must be traced back to x_0 and x_1, respectively, do not overlap when $|G - \Gamma|$ and $|r'|$ are sufficiently large. If $|r'|$ is not large, *i.e.* when the exciting frequency falls into the region of the clearly visible vibrational structure of the electronic absorption band (rigorous resonance), and if $G > \omega_l > \Gamma$, then the relative importance of x_0 (fluorescence) or x_1 (scattering) in the sum $x_0 + x_1$ will change strongly, depending on whether ω_0 coincides with a vibrational maximum or falls in between two maxima. The intensity distribution in the spectra of x_0 and x_1 in Case I can be described qualitatively as follows: the larger G, Γ, $|r'|$, and $\omega' Q_0/\omega_l$ are, the smaller will be the intensity of the total spectra and of all their components. The larger Γ/ω_l, Γ/G and $|r'|$ are, the more intense will be the Rayleigh band relative to the Raman bands in the scattering spectrum (derived from x_1). The fluorescence spectrum (derived from x_0) depends only on the parameters characterizing the oscillator itself and not on G and $|r'|$. If the modulation parameter $\omega' Q_0/\omega_l$ is small the vibrational maximum at $\bar{\omega}$ (corresponding to the pure electronic transition) will be the most intense.

In Case II ($\Gamma = G$) the above conclusions are no longer valid. Here the solution $x(t)$ of the equation of motion of the parametrically excited electronic oscillator does not have the form $x_0(t) + x_1(t)$ as in Case I. Rather one gets a solution $x_a(t) + x_b(t) + x_c(t)$, each term again representing a sum. The non-periodic time dependence of the individual terms in x_b and x_c is again described by the factor $e^{-\frac{1}{2}Gt} = e^{-\frac{1}{2}\Gamma t}$, but that of the terms in x_a by $te^{-\frac{1}{2}Gt}$. If $\omega_0 \approx \bar{\omega} + r'\omega_l$ with very small $|r'|$ (rigorous resonance excitation upon the vibrational maximum r'), this term outweighs the other two terms. The secondary radiation then resembles typical luminescence, for the factor $te^{-\frac{1}{2}Gt}$ starting from $t = 0$ first increases and then again diminishes and thus describes afterglow. When $|r'| > \omega' Q_0/(\omega_l \Gamma)^{\frac{1}{2}}$ the relative importance of x_a is reduced and the secondary radiation becomes similar to non-resonance scattering in which finally, for very large $|r'|$, the Rayleigh band again acquires the highest intensity. For $G \approx \Gamma \gg \omega_l$ the emitted spectrum looks like a very broad structureless band in the region of the exciting frequency, but for $G \approx \Gamma \ll \omega_l$ and small $|r'|$ it has a clear vibrational structure similar to that of the absorption spectrum.

The field function (95) could also be replaced by a coherent pulse function

$$E(t) = E_0 e^{-\frac{1}{2}\Delta^2 t^2} \cos(\omega_0 t + \varphi_0) \tag{100}$$

of Gaussian form. Here Δ plays a similar role to G in (95); the lifetime $1/G$ of the wave train corresponds to the pulse length $1/\Delta$. Regarding Γ and ω^2 in the equation of motion as constants (which means restriction to non-displaced radiation), this possibility was discussed for the case $\Delta \gg \Gamma$ by Penney and Silverstein [50] and supplemented by analogous quantum mechani-

[50] C. M. Penney and S. D. Silverstein, General Electric Information Series Report No. 72CRD150, May 1972.

cal arguments. The results, despite the neglect of nuclear vibrations modulating the electronic motion, qualitatively agree with those of Shorygin and Andreyev: when the difference $|\omega_0 - \omega|$ of the exciting angular frequency and the electronic eigenfrequency is $> \Delta - \Gamma$, the secondary radiation has nearly the same characteristics (*e.g.* the same spectral width Δ) as the incident wave; if $|\omega_0 - \omega| \approx \Delta - \Gamma$ it becomes a mixture of two waves with widths Δ and Γ, and if it is ≈ 0 (rigorous resonance) the width of the emitted line will be approximately equal to the spectral width Γ of the free electronic vibration, *i.e.* we then have fluorescence. Summing up we can see that the secondary radiation, depending on the frequency and the width of the exciting radiation, can show the characteristic features of fluorescence (luminescence), of scattering, or of both (in different spectral regions), and in some cases it can represent a non-discernible mixture of both.

The classical theory sketchily reviewed above gives information on the temporal as well as on the spectral nature of the motion of the radiating electron and consequently of the emitted radiation. Such classical calculations are particularly useful for the study of phase correlations between the individual parts of the interacting system. Because of the usual quantization of radiation fields this would be much more difficult in quantum theory. The classical method also yields insight into the exchange of energy between nuclei and electrons and the different mechanisms of energy balance and loss. Thus the classical theory of the RRE and related phenomena at least heuristically can prepare the way for quantum mechanical investigations.

4 Quantum Mechanical Theories of the Resonance Raman Effect: General Aspects

In what follows, only a very abridged survey of the existing quantum mechanical theories of the RRE and related phenomena can be given. We first in this section consider the completely general theory without introducing any special cases whatsoever. Later, in Section 5, the special theory referring to different systems showing the RRE will be reviewed. In comparison to classical theory the quantum mechanical treatment of the RRE gives better insight into the spectral nature of light–matter interaction. It is only in this way that reliable information about scattered intensities and selection rules can be obtained.

A. Basic Methods for the Description of Two-photon Processes.—The methods used for the description of the RRE and related phenomena all turn out to be more or less refined modifications of Dirac's time-dependent perturbation theory [51, 52] (*cf.* ref. 20, p. 136 ff.) which mathematically is a method of variation of constants. In particular, Dirac's theory can be represented in

[51] P. A. M. Dirac, *Proc. Roy. Soc.*, 1926, **A112**, 661.
[52] P. A. M. Dirac, 'The Principles of Quantum Mechanics', Clarendon Press, Oxford, 3rd edn., 1947, p. 167 ff.

the version of $\hat{U}$-matrix theory (*cf. e.g.* ref. 53, p. 316 ff., ref. 54, p. 183 ff.) in which by a time-displacement operator $\hat{U}(t,t_0)$ the wave function $\Psi_{\mathrm{I}}(t_0)$ describing the state of the total system (including all mutually interacting partial systems) at time t_0 is transformed into the state function $\Psi_{\mathrm{I}}(t)$ at a later time t (the subscript I refers to the so-called interaction picture)

$$\Psi_{\mathrm{I}}(t) = \hat{U}(t,t_0)\Psi_{\mathrm{I}}(t_0) \qquad (101)$$

$\hat{U}(t,t_0)$ can be obtained in the form of a series of successive approximations

$$\hat{U}(t,t_0) = \sum_{\nu=0}^{\infty} \hat{U}^{(\nu)}(t,t_0) \qquad (102)$$

by using an iteration procedure for solving the Schrödinger equation in the interaction picture

$$\mathrm{i}\hbar\frac{\partial}{\partial t}\Psi_{\mathrm{I}}(t) = \hat{H}\Psi_{\mathrm{I}}(t) \qquad (103)$$

with

$$\Psi_{\mathrm{I}}(t) = \mathrm{e}^{\mathrm{i}\hat{H}_0 t/\hbar}\Psi_{\mathrm{S}}(t) \qquad (104)$$

and

$$\hat{H} \equiv \hat{H}_{\mathrm{I}}(t) = \mathrm{e}^{\mathrm{i}\hat{H}_0 t/\hbar}\hat{H}_{\mathrm{S}}\mathrm{e}^{-\mathrm{i}\hat{H}_0 t/\hbar} \qquad (105)$$

The subscript S refers to the Schrödinger picture. $\hat{H}$ is the Hamiltonian describing the mutual interactions of the different parts of the total system.

$$\hat{H}_0 = \hat{H}_{0\mathrm{I}} = \hat{H}_{0\mathrm{S}} \qquad (106)$$

is the Hamiltonian of the 'unperturbed' total system (*i.e.* the system with all interactions switched off). $\hat{H}_{\mathrm{S}}$ usually is time-independent. The individual approximation terms in $\hat{U}(t,t_0)$ have the form

$$\hat{U}^{(0)}(t,t_0) = 1, \ldots,$$

$$\hat{U}^{(\nu)}(t,t_0) = \frac{1}{(\mathrm{i}\hbar)^{\nu}}\int_{t_0}^{t}\mathrm{d}\tau_1\int_{t_0}^{\tau_1}\mathrm{d}\tau_2\ldots\int_{t_0}^{\tau_{\nu-1}}\mathrm{d}\tau_{\nu}\,\hat{H}(\tau_1)\hat{H}(\tau_2)\ldots\hat{H}(\tau_{\nu}), \ldots \qquad (107)$$

It is seen that $\hat{U}^{(\nu)}$ contains in the integrand exactly ν factors $\hat{H}$.

In the limit $t_0 \to -\infty$, $t \to +\infty$ the matrix $\hat{U}(t,t_0)$ goes over into the so-called $\hat{S}$-matrix

$$\hat{S} \equiv \hat{U}(+\infty, -\infty) \qquad (108)$$

which provides another tool frequently used for scattering problems in general and for light scattering in particular. However, the $\hat{S}$-matrix method is not always completely equivalent to the $\hat{U}$-matrix method. This can be seen as follows. As $t_0 \to -\infty$ all lower integration limits in (107) must be replaced by $-\infty$. The convergence of the individual integrals at this lower limit is not assured in every case, however, and besides, as the partial integrals must be

[53] S. S. Schweber, 'An Introduction to Relativistic Quantum Field Theory', Harper and Row, New York, 1964.
[54] J. J. Sakurai, 'Advanced Quantum Mechanics', Addison-Wesley, London, 1967.

calculated successively, a prescription must be given about the order in which the limits $t_0 \rightarrow -\infty$ in $U^{(\nu)}$, $\nu > 1$ are to be carried out. These difficulties can be removed by a trick first used by Dyson in 1951 (see ref. 53, p. 416) which consists in the introduction of artificial convergence factors $e^{\varepsilon\tau\nu}$, ε real and >0, in those partial integrals $\int_{-\infty} \ldots \, d\tau$ which otherwise would be divergent. After the integration $\varepsilon \rightarrow 0$ is taken. But this procedure, which is standard for $\hat{S}$-matrix theory, changes the physical content of the original Schrödinger equation. It removes transient effects due to the 'switching on' of the interaction at the initial time t_0. The interaction is said to be introduced 'adiabatically'. Perhaps the expression 'switching on' which is frequently found in the literature is somewhat misleading because it belongs to the vocabulary of classical physics. In quantum mechanics it does not matter whether the interaction $\hat{H}$ is present or absent at $t < t_0$. The continuous connection with the past ($t < t_0$) of the state function is destroyed at t_0 simply by the fact that at t_0 an experiment is performed identifying the state as a *pure* state M, *i.e.* a state of the unperturbed system. By this measurement the wavefunction $\Psi(t)$, $t < t_0$ is abruptly and radically changed to coincide with $\Psi_M(t)$ at $t = t_0$. [This is independent of the picture (I or S) used. We designate the pure states of the *total* system by capital Latin letters such as A, B, ...; in particular M is the initial, R or R_ρ (see later) an intermediate, and N the final state.] On the other hand, since the explicit form of $\Psi(t)$ for $t < t_0$ is without any importance for the quantum mechanical problem, there is nothing against understanding the term 'switching on the interaction at t_0' in the sense of the correspondence principle. Now, classically, the switching on of a periodic force acting upon a damped oscillator causes a transient vibration of the oscillator with its eigenfrequency. Such transient effects in classical physics are not less real than stationary phenomena. This suggests that, by virtue of the correspondence principle, one should be cautious in discarding transient effects in quantum mechanics simply for mathematical reasons. In fact this constitutes a remarkable physical difference between Dirac's perturbation theory (or the equivalent $\hat{U}$-matrix method) on the one hand and the $\hat{S}$-matrix method on the other. Let us briefly show the equivalence of Dirac's perturbation theory and the $\hat{U}$-matrix formalism.

Dirac's theory starts from the Schrödinger equation in the Schrödinger picture

$$i\hbar \frac{\partial}{\partial t} \Psi_S(t) = (\hat{H}_0 + \hat{H}_S)\Psi_S(t) \tag{109}$$

Equation (109) by (105) and (106) is equivalent to (103). Let us temporarily omit the subscript S. Equation (109) is solved by putting

$$\Psi(t) = \sum_A b_A(t)\Psi_A(t) \tag{110}$$

where

$$\Psi_A(t) = \psi_A e^{-iE_A t/\hbar} \tag{111}$$

is the wavefunction of the pure state A with energy E_A of the unperturbed total system. [In other words, $\Psi_A(t)$ is an eigenfunction belonging to the eigenvalue E_A of equation (109) with $\hat{H}_S$ omitted.]

$$\psi_A \equiv \Psi_A(0) \tag{112}$$

is time-independent. We consider all wavefunctions as orthonormalized, *i.e.*

$$\underset{q}{\mathrm{S}}\Psi_B^{\dagger}(t)\Psi_A(t) = \underset{q}{\mathrm{S}}\psi_B^{\dagger}\psi_A = \delta_{BA} \tag{113}$$

Here $\underset{q}{\mathrm{S}}$ is a generalized summation plus integration symbol, q denoting the time-independent variables on which the state functions Ψ or ψ depend. † means Hermitian or, in the case of scalar wavefunctions, complex conjugation.

Inserting (110) into (109) one finds for the 'probability amplitudes' $b_A(t)$ the system of differential equations

$$\mathrm{i}\hbar\dot{b}_A = \sum_B b_B H_{AB} \mathrm{e}^{\mathrm{i}(E_A - E_B)t/\hbar}, \text{ all } A \tag{114}$$

Here $$H_{AB} \equiv \underset{q}{\mathrm{S}}\psi_A^{\dagger}\hat{H}_S\psi_B \tag{115}$$

is the matrix element of $\hat{H}_S$ for the transition $A \leftarrow B$. Equation (114) can be solved by successive approximations if, after inserting $\lambda\hat{H}_S$ instead of $\hat{H}_S$ and $b_A(\lambda) = \sum_{\nu=0}^{\infty} \lambda^{\nu} b_A^{(\nu)}$ instead of b_A, comparison of equal powers of λ is made. The solution b_A of (114) then is equal to $b_A(\lambda)$ for $\lambda = 1$, *i.e.*

$$b_A(t) = \sum_{\nu=0}^{\infty} b_A^{(\nu)}(t) \tag{116}$$

The approximation terms $b_A^{(\nu)}(t)$, $\nu = 0, 1, \ldots$ for $b_A(t)$ are given by the recursion formulae

$$\dot{b}_A^{(0)}(t) = 0, \ldots, \mathrm{i}\hbar\dot{b}_A^{(\nu)}(t) = \sum_B b_B^{(\nu-1)}(t)\, H_{AB} \mathrm{e}^{\mathrm{i}(E_A - E_B)t/\hbar} \tag{117}$$

Now assuming that the initial ($t = 0$) state of the total system is M, (110) for $t = 0$ must have the form

$$\Psi(0) = \Psi_M(0) = \psi_M \tag{118}$$

from which follows the initial condition for the system (117)

$$b_A(0) = \delta_{MA}, \text{ all } A \tag{119}$$

Equation (117) with (119) can be solved in a straightforward manner. The physical meaning of $b_A(t)$ is revealed as follows: multiplying (110) by $\Psi_B^{\dagger}(t)$, integrating over q, and using the orthonormalization condition (113) one finds

$$\underset{q}{\mathrm{S}}\Psi_B^{\dagger}(t)\Psi(t) = \sum_A b_A(t)\underset{q}{\mathrm{S}}\Psi_B^{\dagger}(t)\Psi_A(t) = \sum_A b_A(t)\delta_{BA} = b_B(t) \tag{120}$$

The absolute square of the left side represents the probability of finding the system in state B when a measurement is performed at time t. Thus, if $b_A(t)$

is calculated by (117) under the initial condition (119), $b_A(t)$ by (120) (simply replace B by A) is the amplitude of the probability that the system experimentally can be found in state A at $t \geqslant 0$ if an experiment at $t_0 = 0$ showed its state to be M.

The orthonormality condition (113) being independent of the quantum mechanical picture chosen, (120) is equally valid for the state vectors taken in the interaction picture. However, for these we have from (101) for $t_0 = 0$, using (118)

$$\Psi_{\mathrm{I}}(t) = \hat{U}(t,0)\Psi_{\mathrm{I}}(0) = \hat{U}(t,0)\Psi_{M\mathrm{I}}(0) \tag{121}$$

and therefore

$$\underset{q}{\mathrm{S}}\Psi^{\dagger}_{B\mathrm{I}}(t)\Psi_{\mathrm{I}}(t) = \underset{q}{\mathrm{S}}\Psi^{\dagger}_{B\mathrm{I}}(t)\hat{U}(t,0)\Psi_{M\mathrm{I}}(0) \equiv U_{BM}(t,0) \tag{122}$$

Consequently, by comparison of (120) and (122)

$$b_B(t) \equiv U_{BM}(t,0) \tag{123}$$

This relation demonstrates the equivalence of the elementary perturbation theory and the $\hat{U}$-matrix method. Equation (122) is also valid for the individual approximations separately. (This is seen from the fact that the νth approximation term in both cases contains exactly ν matrix-element factors of the type H_{AB}.) Therefore, with (107)

$$b_B^{(\nu)}(t) \equiv U_{BM}^{(\nu)}(t,0) = \frac{1}{(\mathrm{i}\hbar)^{\nu}} \int_0^t \mathrm{d}\tau_1 \int_0^{\tau_1} \mathrm{d}\tau_2 \ldots \int_0^{\tau_{\nu-1}} \mathrm{d}\tau_{\nu}\, [\hat{H}_{\mathrm{I}}(\tau_1)\hat{H}_{\mathrm{I}}(\tau_2) \ldots \hat{H}_{\mathrm{I}}(\tau_{\nu})]_{BM} \tag{124}$$

where the νth order matrix element in the brackets can be expanded according to

$$[\hat{H}(\tau_1)\hat{H}(\tau_2) \ldots \hat{H}(\tau_{\nu})]_{BM} = \sum_{R_{\nu-1},\ldots,R_1} H_{BR_{\nu-1}}(\tau_1)H_{R_{\nu-1}R_{\nu-2}}(\tau_2) \ldots H_{R_1M}(\tau_{\nu}) \tag{125}$$

(We have omitted here the subscripts I.)

For the theory of the RRE only very low-order approximation terms are needed. Which order of approximation is required depends on the choice of the interaction Hamiltonian $\hat{H}$. Usually the most convenient way of choosing $\hat{H}$ is the inclusion in $\hat{H}$ of only the mutual interaction between the radiation field and the particle system. In this case and if only two-photon processes are considered only the first- and second-order approximation terms ($\nu = 1, 2$) are needed. Using the standard quantization procedure of the electromagnetic field (see *e.g.* ref. 20, pp. 38 ff., 110, 125) $\hat{H}$ in the non-relativistic approximation consists of two parts

$$\hat{H} = \hat{H}^{(\mathrm{I})} + \hat{H}^{(\mathrm{II})} \tag{126}$$

where

$$\hat{H}^{(\mathrm{I})} = -\sum_k \frac{e_k}{m_k} \hat{\boldsymbol{p}}_k \cdot \hat{\boldsymbol{A}}(\boldsymbol{r}_k), \quad \hat{H}^{(\mathrm{II})} = \sum_k \frac{e_k^2}{2m_kc^2} \hat{\boldsymbol{A}}^2(\boldsymbol{r}_k) \tag{127}$$

with e_k, m_k, and $\hat{\boldsymbol{p}}_k$ the charge, mass, and momentum operator, respectively, of particle k, and $\hat{A}(\boldsymbol{r}_k)$ the vector potential operator of the field at the position $\boldsymbol{r}_k$ of particle k. An additional (already relativistic) term representing the interaction of the spins of the particles with the magnetic field strength of the radiation field must be included in (126) when magnetism plays a role. The operator $\hat{A}(\boldsymbol{r}_k)$ is a sum of terms which contain the photon-creation and -annihilation operators $\hat{c}_\lambda^\dagger$, $\hat{c}_\lambda$ for each field mode (= photon sort) λ linearly. A mode λ is characterized by the wave vector $\boldsymbol{\kappa}_\lambda \equiv \boldsymbol{k}_\lambda/(\hbar c)$, the photon energy $k_\lambda = |\boldsymbol{k}_\lambda| = \hbar\omega_\lambda$, and the polarization unit vector $\boldsymbol{e}_\lambda$ perpendicular to $\boldsymbol{\kappa}_\lambda$. The operator $\hat{A}^2(\boldsymbol{r}_k)$ consequently contains the products of $\hat{c}_\lambda^\dagger$, $\hat{c}_\lambda$ with $\hat{c}_\mu^\dagger$, $\hat{c}_\mu$ of any two modes λ and $\mu \mp \lambda$. $\hat{A}(\boldsymbol{r}_k)$ thus can be called a 'one-photon operator', and $\hat{A}^2(\boldsymbol{r}_k)$ a 'two-photon operator'.

Now consider a light-scattering transition from an initial state M to a final state $N \neq M$. As the elementary process of scattering involves (in unknown succession) the annihilation of one photon of sort a (a indicating 'absorption') and the creation of one photon of sort ε (ε indicating 'emission') the probability amplitude $b_N(t) \equiv U_{NM}(t,0)$ for this transition $N \leftarrow M$ must be the sum of a term $b_N^{(1)}(t)$ containing the matrix element of $\hat{H}(\tau_1) \equiv \hat{H}^{(\mathrm{II})}(\tau_1)$ ('direct transition') and of a term $b_N^{(2)}(t)$ containing the matrix element of $\hat{H}(\tau_1)\hat{H}(\tau_2) \equiv \hat{H}^{(\mathrm{I})}(\tau_1)\hat{H}^{(\mathrm{I})}(\tau_2)$ ('indirect transition' or 'transition *via* an intermediate state R'). The explicit calculation with (124) and (105) gives [we leave off the superscript on $b_N(t)$]:

$$b_N(t) \equiv U_{NM}(t,0) = H_{NM}^{(\mathrm{II})} \frac{1 - \mathrm{e}^{\mathrm{i}(E_N - E_M)t/\hbar}}{E_N - E_M}$$

$$+ \sum_R \frac{H_{NR}^{(\mathrm{I})} H_{RM}^{(\mathrm{I})}}{E_R - E_M} \left[\frac{1 - \mathrm{e}^{\mathrm{i}(E_N - E_R)t/\hbar}}{E_N - E_R} - \frac{1 - \mathrm{e}^{\mathrm{i}(E_N - E_M)t/\hbar}}{E_N - E_M} \right] \qquad (128)$$

$\hat{S}$-matrix theory and similar methods of adiabatic approximation give expressions equivalent to (128) but with the first term in the brackets missing. It is this term which is thought to cause only transient (non-stationary, decaying) effects and whose artificial elimination therefore seems to be justified. Even many authors using Dirac's perturbation theory or the $\hat{U}$-matrix method argued for the irrelevancy of this term [see ref. 52, p. 180, ref. 9, p. 219 (footnote), ref. 20, p. 140, ref. 55, p. 464, ref. 56, p. 279 and ref. 57, p. 470], but the reasons disclosed are not conclusive for the case of resonance. In fact, already Dirac in 1927 (ref. 17, p. 725) and Wentzel in 1933 (ref. 58, p. 748) had emphasized the importance of this term in resonance cases. Sushchinsky (ref. 59, p. 53) even considers it as the sole source of RF.

[55] V. Weisskopf. *Z. Physik*, 1933, **85,** 451.
[56] M. Goeppert-Mayer, *Ann. Physik*, 1931, **9**, 273.
[57] E. Merzbacher, 'Quantum Mechanics', Wiley, New York, 1971.
[58] G. Wentzel, 'Wellenmechanik der Stoss- und Strahlungs-vorgänge', in 'Handbuch der Physik', ed. H. Geiger and K. Scheel, Springer, Berlin, 1933, Vol. XXIV, Part 1, pp. 695—784.
[59] M. M. Sushchinsky, 'Combination Scattering Spectra of Molecules and Crystals' (in Russian), Izdatelstvo Nauka, Moscow, 1969; English translation: 'Raman Spectra of Molecules and Crystals', Israel Program for Scientific Translations, New York (distributor Wiley), 1972.

With this background, the author of this Report believes he has good arguments[60] which prove that it is indeed the presence of this term which, in the case of resonance excitation and under certain additional conditions (*cf.* the end of Section 2), may bring about a separation of the two-photon scattering process into two statistically independent one-photon processes (a real absorption followed by a real emission). Consequently, the riddle of the interrelation of RRE and RF cannot be solved without paying attention to this term. The present author does not think, however, that the individual terms in (128) separately should be assigned to the RRE or to RF as in Sushchinsky's book. The term $b_N(t)$ refers to one definite transition $N \leftarrow M$ and this transition cannot be divided between RR scattering and RF. Depending on the conditions in operation it refers in its entirety either to a RR or to a RF transition. To avoid any misunderstanding it must be emphasized however that for non-rigorous resonance scattering and also for non-resonance scattering the first term in the brackets in (128) in fact is negligible.

In the literature, Dirac's perturbation theory either in the original formulation or in the $\hat{U}$-matrix version was used some time ago for explaining the ordinary (non-resonance) RE (see *e.g.* ref. 51, ref. 61, p. 240, 404, ref. 9, p. 216 ff., ref. 58, p. 747 and ref. 20, p. 189 ff.). In most of these early works only the electric dipole moment approximation for $\hat{H}^{(\mathrm{I})}$ (see Section 4E) was used and $H^{(\mathrm{II})}$ was neglected. This method is also used in recent publications on RR theory (*cf. e.g.* ref. 62, ref. 63, p. 45 ff., ref. 64, ref. 65, ref. 66, p. 27 ff., ref. 59, p. 42 ff.). The $\hat{S}$-matrix theory was employed in ref. 67, p. 521, ref. 68, ref. 69, p. 24, ref. 70. When the interaction Hamiltonian $\hat{H}$ is chosen such that it involves not only the interactions $\hat{H}^{(\mathrm{I})}$ and $\hat{H}^{(\mathrm{II})}$ between the radiation field and the particle system but *e.g.* also internal interactions between the particles, such as the Coulomb interaction among electrons and nuclei, higher-order approximation terms $b_N^{(\nu)}(t) = U_{NM}^{(\nu)}(t,0)$, $\nu > 2$, must be employed. For certain systems (*i.e.* crystals) the convenience of this procedure cannot be denied. However, a posterior introduction of intrinsic interactions among the constituents of the partial systems in the second-order approximation

60 J. Behringer, unpublished work.
61 M. Born and P. Jordan, 'Elementare Quantenmechanik', Springer, Berlin, 1930.
62 M. M. Sushchinsky and V. A. Zubov, *Optika i Spektroskopiya*, 1962, **13**, 766.
63 V. A. Zubov, 'Investigation of the Connection Between Combination Scattering Spectra and Electronic Absorption' (in Russian), Thesis, Moscow, published in *Trudy fiz. Inst. Lebedev Akad. Nauk S.S.S.R.*, Moscow, 1964, **30**, 65 pp.
64 V. A. Morozov and P. P. Shorygin, *Optika i Spektroskopiya*, 1966, **20**, 214.
65 E. M. Verlan, *Optika i Spektroskopiya*, 1966, **20**, 1003.
66 J. Behringer, 'Theorie der molekularen Lichtstreuung', Sektion Physik der Universität, Munich, 1967.
67 M. Jacon, R. Germinet, M. Berjot, and L. Bernard, *J. Phys. Radium*, 1971, **32**, 517.
68 M. Jacon, *Annales de l'Université et de l'A.R.E.R.S., Reims*, 1972, **10**, 39.
69 M. Jacon, 'Etude théorique de l'effet Raman et de la fluorescence de résonance', Thesis, Reims, 1973, 145 pp.
70 M. Jacon, 'Theoretical Study of Resonance Raman Scattering by the Methods of Quantum Electronics', in 'Advances in Raman Spectroscopy', ed. J.-P. Mathieu, Heyden and Son, London, 1973, Vol. 1, p. 325.

(128) is always possible. The resulting equations agree with those obtained by higher-order perturbation theory. For this reason and also because the final expressions would be rather voluminous and intricate we desist here from discussing in detail higher-order matrix elements. Applications of higher- (3rd and 4th) order perturbation theory have been made (ref. 71, ref. 72, p. 325 ff., and ref. 73). An example will be discussed in Section 5C.

B. Inclusion of Damping.—In RR scattering theory the damping of states plays an essential role. If damping were excluded the probability amplitudes would take infinite values ('resonance catastrophe'). The damping of a state is always due to either radiative or non-radiative transitions from this state to other states which shorten its lifetime (which is only definable statistically). In this connection 'state' should be understood as 'state of the total system'. The lifetime of the stationary states of the unperturbed system by definition being infinite, it is clear that it is only the interaction between the parts of the total system which is responsible for damping. Now in a real system there usually are so many different types of interaction among its parts that it is plainly impossible to represent all damping mechanisms explicitly in a theory of radiation. The customary procedures for taking account of damping therefore are either to restrict oneself to radiation damping (which in perturbation theory formally amounts to radiative return processes $A \leftarrow A$) or to allow for damping in a somewhat phenomenological, indiscriminate manner. The first method seems to be superior to the second one because of its clear consistency with the principles of quantum mechanics and its logical nature. However when going to higher approximations for the amplitudes for the radiative return process $A \leftarrow A$, the calculation quickly becomes immensely complicated and there will be the well-known convergence difficulties calling for renormalization theory. The second method, despite its somewhat semi-classical character, has the advantage of being much simpler and more generally applicable. Moreover the fundamental assumption of this method (introduced in its essentials by Weisskopf and Wigner in 1930 [74] in regard to the correspondence principle) that any state A of the system would decay exponentially according to

$$b_A(t) = \mathrm{e}^{-\frac{1}{2}\Gamma_A t}, \; t \geqslant 0 \tag{129}$$

if it were chosen for the initial state ($t = 0$), independently of the comparison with classical damped motions, was shown to be consistent with quantum theory at least in an approximation fully sufficient for optical spectroscopy (see ref. 20, p. 163 ff.). In (129) Γ_A, given by

$$\Gamma_A \equiv \Gamma'_A + \mathrm{i}\Gamma''_A \tag{130}$$

[71] R. Loudon, *Proc. Roy. Soc.*, 1963, **A275**, 218; *Adv. Phys.*, 1964, **13**, 423; *J. Phys. Radium*, 1965, **26**, 677.

[72] D. Marcuse, 'Engineering Quantum Electrodynamics', Harcourt, Brace, and World, New York, 1970.

[73] W. L. Peticolas, L. A. Nafie, P. Stein, and B. Fanconi, *J. Chem. Phys.*, 1970, **52**, 1576.

[74] V. Weisskopf and E. Wigner, *Z. Physik*, 1930, **63**, 54; 1930, **65**, 18.

is a complex constant. Its real and imaginary parts in a first approximation are defined by

$$\Gamma'_A \equiv \mathrm{Re}\Gamma_A = \frac{2\pi}{\hbar}\,[|H_{BA}|^2\rho_B]_{E_B=E_A} \tag{131}$$

where $\rho_B \mathrm{d}E_B$ is the number of states B in an energy interval $\mathrm{d}E_B$ about E_B, and

$$\Gamma''_A \equiv \mathrm{Im}\Gamma_A = \frac{2}{\hbar}\,\mathscr{P}_{(E_A)} \sum_B \frac{|H_{BA}|^2}{E_B - E_A} \tag{132}$$

Equation (131) comprises only radiation damping if $\hat{H}$ is defined by (126); this restriction can be avoided by also including in (131) [either in Γ'_A collectively or by a redefinition of $\hat{H}$ appearing in (131)] non-radiative damping interactions. If the time derivative of (129)

$$\dot{b}_A + \tfrac{1}{2}\Gamma_A b_A \equiv \mathrm{e}^{-\frac{1}{2}\Gamma_A t}\frac{\mathrm{d}}{\mathrm{d}t}(b_A \mathrm{e}^{\frac{1}{2}\Gamma_A t}) = 0,\ t \geqslant 0 \tag{133}$$

which is independent of any initial ($t = 0$) condition, is inserted in (114) instead of $\dot{b}_A$, and simultaneously there is introduced the condition $H_{AA} = 0$ for all A [direct and indirect return processes are already settled by (129)], the equation system

$$\mathrm{i}\hbar\dot{\bar{b}}_A = \sum_{B \neq A} \bar{b}_B H_{AB} \mathrm{e}^{\mathrm{i}(\bar{E}_A - \bar{E}_B)t/\hbar},\ t \geqslant 0 \tag{134}$$

with

$$\bar{b}_A \equiv b_A \mathrm{e}^{\frac{1}{2}\Gamma_A t},\ \bar{E}_A \equiv E_A - \tfrac{1}{2}\mathrm{i}\hbar\Gamma_A,\quad \text{all } A \tag{135}$$

is obtained which, apart from the bars and the exclusion of A from the summation index, is formally equal to (114). Therefore (128) and similar equations derived from (114) can at once be formally transformed into corresponding equations containing the damping terms of all states simply by replacing b_A by $\bar{b}_A$ and E_A by $\bar{E}_A$ defined by (135) and setting $H_{AA} = 0$ for all A occurring in the respective equation. The last requisite is no extra condition because already in the original equations it can be taken to be satisfied by appropriately defining the interaction Hamiltonian. We abstain from reproducing here the final formula analogous to (128) and containing explicitly all damping constants although it is basic for the RRE. [The reader using (128) and (135) will easily find the expression himself.] We also only mention that this result can also be obtained by directly applying the method of Heitler and Ma described in ref. 20, p. 163 ff. (see refs. 62, 63, and 66) or the essentially equivalent resolvant formalism (refs. 67—70). The method described above was used in Sushchinsky's book (ref. 59, p. 42 ff.). There the damping term $-\frac{1}{2}\mathrm{i}\hbar\Gamma_A$ is considered as the diagonal element H^{D}_{AA} of an antihermitian 'damping operator' $\hat{H}^{\mathrm{D}}$ representing an additive part of the interaction Hamiltonian $\hat{H}$. This latter conception in spite of its formal correctness is physically problematic because by a fundamental postulate of quantum mechanics all operators representing observables must be Hermitian. Perhaps, though, the incorporation of damping terms in quantum mechanical equations is never rigorously justifiable by quantum mechanics alone (*cf.* ref. 57, p. 474 and ref. 72, p. 424 ff.)!

By (129) we have

$$|b_A(t)|^2 = e^{-\Gamma'_A t} \tag{136}$$

and by (135)

$$\bar{E}_A = E_A + \tfrac{1}{2}\hbar\Gamma''_A - \tfrac{1}{2}i\hbar\Gamma'_A \tag{137}$$

This shows that Γ'_A is the reciprocal of the lifetime T_A of state A, which is defined as the time interval during which the probability for the system to be found in state A (A being presumed as initial state) diminishes by the factor e. Γ''_A, multiplied by $\frac{1}{2}\hbar$, appears as a positive addition to the energy E_A. This energy displacement usually is extremely small and can be neglected in optical spectroscopy, so that Γ_A is almost real ($\approx \Gamma'_A$). In what follows we shall assume $\Gamma''_A = 0$, for all A. A rather elegant, though for the purpose of non-stimulated Raman scattering somewhat pretentious method of treating damping phenomena makes use of the density matrix (see *e.g.* ref. 72, pp. 270 ff., 423 ff.). Applications of this method to ordinary Raman scattering can be found in ref. 75 and to RR scattering in ref. 69, p. 107 ff. and ref. 70, p. 334 ff. Bogoliubov has given an averaging method [76,77] which can be used for solving (114) instead of the perturbation expansion and leads to solutions free of damping constants. This method was applied to different resonance problems in optics [78] and also to the RRE [79] but was subsequently abandoned because of its semi-classical character and peculiar difficulties arising for experimental control.

C. Specification of States.—In Sections 4A and 4B we only used the states A (designated by capital Latin letters) of the *total* system (electromagnetic field plus particle system). A can be understood as an abbreviation for a complete set of quantum numbers characterizing the respective state of the total system. These quantum numbers comprise the quantum numbers of the states belonging to A of all partial systems contained in the total system. We designate the states of the *partial* systems by lower case Latin letters (eventually with subscripts) of the same species as the letter used for the respective total state. Considering the scattering of light by a molecule

[75] L. Knöll, *Ann. Physik*, 1971, **26,** 121, 129, 140.
[76] N. N. Bogoliubov and D. V. Shirkov, 'Introduction to the Theory of Quantized Fields', Interscience, New York, 1959, p. 484.
[77] N. N. Bogolioubov and J. Mitropolsky, 'Les Méthodes Asymptotiques dans la Théorie des Oscillations Non-linéaires', Gauthier-Villars, Paris, 1963.
[78] G. Lochak and M. Thiounn, *Compt. rend.*, 1967, **264,** *B*, 407, 1533; 1967, **265,** *B*, 1; 1968, **266,** *B*, 825; 1969, **268,** *B*, 897, 1452; *J. Phys. Radium*, 1969, **30,** 482.
[79] M. Jacon, M. Berjot, and L. Bernard, *Compt. rend.*, 1968, **266,** *B*, 681; *J. Phys. Radium*, 1969, **30,** 350.

and using the 'occupation number representation' for the light field, every total state A can be broken up as

$$A = (a; \ldots, a_\lambda, \ldots) \tag{138}$$

with a designating the corresponding molecular state (a comprises all molecular quantum numbers such as electronic, vibrational, rotational, *etc.*) and a_λ the number of photons present in field mode λ. From (138) follows (A is a state of the unperturbed total system)

$$E_A = E_a + \sum_\lambda E_{a\lambda} = E_a + \sum_\lambda (a_\lambda + \tfrac{1}{2})k_\lambda \tag{139}$$

so that abbreviating

$$E_a - E_b \equiv k_{ab},\ E_{a\lambda} - E_{b\lambda} \equiv k_{a\lambda b\lambda} = (a_\lambda - b_\lambda)k_\lambda \tag{140}$$

a typical energy difference of 'total states' is

$$E_A - E_B = k_{ab} + \sum_\lambda (a_\lambda - b_\lambda)k_\lambda \tag{141}$$

By (135) or (137) the damping constants Γ_A appear in additive terms to the energies E_A. It is therefore reasonable (this can also be justified in other ways) to postulate for Γ_A an equation similar to (139) separating Γ_A in parts $\Gamma_a, \ldots, \Gamma_{a\lambda}, \ldots$. However, we have adopted the customary field quantization using plane waves (*cf.* ref. 20, pp. 38 ff., 54 ff.). Therefore $\Gamma_{a\lambda} = 0$, for all λ, and we have

$$\Gamma_A = \Gamma_a \tag{142}$$

Equation (142) expresses the fact that we consider only the molecular states as being damped.

Now formula (128), after being modified by the inclusion of damping (Section 4B), can be explicitly specialized for two-photon (scattering and related) processes. We give the rather complicated result because it is fundamental for everything which follows:

$$\begin{aligned} b_N(t) = {} & H^{\alpha\varepsilon}_{n,m_\alpha - 1,\, m_\varepsilon + 1;\, m,\, m_\alpha,\, m_\varepsilon}\, e^{-\frac{1}{2}\Gamma_n t}\, \frac{1 - e^{t[i(k_{nm} - k_\alpha + k_\varepsilon)/\hbar + \frac{1}{2}(\Gamma_n - \Gamma_m)]}}{k_{nm} - k_\alpha + k_\varepsilon - \frac{1}{2}i\hbar(\Gamma_n - \Gamma_m)} \\ & + \sum_r \Bigg\{ \frac{H^\alpha_{n,\, m_\alpha - 1;\, r,\, m_\alpha} H^\varepsilon_{r,\, m_\varepsilon + 1;\, m,\, m_\varepsilon}}{k_{rm} + k_\varepsilon - \frac{1}{2}i\hbar(\Gamma_r - \Gamma_m)} \Bigg[e^{-\frac{1}{2}\Gamma_n t} \frac{1 - e^{t[i(k_{nr} - k_\alpha)/\hbar + \frac{1}{2}(\Gamma_n - \Gamma_r)]}}{k_{nr} - k_\alpha - \frac{1}{2}i\hbar(\Gamma_n - \Gamma_r)} \\ & \qquad - e^{-\frac{1}{2}\Gamma_n t} \frac{1 - e^{t[i(k_{nm} - k_\alpha + k_\varepsilon)/\hbar + \frac{1}{2}(\Gamma_n - \Gamma_m)]}}{k_{nm} - k_\alpha + k_\varepsilon - \frac{1}{2}i\hbar(\Gamma_n - \Gamma_m)} \Bigg] \\ & + \frac{H^\varepsilon_{n,\, m_\varepsilon + 1;\, r,\, m_\varepsilon} H^\alpha_{r,\, m_\alpha - 1;\, m,\, m_\alpha}}{k_{rm} - k_\alpha - \frac{1}{2}i\hbar(\Gamma_r - \Gamma_m)} \Bigg[e^{-\frac{1}{2}\Gamma_n t} \frac{1 - e^{t[i(k_{nr} + k_\varepsilon)/\hbar + \frac{1}{2}(\Gamma_n - \Gamma_r)]}}{k_{nr} + k_\varepsilon - \frac{1}{2}i\hbar(\Gamma_n - \Gamma_r)} \\ & \qquad - e^{-\frac{1}{2}\Gamma_n t} \frac{1 - e^{t[i(k_{nm} - k_\alpha + k_\varepsilon)/\hbar + \frac{1}{2}(\Gamma_n - \Gamma_m)]}}{k_{nm} - k_\alpha + k_\varepsilon - \frac{1}{2}i\hbar(\Gamma_n - \Gamma_m)} \Bigg] \Bigg\} \end{aligned} \tag{143}$$

Equation (143) refers to the two-photon scattering (inclusive RF) transition

$$N = (n; \ldots, n_\alpha, \ldots, n_\varepsilon, \ldots, n_\rho, \ldots) = (n; \ldots, m_\alpha - 1, \ldots, m_\varepsilon + 1, \ldots, m_\rho, \ldots) \leftarrow M = (m; \ldots, m_\alpha, \ldots, m_\varepsilon, \ldots, m_\rho, \ldots)$$

α, ε denote the modes λ whose occupation number m_α,m_ε is changed (for α, diminished; for ε, augmented) by 1 during the transition. $\hat{H}^{\alpha\varepsilon}$ is that term in $\hat{H}^{(\mathrm{II})}$ which annihilates one photon of sort α and creates one photon of sort ε. $\hat{H}^{\alpha}$ (or $\hat{H}^{\varepsilon}$, respectively) is that term in $\hat{H}^{(\mathrm{I})}$ which annihilates a photon α (or creates a photon ε). Equation (143) implicitly contains all information for non-resonance and resonance scattering (Rayleigh, Brillouin, Raman) and RF. (It does not cover, however, ν-photon process, $\nu > 2$, such as stimulated or hyper-Raman scattering. Moreover photon–spin interactions are neglected.) It is only necessary to specialize (and simplify) (143) for different cases.

We must desist here from discussing (143) and its special forms in detail and, apart from one special case (see below), we shall give only some general comments about the possibilities of applying (143). Let us denote the five typical consecutive terms occurring in (143) by I—V respectively (II—V within the sum over r). The terms II and IV [corresponding to the first term in the brackets in (128)] can be omitted for non-resonance and pre-resonance excitation and for excitation in a continuum of states r. I is only important for Rayleigh scattering. In the ordinary RE (including ordinary Rayleigh scattering) all damping constants can be considered as vanishing. In the rigorous non-continuum RRE (including resonance Rayleigh scattering and RF) either the terms I, II, III ('resonance with an emission frequency of the final state'; ref. 9, p. 247 ff.) or I, IV, V ('resonance with an absorption frequency of the initial state') must be retained. In almost all experiments the resonance state(s) is (are) higher in energy than the initial and final states so that only the latter case (with I, IV, V) is of practical interest.

Equation (143) in the special case of a continuum RRE ($k_\alpha > E_\mathrm{D} - E_m$, E_D = dissociation or ionization energy), omitting Rayleigh scattering, and taking $\Gamma_m = \Gamma_n = 0$, can be reduced to

$$b_N(t) = \left(-\sum_r \frac{H^{\varepsilon}_{n,\, m_\varepsilon + 1;\, r,\, m_\varepsilon} H^{\alpha}_{r,\, m_\alpha - 1;\, m,\, m_\alpha}}{k_{rm} - k_\alpha - \frac{1}{2}\mathrm{i}\hbar\Gamma_r} \right) \frac{1 - \mathrm{e}^{\mathrm{i}(k_{nm} - k_\alpha + k_\varepsilon)t/\hbar}}{k_{nm} - k_\alpha + k_\varepsilon} \tag{144}$$

Only those states r contribute to the sum for which $|E_r - (E_m + k_\alpha)| \equiv |k_{rm} - k_\alpha|$ is not essentially larger than $\frac{1}{2}\hbar\Gamma_r$. Strictly speaking, the sum over r must be understood as a sum over the discrete states ($E_r < E_\mathrm{D}$) plus an integral $\int_{E_\mathrm{D}}^{\infty} \ldots \rho(E_r)\mathrm{d}E_r$ over the continuum states ($E_r \geqslant E_\mathrm{D}$), where $\rho(E_r)$ is the density of states r near E_r. If Γ_r is sufficiently small and $|k_\alpha - (E_\mathrm{D} - E_m)|$ sufficiently large, the sum over r can be approximated by

$$S \equiv \lim_{\Gamma_r \to 0} \int_{-\infty}^{+\infty} \frac{H^{\varepsilon}_{\dots} H^{\alpha}_{\dots}}{k_{rm} - k_{\alpha} - \frac{1}{2}\mathrm{i}\hbar\Gamma_r} \rho(E_r)\mathrm{d}E_r$$

$$= \mathscr{P} \int_{-\infty}^{+\infty} \frac{H^{\varepsilon}_{\dots} H^{\alpha}_{\dots} \rho(E_r)}{E_r - E_m - k_{\alpha} - \frac{1}{2}\mathrm{i}\hbar\Gamma_r} \mathrm{d}E_r + \mathrm{i}\pi[H^{\varepsilon}_{\dots} H^{\alpha}_{\dots} \rho(E_r)]_{E_r = E_m + k_\alpha} \tag{145}$$

For large t the absolute square of (144) can be written [Fermi's 'Golden Rule', *cf. e.g.* ref. 20, p. 139, eq. (8)]

$$|b_N(t)|^2 = \frac{2\pi}{\hbar} t |S|^2 \delta(k_{nm} - k_{\alpha} + k_{\varepsilon}) \tag{146}$$

Equation (146) with the first term in (145) neglected and the square of the density factor $\rho(E_r)$ inadvertently omitted was first derived by the Reporter.[80] The omission of the density factor was justly criticized.[81] Comparison of numerical calculations with experimental data obtained from gaseous iodine[82-84] shows that the justification for neglecting the integral in the RHS of (145) must be carefully examined in individual cases. (Abstractly, however, $\mathscr{P} \int_{-\infty}^{+\infty} \frac{\mathrm{d}x}{x} = 0$.) In particular, the states r in the discrete part of the energy spectrum ($E_r < E_D$) may give a contribution to the sum over r for which the approximation for the lower part ($-\infty < E_r < E_D$) of the integral in S may prove unrealistic.

D. Resonance Raman Effect and Resonance Fluorescence.—It can be shown[60] that the probability $|b_N(t)|^2$ for a scattering transition to take place within the time interval $(0,t)$ can be interpreted as the probability for two statistically independent, consecutive, real, one-photon processes of absorption and emission, in the special case of strictly monochromatic resonance excitation upon one non-degenerate discrete level with zero width. Because of this separation of the scattering process the phenomenon of light transformation should in this case, according to the proposals made at the end of Section 2, be termed RF instead of RR scattering. When non-monochromatic excitation [say with a Lorentzian intensity distribution $I_p(\omega)$ — p for 'primary'— of half-width G and maximum ω_0] is used and the energy $E_{r'}$ of the discrete non-degenerate resonance state r' is associated with a finite damping constant $\Gamma_{r'}$, the more complicated quantum mechanical calculation essentially

[80] J. Behringer, *Z. Physik*, 1969, **229**, 209.
[81] O. S. Mortensen, *Mol. Phys.*, 1971, **22**, 179.
[82] M. Berjot, M. Jacon, and L. Bernard, *Opt. Comm.*, 1971, **4**, 117, 246.
[83] M. Jacon, M. Berjot, and L. Bernard, *Compt. rend.*, 1971, **273**, *B*, 595.
[84] M. Berjot, 'Etude Expérimentale de la Diffusion Résonnante de la Lumière, avec Changement de Fréquence, par les Molécules d'Iode et de Brome', Thesis, Reims, 1973, pp. 27, 86ff.

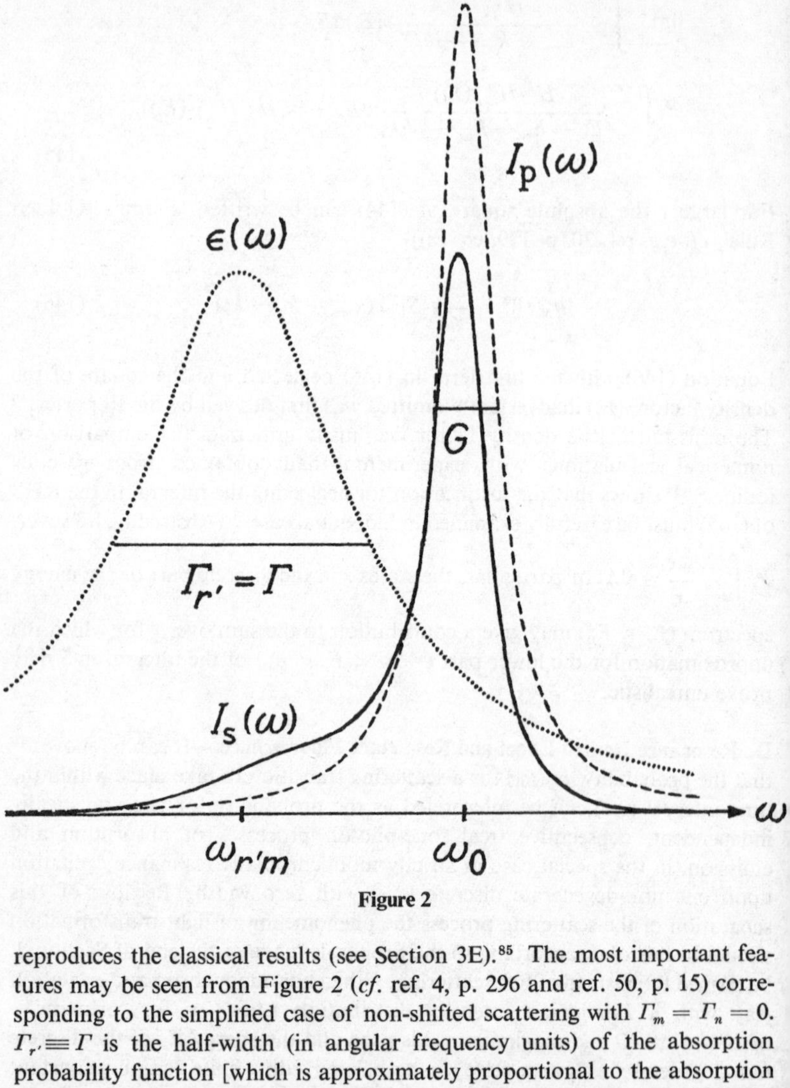

Figure 2

reproduces the classical results (see Section 3E).[85] The most important features may be seen from Figure 2 (*cf.* ref. 4, p. 296 and ref. 50, p. 15) corresponding to the simplified case of non-shifted scattering with $\Gamma_m = \Gamma_n = 0$. $\Gamma_{r'} \equiv \Gamma$ is the half-width (in angular frequency units) of the absorption probability function [which is approximately proportional to the absorption coefficient $\varepsilon(\omega)$] for the transition $r' \leftarrow m$. The secondary intensity $I_s(\omega)$ turns out to be approximately proportional to $\varepsilon(\omega)I_p(\omega)$ so that in the case $|\omega_0 - \omega_{r'm}| \gtrsim |G - \Gamma| \gg 0$ two maxima will appear with half-widths more (>) or less (~) similar to Γ and G respectively. If $|\omega_0 - \omega_{r'm}| \gg |G - \Gamma| \gg 0$ the two maxima are clearly separable and the two peaks can be assigned to RF and Raman scattering. If, however, $|\omega_0 - \omega_{r'm}| < |G - \Gamma|$

[85] *Cf.* D. L. Huber, *Phys. Rev.*, 1969, **178**, 93; 1969, **187**, 392.

the two peaks coalesce into one peak, which in the case $|G - \Gamma| \gg 0$ has a width approximately equal to the smaller one of the two quantities G, Γ, so that the secondary radiation resembles either scattering or fluorescence. If $G \approx \Gamma$ these conclusions must be revised (*cf.* Section 3E, Case II).

In order to determine lineshapes and linewidths explicitly the Fourier transform of $|b_N(t)|^2$ must be calculated. This relates to any processes $N \leftarrow M$ whatsoever (absorption or emission, as well as scattering). The calculation can be performed in a straightforward manner by complex integration. The formulae obtained for RR lineshapes, at least if calculated in full generality, are very complicated and therefore will not be reproduced here.[60]

E. Electric Dipole and Higher Approximations of the Interaction Hamiltonian.— The vector potential operator $\hat{A}(\boldsymbol{r}_k)$ (127) in its explicit form contains terms with factors $e^{\pm i\boldsymbol{\kappa}_\lambda \cdot \boldsymbol{r}_k}$. These factors can be expanded in a Taylor's series:

$$e^{\pm i\boldsymbol{\kappa}_\lambda \cdot \boldsymbol{r}_k} = \sum_\rho \frac{1}{\rho!}(\pm i\boldsymbol{\kappa}_\lambda \cdot \boldsymbol{r}_k)^\rho = 1 \pm i\boldsymbol{\kappa}_\lambda \cdot \boldsymbol{r}_k + \ldots \tag{147}$$

By eliminating the translational motion of the scattering system, $\boldsymbol{r}_k$ can be replaced by $\boldsymbol{r}_k' = \boldsymbol{r}_k - \boldsymbol{R}$ where $\boldsymbol{R}$ is the position vector of the centre of mass of the system. If then

$$|\boldsymbol{\kappa}_\lambda \cdot \boldsymbol{r}_k| \ll 1$$

or, a little more restrictively,

$$|\boldsymbol{\kappa}_\lambda||\boldsymbol{r}_k'| = \frac{\omega_\lambda}{c}|\boldsymbol{r}_k'| = \frac{1}{\lambda_\lambda}|\boldsymbol{r}_k'| \ll 1 \tag{148}$$

i.e. if the dimensions of the scattering particle system are small relative to the incident and scattered wavelengths, all expansion terms in (147) except the first term 1 or, in a more refined approximation, except the first two terms $1 \pm i\boldsymbol{\kappa}_\lambda \cdot \boldsymbol{r}_k$ can be discarded. [In macroscopic crystals (148) is formally never satisfied for optical wavelengths, but the translational symmetry permits the application of (148) to the particles within only one primitive cell.] The first term ($\rho = 0$) in (147) leads to the so-called 'electric dipole approximation (E1)', the second one ($\rho = 1$) to the 'magnetic dipole inclusive electric quadrupole approximation (M1 + E2)'. [These concepts should not be confused with what was called by Placzek (ref. 9, p. 232 ff.) quadrupole or magnetic dipole scattering.]

In the literature on Raman spectroscopy usually only E1 is applied. Just as coherent Rayleigh scattering in E1 is responsible for the elementary optical phenomena of reflection, refraction, and diffraction, in M1 + E2 it causes natural optical activity and (one part of) Faraday rotation. (For the other part of Faraday rotation, due to paramagnetism, the spin–magnetic field interaction in $\hat{H}$ must be included.) Now Rayleigh scattering is only a special case of Raman scattering. It can be expected that in the M1 + E2 approximation the RRE will acquire certain features analogous to those of optical activity (*cf.* ref. 9, p. 232, footnote 2). Early attempts to demonstrate such features in the ordinary RE of optically active compounds failed (see ref. 13, p. 242).* By generalizing the customary quantum mechanical theory of optical activity (ref. 61, p. 250 ff., ref. 86, p. 61 ff., ref. 87, p. 369 ff., ref. 88, p. 51 ff.) some basic formulae for the RRE in the M1 + E2 approximation were derived in ref. 66, p. 86 ff. Non-resonance Rayleigh and Raman scattering from optically active molecules and from molecules in magnetic fields (including symmetry rules) have been extensively discussed in a series of papers by Atkins, Barron, and Buckingham.[89-91]

F. Phase Correlations.—In the usual quantum mechanical occupation number representation of the radiation field, in contrast to classical theory, all knowledge of field component phases is excluded by the uncertainty relation for occupation numbers and phases. Therefore there is no definite phase correlation of the photons absorbed or scattered in an elementary scattering process with the photons still or already present in the respective field modes. Although the intrinsic coherence of Rayleigh scattering from a non-degenerate level can be shown in the well-known way (see *e.g.* ref. 18, p. 55 ff., ref. 20, p. 193 and ref. 72, p. 361 ff.) the full demonstration of the correspondence between the classical and quantum mechanical descriptions of the RRE (including related phenomena) would necessitate the use of coherent field states associated with definite phases. Then the reaction of the scattering system to an exciting wavepacket with exactly defined phases of its monochromatic components could also be studied quantum mechanically. But obviously this aspect of the scattering problem has not so far been investigated systematically although some important preliminary studies were done by Morozov [92] (*cf.* ref. 4, p. 312 ff.).

5 Quantum Mechanical Theories of the Resonance Raman Effect: Special Systems

In this section we review the theoretical methods devoted to the investigation of special systems involved in the scattering process. These systems can be roughly divided as follows: (i) atoms and mononuclear ions (either single or in larger numbers, but in the latter case individually independent in position and orientation), (ii) molecules and polynuclear ions (again either single

* The author is obliged to Prof. B. Schrader, Dortmund, for this notice.

[86] J.-P. Mathieu, 'Les Théories Moléculaires du Pouvoir Rotatoire Naturel', CNRS (Gauthier-Villars), Paris, 1946.

[87] J.-P. Mathieu, 'Activité Optique Naturelle', in 'Encyclopedia of Physics' ed. S. Flügge, Springer, Berlin, 1957, Vol. 28, p. 333–431.

[88] D. J. Caldwell and H. Eyring, 'The Theory of Optical Activity', Wiley, New York, 1971.

[89] P. W. Atkins and L. D. Barron, *Mol. Phys.*, 1969, **16**, 453.

[90] L. D. Barron, *J. Chem. Soc.* (*A*), 1971, 2899.

[91] L. D. Barron and A, D. Buckingham, *Mol. Phys.*, 1970, **20**, 1111; 1972, **23**, 145.

[92] V. A. Morozov, *Optika i Spektroskopiya*, 1966, **20**, 491; 1967, **23**, 3.

or, if not, individually free in space), and (iii) condensed systems (all states of particle aggregation due to interatomic or intermolecular forces, *e.g.* gases under higher pressures, liquids, crystals, liquid crystals, layers and films, disperse systems such as solutions, alloys, colloids, powders, polycrystalline material, *etc.*). Because of the variety of forces which can hold atoms together there is no clear distinction between molecules and condensed matter; *e.g.* crystals in certain respects can be conceived as large molecules and conversely macromolecules such as polymers resemble crystals. Quantum mechanically all of these systems differ only in the types of their state manifolds. Whereas an atom (with neglect of the nuclear states) has only translational and electronic states, a molecule in addition shows rotational and vibrational states and a crystal optical phonon states (the acoustical phonons correspond to the translations of free particles) and eventually still other characteristic types of excitation. It is the aim of the special theories of the RRE to find out which transitions between these states are allowed in RR scattering (selection rules) and to explain as far as possible the measured quantities (wavenumbers, intensities, depolarization ratios, lineshapes, *etc.*) characterizing these transitions from structural principles. Applying the general theory to special systems and special conditions permits certain simplifications of the general formulae, but also necessitates the admission of special parameters characterizing the system in question. In non-stimulated Raman scattering the scattered intensity is proportional to the intensity of the incident light and to the number of scattering particles or particle systems. Experimentally, therefore, always very large numbers of atoms or molecules, *i.e.* only more or less tightly condensed systems, are investigated. The real behaviour of single atoms or molecules must be extrapolated from the measurements on accumulations of numerous atoms or molecules with as low a density as possible.

A. Atoms and Mononuclear Ions.—In atoms (including in this word mononuclear ions) we have only the electronic RRE. The quantum numbers m, r, and n appearing in (143) must be understood as comprising the momentum eigenvalue (as translational quantum number) and a complete set of electronic quantum numbers characterizing an atomic state (see literature on atomic spectroscopy, *e.g.* ref. 27, Ch. 1, [93–95]). It should be noted that only those intermediate states r play a role for which the matrix elements of $\hat{H}^{\varepsilon}$ and $\hat{H}^{\alpha}$ appearing in (143) do not vanish. In particular, occupied atomic states, because of the Pauli principle, can be left out of consideration in (143).

Atoms occur either free or bound in molecules or crystals. In this section we are only interested in free or almost free atoms, *i.e.* in atoms with energy

[93] G. Herzberg, 'Atomic Spectra and Atomic Structure', Dover, New York, 1945.
[94] H. G. Kuhn, 'Atomic Spectra', Longman, London, 2nd edn., 1969.
[95] E. U. Condon and G. H. Shortley, 'The Theory of Atomic Spectra', Cambridge University Press, 1964.

levels at most insignificantly shifted by interatomic interactions or by external fields so that the usual atomic quantum numbers can still be regarded as good quantum numbers.

Free Atoms. Free atoms by definition are freely movable in space. (The free orientability in space is automatically implied by the spherical symmetry of atoms.) Free mobility and orientability lead to certain consequences equally valid for other systems possessing these properties. The translational motion relative to a space-fixed frame of reference causes a Doppler shift of the scattered radiation observed by a spectrograph at rest with respect to this frame of reference. The explicit formula describing the Doppler shift of Raman lines is found by performing the partial summation (here integration) over the translational quantum number (= eigenvalue of the total momentum operator) contained in the sum over r in (143) (see *e.g.* ref. 58, p. 749 and ref. 66, p. 73 ff.). Owing to the free orientability in space the total angular momentum (inclusive electronic and nuclear spins) is a constant of motion. Therefore, leaving out of consideration the nuclear spin, the quantum number J refers to a $(2J + 1)$-fold degenerate energy level, the states with the magnetic quantum numbers $M, -J \leqslant M \leqslant J$, coinciding. Consequently the brackets and the denominators in (143) do not depend on $M \equiv M_r$ (the subscript r indicating 'intermediate state') and the partial summation over M_r contained in r in (143) (which concerns only the product of matrix elements) can be performed without impediments. The scattered intensities and the depolarization ratios depend only on the J values J_m, J_r, J_n of the initial, intermediate, and final atomic states. However, as the matrix elements occurring in (143) contain the polarization vectors $\boldsymbol{e}_\alpha$, $\boldsymbol{e}_\varepsilon$ of the incident and scattered waves [see the remark after (127)] this sum over M_r will be different for different choices of $\boldsymbol{e}_\alpha$ and $\boldsymbol{e}_\varepsilon$. Although the explicit calculations had already been performed in a rather cumbersome way by Placzek and Teller [96] (*cf.* Section 6 of ref. 9) it was certainly valuable and stimulating for work on more complex problems that some authors repeated them in the more modern and elegant way using vector coupling coefficients;[97-101] see also ref. 75, p. 145 ff.

In the rigorous RRE (inclusive RF) observed by excitation of only one discrete resonance level with quantum number J_r the sum in (143) can be drastically reduced to the one resonance term containing J_r among the quantum numbers collectively denoted by r. For the *partial* transitions $J_r \leftarrow J_m$, $J_n \leftarrow J_r$ constituting the resonance scattering process (Figure 3) the ordinary selection rules $\Delta J = 0, \pm 1$ (for one-photon absorption and

[96] G. Placzek and E. Teller, *Z. Physik*, 1933, **81**, 209.
[97] J. A. Koningstein and O. S. Mortensen, *Phys. Rev.*, 1968, **168**, 75; O. S. Mortensen and J. A. Koningstein, *J. Chem. Phys.*, 1968, **48**, 3971.
[98] C. M. Penney, *J. Opt. Soc. Amer.*, 1969, **59**, 34.
[99] O. S. Mortensen, *Chem. Phys. Letters*, 1970, **5**, 515.
[100] Y. N. Chiu, *J. Opt. Soc. Amer.*, 1970, **60**, 607.
[101] K. Altmann and G. Strey, *J. Mol. Spectroscopy*, 1972, **44**, 571.

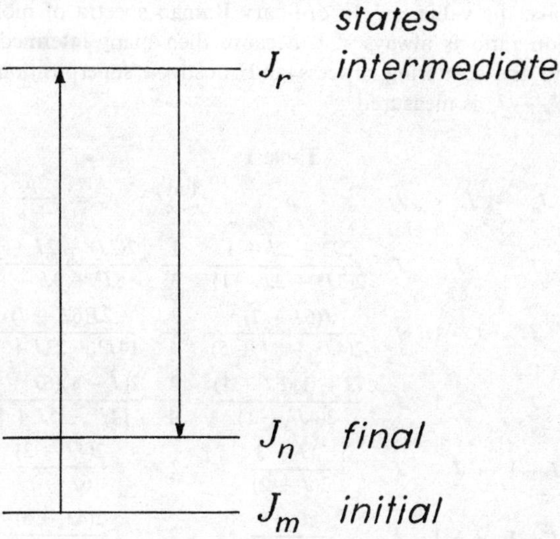

Figure 3

emission) hold, so that nine different types of rigorous RR transition (with $J_n - J_m = 0, \pm 1, \pm 2$) are possible. For these transitions the depolarization ratios ρ (for linearly polarized excitation and 90° observation) and ρ_n (for natural excitation and 90° observation) are reproduced in Table 1 [$J_m \equiv J, J_r, J_n$ referring to the initial, intermediate (resonance), and final states, respectively]. The values for ρ were taken from Mortensen [99] (with one printing error corrected); some of these values were already given explicitly by Placzek [ref. 96, ref. 9, p. 248 (2), p. 371 (2*a*, *b*)] and Morozov.[102]* It can be seen that in some cases ρ and ρ_n can be larger than 1, so that small values of ρ are not characteristic for the RRE although this is widely thought to be the case (*cf.* ref. 29, p. 400 and refs. 103, 104). There may be even very large values of ρ in the RRE of atoms as well as of molecules and other systems. A striking example of this kind discussed already by Placzek is the RR transition shown in Figure 4 in atomic thallium (ref. 9, p. 263; *cf.* also ref. 6, p. 33 ff.). Neglecting the nuclear spin we have for this transition from Table 1 the theoretical value $\rho = 7$. It is easily seen from Table 1 that in some cases ρ

* The values given by Placzek and Morozov refer to diatomic molecules with Σ electronic ground and intermediate states; this case is equivalent to atoms (see Section 5C).

[102] V. A. Morozov, *Optika i Spektroskopiya*, 1965, **18**, 198; 1965, **19**, 35.
[103] I. R. Beattie, G. A. Ozin, and R. O. Perry, *J. Chem. Soc.* (*A*), 1970, 2071.
[104] M. Jacon, M. Berjot, and L. Bernard, *Compt. rend.*, 1971, **273**, *B*, 956; M. Berjot, M. Jacon, and L. Bernard, *Canad. J. Spectroscopy*, 1972, **17**, 60; M. Berjot and L. Bernard, 'Passage Continu de la Fluorescence de Résonance à l'Effet Raman de Résonance', in 'Advances in Raman Spectroscopy' ed. J.-P. Mathieu, Heyden and Son, London, 1973, Vol. 1, p. 343.

may even take the value ∞. In ordinary Raman spectra of molecules the depolarization ratio is always $\leqslant \frac{6}{7}$ because then many intermediate states contribute to the scattering process and mostly a superposition of many transitions $J_n \leftarrow J_m$ is measured.

Table 1

$J_n - J_m$	J_n	$\leftarrow J_r$	$\leftarrow J_m$	ρ	$\lim_{J\to\infty} \rho$	$\rho_n = \frac{2\rho}{1+\rho}$	$\lim_{J\to\infty} \rho_n$
0	J	J	J	$\frac{2J^2+2J+1}{2(3J^2+3J-1)}$	$\frac{1}{3}$	$\frac{2(2J^2+2J+1)}{8J^2+8J-1}$	$\frac{1}{2}$
0	J	$J+1$	J	$\frac{J(6J+7)}{2(4J^2+8J+5)}$	$\frac{3}{4}$	$\frac{2J(6J+7)}{14J^2+23J+10}$	$\frac{6}{7}$
0	J	$J-1$	J	$\frac{(J-1)(6J+1)}{2(4J^2+1)}$	$\frac{3}{4}$	$\frac{2(J-1)(6J+1)}{14J^2-5J+1}$	$\frac{6}{7}$
+1	$J+1$	J	J	$\frac{4J+3}{2(J+2)}$	2	$\frac{2(4J+3)}{6J+7}$	$\frac{4}{3}$
+1	$J+1$	$J+1$	J	$\frac{4J+5}{2J}$	2	$\frac{2(4J+5)}{6J+5}$	$\frac{4}{3}$
−1	$J-1$	J	J	$\frac{4J+1}{2(J-1)}$	2	$\frac{2(4J+1)}{6J-1}$	$\frac{4}{3}$
−1	$J-1$	$J-1$	J	$\frac{4J-1}{2(J+1)}$	2	$\frac{2(4J-1)}{6J+1}$	$\frac{4}{3}$
+2	$J+2$	$J+1$	J	$\frac{3}{4}$	$\frac{3}{4}$	$\frac{6}{7}$	$\frac{6}{7}$
−2	$J-2$	$J-1$	J	$\frac{3}{4}$	$\frac{3}{4}$	$\frac{6}{7}$	$\frac{6}{7}$

The example of Figure 4 can be used for illustrating also two other important features of the rigorous RRE:

(*a*) In this example we have $J_n - J_m = 1$ and a detailed analysis of the dipole approximation to (143) shows that in this case also the *antisymmetric* part (with respect to an inversion of space co-ordinates) of the scattering tensor $\boldsymbol{a}_{nm}$ (36) [which results from (143) by an extraction of $\boldsymbol{e}_\alpha$ and $\boldsymbol{e}_\varepsilon$] contributes to scattering. In Placzek's terminology we have here 'quadrupole scattering' (with a symmetric tensor) plus 'magnetic dipole scattering' (antisymmetric tensor), but no 'spur (or trace) scattering' (symmetric tensor). This can be seen from Placzek, ref. 9, p. 244, Table 1 (ΔJ there is our $J_n - J_m$). In the original (not modified as in Section 3) classical theory of the RE and in Placzek's polarizability theory (see ref. 9, p. 269) the scattering tensor is always considered as symmetric and real. This can be understood very easily by remembering that both of these theories start from coherent Rayleigh scattering (*cf.* ref. 9, p. 265) and only expand the tensor of this Rayleigh scattering in a Taylor's series with respect to nuclear vibrational co-ordinates.

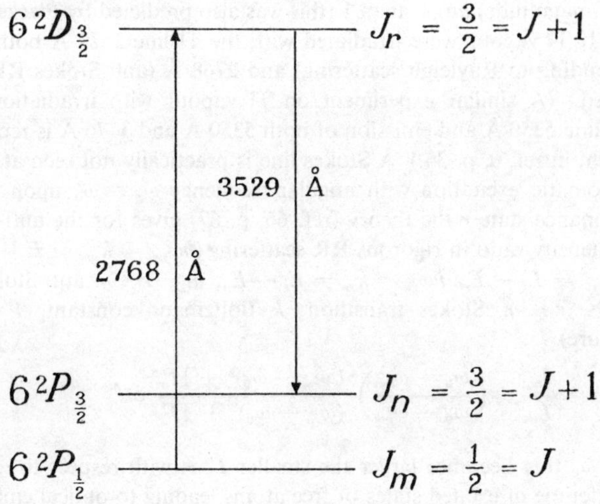

Figure 4

Now the tensor of coherent Rayleigh scattering is for the non-resonance case usually (*i.e.* when the interaction Hamiltonian $\hat{H}$ is not only Hermitian but *real*, which is the case for absent magnetic fields and spin interaction) symmetric and real. The symmetry follows from the reality and *vice versa*, which can be seen as follows from (36) taking $m = n$, $\Gamma_r = 0$: from

$$\boldsymbol{a}_{mm} = \frac{1}{\hbar}\sum_r \left(\frac{\boldsymbol{\mu}_{rm}\,\boldsymbol{\mu}_{mr}}{\omega_{rm} - \omega_0} + \frac{\boldsymbol{\mu}_{mr}\,\boldsymbol{\mu}_{rm}}{\omega_{rm} + \omega_0} \right) \tag{149}$$

there follows (˜ transposition, * complex conjugation) by interchanging the vectors in the dyadic products of the numerators and then using $\boldsymbol{\mu}_{ab} = \boldsymbol{\mu}^*_{ba}$:

$$\tilde{\boldsymbol{a}}_{mm} = \frac{1}{\hbar}\sum_r \left(\frac{\boldsymbol{\mu}_{mr}\,\boldsymbol{\mu}_{rm}}{\omega_{rm} - \omega_0} + \frac{\boldsymbol{\mu}_{rm}\,\boldsymbol{\mu}_{mr}}{\omega_{rm} + \omega_0} \right)$$
$$= \frac{1}{\hbar}\sum_r \left(\frac{\boldsymbol{\mu}^*_{rm}\,\boldsymbol{\mu}^*_{mr}}{\omega_{rm} - \omega_0} + \frac{\boldsymbol{\mu}^*_{mr}\,\boldsymbol{\mu}^*_{rm}}{\omega_{rm} + \omega_0} \right) = \boldsymbol{a}^*_{mm} \tag{150}$$

so that $\boldsymbol{a}_{mm} = \tilde{\boldsymbol{a}}_{mm}$ implies $\boldsymbol{a}_{mm} = \boldsymbol{a}^*_{mm}$ and *vice versa.* In the resonance case (when $\Gamma_r \neq 0$ cannot be neglected) $\boldsymbol{a}_{mm}$ necessarily becomes complex and consequently (even for real $\hat{H}$) acquires also an antisymmetric part. Independently of this, $\boldsymbol{a}_{nm}$, $n \neq m$, in the non-resonance case (even for real $\hat{H}$ and real matrix elements $\boldsymbol{\mu}_{ab}$) is not necessarily symmetric. The antisymmetric part always leads exactly to what was called by Placzek 'magnetic dipole scattering'.

(*b*) The example shown in Figure 4 moreover illustrates the fact that the intensity ratio of the anti-Stokes and Stokes lines can become (even by many orders of magnitude) *larger* than 1 (this was also predicted by Placzek, ref. 9, p. 253). If Tl vapour were irradiated with the Tl line 3529 Å both 3529 Å (corresponding to Rayleigh scattering) and 2768 Å (anti-Stokes RE) would be emitted. (A similar experiment on Tl vapour with irradiation of the green Tl line 5350 Å and emission of both 5350 Å and 3776 Å is reported by Pringsheim in ref. 6, p. 34.) A Stokes line is practically not seen at all. For monochromatic excitation with angular frequency $\omega_\alpha \equiv \omega_0$ upon one discrete resonance state r the theory (ref. 66, p. 67) gives for the anti-Stokes : Stokes intensity ratio in rigorous RR scattering ($\hbar\omega_{nm} = k_{nm} = E_n - E_m > 0$, $\hbar\omega_{rm} = k_{rm} = E_r - E_m$, $\hbar\omega_{rn} = k_{rn} = E_r - E_n$; $m \leftarrow r \leftarrow n$ anti-Stokes transition, $n \leftarrow r \leftarrow m$ Stokes transition; k Boltzmann constant, T absolute temperature)

$$\frac{I_{mn}}{I_{nm}} = \left(\frac{\omega_0 + \omega_{nm}}{\omega_0 - \omega_{nm}}\right)^4 \frac{(\omega_{rm} - \omega_0)^2 + \frac{1}{4}\Gamma_r^2}{(\omega_{rn} - \omega_0)^2 + \frac{1}{4}\Gamma_r^2} \mathrm{e}^{-\hbar\omega_{nm}/kT} \tag{151}$$

If $\omega_{rn} = \omega_0$ this becomes larger the smaller Γ_r is with respect to ω_{nm}. The natural lifetime of excited states of free atoms leading to optical emissions is roughly some 10^{-8} s (*cf.* ref. 14, p. 428, ref. 94, p. 129) which corresponds to about 10^{-2} cm^{-1}. The Raman shifts $\tilde{\nu}_{nm} = \omega_{nm}/(2\pi c)$ observed in electronic rigorous RR spectra are much larger so that anti-Stokes : Stokes ratios which differ considerably from the non-resonance ratio (see ref. 9, p. 253) must be expected for free atoms.

If *displaced* resonance scattering cannot occur because $m = n$ is the only state with an energy level below that of the single discrete resonance state r, and if r is only radiation damped (which can be assumed for free atoms) the scattering cross-section (total energy scattered by one atom per unit of time and per unit exciting energy flux) for near resonance excitation ($\omega_\alpha \equiv \omega_0 \approx \omega_{rm}$) is (*cf.* ref. 49)

$$Q_{nm} \equiv Q_{mm} = \frac{\pi c^2}{2} \frac{\omega_0^4}{\omega_{rm}^6} \frac{2J_r + 1}{2J_m + 1} \frac{\Gamma_r^2}{(\omega_{rm} - \omega_0)^2 + \frac{1}{4}\Gamma_r^2} \tag{152}$$

Here ω_{rm} can be approximated by ω_0. For exact resonance excitation ($\omega_0 = \omega_{rm}$, 'resonance line') (152) becomes independent of the damping constant [*cf.* ref. 9, p. 248, eq. (3)]

$$Q_{mm} = \frac{2\pi c^2}{\omega_0^2} \frac{2J_r + 1}{2J_m + 1} = \frac{\lambda_0^2}{2\pi} \frac{2J_r + 1}{2J_m + 1} \tag{153}$$

The nuclear analogues to (152) and (153) (with the nuclear-spin quantum numbers I instead of the atomic total angular momentum quantum numbers J) are fundamental for Mössbauer spectroscopy.[105] In this case (153) is usually supplemented by a 'conversion factor' which takes into account certain additional energy losses.

[105] *Cf.* N. N. Greenwood and T. C. Gibb, 'Mössbauer Spectroscopy', Chapman and Hall, London, 1971, p. 12.

Atoms Influenced by Fields. For atoms exposed to constant or varying external fields (either artificially applied or generated by neighbouring particles in dense gases or condensed phases) the degeneracy of energy levels due to the free orientability can be lifted. By this splitting of atomic levels some of the properties of scattering described in the preceding section may be changed. (This does not apply, however, to all properties; *e.g.* the anomalous anti-Stokes : Stokes intensity ratio and the antisymmetry of the scattering tensor may still be true.) There is an immense number of possible ways of influencing atoms (and likewise molecules) by fields. We consider, only briefly, two important cases:

(*a*) In an externally applied magnetic field the $2J + 1$ states with different magnetic quantum numbers M belonging to a definite value of J acquire different energies. This Zeeman splitting is proportional to the field strength. The denominators in (143) for different M_r in the set $-J_r \leqslant M_r \leqslant J_r$ become different. If the damping constants are small enough and rigorous resonance excitation upon one discrete magnetic level $J_{r'}$, $M_{r'}$ is used, all other levels ($J_r \neq J_{r'}$, M_r arbitrary; $J_r = J_{r'}$, $M_r \neq M_{r'}$) because of the relative largeness of their denominators in (143) will not contribute much to the scattering probability. In contrast to the case of free atoms the terms with different M_r, $-J_{r'} \leqslant M_r \leqslant J_{r'}$, no longer 'interfere' with one another (because only the term with $M_r = M_{r'}$ is important and those with $M_r \neq M_{r'}$ can be neglected) and consequently the depolarization ratios in Table 1 are no longer applicable to this case (they depend also on M).[60] Similar arguments apply to the case of an external electric field (Stark splitting of levels). A further discussion of these cases would lead into the vast territory of level-crossing (double resonance) spectroscopy (including the Hanle effect; for literature *cf.* ref. 6, p. 63 ff., ref. 106, p. 139 ff. and refs. 107, 108). Apparently so far level-crossing experiments have been devoted only to the non-shifted 'resonance radiation', but perhaps now laser technology could also yield for the shifted RR lines sufficient intensities for analogous measurements of depolarization ratios *etc.* Placzek (ref. 9, p. 249) was already encouraging such experiments in 1934!

(*b*) If a mononuclear ion is embedded in a solid (either as a constituent ion in a pure ionic crystal or as an impurity ion in a dilute crystal or in non-crystalline material) the energies of its (electronic) states are varied by smaller or larger amounts as a consequence of the distortion of the electronic cloud by the crystal or quasi-crystal field. This implies splitting of levels degenerate in the free ion. An immense literature exists giving information about the electronic states of mononuclear (and also complex) ions in crystals.[109] Since

[106] W. Hanle and K. Larché, 'Anregung und Ionisierung', in 'Handbuch der Radiologie', ed. E. Marx, Akad. Verlagsges., Leipzig, 1933, Vol. VI, Part I, Chap. 2, pp. 115–225.
[107] R. Bernheim, 'Optical Pumping', Benjamin, New York, 1965.
[108] W. Hanle and R. Pepperl, *Phys. Blätter*, 1971, **27**, 19; H. Bucka, *J. Phys. Radium*, 1969, **30**, C1.
[109] See *e.g.* D. S. McClure, 'Electronic Spectra of Molecules and Ions in Crystals', Academic Press, New York, 1959; A. B. P. Lever, 'Inorganic Electronic Spectroscopy', Elsevier, Amsterdam, 1968; K. H. Hellwege, 'Einführung in die Festkörperphysik', Part II, Springer, Berlin, 1970.

1963 an increasing number of observations of the ordinary electronic RE from such ions has been reported. However, the RRE of such systems has only recently started to become the object of experimental investigations.[110] The reason why, in contrast to free atoms, both RR scattering and also ordinary Raman scattering can relatively easily be observed from ions embedded in crystals at sufficiently low temperatures must be seen in the favourable (*i.e.* not too small) width of the electronic levels combined by the Raman transition. At ambient temperatures the levels, because of the increased perturbations of the ions in the crystal (*e.g.* by the thermally excited mutual vibrational motion of all particles), usually become very broad, which is detrimental for the observation of the ordinary as well as the resonance RE. The energy separation of the state manifolds with different J of ions in crystals is, because of the smaller moment of inertia, much larger than in molecules. For example in $PrCl_3$ crystals the $^3H_4$, $^3H_5$, $^3F_2$ states of the Pr^{3+} ion in this order are separated by about 2100 and 2700 cm^{-1},[111] whereas even in the lightest molecule H_2 the distance between the levels $J = 0$ and $J = 1$ is only 121 cm^{-1}. In ordinary vibrational Raman spectroscopy the molecules considered are so large and heavy that the rotational structure in most cases cannot be resolved. This shows in connection with the discussion of free atoms on p. 150 ff. that even in the non-resonance electronic RE of ions in solids, in contrast to ordinary vibrational Raman spectroscopy, magnetic dipole scattering may play a role. This is all the more true for the electronic RRE. The importance of the antisymmetric part of the scattering tensor for the (non-resonance) electronic RE was discussed theoretically.[97, 112]

The splitting of the electronic levels of ions in crystals is caused essentially by an inhomogeneous electrical field having the site symmetry of the ion in question. The resulting Stark levels either are non-degenerate (Kramers singlets) or doubly degenerate (Kramers doublets). The latter Kramers degeneracy can be lifted by an additional magnetic field. Because of the inhomogeneity of the crystal field the magnetic quantum number M is in general no more defined and must be replaced by a 'crystal quantum number' μ referring to the principal symmetry axis through the ion and by additional quantum numbers. J, however, remains a good quantum number. If J is integral (half-integral), a single- (double-) valued representation of the site group of the ion must be used for classifying the symmetry of the split levels.[113, 114]

B. Molecules and Polynuclear Ions.—In contrast to atoms the essential new feature of molecules (including di- and poly-nuclear ions) would seem to be

[110] R. L. Wadsack and R. K. Chang, *Solid State Comm.*, 1972, **10**, 45.
[111] J. T. Hougen and S. Singh, *Phys. Rev. Letters*, 1963, **10**, 406; *Proc. Roy. Soc.*, 1964, **A277**, 193.
[112] J. A. Koningstein and O. S. Mortensen, *J. Opt. Soc. Amer.*, 1968, **58**, 1208.
[113] *Cf.* M. Hamermesh, 'Group Theory and its Applications to Physical Problems', Addison-Wesley, Reading, Mass., 1962, p. 337, 357.
[114] J. A. Koningstein, *Phys. Rev.*, 1968, **174**, 477; *Appl. Spectroscopy*, 1968, **22**, 438.

the possibility of vibrations. Vibrations, however, presuppose the existence of a stable equilibrium configuration of the nuclei. It is well known that molecules often when in excited electronic states are not stable but can dissociate (this includes ionization). Such unstable systems of nuclei and electrons with electronic 'repulsion states' are still referred to as molecules. Thus we must stop considering vibrations (in the sense of a periodic change of nuclear distances) as essential for molecules. Rather we must also regard non-periodic changes of the nuclear distances (let us call them 'pseudo-vibrations') as characteristic of molecules in this wider sense. Quantum mechanically, periodic (non-periodic) motions lead to discrete (continuous) energy eigenvalues. The typical potential models corresponding to these cases are, in rough simplification, a well and a wall, respectively. It is clear that the idea of nuclei moving in a potential generated by the electrons is the result of a simplified view of the nuclear–electronic interaction taking account of the large nuclear : electronic mass ratio. Nevertheless this view, which is basic to the Born–Oppenheimer approximation, must be accepted if a theory of RR scattering generally valid for all polynuclear systems of at least slowly transient stability (with respect to electronic motion) is desired. Using this approximation a state of a free molecule can be identified by the quantum numbers of translation (total momentum eigenvalue), rotation, (pseudo-) vibration, electronic motion, and nuclear spin. The centre of mass translation is easily separated as in atoms (see the section on free atoms). The nuclear-spin splitting of molecular levels is extremely small; more important for molecular spectra is the determination of the average populations of symmetric and antisymmetric (with respect to an exchange of identical nuclei) rotational levels by the nuclear spin (*cf.* ref. 27, p. 138 ff., ref. 9, p. 340). In rigorous rotational RR scattering this effect must be taken account of. The electronic RE in free molecules has only been observed in NO with non-resonance excitation,[115] and probably also for future RR spectroscopy it will remain a curiosity. These remarks justify a restriction of the following review to the vibrational and rotational RRE on molecules.

Vibrational Resonance Raman Effect. Generalities. Most of the theoretical work performed in recent years on the RRE was devoted to diatomic molecules. The existence of only one normal vibration favours the concentration of theoretical efforts upon the central questions of the RR scattering mechanism. However, it is clear that the results obtained for diatomic molecules cannot answer, for example, the important question as to why the Raman intensities of different normal vibrations of polyatomic molecules behave differently when the exciting frequency is changed.

The starting-point for every theory of the vibrational RRE is the scattering tensor given in (36) in E1 approximation. The molecular state designations *m* (initial), *r* (intermediate), and *n* (final) more explicitly can be understood

[115] H. Fast, H. L. Welsh, and D. W. Lepard, *Canad. J. Phys.*, 1969, **47**, 2879; D. W. Lepard, *ibid.*, 1970, **48**, 1664.

as abbreviations for complete sets of quantum numbers characterizing these states. As in the section on free atoms, we shall use *m*, *r*, *n* also as subscripts on special quantum numbers in order to identify the states to which they belong. In order to simplify the formulae it is desirable to restrict these quantum-number sets to as few quantum numbers as possible. In the vibrational RRE usually only the electronic quantum numbers (collected in the letter *e*) and the vibrational quantum numbers (collected in *v*) are retained, but it should be kept in mind that the elimination of the rest of the quantum numbers (translational, nuclear spin, rotational) is only possible under certain conditions and eventually leads to modifications (*e.g.* by introduction of statistical weight factors) of the scattering tensor. The chief condition is that no individual rotational, nuclear-spin, or even translational state appearing in (36) is favoured owing to sharp resonance excitation. Because of the insufficient monochromaticity (= relative large fractional linewidths) of optical rays there is hardly danger for resonance in the last two cases (this is the field for Mössbauer spectroscopy). However, the omission of the rotational quantum numbers needs vindication. Their elimination was discussed explicitly by Placzek for non-resonance Raman scattering (see Section 15 of ref. 9) and this procedure can be adopted also for vibrational RR scattering when sharp rotational resonance cannot occur. This is the case for sufficiently broad exciting radiation or for strongly broadened (and therefore numerously overlapping) rotational states, *e.g.* in liquids. On this occasion we mention that the recently developed very successful methods of investigating the orientational and vibrational relaxation of molecules in liquids by an analysis of their ordinary Raman vibrational band shapes are based on the polarizability theory and therefore also exclude rotational resonance.[116]

Thus under the above conditions the Raman tensor for the vibronic transition $(e_n,v_n) \leftarrow (e_m,v_m)$ through the intermediate states $(e_r,v_r) \equiv (e,v)$ can be written

$$\boldsymbol{\alpha}^{v_n v_m}_{e_n e_m} = \frac{1}{\hbar}\sum_e \sum_v \left(\frac{\boldsymbol{\mu}^{v\,v_n}_{e\,e_n}\,\boldsymbol{\mu}^{v_m v}_{e_m e}}{\omega^{v\,v_m}_{e\,e_m} - \omega_0 + \tfrac{1}{2}\mathrm{i}\Gamma^v_e} + \frac{\boldsymbol{\mu}^{v_m v}_{e_m e}\,\boldsymbol{\mu}^{v\,v_n}_{e\,e_n}}{\omega^{v\,v_n}_{e\,e_n} + \omega_0 + \tfrac{1}{2}\mathrm{i}\Gamma^v_e} \right) \qquad (154)$$

In normal pure vibrational Raman spectroscopy the initial and final electronic states coincide with the electronic ground state: $e_m = e_n = g$. In this case of 'scattering in the ground state' g the 'infrared term' with $e = g$ in (154) can be neglected when the exciting angular frequency ω_0 is larger than all nuclear vibrational angular frequencies $\omega^{vv_m}_{gg}$, $\omega^{vv_n}_{gg}$. Then (154) reads

$$\boldsymbol{\alpha}^{v_n v_m}_{g\ g} = \frac{1}{\hbar}\sum_{e \neq g} \sum_v \left(\frac{\boldsymbol{\mu}^{v\,v_n}_{eg}\,\boldsymbol{\mu}^{v_m v}_{g\ e}}{\omega^{v\,v_m}_{eg} - \omega_0 + \tfrac{1}{2}\mathrm{i}\Gamma^v_e} + \frac{\boldsymbol{\mu}^{v_m v}_{g\ e}\,\boldsymbol{\mu}^{v\,v_n}_{eg}}{\omega^{v\,v_n}_{eg} + \omega_0 + \tfrac{1}{2}\mathrm{i}\Gamma^v_e} \right) \qquad (155)$$

[116] *Cf.* A. V. Rakov, *Optika i Spektroskopiya*, 1959, **7**, 202; 1961, **10**, 713; 1962, **13**, 369; R. G. Gordon, *J. Chem. Phys.*, 1964, **40**, 1973; 1965, **42**, 3658; 1965, **43**, 1307; F. J. Bartoli and T. A. Litovitz, *ibid.*, 1972, **56**, 404, 413; J. H. R. Clarke and S. Miller, *Chem. Phys. Letters*, 1972, **13**, 97; S. Bratos and E. Maréchal, *Phys. Rev.* (*A*), 1971, **4**, 1078; E. Brindeau, S. Bratos, and J. C. Leicknam, *ibid.*, 1972, **6**, 2007; J. P. Perchard, W. F. Murphy, and H. J. Bernstein, *J. Mol. Phys.*, 1972, **3**, 499, 519, 535.

The vibrational quantum numbers v_m, v, v_n in these expressions in N_j-atomic molecules, $N_j > 2$, are themselves sets of quantum numbers referring to the individual normal vibrations. The conventional numbering of these normal vibrations for specific molecules can be seen from ref. 117, p. 580 ff. Equation (155) is the basis for almost all quantum mechanical theories of the vibrational RRE developed in recent years. [The classical theories and their relation to (155) have already been discussed in Section 3.] Expressions analogous to (155) can also be obtained for 'optically active' RR scattering using the higher M1 + E2 approximation (see Section 4E; *cf.* ref. 118).

The major difficulties in evaluating (155) are caused by the fact that the normal vibrations of a polyatomic molecule in the different electronic states can be strongly diverging or even missing. The electronic states can correspond to stable or unstable nuclear configurations. In the first case the nuclear equilibrium configurations in different electronic states in general will be different. It is even not unusual that a molecule, on electronic excitation, changes its symmetry. In the second case the 'repulsive' electronic state is associated with a continuous distribution of pseudo-vibrational levels. The definition of normal co-ordinates presupposes the existence of a potential minimum for a set of *finite* nuclear co-ordinates and moreover depends on the spatial symmetry of the potential function which, for its part, agrees with the symmetry of the nuclear configuration. These very formidable complications arising for the general vibrational RRE of polyatomic molecules can only be overcome by considering special cases and introducing convenient simplifications. First of all, separate treatment of the vibrational RRE by irradiation into a continuum and of that by irradiation into a discrete distribution of vibronic levels seems to be imperative. We note, however, that both effects can occur simultaneously (with certain probabilities) for the same molecule when discrete and continuous states exist side by side with energies equal or comparable to the incident-light quantum (*cf.* the cases of predissociation, Auger process, intersecting potential surfaces, *etc.*); the double sum in (155) then even in rigorous resonance keeps (at least) two terms, one of them (for the repulsive state) transformable in an integral as in Section 4C. Fortunately the RRE by irradiation into a continuum (being always a rigorous RRE) can be handled relatively easily (see Section 4C) although even then a knowledge of the pseudo-vibrational wavefunction(s) of the intermediate resonance state(s) is (are) necessary [*cf.* equation (145)]. Despite this separation of the continuum RRE there remain many problems for RR scattering which can only be solved gradually by attacking first simpler, special cases.

Diatomic molecules. A radical simplification can be achieved in a first approach to the general vibrational RR scattering problem by considering

[117] G. Herzberg, 'Molecular Spectra and Molecular Structure', Vol. III: 'Electronic Spectra and Electronic Structure of Polyatomic Molecules', Van Nostrand, New York, 1966.

[118] H. F. Hameka, *J. Chem. Phys.*, 1964, **41**, 3612.

only diatomic molecules. Diatomic molecules possess only one normal vibration and consequently only one normal co-ordinate Q. The quantum numbers v_m, v, v_n occurring in (155) are simple integers $\geqslant 0$. The energy dependence of the electronic states e on the nuclear distance is represented by potential curves instead of potential surfaces. It is important to note that nevertheless the problem of introducing normal co-ordinates must be solved independently for every electronic state. The normal co-ordinates Q_e of the individual electronic states e (including g) in general are not the same nor are the vibrational eigenfunctions $u_{ev}(Q_e)$ even in the harmonic approximation of the (non-repulsive) potential curves. An exact solution of this problem can be found in ref. 66, p. 93 ff. However, there is no difficulty in correlating different Q_e. In the harmonic approximation all Q_e can be expressed as linear functions of the ground-state normal co-ordinate $Q \equiv Q_g$, which then can be used as a universal nuclear co-ordinate. We desist here from repeating the well-known procedure of introducing in (155) the Born–Oppenheimer approximation for the molecular vibronic wavefunctions and of expanding the 'pure electronic transition moments' $\boldsymbol{\mu}_{eg}$ and $\boldsymbol{\mu}_{ge}$ with respect to the nuclear co-ordinate Q (see ref. 119, ref. 66, p. 95 ff., ref. 4, p. 299 ff., and references given therein). The expansion of $\boldsymbol{\mu}_{eg}$ with respect to the normal co-ordinate(s) was already proposed by Placzek [ref. 9, p. 367, equation (1)]. In the non-resonance case the denominators in (155) can be regarded as approximately independent of v and the sum over v can be extended over the numerators alone using Van Vleck's sum rule. The result is equivalent to Placzek's polarizability theory. In the pre-resonance case [*i.e.* when ω_0 for one or several electronic 'resonance' state(s) e approaches from below so closely to ω_{eg}^{00} that the difference of the first denominators in (155) with respect to varying v becomes marked, although still $\omega_0 < \omega_{eg}^{00}$] the variation of the respective first fraction(s) with v must be taken into account (the second fractions can usually be neglected). This is performed to a very good approximation by Shorygin's semi-classical theory, which can be justified by a strict quantum mechanical analysis. (For a short introduction to this theory, including older references, see ref. 1, p. 650 or ref. 2, p. 187; other, more recent Russian papers on this subject have also appeared.[4, 23, 38, 40, 120, 121]) Let us emphasize one very important point inherent in Shorygin's theory as far as the Raman fundamental is concerned. In this case the procedure applied by Shorygin for $v_m = 0$, $v_n = 1$ gives [see ref. 2, p. 190; in (155) only one non-degenerate resonance electronic state is assumed and all small terms and also the damping constants are omitted]:

119 J. Behringer, *Z. Elektrochem.*, 1958, **62**, 906.

120 L. L. Krushinsky and P. P. Shorygin, *Optika i Spektroskopiya*, 1961, **11**, 24, 151; *Izvest. Akad. Nauk S.S.S.R., Ser. fiz.*, 1963, **27**, 497.

121 P. P. Shorygin and L. L. Krushinsky, *Doklady Akad. Nauk S.S.S.R.*, 1960, **133**, 337; 1961, **136**, 577; 1964, **154**, 571; *Optika i Spektroskopiya*, 1964, **17**, 551; *Ber. Bunsengesellschaft phys. Chem.*, 1968, **72**, 495.

$$\alpha_{gg}^{10} = \left(\frac{\mathrm{d}}{\mathrm{d}Q}\,[f(Q)\mu_{eg}(Q)\mu_{ge}(Q)]\right)_0 Q_0$$

$$= \mu_{eg}(0)\mu_{ge}(0)\left(\frac{\mathrm{d}f}{\mathrm{d}Q}\right)_0 Q_0 + f(0)\left(\frac{\mathrm{d}}{\mathrm{d}Q}\,[\mu_{eg}(Q)\mu_{ge}(Q)]\right)_0 Q_0 \qquad (156)$$

with

$$Q_0 \equiv \langle g1|Q|g0\rangle = [\hbar/(2\omega_{gg}^{10})]^{\frac{1}{2}} \text{ (zero-point amplitude)} \qquad (157)$$

and

$$f(Q) \equiv \frac{2}{\hbar}\,\frac{\omega_{eg}(Q)}{\omega_{eg}^2(Q) - \omega_0^2} \qquad (158)$$

Equation (156) consists of two terms: the first one reflects the Q- (or v-) dependence of the resonance denominators and the second one that of the numerators in (155). Approaching resonance the importance of the first term becomes predominant, provided of course that $\mu_{eg}(0) \neq 0$, *i.e.* that the pure electronic dipole transiton $e \leftarrow g$ is allowed. This was already recognized by Placzek (ref. 9, p. 368), who wrote that the constant term of the expansion of μ_{eg} [with respect to the nuclear co-ordinate(s)] in the immediate proximity of resonance is responsible for the main part of the scattered radiation. It can immediately be seen from Figure 5 and confirmed by a simple calculation that for given shapes of the two potential curves $\left|\left(\frac{\mathrm{d}\omega_{eg}}{\mathrm{d}Q}\right)_0\right|$, and consequently also $\left|\left(\frac{\mathrm{d}f}{\mathrm{d}Q}\right)_0\right|$, increases with increasing slope —tan α of the upper potential curve at point R (*i.e.* vertically above the lower minimum) or, qualitatively in case of stable potential curves, with increasing distance $|b|$ of the potential minima. This rule can also readily be generalized for polyatomic molecules: provided that the pure electronic transition from the electronic ground state to the electronic state in near resonance is allowed, the slope of the upper potential surface roughly vertically above the lower minimum is all important for pre-resonance Raman intensities. We have already found this result by classical theory (Section 3A). If this is true, the higher terms of the expansion of μ_{eg} with respect to Q (or, in the case of polyatomic molecules, to all normal co-ordinates Q_l) only play a significant role if the pure electronic transition is forbidden. The reader not fully acquainted with Shorygin's theory can also see this in the following way (ref. 66, p. 99). Cancel from (155) all terms with relatively large denominators under near resonance conditions ($\omega_0 \rightarrow \omega_{eg}^{00}$). Expand $\mu_{eg}(Q)$ about the lower minimum ($Q = 0$). Then in zeroth approximation, with $\omega_{eg}^{v0} \approx \omega_{eg}^{v1}$,

$$\alpha_{gg}^{10} \approx \frac{1}{\hbar}\sum_e{}' \mu_{eg}(0)\mu_{ge}(0) \sum_v \frac{2\omega_{eg}^{v0}}{(\omega_{eg}^{v0})^2 - \omega_0^2}\,\langle ev|g1\rangle\langle g0|ev\rangle \qquad (159)$$

The prime on the sum over e indicates that the sum is extended only over the electronic states near resonance [in (156) and Figure 5 only one such state was assumed]. Since $\omega_0 \rightarrow \omega_{eg}^{v0}$, the fraction in (159) can also be approximated

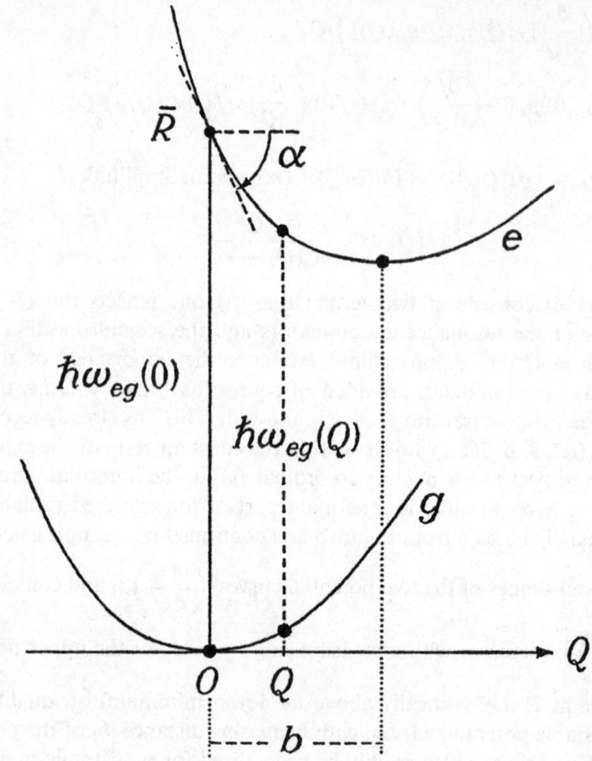

Figure 5

roughly by $1/(\omega_{eg}^{v0} - \omega_0)$ which corresponds to reducing (155) to the first terms only. Let us consider only one *e*-term. If the fraction in (159) were independent of *v*, the sum over *v* could be extended over the overlap integral product $\langle ev|g1\rangle\langle g0|ev\rangle$ alone, giving $\langle g0|g1\rangle$ by Van Vleck's sum rule. But this simple elimination of the influence of the *ev*-states is frustrated when under pre-resonance conditions the dependence of the fraction on *v* becomes marked. In the example shown in Figure 6, where only the terminal maxima of the vibrational eigenfunctions are drawn, the product $\langle ev|g1\rangle$ $\langle g0|ev\rangle$ for $v = 0, 1, 2$ becomes < 0 and for $v = 3, 4$ approximately ≈ 0, but >0 for $v = 5, 6, 7$. Now for increasing *v* the fraction in (159) gradually diminishes. Therefore the negative terms (with $v < 2$) in the sum outweigh the positive terms ($v > 5$) and it can easily be seen in a qualitative manner that this unbalance is enhanced when ω_0 approaches ω_{eg}^{00}. Moreover, it is clear that this effect is all the more pronounced the steeper the slope of the upper potential curve above the lower minimum. This again (for comparable potential shapes) depends on the inter-minimum distance *b*.

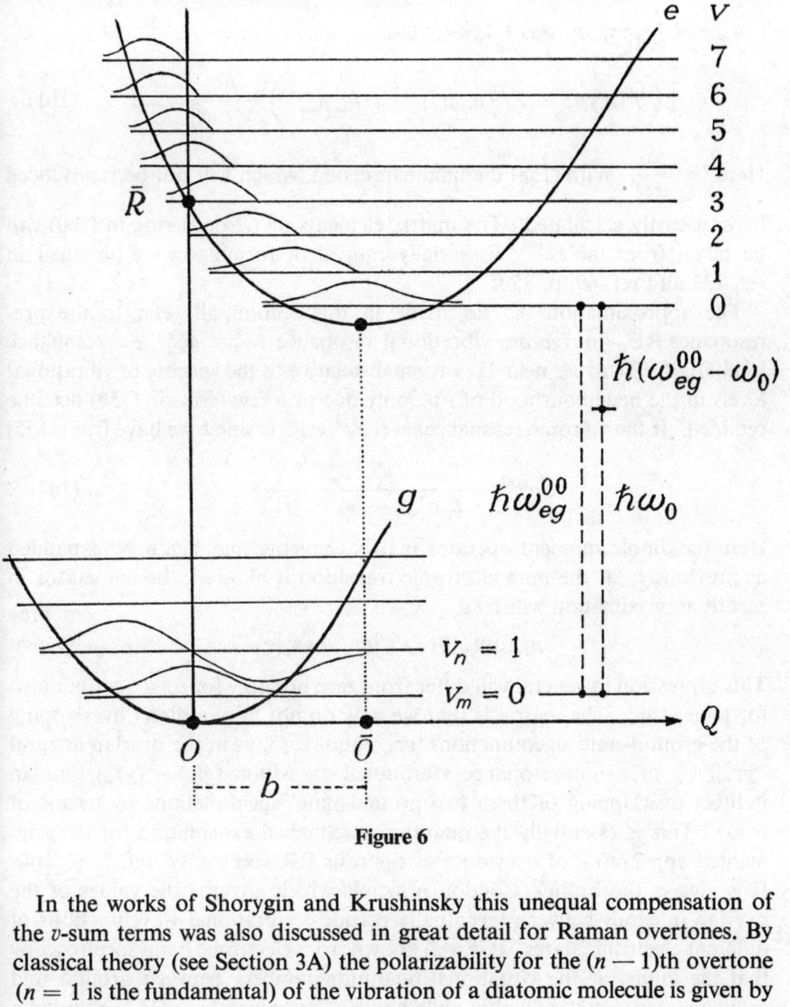

Figure 6

In the works of Shorygin and Krushinsky this unequal compensation of the v-sum terms was also discussed in great detail for Raman overtones. By classical theory (see Section 3A) the polarizability for the $(n - 1)$th overtone ($n = 1$ is the fundamental) of the vibration of a diatomic molecule is given by $\frac{1}{n!}\left(\frac{d^n\alpha}{dQ^n}\right)_0 Q^n$ [the additional factor $1/2^n$ in (11) arises from the factor $\frac{1}{2}$ in (6)]. Using (158) this can be translated at once into the quantum mechanical formalism [Stokes vibrational transition $(gv') \leftarrow (gv'')$, $v' > v''$]

$$\alpha_{gg}^{v'v''} = \frac{1}{(v' - v'')!}\left(\frac{d^{v'-v''}}{dQ^{v'-v''}}[f(Q)\mu_{eg}(Q)\mu_{ge}(Q)]\right)_0 \langle gv'|Q^n|gv''\rangle \quad (160)$$

E.g. for $v'v'' = 10$ (fundamental) (156) is obtained again and for $v'v'' = 20$ (first overtone) we have

$$\alpha_{gg}^{20} = \tfrac{1}{2}\left(\frac{d^2}{dQ^2}[f\mu_{eg}\,\mu_{ge}]\right)_0 \langle g2|Q^2|g0\rangle$$

$$= \tfrac{1}{2}\left(f''\mu_{eg}\,\mu_{ge} + 2f'(\mu_{eg}\,\mu_{ge})' + f(\mu_{eg}\,\mu_{ge})''\right)_0 \frac{\hbar}{\sqrt{2}\,\omega_{gg}^{10}} \qquad (161)$$

Here $' \equiv \frac{d}{dQ}$. With (158) the final expression, which will not be reproduced here, is easily calculated. The matrix elements of Q^n occurring in (160) can be taken from tables.[122] Essentially equivalent formulae were obtained in ref. 123 and ref. 69, p. 85 ff.

The approximations so far made in this section all refer to the pre-resonance RE. In *rigorous* vibrational resonance [$\omega_0 \approx \omega_{e'g}^{v'v_m}$, $e'v'$ resonance level(s)], if Γ_e^v (for E_{ev} near $E_{e'v'}$) is small relative to the spacing of vibrational levels in the neighbourhood of $e'v'$, only one or a few terms in (155) need be retained. If the vibronic resonance level $e'v'$, *e.g.*, is single we have from (155)

$$\alpha_{gg}^{v_n v_m} = \frac{1}{\hbar}\,\frac{\mu_{e'g}^{v'v_n}\,\mu_{ge'}^{v_m v'}}{\omega_{e'g}^{v'v_m} - \omega_0 + \tfrac{1}{2}i\Gamma_e^v} \qquad (162)$$

Here the dipole moment operator matrix elements may again be expanded as previously. If the pure electronic transition is allowed, the numerator in zeroth approximation will read

$$\mu_{e'g}(0)\mu_{ge'}(0)\,\langle e'v'|gv_n\rangle\langle gv_m|e'v'\rangle$$

This expression in general will differ from zero not only for $\Delta v = \pm 1$ but also for larger $|\Delta v|$. The reason is that we now do not have a direct overlapping of the ground-state eigenfunctions $|gv_m\rangle$ and $|gv_n\rangle$, as in the overlap integral $\langle gv_m|gv_n\rangle$ of a non-resonance vibrational transition $(gv_n) \leftarrow (gv_m)$, but an indirect overlapping of these two ground-state eigenfunctions by means of $|e'v'\rangle$. This is essentially the quantum mechanical explanation for the augmented appearance of overtones in rigorous RR spectra (*cf.* ref. 2, p. 214). It is always the Franck–Condon principle which governs the values of the overlap integrals between (eventually pseudo-) vibrational wavefunctions of different electronic states. It is well known from electronic band spectroscopy that the jumps of the vibrational quantum numbers between ground and excited states may vary within wide limits. Consequently in (162) even large jumps $\Delta v = v_n - v_m$ of the ground-state vibrational quantum number may be made possible by a favourably mediating resonance sublevel v' of the excited electronic state e'. This situation is essentially identical to that which occurs in the resonance spectra of fluorescence, although under the usual conditions of observation for these spectra, owing to the insignificant broadening of the rotational sublevels and with narrow exciting lines, usually only very few rotational lines (singlets, doublets, or superpositions) of a vibrational band

[122] *E.g.* E. B. Wilson, jun., J. C. Decius, and P. C. Cross, 'Molecular Vibrations', McGraw-Hill, New York, 1955, p. 290.
[123] M. Berjot, M. Jacon, and L. Bernard, *Opt. Comm.*, 1972, **5**, 94.

appear. A peculiar situation arises for RR spectra excited by irradiation into an absorption continuum. Here there are practically no opportunities for fluorescence, and the RRE is always rigorous because there is always a pseudovibronic level $e'v'$ bridging the transition $(gv_n) \leftarrow (gv_m)$. In a more precise sense, this is also true concerning rovibronic transitions. The consequence is that the vibrational RR band, just as in the ordinary RE, will display a completely developed rotational structure. There is no restriction to only a few rotational lines (singlets or doublets); rather we get all the branches allowed by the symmetry selection rules.

Table 2 *Depolarization ratios for diatomic molecules with Σ electronic ground state*

	Lin. polarized excitation	*Unpolarized excitation*
Total vibrational band	$\rho = \dfrac{3\beta^2}{45\alpha^2 + 4\beta^2}$	$\rho_n = \dfrac{6\beta^2}{45\alpha^2 + 7\beta^2}$
O,S-branch	$\rho_O = \rho_S = \dfrac{3}{4}$	$\rho_{nO} = \rho_{nS} = \dfrac{6}{7}$
Q-branch	$\rho_Q = \dfrac{3\beta^2}{180\alpha^2 + 4\beta^2}$	$\rho_{nQ} = \dfrac{6\beta^2}{180\alpha^2 + 7\beta^2}$

If this knowledge is used for the calculation of depolarization ratios, for diatomic molecules with Σ ground state the values given in Table 2 are found. These formulae are also valid for non-resonance excitation. z is the direction of the molecular axis. In the non-resonance case $\alpha \equiv \frac{1}{3}(2\alpha'_x + \alpha'_z)$ is the spherical part of the derived polarizability and $\beta \equiv \alpha'_z - \alpha'_x$ its anisotropy (*cf.* ref. 9, p. 359 ff.). In the resonance case this interpretation of α and β must be modified. Then α and β are the absolute squares of the transition moments (= dipole moment matrix elements) along or across the molecular axis, respectively.[124] The exact meaning of α and β can only be understood by considering the theoretical details. We only note here that in the case of resonance α and β take special values depending on the classification of the electronic resonance state e'. If e' is a Σ state, we have $\alpha = \frac{1}{3}\beta$; if it is a Π state, $\alpha = -\frac{2}{3}\beta$. With these special values of α and β the depolariza-

Table 3 *Diatomic molecules with Σ ground state*

Resonance state	Σ	Π
Lin. polarized excitation		
Total band ρ	$\frac{1}{3} = 0.333$	$\frac{1}{8} = 0.125$
O,S-branch $\rho_O = \rho_S$	$\frac{3}{4} = 0.75$	$\frac{3}{4} = 0.75$
Q-branch ρ_Q	$\frac{1}{8} = 0.125$	$\frac{1}{28} = 0.036$
Unpolarized excitation		
Total band ρ_n	$\frac{1}{2} = 0.5$	$\frac{2}{9} = 0.222$
O,S-branch $\rho_{nO} = \rho_{nS}$	$\frac{6}{7} = 0.857$	$\frac{6}{7} = 0.857$
Q-branch ρ_{nQ}	$\frac{2}{9} = 0.222$	$\frac{2}{29} = 0.069$

[124] C. Manneback, *Z. Physik*, 1930, **62**, 224 [equations (26) and (26′)].

tion ratios of Table 3 are found. The two values for total band ρ (with A_1, E instead of Σ, Π) were given in ref. 125. In the pre-resonance RE and always when several intermediate electronic states of different angular momenta contribute to scattering the depolarization ratio will be some average of these values. Incidentally the assumptions implicit in the values given in Tables 2 and 3 are general enough to admit application also to certain quasi-linear polyatomic molecules such as chain molecules with conjugated double bonds. If the exciting frequency approaches an electronic absorption band with a transition moment directed either along or across the molecular axis, the depolarization ratios ρ_n for a RR band are supposed to approach $\frac{1}{2}$ or $\frac{2}{9}$ respectively. Data confirming these limiting values are quoted, *e.g.* in ref. 2, p. 210.*

Polyatomic molecules. In linear (non-linear) polyatomic molecules or complex ions we have $3N_j - 5$ ($3N_j - 6$) normal vibrations l. This number of normal co-ordinates must be assumed for every electronic state e (including the ground state g). As already mentioned (p. 157), in different stable electronic states in general the nuclear equilibrium configurations and often even their symmetries are different. In the cases of Renner–Teller and Jahn–Teller degeneracy there may be several completely equivalent but different equilibrium configurations. A linear molecule may become non-linear when excited. Circumstances like these cause immense complications in defining and correlating the normal co-ordinates $Q_l^{(e)}$ of different e. However, it is clear that these problems must be solved before seriously attempting an analysis of RR spectra of polyatomic molecules. The example of diatomic molecules shows that the RRE (in particular in rigorous resonance) depends on the properties of the intermediate vibronic level(s). (This dependence of course guarantees a gain of information about excited electronic states by means of RR studies.) In order to describe these intermediate states properly, however, explicit knowledge of the vibrational wavefunctions of excited electronic states e in terms of their own normal co-ordinates $Q_l^{(e)}$ is indispensable. In the literature surprisingly little effort seems to have been spent on the exact solution of this normal co-ordinate problem in polyatomic molecules, which is basic also for calculating Franck–Condon factors in electronic band spectroscopy (*cf.* ref. 117, p. 142 ff.). Attempts to correlate normal co-ordinates of different electronic states were made in ref. 66, p. 117 ff. and ref. 126. In almost all other papers on the RRE in polyatomic molecules a coincidence of the normal co-ordinates of the electronic excited states with those of the ground state is tacitly assumed. This rather unrealistic assumption (which is, however, justifiable for non-resonance scattering) is for example made in a theory of Raman intensities developed by Albrecht and

* Ref. 2, p. 209, equation (32) contains an erratum: the last line (referring to $\rho = \frac{8}{11}$) should read $\alpha'_{xx} = \alpha'_{yy} = -\alpha'_{zz}$.

[125] O. S. Mortensen, *Chem. Phys. Letters*, 1969, **3**, 4.
[126] E. M. Verlan, *Optika i Spektroskopiya*, 1966, **20**, 605, 802, 916; T. E. Sharp and H. M. Rosenstock, *J. Chem. Phys.*, 1964, **41**, 3453.

Tang [127–130] which takes account of the vibronic coupling of different electronic states. Owing to space limitation it is not possible to discuss here this well-known theory in detail. We mention only some important points. In the original version of this theory [127] the expansion coefficients of the first-order terms in the $\boldsymbol{\mu}_{eg}(Q)$ [and its complex conjugate $\boldsymbol{\mu}_{ge}(Q)$] expansion with respect to the normal co-ordinates Q_l are transformed by means of the Herzberg–Teller [131] (*cf.* ref. 117, p. 166 ff.) perturbation theory for vibronic interactions. The transformation applied only takes account of the vibronic mixing of *e* (not *g*) with other electronic excited states *s*. The net result for the RRE is that (155) (with the second terms neglected) splits into two terms called A''' and B''', A''' [which contains only the zeroth-order approximation $\boldsymbol{\mu}_{eg}(\mathbf{0})$] being identical with (159) (apart from unessential details) and B''' showing the mixing of states *e* and *s*. Later [ref. 129, p. 37, equations (6) and (7)] the theory was amplified by introducing the vibronic mixing with excited states *t* also for *g*. This generates a third term *C* additional to A''' and B''', which now are called *A* and *B*. Summing up, the whole procedure consists only in transforming the numerators in the basic formula (155) but does not change anything in the denominators. Clearly this yields neither an advantage nor a disadvantage for a description of the (pre-)resonance RE which is connected with the variation of the denominators in (155) owing to ω_0 moving towards the vibronic absorption frequencies. Also, the appearance of additional factors $\omega_s - \omega_e$ or $\omega_t - \omega_g$ in the denominators of *B* or *C*, respectively, does not seem to improve the understanding of the RR mechanism because the corresponding terms in their numerators simultaneously contain both $\boldsymbol{\mu}_{gs}(\mathbf{0})$ *and* $\boldsymbol{\mu}_{ge}(\mathbf{0})$ (in *B*) or $\boldsymbol{\mu}_{gt}(\mathbf{0})$ *and* $\boldsymbol{\mu}_{ge}(\mathbf{0})$ (in *C*), respectively, as factors and consequently are only non-zero when *both* of these zeroth-order, purely electronic transition moments are non-zero. But then also in *A* [essentially equivalent to (159)] the terms with these matrix elements do not vanish and we come back to the question already discussed in connection with Figure 6, *i.e.* whether the invalidation of Van Vleck's sum rule due to the variation of the resonance denominators in *A* will give non-vanishing Raman ($v_n \neq v_m$) intensities despite their vanishing far from resonance (contrary to the Rayleigh band $v_n = v_m$). The merits of Albrecht and Tang's theory for *non-resonance* conditions will not be contested, but it is hard to understand the claim that 'as resonance is approached the intensity of the very vibration that is mixing the two electronic states should increase relative to all other Raman lines' (ref. 127, p. 1480). In Albrecht's term B''' [ref. 127, equation (15*b*)] the only quantity sensitive to a change of the exciting frequency ν_0 is $\nu_{ev,gi} - \nu_0 + i\gamma'_e$ [or in our notation $\frac{1}{2\pi}(\omega_{eg}^{vvm} - \omega_0 + \frac{1}{2}i\Gamma_e^v)$]. In this factor the

[127] A. C. Albrecht, *J. Chem. Phys.*, 1961, **34**, 1476.
[128] J. Tang and A. C. Albrecht, *J. Chem. Phys.*, 1968, **49**, 1144.
[129] J. Tang and A. C. Albrecht, 'Developments in the Theory of Vibrational Raman Intensities', in 'Raman Spectroscopy Theory and Practice', ed. H. A. Szymanski, Plenum Press, New York, 1970, Vol. 2, pp. 33—67.
[130] A. C. Albrecht and M. C. Hutley, *J. Chem. Phys.*, 1971, **55**, 4438.
[131] G. Herzberg and E. Teller, *Z. phys. Chem. (Leipzig)*, 1933, **B21**, 410.

vibration mixing e and s is in no wise distinguished from other vibrations. Incidentally, this term B''' of course is not absent in Shorygin's theory; it appears in addition to (159) when the first-order term in the expansion of $\mu_{eg}(Q)$ is also taken account of. Moreover, Shorygin's theory also does not exclude the possibility that more than one electronic state contributes to resonance scattering. In this case of course the much simplified version expressed by (156) and (160) of Shorygin's theory, which version specifically assumes the 'RR activity' of only one non-degenerate electronic state, must be given up and replaced by formulations taking account of two or more electronic resonance states. For the RRE the application of the Herzberg–Teller theory will doubtless be useful or even necessary if and only if (i) there are at least two such electronic resonance states and (ii) these states are vibronically perturbed. This does not, however, seem to be the normal case (*cf.* ref. 117, p. 66 ff.; see also below). Moreover, if there are indeed such vibronically perturbed electronic resonance states, a diminution of the energy of the lowest perturbed electronic state only by the perturbing vibration should show up in the resonance denominator containing the exciting frequency. It is only in this way that the intensity of the perturbing vibration can be shown to be preferentially enhanced in the pre-RRE. The benefit of the theory of Albrecht and Tang in its present form for the explanation of Raman intensities seems to be evident only in the non-resonance case when Van Vleck's sum rule becomes applicable. In Albrecht's [127] terms B, C (or B', B'', B''') the coefficient $\left(\frac{\partial \mu_{ge}}{\partial Q_l}\right)_0$ of the linear term in the expansion of $\mu_{ge}(Q)$ takes the explicit form $\sum_s \frac{h^l_{es}}{E^0_s - E^0_e}\mu_{gs}(0)$, which contains again the zeroth-order approximations $\mu_{gs}(0)$ of the purely electronic transition moments. Particularly when the individual Cartesian components of the scattering tensor are considered, this formulation should provide a good way of judging the influence of purely electronic transitions and particularly of their polarization on non-resonant Raman intensities.

Another way of taking account of vibronic coupling of electronic states in Raman spectroscopy is to start from third-order perturbation theory instead of the second-order formula (155) and to add the vibronic coupling Hamiltonian from the very beginning to the perturbing Hamiltonian $\hat{H}$ (126) (interaction radiation–molecule). This was done by Peticolas *et al.*[73] (*cf.* also ref. 132). The comments given on Albrecht's theory apply also here. We further note that the treatment of overtones in these theories, although possible in principle, would be almost hopelessly complicated in practice because second- and higher-order vibronic interaction terms must be used.

In the rigorous vibrational RRE the selection rules due to symmetry must be modified in a way which takes account of the influence of the intermediate

[132] L. A. Nafie, P. Stein, and W. L. Peticolas, *Chem. Phys. Letters*, 1971, **12**, 131.

resonance state(s) which can no longer be neglected (ref. 66, p. 104 ff.). This can be seen from (162), where both matrix elements in the numerator must be non-zero for the transition $(gv_n) \leftarrow (gv_m)$ not to be forbidden. By group theory then the totally symmetric representation Γ_0 of the molecular point group must be contained in the direct products $\Gamma(\boldsymbol{\mu}) \times \Gamma(e'v') \times \Gamma(gv_n)$ and (independently thereof) $\Gamma(\boldsymbol{\mu}) \times \Gamma(e'v') \times \Gamma(gv_m)$. From this it follows that totally symmetric fundamentals (in contrast to non-totally symmetric ones) are always allowed in RR scattering. However, this does not give any information about the intensities of allowed vibrations.

There are two fundamental problems arising from observation of Raman and RR scattering on polyatomic molecules for which a fully satisfactory solution is still lacking: (i) why do vibrational bands in one Raman spectrum (excited with a fixed frequency) show different intensities? and (ii) why do different vibrational bands in one Raman spectrum change their intensities differently (even if they belong to the same irreducible representation of the molecular point group) when the exciting frequency is shifted? These problems were tentatively attacked in ref. 66, pp. 114–124, but this author is conscious of the fact that much more work is necessary in order to obtain a quantitatively satisfactory explanation of the observed phenomena. It cannot be excluded that the idea of vibronic interactions among different electronic states offers some clues in certain cases. It is more likely, however, that all the intricacies existing for potential surfaces of polyatomic molecules which are described, *e.g.* in Herzberg's book,[117] must be taken into consideration. In particular the Renner–Teller and Jahn–Teller effects of potential deformation and level splitting together with the Franck–Condon principle will play an essential role in finding answers to the above questions (i) and (ii). Vibronic interactions seem to be much more important when degenerate excited electronic states are involved rather than different (also by symmetry) non-degenerate electronic states. There is no doubt that the slopes of the potential surfaces of the excited electronic states roughly vertically above the ground-state minimum and (in the case of stable excited states) the relative situation of the (single or multiple) minima of the excited- and ground-state surfaces in normal co-ordinate space are all important for RR intensities. This is a result of quantum mechanical as well as of classical theory. For certain vibrations these slopes and the potential minima are strongly influenced by the Renner–Teller or the Jahn–Teller effect. In ref. 133 and ref. 66, p. 115 some examples are discussed which qualitatively demonstrate the importance of the Jahn–Teller effect for Raman intensities.

Rotational Resonance Raman Effect. The term 'rotational RRE' means a RE excited by sufficiently monochromatic radiation whose frequency falls within the well-resolved rovibronic structure of the i.r., visible, or u.v. absorption spectrum of the molecule. There is also an i.r.-excited pure rota-

[133] I. I. Kondilenko and V. L. Strizhevsky, *Optika i Spektroskopiya*, 1964, **17**, 528.

tional ordinary or resonance RE. The rigorous RRE, in this case under the name of resonance radiation, is used in i.r. molecular laser physics.

Rotational RR spectroscopy has been concerned so far only with diatomic molecules. A diatomic molecule as a system of nuclei plus electrons is a symmetric top (*cf.* ref. 27, p. 115). For classifying the rotational levels three quantum numbers J, Λ (or K), and M are necessary. The case $\Lambda = 0$ is equivalent to the case of atoms (Section 5A). An explicit theory for the rotational RRE in diatomic molecules with a Σ electronic ground state and a Σ electronic intermediate (resonance) state was developed by Morozov.[102] Also the case of a Σ ground and a Π resonance state was studied.[60] More general cases do not seem to have been discussed in detail. A general theory of the rotational RRE in diatomic molecules would be very voluminous because of the numerous complexities existing for the rotational structure of their electronic bands.[27, 134] The Placzek–Teller coefficients $b^{JK}_{J'K'}$, (ref. 9, p. 338, ref. 96, p. 225) cannot be used immediately in rotational RR theory because in these coefficients, when the summation over the intermediate state J_r is performed, information is lost which is needed for sharp resonance with a definite J_r state.

To the knowledge of this author there have been no systematic investigations of the rotational RRE on polyatomic molecules either in theory or in experiment. The elaboration of a theory of this effect would be straightforward because in comparison with Section 5A (atoms) and the statements above concerning diatomic molecules there is nothing essentially new except the much greater complexity of rovibronic states found in spherical, symmetric-, or asymmetric-top molecules (*cf.* ref. 117, Chaps. I.3.4. and II.3.4, ref. 6, p. 244 ff.). It must be doubted, however, whether it would be worthwhile to work out a rotational RR theory for polyatomic molecules because the experimental situation (with the exception of a few small molecules) is rather discouraging. Sponer and Teller [135] wrote in 1941: '. . . the rotational structure in spectra of polyatomic molecules does not have the same great practical importance as the rotational structure for diatomic molecules. With the exception of rather simple molecules the resolution of rotational structure is all but impossible.' The hope expressed by these authors in the immediately following text that better instrumentation will allow better resolution and 'eventually yield valuable information' can be repeated to-day for rotational RR spectroscopy.

C. Condensed Systems.—Explicit theories of RR scattering in condensed systems have been developed only for crystals, which show the greatest order in the arrangement of nuclei and electrons, so that relatively uncomplicated methods suffice. In non-crystalline condensed systems the disturbed spatial

[134] J. E. Wollrab, 'Rotational Spectra and Molecular Structure', Academic Press, New York, 1967; I. Kovacs, 'Rotational Structure in the Spectra of Diatomic Molecules', Hilger, London, 1969.

[135] H. Sponer and E. Teller, *Rev. Mod. Phys.*, 1941, **13**, 75.

and temporal order will considerably influence light scattering and it is only by statistical methods that the different types of disorder can be studied quantitatively. Several authors have discussed this topic.[136] We shall only very briefly review RR theories on crystals. The importance and length of the material would justify an extra Report.

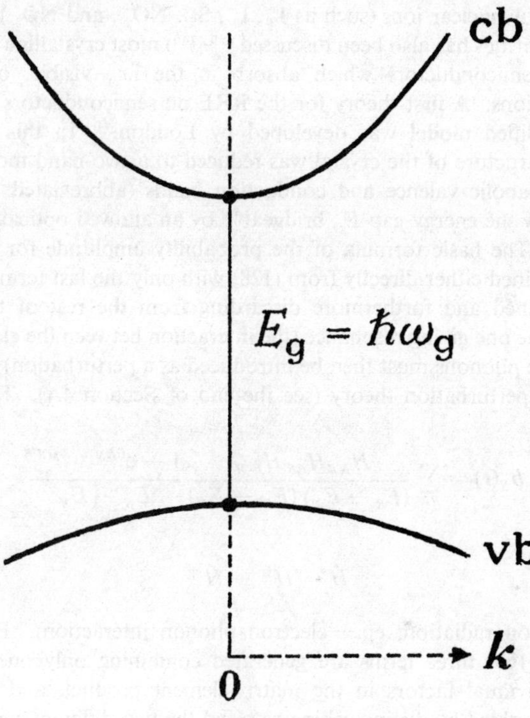

Figure 7

Theories on ordinary Raman scattering of crystals are as old as Raman spectroscopy itself but it is only since the advent of lasers that the most important basic quantum mechanical processes could be revealed under experimental control (*cf.* ref. 59, p. 366 ff., ref. 71, ref. 137, p. 201 ff., 367 ff.

[136] B. J. Berne, 'Time-dependent Properties of Condensed Media', in 'Physical Chemistry', ed. H. Eyring, D. Henderson, and W. Jost, Academic Press, New York, 1971, Vol. VIIIB; F. Kohler, 'The Liquid State', Verlag Chemie, Weinheim, 1972; N. F. Mott and E. A. Davis, 'Electronic Processes in Non-Crystalline Materials', Clarendon Press, Oxford, 1971; H. E. Stanley, 'Introduction to Phase Transitions and Critical Phenomena', Clarendon Press, Oxford, 1971.

[137] M. Born and K. Huang, 'Dynamical Theory of Crystal Lattices', Clarendon Press, Oxford, 1962.

and ref. 138). The objects of ordinary RE studies on crystals are lattice vibrations (one- and two-phonon scattering) including a number of peculiar crystal phenomena such as polaritons, anharmonic phonon-coupling, crystal defects and impurities, localized modes, excitons, plasmons, and magnons. Apart from impressive RREs observed on crystals with *F*-centres and doped with mononuclear (see the section on atoms influenced by fields) and especially polynuclear ions (such as I_2^-, I_3^-, S_2^-, NO_2^-, and NO_3^-) (the RE on crystal impurities has also been discussed [139, 140]) most crystalline RREs were found in semiconductors which absorb in the i.r., visible, or near-u.v. spectral regions. A first theory for the RRE on semiconductors based on a much simplified model was developed by Loudon.[71] In this theory the electronic structure of the crystal was reduced to a two-band model (Figure 7) with parabolic valence and conduction bands (abbreviated vb and cb) separated by the energy gap E_g, bridgeable by an allowed optical absorption transition. The basic formula of the probability amplitude for this theory can be obtained either directly from (128) with only the last term in the sum over *R* retained and furthermore discarding from the rest of the sum all terms but the one giving resonance (the interaction between the electron–hole pair and the phonons must then be introduced as a perturbation) or by using third-order perturbation theory (see the end of Section 4A). The formula then reads

$$b_N(t) = \sum_{RR'} \frac{H_{NR}H_{RR'}H_{R'M}}{(E_R - E_M)(E_{R'} - E_M)} \frac{1 - e^{(iE_N - E_M)t/\hbar}}{E_N - E_M} \tag{163}$$

where now

$$\hat{H} = \hat{H}^{\text{er}} + \hat{H}^{\text{ep}} \tag{164}$$

(er = electron–radiation, ep = electron–phonon interaction). By inserting (164) into (163) three terms are generated containing only one of $\hat{H}^{\text{er}}$ or $\hat{H}^{\text{ep}}$ as individual factors in the matrix-element product, and these three terms are doubled by distinguishing as usual the two different successions of (virtual) absorption and emission. In the decisive resonance term (corresponding to a hole in the vb and an electron in the cb), which contains the fraction

$$\frac{H^{\text{er}}_{NR}H^{\text{ep}}_{RR'}H^{\text{er}}_{R'M}}{(E_R - E_M)(E_{R'} - E_M)} \tag{165}$$

138 J.-P. Mathieu, 'Spectres de Vibration et Symétrie des Molécules et des Cristaux', Hermann, Paris, 1945; H. Poulet and J.-P. Mathieu, 'Spectres de Vibration et Symétrie des Cristaux', Gordon & Breach, Paris–London, 1970; G. Turrell, 'Infrared and Raman Spectra of Crystals', Academic Press, New York, 1972; R. A. Cowley, 'The Theory of Raman Scattering from Crystals', in 'The Raman Effect', ed. A. Anderson, Dekker, New York, 1971, Vol. I, pp. 95—181.

139 *Cf.* K. K. Rebane, 'Impurity Spectra of Solids', Plenum Press, New York, 1970, p. 163ff.

140 *Cf.* 'Light Scattering in Solids', ed. M. Balkanski (Conference, Paris, July 1971), Flammarion Sciences, Paris, 1971, pp. 57—79.

the energy differences

$$E_R - E_M = \hbar(\omega_g + \frac{\hbar k^2}{2\mu} - \omega_\alpha + \omega_k), E_{R'} - E_M = \hbar(\omega_g + \frac{\hbar k^2}{2\mu} - \omega_\alpha) \quad (166)$$

($\hbar\omega_g$ is the energy gap; $\boldsymbol{k}$ and $\hbar\omega_k$ the phonon wave vector and energy respectively, μ the reduced effective mass, and $\hbar\omega_\alpha$ and $\hbar\omega_\varepsilon = \hbar(\omega_\alpha - \omega_k)$ the exciting and scattered photon energies respectively) are inserted. Integration over $\boldsymbol{k}$ (the matrix elements are considered as independent on $\boldsymbol{k}$) following Loudon gives a proportionality of $b_N(t)$ to

$$(\omega_g + \omega_k - \omega_\alpha)^{\frac{1}{2}} - (\omega_g - \omega_\alpha)^{\frac{1}{2}} \quad (167)$$

The scattered intensity then is proportional to the absolute square of this expression. When ω_α is increased starting from small values, the intensity first (for $\omega_\alpha < \omega_g$) increases, reaching a sharp maximum at $\omega_\alpha = \omega_g$, and then suddenly falls to low values. If the concurring absorption losses of the scattered radiation in a practical experiment are taken account of, it is seen that the most favourable exciting frequency yielding an optimum scattering intensity for observation lies some distance in the long-wave direction from the absolute maximum of (167).

Loudon himself pointed to the deficiencies of this model, indicating that in real crystals the absorptivity at the flank of the gap absorption can be influenced by exciton effects. In fact subsequently developed theories taking account of exciton states and also of the interaction with short-range (deformation) and long-range (Fröhlich) fields could considerably improve the agreement of theoretical predictions with experimental results. It is clear that long-range Coulomb fields will play an important role for the RE in non-centrosymmetric (piezoelectric) semi-conducting crystals. These theories being far too involved to be discussed comprehensively in a short Report, the reader must be referred to the original literature.[141] This must also be the case for another group of theories considering also the direct interaction of radiation with transverse optical phonons or with discrete excitons (polariton model) and/or with plasmons (ref. 140, p. 19 ff., ref. 142).

Exceptionally under certain conditions (TO-modes) an intensity decrease or even cancellation under resonance conditions was observed ('anti-

[141] A. K. Ganguly and J. L. Birman, *Phys. Rev. Letters*, 1966, **17**, 647; *Phys. Rev.*, 1967, **162,** 806; P. J. Colwell and M. V. Klein, *Solid State Comm.*, 1970, **8,** 2095; C. Mavroyannis, *J. Math. Phys.*, 1967, **8,** 1515, 1522; J. F. Scott, R. C. C. Leite, and T. C. Damen, *Phys. Rev.*, 1969, **188,** 1285; E. Mulazzi, *Phys. Rev. Letters*, 1970, **25,** 228.

[142] E. Burstein, D. L. Mills, A. Pinczuk, and S. Ushioda, *Phys. Rev. Letters*, 1969, **22,** 348; D. L. Mills and E. Burstein, *Phys. Rev.*, 1969, **188,** 1465; J. J. Hopfield, *ibid.*, 1969, **182,** 945; B. Bendow and J. L. Birman, 'Theory of Interaction of Light with Insulating Crystals', in 'Light Scattering Spectra of Solids', ed. G. B. Wright, Springer, Berlin, 1969, p. 381; B. Bendow, J. L. Birman, A. K. Ganguly, T. C. Damen, R. C. C. Leite, and J. F. Scott, *Opt. Comm.*, 1970, **1,** 267; B. Bendow and J. L. Birman, *Phys. Rev.* (*B*), 1970, **1,** 1678; 1971, **4,** 569; B. Bendow, *ibid.*, 1970, **2,** 5051; 1971, **4,** 552; R. Zeyher, J. L. Birman, and W. Brenig, *ibid.*, 1972, **6,** 4613.

resonance'). Theoretically this is explained by 'interference effects' of opposite-sign contributions to the scattering tensor.[143]

A particularly interesting and probably prolific subject for future research is RR scattering induced by strong electric [144] and magnetic fields. In view of the large amount of knowledge about magneto-optical properties of solids [145] it seems somewhat surprising that only one solid-state magnetic RRE has been reported.[146]

The author would like to thank Prof. D. A. Long, Bradford, for kindly revising the English text and Prof. J. Brandmüller, Dr. H.-W. Schrötter, and Dr. G. Strey, Munich, for useful discussions and for help in providing the literature.

[143] J. M. Ralston, R. L. Wadsack, and R. K. Chang, *Phys. Rev. Letters*, 1970, **25**, 815; J. L. Lewis, R. C. Wadsack, and R. K. Chang, ref. 140, p. 41; J. Ruvalds and A. Zawadowski, ref. 140, p. 29.
[144] J. F. Gay, J. D. Dow, E. Burstein, and A. Pinczuk, ref. 140, p. 33.
[145] *Cf.*, for example, J. G. Mavroides, 'Magneto-Optical Properties', in 'Optical Properties of Solids', ed. F. Abelès, North-Holland, Amsterdam, 1972, 351—528.
[146] J. F. Scott, 'Double Spin-Flip Resonant Raman Effect', in 'Advances in Raman Spectroscopy', ed. J.-P. Mathieu, Heyden and Son, London, 1973, Vol. 1, p. 353.

3

Infrared and Raman Studies of Molecular Motion

BY R. T. BAILEY

1 Introduction

A variety of experimental techniques may be used to probe the dynamics of molecular motion in solids, liquids, and gases. Lineshape studies in dielectric relaxation,[1] electronic,[2] and infrared absorption and Raman scattering [3, 4] all yield detailed information on the dynamics of molecular collisions. Other techniques which yield similar information include studies of the shape of the spectral distribution function obtained from light- and neutron-scattering experiments,[5–7] spin relaxation,[8] and acoustic attenuation measurements.[9]

I.r. and Raman bandshapes contain considerable information on molecular motion in condensed systems. Information on the intermolecular forces and torques which govern the reorientational motion as well as the motion itself may be recovered from the spectra. In the conventional Schrödinger picture attention is focused on the energy levels of the system rather than on its time development. Molecules in the gas phase, at low pressures, undergo almost free rotation between collisions, and information about the rotational energy levels is contained in the resolvable fine structure. In high-pressure gases, liquids, and many solids, however, the intermolecular potentials between neighbouring molecules damp the rotational and translational motion, resulting in a broadening of the transitions and a loss of discrete rotational fine structure.

In the Heisenberg picture, attention is focused instead on the time development of the system rather than on its quantum states. In this way, information about the nature of rotational and translational motion can be extracted from the broadened contour of a pure rotational or rotational–vibrational band. In addition, a classical correspondence exists which may be exploited for those systems which approach classical behaviour.

1 T. A. Litovitz and D. Settle, *J. Chem. Phys.*, 1953, **21**, 17.
2 Y. Toyazawa, *Prog. Theor. Phys.* (*Kyoto*), 1958, **53**, 20.
3 G. Ewing, *Accounts Chem. Res.*, 1969, **2**, 168.
4 R. G. Breene, 'The Shift and Shape of Spectral Lines', Pergamon Press, New York, 1961.
5 A. Sjolander, in 'Phonons and Phonon Interactions', ed. T. A. Bak, Benjamin, New York, 1964.
6 P. A. Egelstaff, 'Thermal Neutron Scattering', Academic Press, New York, 1965.
7 D. Pines and P. Nozieres, 'The Theory of Quantum Liquids', Benjamin, New York, 1966.
8 S. Abragam, 'The Principles of Nuclear Magnetism', Oxford, London, 1961.
9 K. F. Herzfeld and T. A. Litovitz, 'Absorption and Dispersion of Ultrasonic Waves', Academic Press, New York, 1959.

In the Heisenberg approach, the spectrum is considered as the Fourier transform of an appropiate time correlation function. Inversion of this relationship gives the correlation function as a Fourier integral over a complete experimental frequency spectrum. This allows the short-time and the long-time behaviour of the correlation function to be treated separately. The short-time behaviour can be discussed fairly rigorously in terms of the dynamics of the many-molecule system, whereas statistical arguments are usually employed to establish the form of the correlation function at long times.

In this Report, an account will be given of the application of i.r. and Raman bandshape studies to molecular motion. Emphasis will be placed on the time correlation function approach which has developed rapidly since Gordon and others first drew attention to the advantages for molecular spectroscopy of the Heisenberg approach. More conventional bandshape studies will also be considered, however, where appropriate.

Molecular-motion studies obviously cover a very wide field. In this Report attention will be focused primarily on reorientational motions, such as rotational diffusion and librational motions. Vibrational motions will only be considered in the context of their interaction with reorientational motions and their contribution to the observed band profile. Lattice phonons in molecular crystals will not be considered here. Molecular motions in polymeric molecules will also be excluded.

2 Background Molecular Theory

Correlation Function Approach.—According to linear response theory, whenever two systems are weakly coupled, for instance when radiation is weakly coupled to matter or molecular vibrations are weakly coupled to molecular motion, it is only necessary to know how both systems behave in the absence of coupling in order to deduce the way in which one system responds to the other. The response of one system to another can be completely described in terms of time correlation functions of the appropriate dynamical properties.

Time-dependent correlation functions have been well known in some fields since the early work of Bloembergen *et al.*[8] In recent years, they have found increasing application in many areas of statistical physics and spectroscopy. Correlation functions provide a concise method of expressing the degree to which two dynamical properties are correlated as a function of time. In other words, they describe quantitatively how long some property of the system persists until it is averaged out by molecular motion in the system.

The correlation function $C_{\alpha\beta}(t)$ for two dynamical properties α and β is defined as

$$C_{\alpha\beta}(t) = \langle \alpha(0) \cdot \beta(t) \rangle_0$$

When α and β are different properties of the system, $C_{\alpha\beta}$ is called a cross-correlation function, and when they are identical, it is called an auto-correlation function. The time dependence is that produced by the natural molecular motion in the system. The average $\langle\ \rangle_0$ is over an ensemble of systems at the reference time 0. Normally, this ensemble is a canonical Boltzmann distribution, characteristic of systems in thermal equilibrium. When, however, the system is subject to some special preparation or constraint, such as in fluorescence measurements, this information must be built into the ensemble. For the purpose of this Report, only auto-correlation functions, defined as

$$C(t) = \langle a(0) \cdot a(t) \rangle_0$$

will be considered. The dynamical quantities $a(t)$ are usually defined in such a way that the ensemble average of a is 0. Then if the system is ergodic (*i.e.* if time averages are the same as averages over the initial ensemble), $C(t)$ will approach zero as the time becomes large (as $t \rightarrow \infty$).

Because the response of the system to a weak, specific probe is related directly to a particular correlation function, many experiments have been devised to determine specific correlation functions. Only i.r. and Raman correlation functions will be considered in detail in this Report. Information on other specific correlation functions can be found in a number of excellent reviews.[10–18] The shape of vibration–rotation bands in i.r. absorption and Raman-scattering experiments [19] on linear molecules can be used to determine the auto-correlation functions $\langle \boldsymbol{u}(0) \cdot \boldsymbol{u}(t) \rangle$ and $\langle P_2[\boldsymbol{u}(0) \cdot \boldsymbol{u}(t)] \rangle$, where $\boldsymbol{u}$ is a unit vector pointing along the molecular axis and $P_2(x)$ is the Legendre polynomial of index 2. These correlation functions measure the rate of reorientational motion of the molecules in a particular molecular environment and have the advantage over other techniques (such as n.m.r.) that the form of $C(t)$ is obtained, and not just the correlation time.

Orientational motion in a molecular system occurs in response to interaction with the molecular environment and so can be used to probe the overall structural dynamics occurring on the same characteristic time-scale (10^{-10}—10^{-13} s). Reorientational motion is determined largely by the nature of the interactions between the molecule and its immediate environment. For instance, in the case of strong interactions orientational motions will be slow compared with those of a freely rotating molecule and rotational

10 R. Kubo, 'Lectures in Theoretical Physics', Interscience, New York, 1961, Vol. 1, p. 120.
11 R. G. Gordon, *Adv. Magn. Resonance*, 1968, **3**, 1.
12 R. Zwanzig, *Ann. Rev. Phys. Chem.*, 1965, **16**, 67.
13 R. Kubo, *J. Phys. Soc. Japan*, 1957, **12**, 570.
14 R. D. Mountain, 'Critical Reviews, Solid State Sciences', Chemical Rubber Co., Cleveland, Ohio, 1970, Vol. 1, p. 5.
15 B. J. Berne and G. D. Harp, *Adv. Chem. Phys.*, 1970, **17**, 63.
16 G. D. Harp and B. J. Berne, *Phys. Rev.* (*A*), 1970, **2**, 975.
17 J. M. Deutch and I. Oppenheim, *Adv. Magn. Resonance*, 1968, **3**, 43.
18 G. Williams, *Chem. Rev.*, 1972, **72**, 55.
19 R. G. Gordon, *J. Chem. Phys.*, 1965, **43**, 1307.

diffusion probably occurs by a jump mechanism. Where the interactions are weak, however, rotational diffusion may occur by small-angle Debye-type motions.

Theoretical work has developed rapidly since the early work of Gordon and of Shimizu.[20] Three general groups of theories which relate spectral bandshapes to reorientational correlation functions can be distinguished. In the first group of theories following Gordon,[11] the basic Fourier relationship between the spectral density and the appropriate correlation function is established. No assumptions regarding the motion in the molecular system are made. In the second group of theories following Sobelman [21] random orientational motions are assumed to represent the basic broadening mechanism. Detailed information on the nature of the molecular motion may be obtained by these methods but a suitable model must be established in this case. The third group of theories involves the calculation of some simple properties of the spectral bandshapes such as their frequency moments.[22–24] Information regarding intermolecular forces and torques can be recovered from the spectra using these techniques.

Much of the recent theoretical work has been aimed at extending the original theories, which were limited essentially to diatomic and linear molecules, to more complicated molecular systems. It is therefore appropriate to review briefly the relationship between the spectral bandshape and the correlation function derived by Gordon and then to introduce the subsequent developments in a logical fashion. Methods of relating the correlation function to the detailed molecular motion in the system will then be discussed.

Infrared Absorption. If we consider a group of interacting molecules in a quantum state described by a vector $| i\rangle$ and we induce transitions to other quantum states $|f\rangle$ by irradiation with light at frequency ω, the probability per unit time of a transition is given by [11, 19]

$$P_{f\leftarrow i}(\omega) = \frac{\pi}{2\hbar^2} | \langle f|\boldsymbol{E}_0 \cdot \boldsymbol{u}|i\rangle |^2 \{\delta(\omega_{fi} - \omega) + \delta(\omega_{fi} + \omega)\} \qquad (1)$$

here E_0 is the amplitude of the electric field in the radiation and $\boldsymbol{u}$ is the total dipole moment operator for molecules in the system. Usually, one measures the rate of energy loss $-\dot{E}_{\rm rad}$ in an absorption experiment. This is given by

$$\begin{aligned} -\dot{E}_{\rm rad} &= \Sigma\hbar\omega_{fi} P_{f\leftarrow i} \\ &= \frac{\pi}{2\hbar} \sum_f \sum_i \omega_{fi}\, \rho_i\, | \langle f|\boldsymbol{E}_0 \cdot \boldsymbol{u}|i\rangle |^2 \{\delta(\omega_{fi} - \omega) + \delta(\omega_{fi} + \omega)\} \qquad (2) \end{aligned}$$

[20] H. Shimizu, *J. Chem. Phys.*, 1965, **43**, 2453.
[21] I. I. Sobelman, *Izvest. Akad. Nauk S.S.S.R., Ser. fiz.*, 1953, **57**, 554.
[22] R. G. Gordon, *J. Chem. Phys.*, 1963, **38**, 1724.
[23] R. G. Gordon, *J. Chem. Phys.*, 1963, **38**, 2788.
[24] R. G. Gordon, *J. Chem. Phys.*, 1964, **41**, 1819.

where ρ_i is the probability of finding the molecules in the ith state in the initial ensemble. Since the sums i and f run over all the quantum states of the molecules, we may interchange these indices in the terms coming from the second δ function, giving

$$-\dot{E}_{\mathrm{rad}} = \left(\frac{\pi}{2\hbar}\right) \sum_f \sum_i \omega_{fi}(\rho_i - \rho_f) |\langle f|\boldsymbol{E}_0 \cdot \boldsymbol{u}|i\rangle|^2 \,\delta(\omega_{fi} - \omega) \tag{3}$$

If the initial ensemble of states is Boltzman, then

$$\rho_f = \rho_i \exp(-\hbar\omega_{fi}/kT)$$

and (3) becomes

$$-\dot{E}_{\mathrm{rad}} = \left(\frac{\pi}{2\hbar}\right)[1 - \exp(-\hbar\omega/kT)] \sum_i \sum_f \rho_i |\langle f|\boldsymbol{E}_0 \cdot \boldsymbol{u}|i\rangle|^2 \,\delta(\omega_{fi} - \omega) \tag{4}$$

where ω_{fi} has been replaced by ω because of the energy-conserving δ functions. The imaginary part of the dielectric constant $\varepsilon''(\omega)$ is obtained by dividing by ω and the average density in the radiation field,

$$\bar{E}_{\mathrm{rad}} = E_0^2/8\pi \tag{5}$$

so that

$$\varepsilon''(\omega) = -\dot{E}_{\mathrm{rad}}/\omega\bar{E}_{\mathrm{rad}} \tag{6}$$

$$= \left(\frac{4\pi^2}{\hbar}\right)\{1 - \exp(-\hbar\omega/kT)\} \sum_f \sum_i \rho_i |\langle f|\hat{\epsilon} \cdot \boldsymbol{u}|i\rangle|^2 \,\delta(\omega_{fi} - \omega) \tag{7}$$

where $\hat{\epsilon}$ is a unit vector along the electric field of the radiation. Thus we obtain the usual expression for an absorption bandshape in terms of transitions between the quantum states (Bohr–Schrödinger), *i.e.*

$$I(\omega) = \frac{3\hbar\varepsilon''(\omega)}{4\pi^2[1 - \exp(-\hbar\omega/kT)]} = 3\sum_i \sum_f |\langle f|\hat{\epsilon} \cdot \boldsymbol{u}|i\rangle|^2 \,\delta(\omega_{fi} - \omega) \tag{8}$$

The transformation of this expression into the Heisenberg picture involves representing the δ function by its Fourier integral,

$$\delta(\omega) = \frac{1}{2\pi}\int_{-\infty}^{+\infty} e^{i\omega t}\,dt \tag{9}$$

giving

$$I(\omega) = \frac{3}{2\pi}\sum_{if} \rho_i \langle i|\hat{\epsilon} \cdot \boldsymbol{u}|f\rangle\langle f|\hat{\epsilon} \cdot \boldsymbol{u}|i\rangle$$

$$\times \int_{-\infty}^{+\infty} dt \exp i\{(E_f - E_i)/(\hbar - \omega)\}t \tag{10}$$

The energy eigenvalues E_f and E_i may be expressed in terms of the Hamiltonian operator H for the rotational–translational motion, giving

$$I(\omega) = \frac{3}{2\pi}\int_{-\infty}^{+\infty} dt\, e^{-i\omega t} \sum_{if} \rho_i \langle i|\hat{\epsilon} \cdot \boldsymbol{u}(0)|f\rangle\langle f|\hat{\epsilon} \cdot \boldsymbol{u}(t) i\rangle \tag{11}$$

where the time-dependent Heisenberg operator $\boldsymbol{u}(t)$ has been used for the dipole at time t, defined by

$$\boldsymbol{u}(t) = \mathrm{e}^{\mathrm{i}Ht/\hbar}\,\boldsymbol{u}(0)\,\mathrm{e}^{-\mathrm{i}Ht/\hbar} \tag{12}$$

Performing the sum over the complete set of final states $|f\rangle$ we have,

$$I(\omega) = \frac{3}{2\pi}\int_{-\infty}^{+\infty} \mathrm{d}t\,\mathrm{e}^{-\mathrm{i}\omega t}\,\Sigma\,\rho_i\langle i|\hat{\epsilon}\cdot\boldsymbol{u}(0)\hat{\epsilon}\cdot\boldsymbol{u}(t)|i\rangle \tag{13}$$

The sum over initial states i, weighted by ρ_i is simply an equilibrium average, $\langle\ \rangle_0$:

$$I(\omega) = \frac{3}{2\pi}\int_{-\infty}^{+\infty} \mathrm{d}t\,\mathrm{e}^{-\mathrm{i}\omega t}\,\langle\hat{\epsilon}\cdot\boldsymbol{u}(0)\;\hat{\epsilon}\cdot\boldsymbol{u}(t)\rangle_0 \tag{14}$$

If the system is isotropic, *i.e.* gas, liquid, or glass, the same result is obtained for any direction of polarization $\hat{\epsilon}$, so that

$$I(\omega) = \frac{1}{2\pi}\int_{-\infty}^{+\infty} \mathrm{d}t\,\mathrm{e}^{-\mathrm{i}\omega t}\,\langle\boldsymbol{u}(0)\cdot\boldsymbol{u}(t)\rangle_0 \tag{15}$$

Thus in the Heisenberg picture the spectral distribution is given by the Fourier transform of a correlation function of the dipole moment operator of the absorbing molecules. In general, the dipole moment of the whole group of interacting molecules must be considered. Thus in every product of $\boldsymbol{u}(0)\cdot\boldsymbol{u}(t)$ there will be cross-terms between the dipole moments of different molecules $\boldsymbol{u}_i(0)\cdot\boldsymbol{u}_j(t)$ and the total dipole moment will be a sum of molecular dipole moments. Because of these cross-terms the dipole correlation function cannot be interpreted simply in terms of the reorientation of a single molecule when the concentration of dipolar molecules is high.

Often, however, in the case of rotation–vibration bands, it is a good approximation to consider the internal vibrational motion of the molecules as separable from the rotational and translational motion. In this approximation, a vibrational excitation is localized on a single molecule, and the correlation function may be considered to be that of the vibrational transition dipole moment of a typical molecule. In this case, the angular frequency ω is measured as a displacement from the shifted vibrational frequency, so taking into consideration the oscillations of the transition vibration dipole, which may then be considered to change with time only with the orientational motion of the molecule. A correlation function of this form describes the average projection of a molecule's vibrational transition moment on to the direction the dipole had a time t earlier.

The dipole correlation function can be obtained from a rotation–vibration bandshape by Fourier inversion, that is

$$\langle\boldsymbol{u}(0)\cdot\boldsymbol{u}(t)\rangle = \int_{\text{band}} \mathrm{d}\omega\,\mathrm{e}^{\mathrm{i}\omega t}\,I(\omega) \tag{16}$$

Usually, it is convenient to normalize the observed spectrum according to

$$\hat{I}(\omega) = I(\omega) \Big/ \int_{\text{band}} I(\omega)\,\mathrm{d}\omega \tag{17}$$

so that

$$\langle \boldsymbol{u}(0) \cdot \boldsymbol{u}(t) \rangle = \int_{\text{band}} \mathrm{d}\omega\, \hat{I}(\omega) \exp(\mathrm{i}\omega t) \tag{18}$$

where $\boldsymbol{u}(t)$ is a unit vector along the direction of the transition dipole. The use of a normalized spectrum largely eliminates dielectric effects on the local electric field due to the radiation. Only fluctuations above the average local field will effect the normalized spectrum. Vibrational perturbations (frequency shifts) were neglected in this treatment. If necessary, these effects can be included by allowing the intermolecular potential energy for a vibrationally excited molecule to differ from that of a molecule in its ground state. However, when frequency shifts are significant, the simple physical interpretation of the correlation function is obscured. This difficulty is reduced by removing the mean frequency shift from the vibrational band origin; then only fluctuations of the frequency shift distort the bandshape. Generally the effects of frequency shifts are more marked for larger and heavier molecules. Vibrational relaxation is also neglected in Gordon's treatment since the final vibrational state was assumed to be sharp, and the density of final states was written in terms of a delta function expressing energy conservation between the initial and final states of the system.

Recently, Nafie and Peticolas [25] have extended their treatment of vibrational Raman line broadening to i.r. absorption. Their Raman treatment will be discussed in the following section. For a randomly oriented isotropic system they obtained an expression for the intensity distribution of an i.r. absorption band, $I(\omega)$

$$I(\omega) = \frac{B}{2\pi} \int_{-\infty}^{+\infty} \tfrac{1}{3} \langle \boldsymbol{m}^{\mathrm{v}}(0) \cdot \boldsymbol{m}^{\mathrm{v}}(t) \rangle_{\mathrm{tr}} \times \langle \hat{Q}^{\mathrm{v}}(0)\, \hat{Q}^{\mathrm{v}}(t) \rangle_{\mathrm{vib}} \exp(\mathrm{i}\omega t)\,\mathrm{d}t \tag{19}$$

where B is a constant. On normalizing we obtain

$$I(\omega) = \frac{1}{2\pi} \int_{-\infty}^{+\infty} \langle \boldsymbol{u}(0) \cdot \boldsymbol{u}(t) \rangle_{\mathrm{tr}} \times \langle \hat{Q}^{\mathrm{v}}(0)\, \hat{Q}^{\mathrm{v}}(t) \rangle_{\mathrm{vib}} \exp(\mathrm{i}\omega t)\,\mathrm{d}t \tag{20}$$

where $\boldsymbol{u}$ is a unit vector along the molecular axis which is parallel to the i.r. transition moment for the normal co-ordinate of vibration Q. In this expression, the contributions from orientational and vibrational relaxation are separated. The vibrational relaxation correlation function has the same form in both i.r. and Raman spectra but the orientational contributions are

[25] L. A. Nafie and W. L. Peticolas, *J. Chem. Phys.*, 1972, **57**, 3145.

different. An *exact* separation of the orientational contribution from the vibrational relaxation correlation function is possible only for the i.r.-active totally symmetric modes $C_n(A)$ and $C_{n\nu}(A)$ where one has also obtained $\langle \hat{Q}^{\nu}(0) \cdot \hat{Q}^{\nu}(t) \rangle$ from analysis of the corresponding Raman line. If the assumption is made, however, that all the vibrational modes of a molecule have the same vibrational relaxation correlation function, the separation can be achieved using

$$\langle \boldsymbol{u}(0) \cdot \boldsymbol{u}(t) \rangle_{\text{tr}} = \int_{-\infty}^{+\infty} \hat{I}_{\text{ir}}(\omega) \exp(-i\omega t)\, d\omega \Big/ \int_{-\infty}^{+\infty} \hat{I}_{\text{vib}}(\omega) \exp(-i\omega t)\, d\omega \qquad (21)$$

This assumption should hold to better than an order of magnitude in most cases for all the vibrational relaxation correlation functions of a given molecule. In addition to contributions from this source, larger molecules have increasingly greater contributions from vibrational relaxation owing to a large number of vibrational manifolds and smaller contributions from molecular reorientation owing to larger moments of inertia.

The effect of vibrational relaxation on the decay of the dipole correlation function was investigated by Morawitz and Eisenthal.[26] In their theory, the coupling between the i.r. vibrational mode and translational–rotational degrees of freedom was treated. Many-body perturbation theory was used to account for collision-induced de-excitation of the vibration, which leads to a frequency shift and line broadening. The additional time dependence of the dipole correlation function was a damped exponential resulting in a modified lineshape. The inclusion of a damping term also changed the interpretation of the first few terms in the short-time expansion of the correlation function, which are related to the moment of inertia and the mean square torque.

Raman Scattering. Following Placzek,[27] the polarizability formula for non-resonant Raman scattering is

$$\bar{\lambda}^4 \frac{d\sigma}{d\Omega} = \sum_i \rho_i \sum_f | \langle i | \boldsymbol{\epsilon}^0 \cdot \boldsymbol{a}^{\nu} \cdot \boldsymbol{\epsilon}^s | f \rangle |^2 \, \delta(\omega^s - \omega^0 + \omega_\nu + \omega_f - \omega_i) \qquad (22)$$

where $2\pi\bar{\lambda}$ is the wavelength of the scattered light, $d\sigma/d\Omega$ is the differential light scattering cross-section per molecule, for the transition from the ground vibrational state of a molecule to the vibrational state ν with vibrational energy $h\omega^{\nu}$, $\boldsymbol{a}^{\nu}$ is the off-diagonal matrix element of the polarizability tensor between the ground vibrational state of the molecule and the vibrational state ν, ω^0 is the frequency of the incident photon with its electric vector polarized along the unit vector $\boldsymbol{\epsilon}^0$ and ω^s and $\boldsymbol{\epsilon}^s$ are the analogous quantities for the scattered photon. This expression is analogous to the corresponding relationship for the absorption coefficient [equation (8)] and we employ similar arguments to the i.r. absorption case.

26 H. Morawitz and K. B. Eisenthal, *J. Chem. Phys.*, 1971, **55**, 887.
27 G. Placzek, *Z. Physik*, 1931, **70**, 84.

Again thermal equilibrium is assumed before each scattering process:

$$\rho_i = \exp(-\hbar\omega_i/kT) \Big/ \sum_i \exp(-\hbar\omega_i/kT) \tag{23}$$

Conservation of energy in the scattering act requires that

$$\omega^0 + \omega_i = \omega^s + \omega_v + \omega_f \tag{24}$$

Using the usual Fourier representation of a delta function [equation (9)] we obtain

$$\lambda^4 \frac{d\sigma}{d\Omega} = \frac{1}{2\pi} \sum_i \rho_i \sum_f \langle i|\boldsymbol{\epsilon}^0 \cdot \boldsymbol{\alpha}^v \cdot \boldsymbol{\epsilon}^s|f\rangle\langle f|\boldsymbol{\epsilon}^0 \cdot \boldsymbol{\alpha}^v \cdot \boldsymbol{\epsilon}^s|i\rangle$$
$$\times \int_{-\infty}^{+\infty} \exp[i(\omega^s - \omega^0 + \omega_v)t] \exp(i\omega_f t) \exp(-i\omega_i t)\, dt \tag{25}$$

Expressing the rotation–translation energies $h\omega_i$ and $h\omega_f$ as eigenvalues of the Hamiltonian acting on states $|i\rangle$ and $|f\rangle$ and summing over the complete set of states $|f\rangle$, we arrive at

$$\lambda^4 \frac{d\sigma}{d\Omega} = \frac{1}{2\pi} \int_{-\infty}^{+\infty} \langle[\boldsymbol{\epsilon}^0 \cdot \boldsymbol{\alpha}^v(0) \cdot \boldsymbol{\epsilon}^s][\boldsymbol{\epsilon}^0 \cdot \boldsymbol{\alpha}^v(t) \cdot \boldsymbol{\epsilon}^s]\rangle \exp(-i\omega t)\, dt \tag{26}$$

where ω is the frequency displacement from the centre of the band, *i.e.* $\omega = \omega^0 - \omega^s - \omega_v$. The equilibrium statistical average is denoted by $\langle\ \rangle$. If this average is interpreted as a classical average, one regains the classical description of light scattering.

For spatially isotropic systems, and for scattering where the incident and scattered electric vectors are perpendicular, the cross-section [equation (26)] may be written in the form [28, 29]

$$\lambda^4 \left.\frac{d\sigma}{d\Omega}\right|_\perp = \frac{1}{60\pi} \int_{-\infty}^{+\infty} \langle 3\mathrm{Tr}\, \boldsymbol{\alpha}^v(0) \cdot \boldsymbol{\alpha}^v(t)$$
$$- [\mathrm{Tr}\, \boldsymbol{\alpha}^v(0)][\mathrm{Tr}\, \boldsymbol{\alpha}^v(t)]\rangle \exp(-i\omega t)\, dt \tag{27}$$

where the traces Tr are taken over the spatial indices of $\boldsymbol{\alpha}^v$. This can be simplified by separating the transition polarizability $\boldsymbol{\alpha}^v$, into its spherical ($\bar{\alpha}^v$) and traceless ($\boldsymbol{\beta}^v$) parts,

$$i.e. \quad \boldsymbol{\alpha}^v = \bar{\alpha} + \boldsymbol{\beta}^v; \quad \mathrm{Tr}\, \boldsymbol{\beta}^v = 0$$

$$\lambda^4 \left.\frac{d\sigma}{d\Omega}\right|_\perp = \frac{1}{2\pi} \int_{-\infty}^{+\infty} \tfrac{1}{10} \langle \mathrm{Tr}\, \boldsymbol{\beta}^v(0) \cdot \boldsymbol{\beta}^v(t)\rangle \exp(-i\omega t)\, dt \tag{28}$$

The Fourier inversion of this gives the correlation function in terms of the scattering cross-section:

$$\tfrac{1}{10} \langle \mathrm{Tr}\, \boldsymbol{\beta}^v(0) \cdot \boldsymbol{\beta}^v(t)\rangle = \int_{-\infty}^{+\infty} \lambda^4 \left.\frac{d\sigma}{d\Omega}\right|_\perp \exp(i\omega t)\, d\omega \tag{29}$$

[28] R. G. Gordon, *J. Chem. Phys.*, 1964, **40**, 1973.
[29] R. G. Gordon, *J. Chem. Phys.*, 1965, **42**, 3658.

By normalizing the intensity distribution $I(\omega)$ to unit integrated band intensity,

$$I(\omega) = \lambda^4 \left.\frac{d\sigma}{d\Omega}\right|_{\perp} \Big/ \int_{-\infty}^{+\infty} \lambda^4 \left.\frac{d\sigma}{d\Omega}\right|_{\perp} d\omega \qquad (30)$$

a normalized correlation function $\hat{C}(t)$ is obtained which decays from unity at $t = 0$:

$$\hat{C}(t) = \frac{\langle \mathrm{Tr}\, \boldsymbol{\beta}^{\nu}(0) \cdot \boldsymbol{\beta}^{\nu}(t) \rangle}{\langle \mathrm{Tr}\, \boldsymbol{\beta}^{\nu}(0) \cdot \boldsymbol{\beta}^{\nu}(0) \rangle} \qquad (31)$$

Equations (25) and (26) then become

$$\hat{I}(\omega) = \frac{1}{2\pi} \int_{-\infty}^{+\infty} \hat{C}(t) \exp(-i\omega t)\, dt \qquad (32)$$

and

$$\hat{C}(t) = \int_{-\infty}^{+\infty} I(\omega) \exp(i\omega t)\, d\omega \qquad (33)$$

In the specific case of a totally symmetric vibration in a linear molecule or a symmetric top, the anisotropic part of the transition polarizability has the form

$$\beta_{ij} = \text{const} \times (\mu_i^z \mu_j^z - \tfrac{1}{3}\delta_{ij}) \qquad (34)$$

where μ_i^z is the ith component of a unit vector fixed along the symmetry axis (z) of the molecule and δ_{ij} is the Kronecker delta. Thus,

$$\begin{aligned}\mathrm{Tr}\, \boldsymbol{\beta}(0) \cdot \boldsymbol{\beta}(t) &\propto \sum_{i,j=1}^{3} [\mu_i^z(0)\, \mu_j^z(0) - \tfrac{1}{3}\delta_{ij}][\mu_i^z(t)\, \mu_j^z(t) - \tfrac{1}{3}\delta_{ij}] \\ &= \sum_{i,j=1}^{3} [\mu_i^z(0)\, \mu_j^z(0)\, \mu_i^z(t)\, \mu_j^z(t) - \tfrac{1}{3}] \end{aligned} \qquad (35)$$

and the normalized correlation function [equation (31)] becomes

$$\hat{C}(t) = \tfrac{1}{2} \langle 3 \sum_{i,j=1}^{3} \mu_i^z(0)\, \mu_j^z(0)\, \mu_i^z(t)\, \mu_j^z(t) - 1 \rangle \qquad (36)$$

In the classical limit, $\mu(0)$ and $\mu(t)$ commute, so that we have

$$\hat{C}(t) = \langle P_2[\boldsymbol{u}^z(0) \cdot \boldsymbol{u}^z(t)] \rangle \qquad (37)$$

where $P_2(x) = \frac{1}{2}(3x^2 - 1)$

The same expression is obtained for a degenerate vibration of a tetrahedral molecule,

$$\hat{C}(t) = \langle P_2[\boldsymbol{u}^{(2)}(0) \cdot \boldsymbol{u}^{(2)}(t)] \rangle \qquad (38)$$

where $\boldsymbol{u}^{(2)}$ is a unit vector along any of the twofold axes.[22] In the absence of Coriolis coupling, the triply degenerate vibrations of a tetrahedral molecule also have the same correlation function. Because of interactions between the degenerate components, however, these vibrations usually exhibit a large Coriolis effect.

In this treatment,[19, 28, 29] the final vibrational state was assumed to be sharp, thus neglecting the effects of vibrational relaxation, which may be significant, especially for larger molecules. Several authors have subsequently included vibrational relaxation in their treatments of Raman bandshapes. Both Bartoli and Litovitz [30, 31] and Nafie and Peticolas [25] have used a somewhat similar approach in their treatments of vibrational Raman bandshapes. Basically, whereas Bartoli and Litovitz used the semi-classical expression for the expansion of the electronic polarizability in terms of the normal co-ordinates,

$$\alpha_{ij}(t) = \alpha_{ij}^0(t) + \sum_v [\partial\alpha_{ij}(t)/\partial Q^v(t)]_{Q=0} \quad Q^v(t) + \ldots \tag{39}$$

Nafie and Peticolas [25] used the corresponding quantum mechanical expression for Raman intensities which they had derived previously.[32] Also, the former authors [30, 31] used the characteristic rotation matrices $D_{mn}^J(\Omega)$ to perform the averaging whereas the latter workers employed the direction cosine method.

The relevant expressions derived for the parallel and perpendicular scattered Raman intensities were

$$I_{\|}(\omega) = A(\omega_1 - \Omega_v)^4 \frac{1}{2\pi} \int_{-\infty}^{+\infty} \langle(\alpha^v)^2 + \tfrac{2}{15}\,\mathrm{Tr}\,[\boldsymbol{\beta}^v(0)\cdot\boldsymbol{\beta}^v(t)]_{\mathrm{tr}}\rangle \times \langle Q^v(0)\cdot Q^v(t)\rangle_{\mathrm{vib}} \exp(i\omega t)\,\mathrm{d}t \tag{40}$$

and

$$I_{\perp}(\omega) = A(\omega_1 - \Omega_v)^4 \frac{1}{2\pi} \int_{-\infty}^{+\infty} \langle\tfrac{1}{10}\,\mathrm{Tr}\,[\boldsymbol{\beta}^v(0)\cdot\boldsymbol{\beta}^v(t)]_{\mathrm{tr}}\rangle \times \langle Q^v(0)\cdot Q^v(t)\rangle_{\mathrm{vib}} \exp(i\omega t)\,\mathrm{d}t \tag{41}$$

where ω_1 is the incident and ω_2 the scattered photon frequency and Ω_v the frequency of the vth normal mode. The other quantities are in their usual notation. In order to separate the translational–rotational average from the vibrational average we combine (40) and (41) and write

$$I_{\|}(\omega) - \tfrac{4}{3} I_{\perp}(\omega) = A(\omega_1 - \Omega_v)^4 \frac{(\alpha^v)^2}{2\pi} \int_{-\infty}^{+\infty} \langle Q^v(0)\cdot Q^v(t)\rangle_{\mathrm{vib}} \times \exp(i\omega t)\,\mathrm{d}t \tag{42}$$

Normalizing this expression,

$$\hat{I}_{\mathrm{vib}}(\omega) \equiv I_{\|}(\omega) - \tfrac{4}{3} I_{\perp}(\omega) \Big/ \int_{-\infty}^{+\infty} [I_{\|}(\omega) - \tfrac{4}{3} I_{\perp}(\omega)]\,\mathrm{d}\omega$$

$$= (2\pi)^{-1} \int_{-\infty}^{+\infty} \langle Q^v(0)\cdot Q^v(t)\rangle_{\mathrm{vib}} \exp(i\omega t)\,\mathrm{d}t \Big/ \langle Q^v(0)\cdot Q^v(0)\rangle_{\mathrm{vib}}$$

$$= (2\pi)^{-1} \int_{-\infty}^{+\infty} \langle \hat{Q}^v(0)\cdot \hat{Q}^v(t)\rangle_{\mathrm{vib}} \exp(i\omega t)\,\mathrm{d}t \tag{43}$$

30 F. J. Bartoli and T. A. Litovitz, *J. Chem. Phys.*, 1972, **56**, 404.
31 F. J. Bartoli and T. A. Litovitz, *J. Chem. Phys.*, 1972, **56**, 413.
32 W. L. Peticolas, L. Nafie, P. Stein, and B. Fanconi, *J. Chem. Phys.*, 1970, **52**, 1576.

The corresponding expression for $I_{\perp}(\omega)$ can be similarly written:

$$\hat{I}_{\perp}(\omega) \equiv I_{\perp}(\omega) \Big/ \int_{-\infty}^{+\infty} I_{\perp}(\omega)\, d\omega$$
$$= (2\pi)^{-1} \int_{-\infty}^{+\infty} \langle \mathrm{Tr}\, \boldsymbol{\beta}^{v}(0) \cdot \boldsymbol{\beta}^{v}(t) \rangle_{tr} \langle \hat{Q}^{v}(0) \cdot \hat{Q}^{v}(t) \rangle_{vib} \times \exp(i\omega t)\, dt \qquad (44)$$

Fourier transformation of these expressions then yields

$$\langle \hat{Q}^{v}(0) \cdot \hat{Q}^{v}(t) \rangle_{vib} = \int_{-\infty}^{+\infty} \hat{I}_{vib}(\omega) \exp(-i\omega t)\, d\omega \qquad (45)$$

$$\langle \mathrm{Tr}\, \boldsymbol{\beta}^{v}(0) \cdot \boldsymbol{\beta}^{v}(t) \rangle_{tr} \langle \hat{Q}^{v}(0) \cdot \hat{Q}^{v}(t) \rangle_{vib} = \int_{-\infty}^{+\infty} \hat{I}_{\perp}(\omega) \exp(-i\omega t)\, d\omega \qquad (46)$$

and finally

$$\langle \mathrm{Tr}\, \boldsymbol{\beta}^{v}(0) \cdot \boldsymbol{\beta}^{v}(t) \rangle_{tr} = \int_{-\infty}^{+\infty} \hat{I}_{\perp}(\omega) \exp(-i\omega t)\, d\omega \Big/ \int_{-\infty}^{+\infty} \hat{I}_{vib}(\omega) \times \exp(-i\omega t)\, d\omega \qquad (47)$$

Equations (45) and (47) represent the correlation functions associated solely with vibrational relaxation and reorientation of the anisotropy. This separation depends upon the trace of the tensor $\boldsymbol{\alpha}^{v}$ being non-zero and is therefore only strictly applicable to vibrational modes belonging to the totally symmetric representations. Assuming, however, that the vibrational correlation function is the same for all the modes in a given molecule, separation may be accomplished using (47), where $\hat{I}_{\perp}(\omega)$ is obtained from any Raman band and $\hat{I}_{vib}(\omega)$ is obtained from a totally symmetric band.

Nafie and Peticolas [25] also catalogued the formulae for all the orientation correlation functions for most of the common irreducible representations of molecular point groups. These results were previously only available for totally symmetric vibrations of symmetric-top and linear molecules, and for the doubly degenerate mode of spherical-top molecules. These authors [25] also extended their treatment of Raman band contours to include i.r. lineshapes and hyper-Raman scattering. For the same normal mode, the vibrational correlation function [equation (45)] has the same form in Raman, hyper-Raman, and i.r. spectroscopy.

The Calculation of Correlation Functions.—The relationship between correlation functions and spectral lineshapes in i.r. absorption and Raman scattering discussed in the previous section furnishes only the formal framework for a comparison between experiment and theory. The evaluation of the relevant correlation function and spectral densities from a molecular description of the system is much more difficult. Two general types of approach have been applied to the theoretical evaluation of correlation functions, a dynamical approach and a stochastic technique. The quantum mechanical or classical equations of motion are solved directly in the first approach. Binary collisions in a gas can be treated in this way, using either classical or quantum methods, but in liquids only classical equations are tractable. In the stochastic approach, the difficult quantum expressions are replaced by stochastic equations such as a diffusion equation, Langevin equation, or Fokker–Planck equations. These stochastic equations have fewer variables and are usually easier to handle than the mechanical expressions. Both approaches have been employed in the computation of theoretical correlation functions.

It is appropriate at this point to consider some of the basic properties of the correlation functions

$$\hat{C}(t) = \langle \alpha(0) \cdot \alpha(t) \rangle_0$$

In quantum mechanics $\alpha(0)$ and $\alpha(t)$ do not generally commute, so that a quantum mechanical correlation function is a complex quantity, with its real part an even function of time and its imaginary part an odd function of time. The real and imaginary parts are related by the expression

$$\text{Im } C(t) = -\tan\left[\left(\frac{\hbar}{2kT}\right)\frac{\partial}{\partial t}\right] \text{Re } C(t) \tag{48}$$

For the nearly classical case $\left(\frac{\hbar}{2kT} \text{ small}\right)$ we have

$$\text{Im } C(t) \approx -\frac{\hbar}{2kT}\frac{\partial}{\partial t} \text{Re } C(t) \tag{49}$$

Thus Im $C(t)$ is only a small quantum mechanical term which goes to zero in the classical limit ($\hbar \to 0$).

In considering theoretical models of orientational correlation functions, it is sometimes convenient to separate the short- and long-time behaviour of the molecular system. This separation occurs naturally in the Fourier analysis of a band contour. Since the classical correlation function is a real, even function of time, it may be approximated at short times by an even power series in time: [11, 19]

$$\langle \alpha(0) \cdot \alpha(t) \rangle = \sum_{n=0}^{\infty} \left(\frac{t^n}{n!}\right)\left[\frac{d^n}{dt^n} \langle \alpha(0) \cdot \alpha(t) \rangle_{t=0}\right] \tag{50}$$

The time derivatives can be evaluated by repeated application of the Heisenberg equation of motion for $\alpha(t)$,

$$\frac{d\alpha}{dt} = \left(\frac{i}{\hbar}\right)[H, \alpha] \tag{51}$$

giving the general expression

$$\langle \alpha(0) \cdot \alpha(t) \rangle = \sum_{n=0}^{\infty} \left[\frac{(it)^n}{n!}\right] \hbar^{-n} \langle \alpha(0)[H,[H,[H \ldots [H,\alpha(0)] \ldots]]] \rangle \tag{52}$$

The factors governing the convergence of this power-series expansion have been discussed by Gordon.[11] He showed that the coefficients in this time series may be identified with the frequency moments of the spectrum.[24, 28] Remembering the relationship

$$C(t) = \int_{-\infty}^{+\infty} I(\omega) \exp(i\omega t)\, d\omega \tag{53}$$

which relates $C(t)$ to an integral over the spectral density, expanding the exponential factor in a power series, and interchanging the order of summation and integration, we obtain

$$\begin{aligned} C(t) &= \sum_{k=0}^{\infty} \frac{(it)^k}{k!} \int_{-\infty}^{+\infty} \omega^k I(\omega)\, d\omega \\ &= \sum_{k=0}^{\infty} \frac{(it)^k}{k!} M(k) \end{aligned} \tag{54}$$

where $i = \sqrt{(-1)}$ and the $M(k)$ are the frequency moments of the spectrum, defined by

$$M(k) = \int_{-\infty}^{+\infty} \omega^k I(\omega)\, d\omega \tag{55}$$

Thus the frequency moments are the coefficients of a power-series expansion in time for $C(t)$, and hence are time derivatives of $C(t)$ evaluated at $t = 0$:

$$M(k) = (-i)^k \left[\frac{d^k C(t)}{dt^k}\right]_{t=0} \tag{56}$$

For a classical system only even values of k are considered. Normally the power series [equation (54)] converges slowly except at short times, so it is not directly useful for evaluating spectral densities. Each coefficient in the time series is an equilibrium property of the molecular system and can be evaluated without solving any equations of motion. The evaluation of these coefficients will be discussed in the section on moment analysis. As an example, for the dipole correlation function for a parallel transition of a classical linear rotor,[23, 24, 28] we have

$$\begin{aligned} M(2) &= \frac{2kT}{I} \\ M(4) &= [8(kT)^2 + \langle(\boldsymbol{O}V)^2\rangle]/I^2 \end{aligned} \tag{57}$$

where $-\boldsymbol{O}V$ is the torque on a molecule due to its neighbours. Thus the power-series expansion begins

$$\begin{aligned} \langle \boldsymbol{u}(0) \cdot \boldsymbol{u}(t) \rangle = 1 &- \left(\frac{kT}{I}\right) t^2 \\ &+ \left[\frac{1}{3}\left(\frac{kT}{I}\right)^2 + \frac{1}{24I^2}\langle(\boldsymbol{O}V)^2\rangle\right] t^4 + O(t^6) \end{aligned} \tag{58}$$

Quantum corrections will add odd powers of t to this expansion. Similarly, the moments of a Raman spectrum may be used to compute the correlation function,

$$\langle P_2[u(0) \cdot u(t)] \rangle = 1 - \left(\frac{3kT}{I}\right) t^2 + \left[4\left(\frac{kT}{I}\right)^2 + \frac{1}{8I^2}\langle (OV)^2 \rangle\right] t^4 + O(t^6) \quad (59)$$

for a classical linear molecule. These expansions apply to solids, liquids, and gases provided the appropriate equilibrium averages $\langle (OV)^2 \rangle$ are performed.

The initial curvatures (second moments) of these correlation functions depend only on the temperature of the system and the molecular moment of inertia. This occurs because the second moment is entirely a kinetic energy effect which in the classical limit is determined by the temperature alone. Thus the intermolecular forces have no effect on the initial decay of $C(t)$. The intermolecular torques come into play with the t^4 term and have the effect of decreasing the decay of $C(t)$ with time. Thus the orientational correlation function for interacting molecules initially will lie above that for free molecules.

The classical invariance of the second moment provides a method of checking if the spectral distribution was measured over a sufficiently wide range of frequencies. If $M(2)$ is far below its classical value the measurements may not have been carried sufficiently far into the wings of the band. An experimental second moment which considerably exceeds the classical value, however, suggests that the fluctuations of the vibrational frequency shift are a significant broadening feature in the spectrum. The theoretical second moment forms the basis of a method for correcting the observed spectral bandshape which will be discussed in the experimental section.

The lower moments of the frequency spectrum are essentially simple properties of the system primarily because they describe the short-time behaviour of the system. The motion at longer times, however, is more complicated and requires many higher moments in the power series. These higher moments are more complex equilibrium properties and become more difficult to evaluate as one goes to higher powers of t. Thus, even if the power series converges at longer times computational difficulties limit the usefulness of the expansion. Alternative techniques must therefore be used for evaluating $C(t)$ at longer times. One approach outlined by Gordon [33] for calculating spectral densities is based on the rigorous equations of motion. No attempt is made, however, to solve these equations of motion directly; instead, a spectral density is regarded as some function of unknown form, certain features regarding which can be calculated directly from the equations of motion. These features are the moments and also averages of powers of

[33] R. G. Gordon, *Adv. Chem. Phys.*, 1969, **15**, 79.

the frequency over the spectral density. Error bounds were constructed to determine the limits of certain averages of the spectral distribution.

Models of Molecular Motion. To make progress in the theoretical prediction of bandshapes at this point, a simplified model for the orientational motion must be constructed. Three approaches have been particularly useful in molecular reorientation studies. In the first of these, the molecules reorient by means of free, classical rotations; the molecules undergo frequent intermolecular collisions of short duration and their angular momenta are randomized by the impulsive torques associated with the collisions. In this model, if the time between collisions is short compared with the mean free rotational period, then the motion is well-described by rotational diffusion.[34] If, however, the mean free rotational period is comparable to the time between collisions, the reorientational motion is rather more complex.[35] The second approach, used principally for liquids, pictures the fluid as a pseudo-solid and is exemplified by the various cell models.[36] A molecule is pictured as undergoing solid-like torsional oscillations in a potential well formed by the other molecules and can reorient appreciably only if (*a*) it can tunnel through the barrier (an unlikely event), (*b*) can acquire sufficient activation energy through collisions to surpass the potential barrier, or (*c*) if the barrier itself dissipates because molecular motions open up the cell and leave some free volume, allowing the molecule to rotate freely. The third approach is based on the van Vleck–Weisskopf [37] collision-broadening theory, which is extended to include molecular reorientation.[38] The molecule in this picture reorients instantaneously and randomly upon collision.

The most frequently employed model is the rotational diffusion model proposed by Debye [34] to describe dielectric relaxation phenomena. In this model, the orientational motion of the molecules is assumed to follow a rotational diffusion equation with the rotational diffusion coefficient given by a Stokes–Einstein relationship. More elaborate treatments of the rotational diffusion equation approach have been proposed,[39, 40] but generally an adequate description of the reorientational process is obtained only if each diffusive step is limited to small angles. The Debye model is particularly inadequate in dealing with rotational diffusion in liquids composed of spherical molecules [41–44] and in gases where the molecules rotate through large angles between collisions.

34 P. Debye, 'Polar Molecules', Reinhold, New York, 1929.
35 W. A. Steele, *J. Chem. Phys.*, 1963, **38**, 2404, 2411.
36 J. E. Anderson, *J. Chem. Phys.*, 1967, **47**, 4879; S. H. Glarum and J. H. Marshall, *ibid.*, 1967, **46**, 55.
37 J. H. Van Vleck and V. F. Weisskopf, *Rev. Mod. Phys.*, 1945, **17**, 227.
38 W. A. Steele, in 'Transport Phenomena in Fluids', ed. N. J. M. Hanley, Marcel-Dekker, New York, 1969, Ch. 8, p. 296.
39 W. H. Furry, *Phys. Rev.*, 1957, **107**, 7.
40 L. D. Favro, *Phys. Rev.*, 1960, **119**, 53.
41 J. H. Rugheimer and P. S. Hubbard, *J. Chem. Phys.*, 1963, **39**, 552.
42 W. R. Hackelman and P. S. Hubbard, *J. Chem. Phys.*, 1963, **39**, 2688.
43 P. Rigny and J. Virlet, *J. Chem. Phys.*, 1967, **47**, 4645.
44 M. Bloom, F. Bridges, and W. N. Hardy, *Canad. J. Phys.*, 1967, **45**, 3533.

Gordon [45] has generalized the Debye model of rotational diffusion to allow reorientation through angular steps of arbitrarily large size. In this extended diffusion model, the molecules are assumed to rotate freely in the intervals between collisions with other molecules. A 'collision' in this context is taken to mean an event by which the rotational angular momentum of the molecule is changed. Two limiting cases of the extended diffusion model were considered by Gordon, the *J*-diffusion and the *M*-diffusion models. Both of these models allow angular diffusion steps of arbitrary size. In *J*-diffusion, both the magnitude and direction of the angular momentum vector of the molecule are randomized into a Boltzmann distribution by a collisional process. The *M*-diffusion model terminates a free rotational step by randomizing the orientation of the angular momentum vector, the magnitude of this vector remaining unchanged. The time between collisions is a random variable characterized by its mean, the correlation time for the rotational angular momentum, τ_J, which appears as an adjustable parameter in the extended diffusion model. In Gordon's [45] original treatment only linear molecules were considered. The general diffusion equation was found to obey exact classical mechanics at short times. Thus the power-series expansion at short times begins

$$\langle \boldsymbol{u}(0) \cdot \boldsymbol{u}(t) \rangle = 1 - t^2kT/I + O(t^3) \tag{60}$$

which is the correct classical result. This shows that the initial curvature of a classical dipole correlation function is independent of the intermolecular interactions. The usual Debye exponential formulae for small-angle diffusion do not have the correct initial time dependence. At longer times, the form of the dipole correlation function is determined explicitly by the molecular motion which is specified in Gordon's model by giving the relation between the rotation frequencies, ω_R, during successive diffusion steps. The *M*-diffusion model specifies the ω_R to be unchanged in successive steps whereas in the *J*-diffusion model the rotational frequencies ω_R are completely uncorrelated in successive steps so that the ω_R are distributed independently over a Boltzmann distribution. The dipole correlation function in the *M*-diffusion limit is given by

$$\langle \boldsymbol{u}(0) \cdot \boldsymbol{u}(t) \rangle = \exp(-t/\tau) \sum_{n=0}^{\infty} (n+1)(t/2\tau)^n \times \sum_{k=0}^{(n+1)/2} F_k(t)/2^k k!\,(n+1-2k)! \tag{61}$$

where $F_0(t)$, the correlation function for a free linear molecule, is given by

$$F_0(t) = \langle \cos \omega t \rangle = \int_{-\infty}^{+\infty} \cos(\omega t)\omega \exp(-\tfrac{1}{2}\omega^2)\, d\omega \tag{62}$$

[45] R. G. Gordon, *J. Chem. Phys.*, 1966, **44**, 1830.

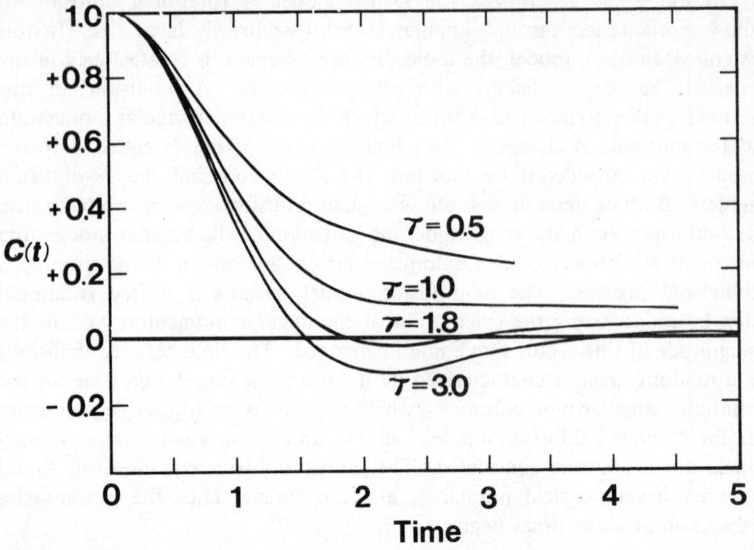

Figure 1 *Dipole correlation functions for the M-diffusion model, with* $\tau = 3.0, 1.8, 1.0,$ *and* 0.5 *(from bottom to top). The time-scale is in the reduced units* $(I/kT)^{\frac{1}{2}}$ *in all the Figures* 1—6
(Reproduced by permission from *J. Chem. Phys.*, 1966, **44**, 1830)

The results of the calculations for some typical parameters are shown in Figure 1. These curves sometimes become negative, suggesting that after a certain time it is more probable to find the dipole in the half of the sphere opposite to that in which it began at a time t earlier. Molecules with the most probable thermal angular momentum will have rotated through about 180° at the negative minima. Large-angle reorientations cannot occur in the Debye treatment. This model was also generalized by allowing the rate $(1/\tau)$ at which diffusion steps end to depend on the rotation frequency ω. A frequency dependence of the angular momentum relaxation time is suggested by several simple physical models. In this modification, the rate at which the *direction* of the angular moment diffuses decreases with increasing rotation frequency. Thus,

$$1/\tau = (1/\tau_0)f(\omega) \tag{63}$$

where $f(\omega)$ is a function which decreases at large ω. The dipole correlation function was integrated numerically for a number of dispersion functions $f(\omega)$. Typical families of such curves are shown in Figure 2. Quantitatively, these dispersion functions produce a secondary maximum in the correlation function and show a damped oscillatory approach to rotational equilibrium, rather than the monotonic decay usually assumed in statistical mechanics.

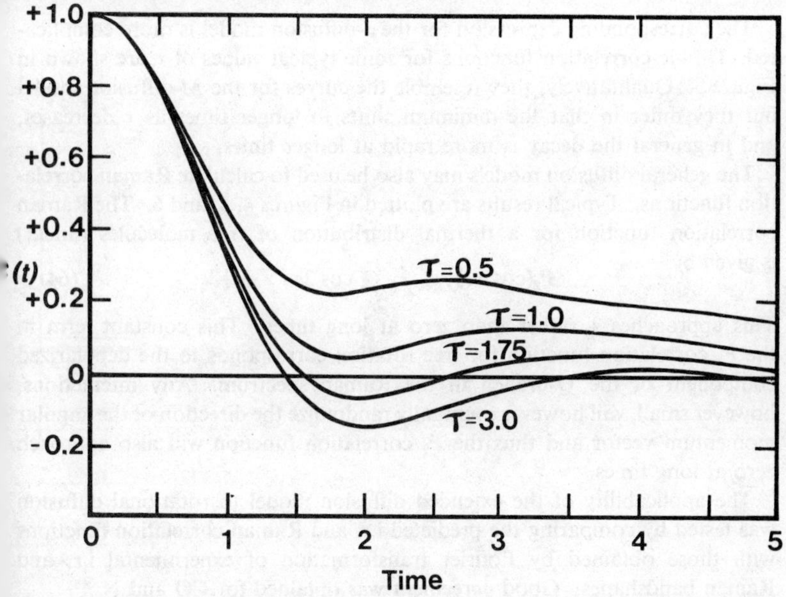

Figure 2 *Dipole correlation functions for the M-diffusion model, including a distribution of angular momentum relaxation times* $\tau(1 + 0.4\,\omega^4)$, *with* $\tau = 3.0, 1.75, 1.0,$ *and* 0.5 (Reproduced by permission from *J. Chem. Phys.*, 1966, **44,** 1830)

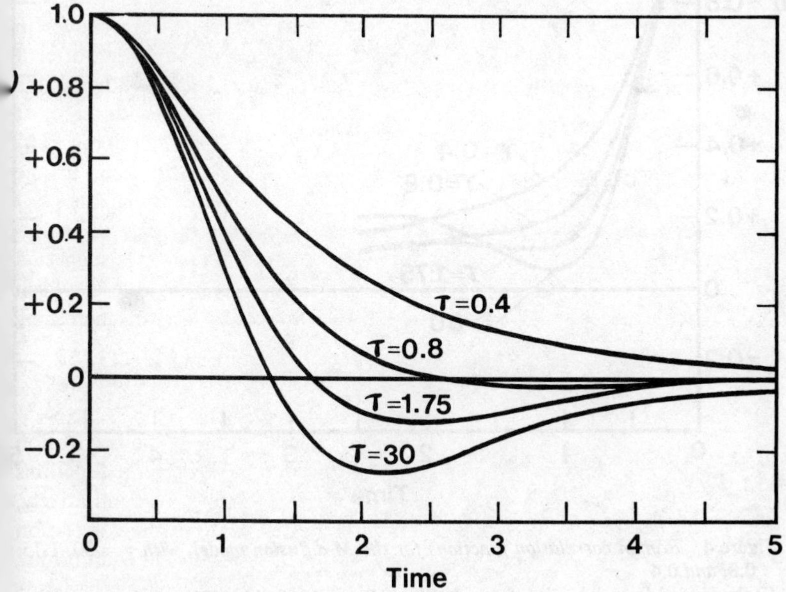

Figure 3 *Dipole correlation functions for the J-diffusion model, with* $\tau = 30, 1.75,$ 0.8, *and* 0.4 (Reproduced by permission from *J. Chem. Phys.*, 1966, **44,** 1830)

The corresponding expression for the *J*-diffusion model is more complicated. Dipole correlation functions for some typical values of τ are shown in Figure 3. Qualitatively, they resemble the curves for the *M*-diffusion model but they differ in that the minimum shifts to longer times as τ decreases, and in general the decay is more rapid at longer times.

The general diffusion models may also be used to calculate Raman correlation functions. Typical results are plotted in Figures 4, 5, and 6. The Raman correlation function for a thermal distribution of free molecules (linear) is given by

$$\langle P_2(\cos \omega t)\rangle_{\text{free}} = \langle \tfrac{3}{4} \cos 2\omega t\rangle + \tfrac{1}{4} \quad (64)$$

This approaches $\frac{1}{4}$ rather than zero at long times. This constant term in the P_2 correlation function for free rotation corresponds to the depolarized component of the *Q*-branch in the Raman spectrum. Any interactions, however small, will however eventually randomize the direction of the angular momentum vector and thus the P_2 correlation function will also approach zero at long times.

The applicability of the extended diffusion model to rotational diffusion was tested by comparing the predicted i.r. and Raman correlation functions with those obtained by Fourier transformation of experimental i.r. and Raman bandshapes. Good agreement was obtained for CO and N_2.[45]

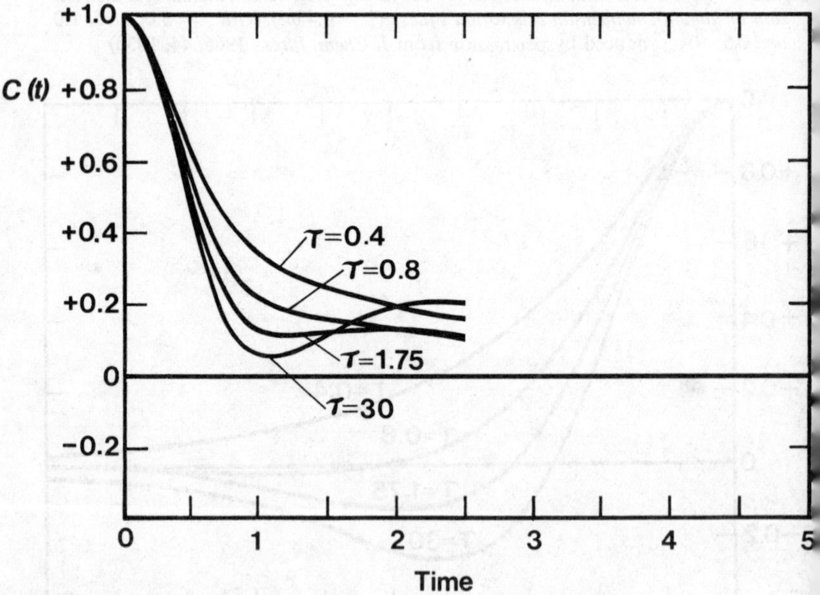

Figure 4 *Raman correlation functions for the M-diffusion model, with* τ = 30, 1.75, 0.8, *and* 0.4
(Reproduced by permission from *J. Chem. Phys.*, 1966, **44**, 1830)

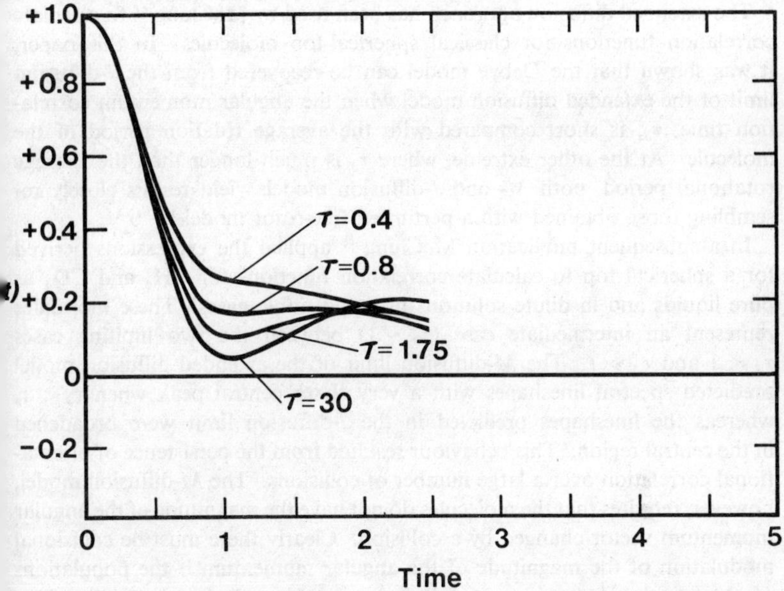

Figure 5 *Raman correlation functions for the M-diffusion model, including a distribution of angular-momentum relaxation times* $\tau(1 + 0.4\ \omega^4)$, *with* $\tau = 30$, 1.75, 0.8, *and* 0.4 (Reproduced by permission from *J. Chem. Phys.*, 1966, **44**, 1830)

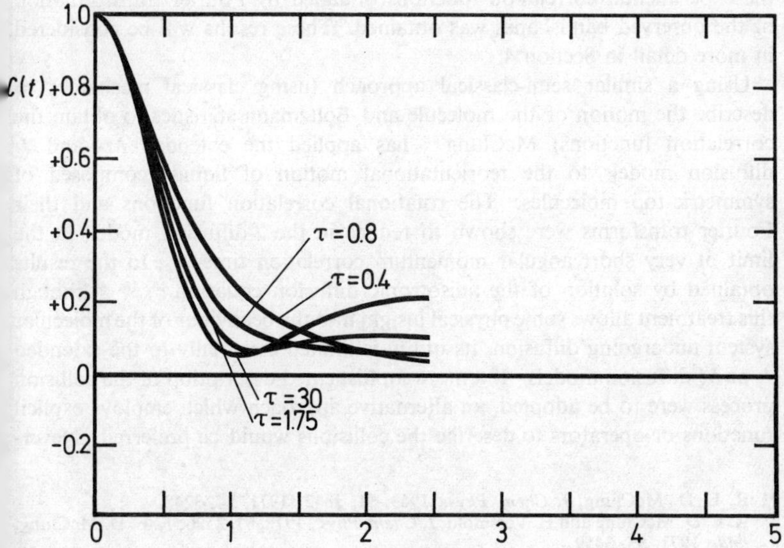

Figure 6 *Raman correlation functions for the J-diffusion model, with* $\tau = 30$, 1.75, 0.8, *and* 0.4 (Reproduced by permission from *J. Chem. Phys.*, 1966, **44**, 1830)

The extended diffusion approach has been used by McClung [46] to calculate correlation functions for classical spherical-top molecules. In this paper, it was shown that the Debye model can be recovered from the *J*-diffusion limit of the extended diffusion model when the angular momentum correlation time, τ_J, is short compared with the average rotation period of the molecule. At the other extreme, where τ_J is much longer than the average rotational period, both *M*- and *J*-diffusion models yield results closely resembling those obtained with a perturbed free-rotor model.[13, 34, 44]

In a subsequent publication McClung [47] applied the expressions derived for a spherical top to calculate correlation functions for CH_4 and CD_4 as pure liquids and in dilute solutions in liquid noble gases. These molecules represent an intermediate case ($\tau_J \sim 1$) between the two limiting cases $\tau_J \ll 1$ and $\tau_J \gg 1$. The *M*-diffusion limit of the extended diffusion model predicted spectral lineshapes with a very sharp central peak when $\tau_J \leqslant 1$, whereas the lineshapes predicted in the *J*-diffusion limit were broadened in the central region. This behaviour resulted from the persistence of orientational correlation over a large number of collisions. The *M*-diffusion model, however, requires that the molecules do not have the magnitude of the angular momentum vector changed by a collision. Clearly there must be collisional modulation of the magnitude of the angular momentum if the populations of the rotational states remain Boltzmann. In real situations, therefore, the *M*-diffusion model should be applied with caution in bandshape analysis, especially near the centre of the band and in the analysis of correlation function at very long times. Satisfactory agreement between the predicted and the experimental correlation functions obtained by Fourier transformation of the observed bandshapes was obtained. These results will be considered in more detail in Section 4.

Using a similar semi-classical approach (using classical mechanics to describe the motion of the molecule and Boltzmann statistics to obtain the correlation functions) McClung [48] has applied the extended *M*- and *J*-diffusion models to the reorientational motion of liquids composed of symmetric-top molecules. The rotational correlation functions and their Fourier transforms were shown to reduce in the *J*-diffusion model, in the limit of very short angular momentum correlation times τ_J, to the results obtained by solution of the anisotropic diffusion equation.[40, 49] Although this treatment allows some physical insight into the behaviour of the molecular system undergoing diffusion, its utility is limited essentially to the extended *J*- or *M*-diffusion models. If a more sophisticated description of the collision process were to be adopted, an alternative approach which employs explicit functions or operators to describe the collisions would be preferred. Never-

[46] R. E. D. McClung, *J. Chem. Phys.*, 1969, **51**, 3842; 1971, **54**, 3248.
[47] R. E. D. McClung and H. Versmold, *J. Chem. Phys.*, 1972, **57**, 2596; R. E. D. McClung, *ibid.*, 1971, **55**, 3459.
[48] R. E. D. McClung, *J. Chem. Phys.*, 1972, **57**, 5478.
[49] F. Perrin, *J. Phys. Radium.*, 1936, **7**, 1.

theless, in spite of its deficiencies, the extended diffusion model, although based on a greatly oversimplified picture of the rotational motion in fluids, does seem to be a valid approximation for many systems of linear, spherical-top, and symmetric-top molecules. It seems unlikely that this model will be developed for asymmetric-top molecules owing to the complexity of the mathematical treatment of the motion. For some molecules, however, the expressions for the symmetric rotor will provide a reasonable approximation.

Bliot and Constant [50] recently pointed out that expressions for the dipole correlation function equivalent to those of the *M*- and *J*-diffusion models could be obtained from the memory functions introduced by Berne and Harp.[15] The results were valid for linear, spherical-top, and symmetric-top molecules. The main features of the theory were illustrated by calculations for CH_3F.

An alternative and more general approach to the problem of rotational motions in fluids has been developed by Fixman and Rider [51] and by St. Pierre and Steele.[52, 53] These workers derived expressions for the $k = 0$ reorientational correlation functions and spectral densities. The approach of Fixman and Rider [51] was based on the assumption that the rotational motion could be treated as a Markoffian process (as are the extended diffusion models) and was able to derive a general expression for the time-dependent orientational and angular momentum distribution function. The free motion of the molecule was described by a streaming operator $\mathscr{S}$ and the effects of the intermolecular interactions by a collisional operator $\mathscr{L}$. Construction of matrix representations of these operators in a suitable Hilbert space enabled these authors to obtain expressions for the $k = 0$ reorientational correlation functions and spectral densities for a general Markoff process. Changes in both orientation and angular momentum could be included in the specification of the collision. In the application of their general theory to specific collisional models, no changes in the orientation of the molecules during collisions were considered. Their relaxation model was thus essentially a modification of the *M*- and *J*-diffusion models. As well as relaxation models, a Langavin model was also considered in which the angular momentum distribution function followed a Fokker–Planck equation. In the limit of small τ_J, this model predicted rotational correlation functions in agreement with those from the rotational diffusion equation theory.

St. Pierre and Steele [52, 53] approached the problem from the standpoint of the collision-interrupted free-rotation model and calculated the conditional phase space distribution functions for symmetric-top molecules undergoing *J*- and *M*-diffusion. Starting from explicit distribution functions obtained for classical symmetric tops and linear rotors derived in a previous paper,[52] they showed how collisional events that change the angular moment

[50] P. Bliot and E. Constant, *Chem. Phys. Letters*, 1973, **18**, 253.
[51] M. Fixman and K. Rider, *J. Chem. Phys.*, 1969, **51**, 2425.
[52] A. G. St. Pierre and W. A. Steele, *Phys. Rev.*, 1969, **184**, 172.
[53] A. G. St. Pierre and W. A. Steele, *J. Chem. Phys.*, 1972, **57**, 4638.

can be accounted for. The collisional distribution function was represented as a sum of contributions from those molecules undergoing a particular number of collisions. As in Gordon's treatment of the linear molecule the *n*th of these terms turns out to be an $(n - 1)$-fold integral. These integrals were computed in general form and for the specific case of a symmetric top. Using Fourier transform techniques the integrals were expressed as convolutions which were analogous to the expressions derived by Fixman and Rider.[51] The results obtained indicated that the *M*- and *J*-diffusion models provided plausible predictions for experimental spectra and correlation functions over a wide range of collision parameters.

A modification of the *M*-diffusion model in which fewer implicit assumptions and restrictions were made has been described by Frenkel, Wegdam, and van der Elsken.[54] A matrix description of the rotational motion of a dipolar molecule undergoing frequent collisions was given. The probability that the molecule has undergone an *n*-collisional process is contained explicitly in the matrix. By taking the ensemble average, an analytical expression for the dipole moment and angular momentum correlation functions was obtained. Other collisional distributions, apart from the usual Poisson distribution, can be employed. Thus the model can be used to describe systems where strong correlation exists between successive collisions. Librational motions are naturally included in this model.

The Debye diffusion model has been generalized to include arbitrary reorientational angles by Cukier and Lakatos-Lindenburg.[55] Any reorientational process, including large-angle reorientations, reorientations between distinct sites appropriate to continuous random walks, and the usual diffusion equation approximation, can be described by this model. Inertial effects were neglected. The form of the correlation functions for i.r. and Raman spectroscopy was found to depend only on the medium isotropy and molecular geometry. Rotational motion in non-isotropic systems such as liquid crystals was also considered.

A somewhat different model of orientational motion in liquids has been developed by Kivelson and Keyes.[56] This theory, which was developed in terms of Mori's [57] formalism for generalized hydrodynamics, was developed for three orientational variables interrelated by three coupled linear transport equations. In appropriate limits, the theory reduces to that for rotational diffusion, gas-like extended rotational diffusion, and solid-like oscillatory rotations or cell model motions. This theory therefore gives a unified picture encompassing all limits of molecular motion in liquids. Most of the interesting features of the rotational diffusion theory, Gordon's extended rotational diffusion theory, Steele's inertial effects, and Ivanov's jump theory [58] are

[54] D. Frenkel, G. H. Wegdam, and J. van der Elsken, *J. Chem. Phys.*, 1972, **57**, 2691.
[55] R. I. Cukier and K. Lakatos-Lindenberg, *J. Chem. Phys.*, 1972, **57**, 3427.
[56] D. Kivelson and T. Keyes, *J. Chem Phys.*, 1972, **57**, 4599.
[57] H. Mori, *Progr. Theor. Phys.*, 1965, **33**, 423.
[58] E. N. Ivanov, *Zhur. eksp. teor. Fiz.*, 1963, **45**, 1509.

shown. Isotropic rotational motion was assumed and all coupling between rotational motions and collective hydrodynamic transport modes neglected, This three-variable theory is, however, only accurate at relatively long times. so that the results may not be reliable at very short times corresponding to the wings of the spectral band. Kometani and Shimizu [59] have also presented a three-variable theory and applied it to a discussion of correlation times τ_θ, primarily in the rotational diffusion limit. They were able to express τ_θ in terms of torques and angular momenta.

Shimizu [20] obtained essentially the same relationship between the correlation function and spectral density as Gordon, using a somewhat different approach. The correlation function for rotational Brownian motion of a molecular system was considered theoretically. Two limits of Brownian motion, the inertial limit and the Debye limit, were assumed. A theory of i.r. bandshapes in terms of the correlation function was subsequently developed. More recently, a stochastic-type theory for diatomic molecules in inert solutions was developed by Bratos *et al.*[60–62] This semi-classical theory was limited to the intermediate temperature range. The model consisted of one active molecule isolated in a bath of non-absorbing inert solvent molecules. The active molecule was assumed to execute anharmonic vibrations modulated by a stochastic potential $V_s(r,t)$ due to the solvent. The active molecule also executed stochastic reorientations which were not correlated with the vibrations. Quantum theory was used to describe the vibrations whereas the reorientations were treated classically. According to the fluctuation-dissipation theorem,

$$I(\omega) \sim \int_{-\infty}^{+\infty} \langle \boldsymbol{M}^{\mathrm{H}}(t) \cdot \boldsymbol{M}^{\mathrm{H}}(0) \rangle \exp(\mathrm{i}\omega t)\, \mathrm{d}t$$

$$= \int_{-\infty}^{+\infty} G(t) \exp(\mathrm{i}\omega t)\, \mathrm{d}t \qquad (65)$$

where $\boldsymbol{M}^{\mathrm{H}}(t)$ represents the Heisenberg operator for $\boldsymbol{M}$. The function $G(t)$, called the relaxation function, contains all the information needed to relate the shape of an i.r. band to the nature of the random modulation processes. Once $G(t)$ is determined the band contour can be calculated by Fourier inversion. The relaxation function $G(t)$ was split into two contributions, the vibrational relaxation term, $G_v(t)$, and the reorientational relaxation term, $G_R(t)$. The vibrational relaxation process was determined basically by the same statistical theories used in lineshape studies.[63, 64] The bandshapes predicted by this theory were basically asymmetric with limiting symmetric

[59] K. Kometani and H. Shimizu, *J. Phys. Soc. Japan*, 1971, **30,** 1036.
[60] S. Bratos and J. Rios, *Compt. rend.*, 1969, **269,** *B*, 90.
[61] Y. Guissani, S. Bratos, and J. C. Leicknam, *Compt. rend.*, 1969, **269,** *B*, 137.
[62] S. Bratos, J. Rios, and Y. Guissani, *J. Chem. Phys.*, 1970, **52,** 439.
[63] A. D. Buckingham, *Proc. Roy. Soc.*, 1960, **A255,** 32.
[64] K. A. Valiev, *Optika i Spektroskopiya*, 1960, **11,** 465.

cases. A continuous sequence of band forms was predicted, including Lorentzian, Gaussian, and Voight profiles as well as several forms of asymmetric profile. The form of a particular band was shown to depend on the nature of the dominating relaxation process.

Recently, Bratos has extended his theory of i.r. bandshapes of diatomic molecules to Raman band profiles.[65, 66] The theory proposed was again a semi-classical theory of the stochastic type similar to that used for the i.r. case. The scattered intensity $I(\omega)$ was given by the fluctuation-dissipation theorem as

$$I(\omega) = \int_{-\infty}^{+\infty} G(t) \cdot \exp(i\omega t)\, dt \qquad (66)$$

where $G(t)$ is called the relaxation function. It was shown that

$$G(t) = A \cdot G_i(t) + B \cdot G_a(t) \qquad (67)$$

where the coefficients A and B depend on the geometry of the scattering arrangement. $G_i(t)$, which represents the scattering associated with the isotropic component of the polarizability tensor, is orientation independent and describes the relaxation of the vibrational energy; $G_a(t)$ describes the relaxation of both the vibrational and the rotational energy associated with the anisotropic part of the tensor. Thus, the profile of the isotropic scattering is determined only by the vibrational relaxation processes and can be described in terms of the correlation function $G_v(t)$,

$$G_v(t) = \int_{-\infty}^{+\infty} \lambda^4 I_{iso}(\omega) \cdot \exp(i\omega t)\, d\omega \qquad (68)$$

The profile of the perpendicular components, associated with the anisotropic part of the tensor, depends on both rotational and vibrational relaxation mechanisms. If rotational and vibrational motions are not correlated, we can write

$$G_v(t) \cdot G_r(t) = \int_{-\infty}^{+\infty} \lambda^4 I_{aniso}(\omega) \cdot \exp(-i\omega t)\, d\omega \qquad (69)$$

This relationship should be used where vibrational broadening is important. Thus Raman spectroscopy, in contrast to i.r., allows us to separate the contributions from the vibrational and rotational relaxation processes. In his paper, Bratos [65] goes on to predict theoretical band profiles for various contributions from vibrational and rotational relaxation. Bands associated purely with vibrational relaxation were all asymmetric although the asymmetry was small in certain cases. The Raman profile in this case was basically dependent on whether the solute–solvent interaction was one to one (or specific), yielding an asymmetric band contour, or a non-specific collective

[65] S. Bratos and E. Marechal, *Phys. Rev. (A)*, 1971, **4**, 1078.
[66] E. Marechal, *Ber. Bunsengesellschaft. phys. Chem.*, 1971, **75**, 343.

interaction which produced a symmetric (Gaussian) contour. For anisotropic scattering, the situation was more complex because of the interaction between vibrational and rotational relaxation. A continuous sequence of band contours was predicted which, depending on the relative contributions from the two mechanisms, ranged through an *O–Q–S*-type structure, a Lorentzian, a Gaussian, a Voight profile, and several asymmetric profiles. For the two limiting types, essentially free rotation and at the other extreme rotational diffusion, two wings, corresponding to the *O*- and *S*-branches of the vapour spectrum, and a broad symmetric band respectively were predicted.

Lassier and Brot [67] considered a model in which a molecular rotor could undergo librations in a potential well determined by neighbouring molecules, and under the influence of strong collisions might reorient from one well to another. Two primary processes were predicted at low temperatures, a small high-frequency librational motion and a much larger reorientational process. At higher temperatures, the two processes tended to merge. The absorption coefficient $\alpha(\omega)$ was found to peak at a specific frequency and also tended to zero as $\omega \rightarrow \infty$, *i.e.* the system becomes transparent at high frequencies. This behaviour has also been discussed by LeRoy *et al.*[68, 69] and by Constant *et al.*[70] Various correlation functions have been calculated numerically by Lassier and Brot [71] for a classical linear rotor hindered by a two-well potential and submitted to random torque impulses. The potential has the form

$$W(\theta) = V \sin^2 \theta$$

where θ is the polar angle. The equations of motion were integrated numerically with the random numbers required for determining the times and the parameters of the impacts generated by a computer programme. When $V = 0$ (the zero-impact case) the correlation function approached that of the free rotor for very rare impacts. For more numerous impacts, the correlation functions were identical with those computed by Gordon for his analogous *J*-diffusion model. In the case of very frequent collisions, $C(t)$ turned out to be essentially exponential, corresponding to the Debye picture of small-step diffusion. The correct behaviour at very short times was also obtained.

Calculations were also performed for $V = 3$—$7kT$. Their results for $V = 3kT$ indicated that, for infrequent impacts, the correlation function decays initially in an oscillatory fashion, and at longer times the decay becomes exponential in time. This implies in physical terms that the dipole librates in a potential minimum and occasionally crosses the barrier between minima. For more frequent collisions the short-time behaviour is retained,

[67] B. Lassier and C. Brot, *Chem. Phys. Letters*, 1968, **1**, 581.
[68] Y. LeRoy, E. Constant, and P. Desplanques, *J. Chim. phys.*, 1967, **64**, 1499.
[69] Y. LeRoy, E. Constant, C. Abbar, and P. Desplanques, *Adv. Mol. Relaxation Processes*, 1968, **1**, 273.
[70] E. Constant, L. Galatry, Y. LeRoy, and D. Robert, *J. Chim. phys.*, 1968, **65**, 1022.
[71] B. Lassier and C. Brot, *Discuss. Faraday Soc.*, 1969, **48**, 39.

but this changes over to an exponential decay sooner than for the infrequent-collision case. The plots of $\alpha(\omega)$ against frequency ω shown in Figure 7, for $V = 3kT$, show that the absorption coefficient α_{max} increases as the number of impacts decreases. This absorption corresponds to librational motion of the rotor in the potential wells. The deeper these wells, the better separated are the relaxational domain and the librational band. Also, the rarer the impacts, the narrower the librational band, as Figure 7 shows. The transparency at high frequencies is again a feature of this model and can be compared with the earlier analytical work of Lassier and Brot.[67]

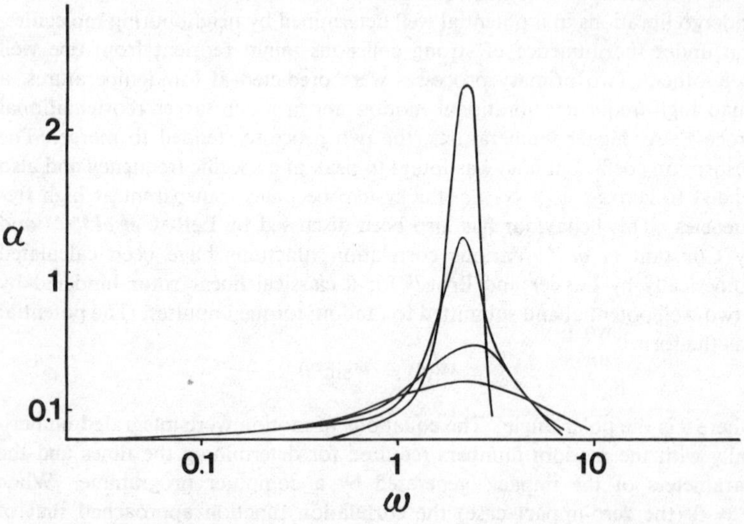

Figure 7 *Absorption per unit length for* $V = 3kT$. *From top to bottom at maximum absorption* $\tau_i = 5$, 0.5, 0.2, *and* 0.1 $(I/kT)^{\frac{1}{2}}$
(Reproduced by permission from *Discuss. Faraday Soc.*, 1969, **48**, 39)

The problems arising from the variation of refractive index with frequency in regions of absorption has been considered by Fulton.[72] Strictly, it is the product of refractive index and absorption coefficient which should be Fourier-inverted, not the absorption coefficient alone as is the usual practice. The error introduced into the computed dipole correlation function could be significant if the refractive index varies substantially through a strong absorption band. For weaker bands, however, the error introduced is generally small.

Molecular Dynamics Calculations. Considerable interest has been shown recently in the numerical evaluation of various correlation functions. This technique, usually called 'molecular dynamics', involves solving numerically

[72] R. L. Fulton, *J. Chem. Phys.*, 1971, **55**, 1386; 1969, **48**, 39.

the equations of motion for a model system. Alder and Wainwright [73] first used this technique to study fluids consisting of 'hard-sphere' atoms and 'square-well' atoms. More realistic dynamical studies were carried out by Rahman [74] for liquid argon using two-body interaction potentials of the Lennard-Jones type. Recently, Berne and Harp [15] have carried out computer simulations on liquids consisting of molecules such as CO and N_2 with internal degrees of freedom.

The use of molecular dynamics calculations has enabled a realistic and detailed picture of molecular rotation and translation to be obtained. The effect of the different intermolecular potentials and the various stochastic assumptions on the molecular motion can then be readily assessed. A discussion of the techniques used in computer experiments is contained in a review by Berne and Forster.[75]

Berne and Harp [15] performed their molecular dynamics calculations for CO and N_2 using three different models:

(*a*) A Stockmayer potential which used a spherically symmetric Lennard-Jones 6–12 potential between the centres of mass of two molecules, together with a dipole–dipole interaction term, $\mu = 0.1172$ D. The latter term contains the orientation dependence.

(*b*) A Stockmayer potential similar to that already described but with $\mu = 1.172$ D.

(*c*) A modified Stockmayer potential again similar to that already described but with the addition of quadrupole–quadrupole and quadrupole–dipole terms. In this case, the orientational part of the potential was derived from a multipole expansion of the electrostatic interaction between the charge distributions on two different molecules and only permanent (not induced) multipoles were considered. For CO, a quadrupole moment $Q = 2.43 \times 10^{-26}$ esu was used, and for N_2, a value for $Q = 2.05 \times 10^{-26}$ esu was employed in the simulation. A temperature of 68 K was assumed. Most simulations were performed for 216 and 512 molecules. The computation time rises dramatically with larger numbers of molecules. The results of the simulation for CO using potential (*a*) were nearly the same as the free-rotor correlation function, suggesting that the small dipole–dipole term is not important in this case. The results of the simulation for CO using potential (*b*) and for N_2 using potential (*c*) are shown in Figure 8. The quadrupole–quadrupole term was found strongly to influence the decay of the correlation function. Similar molecular dynamics calculations were carried out by Quantrec and Brot.[76] They calculated the dipole correlation function for a two-dimensional system of rigid diatomics with their centres fixed on the sites of a square lattice, so restricting the molecules to one (angular) degree of freedom. A Lennard-Jones 6–12 potential between the non-bonded atoms was used

[73] B. J. Alder and T. E. Wainwright, *J. Chem. Phys.*, 1959, **31**, 459; 1960, **33**, 1439.
[74] A. Rahman, *Phys. Rev. (A)*, 1964, **136**, 405.
[75] B. J. Berne and D. Forster, *Ann. Rev. Phys. Chem.*, 1971, **22**, 563.
[76] B. Quantrec and C. Brot, *J. Chem. Phys.*, 1971, **54**, 3655.

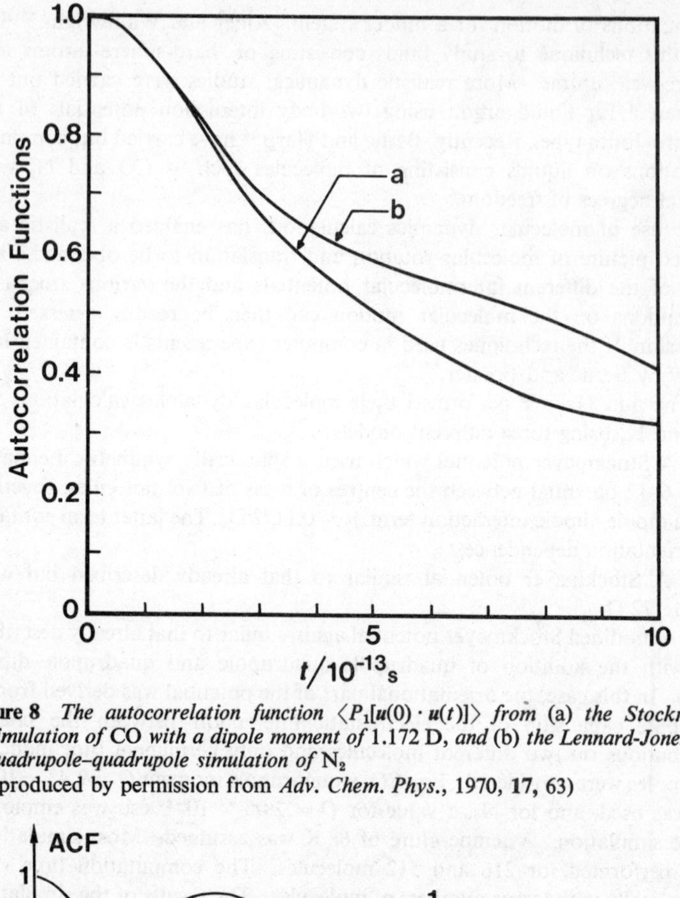

Figure 8 *The autocorrelation function* $\langle P_1[u(0) \cdot u(t)]\rangle$ *from* (a) *the Stockmayer simulation of* CO *with a dipole moment of* 1.172 D, *and* (b) *the Lennard-Jones plus quadrupole–quadrupole simulation of* N_2
(Reproduced by permission from *Adv. Chem. Phys.*, 1970, **17**, 63)

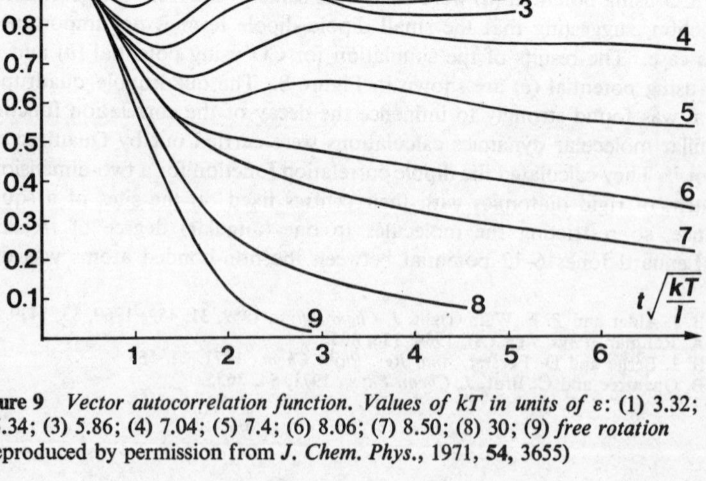

Figure 9 *Vector autocorrelation function. Values of kT in units of* ε: (1) 3.32; (2) 5.34; (3) 5.86; (4) 7.04; (5) 7.4; (6) 8.06; (7) 8.50; (8) 30; (9) *free rotation*
(Reproduced by permission from *J. Chem. Phys.*, 1971, **54**, 3655)

and the simulation was performed for 100 molecules. At low temperatures, the dipole correlation function was found to undergo a few damped oscillations and then slowly decrease with time (see Figure 9). This was interpreted in terms of a librational motion with changing amplitudes and slightly varying periods about a stable orientation, with rare end-to-end reorientations. At high temperatures the motion was non-uniform but with more or less continuous rotation with $C(t)$ approaching the free-rotor result. They conclude that the atom–atom potential has a far greater effect on the decay of $C(t)$ than does dipolar interaction even at high temperatures.

Further studies on the problem of molecular motion in dense argon were recently reported.[77] Using the molecular dynamics method, time correlation functions for depolarized light scattering were computed for several models of $\beta(R)$, the optical anisotropy, and compared with experimental values. No model was found which was consistent with all the results. Correlated reorientation of pairs of argon atoms was suggested to explain the observed discrepancy. In self-diffusion studies in liquid argon, O'Reilly [78] constructed

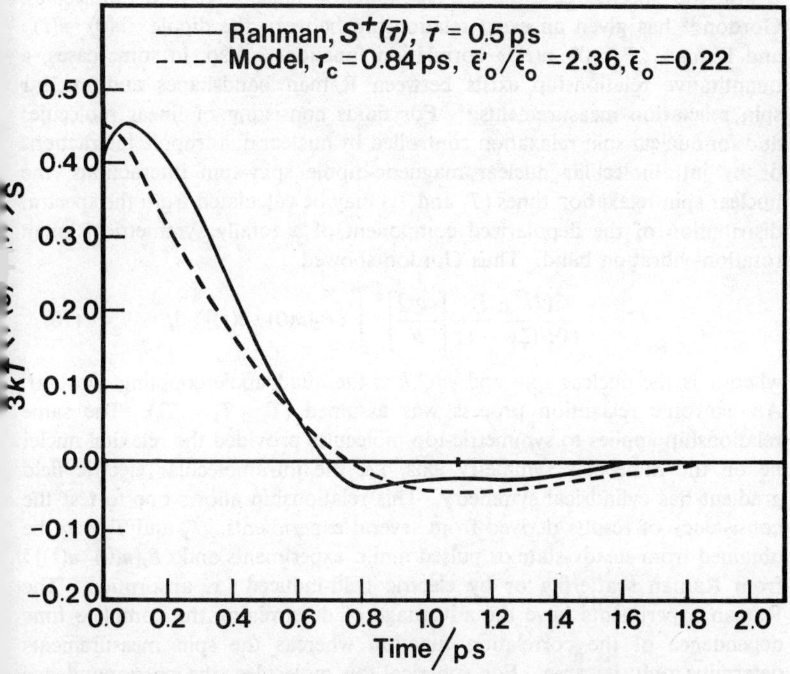

Figure 10 *Sliding velocity autocorrelation function for argon at* 94 K *computed by Rahman* (——) *and computed from model* (----) *with parameters given*
(Reproduced by permission from *J. Chem. Phys.*, 1971, **55**, 2876)

[77] B. J. Berne, M. Bishop, and A. Rahman, *J. Chem. Phys.*, 1973, **58**, 2696.
[78] D. E. O'Reilly, *J. Chem. Phys.*, 1971, **55**, 2876.

a general velocity autocorrelation function based on the quasi-lattice model of the liquid state.[79] The computed correlation functions were compared with the function $(m/3kT)\langle V(0) \cdot V(\tau)\rangle$ computed by Rahman [74] from molecular dynamics studies on liquid argon. Using three adjustable parameters, the mean time between 'hard' collisions, and the mean barrier heights for the creation of an interstitial molecule in the absence and in the presence of a nearest-neighbour vacancy, good agreement was obtained. The relevant functions are compared in Figure 10.

Relationship between some Correlation Functions. The dipole and polarizability correlation functions are only two of many correlation functions which describe molecular motion. These different correlation functions, such as, for example, those associated with depolarization of fluorescence and certain nuclear magnetic resonance experiments, explore different aspects of the molecular motion. It is important, therefore, to relate the various correlation functions in order to correlate the information derived from the various experimental techniques.

For the specific case of an ensemble of linear free-rotor molecules, Gordon [45] has given an exact relationship between the dipole $\langle \boldsymbol{u}(0) \cdot \boldsymbol{u}(t)\rangle$ and Raman $\langle P_2[\boldsymbol{u}(0) \cdot \boldsymbol{u}(t)]\rangle$ correlation functions. Also, in some cases, a quantitative relationship exists between Raman bandshapes and nuclear spin relaxation measurements.[29] For fluids consisting of linear molecules and for nuclear spin relaxation controlled by nuclear quadrupole interactions or by intramolecular nuclear magnetic-dipole spin–spin interactions, the nuclear spin relaxation times (T_1 and T_2) may be calculated from the spectral distribution of the depolarized component of a totally symmetric Raman rotation–vibration band. Thus Gordon showed

$$T^{-1} = \frac{3(2I+3)}{80I^2(2I-1)} \left[\frac{eqQ}{\hbar}\right]^2 \int_{-\infty}^{+\infty} \langle P_2[\boldsymbol{u}(0) \cdot \boldsymbol{u}(t)]\rangle \, \mathrm{d}t \qquad (70)$$

where I is the nuclear spin and $eqQ/\hbar$ is the quadrupole coupling constant. An isotropic relaxation process was assumed ($T = T_1 = T_2$). The same relationship applies to symmetric-top molecules provided the relaxing nuclei lie on the molecular symmetry axis, *i.e.* the intramolecular electric field gradient has cylindrical symmetry. This relationship allows one to test the consistency of results derived from several experiments. T_1 and T_2 may be obtained from steady-state or pulsed n.m.r. experiments and $\langle P_2[\boldsymbol{u}(0) \cdot \boldsymbol{u}(t)]\rangle$ from Raman scattering or by electric field-induced i.r. absorption. The Raman experiments have the advantage of determining the complete time dependence of the correlation function whereas the spin measurements determine only its area. For spherical-top molecules, the correspondence between light scattering and spin relaxation is not exact but is close enough for practical comparison, since the two correlation functions become identical

[79] J. O. Hirschfelder, C. F. Curtiss, and R. B. Bird, 'Molecular Theory of Gases and Liquids', Wiley, New York, 1964, p. 276.

in the opposing limits of rotational diffusion and free rotation. These comparisons do *not* depend on dynamical models, such as models of rotational diffusion, and they hold for both classical and quantum systems.

A relationship between dielectric dispersion and the dipole correlation function has been obtained by Keller, Ebersold, and Kneubühl.[80] The connection between i.r. and Raman bandshapes has been discussed by Berne, Pechukas, and Harp.[81] They outlined an approximate method whereby a knowledge of the dipole correlation function $\langle \boldsymbol{u}(0) \cdot \boldsymbol{u}(t) \rangle$ would enable a reasonable prediction of $\langle P_2[\boldsymbol{u}(0) \cdot \boldsymbol{u}(t)] \rangle$ to be made and vice versa. Two procedures were used for estimating the distribution of $\langle \boldsymbol{u}(0) \cdot \boldsymbol{u}(t) \rangle$ and thereby predicting Raman bandshapes. In the first of these, the information entropy of the distribution $P(\theta,\varphi,t)$ under the constraint imposed by the known $\langle \boldsymbol{u}(0) \cdot \boldsymbol{u}(t) \rangle$ was maximized; the second maximized the mean-square difference between the distribution and the equilibrium ($t \rightarrow \infty$) distribution under the same constraint. Using the first of these methods the approximate relations (71) and (72) were obtained, where $\beta(t)$ is a Lagrange undetermined

$$\langle \boldsymbol{u}(0) \cdot \boldsymbol{u}(t) \rangle \sim [\coth \beta(t) - 1/\beta(t)] \tag{71}$$

$$\langle P_2[\boldsymbol{u}(0) \cdot \boldsymbol{u}(t)] \rangle \sim 1 - 3/\beta(t)[\coth \beta(t) - 1/\beta(t)] \tag{72}$$

multiplier. Thus if $\langle \boldsymbol{u}(0) \cdot \boldsymbol{u}(t) \rangle$ is known at time t, $\beta(t)$ and hence $\langle P_2[\boldsymbol{u}(0) \cdot \boldsymbol{u}(t)] \rangle$ can be obtained. The P_2 correlation function for CO obtained by this procedure is compared with that obtained by molecular dynamics computer simulation in Figure 11. Also included is the correlation function predicted by the least-mean-square theory. Good agreement was obtained with the

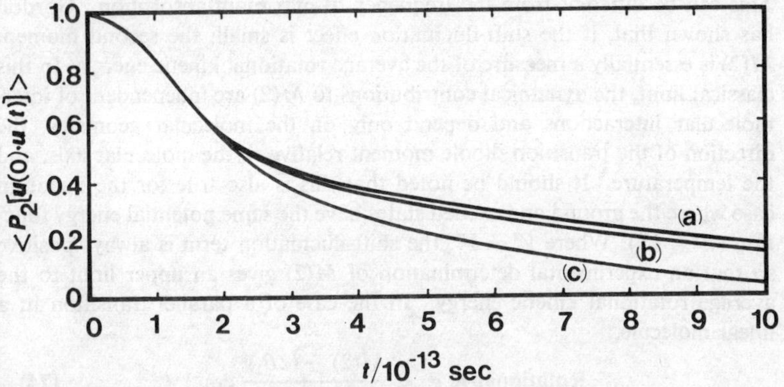

Figure 11 (a) *Molecular dynamics calculation of* $\langle P_2[\boldsymbol{u}(0) \cdot \boldsymbol{u}(t)] \rangle$; (b) *predictions from information theory* [*equation* (72)]; (c) *predictions from least-mean-square theory*
(Reproduced by permission from *J. Chem. Phys.*, 1968, **49**, 3125)

[80] B. Keller, P. Ebersold, and F, Kneubühl, *Proc. Phys. Soc.* (*Atomic Mol. Phys.*), 1970, **3**, 688.
[81] B. J. Berne, P. Pechukas, and G. D. Harp, *J. Chem. Phys.*, 1968, **49**, 3125.

best results obtained from information theory. There is obvious need to extend this type of approach to include other correlation functions.

Moment Analysis.—Important information regarding the non-spherical part of the intermolecular potential which governs molecular motion can be deduced from spectral moment calculations. The *n*th moment of a spectrum is defined as

$$M(n) = \int (\omega - \omega_0)^n I(\omega) \, d\omega \qquad (73)$$

where $I(\omega)$ is the intensity at frequency ω and ω_0 is some suitably chosen band origin. Generally, $I(\omega)$ is normalized so that $M(0) = 1$. The lower moments depend only on low-order molecular distribution functions, *i.e.* they are determined by simple molecular properties of the system such as moment of inertia and mean square torque. The higher moments are more complex properties of the system. The intensity distribution $I(\omega)$ can usually be expanded in a complete set of polynomials in ω. The coefficients in this expansion are then simply linear combinations of the moments of the spectrum, so that the spectrum is completely characterized by its moments. Properties such as I_{max}, the half-width of the band, or $I(\omega)$, depend in general on *all* the moments and thus on *all* the molecular distribution functions.

Gordon [23] has analysed molecular absorption and emission spectra (for vibrational transitions or of a single component of an electronic transition) in terms of moments of their intensity distributions. From the expression for the first moment the true frequency of the molecular transition can be located. This can be different from the frequency of maximum absorption. Gordon has shown that, if the shift-fluctuation effect is small, the second moment $M(2)$ is essentially a measure of the average rotational kinetic energy. In this classical limit, the dynamical contributions to $M(2)$ are independent of intermolecular interactions and depend only on the molecular geometry, the direction of the transition dipole moment relative to the molecular axis, and the temperature. It should be noted that this is also true for the quantum case where the ground and excited states have the same potential energy function ($V' = V^0$). Where $V' \neq V^0$, the shift-fluctuation term is always positive so that an experimental determination of $M(2)$ gives an upper limit to the average rotational kinetic energy. In the case of a parallel transition in a linear molecule,

$$\text{Rotational K.E.} \leqslant \frac{M(2) - (2B_1)^2}{4(3B_1 - 2B_0)} \text{ cm}^{-1} \qquad (74)$$

For i.r. absorption, the dynamical width generally dominates the shift-fluctuation effect. For instance, the values of $M(2)$ for CO in N_2 solution and for liquid HCl [83] are in satisfactory agreement with classical values.

[82] G. E. Ewing *J. Chem. Phys.*, 1962, **37**, 2250.
[83] W. West, *J. Chem. Phys.*, 1939, **7**, 795.

In the classical approximation, the odd moments of a band are zero, so that the asymmetry in an i.r. band is essentially a quantum effect. The third moment $M(3)$ of a band is a measure of its asymmetry,

$$M(3) = \int_{\text{band}} (\omega - \omega_s)^3 \hat{I}(\omega)\, d\omega \tag{75}$$

Hindering of the molecular motion always increases the value of $M(3)$, giving rise to skewness on the high-frequency side of the absorption band. The fourth moment of a band is the most useful for deriving information on molecular torques. Expressions for the frequency moments for a parallel transition in a classical linear rotor have been given by Gordon.[24] Corrections for quantum effects and for rotation–vibration interactions have been included in these expressions. The first quantum corrections to the second moments were given as

$$M(2) = M(2)_{\text{class}} + \frac{\hbar^2 M(4)}{12(kT)^2} + O(h^4) \tag{76}$$

Higher-order quantum corrections were also included. For the case of CO at room temperature, in 1000 atm of argon, the first quantum correction to $M(2)$ was only 0.78% above the classical value. Quantum corrections may therefore generally be neglected at ordinary temperatures.

The dominant effects of rotation–vibration interactions can be classified as frequency perturbations and intensity perturbations. The largest of the frequency perturbations arises from the fact that the moment of inertia of the excited vibrational state differs from that of the ground state. Again for the special case of a parallel transition in a linear molecule, the corrected expressions for $M(1)$ and $M(2)$ become

$$M(1) \sim 2B_1 + \left(\frac{kT}{hc}\right)\Delta \tag{77}$$

and

$$M(2) \sim 4\left(\frac{kT}{hc}\right)B_1 + 8\left(\frac{kT}{hc}\right)B_1\Delta + 2\left(\frac{kT}{hc}\right)^2\Delta^2 \tag{78}$$

where B_0 and B_1 are the rotational constants for the ground and excited states, respectively, and $\Delta = (B_1 - B_0)/B_0$.

The intensity perturbations result from the variation in moment of inertia during vibrational motion, giving rise to an oscillatory component of dipole moment perpendicular to both the axis and direction of the angular momentum. For even frequency moments the correction is generally small but for odd moments it may be significant. The corresponding expressions for $M(3)$ and $M(4)$ become

$$M(3) \simeq (4kT/hc)B_1^2(4 + 7\Delta) + 12B_1(kT/hc)^2\Delta(2 + 3\Delta) + 3(kT/hc)^3\Delta^3 + 2B_0^2(1 + \tfrac{3}{2}\Delta)\langle O^2V\rangle \tag{79}$$

and

$$M(4) \simeq 32B_1^2(kT/hc)^2(1+6\Delta) + 48B_1(kT/hc)^3\Delta^2(3+4\Delta) + 24(kT/hc)^4\Delta^4 + 4B_0^2[1+4\Delta+O(\Delta^2)]\langle(OV)^2\rangle \quad (80)$$

In most cases, (79) and (80) can be simplified to

$$M(3) \simeq 16B_1^2(kT/hc) + 2B_0^2\langle O^2V\rangle \quad (81)$$

$$M(4) \simeq 32B_1^2(kT/hc)^2 + 4B_0^2\langle(OV)^2\rangle \quad (82)$$

so that the mean square torque $\langle(OV)^2\rangle$ can be written in terms of $M(2)$ and $M(4)$,

$$\langle(OV)^2\rangle \simeq \{M(4) - 2[M(2)]^2\}/4B_0^2 \quad (83)$$

The corresponding relationship for a tetrahedral spherical-top molecule has also been derived.[84]

Moment analysis has also been applied by Gordon[28] to pure rotational Raman bandshapes. If it is assumed that the sample is isotropic, that there is no angular correlation between molecules separated by a distance corresponding to approximately the wavelength of the scattered light, and that the polarizability of an ensemble of molecules is a simple sum of the polarizability tensors of the individual molecules, the ratio between any two moments is independent of the scattering geometry. In the classical limit the lower moments are related by the expression[28]

$$M(2)/M(1) = M(4)/M(3) = 2kT/hc \quad (84)$$

For rotation–vibration Raman scattering the second moment of the Raman band of a linear molecule is three times that for absorption, and the relation between the second and fourth moments for scattering is[84]

$$M(4) = 8/3[M(2)]^2 + 12B_0^2\langle(OV)^2\rangle \quad (85)$$

Brindeau, Bratos, and Leicknam[85] calculated the spectral moments of diatomic molecules in inert solutions by using the stochastic theories of i.r. and Raman band profiles discussed previously. The spectral moments (M_n) of a band were related to the correlation function $G(t)$. Writing

$$\omega^n e^{i\omega t} = (-i)^n \frac{d^n}{dt^n}(e^{i\omega t}) \quad (86)$$

and integrating by parts,

$$(M_n)_0 = \frac{\int I(\omega)\omega^n\, d\omega}{\int I(\omega)\, d\omega} = \frac{\int d\omega\, \omega^n \left[\int_{-\infty}^{+\infty} G(t)\, e^{i\omega t}\, dt\right]}{\int d\omega \left[\int_{-\infty}^{+\infty} G(t)\, e^{i\omega t}\, dt\right]} = \frac{i^n G^{(n)}(0)}{G(0)} \quad (87)$$

[84] R. L. Armstrong, S. M. Blumenfeld, and C. G. Gray, *Canad. J. Phys.*, 1968, **46**, 1331.
[85] E. Brindeau, S. Bratos, and J. C. Leicknam, *Phys. Rev. (A)*, 1972, **6**, 2007.

$$(M_n) = \frac{\int I(\omega)(\omega - \omega')^n \, d\omega}{\int I(\omega) \, d\omega}$$

$$= \sum_{p=0}^{n} (-1)^p \, C_n^p (M_{n-p})_0 \, \omega'^p \tag{88}$$

where $I(\omega)$ is the intensity at frequency ω and C_n^p are the binomial coefficients. Thus calculating spectral moments essentially reduces to the differentiation of correlation functions. The spectral moments $(M_n)_{\omega'}$ of an i.r. band can be written in terms of pure vibrational $(M_n)_{\omega'}^{\mathrm{v}}$ and pure rotational $(M_n)_0^{\mathrm{r}}$ moments. These are the band moments associated with pure vibrational and pure rotational relaxation respectively. The vibrational moments $(M_n)_{\omega'}^{\mathrm{v}}$ were calculated using (87) and (88) with the vibrational correlation function. The purely rotational classical spectral moments $(M_n)_0^{\mathrm{r}}$ were determined using (87) with the classical rotational correlation function. The results were

$$(M_1)_0^{\mathrm{r}} = \hbar/I \tag{89}$$

$$(M_2)_0^{\mathrm{r}} = 2kT/I \tag{90}$$

$$(M_3)_0^{\mathrm{r}} = \frac{4\hbar kT}{I^2} + \frac{\hbar}{2kT}\left\langle\left(\frac{\partial\omega}{\partial t}\right)^2\right\rangle \tag{91}$$

$$(M_4)_0^{\mathrm{r}} = \frac{8k^2T^2}{I^2} + \left\langle\left(\frac{\partial\omega}{\partial t}\right)^2\right\rangle \tag{92}$$

up to terms linear in $\hbar$. The complete expressions for the rotation–vibration spectral moments can then be expressed in terms of $(M_n)_{\omega'}^{\mathrm{v}}$ and $(M_n)_0^{\mathrm{r}}$.

Similar arguments apply in the case of Raman bandshapes except that here the vibrational and rotation effects can be separated. This is not possible with i.r. absorption. The pure rotational contributions are given by

$$(M_1)_0^{\mathrm{r}} = 3\hbar/I \tag{93}$$

$$(M_2)_0^{\mathrm{r}} = 6kT/I \tag{94}$$

$$(M_3)_0^{\mathrm{r}} = \frac{48\hbar kT}{I^2} + \frac{3\hbar}{2kT}\left\langle\left(\frac{\partial\omega}{\partial t}\right)^2\right\rangle \tag{95}$$

$$(M_4)_0^{\mathrm{r}} = \frac{96(kT)^2}{I^2} + 3\left\langle\left(\frac{\partial\omega}{\partial t}\right)^2\right\rangle \tag{96}$$

In principle, all moments of an anisotropic spectrum differ from the corresponding moments of an i.r. spectrum, since the rotational contributions are different in the two cases. However, if the vibrational relaxation processes dominate the spectrum, the differences may not be significant. The results of the stochastic theory closely resemble the earlier theories of Gordon [23, 24, 28] for i.r. and rotational Raman bands. The expressions for the spectral moments of isotropic and for vibrational–rotational anisotropic Raman bands have not been derived previously.

3 Experimental Techniques

In order to obtain meaningful information on molecular motion from spectral bandshapes very accurate measurements of the spectral distribution $I(\omega)$ must be carried out. Particular care must be taken to eliminate instrumental effects and to perform the measurements sufficiently far into the wings of the band. Since the experimental requirements of i.r. absorption and Raman scattering are somewhat different they will be considered separately.

Infrared Absorption.—*Bandshape Measurements.* The suitability of an i.r. absorption band for molecular motion studies is governed by a number of criteria. The first and most important requirement is that the band, especially in the wings, be free from extraneous absorptions arising from weak overtones and combination bands, hot bands, and overlapping fundamentals. These requirements are sometimes difficult to meet especially in fluids composed of fairly complex molecules, and artificial curve-resolving methods have to be resorted to. The effect of these techniques on the bandshape, however, is difficult to assess, especially in the wings of the band, and they should be avoided if possible. Bands of moderate intensity should be selected. If the band is very intense, the transmission must be measured through thin films, which results in distortion from interference effects within the films and from phase changes at the boundaries. Conventional double-beam spectrometry is not applicable in this case and many investigators have turned to methods based on interferometry, simple reflection, or attenuated total reflection (ATR). In addition, changes in refractive index through an absorption band could lead to considerable inaccuracies with strong bands measured by conventional techniques.[72] ATR measurements, however, overcome these problems, but the experiments are difficult to perform, requiring very accurate angular measurements and precise collimation if meaningful data are to be obtained. Detailed measurements on liquids using the ATR technique have been made by Irons and Thompson[86] and by Crawford and co-workers.[87] The problems associated with thin-film measurements for strongly absorbing liquids have recently been considered by Young and Jones.[88] These authors have also recently reviewed the characteristics and suitability of various spectroscopic techniques for recording accurate bandshapes.[89]

With modern high-resolution double-beam spectrometers, the effect of instrumental factors on the absorption profile should be small provided the spectral slit width is small compared with the half-width of the band. Various

[86] G. M. Irons and H. W. Thompson, *Proc. Roy. Soc.*, 1967, **A298,** 160.
[87] (*a*) T. Fujiyama and B. Crawford, *J. Phys. Chem.*, 1968, **72,** 2174; (*b*) C. E. Favelukes, A. A. Clifford, and B. Crawford, *J. Phys. Chem.*, 1968, **72,** 962.
[88] R. P. Young and R. N. Jones, *Canad. J. Chem.*, 1969, **47,** 3463.
[89] R. P. Young and R. N. Jones, *Chem. Rev.*, 1971, **71,** 219.

techniques are available, however, for recovering the true bandshape from the observed profile. These techniques were reviewed by Seshadri and Jones [90] in 1963 but since that date the availability of computers has simplified all aspects of data processing, including initial smoothing of raw data to sophisticated methods of bandshape analysis and curve-fitting.[91, 92] Some care is advisable in the use of certain techniques, however, notably those involving deconvolution, since these can lead to the introduction of oscillatory features in the wings of the band.[93]

Computation of Dipole Correlation Functions. The experimental dipole correlation function is computed from the observed spectral density $I(\omega)$ according to equation (18). Normally only the real part of $C(t)$ is calculated, the imaginary part contributing only a small quantum correction. Thus we obtain

$$\hat{C}(t) = \int_{\text{band}} \hat{I}(\omega) \cos[(\omega - \omega_s)t]\mathrm{d}\omega \tag{97}$$

where ω_s is the shifted band centre which is obtained from

$$\omega_s = \int_{\text{band}} \omega\hat{I}(\omega)\,\mathrm{d}\omega - M(1) \tag{98}$$

in which $M(1)$ is the first moment of the band. The spectral density $I(\omega)$ should strictly be calculated from the absorption coefficient $\sigma(\omega)$, which is given by

$$I(\omega) = \frac{\sigma(\omega)}{1 - \exp(-\hbar\omega/kT)} \tag{99}$$

The Boltzmann factor in the denominator corrects for induced emission. It is generally not necessary to use this expression, however, since the absorbance, $I(\omega) = \ln(I_0/I)$, is a sufficiently good approximation. The integration is performed over the whole band. Since it is not always obvious to what extent the measurements should be carried into the wings, Rothschild [94] has developed a procedure for adjusting the integration limits to obtain the theoretical value of the second moment. This was accomplished by parallel shifting the zero-absorption line up or down so that its intersection with the absorption curve moved toward or away from the band centre. The physical implications of this procedure are not clear at present since other factors such as shift fluctuations can also influence the second moment.

[90] K. S. Seshadri and R. N. Jones, *Spectrochim. Acta*, 1963, **19**, 1013.
[91] R. N. Jones, *Pure Appl. Chem.*, 1969, **18**, 303.
[92] R. N. Jones, *Appl. Optics*, 1969, **8**, 597.
[93] A. F. Jones and D. S. Misell, *J. Phys.* (*A*), 1970, **3A**, 462; V. G. Cooper, *Appl. Optics*, 1971, **10**, 525.
[94] W. G. Rothschild, *J. Chem. Phys.*, 1970, **53**, 990.

The Fourier transformation of $I(\omega)$ is best carried out by numerical integration using a digital computer. It is advantageous if the intensity data can be produced in digital form, so allowing direct accurate processing to be performed. Where a large number of data points, $I(\omega)$, are to be processed it is sometimes an advantage to use a specially developed algorithm, the so-called Fast Fourier Transform.[95, 96] This algorithm is particularly useful for deconvolution procedures and molecular dynamics [97] calculations. Most calculations are carried out using time intervals of about 0.05—0.1 $\times 10^{-12}$ s. The extent to which the decay of $C(t)$ can be followed is limited by the resolution of the measurement. Thus the curves will become unreliable at times longer than about $1/\Delta\omega_r$, where $\Delta\omega_r$ is the experimental resolution. This corresponds to about 5 ps at 1 cm^{-1} resolution.

Raman Scattering.—*Bandshape Measurements.* Most of the requirements for i.r. bandshape work also apply to Raman scattering. With modern double or triple Raman monochromators and high-intensity, stable laser sources, accurate band contours relatively free from instrumental distortion can be obtained. In addition, with photon-counting detection systems, the intensity data are already in digital form, greatly facilitating data processing. At the same time, increased accuracy can be obtained. Ideally, the intensity data $I(\omega)$, would be accumulated at predetermined frequency intervals across the band. The time interval over which the counts are accumulated would be determined by the accuracy required. The statistics of the counting process are of course well known so that the data are well characterized statistically. A small dedicated computer is an advantage for this work.

Unlike i.r. absorption, we have seen that the orientational and vibrational relaxation processes can be separated in the Raman case. To accomplish this, however, accurate measurements of the isotropic and anisotropic Raman scattering must be performed. The theoretical arguments underlying the separation of vibrational relaxation and reorientation of anisotropy have already been discussed. The relevant expressions for the vibrational and reorientational correlation functions are given by equations (45) and (47).

The study of strongly polarized lines requires an optical system capable of isolating the strong isotropic component from the depolarized spectrum. A Glan–Thompson polarizing prism is useful to ensure that only light of the correct polarization enters the cell. A polarization rotator placed after the polarizer enables the plane of polarization to be rotated through 90°. An analyser is generally used to select the appropriate polarization of the scattered light. A quartz wedge polarization scrambler placed before the entrance slit is necessary if corrections for grating effects are to be avoided. Generally, the instrument functions are Gaussian and both the intrinsic

[95] J. W. Cooley and J. W. Tukey, *Math. Comput.*, 1965, **19**, 297.
[96] W. T. Cochran, J. W. Cooley, D. L. Favin, H. D. Helms, R. A. Kaenel, W. W. Lang, G. C. Maling, D. E. Nelson, C. M. Rader, and P. D. Welch, *Proc. I.E.E.E.*, 1967, **55**, 1664.
[97] R. P. Futrelle and D. J. McGinty, *Chem. Phys. Letters*, 1971, **12**, 285.

lineshape and orientational spectrum are Lorentzian, so that instrumental effects can readily be removed where necessary by the application of closed expressions.[30]

A second method useful for obtaining the orientational linewidths and associated correlation times for weakly polarized Raman bands has been described by Rakov [98] and more recently revived by Bartoli and Litovitz.[30] The depolarized Raman linewidth $\omega(T)$ is measured over a wide temperature range so that at higher temperatures $\omega(T)$ has a large orientational contribution whereas at low temperatures $\omega(T)$ is approximately equal to the vibrational width. After removal of instrumental effects, the vibrational bandshapes of six liquids shown in Table 1 were approximately Lorentzian.

Table 1 *Molecular reorientational times for some liquid systems from depolarized Raman linewidths* $\omega(T)^a$

Liquid	*Line (wavenumber/* cm^{-1})	*Point group*	*Assignment*	τ_{vib}/cm^{-1}	τ_{or}/cm^{-1}
CCl_4	217	T_d	CCl_2 deformation	2.3	3.0 ± 0.5
C_6H_{12}	1027	D_{3h}	CC stretch	3.0	3.1 ± 0.5
C_6H_{12}	1266	D_{3h}	CH_2 twist	3.0	3.1 ± 0.5
CH_3OH	1030	C_s	CO stretch	9.0	6.0 ± 0.5
C_4H_9OH	1300	C_s	CH_2 twist	5.3	2.6 ± 0.5
$C_8H_{17}OH$	1300	C_s	CH_2 twist	8.0	1.6 ± 0.5

[a] Ref. 30.

Assuming that the orientational spectrum is also Lorentzian, the observed spectrum will also be Lorentzian whose width will be the sum of the orientational width $\omega_{or}(T)$ and the vibrational width ω_{vib}. Thus, we have

$$\omega(T) = \omega_{vib} + \omega' \exp(-A/T) \qquad (100)$$

in which it is assumed that the reorientational linewidth has an Arrhenius temperature dependence, *i.e.* $\omega_{or} = \omega' \exp(-A/T)$, where ω' and A are arbitrary constants. The vibrational linewidth ω_{vib} was further assumed to be independent of temperature. ω_{vib}, ω', and A can be obtained by fitting the measured Raman linewidth to equation (100). This technique is particularly useful for liquids which form plastic crystals, *i.e.* where rotational freedom persists in the solid phase. I.r. bandshapes can also be analysed by this method.

The use of linewidth instead of the complete correlation function is a useful technique, but it gives no information regarding the very short-time (the first fraction of a picosecond) behaviour of the system. This particular time interval is sacrificed in the interests of expediency. It should be remembered, however, that the data relating to this part of the rotational motion are contained in the far wings of the band contour, which is the part of the profile known with least accuracy.

98 A. V. Rakov, *Trudy Fiz. Inst. Akad. Nauk S.S.S.R.*, 1964, **27**, 111; *Optika i Spectroskopiya*, 1959, **7**, 202 (*Optics and Spectroscopy*, 1959, **7**, 128).

4 Applications

Infrared Studies.—Most of the bandshape work that has been published to date, especially the earlier work, considers rotational relaxation as the dominant broadening effect and completely neglects other energy relaxation processes which may contribute to the observed band profile. In some cases this procedure is justified but in others it is not, depending on the nature of the particular system under consideration. Care must therefore be exercised

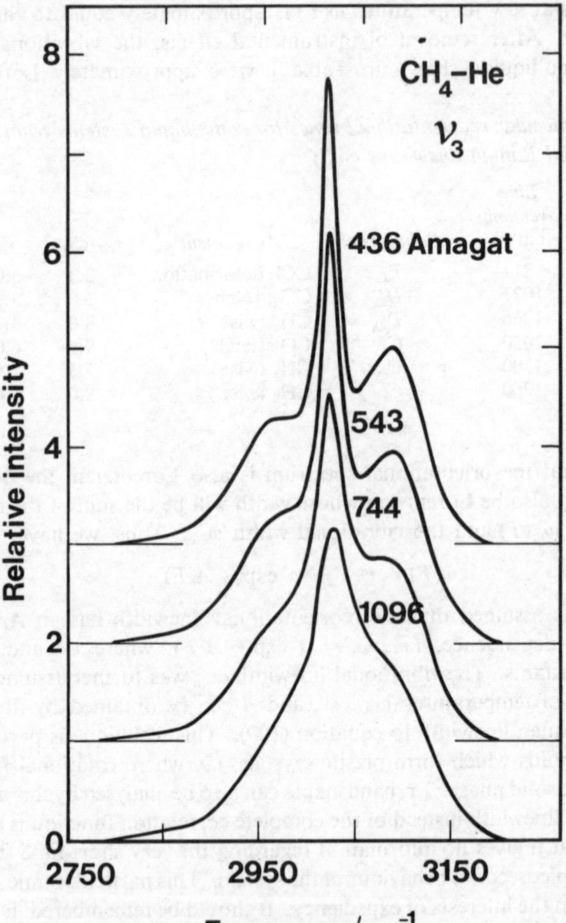

Figure 12 *The effect of various densities of helium on the* ν_3 *band of* CH_4. *The individual profiles have been normalized to show equal areas; the relative intensity scales have been displaced for clarity*

(Reproduced by permission from *Canad. J. Phys.*, 1968, **46**, 1331)

where the quantitative interpretation of the orientational motion is being undertaken, particularly with larger, more complex systems where vibrational relaxation may be the dominant process. More quantitative work, particularly Raman measurements, is needed to clarify the situation.

Gaseous Phase. Relatively little work on gaseous molecular systems has been published compared with the extensive liquid-phase studies. One of

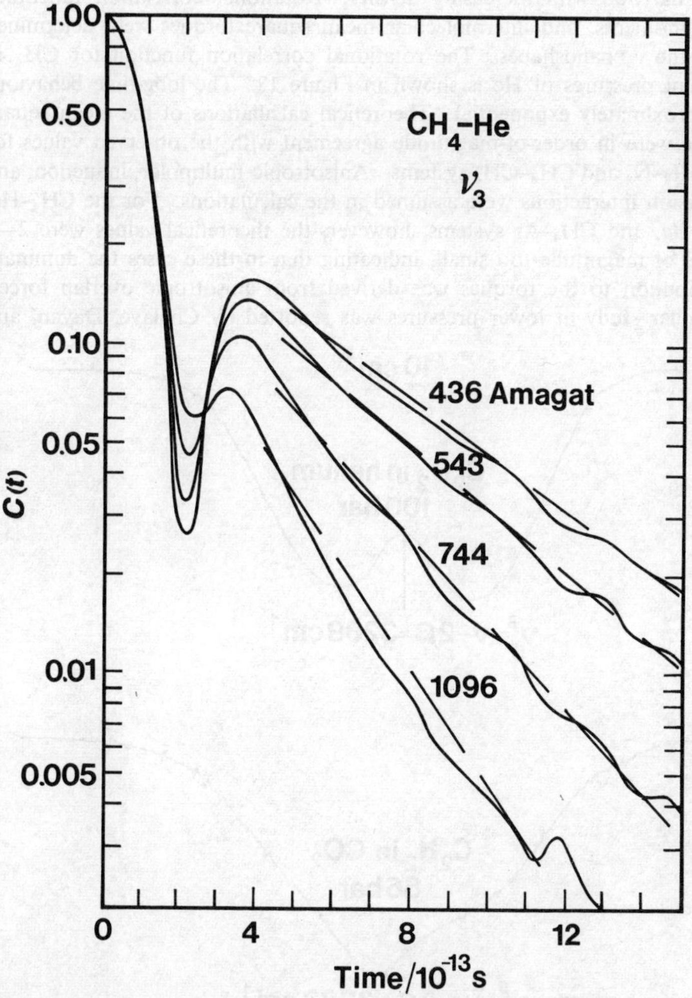

Figure 13 *The effect of various densities of helium on the rotational correlation function of* CH_4
(Reproduced by permission from *Canad. J. Phys.*, 1968, **46,** 1331)

the first quantitative studies by Armstrong, Blumenfeld, and Gray [84] was aimed at an understanding of the intermolecular interactions in liquid methane. Extensive bandshape measurements on the ν_3 and ν_4 bands in CH_4–He, and on the ν_3 band in CH_4–He, CH_4–N_2, and CD_4–He mixtures were carried out at pressures up to 3000 atm. The effect of different He pressures on the profile of the ν_3 band is shown in Figure 12. The distinction between the *P*-, *Q*-, and *R*-branches becomes less apparent and the total band narrows with increasing density. Rotational correlation functions, band moments, and intermolecular mean square torques were determined from the ν_3 bandshapes. The rotational correlation function for CH_4 in different pressures of He is shown in Figure 13. The long-time behaviour is approximately exponential. Theoretical calculations of the mean square torque were in order-of-magnitude agreement with the observed values for the CH_4–N_2 and CH_4–CH_4 systems. Anisotropic multipolar, induction, and dispersion interactions were assumed in the calculations. For the CH_4–He, CD_4–He, and CH_4–Ar systems, however, the theoretical values were 2—3 orders of magnitude too small, indicating that in these cases the dominant contribution to the torques was derived from anisotropic overlap forces. A similar study at lower pressures was reported by Chalaye, Dayan, and

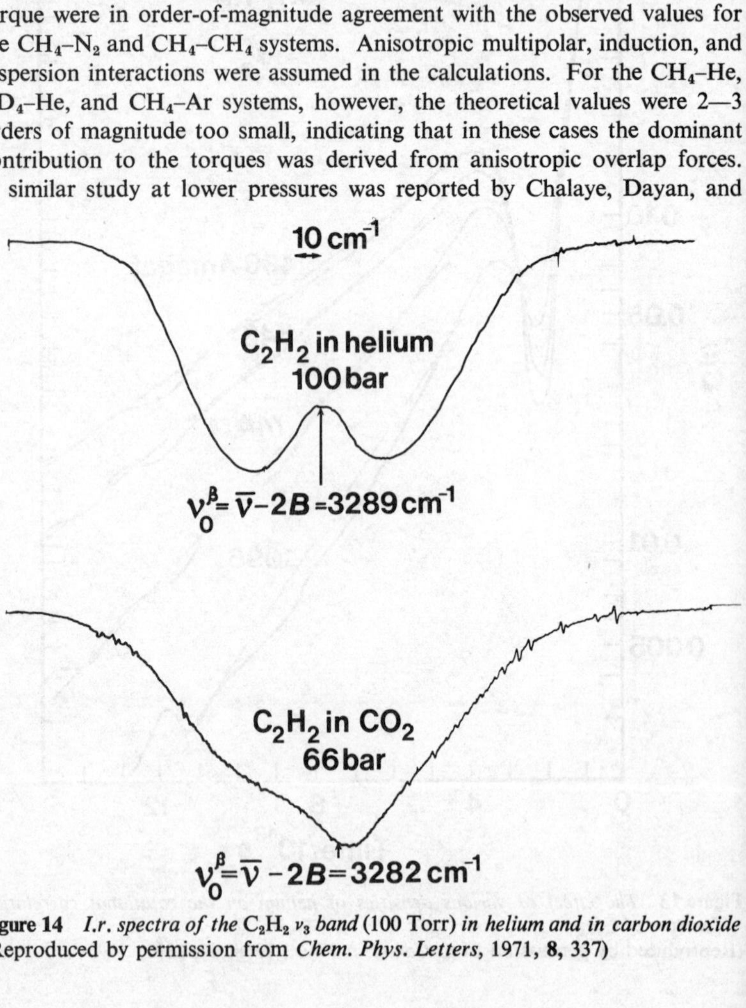

Figure 14 *I.r. spectra of the* C_2H_2 ν_3 *band* (100 Torr) *in helium and in carbon dioxide* (Reproduced by permission from *Chem. Phys. Letters*, 1971, **8**, 337)

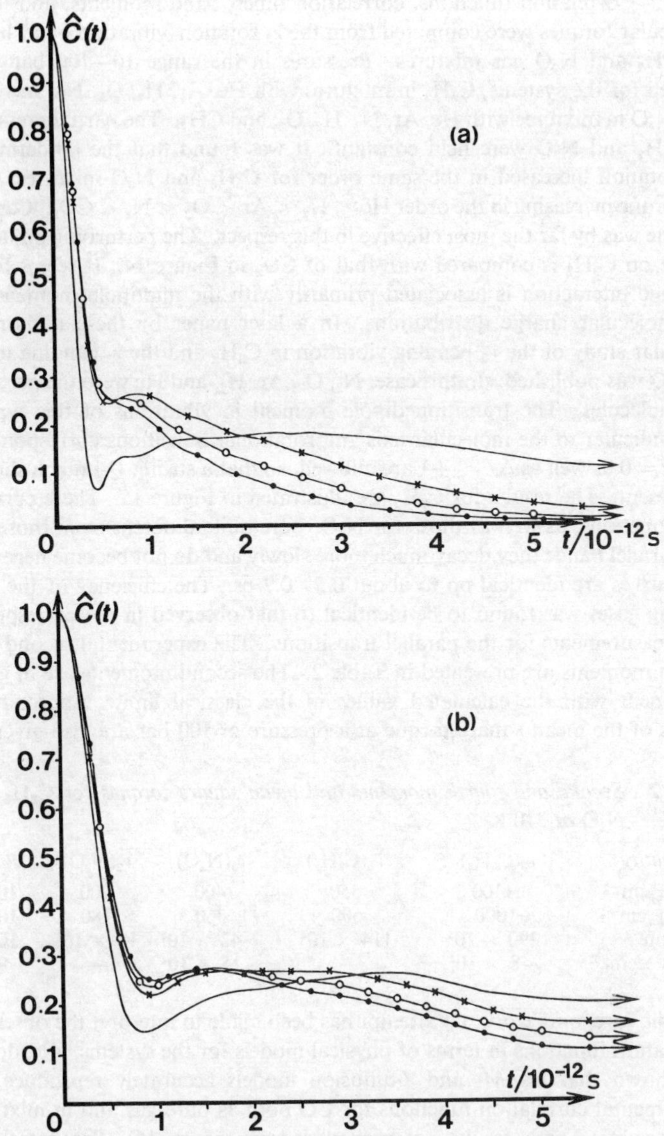

Figure 15 *Correlation functions in various gas mixtures*: (a) ν_5 *for* C_2H_2 *in* N_2: —•—•— (100 bar); —o—o— (70 bar); —×—×— (40 bar); ——— pure (20 Torr); (b) ν_2 *for* N_2O *in* He: —•—•— (100 bar); —o—o— (70 bar); —×—×— (40 bar); ——— pure (160 Torr)
(Reproduced by permission from *Chem. Phys. Letters*, 1972, **12**, 462)

Levi.[99] Correlation functions, correlation times, band moments, and intermolecular torques were computed from the ν_3 rotation–vibration bandshapes in C_2H_2 and N_2O gas mixtures. Pressures in the range 10—100 bar were studied for the systems, C_2H_2 in mixtures with He, Ar, H_2, O_2, N_2, and CO_2 and N_2O in mixtures with He, Ar, N_2, H_2, O_2, and CH_4. The partial pressures of C_2H_2 and N_2O were held constant. It was found that the hindering of the rotation increased in the same order for C_2H_2 and N_2O mixtures, with the torque increasing in the order He $< H_2 <$ Ar $< O_2 < N_2 \leqslant CO_2$. Carbon dioxide was by far the most effective in this respect. The perturbing influence of He on C_2H_2 is compared with that of CO_2 in Figure 14. It seems likely that the interaction is associated primarily with the multipole moments of the molecular charge distributions. In a later paper by these authors,[100] a similar study of the ν_5 bending vibration in C_2H_2 and the ν_2 bending mode in N_2O was published. In this case, N_2, O_2, Ar, H_2, and He were the perturbing molecules. The transition dipole moment in vibrations of this type is perpendicular to the molecular axis and rotational transitions corresponding to $\Delta J = 0$ as well as $\Delta J = \pm 1$ are allowed, so that a strong Q-branch should be present. The results for C_2H_2 are illustrated in Figure 15. These correlation functions, as well as those for N_2O, were quite different from those for the parallel band; they decay much more slowly and do not become negative. The curves are identical up to about 0.5—0.7 ps. The efficiency of the perturbing gases was found to be identical to that observed in the corresponding measurements for the parallel transitions. The experimental second and fourth moments are presented in Table 2. The second moments are in good agreement with the calculated values in the classical limit. Approximate values of the mean square torque at a pressure of 100 bar are also given.

Table 2 *Second and fourth moments and mean square torques for* C_2H_2 *and* N_2O *at* 310 K

Quantity	$\nu_3(C_2H_2)$	$\nu_5(C_2H_2)$	$\nu_3(N_2O)$	$\nu_2(N_2O)$	*Ref.*
$M(2)_{exp}/cm^{-1}$	1100	550	400	200	100
$M(2)_{calc}/cm^{-1}$	1000	500	360	180	100
$M(4)/cm^{-4}$	290×10^4	114×10^4	42×10^4	17×10^4	100
$\langle (OV)^2 \rangle/cm^{-2}$	$\sim 8 \times 10^4$	—	$\sim 15 \times 10^4$	—	99

In the foregoing work, no attempt has been made to interpret the observed correlation functions in terms of physical models for the systems. Gordon[45] has shown that his M- and J-diffusion models accurately reproduce the experimental correlation functions for CO both as pure gas and in mixtures with argon. These results are reproduced in Figure 16. The rotational correlation functions of CO and the parallel transition in N_2O have also been interpreted using the M-diffusion model.[101] Mixtures with N_2, O_2, H_2,

[99] M. Chalaye, E. Dayan, and G. Levi, *Chem. Phys. Letters*, 1971, **8**, 337.
[100] G. Levi, M. Chalaye, and E. Dayan, *Chem. Phys. Letters*, 1972, **12**, 462.
[101] G. Levi, M. Chalaye, and E. Dayan, *Compt. rend.*, 1972, **274**, *B*, 50.

and He at pressures of between 60 and 100 bar were used. Good agreement with the experimental data was obtained. Recently, Levi and Chalaye[102] studied rotational diffusion in compressed CH_3D in mixtures with He, Ne, Ar, H_2, O_2, N_2, and CH_4 at pressures of up to 150 bar. The $\nu_2(A_1)$ parallel band at 2204 cm^{-1} was chosen and the computed rotational correlation function compared with the predictions of the *M*- and *J*-diffusion models.

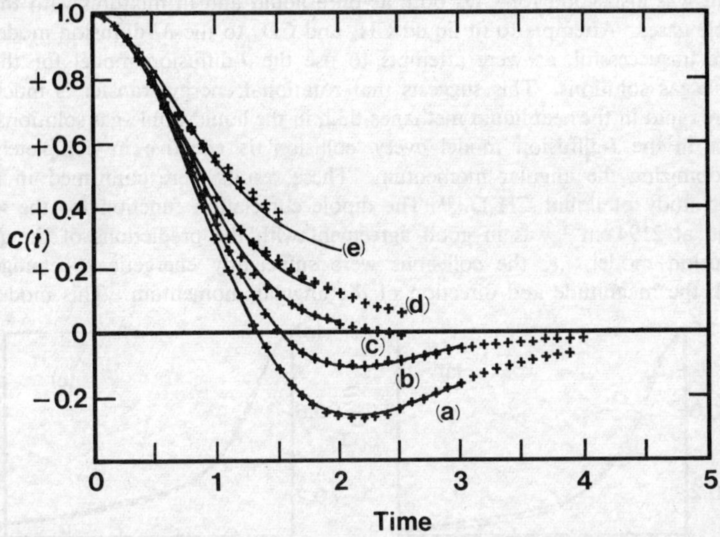

Figure 16 *I.r. correlation functions for carbon monoxide gas at room temperature* (*with argon at* 270 *and* 850 amagat) *in the gas just above the critical point* (*at* 270 *and* 1520 atm) *and in the liquid. The diffusion parameters for the theoretical curves are* $\tau = 30(1 + 0.05\omega^4)$, $3.1(1 + 0.05\omega^4)$, $1.5(1 + 0.1\omega^4)$, $1.0(1 + 0.1\omega^4)$, *and* 0.6
(Reproduced by permission from *J. Chem. Phys.*, 1966, **44,** 1830)

Vibrational relaxation was neglected in this treatment since the half-width of the corresponding isotropic Raman band was only 2 cm^{-1} compared with the width of the i.r. band which was >300 cm^{-1}. The *J*-diffusion model was found to give the best fit to the experimental data.

Condensed Phase. Extensive measurements of the i.r. absorption spectra of the spherical-top molecule methane, both as the pure liquid and dissolved in liquid Ar, Kr, and Xe, have been carried out by Cabana, Bardoux, and Chamberland.[103] Reorientational correlation functions for the ν_3 and ν_4 bands of CH_4 and CD_4 were computed. From their results they concluded that free rotational effects dominated the short-time dependence of the reorientational correlation functions whereas intermolecular effects became important at long times. Band moments and approximate intermolecular

[102] G. Levi and M. Chalaye, *Chem. Phys. Letters*, 1973, **19**, 263.
[103] A. Cabana, R. Bardoux, and A. Chamberland, *Canad. J. Chem.*, 1969, **47**, 2915.

torques were also computed. In a later publication, McClung[47] used the *M*- and *J*-diffusion model to calculate the reorientational correlation functions and spectral densities for CH_4 and CD_4 and compared them with the experimental values of Cabana *et al.*[103] The agreement was satisfactory for liquid CH_4 in the *J*-diffusion limit and for liquid Ar, Xe, and Kr solutions of CH_4 in the *M*-diffusion limit. These results are compared in Figure 17. Agreement was also good for CD_4 both as pure liquid and in mixtures with the noble gases. Attempts to fit liquid CH_4 and CD_4 to the *M*-diffusion model were unsuccessful, as were attempts to use the *J*-diffusion model for the noble-gas solutions. This suggests that rotational energy transfer is much more rapid in the neat liquid methanes than in the liquid noble-gas solutions, since in the *J*-diffusion model every 'collision' is effective in completely randomizing the angular momentum. These results were confirmed in a later study of liquid CH_3D.[104] The dipole correlation function for the ν_1 band at 2194 cm^{-1} was in good agreement with the predictions of the *J*-diffusion model, *i.e.* the collisions were sufficiently energetic to change both the magnitude and direction of the angular momentum. This model

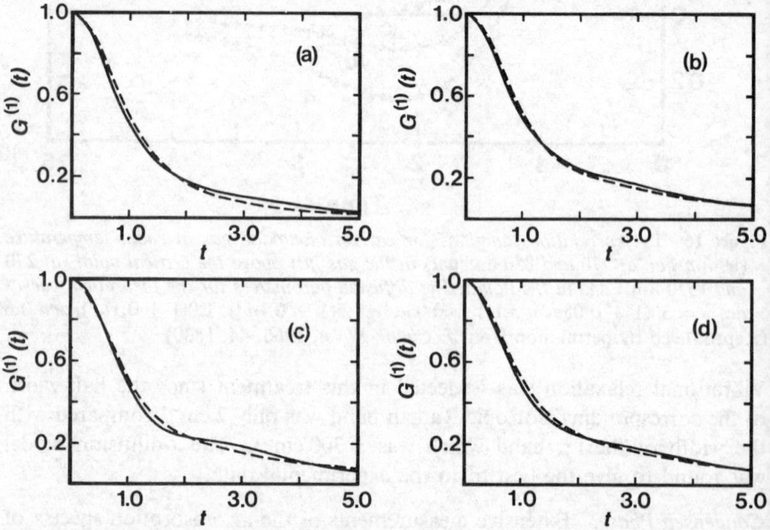

Figure 17 *Dipole correlation functions for the* ν_3 *band of* CH_4. (a) ——: *liquid* CH_4 (117 K), ----: *calculated J-diffusion (reduced* $\tau_j = 0.80$); (b) ——: 0.15% CH_4 *in liquid argon* (111 K), ----: *calculated M-diffusion model (reduced* $\tau_j = 0.80$); (c) ——: 0.15% CH_4 *in liquid xenon* (163 K), ----: *calculated M-diffusion (reduced* $\tau_j = 0.75$); (d) ——: 0.15% CH_4 *in liquid krypton* (135 K), ----: *calculated M-diffusion (reduced* $\tau_j = 0.70$)
(Reproduced by permission from *J. Chem. Phys.*, 1971, **55**, 3459)

[104] F. Marsault-Herail, J. P. Marsault, M. Chalaye, J. L. Saulnier, and G. Levi, *Compt. rend.*, 1972, **275**, *B*, 307.

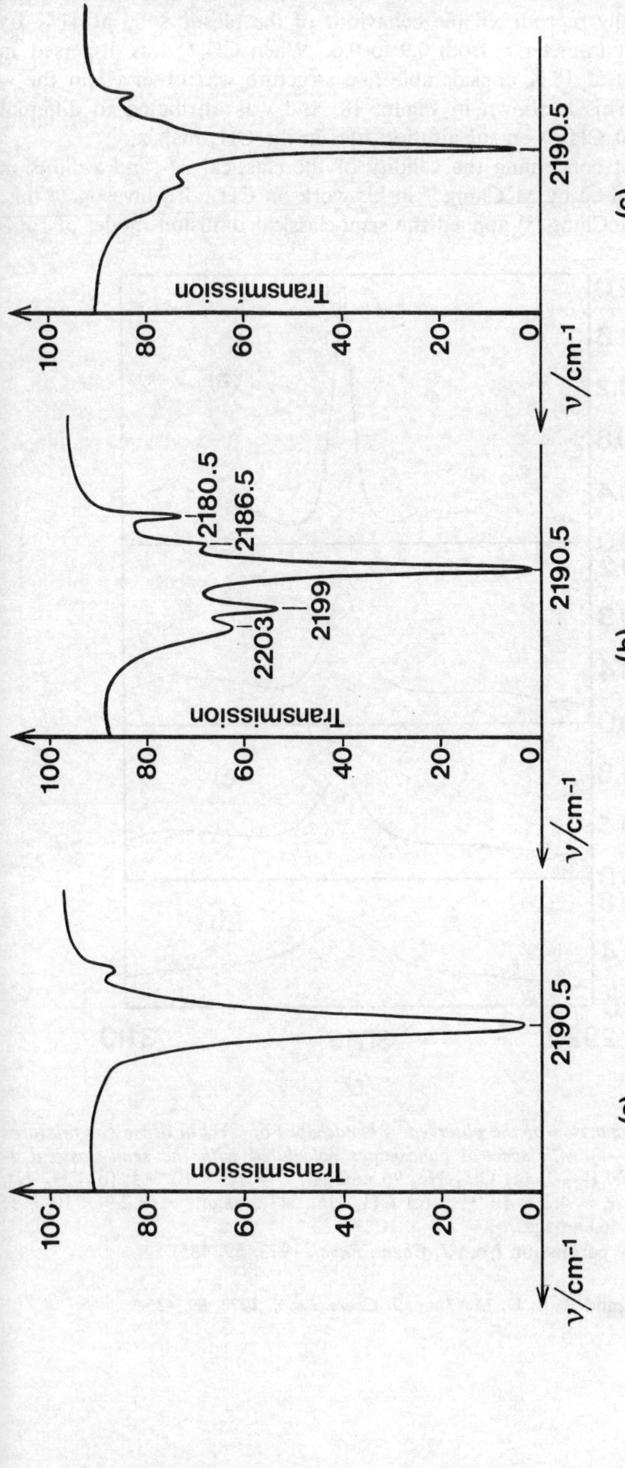

Figure 18 *Band profiles for* CH_3D *at 18 K: (a) pure solid; (b) 3% solution in* CH_4; *(c) 0.5% solution in* CH_4
(Reproduced by permission from *Compt. rend.*, 1972, **275**, *B*, 307)

also successfully reproduced the behaviour of the plastic solid at 41 K by changing the parameter τ_j from 0.9 to 0.6. When CH_3D was dispersed in a CH_4 matrix at 18 K considerable fine structure was observed in the ν_1 absorption. This is shown in Figure 18, and was attributed to different orientations of CH_3D on substitution sites in the CH_4 matrix.

Some doubt concerning the validity of the classical *M*- and *J*-diffusion models was raised by McClung [47] in his work on CH_4. To investigate this, Eagles and McClung [105] applied the semi-classical diffusion model of rota-

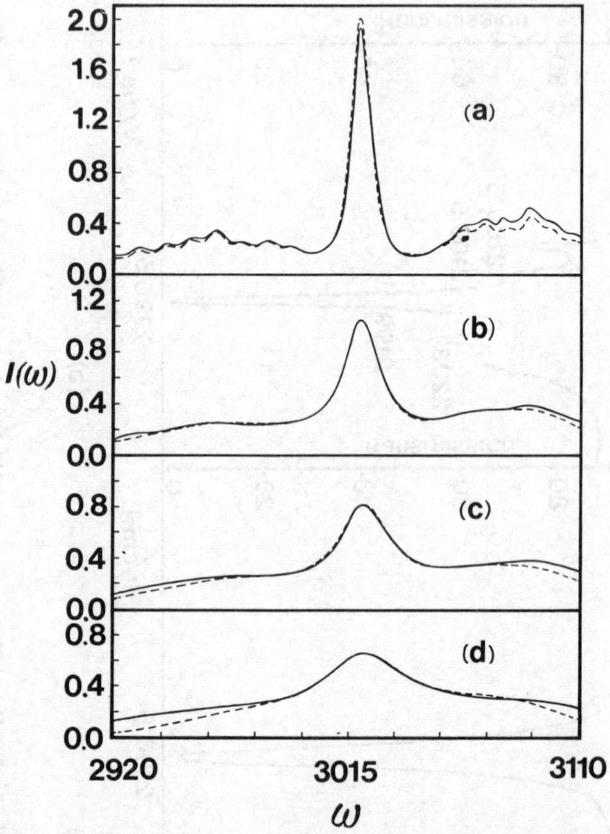

Figure 19 *Comparison of the observed* ν_3 *bandshapes of* CH_4 *in dense gas mixtures at* 295 K (——) *with spectral bandshapes calculated with the semi-classical J-diffusion model* (----): (a) CH_4–He, 96 amagat, $\tau_j = 12 \times 10^{-13}$ s; (b) CH_4–N_2, 186 amagat, $\tau_j = 4.2 \times 10^{-13}$ s; (c) CH_4–He, 543 amagat, $\tau_j = 2.9 \times 10^{-13}$ s; (d) CH_4–N_2, 463 amagat, $\tau_j = 1.8 \times 10^{-13}$ s
(Reproduced by permission from *J. Chem. Phys.*, 1973, **59**, 435)

[105] T. E. Eagles and R. E. D. McClung, *J. Chem. Phys.*, 1973, **59**, 435.

tional diffusion, developed by Gordon[11, 106] to interpret variations in the i.r. bandshape of CO, to calculate the ν_3 band contour of methane. This model, which is based on the transfer of intensity between vibration–rotation lines by collisions, was used to set up semi-classical analogues of the *M*- and *J*-diffusion models. The bandshapes predicted by these models were compared with the ν_3 contours of CH_4 in liquid methane, in dilute solutions of liquid noble gases, and in high-pressure gas mixtures with He and N_2. The observed ν_3 bandshapes of compressed CH_4 mixtures were compared with the spectral bandshapes computed with the semi-classical *J*-diffusion model in Figure 19. Equally satisfactory fits were obtained with both the *M*- an *J*-diffusion models. The best agreements were obtained with the *J*-diffusion model for pure liquid CH_4 and with the *M*-diffusion model for the liquid noble-gas solutions. This was in accord with the earlier results of McClung[47] using the classical diffusion models. Also, the bandshape computed from the classical model was compared with the symmetric part of the bandshape computed from the semi-classical model. The two models give similar results for correlation times greater than 2×10^{-13} s but for shorter correlation times the semi-classical spectra begin to narrow much more rapidly than the classical spectra. This implies that the asymmetry of a band is essentially quantum mechanical in origin, and a classical analysis of the symmetric part of the band led to the same dynamical results as the quantum models.

Several i.r. studies of the molecular reorientation of small molecules in solution have been published recently. Measurements have been made of the half-widths and band profiles of the three fundamentals and some harmonics of N_2O dissolved in a variety of solvents.[107] In the case of non-polar solvents the widths of the bands were consistent with a rotational relaxational mechanism, but in other cases variations in the intermolecular potential were important. Spectral moments and dipole correlation functions were calculated and interpreted in terms of rotational diffusion. Approximate values of the intermolecular torques were also calculated. Dipole correlation functions for the ν_3 and ν_5 bands of C_2H_2 dissolved in several polar solvents were also calculated.[108] Again rotational relaxation was assumed to be the dominant broadening mechanism.

The molecular motion of the condensed phases of the hydrogen halides have always held particular interest for the spectroscopist owing to the possibility of relatively unhindered molecular reorientation. Much of the work on HCl has been carried out by the French group at Bordeaux. Huong, Couzi, and Perrot[109] studied the vibration–rotation i.r. spectra of HF, DF, and HCl dissolved in liquid SF_6. The fine structure normally present only in the gas phase was found to persist in the SF_6 solutions. The HF and DF

[106] R. G. Gordon, *J. Chem. Phys.*, 1966, **45**, 1649.
[107] J. Vincent-Geisse, J. Soussen-Jacob, Nguyen-Tan Tai, and D. Descont, *Canad. J. Chem.*, 1970, **48**, 3918.
[108] G. Levi and M. Chalaye, *Compt. rend.*, 1970, **271**, *B*, 1093.
[109] P. V. Huong, M. Couzi, and M. Perrot, *Chem. Phys. Letters*, 1970, **7**, 189.

spectra are illustrated in Figure 20. These spectral features were interpreted in terms of rotational motions in solution. Birnbaum and Ho [110] recently found rotational lines in the far-i.r. spectrum of HCl dissolved in liquid SF_6.

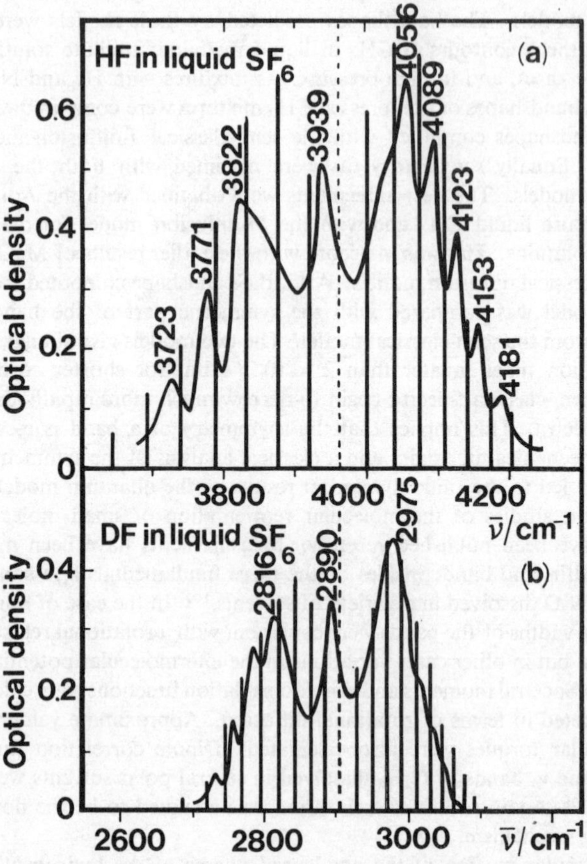

Figure 20 *I.r. spectra of solutions of* HF *and* DF *in liquid* SF_6
(Reproduced by permission from *Chem. Phys. Letters*, 1970, 7, 189)

A theoretical interpretation of the spectrum was proposed by Galatry and Robert.[111] Subsequently the work on SF_6 was extended by Perrot, Huong, and Lascombe [112] to include other solvents. Absorption studies of HCl dissolved in a variety of solvents were carried out in the far-, mid-, and near-i.r. regions to observe pure rotation, rotation–vibration, and second and third

[110] G. Birnbaum and W. Ho, *Chem. Phys. Letters*, 1970, **5**, 334.
[111] L. Galatry and D. Robert, *Chem. Phys. Letters*, 1970, **5**, 120.
[112] M. Perrot, P. V. Huong, and J. Lascombe, *J. Chim. phys.*, 1971, **68**, 614.

overtones respectively. The vibrational Raman bandshape of HCl in CCl_4 was also investigated. By hydrogen/deuterium substitution these workers were able to show that rotational relaxation accounted for most of the pure rotational band profile and was also the dominant mechanism in the pure and overtone vibrational absorptions. Nevertheless, a careful comparison of the profiles observed in the various spectral regions suggested that vibrational relaxation also contributed to the vibrational band contours. This relaxation was interpreted as arising from fluctuations in the environment of HCl molecules and also from rotation–vibration interactions. The analysis of the Raman profile supported this conclusion.

The influence of rotation–vibration interactions on the band profile was subsequently investigated by Perrot, Caloine, and Lascombe.[113] A technique for correcting the observed profile for these effects was developed and applied to the $0 \rightarrow 2$ absorption of HCl dissolved in liquid SF_6. Later i.r. studies on the HCl system by the same group [114] have shown, in addition to fluctuations of orientational states, the presence of a distribution of vibrational frequencies. The isotropic profiles of Raman bands were also found to originate essentially from the frequency distribution found in the i.r.

The dipole correlation function approach has also been used to investigate rotational motion in the hydrogen halides. Rothschild [115] used previously published extrapolated far-i.r. data [116] to compute correlation functions for HCl, DCl, and HF dissolved in cyclohexane. The results indicated that the dissolved halides undergo fast, large-angle rotational diffusion jumps. A quantitative estimate of collision-induced absorption between the HF and cyclohexane molecules indicated that dipole–induced dipole and quadrupole–induced dipole absorption was appreciable. The absorption band was corrected for quadrupole–induced dipole absorption in order to obtain the pure dipolar absorption bandshape. The i.r. and Raman band profiles of HCl and CO dissolved in CCl_4 were recently analysed by Guissani and Leicknam [117] using previously developed stochastic theories. The $0 \rightarrow 1$ absorption was studied for CO whereas for HCl the $0 \rightarrow 1$, $0 \rightarrow 2$, and $0 \rightarrow 3$ transitions were investigated. Both rotational and vibrational relaxation were found to be significant. Experimental dipole correlation functions were computed by Keller, Ebersold, and Kneubuhl [80] for CH_3Cl, HCl, and DCl dissolved in some Group IV tetrachlorides. By using these molecules as probes, the dynamics of the tetrahalides were investigated. CH_3Cl has a shape and size similar to those of the host molecules, whereas the hydrogen chlorides can enter the interstices between the host molecules. Dipole–dipole interactions were negligible in the case of the HCl and DCl molecules but important for $CHCl_3$. Relatively simple cavity models were sufficient to explain the main features of the HCl and DCl absorptions.

[113] M. Perrot, P. B. Caloine, and J. Lascombe, *Compt. rend.*, 1972, **274,** *C*, 104.
[114] M. Perrot and J. Lascombe, *J. Chim. phys.*, 1973, **70,** 5.
[115] W. G. Rothschild, *J. Chem. Phys.*, 1968, **49,** 2250.
[116] P. Datta and G. M. Barrow, *J. Chem. Phys.*, 1965, **43** 2137.
[117] Y. Guissani and J. L. Leicknam, *Canad. J. Phys.*, 1973, **51,** 938.

Considerable interest has recently been generated in the molecular dynamics of substituted methanes. This is due in part to the suitability of these small and relatively simple molecules of differing molecular symmetries for testing theories of rotational and vibrational relaxation. Since the pioneering work of Favelukes, Clifford, and Crawford [87b] on CH_3I a number of studies on substituted methanes have appeared in the literature. In Crawford's work, dipole correlation functions for the 885 cm^{-1} perpendicular band in CH_3I were computed from attenuated total reflectance measurements. It was concluded that on average the CH_3I molecules experience essentially free rotation over periods of about 0.2 ps, after which collisional effects begin to randomize the motion and rotational diffusion sets in. Substitution of D for H produced a slowing down of the diffusion rate. Essentially the same correlation function was obtained by Rothschild [118] using conventional transmission measurements. These data were consistent with a free rotational angle of 60° for the CH_3I molecule. Dipole correlation functions have also been computed from the ν_1 bands of $CHCl_3$, $CHBr_3$, and CHI_3 dissolved in CCl_4 and CS_2, and for liquid and gaseous $CHCl_3$ and $CHBr_3$.[119] CO dissolved in CCl_4 was also studied. The CO results were interpreted principally in terms of rotational fluctuations. Vibrational relaxation was found to be important for some of the haloform systems. Spectral moments and intermolecular mean square torques were also tabulated. The experimental moments $M(2)$ and $M(4)$ changed dramatically with different solvents for the CHX_3 molecules.

The bandshapes of three fundamentals of methylene chloride of different symmetry were recorded by Rothschild.[94] These were 283 (a_1), 895 (b_1), and 1266 cm^{-1} (b_2). Some overtone and combination bands were also investigated. These are included in Table 3. A primary aim of this work was to determine if the correlation functions produced from different modes of the same symmetry in CH_2Cl_2 were identical within experimental error. The dipole correlation functions of all the absorption bands were computed

Table 3 *Wavenumbers and assignments of some vibrational transitions of* CH_2Cl_2

Wavenumber/cm^{-1}	*Mode*	*Species*	*Assignment*[a]
283	ν_4	a_1	CCl_2 deformation
452	$\nu_9 - \nu_4$	b_2	
895	ν_7	b_1	CCl_2 rocking
1266	ν_8	b_2	CH_2 rocking
2308	$\nu_2 - \nu_7$	b_1	
2414	$\nu_5 + \nu_8$	b_1	
2525	$2\nu_8$	a_1	
2688	$\nu_2 + \nu_8$	b_2	

[a] Ref. 94.

[118] W. G. Rothschild, *J. Chem. Phys.*, 1969, **51**, 5187; 1970, **52**, 6453; 1972, **56**, 4722.
[119] I. Rossi-Sonnichsen, J. P. Bouanich, and N. V. Thanh, *Compt. rend.*, 1971, **273**, *C*, 19.

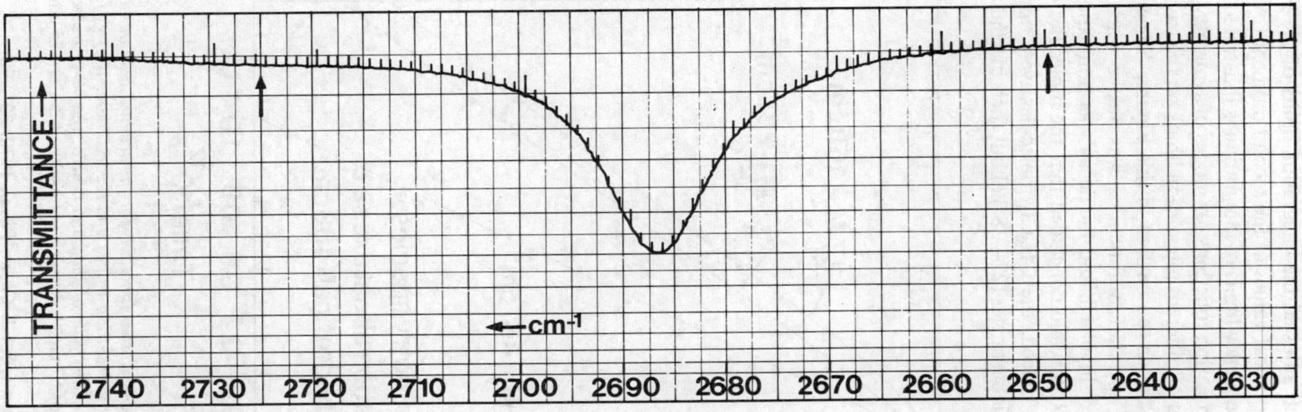

Figure 21 *Band contour of the $\nu_2 + \nu_8$ (b_2) i.r. combination. The vertical arrows show the frequency limits which yield the theoretical value of the second moment*
(Reproduced by permission from *J. Chem Phys.*, 1970, **53**, 990)

using a technique which adjusted the integration limits of the Fourier transform to obtain the theoretical values of the second moment. A typical band contour showing the frequency limits of integration is illustrated in Figure 21. Some experimental correlation functions for modes of symmetry a_1 are shown in Figure 22. The effect of different integration limits on the second moments and the correlation functions can be seen in this Figure. Since the observed rate of decay of the correlation functions was faster when the smaller moments of inertia were involved, and the decay was non-exponential for a certain initial period, Rothschild concluded that the molecules did not obey Debye-type diffusion. Indeed, the CH_2Cl_2 molecules were apparently able to turn through appreciable angles (27—38°) about their inertial axes. Agree-

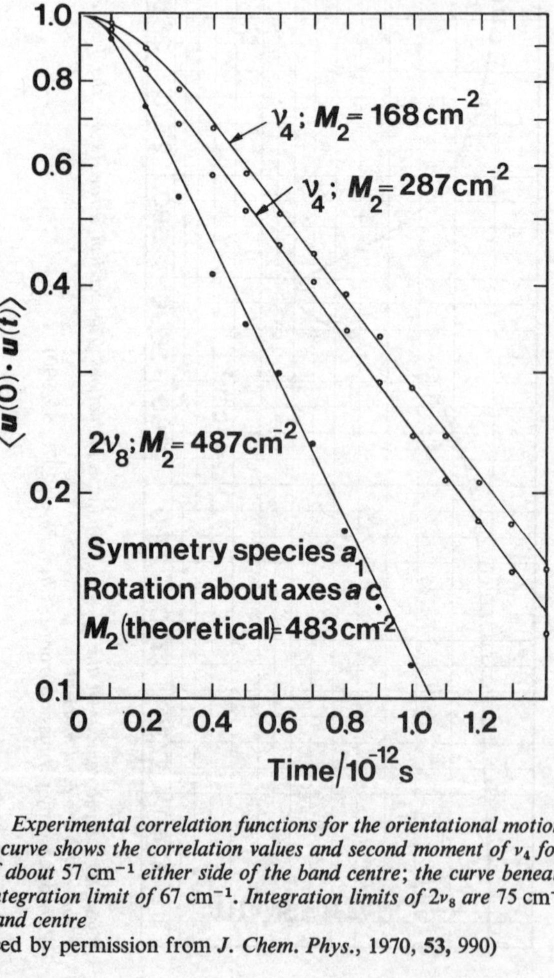

Figure 22 *Experimental correlation functions for the orientational motion of* CH_2Cl_2: *The top curve shows the correlation values and second moment of* ν_4 *for integration limits of about* 57 cm^{-1} *either side of the band centre; the curve beneath shows this for an integration limit of* 67 cm^{-1}. *Integration limits of* $2\nu_8$ *are* 75 cm^{-1} *either side of the band centre*

(Reproduced by permission from *J. Chem. Phys.*, 1970, **53**, 990)

ment within experimental error was obtained between the correlation functions computed from different vibrational bands of the same species. The anisotropy of the rotational motion in the liquid was found to resemble that of freely rotating CH_2Cl_2 molecules, indicating that the reorientational motion was not sensitive to intermolecular interactions. This behaviour is similar to that observed with CH_3I. Approximate values of the intermolecular torques were calculated from the observed $M(4)$, using the expression

$$\tfrac{1}{4}M(4)^a = (kT)^2[B^2(1 + C/A + A/C) + C^2(1 + B/A + A/B) - A(B + C) + 6BC] + \langle(OV)_b^2\rangle B^2 + \langle(OV)_c^2\rangle C^2 \quad (101)$$

which was derived from Gordon's general result. A, B, and C are rotational constants in this expression. Equation (101) describes the intermolecular effects influencing the motion of the inertial axis a (the relations for the other two axes are found by cyclic permutation of the symbols). $\boldsymbol{O}$ is an operator whose components O_a, O_b, and O_c along the axes, a, b, and c represent derivatives with respect to angles of rotation about a, b, and c; the quantity V denotes the angle-dependent part of the intermolecular potential. The values obtained for the mean square torque were $\langle(OV)_a^2\rangle^{\frac{1}{2}} \sim 280\ \mathrm{cm^{-1}\ rad^{-1}}$ and $\langle(OV)_b^2\rangle^{\frac{1}{2}} \sim \langle(OV)_c^2\rangle^{\frac{1}{2}} \sim 440 \pm 270\ \mathrm{cm^{-1}\ rad^{-1}}$, indicating that the torques in CH_2Cl_2 are not large and not very anisotropic.

The dynamical behaviour of acetonitrile CH_3CN is highly anisotropic owing to its molecular geometry. Because of the anisotropy of this elongated symmetric-top molecule, it provides an interesting molecule for testing theories of rotational diffusion.[54, 120] Frenkel *et al.*[54] have used CH_3CN as a test molecule for their variable-collision, rotational-diffusion model. Using a Poisson collisional distribution, the simulated absorption spectrum peaked at 25 cm^{-1} compared with the observed absorption maximum of 60 cm^{-1}. The use of a modified distribution, however, enabled the observed absorption to be closely reproduced. The peak in the collision distribution that was used corresponded approximately to a frequency of 200 cm^{-1}, indicating that there was considerable coupling between the internal vibrations and the rotational motions.

The size of the rotational and translational diffusion steps in several simple liquids was the subject of an article by Rothschild.[121] By using the dipole correlation function approach, he was able to show that the molecular reorientation in CH_2Cl_2, CH_3I, CH_3Cl, and C_6H_{12} consisted of angular jumps of the order of 20—60°. During a jump, which takes place in about 0.4 ps, the molecules behave as free rotors, regardless of the size of the molecule or the intermolecular forces. Furthermore, it transpired that the translational diffusion occurred by a jump mechanism, with a molecular

[120] J. E. Griffiths, *J. Chem. Phys.*, 1973, **59**, 751.
[121] W. G. Rothschild, *J. Chem. Phys.*, 1970, **53**, 3265.

trapping time of the order of 10^{-11}—10^{-10} s and jumps of the order of a molecular diameter. Large rotational and translational jumps of CH_2Cl_2 were even observed in a very viscous solution of polystyrene.[122] The macroscopic viscosity of the solutions appeared to have little effect on the rotational diffusion of the CH_2Cl_2 molecules. This type of large-angle rotational diffusion seems to be quite general for fluids composed of fairly small molecules where there are no specific strong interactions such as hydrogen-bond formation.

The effect of weak complex formation on the reorientational motion of liquid $CHCl_3$ has also been discussed by Rothschild.[123, 124] The dipole correlation function and proton nuclear magnetic relaxation techniques were combined to obtain details of the rotational motion and of the translation motion in the $CHCl_3$–C_6H_6 system. The contours of the 362 cm^{-1} parallel band in $CDCl_3$ and of the 1219 cm^{-1} perpendicular band in $CHCl_3$ were obtained in C_6H_6 solutions and Fourier-transformed. The resulting dipole correlation functions indicated that the two components forming the weak complex behaved essentially as random rotors, performing gas-like orientational jumps through average angles of 0.4—0.7 rad around their inertial axes. There was no evidence of any specific motion associated with the whole complex or of any directional intermolecular interaction. Similarly, the translational motion appeared to be characterized mainly by large-distance jumps (*ca.* 5 Å) of the individual moieties and not by jumps of the complex as a whole. It seems that the intermolecular interactions only exist during the lifetime of a molecular collision. This aspect will be discussed more fully in the section on low-frequency studies. Similar results have been obtained for the C_6H_6–C_6F_6 and other weakly complexed systems.[125] This picture is probably correct for most *weakly* interacting systems such as weak complexes and non-ideal liquids and solutions.

Both i.r. and Raman bands in water are unusually broad and asymmetric. The shapes of these bands are similar irrespective of their vibrational origin, indicating that their broadening and tailing is caused by similar perturbations. Time correlation functions were computed from Raman bands arising from both OH and OD stretching vibrations of HDO in liquid water.[128] As shown in Figure 23 both these functions rapidly decayed during the first fraction of a picosecond, faster in fact than the free-rotor correlation functions. Klier [127] observed similar behaviour for the dipole correlation functions of liquid water and of water adsorbed on a silica surface, which he interpreted as arising from strong vibrational perturbations of the rotational motion of the water molecules. This results in a periodic acceleration and slowing down (modulation) of the rotary motion. The application of first-

122 W. G. Rothschild, *Macromolecules*, 1968, **1**, 43.
123 W. G. Rothschild, *Chem. Phys. Letters*, 1971, **9**, 149.
124 W. G. Rothschild, *J. Chem. Phys.*, 1971, **55**, 1402.
125 R. T. Bailey and R. Ferri, unpublished work.
126 T. L. Wall, *J. Chem. Phys.*, 1969, **51**, 113.
127 K. Klier, *J. Chem. Phys.*, 1973, **55**, 737.

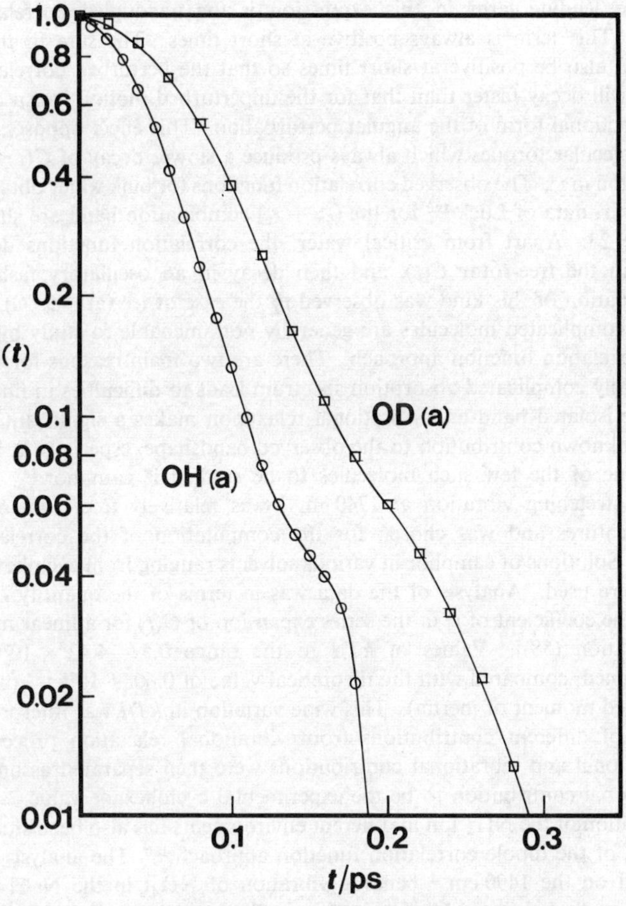

Figure 23 *The Fourier transforms for the* O—H (a) *and* O—D (b) *bands* (Reproduced by permission from *J. Chem. Phys.*, 1969, **51**, 113)

order perturbation theory led to a correlation function which differed from the usual expression for uncoupled vibrations and rotations by the modulating factor $\exp[i(\Delta\omega_l - \Delta\omega_m)t]$ which gives rise to the observed modulation. Thus the frequency of the modulation is determined by the difference between the average perturbation of the excited- and ground-state vibrational levels. Expanding the modulating factor as a function of time gives

$$\begin{aligned} C(t) = {} & \sum_l \rho_l \langle r_l|\boldsymbol{u}(0)\cdot\boldsymbol{u}^*(t)|r_l\rangle \\ & + it\sum_l \rho_l(\Delta\omega_l - \Delta\omega_m)\langle r_l|\boldsymbol{u}(0)\cdot\boldsymbol{u}^*(t)|r_l\rangle \\ & - \tfrac{1}{2}t^2\sum_l \rho_l(\Delta\omega_l - \Delta\omega_m)^2\langle r_l|\boldsymbol{u}(0)\cdot\boldsymbol{u}^*(t)|r_l\rangle \end{aligned} \tag{102}$$

where the leading term in this expression is the uncoupled correlation function. This term is always positive at short times. The sum in the t^2 term must also be positive at short times so that the perturbed correlation function will decay faster than that for the unperturbed motion irrespective of the functional form of the angular perturbation. This effect opposes that of the molecular torques which always produce a slower decay of $C(t)$ by a contribution in t^4. The observed correlation functions for bulk water obtained from the i.r. data of Luck [128] for the $(\nu_3 + \nu_2)$ combination band are shown in Figure 23. Apart from critical water, the correlation functions decay faster than the free-rotor $C(t)$, and then decay in an oscillatory fashion. No modulation of this kind was observed in the case of ice (at −9 °C).

Large, complicated molecules are generally not amenable to study by the dipole correlation function approach. There are two main reasons for this: the generally complicated absorption spectrum leads to difficulties in finding a suitable isolated band and vibrational relaxation makes a significant and usually unknown contribution to the observed bandshape, especially at long times. One of the few such molecules to be studied is camphor.[129] The carbonyl stretching vibration at 1740 cm^{-1} was relatively free from overlapping features and was chosen for the computation of the correlation function. Solutions of camphor in various solvents ranging from cyclohexane to CS_2 were used. Analysis of the data was in terms of the quantity kT/I, which is the coefficient of t^2 in the series expansion of $C(t)$ for a linear molecule [equation (58)]. Values of kT/I in the range 0.52—$4.42 \times 10^{24}\ s^{-2}$ were obtained, compared with the theoretical value of $0.60 \times 10^{24}\ s^{-2}$ (using an averaged moment of inertia). This wide variation in kT/I was interpreted in terms of different contributions from vibrational relaxation processes. The rotational and vibrational contributions were then separated assuming the rotational contribution to be the experimental cyclohexane value.

The motion of the NH_4^+ ion in different environments has also been studied by means of the dipole correlation function approach.[130] The analysis was performed on the 1400 cm^{-1} bending vibration of NH_4I in the NaCl-type lattice at room and liquid-N_2 temperatures. The resulting correlation function was in fair agreement with a model consisting of a torsional oscillator, interrupted by collisions with phonons in the lattice.

Raman Studies.—Raman scattering has only emerged recently as a tool for elucidating the details of rotational dynamics in molecular systems. Although Rakov and others have studied the relation between spectral bandshapes and rotational motions for many years and Gordon has provided the necessary framework for the correlation function approach, only recently has the detailed relationship between rotational relaxation and other relaxational processes been treated.

[128] W. A. P. Luck, *Ber. Bunsengesellschaft phys. Chem.*, 1965, **69**, 626.
[129] M. Jauquet and P. Laszlo, *Chem. Phys. Letters*, 1972, **15**, 600.
[130] B. Borstnick and A. Azman, *Spectrochim. Acta*, 1972, **28A**, 188.

Gaseous Phase. A recent study of Raman scattering on gaseous systems investigated the self-broadening of HCl, DCl, HBr, and DBr and their broadening by gaseous SF_6 and C_2F_6.[131] The total scattered intensity for HCl at pressures between 1 and 44 atm is shown in Figure 24. The results

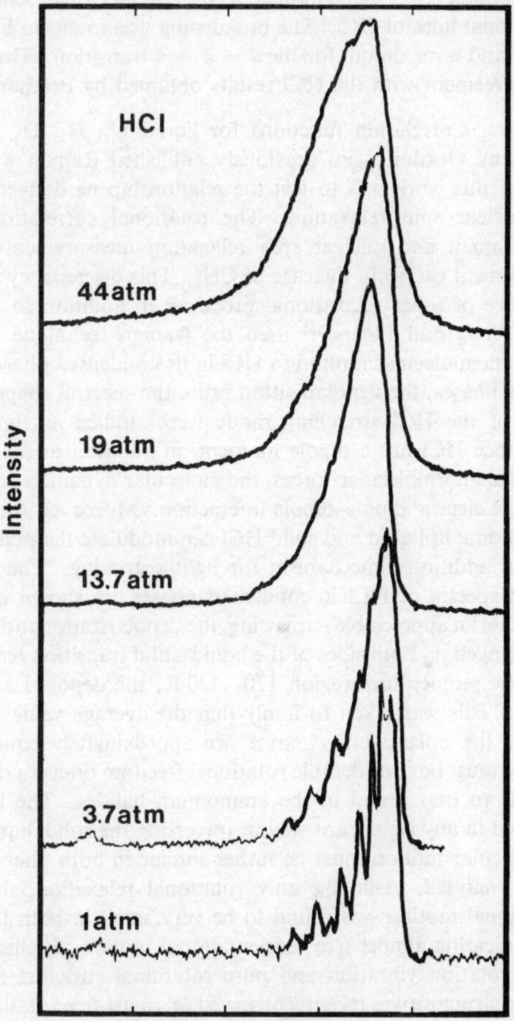

Figure 24 *Total scattering intensity for the Raman band of* HCl *gas between* 1 *and* 44 atm (Reproduced by permission from *Mol. Phys.*, 1972, **23**, 535)

[131] J. P. Perchard, W. F. Murphy, and H. J. Bernstein, *Mol. Phys.*, 1972, **23**, 535.

were compared with calculations using van Kranendonck's theory of line broadening.[132] It was shown that the self-broadening of the isotropic *Q*-branch does not follow the theory whereas broadening by a noble gas does. Optical cross-sections for multipolar forces were derived by Gray [133] from the general collision theory [132, 134] and applied to the special case of self-broadening of the rotational lines of HCl. The broadening was found to be strongly dependent on J and a maximum for the $J = 2 \rightarrow 4$ transition. These results were in good agreement with the HCl results obtained by Perchard *et al.*[131]

Condensed Phase. Correlation functions for liquid N_2, H_2, D_2, and CH_4 were calculated by Gordon from previously published Raman work. The main purpose of this work was to test the relationship he derived between Raman and nuclear spin relaxation. The rotational correlational times derived from Raman and nuclear spin relaxation measurements were in reasonable agreement except in the case of CH_4. This discrepancy may arise from the presence of other relaxational processes in addition to molecular reorientation. Wang and Fleury [135] used the Raman technique to obtain information on the molecular motion in HCl in its condensed phases. In the liquid and cubic phases, the depolarization ratio, the spectral shape, and the spectral width of the HCl stretching mode were studied as functions of temperature. Since HCl has a dipole moment, in addition to the influence of the short-range intermolecular forces, the molecular dynamics are effected by the long-range electric dipole–dipole interaction. Moreover, the presence of hydrogen-bonding in liquid and solid HCl can modulate the polarizability and provides an additional mechanism for light scattering. The polarized and depolarized spectra of HCl in condensed phases are shown in Figures 25 and 26. Despite an appreciable narrowing, the depolarization ratio remains essentially unchanged on both sides of the liquid–solid transition temperature (158.9 K). In the temperature region 170—120 K, the depolarization ratio was about 0.06. This was taken to imply that the average value α and the anisotropy β of the polarizability tensor are approximately equal. Thus, in the solid there must be considerable rotational freedom or else a disordered structure similar to that found in the ammonium halides. The lineshapes were not changed in any significant way in traversing the solid–liquid transition so the molecular motion must be rather similar in both phases. When the data were analysed, assuming only rotational relaxation, the barrier hindering rotational motion was found to be very small in both liquid and solid phases, indicating almost free reorientational motion. If this were the case, however, rotation–vibration and pure rotational structure should be observed.[65] Fine structure was recently observed for quasi-free rotation of HCl dissolved in liquid SF_6 and C_2F_6.[136] This would seem to indicate that other

[132] J. van Kranendonk, *Canad. J. Phys.*, 1963, **41**, 430.
[133] C. G. Gray, *Chem. Phys. Letters*, 1971, **8**, 527.
[134] C. G. Gray and J. van Kranendonk, *Canad. J. Phys.*, 1966, **44**, 2411.
[135] C. H. Wang and P. A. Fleury, *J. Chem. Phys.*, 1970, **53**, 2243.
[136] J. P. Perchard, W. F. Murphy, and H. J. Bernstein, *Mol. Phys.*, 1971, **23**, 519.

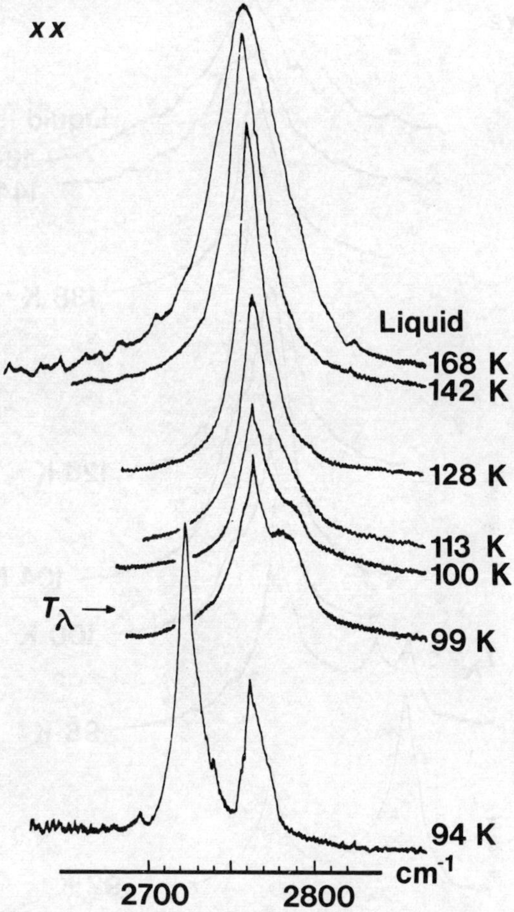

Figure 25 *Temperature variation of the polarized Raman spectrum of the* HCl *internal stretching mode*
(Reproduced by permission from *J. Chem. Phys.*, 1970, **53**, 2243)

relaxation effects were contributing to the observed band profile in this case.

In an attempt to resolve this problem, the combined techniques of Raman and Rayleigh scattering were used by Perchard, Murphy, and Bernstein [137] to obtain information on molecular reorientation in liquid HCl, DCl, and HBr systems. The results were discussed in terms of existing theories relating bandshapes and molecular motion.[138–140] The comparison of the Rayleigh and

[137] J. P. Perchard, W. F. Murphy, and H. J. Bernstein, *Mol. Phys.*, 1972, **23**, 499.
[138] J. Fiutak and J. van Kranendonk, *Canad. J. Phys.*, 1962, **40**, 1085.
[139] J. Fiutak and J. van Kranendonk, *Canad. J. Phys.*, 1963, **41**, 21.
[140] J. van Kranendonk, *Canad. J. Phys.*, 1963, **41**, 433.

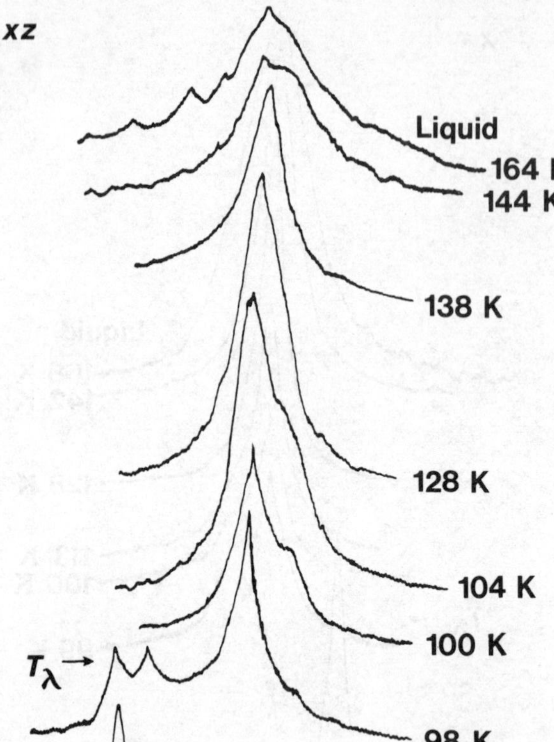

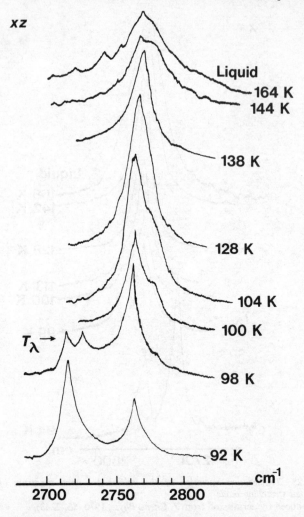

Figure 26 *Temperature variation of the depolarized Raman spectrum of the* HCl *internal stretching mode*
(Reproduced by permission from *J. Chem. Phys.*, 1970, **53**, 2243)

Raman scattering results showed that the profile of the anisotropic Raman band could not be explained simply in terms of the molecular reorientational processes. Thus, the molecular motion was analysed in terms of both the rotational $G_r(t)$ and vibrational $G_v(t)$ correlation functions in the usual way. The results for HCl are presented in Figure 27. The resulting values of the rotational correlation time τ_r, shown in Table 4, were found to be in good

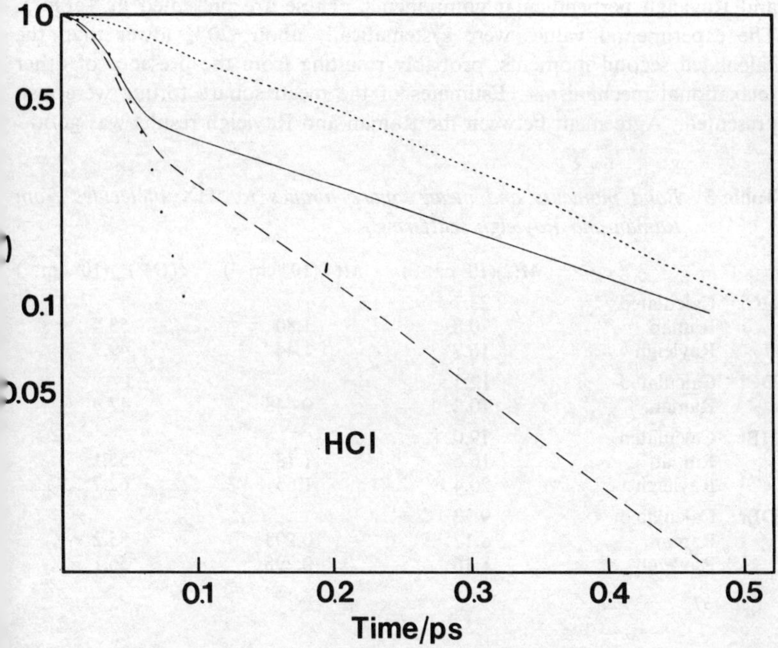

Figure 27 HCl *correlation functions*: ------ $G_v(t)$; —— $G_r(t)$; — — — $G_v(t) \times G_r(t)$; .... *correlation function for* HCl *gas*
(Reproduced by permission from *Mol. Phys.*, 1972, **23**, 499)

agreement with the Rayleigh scattering results. Three basic mechanisms for vibrational broadening were considered, *viz.* translational motion, resonance broadening, and vibration–rotation coupling. Most of the broadening was found to be due to translational movement, but self-broadening could also contribute significantly. After rotation through angles of *ca.* 30—40°, the motion becomes Brownian with a correlation time of the order of 0.25—0.3 ps. This correlation time was isotopically independent, in agreement with Debye's [34] relaxation theory and the results of rotational Brownian motion theory.[40] The second and fourth moments were calculated for the Raman

Table 4 *Rotational and vibrational correlation times for* HCl, DCl, HBr, *and* DBr[a] ($\tau \times 10^{13}$ s)

Species	*From* $G_v \times G_r$	*From* G_v	*From* G_r	*Rayleigh*
HCl	1.24	2.19	2.77	2.40
DCl	1.53	3.35	2.75	—
HBr	1.64	3.31	3.28	2.90
DBr	2.02	5.05	3.32	2.88

[a] Ref. 137.

and Rayleigh perpendicular components. These are presented in Table 5. The experimental values were systematically about 20% lower than the calculated second moments, probably resulting from the presence of other relaxational mechanisms. Estimates of the mean square torque were also presented. Agreement between the Raman and Rayleigh results was good.

Table 5 *Band moments and mean square torques for* HX *molecules from Raman and Rayleigh scattering*[a]

		$M(2)(10^3\ cm^{-2})$	$M(4)(10^9\ cm^{-1})$	$\langle(OV)^2\rangle(10^4\ cm^{-2})$
HCl	Calculated	23.7		
	Raman	20.8	1.80	53.4
	Rayleigh	16.7	1.44	49.9
DCl	Calculated	12.5		
	Raman	10.2	0.448	47.9
HBr	Calculated	19.0		
	Raman	16.6	1.18	52.1
	Rayleigh	20.4	1.65	62.7
DBr	Calculated	9.92		
	Raman	8.17	0.293	55.2
	Rayleigh	8.16	0.296	56.1

[a] Ref. 137.

In a subsequent paper these authors [136] extended their work to include solutions of hydrogen halides in various solvents. Extensive measurements of vibrational and, in some cases, pure rotational Raman spectra of HCl, DCl, and DBr dissolved in SF_6, C_2F_6, CCl_4, and SO_2 were carried out at room temperature. Band profiles of both the isotropic and anisotropic scattering were analysed and discussed in the light of existing theories of molecular motion in liquids. The change in band profile with solvent was generally comparable for HCl and HBr; the bands became broader as the intermolecular interactions increased. For SF_6 and C_2F_6 solutions, the band profiles of the total intensity are similar to those observed with low-pressure gases or gases pressured with noble gases.[131] The perpendicular components of HCl liquid and HCl dissolved in three solvents are compared in Figure 28. The *Q*-branch asymmetry due to vibration–rotation interactions was clearly observed, but the individual lines could not be resolved as in the case of gases. In the case of the inert solvents (SF_6, C_2F_6), the band contours were explained by an extension of the collisional theory [138] of Raman broadening. The most dramatic result was the observation of rotational fine structure for the vibration–rotation bands of HCl and HBr dissolved in SF_6 (Figure 29). These results can be compared with the pure rotational Raman spectra of the same solutions [141] and with the corresponding vibration–rotation i.r.

[141] J. P. Perchard, W. F. Murphy, and H. J. Bernstein, *Chem. Phys. Letters*, 1971, **8**, 559.

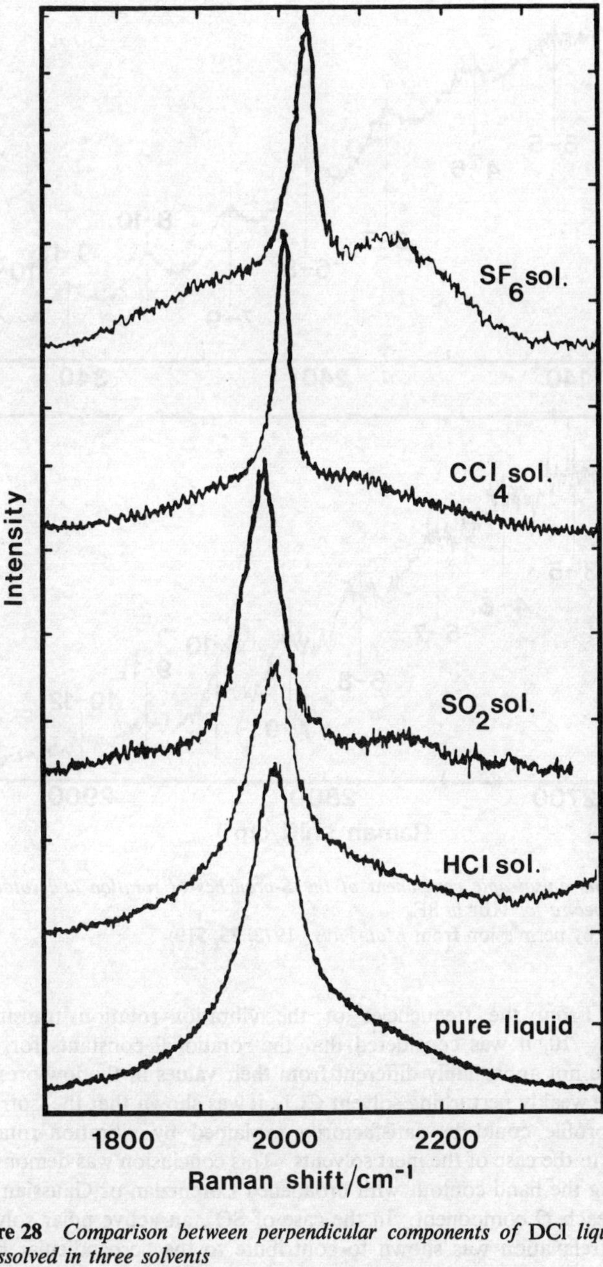

Figure 28 *Comparison between perpendicular components of* DCl *liquid and* DCl *dissolved in three solvents*
(Reproduced by permission from *Mol. Phys.*, 1972, **23**, 519)

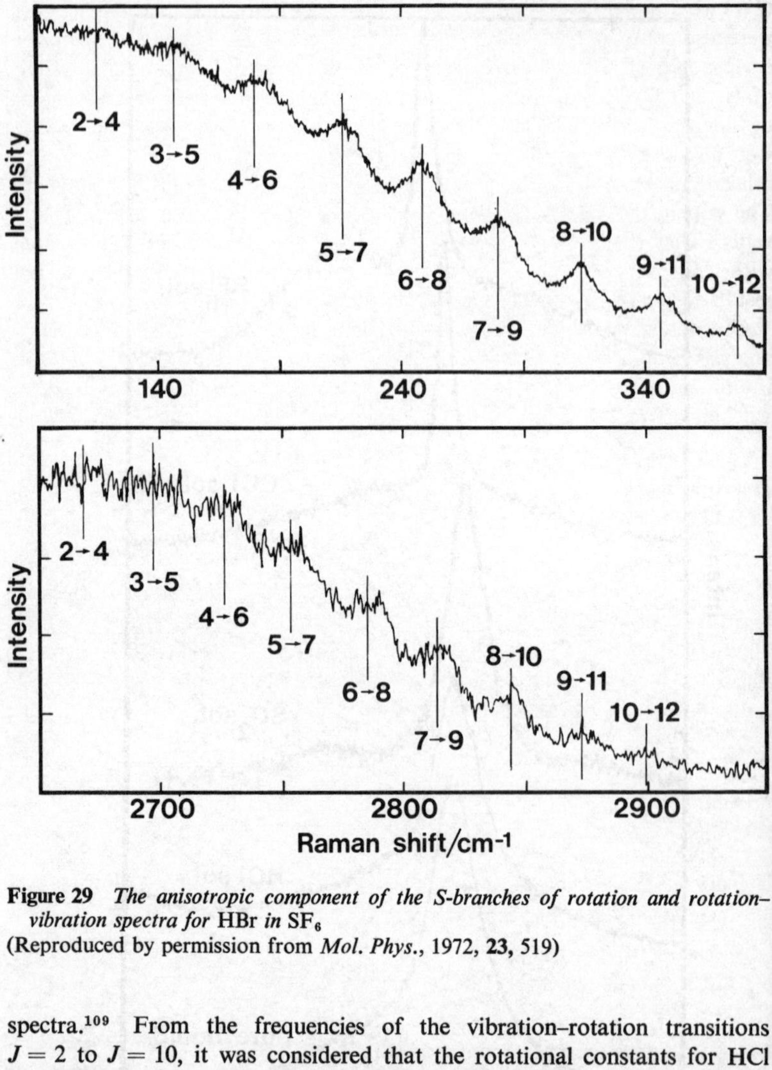

Figure 29 *The anisotropic component of the S-branches of rotation and rotation–vibration spectra for* HBr *in* SF_6
(Reproduced by permission from *Mol. Phys.*, 1972, **23**, 519)

spectra.[109] From the frequencies of the vibration–rotation transitions $J = 2$ to $J = 10$, it was considered that the rotational constants for HCl and HBr are not appreciably different from their values in the low-pressure gas. For the weakly perturbing solvent CCl_4, it was shown that the isotropic scattering profile could be satisfactorily explained by vibration–rotation coupling as in the case of the inert solvents. This conclusion was demonstrated by fitting the band contour with broadened Lorentzian or Gaussian line profiles of each Q component. In the case of SO_2, an active polar solvent, vibrational relaxation was shown to contribute to the perpendicular band profile to an appreciable extent. Both the correlation functions $G_r(t)$ and

$G_v(t)$ were therefore computed. Band moments and intermolecular torques were also computed; the values of the latter were in good agreement with the corresponding i.r. values. In similar work, the Raman spectrum of solutions of H_2 and D_2 in SF_6 and CCl_4 were investigated.[142] Precise measurements of linewidth, position, and intensity were made. The observed spectra were discussed in terms of the contribution made by rotational and vibrational relaxation. Rotational fine structure was observed for H_2 in both solvents. The same group [143] also analysed the half-widths of SO_2 in a variety of solvents. They found that the half-width of the ν_1 symmetric stretch decreased with temperature whereas that of the ν_3 asymmetric stretch increased. The results were interpreted using a relationship developed by Rokov.

Griffiths [120] recently reported isotropic and anisotropic Raman bandshape analysis of the $\nu_1\,(a_1)$ fundamentals in liquid CH_3CN and CD_3CN. From the orientational width $\omega_{or}(T)$, a reorientational correlation time τ_{or} can be calculated from the relation,

$$\tau_{or} = [2\pi c\omega_{or}(T)],$$

and this can then be used to compute a value for the perpendicular component $D_\perp$ of the diffusion tensor by

$$D_\perp = (6\tau_{or})^{-1}$$

Values of $D_\perp$ determined by the Raman method can then be used for comparison with values calculated from other techniques. Also, the experimental values of $D_\perp$ can be used to compare with theoretical values based on specific models. The values of $D_\perp$ computed from Raman bandshapes are compared with those obtained by n.m.r. and dielectric measurements and also with calculated values using the microviscosity theory [144] in Table 6. The agreement was good. I.r. transmission measurements have also been employed to determine $D_\parallel$ and $D_\perp$ for acetonitrile.[145] The results quoted for $D_\parallel$ and $D_\perp$, however, appear to be in error.[120]

Table 6 *Rotational diffusion in acetonitrile from Raman and n.m.r. measurements*[a] ($D \times 10^{-10}$ s^{-1})

Species	*Transition*/cm^{-1}	*Temperature*/°C	$D_\perp$ (*Raman*)	$D_\perp$ (*n.m.r.*)	*D* (*calc.*)
CD_3CN	2112 (ν_1)	25	16.3	14.0	17.6
		1	12.0	—	—
		—11.2	9.6	—	—
		—31	7.5	—	—
CH_3CN	2942 (ν_1)	25	18.3	15.1	18.9
	2249 (ν_2)	25	11.0	—	—
	918 (ν_4)	25	12.6	—	—

[a] See ref. 120 for origin of data.

142 W. Holzer, Y. LeDuff, and R. Ouillon, *Compt. rend.*, 1971, **273**, *B*, 313; Y. LeDuff, W. Holzer, and R. Ouillon, *Adv. Raman Spectroscopy*, 1973, **1**, 423.
143 R. Ouillon and Y. LeDuff, *Adv. Raman Spectroscopy*, 1973, **1**, 628.
144 D. E. Woessner, B. S. Snowden, jun., and E. T. Strom, *Mol. Phys.*, 1968, **14**, 265.
145 J. Yarwood, *Spectroscopy Letters*, 1972, **5**, 193.

Acetonitrile was also included in an i.r. and Raman study of highly polar and strongly associated liquid systems.[146] In this work, the characteristics of the rotational–vibrational relaxation of acetonitrile, benzonitrile, propan-2-ol, and 1,4-dioxan were observed by determining the correlation functions of several i.r. and Raman bands. Benzonitrile and propan-2-ol were also investigated as dilute solutions in CCl_4 and in CS_2, whereas 1,4-dioxan was studied in a variety of polar and non-polar molecules. The short-time behaviour of the correlation function is of primary interest in this study whereas in the previous work [120] the method of analysis dictated that only information regarding the longer-time behaviour could be obtained. As Rothschild pointed out, by concentrating on the short-time behaviour of the system, the effects of vibrational relaxation become less important. The results show that the intermolecular forces need *ca.* 0.1—0.2 $\times 10^{-12}$ s to produce observable effect on the molecular motion and that regardless of the polarity or association of the liquids or solutions, rotational relaxation was the dominant mode of decay for times less than 0.5 $\times 10^{-12}$ s. Benzonitrile and 1,4-dioxan were exceptions, owing to fast vibrational decay. The rotation of CH_3CN was also shown to be strongly hindered in those motions which tend to tilt the permanent dipole moment axis, but it was also shown that the molecules undergo rotational jumps by *ca.* 0.5—1.2 rad about this axis. The long-time decay (beyond 0.8 $\times 10^{-12}$ s), of the C—C stretch in propan-2-ol was mainly governed by vibrational decay, whereas for shorter times rotational relaxation was dominant. Molecular relaxation

Table 7 *Reorientational times for some liquid systems,* (T = 23 °C) *from Raman lineshapes*[a]

Liquid	*Point group*	*Transition/* cm^{-1}	*Symmetry species and assignment*		ω_{or}/cm^{-1}	τ (Raman)/ps
C_6H_{12}	802	D_{3h}	A_{1g}	CC stretch	3.5 ± 0.5	1.5 ± 0.3
	1027		E_u	CC stretch	3.1 ± 0.5	1.7 ± 0.3
	1157		A_{1g}	CH_2 rock	3.5 ± 0.5	1.5 ± 0.3
	1266		E_g	CH_2 twist	3.1 ± 0.5	1.7 ± 0.3
CCl_4	217	T_d	E	degn. defm.	3.0 ± 0.5	1.7 ± 0.3
C_6F_6	558	D_{6h}	A_{1g}	ring breathing	0.8 ± 0.5	7.0 $^{+10}_{-3}$
C_6H_6	992	D_{6h}	A_{1g}	ring breathing	2.0 ± 0.5	2.6 ± 0.3
CS_2	656	$D_{\infty h}$		sym. stretch	3.5 ± 0.5	1.5 ± 0.3
CH_3CN	918	C_{3v}	A_1	CC stretch	4.0 ± 0.5	1.5 ± 0.3
	2249		A_1	CN stretch	3.5 ± 0.5	1.5 ± 0.3
CH_3I	527	C_{3v}	A_1	Cl stretch	3.5 ± 0.5	1.5 ± 0.3
	1251		A_1	CH_3 defm.	3.8 ± 0.5	1.4 ± 0.3
$CHCl_3$	667	C_{3v}	A_1	CCl_3 stretch	3.8 ± 0.5	1.4 ± 0.3
	3019		A_1	CH stretch	3.5 ± 0.5	1.5 ± 0.3
$CHBr_3$	222	C_{3v}	A_1	CBr_3 defm.	1.0 ± 0.3	5.3 ± 2
$CBrCl_3$	247	C_{3v}	A_1	CCl_3 defm.	1.9 ± 0.3	2.8 ± 0.4

[a] Ref. 31.

[146] W. G. Rothschild, *J. Chem. Phys.*, 1972, **57**, 991.

in pure 1,4-dioxan and its solutions was apparently strongly influenced by vibrational relaxation effects.

A systematic study of reorientational motions in ten liquids was undertaken by Bartoli and Litovitz [31] using the linewidth technique.[30] Their results are given in Table 7, which includes the point group, symmetry species, ω_{or}, and τ_{or}. The Raman reorientational correlation times are compared in Table 8 with the corresponding τ values obtained from n.q.r. spectroscopy and Rayleigh scattering. For all liquids, the reorientational times from all three methods show agreement well within experimental uncertainties. The rotational diffusion tensor of benzene was found to be approximately isotropic. This result was at variance with the data of Gillen and Griffiths,[147] who found that reorientational motion in benzene was highly anisotropic.

Table 8 *Comparison of reorientational times for various liquids measured by different techniques* ($T = 25$ °C)[a]

Liquid	τ (*Raman*)/ps	τ (*n.q.r.*)/ps	τ (*Rayleigh*)/ps
CS_2	1.5	1.4	2.0
C_6H_6	2.6	—	2.6
CH_3CN	1.4	1.2	1.8
$CHCl_3$	1.5	1.8	—
CCl_4	1.8	1.7	—
C_6H_{12}	1.7	—	—
CH_3I	1.5	—	1.6

[a] Ref. 31.

The liquids studied were classified according to whether reorientational motion was determined by structural dynamics or by collisional motions and the detailed mechanism of reorientation in liquids was discussed. The plastic crystalline phase was also considered. Very little change in reorientational correlation times was found in traversing the liquid–plastic crystal phase boundary. The temperature dependences of the i.r. and Raman orientational times are shown in Figure 30. This seems to rule out a reorientational mechanism in which reorientation can only occur during a translational jump of the molecule.[148]

As has already been indicated, the recent work on benzene has shown the rotational diffusion tensor of benzene to be highly anisotropic.[147] By analysing Raman bandshapes and ^{2}D n.m.r. relaxation times as a function of temperature, values for the diffusion constants $D_\perp$ and $D_\parallel$ were obtained. Two models were used to interpret the data, the hydrodynamic [149] and the slightly damped free rotor (SDFR).[150] These models represent opposite microdynamic extremes, and predict quite a different temperature dependence

[147] K. T. Gillen and J. E. Griffiths, *Chem. Phys. Letters*, 1972, **17**, 559.
[148] D. E. O'Reilly and G. E. Schachen, *J. Chem. Phys.*, 1963, **39**, 1768; D. E. O'Reilly, *ibid.*, 1968, **49**, 5416.
[149] K. T. Gillen and J. H. Noggle, *J. Chem. Phys.*, 1970, **53**, 801.
[150] W. A. Steele, *J. Chem. Phys.*, 1963, **38**, 2404, 2410; W. B. Moniz, W. A. Steele, and J. A. Dixon, *ibid.*, p. 2418; P. W. Atkins, *Mol. Phys.*, 1969, **17**, 321.

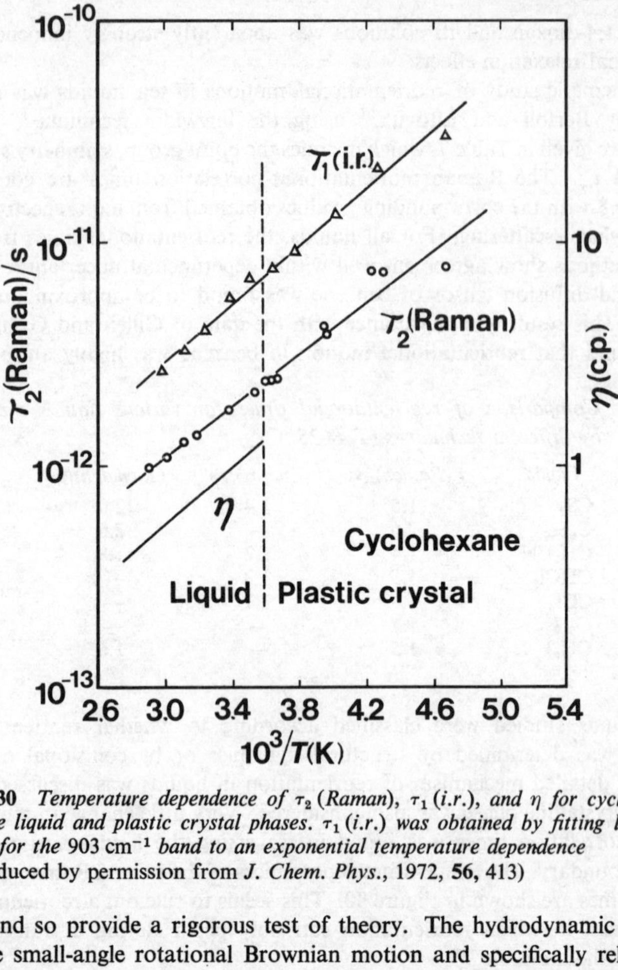

Figure 30 *Temperature dependence of* τ_2 *(Raman),* τ_1 *(i.r.), and* η *for cyclohexane in the liquid and plastic crystal phases.* τ_1 *(i.r.) was obtained by fitting linewidth data for the* 903 cm^{-1} *band to an exponential temperature dependence*
(Reproduced by permission from *J. Chem. Phys.*, 1972, **56**, 413)

of τ, and so provide a rigorous test of theory. The hydrodynamic models assume small-angle rotational Brownian motion and specifically relate the rotational and translational motions, whereas in the SDFR model molecular reorientations approach those in the low-pressure gas phase with large-angle rotational jumps. On comparison with the observed results, it was found that rotational motion about the symmetry axis of benzene was consistent with the SDFR model of diffusion whereas the rotational motion perpendicular to the C_6 axis was best interpreted in terms of small-angle rotational Brownian motion.

A rotational correlation time for benzene has also been obtained from the 992 cm^{-1} band by Clark and Miller,[151] assuming an exponential form for $C(t)$ at short times. They obtained a value of 1.3 ps, which is about a factor of two lower than the currently accepted value.

[151] J. H. R. Clark and S. Miller, *Chem. Phys. Letters*, 1972, **13**, 97.

Considerable interest has recently been generated in Raman studies of the orientational motion in substituted methanes. The explanation for this, as in the i.r. case, is their suitability as model compounds with which to test the validity of current experimental techniques and to seek justification for the theory underlying the separation of the various relaxational processes. In one of the first such investigations, i.r. and Raman correlation times for the C—I vibration in CH_3I and C_2H_5I and the C—Cl vibration in $(CH_3)_3CCl$ were obtained by Constant *et al.*[152] Both pure liquids and solutions in a variety of solvents were investigated. The temperature dependence of τ was also obtained.

Rayleigh scattering can also be used in principle to extract information concerning rotational behaviour. The expressions for G_v and $G_v \times G_r$ are analogous to the corresponding expressions for Raman scattering [equations (45) and (47)] except that the polarizability derivatives are replaced by the equilibrium polarizabilities. Vibrational relaxation does not contribute to these spectra, but other difficulties arise. The most important of these is the extent of the contribution of co-operative scattering between different molecules, but there are also experimental difficulties due to working very near the exciting line. A comparison of both Raman and Rayleigh studies of rotational diffusion in the anisotropic molecule $CHCl_3$ has been reported.[153] Correlation functions were computed from the ν_6 Raman band and the depolarized Rayleigh scattering and compared with theoretical expressions.[19, 35]

A comprehensive Raman study of the molecular interactions and motion in liquid CH_3I has recently appeared.[154] The temperature dependence of the 524 cm^{-1} (a_1) mode and the low-frequency Raman scattering (0—200 cm^{-1}) from room temperature down to its freezing point, and also in the super-cooled liquid, were investigated. The reorientational line broadening was separated from contributions from vibrational and other line-broadening processes. Corrected correlation functions for isotropic and anisotropic scattering for the two temperature extremes are shown in Figure 31. The room-temperature reorientational correlation function G_r is seen to decay much more quickly than the non-reorientational (isotropic) function G_v; at the lowest temperature studied this situation is totally reversed. Thus at room temperature the dominant process is molecular reorientation, whereas as the temperature is decreased this mechanism becomes less important and other mechanisms become dominant. The approximate correlation function for short times included in Figure 31 is given by the truncated Taylor series expansion

$$G_{approx} = 1 - \tfrac{1}{2}(\partial^2 G/\partial \tau^2)\,|_0\, \tau^2 = 1 - [M(2)_{exp}/2]\tau^2$$

The free-rotor correlation function G_{fr} decays much more quickly than

[152] M. Constant, M. Delhaye, and R. Fauquembergue, *Compt. rend.*, 1970, **271**, *B*, 1177.
[153] P. Lallemand, *Compt. rend.*, 1971, **272**, *B*, 429.
[154] R. B. Wright, M. Schwartz, and C. H. Wang, *J. Chem. Phys.*, 1973, **58**, 5125.

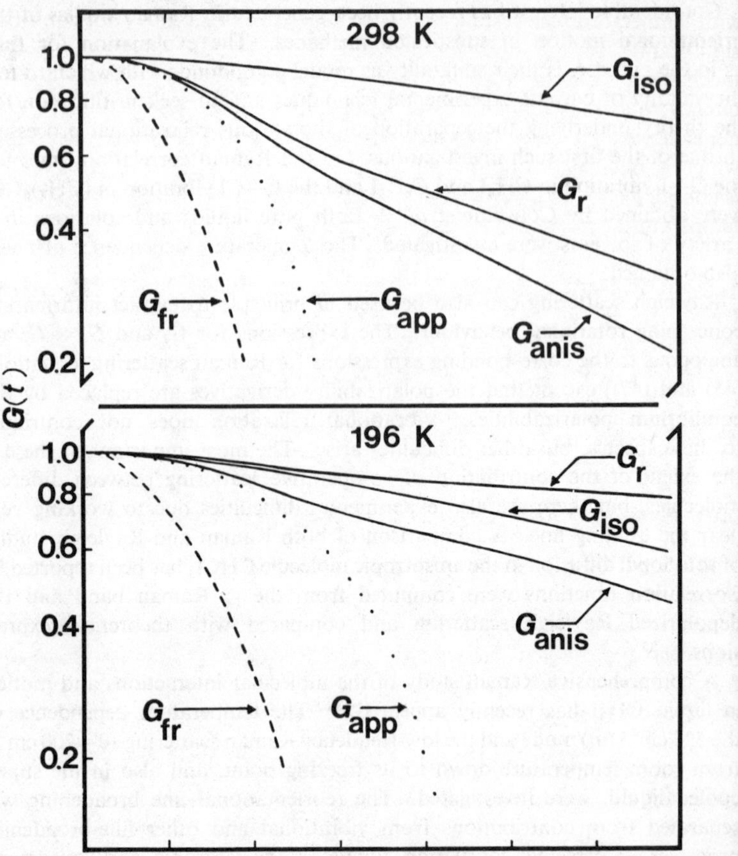

Figure 31 *Real part of the calculated correlation functions for the* 524 cm^{-1} *band of* CH_3I *at* 298 *and* 196 K
(Reproduced by permission from *J. Chem. Phys.*, 1973, **58**, 5125)

G_r, indicating that free rotation of CH_3I molecules is quite restricted, even at very short times. Finally, all three experimental correlation functions become exponential at long times, as predicted by the stochastic nature of the various contributing processes. The reorientation correlation time $\tau_{or} = 1.6$ ps was in good agreement with Bartoli and Litovitz's [31] room-temperature value of 1.5 ps. The temperature dependence of the correlation times associated with the 524 cm^{-1} band is shown in Table 9. It can be seen that the τ_{iso} become larger with increasing temperature. All theories which describe the vibrational relaxation process predict a decrease in τ_{iso}

with temperature, suggesting that vibrational relaxation does not contribute significantly to the observed isotropic broadening. Another important mechanism, collision-induced broadening, often used to explain low-frequency rotational Raman lineshapes, again predicts the wrong temperature dependence. However, this model is based on isolated binary collisions, which is obviously a poor approximation for liquid systems. If the CH_3I is considered to 'bounce' around inside a cage of similar molecules the correct temperature dependence is obtained.

Table 9 *Correlation times associated with the* 524 cm^{-1} CH_3I *Raman band*[a]

T/K	τ_{iso}/ps	τ_{aniso}/ps	τ_r/ps
196	1.82	1.45	7.13
209	1.92	1.55	7.93
228	1.91	1.38	5.03
242	2.08	1.36	3.96
259	2.18	1.24	2.87
298	2.31	0.95	1.61

[a] Ref. 154.

Further Raman work on the substituted methanes CH_3I and $CHCl_3$ was recently reported by Johnson and Drago.[155] This paper also included a general discussion of the correlation function approach and considered in particular the problem of separating the Raman reorientational correlation from the contribution arising from other relaxational processes. Reorientational correlation functions were computed for the totally symmetric C—D stretching mode in $CDCl_3$ and the C—I stretch in CH_3I over a range of temperatures. Correlation times computed from the $\langle P_2[\boldsymbol{u}(0) \cdot \boldsymbol{u}(t)]\rangle$ curves were used to evaluate an activation energy for the reorientational process according to

$$\tau = \tau_0 \exp(-E_a/kT)$$

The values of E_a obtained by this process agreed well with those obtained by n.m.r. relaxation, microviscosity, and dielectric relaxation. These results are compared in Table 10. Diffusion coefficients are also included in this Table.

Virtually all treatments of i.r. and Raman bandshapes, both experimental and theoretical, have neglected any coupling between rotation and vibration. With this in mind, Goldberg and Pershan [156] recently discussed the Raman bandshapes in CH_3I and CD_3I with particular emphasis on rotation–vibration coupling. The rotational and vibrational correlation functions for the a_1 bands were obtained in the usual way. On comparing the linewidths of the a_1 bands, it was found that, within experimental error, the ν_2 width was equal to the ν_3 width in each molecule, but the two ν_1 widths were larger

[155] E. R. Johnson and R. S. Drago, *J. Amer. Chem. Soc.*, 1973, **95**, 1391.
[156] H. Goldberg and P. S. Pershan, *Adv. Raman Spectroscopy*, 1973, **1**, 437.

than the corresponding ν_2 and ν_3 widths. In the absence of rotation–vibration coupling the a_1 rotation–vibration lineshapes (both i.r. and Raman) must be identical. The experimental results therefore indicate the presence of such coupling. Significant A_1–E-type Coriolis coupling was the proposed explanation for this rather unexpected result. The ν_2 and ν_3 modes are unaffected by this type of coupling and therefore provide more reliable information regarding molecular dynamics in the system. The diffusion coefficients $D_\perp$ derived from the ν_2 and ν_3 bands (1.5 and 1.2 × 10^{11} s^{-1} respectively) were in good agreement with $D_\perp$ values obtained from other techniques, but the $D_\perp$ value obtained from the ν_1 band (2.2 × 10^{11} s^{-1}), deviated significantly from other results.

Table 10 *Comparison of experimental values of E_a and D for* $CDCl_3$ *and* CH_3I

Experiment	E_a/kcal mol^{-1}	*D*/s^{-1}
	$CDCl_3$	
Raman	1.61	1.05 × 10^{11}
N.m.r. relaxation[a]	1.6 ± 0.1	
Dielectric relaxation[b]	1.75	0.68 × 10^{11}
	CH_3I	
Raman	1.88	1.1 × 10^{11}
Microviscosity[c]	1.9	1.07 × 10^{11}
Dielectric relaxation[c]	—	1.17 × 10^{11}

[a] W. T. Huntress, *J. Phys. Chem.*, 1969, **73**, 103. [b] S. Mallikarijun and N. E. Hill, *Trans. Faraday Soc.*, 1965, **61**, 1389. [c] P. Debye, 'Polar Molecules', Dover Publications, New York, 1929.

An interesting Raman study of reorientational motion in $CHCl_3$ as a function of pressure has been published by Campbell and Jonas.[157] The a_1 3019 cm^{-1} CH stretching band profile was analysed at pressures of 1, 1000, and 2000 bar at room temperature. The reorientational correlation function was separated from the effects of other relaxational processes in the manner discussed above. The predicted short-time behaviour of this function was obtained using the Gaussian function introduced by Bratos and Marechal.[65] Deuterium spin–lattice relaxation times (T_1) were also obtained as a function of pressure and used to compute rotational correlation times. Both n.m.r. and Raman experiments observe the same motion in $CHCl_3$, the reorientational motion about an axis perpendicular to the symmetry axis. The results from both sets of experiments are listed in Table 11. Theoretical values of τ_{or} were also calculated using the multiviscosity approach of Gierer and Wirtz [158] and Bridgeman's high-pressure viscosity results.[159] These are also included in Table 11. Although the theoretical values of τ_{or} were about 1.3

[157] J. H. Campbell and J. Jonas, *Chem. Phys. Letters*, 1973, **18**, 441.
[158] A. Gierer and K. Wirtz, *Z. Naturforsch.*, 1953, **8a**, 532.
[159] P. W. Bridgeman, *Proc. Amer. Acad. Arts Sci.*, 1926, **61**, 57.

times larger than the Raman results they both showed approximately the same temperature dependence. This behaviour was not observed in the n.m.r. times. Mean square torques were also computed and found to be of the same order as those reported for CH_3CN in n-heptane.[54]

Table 11 *Reorientational correlation times for* $CHCl_3$ *as a function of pressure* ($\tau_{or} \times 10^{12}$ s)[a]

		τ_{or}		
Pressure/bar	T_1/s	*Theoretical*	*Raman* (23 °C)	*N.m.r.* (25 °C)
1	1.39	2.60	1.97	1.74
1000	0.92	4.26	3.14	2.60
2000	0.77	6.36	5.42	3.10

[a] Ref. 157.

The librational motion of the NH_4^+ ion in NH_4Br below T_λ has been studied by Wang and Wright.[160] The frequency, intensity, spectral bandwidth, and Raman polarizability correlation function associated with this motion were obtained as a function of temperature in the range 90—200 K. At temperatures above 170 K, all correlation functions appeared to decay rapidly in the 0.4—1.5 ps region, and then decay much more slowly at times greater than 1.5 ps. Below 150 K, the decay was slow with no change in the decay rate. The two decay rates above 170 K indicated that two competitive intermolecular processes were probably present. Since the second moment of the band $M(2)$ is equal to the second derivative of the correlation function, the study of $M(2)$ as a function of temperature provides information regarding G_{lib} at short times and thus reveals the nature of the fast interaction process. The experimental second moments were found to increase exponentially with temperature, a result contrary to the behaviour expected if reorientational processes were dominant; $M(2)$ should increase linearly with temperature if this were the case. The magnitude of the observed $M(2)$ values was also inconsistent with a reorientational model. Correlation times were computed from the area of the correlation function up to 10 ps. Both these and the spectral linewidths were found to exhibit an Arrhenius-type temperature dependence. These data were consistent with a thermally induced hydrogen-bond-breaking process as the dominant interaction in the modulation of the librational mode.

Low-frequency Studies.—Low-frequency Raman and far-i.r. studies of molecular motion have developed substantially in the past few years. Advances in experimental techniques have contributed in no small part to the increased exploitation of this important spectral region. It is now possible to investigate transitions down to a few wavenumbers both in absorption and light scattering.

[160] C. H. Wang and R. B. Wright, *J. Chem. Phys.*, 1973, **58**, 2934.

Molecular dynamics studies in the low-frequency region can be broadly classified under the headings of rotational and translational motions. Since, however, in many systems the frequencies of the two motions are comparable it is difficult in some cases to make this distinction. No attempt will be made here to consider rotational and translational motions separately.

Far-infrared Studies. There have been a number of recent studies of the far-i.r. rotational absorption of compressed non-polar [161–168] and polar [168–170] gases and of the effect of inert gases on pure rotational lines.[171] Collision-induced spectra in the far-i.r. region yield valuable information on molecular motion and intermolecular forces. In the case of linear molecules, these spectra arise from a transient dipole moment induced in a pair of colliding molecules by the quadrupolar electric field. Tetrahedral molecules have no quadrupole moment but collision-induced spectra can arise from the transient dipole moment induced by the octupolar field and by molecular distortion due to overlap interaction. Molecular quadrupole moments are sensitive to the outer electronic charge distribution in a molecule and so are useful as tests for electronic wavefunctions. Furthermore, the electric fields of quadrupole moments can make significant contributions to the orientation-dependent part of the intermolecular potential.

The theory of pressure-induced spectra of noble-gas atoms and of linear molecules is well established [172, 173] and has been recently extended to include tetrahedral molecules.[167] In the case of tetrahedral molecules, the molecular octupole moment is related to the integrated intensity resulting from the modulation of the octupole-induced dipole moment by the rotational motion of the molecules.[167] Translational motion was not considered. It was shown that for the pure rotational spectrum the selection rules $\Delta J = \pm 1, \pm 2, \pm 3,$ will hold (as well as $\Delta J = 0$). The $\Delta J = 3$ transitions contribute some $\frac{2}{3}$ of the total intensity, thus dominating the shape of the spectrum. The analysis was applied [164–6] to experimental far-i.r. spectra of CH_4, CD_4, and CF_4 and gave octupole moments in good agreement with those obtained by other techniques. The general features of the spectra were also reproduced.

As an extension of their work on tetrahedral molecules Rosenberg and

161 D. R. Bosomworth and H. P. Gush, *Canad. J. Phys.*, 1965, **43**, 751.
162 A. Rosenberg and G. Birnbaum, *J. Chem. Phys.*, 1970, **52**, 683.
163 J. E. Harris, *J. Phys. (B)*, 1970, **3**, 704.
164 S. Weiss, G. Leroi, and R. H. Cole, *J. Chem. Phys.*, 1969, **50**, 2267.
165 A. Rosenberg and G. Birnbaum, *J. Chem. Phys.*, 1968, **48**, 1396.
166 G. Birnbaum and A. Rosenberg, *Phys. Letters (A)*, 1968, **27**, 272; W. Ho, G. Birnbaum, and A. Rosenberg, *J. Chem. Phys.*, 1971, **55**, 1028.
167 I. Ozier and K. Fox, *Phys. Letters (A)*, 1968, **27**, 274; *J. Chem. Phys.*, 1970, **52**, 1416.
168 S. Weiss and R. H. Cole, *J. Chem. Phys.*, 1967, **46**, 644.
169 A. I. Baise, *Chem. Phys. Letters*, 1971, **9**, 627.
170 I. Darmon, A. Gerschel, and C. Brot, *Chem. Phys. Letters*, 1971, **8**, 454.
171 M. E. van Kreveld, R. M. van Aalst, and J. van der Elsken, *Chem. Phys. Letters*, 1970, **4**, 580.
172 J. van Kranendonk and Z. J. Kiss, *Canad. J. Phys.*, 1959, **37**, 1187.
173 J. D. Poll and J. van Kranendonk, *Canad. J. Phys.*, 1961, **39**, 189.

Birnbaum[162] measured the far-i.r. spectra (12—250 cm^{-1}) of SF_6 at pressures from 3 to 19 atm in the gas phase and at 0 and —40 °C in the liquid phase. In view of the O_h molecular symmetry of SF_6 the lowest-order non-vanishing electric moment is the hexadecapole in the ground vibrational state. A number of absorption bands were found in the 12—250 cm^{-1} region. The intensity of one band at 50 cm^{-1} varied as the (density)^2 and so was attributed to collision-induced absorption. Other bands at 94 and 173 cm^{-1} were assigned to difference vibrations. The hexadecapole moment of SF_6 was estimated from the integrated intensity of the 50 cm^{-1} band.

Despite the extensive interest in the collision-induced spectra of non-polar gases, very little work on polar gases has been reported. This is due in the main to the complexity of these systems. In one such investigation Weiss and Cole[168] observed the collision-induced spectra of HCl and DCl in an attempt to separate the effects of dipolar and quadrupolar interactions. Quadrupole–induced dipole interactions lead to one member of a 'collision pair' being raised to a higher level whereas the other molecule remains in its original state. The selection rules for rotational transitions of this kind are $\Delta J = 0, \pm 2$ rather than the usual dipolar ($\Delta J = \pm 1$). By making use of this difference in selection rules, together with the fact that different populations of the molecular rotational states are involved in the two cases, these workers were able to separate contributions arising from dipolar and quadrupole–induced dipole effects. Molecular quadrupole moments calculated from the integrated intensities of the separated bands were $Q_{HCl} = 5.8 \times 10^{-26}$ esu cm^2 and $Q_{HBr} = 5.5 \times 10^{-26}$ esu cm^2.

Pure rotational absorption in the far-i.r. region is proportional to the density of polar molecules whereas induced effects are proportional to the density squared. Baise[169] used this effect to separate the quadrupole–induced dipole and dipole–induced dipole absorption from pure dipole–dipole rotational absorption in compressed N_2O. Since the dipole moment of N_2O is small (0.17 D), the induced absorption becomes larger than the pure rotational absorption at high densities. The far-i.r. spectra (8—100 cm^{-1}) of gaseous N_2O in the range 34—64 bar were reported and an approximate value of the quadrupole moment was computed ($Q \sim 8 \times 10^{-26}$ esu cm^2).

Measurements of line broadening in the pure rotational spectra of HCl in gaseous mixtures with inert gases have revealed that heavy atoms are more effective in broadening the rotational lines than are light atoms.[174] In this work, HCl was studied in the region 15—170 cm^{-1} at pressures of up to 100 atm in mixtures with He and Ar. The results were interpreted in terms of energy and momentum transfer in collisions of short duration. Related work on HCl[175] and the HBr–Ar system[176] has also been published.

The effect of pressure perturbation on the pure rotational spectrum of

[174] M. E. van Kreveld, R. M. van Aalst, and J. van der Elsken, *J. Chem. Phys.*, 1971, **55**, 2853.
[175] M. Giraud and J. Pourcin, *Compt. rend.*, 1968, **266**, *B*, 1593.
[176] J. Pourcin, G. Bachet, and R. Coulon, *Compt. rend.*, 1967, **264**, *B*, 975.

HCl was also investigated by Fabre *et al.*[177] Pure HCl at pressures of up to 35 bar and mixtures with argon in the pressure range 0—275 bar were the subjects of this Raman study. A true linewidth was computed from the experimental spectra and was found to be proportional to the perturber pressure. The results were in good agreement with the predictions of Gray's theory which was based on the effects of anisotropic intermolecular forces.[133, 134]

A number of far-i.r. studies of polar molecules dissolved in non-polar solvents have appeared in the literature.[70, 115, 116, 178, 179] One point of particular interest in these studies involves the question of free rotation in these liquid systems. In most cases, no fine structure indicative of free rotation was found. Only in the cases of HCl dissolved in liquid SF_6 [110] and in liquid Ar and Kr [180] have the remnants of fine structure been observed. The small perturbing effect of SF_6 in this case may be associated with the very high symmetry of this molecule. No fine structure was observed for DCl dissolved in liquid SF_6, which may be the result of the smaller spacing of the rotational transitions.

To obtain more information regarding the intermolecular origin of this fine structure, van Aalst and van der Elsken [180] computed the pure rotational dipole correlation function for HCl dissolved in liquid argon and liquid krypton. Both absorption bands exhibit residual fine structure on the high-frequency side. The dipole correlation function for HCl dissolved in argon is shown in Figure 32. A free-rotor correlation function for HCl is a periodic function with a period of 1.6 ps. At this time all freely rotating molecules should again have the same orientation as at $t = 0$. The experimental $C(t)$ does indeed have a maximum at 1.6 ps, but it falls considerably below this correlation function near $t = 0$. Also, the experimental function oscillates much more rapidly around the secondary maximum. In the *M*- and *J*-diffusion limit,[45] the ratio between the damping constants for HCl in argon and HCl in krypton should be 2 : 7. In fact, this ratio was found to be <1.0. Thus the damping of the rotational motion in HCl must to a certain extent be determined by the specific properties of the host liquid.

Robert and Galatry [181, 182] have presented a detailed quantum mechanical calculation of the far-i.r. absorption spectrum of polar molecules dissolved in non-polar liquid solvents. They used a cell model to represent the solution and calculated the total interaction between the solute and solvent molecules. Using this interaction, they obtained the resulting average deformation of the rotational wavefunctions in lowest order. They compared their predictions

[177] D. Fabre, G. Widerlocher, M. M. Thiery, H. Vie, and B. Voder, *Adv. Raman Spectroscopy*, 1973, **1**, 497.
[178] S. G. Kroon and J. van der Elsken, *Chem. Phys. Letters*, 1967, **1**, 285.
[179] E. Constant, Y. Leroy, J. L. Barois, and P. Desplanques, *Compt. rend.*, 1967, **264**, *B*, 228.
[180] R. M. van Aalst and J. van der Elsken, *Chem. Phys. Letters*, 1972, **13**, 631.
[181] D. Robert and L. Galatry, *Chem. Phys. Letters*, 1968, **1**, 526.
[182] D. Robert and L. Galatry, *J. Chem. Phys.*, 1971, **55**, 2347.

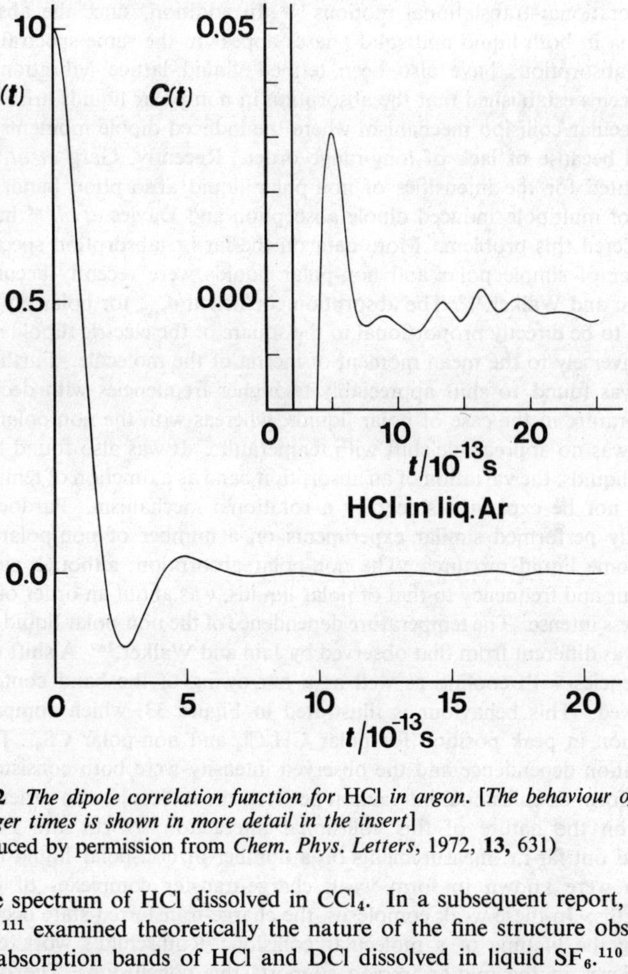

Figure 32 *The dipole correlation function for* HCl *in argon.* [*The behaviour of* $C(t)$ *at longer times is shown in more detail in the insert*]
(Reproduced by permission from *Chem. Phys. Letters*, 1972, **13,** 631)

with the spectrum of HCl dissolved in CCl_4. In a subsequent report, these authors [111] examined theoretically the nature of the fine structure observed on the absorption bands of HCl and DCl dissolved in liquid SF_6.[110] Two factors which govern the shape of an absorption band of a diatomic molecule dissolved in a polar solvent were discussed, the average value of the interaction energy between the dissolved molecule and its immediate environment, and the mean square of the fluctuations of that interaction. The periodic fine structure observed in the HCl solution in SF_6 was explained on this basis.

Far-i.r. absorption of liquids has received increasing attention in recent years. The absorption spectra of a number of polar and non-polar liquids have been reported. Low-frequency absorption in non-polar liquids has been interpreted in various ways, including librational motions in a cage [183]

[183] N. E. Hill, *Proc. Phys. Soc.*, 1963, **82,** 723.

and rotational–translational motions.[178] In addition, since the absorption maxima in both liquid and solid phases appear in the same spectral region, these absorptions have also been termed 'liquid lattice' vibrations.[184] It now seems established that the absorption in non-polar liquids arises from a bimolecular collision mechanism where the induced dipole moments do not cancel because of lack of long-range order. Recently, Garg *et al.*[185] have accounted for the intensities of non-polar liquid absorption bands on the basis of multipole–induced dipole absorption and Davies *et al.*[186] have also considered this problem. More data on the far-i.r. absorption spectra of a number of simple polar and non-polar liquids were recently accumulated by Jain and Walker.[187] The absorption coefficient α_{max} for polar liquids was found to be directly proportional to the square of the electric dipole moment and inversely to the mean moment of inertia of the molecule. Furthermore, α_{max} was found to shift appreciably to higher frequencies with decrease in temperature in the case of polar liquids, whereas with the non-polar liquids there was no appreciable shift with temperature. It was also found that, for polar liquids, the variation of an absorption band as a function of temperature could not be explained solely by a rotational mechanism. Pardoe [188] has recently performed similar experiments on a number of non-polar liquids and some liquid mixtures. The non-polar absorption, although similar in contour and frequency to that of polar liquids, was about an order of magnitude less intense. The temperature dependence of the non-polar liquid absorption was different from that observed by Jain and Walker.[187] A shift to lower frequencies with cooling as well as a narrowing of the band contour was observed. This behaviour is illustrated in Figure 33, which compares the variation in peak position for polar CH_2Cl_2 and non-polar CS_2. The concentration dependence and the observed intensity were both consistent with some form of collisional interaction between the molecules. In order to shed light on the nature of this collisional interaction, North and Parker [189] carried out far-i.r. measurements on a number of non-polar liquid mixtures which were known to form weak charge-transfer complexes of differing strengths. In these weak complexes, the charge-transferred state occurs only during the lifetime of a molecular collision. Rothschild's work on weak complexes in the mid-i.r. region supports this conclusion. The object of employing interacting molecules of this type was to enhance the collisional polarization and thus increase the absorption intensity. Carbon tetrachloride was used as the acceptor and benzene, toluene, *p*-xylene, and mesitylene were the donor molecules. As expected, the collisional polarization was found to be greatest and collisional frequency lowest when the charge-

[184] G. W. Chantry, H. A. Gebbie, B. Lassier, and G. Wyllie, *Nature*, 1967, **214**, 163.
[185] S. K. Garg, J. E. Bertie, H. Kilp, and C. P. Smyth, *J. Chem. Phys.*, 1968, **49**, 2551.
[186] M. Davies, G. W. F. Pardoe, J. E. Chamberlain, and H. E. Gebbie, *Trans. Faraday Soc.*, 1970, **66**, 273.
[187] S. R. Jain and S. Walker, *J. Phys. Chem.*, 1971, **75**, 2942.
[188] G. W. F. Pardoe, *Trans. Faraday Soc.*, 1970, **66**, 2699.
[189] A. M. North and T. G. Parker, *Trans. Faraday Soc.*, 1971, **67**, 2234.

transfer interaction was strongest. It was suggested that the slower 'sticky' collision arises because of a reduction in the slope of the repulsive pair-interaction potential.

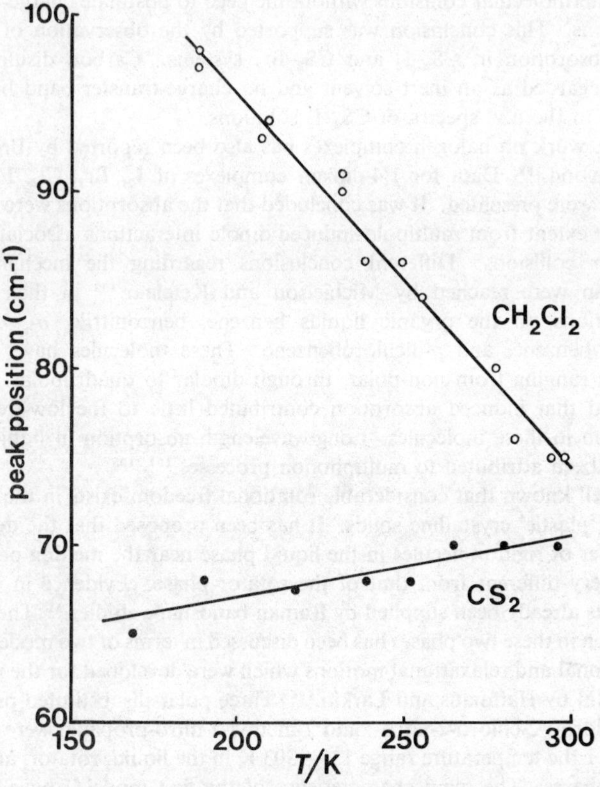

Figure 33 *The dependence of peak position $\bar{\nu}_0$ on temperature for* CS_2 *and* CH_2Cl_2 (Reproduced by permission from *Trans. Faraday Soc.*, 1970, **66**, 2699)

A somewhat similar approach was employed by Kettle and Price [190] to examine the influence of charge-transfer interactions on the absorption of non-polar liquids in the far-i.r. region. Low-frequency absorption measurements on solutions of iodine in benzene, dioxan, and carbon disulphide and on bromine in carbon disulphide were reported. Charge-transfer interactions are known to exist in these systems and dielectric relaxation measurements [191] have shown that the lifetimes of these weak complexes are of the order of 10^{-12} s. It is thus apparent that the excess polarization observed

[190] J. P. Kettle and A. H. Price, *J.C.S. Faraday II*, 1972, **68**, 1306.
[191] R. A. Crump and A. H. Price, *Trans. Faraday Soc.*, 1970, **66**, 22.

in solutions of weak charge-transfer complexes arises from collisional polarization. The intensities of all the low-frequency liquid absorption bands were considerably enhanced by the addition of iodine. The observed absorptions were adequately explained by induced dipole moments produced during intermolecular collisions without the need to postulate charge-transfer interactions. This conclusion was supported by the observation of similar far-i.r. absorption in CS_2–I_2 and CS_2–Br_2 systems. Carbon disulphide is usually regarded as an inert solvent and no charge-transfer band has been observed in the u.v. spectra of CS_2–I_2 solutions.

Far-i.r. work on halogen complexes has also been reported by Brownson and Yarwood.[192] Data for 1,4-dioxan complexes of I_2, Br_2, Cl_2, ICl, IBr, and ICN were presented. It was concluded that the absorptions were derived to a large extent from multipole–induced dipole interactions associated with molecular collisions. Different conclusions regarding the mechanism of absorption were reached by Michielson and Ketelaar[193] in their far-i.r. investigations on the organic liquids benzene, benzonitrile, *o*-, *m*-, and *p*-dicyanobenzene, and *p*-dichlorobenzene. These molecules have varying moments ranging from non-polar, through dipolar to quadrupolar. It was concluded that induced absorption contributed little to the low-frequency absorption in these molecules. Long-wavelength absorption in liquid paraffins has been attributed to multiphonon processes.[194, 195]

It is well known that considerable rotational freedom exists in the rotator phase of 'plastic' crystalline solids. It has been proposed that the degree of local order of rigid molecules in the liquid phase near the melting point will not be very different from that of the rotator phase. Evidence in support of this has already been supplied by Raman bandshape studies.[31] The far-i.r. absorption in these two phases has been discussed in terms of two models[67, 195] of librational and relaxational motions which were developed for the rotator-phase solid by Haffmans and Larkin.[196] Three polar disubstituted propanes (2,2-dichloro-, 2-chloro-2-nitro-, and 2-methyl-2-nitro-propane) were investigated over the temperature range 133—303 K in the liquid, rotator, and non-rotator phases. The main characteristics of the first model[67] have already been discussed. The second model consists basically of a molecule undergoing damped librational motion within a cage of nearest neighbours, the damping arising from lack of rigidity of the cage. The co-operative diffusional reorientation of the cage molecules about their centres of gravity is a function of the moment of inertia of the cage and its damping coefficient. The inclusion of two damping coefficients allows good agreement with the experimental spectra to be obtained for all three compounds for both liquid and plastic phases. A good fit was also obtained with the first model. Thus, although only tran-

[192] G. W. Brownson and J. Yarwood, *Spectroscopy Letters*, 1972, **5**, 185.
[193] C. Hall, J. W. Fleming, G. W. Chantry, and J. A. D. Matthew, *Mol. Phys.*, 1971, **22**, 323.
[194] C. Hall and R. J. Bell, *Mol. Phys.*, 1972, **24**, 511.
[195] G. Wyllie, *J. Phys.* (*C*), 1971, **4**, 564.
[196] R. Haffmans and I. W. Larkin, *J.C.S. Faraday II*, 1972, **68**, 1729.

sient local order is to be expected in the liquid phase of spherical molecules, even at temperatures just above the melting point, the models do give a reasonable representation of the liquid spectra in terms of librational and relaxational motion.

The relationship between rotation–vibration Raman scattering and far-i.r. absorption has been considered by Constant and Fauquembergue.[197] The reorientational correlation function for Raman scattering G_{2R} was computed for the 525 cm^{-1} band of CH_3I and compared with the correlation function G_{1R} obtained from the far-i.r. spectra of dilute solutions of CH_3I in hexane and carbon tetrachloride. The ratio of the correlation times τ_{1R}/τ_{2R} was found to lie in the range 2.3—3, indicating that the reorientational motions followed a rotational diffusion process. Good agreement was also obtained with the expressions derived by Berne, Pechukas, and Harp [81] [equations (71) and (72)]. The far-i.r. absorption of liquid N_2 and of dilute solutions of N_2 in argon has been measured by Buontempo *et al.*[198] The shape of the absorption curve was essentially identical in both cases, implying that quadrupole-induced rotational transitions dominate the absorption. The induced far-i.r. absorption of neon dissolved in liquid argon has also been reported and the corresponding autocorrelation function computed.[199] Intercollisional interference effects were found to be important at low frequencies (wavenumbers below 70 cm^{-1}). The translational absorption spectra of dilute solutions of He and Ne dissolved in liquid argon were also reported by this group.[200] The two spectra consisted of broad bands similar in shape peaked at 70 cm^{-1} (Ne–Ar) and 120 cm^{-1} (He–Ar). They were characterized by a long high-frequency tail and a rapid cut-off at low frequency. The two lineshapes were compared, taking into account the difference in the duration of the collision-induced dipole moment. At high frequencies the two curves were similar to those observed in the gas phase, whereas at low frequencies their departure from the gas-phase spectra was attributed to a negative correlation of the dipole moment induced in successive repulsive interactions. The correlation functions of the induced dipole moment were also computed. The far-i.r. spectra of liquid argon doped with Kr and Xe were found [201] to resemble the absorption of solid mixtures. This result was explained in terms of impurity-induced absorption through collective modes.

Low-frequency Raman Studies. Interest in low-frequency Raman scattering is currently increasing rapidly and this technique is certainly destined to yield a great deal of information on molecular interactions to complement the far-i.r. studies. Weak depolarized Raman scattering at low frequencies in a number of spherical and tetrahedral molecules in dense gases and liquids has also been

[197] M. Constant and R. Fauquembergue, *Compt. rend.*, 1971, **272,** *B*, 1293; *Adv. Raman Spectroscopy*, 1973, **1,** 413.
[198] U. Buontempo, S. Cunsolo, and G. Jacucci, *Mol. Phys.*, 1971, **21,** 381.
[199] U. Buontempo, S. Consolo, and G. Jacucci, *Phys. Letters* (*A*), 1970, **31,** 128.
[200] U. Buontempo, S. Consolo, and G. Jacucci, *Canad. J. Phys.*, 1971, **49,** 2870.
[201] U. Biskup and V. Wagner, *Optics Comm.*, 1969, **1,** 3.

interpreted in terms of collision-induced effects.[178, 202–208] The strong intermolecular interactions in fluids produce a fluctuating distortion of the symmetric polarizability tensors of the isolated molecules, giving rise to low-energy light scattering. These interactions of course also give rise to the excess absorption in the far-i.r. Three basic models of collision-induced light scattering have been proposed. Briefly, they consist of:

(*a*) Dipole–induced dipole. The anisotropy in this case results from the interaction between electric dipoles induced in the molecule by the incident light. The total field acting on a given molecule is then the sum of this external light field and the internal field due to the induced dipoles in its neighbours. Orientational fluctuations in the net induced moment due to the molecular motion yield the depolarized scattering. This is a long-range interaction giving an effective pair polarizability which varies as r^{-3} where r is the distance between the molecules in the colliding pair.

(*b*) Electron overlap. The distortion of the electron clouds during collisions produce this induced polarizability. Both $r^2 \exp(-ar^2)$ and r^{-9} dependences have been proposed.

(*c*) Frame distortion. Frame distortion during close collisions gives a pair polarizability varying as r^{-13}. This is the principle induction mechanism in molecular liquids.

Theoretical treatments of Rayleigh-wing light scattering have published by several authors. Thibeau *et al.*[209] calculated the intensity of depolarized scattering expected in argon gas from binary collisions. A dipole–induced dipole model was assumed with complete neglect of overlap effects. Reasonable agreement with the experimental depolarized scattering was obtained for gaseous argon in the range 0—150 bar. In a later publication, Thibeau and Oksengorn [210] calculated the spectrum of argon gas at 300 K which agreed well with the experimental data.[205] Thus both the intensity and the spectral behaviour in the 0—40 cm^{-1} region of the argon spectrum were adequately explained by assuming that the polarizability distortion was given by a dipole–induced dipole model. A similar calculation of the correlation function confirmed that dipole–induced dipole moments were the main cause of the scattering.[211] An overlap contribution to the polarizability distortion in argon has been detected by Volterra, Bucaro, and Litovitz.[212]

The low-frequency spectra of some liquids also show the same exponential

[202] H. S. Gabelnick and H. L. Strauss, *J. Chem. Phys.*, 1968, **49**, 2334.
[203] W. S. Gornall, H. E. Howard-Lock, and B. P. Stoicheff, *Phys. Rev.* (*A*), 1970, **1**, 1288.
[204] H. B. Levine and G. Birnbaum, *Phys. Rev. Letters*, 1968, **20**, 439.
[205] J. P. McTague and G. Birnbaum, *Phys. Rev. Letters*, 1968, **21**, 661; *Phys. Rev.* (*A*), 1971, **3**, 1376.
[206] J. P. McTague, P. A. Fleury, and D. B. Dupré, *Phys. Rev.*, 1969, **188**, 303.
[207] J. A. Bucaro and T. A. Litovitz, *J. Chem. Phys.*, 1971, **54**, 3846.
[208] J. A. Bucaro and T. A. Litovitz, *J. Chem. Phys.*, 1971, **55**, 3585.
[209] M. Thibeau, B. Oksengorn, and B. Vodar, *J. Phys.*, 1968, **29**, 287.
[210] M. Thibeau and B. Oksengorn, *Mol. Phys.*, 1968, **15**, 579.
[211] P. M. Lallemand, *Phys. Rev. Letters*, 1970, **25**, 1079.
[212] V. Volterra, J. A. Bucaro, and T. A. Litovitz, *Phys. Rev. Letters*, 1971, **26**, 55.

character in the wings as the gas spectra.[203] It was thus concluded that the depolarized Rayleigh spectra for liquids composed of isotropic molecules could be explained by a mechanism similar to that producing collision-induced light scattering in noble gases. At high gas densities, however, the dipole–induced dipole mechanism is suppressed and short-range electron overlap and molecular frame distortion effects may predominate. Bucaro and Litovitz [213] assumed a zero-impact parameter for liquids, a Lennard-Jones potential, and a $1/r^m$ form for the change in polarizability between two atoms, where m is the value appropriate to whether electron-overlap or frame-distortion processes predominate.

McTague *et al.*[206] presented calculations of the argon spectrum based on a second-order Raman process from pairs of overdamped phonons. These phonons polarized one another by a dipole field. The second and fourth moments of the experimental spectra agreed well with the predictions of the second-order Raman picture but the calculated bandwidth was too small. Bucaro and Litovitz [213] proposed that the higher-frequency components in liquid argon arose from very short-range interactions, *e.g.* electron-overlap effects in monatomic liquids and molecular frame distortions in molecular liquids. They carried out light-scattering measurements over the range 5—500 cm^{-1} in CCl_4, C_6H_{12}, C_5H_{12}, CH_3OH, C_2H_5OH, H_2O, NH_3, and $CHCl_3$ and compared these with calculations based on a binary-interaction model employing a Lennard-Jones potential and a short-range electronic overlap distortion. Assuming that molecular frame distortion was proportional to the interaction force, excellent agreement was obtained. The calculations were extended in a subsequent paper to include zero frequency shift.[214] The collision-induced light-scattering spectrum of CCl_4 was also compared with the collision-induced dielectric and far-i.r. absorption spectra. It was proposed that both absorption and scattering in CCl_4 contain contributions from the two processes, a high-frequency process, identical for both spectra, and, at frequencies below 20 cm^{-1}, a second process, characterized by an i.r. absorption which falls well below the scattering. A binary-collision model accounted satisfactorily for the high-frequency behaviour. The difference in the spectra was due to correlation between successive collisions. The low-frequency Raman spectrum of liquid CCl_4 was also reported by Gabelnick and Strauss.[202] The correlation function computed by these workers showed exponential behaviour at long times and yielded a relaxation time in excellent agreement with that of Bucaro and Litovitz.

A comparison of far-i.r. absorption and low-frequency Raman scattering was used by Dardy, Volterra, and Litovitz [215] to obtain information on the molecular dynamics of polar liquids. The far-i.r. absorption maximum in polar liquids shifts to lower frequencies with increasing temperature whereas

[213] J. A. Bucaro and T. A. Litovitz, *J. Chem. Phys.*, 1971, **54**, 3846.
[214] J. A. Bucaro and T. A. Litovitz, *J. Chem. Phys.*, 1971, **55**, 3585.
[215] H. Dardy, V. Volterra, and T. A. Litovitz, *Faraday Symp. Chem. Soc.* 1972, No. 6, p. 71.

in non-polar liquids there is a slight shift to higher frequencies at higher temperatures. The corresponding differences were shown to exist between the Rayleigh wing of highly anisotropic molecules and the pure collisional spectrum of isotropic molecules. The low-frequency regions of both absorption and scattering spectra yield information on the orientational relaxation times while consideration of the high-frequency 'wings' of the spectra can produce information on the details of the orientational motion in the system. By considering collision-induced effects in addition to orientational effects Dardy *et al.*[215] were able to explain the temperature dependence of the far-i.r. absorption maxima and to account for the origin of this peak in both polar and non-polar liquids. They also discussed the relationship between the observed spectra, the angular velocity correlation function, and the time between collisions.

Measurements of low-frequency depolarized Raman scattering of liquid CH_3I, $CHCl_3$, CH_2Cl_2, and CCl_4 have been reported by van Konynenburg and Steele [216] and used to compute polarizability correlation functions. Dipole correlation functions for parallel transitions obtained by previous investigators were corrected for bandshape changes due to variable refractive index, isotope splitting, and hot bands. The Raman correlation function contains contributions from orientational motions and collision-induced changes in polarizability, but it was shown that the collision-induced contributions were only important at short times. The long-time behaviour of the correlation functions was compared with several models of molecular motion in liquids. The best agreement was obtained with the *M*- and *J*-diffusion models.[45] In subsequent work,[217] low-frequency Raman scattering was reported for liquid C_2H_6, C_2H_4, and CO_2. Spectral moments $M(0)$, $M(1)$, and $M(2)$ were computed for both Stokes and anti-Stokes components. The results indicated that all spectra were asymmetric, that is, the Stokes and anti-Stokes components differed appreciably in width and shape. Correlation functions were computed and corrected for quantum effects. The collision-induced effects were neglected at long times. Several current theories of molecular reorientation were used to predict the dynamical behaviour of the fluids. The *J*-diffusion model [45] produced the best agreement between experimental and theoretical correlation functions. Preliminary reports of induced light scattering from molecular liquids have been reported by Howard-Lock and Taylor [218] and by Tabisz *et al.*[219]

[216] P. van Konynenburg and W. A. Steele, *J. Chem. Phys.*, 1972, **56**, 4776.
[217] P. van Konynenburg and W. A. Steele, *Adv. Raman Spectroscopy*, 1973, **1**, 450.
[218] H. E. Howard-Lock and R. S. Taylor, *Adv. Raman Spectroscopy*, 1973, **1**, 460.
[219] G. C. Tabisz, W. R. Wall, D. P. Shelton, and J. Ho, *Adv. Raman Spectroscopy*, 1973, **1**, 466.

5 Summary

The study of i.r. and Raman bandshapes in fluids is a rapidly developing field and is providing theoreticians with the data necessary to test models of the liquid state. The correlation function approach, in particular, provides a direct means for proceeding from a specific model of molecular motion to the predicted spectroscopic behaviour. This is true in spite of the fact that the wings of i.r. and Raman bands are frequently influenced by extraneous features. These include hot bands, weak underlying overtone and combination bands, and displaced bands arising from molecules containing non-dominant isotopic species. As we have seen in Section 4, however, considerable progress has already been made in eliminating the influence of these features from the computed correlation function. These effects will take on increasing significance as bandshape studies are used to provide more rigorous tests of theoretical models of molecular motion. The contribution of other relaxation processes, in addition to reorientation, will also need careful assessment. Again considerable progress has been made toward this end, particularly for Raman scattering, but much remains to be done. It is likely that the various relaxational processes will contribute differently to the correlation functions obtained from various experiments such as n.m.r., dielectric, low-frequency absorption, i.r., and Raman. A careful comparison of the various measurements will probably lead to a better understanding of the relative importance of the relaxation mechanism operative for a particular molecule in the liquid or dense gas. Much more effort needs to be expended in this area in view of its importance in the interpretation of experimental data.

In conclusion, the adoption of the correlation function approach has had the effect of shifting the attention of the spectroscopist from the peak positions of the bands to the wing areas, where the accuracy of most techniques is much lower. A new set of technical problems has been generated in the process. As Young and Jones aptly remark, 'in some respects we appear to be entering a phase in the study of liquid phase physics comparable to that which revolutionized solid state physics over a decade ago'.

4
Infrared Fluorescence Studies

BY R. T. BAILEY AND F. R. CRUICKSHANK

1 Introduction

The mechanism of energy transfer between quantum states is of fundamental importance to our understanding of chemical kinetics and of other non-equilibrium processes. It has recently been shown[1] for example that energy is not transferred intramolecularly infiinitely rapidly on the chemical reaction time-scale. CH_2 was added to the double bond of CF_2—CF—$CF=CF_2$ and

$$CD_2$$

preferentially eliminated CF_2 from the CH_2-containing cyclopropane ring. The recent development of powerful i.r. laser sources has made possible the direct study of these processes with greater precision than has previously been possible. These experiments involve the direct excitation of molecular vibration–rotation energy levels using i.r. laser radiation. The population of these levels and of the other energy levels in the molecule is then monitored by i.r. fluorescence spectroscopy. Two general types of laser experiment can be distinguished. In the first type, measurements are carried out under steady-state conditions, whereas in the second type, time-resolved techniques are employed. In the steady-state experiments a CW laser is employed and the fluorescence spectrum scanned in the usual way. The kinetics of energy transfer are difficult to unravel from such data, but useful analytical information can be obtained. In the second type of experiment, the excitation is in the form of short intense pulses and the fluorescence monitored by fast-response i.r. detection systems. Such data are more readily interpreted kinetically since vibrational lifetimes are obtained directly.

This Report will be concerned with the changes in the vibrational, rotational, and translational energy states that occur during molecular collisions, the molecule remaining in the ground electronic state. Intramolecular energy-transfer processes will not be considered. Also, although attention will be focused on non-reactive systems, *i.e.* those in which no chemical reaction occurs during the energy-transfer processes, brief consideration will be given to laser-induced chemical reactions.

Much of the impetus for the recent growth of interest in vibrational energy-

[1] J. N. Reynbrandt and B. S. Rabinovitch, *J. Chem. Phys.*, 1971, **54**, 2275; *J. Phys. Chem.*, 1971, **75**, 2164.

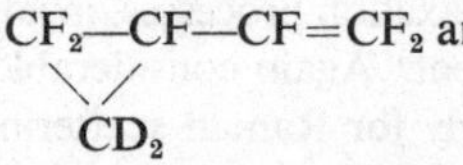

transfer processes has resulted from the need for fundamental data necessary for the development of gas lasers, particularly the CO_2 laser. The theoretical treatment of energy transfer has been hampered by the lack of realistic intermolecular potentials. However, in the past few years the experimental situation has improved considerably and has enabled calculations of energy-transfer cross-sections to be tested. Improved calculations must include the influence of the long-range attractive part of the potential in a proper manner.

Several well-established techniques are available for the study of vibration to translational (*V–T*) transfer. These include ultrasonic, shock-tube, spectrophone, and laser methods.[2–6] The transfer of energy from vibration into rotation (*V–R*) can also be studied in suitable cases by these techniques. More recently, stimulated-Raman and double-resonance techniques have also been used. Attention will be focused in this article on laser-excited i.r. fluorescence techniques, which promise to revolutionize the field, particularly with the recent introduction of tunable i.r. lasers. Several excellent reviews [7–9] dealing with various aspects of laser-excited fluorescence spectroscopy have appeared. For an account of (*V–V*) energy transfer a very recent review by Moore [10] should be consulted. A critical evaluation of current experimental techniques is contained in this article.

Since laser-excited i.r. fluorescence spectroscopy is a relatively recent technique, the experimental aspects will be discussed fairly comprehensively. Particular attention will be paid to the properties of the laser excitation sources and to the fast i.r. detection systems necessary for energy-relaxation studies. Until recently, fluorescence spectroscopy has been the domain of the molecular physicist concerned with the design of efficient molecular laser systems. Consequently, there is a large body of data on molecules suitable for laser systems such as CO_2, HF, HCl, *etc.*, and also on molecules such as SF_6, which are useful as passive *Q*-switches. Energy-transfer studies on many other molecular systems are now appearing at an increasing rate and provide stringent tests of current theories of energy transfer.

In view of the interdisciplinary nature of the energy-transfer field today, it is not possible to be completely comprehensive in coverage. Major contributions to this field have been made by engineers, physicists, and chemists, all with widely differing aims and objectives. In this article we have selected those aspects of the subject which are of chemical and spectroscopic interest.

[2] A. B. Callear and J. D. Lambert, in 'Comprehensive Chemical Kinetics', ed. C. H. Bamford and C. F. H. Tipper, Elsevier, Amsterdam, 1969, vol. 3, p. 182.
[3] T. L. Cottrell and J. C. McCoubrey, 'Molecular Energy Transfer in Gases', Butterworth, London, 1961.
[4] R. G. Gordon, W. Klemperer, and J. I. Steinfeld, *Ann. Rev. Phys. Chem.*, 1968, **19**, 215.
[5] P. Borrell, *Adv. Mol. Relaxation Processes*, 1967, **1**, 69.
[6] 'Transfer and Storage of Energy by Molecules', ed. G. M. Burnett and A. M. North, Wiley–Interscience, New York, 1969, vol. 2, p. 58.
[7] C. B. Moore, *Accounts Chem. Res.*, 1969, **2**, 103.
[8] C. B. Moore, *Ann. Rev. Phys. Chem.*, 1971, **22**, 387.
[9] C. B. Moore in 'Fluorescence', ed. G. G. Guilbault, Dekker, New York, 1967, p. 133.
[10] C. B. Moore, *Adv. Chem. Phys.*, 1973, **23**, 41.

2 Experimental Techniques

Laser Sources.—The development of laser sources has proceeded at a dramatic rate in recent years. New laser frequencies, new techniques, and improvements in existing lasers appear almost daily in the literature. Wavelength-tunable i.r. lasers are a particularly promising development as pumping sources in the field of energy relaxation and transfer. The basic principles of the laser can be found in a number of texts [11–13] and the chemical applications of lasers have also been reviewed.[14–16] Only i.r. laser systems useful for i.r. fluorescence studies will be mentioned in this chapter.

Helium–Neon. Some of the early experiments in energy transfer were carried out using the helium–neon 3.39 μm line. This laser frequency coincides with the asymmetric stretching vibration in methane.[17, 18] Mechanical modulation of the laser radiation was employed, imposing limits on the pulse widths and the energy which could be obtained.

Carbon Dioxide. A more useful laser source for energy-transfer studies is the CO_2 laser. Continuous-wave (CW) operation at 10.6 μm was first announced by Patel [19] in 1964. Since that date, the CO_2 laser has undergone remarkably rapid development.[20] The intense interest in this laser is due partly to the high efficiency and very high CW output obtainable. A variety of commercial devices are now available providing CW outputs from a few watts to several kilowatts. With the addition of a wavelength selector in the cavity the CO_2 laser can be tuned to a large number of vibration–rotation transitions between 908 and 1096 cm^{-1}. These frequencies arise from transitions between the (001–100) and (001–020) levels.

The laser can also be pulsed and *Q*-switched to provide short high-power pulses. Mechanical *Q*-switching can be accomplished by the rotating-mirror technique or by chopping the beam in the cavity with a mechanical shutter. Typically, for a laser tube 2 m long, peak powers of about 10 kW are obtained with a half-width of about 300 ns. Repetition rates of about 200 Hz provide optimum power. Passive *Q*-switching can also be used to obtain short CO_2 pulses. In this method, a saturable absorber such as SF_6 is placed in the cavity.[21] The absorber must have a strong absorption band at CO_2 laser

[11] A. Yariv, 'Quantum Electronics', Wiley, New York, 1967.
[12] D. Ross, 'Lasers: Light Amplifiers and Oscillators', Academic Press, New York, 1969.
[13] 'Lasers', ed. A. K. Levine and A. J. DeMaria, Dekker, New York, 1971, vols. 1, 2, and 3.
[14] A. F. Haught, *Ann. Rev. Phys. Chem.*, 1968, **19**, 343.
[15] W. J. Jones, *Quart. Rev.*, 1969, **23**, 73.
[16] R. M. Wilson, *Ann. New York Acad. Sci.*, 1970, **168**, 615.
[17] J. T. Yardley and C. B. Moore, *J. Chem. Phys.*, 1966, **45**, 1066.
[18] J. T. Yardley and C. B. Moore, *J. Chem. Phys.*, 1968, **49**, 1111.
[19] C. K. N. Patel, W. L. Faust, and R. A. McFarlane, *Bull. Amer. Phys. Soc.*, 1964, **9**, 500.
[20] P. K. Cheo in 'Lasers', ed. A. K. Levine and A. J. DeMaria, Dekker, New York, 1971, vol. 3, p. 111.
[21] O. R. Wood and S. E. Schwarz, *Appl. Phys. Letters*, 1967, **11**, 88.

Table 1 *Operating characteristics of some* CO_2 *saturable absorbers*

Absorber	*Abs. coeff.*/ $Torr^{-1}$ cm^{-1}	*Path length*/m	*Pressure*/ mTorr	*Laser tube length*/m	*Laser transition wavelength*/μm	*Pulse width*/μs	*Pulse separation*/μs	*Peak power*[a]
CH_3F	0.018	1	15—20	1	9.6	1—2	40—100	12 P_{CW}
CO_2–propene	—	0.75	500	1	10.6	1—2	18	—
Formic acid	—	0.2	500	1	~9.2	—	—	P_{CW}
Vinyl chloride	0.01	0.1	2000	3.4	10.6	2	20	4 P_{CW}
BCl_3	0.01—0.02	0.2	1000	6.8	10.4	—	40—100	—
PF_5	0.02—0.12	1	5—70	1	10.4	0.6—1	150—500	50 P_{CW}
SF_6	0.1—0.5	0.1	20—50	3	10.4	0.4—2	250—450	20 P_{CW}
C_2F_3Cl	0.019	0.25	500	2.5	9.4	1.8—5	15—40	10 P_{CW}
CF_2Cl_2	0.076	0.25	20—50	2.5	10.4	0.8—3	15—100	20 P_{CW}

[a] P_{CW} = CW power.

wavelengths and a long-relaxation time for the upper state. Other suitable bleachable absorbers include formic acid vapour, propene, hot CO_2,[22] vinyl chloride,[23] BCl_3,[24] CH_3F and PF_5,[25] and CF_2Cl_2 and C_2F_3Cl.[26] A typical pulse obtained by this technique would be about 0.5—2 μs wide with a peak power of about 20—200 times the CW output. Table 1 lists the characteristics of some saturable absorbers.

To obtain shorter pulse widths from the CO_2 laser, mode-locking techniques must be used. By using an intra-cavity GaAs electro-optic modulator, Bridges and Cheo [27] obtained a chain of 10 or more 20 ns pulses with a total duration of about 400 ns in a single $P(20)$ transition in the 10.4 μm band. With their arrangement it was possible to dump any one of the short pulses from the train out of the cavity through an output coupling polarizer to produce a single pulse of ~10 kW peak power and 20 ns duration. The experimental procedure is, however, somewhat involved. Various other mode-locking techniques are discussed in a comprehensive review by Cheo.[20]

Probably the most significant development in the CO_2 laser field with regard to their use as pumping sources for i.r. fluorescence studies was the introduction of the so-called transversely excited atmospheric 'TEA' laser. As the acronym suggests, this laser operates at atmospheric pressure, thus eliminating vacuum problems and at the same time increasing the power output per unit volume. Furthermore, the pulse length is determined to a large extent by the relaxation times of the excited molecules. As these become shorter at higher pressures, the pulse length decreases proportionally. The peak power is thus proportional to the square of the gas pressure. Higher peak powers and shorter pulses are therefore obtained from a more compact device when the lasing medium is at atmospheric pressure.

It is difficult to achieve uniform excitation of the gas mixture at atmospheric pressure since at about 200 Torr the normal glow discharge constricts to an arc, heating the gas and destroying laser action. Several ingenious techniques have been devised to overcome this problem. Beaulieu [28] used very short 17 kV pulses to excite a transverse discharge between a cathode consisting of a row of pins and a round bar anode. A shower of brush discharges opens out from each pin simultaneously so that the excited volume becomes a linear array of overlapping discharges. This system is discussed in detail by Beaulieu [29] and by Fortin.[30] A helical arrangement designed to improve the uniformity of the discharge is shown in Figure 1.

A number of electrode geometries have been designed to provide a more

[22] P. L. Hanst, J. A. Morreal, and W. J. Henson, *Appl. Phys. Letters*, 1968, **12**, 58.
[23] J. T. Yardley, *Appl. Phys. Letters*, 1968, **12**, 120.
[24] N. V. Karlov, G. P. Kuzmin, Yu. N. Petrov, and A. M. Prokhorov, *J.E.T.P. Letters*, 1968, **7**, 134.
[25] T. Y. Chang, C. H. Wang, and P. K. Cheo, *Appl. Phys. Letters*, 1969, **15**, 157.
[26] S. Marcus, *Appl. Phys. Letters*, 1969, **15**, 217.
[27] T. J. Bridges and P. K. Cheo, *Appl. Phys. Letters*, 1969, **14**, 262.
[28] A. J. Beaulieu, *Appl. Phys. Letters*, 1970, **16**, 504.
[29] A. J. Beaulieu, *Proc. I.E.E.E.*, 1971, **59**, 667.
[30] R. Fortin, *Canad. J. Phys.*, 1971, **49**, 257.

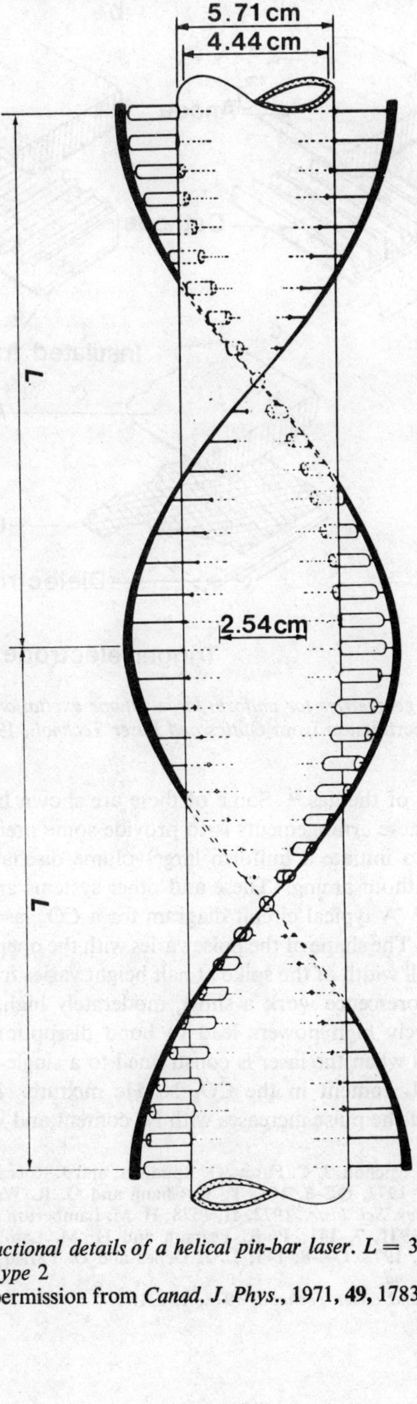

Figure 1 *Constructional details of a helical pin-bar laser.* $L = 36$ cm *for type* 1 *helix and* 18 cm *for type* 2
(Reproduced by permission from *Canad. J. Phys.*, 1971, **49**, 1783)

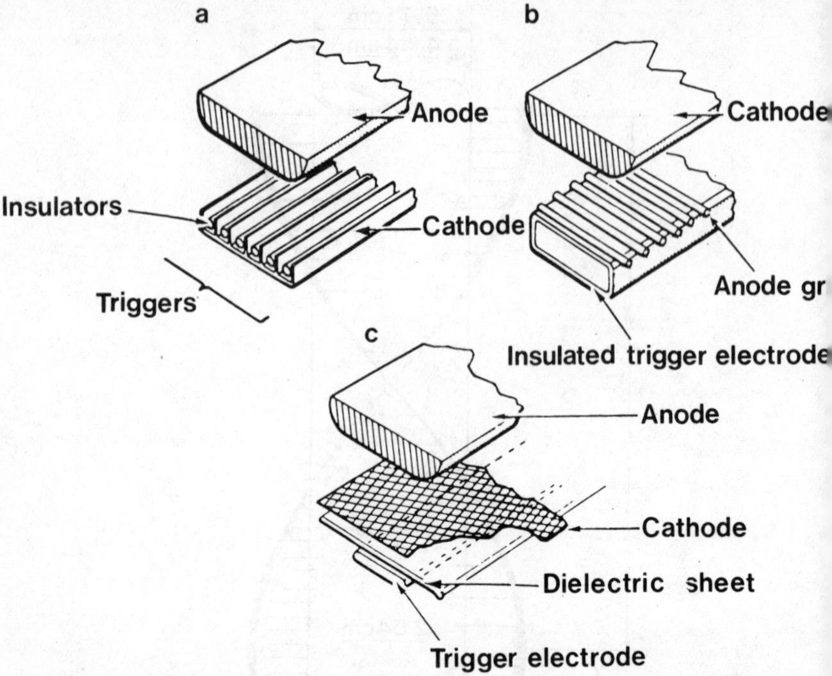

Figure 2 *Electrode geometries for uniform large-volume excitation*
(Reproduced by permission from *Optics and Laser Technol.*, 1972, **4**, 121)

uniform excitation of the gas.[31] Some of these are shown in Figure 2. The main purpose of these arrangements is to provide some preionization of the gas, which helps to initiate a uniform large-volume discharge between the main electrodes without arcing. These and other systems are discussed in a review by Foster.[32] A typical circuit diagram for a CO_2 laser of this type is shown in Figure 3. The shape of the pulse varies with the operating conditions but typically the full width of the spike at half height varies from 70 to 250 ns. For pulsed i.r. fluorescence work a short, moderately high, power pulse is required. Excessively high powers lead to bond disruption. The shortest pulses are obtained when the laser is constrained to a single-mode operation and with a low N_2 content in the CO_2–N_2–He mixture. In addition, the energy in the tail of the pulse increases with N_2 content and with multi-mode

[31] R. Dumanchin, M. Michon, J. C. Farcy, G. Boudinet, and J. Rocca-Serra, *I.E.E.E.-J. Quantum Electron*, 1972, **QE–8**, 163; T. Y. Chang and O. R. Wood, *ibid.*, p. 721; A. K. Laflamme, *Rev. Sci. Instr.*, 1972, **41**, 1578; H. M. Lamberton and P. R. Pearson, *Electron. Letters*, 1971, **7**, 141; P. R. Pearson and H. M. Lamberton, *I.E.E.E.-J. Quantum Electron.*, 1972, **QE–8**, 145; L. J. Denes and O. Farish, *Electron. Letters*, 1971, **7**, 337.

[32] H. Foster, *Optics and Laser Technol.*, 1972, **4**, 121.

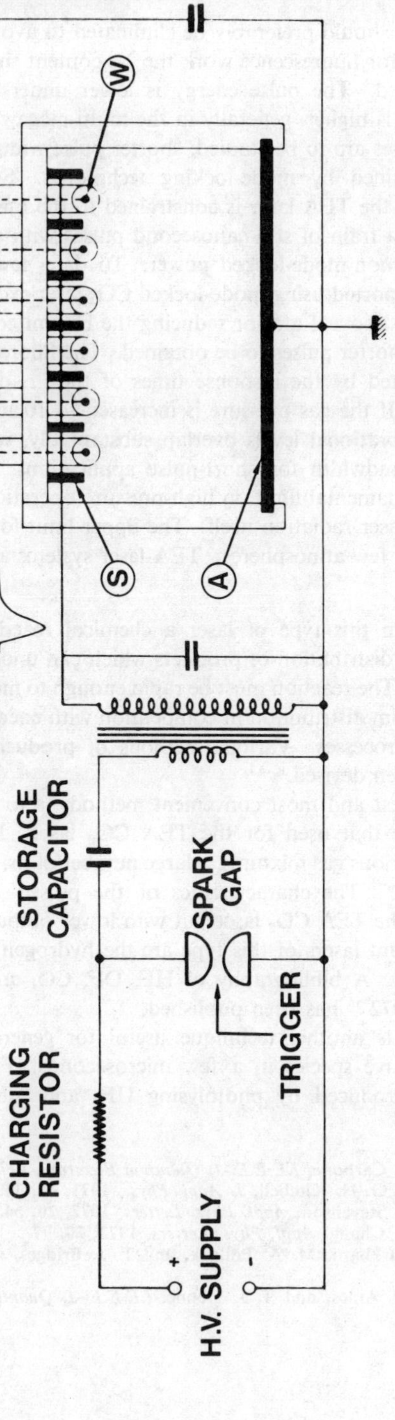

Figure 3 *Electrical pulse circuit for TEA CO_2 laser*
(Reproduced by permission from *I.E.E.E.-J. Quantum Electron.*, 1972, QE-8, 163)

operation. This tail should preferably be eliminated to avoid double-excitation effects, so that for fluorescence work the N_2 content should be small or completely eliminated. The pulse energy is lower under these conditions but the peak power is higher, generally in the multi-megawatt region.

When fast processes are to be studied, shorter pulse widths are necessary. These can be obtained by mode-locking techniques. Self-mode-locking occurs readily when the TEA laser is constrained to the fundamental mode. The output is then a train of sub-nanosecond pulses with a peak power of about 20 times the non-mode-locked power. To date, few energy-transfer studies have been reported using mode-locked CO_2 laser excitation. Increasing the gas pressure above 1 atm or reducing the helium content of the gas should allow even shorter pulses to be obtained. The fluorescence technique would then be limited by the response times of the i.r. detection systems currently available. If the gas pressure is increased to 10 atm the rotational lines of the CO_2 vibrational levels overlap substantially, which offers even greater potential bandwidth for short-pulse applications. However, there appears to be a fundamental limit on high-pressure operation caused by gas breakdown by the laser radiation itself. The upper limit for efficient operation appears to be a few atmospheres. TEA laser systems are now available commercially.

Chemical Lasers. In this type of laser a chemical reaction produces an inverted population distribution of products which can undergo laser action in a suitable cavity. The reaction must be rapid enough to maintain an appreciable non-equilibrium distribution in competition with energy-redistribution and -deactivation processes. Various methods of producing the chemical laser action have been devised.[8, 44]

One of the simplest and most convenient methods is to use a transverse discharge similar to that used for the TEA CO_2 laser. By replacing the CO_2–N_2–He with various gas mixtures, a large number of i.r. laser frequencies can be obtained.[33–37] The characteristics of the pulsed laser output are similar to those of the TEA CO_2 laser but with lower output powers. Some of the more important lasers of this type are the hydrogen halide (HX and DX) and CO lasers. A bibliography of HF, DF, CO, and CO_2 chemical lasers up to June 1972 [38] has been published.

Flash photolysis is another technique useful for generating large concentrations of reactive species in a few microseconds. For instance, HF lasers have been produced by photolysing UF_6 and other fluorides, by

[33] S. Marcus and R. J. Carbone, *I.E.E.E.-J. Quantum Electron.*, 1971, **QE-7**, 493.
[34] T. V. Jacobson and G. H. Kimbell, *J. Appl. Phys.*, 1971, **42**, 3402.
[35] J. Wilson and T. S. Stevenson, *Appl. Phys. Letters*, 1972, **20**, 64.
[36] R. Wood and T. Y. Chang, *Appl. Phys. Letters*, 1972, **20**, 77.
[37] R. Wood, E. G. Burkhard, M. A. Pollack, and T. J. Bridges, *Appl. Phys. Letters*, 1971, **18**, 112.
[38] C. E. Wiswall, D. P. Ames, and T. J. Menne, *I.E.E.E.-J. Quantum Electron.*, 1973, **QE-9**, 181.

the reaction $F + H_2$.[39, 40] Hydrogen-atom abstraction from hydrocarbons and alkanes can also be used to produce laser action.[41] The HCl laser, operating on the reaction $Cl\cdot + HI \rightarrow HCl + I\cdot$, produces pulses 10 μs wide of about 3×10^4 W.[42, 43] The power obtainable from this type of chemical laser is generally lower than can be obtained from the corresponding electric discharge laser. HCl lasers have been made by flash initiation of a $H_2 + Cl_2$ mixture.[45] The HF chain reaction is a particularly useful laser system.[46–48]

The transfer chemical laser (TCL) is another device which makes use of the non-thermal vibrational energy of reaction products (*e.g.* hydrogen halides) in the creation of total population inversion in a second molecule (*e.g.* CO_2). In suitable systems, the TCL is more efficient than the conventional chemical laser and can be operated at much higher pressures. Both CW and pulsed operation of the DF–CO_2 laser have been achieved.[44, 49] At present, these systems do not appear to have any advantage over TEA laser systems for fluorescence work except where exact resonance between the emitter and absorber is required.

It should be mentioned in passing that chemical lasers provide information on the energy distribution in the reaction products. For instance, the quenching of laser action by additives or the increased power in the *Q*-switched mode [50] are a measure of transfer rates of vibrational energy. The competition between the various vibration–rotation transitions is a measure of rotational energy-transfer rates and stimulated emission or other vibrational transfer processes. Parker and Pimentel [51] have developed a technique (equal-gain temperature) for determining the ratios of rate constants in the upper and lower laser levels, N_v/N_{v-1} which, may be calculated from spectroscopic constants and the equal-gain temperature, T_{eq}.

Tunable Lasers. The majority of the infrared laser fluorescence work that has been published to date has employed fixed-frequency laser sources. Excitation has depended on fortuitous overlap between laser frequencies and molecular i.r. absorption bands or on sensitization by an intermediate

39 K. L. Kompa and G. C. Pimentel, *J. Chem. Phys.*, 1967, **47**, 857.
40 K. L. Kompa, J. H. Parker, and G. C. Pimentel, *J. Chem. Phys.*, 1968, **49**, 4257.
41 K. L. Kompa, P. Gensel, and J. Wanner, *Chem. Phys. Letters*, 1969, **3**, 210.
42 J. R. Airey, *I.E.E.E.-J. Quantum Electron.*, 1967, **3**, 208.
43 H. L. Chen, J. C. Stephenson, and C. B. Moore, *Chem. Phys. Letters*, 1968, **2**, 593.
44 T. A. Cool, *I.E.E.E.-J. Quantum Electron.*, 1973, **QE–9**, 72.
45 P. H. Corneil and G. C. Pimentel, *J. Chem. Phys.*, 1968, **49**, 1379.
46 O. M. Batovskii, G. K. Vasilev, E. F. Makarov, and V. L. Tal'Roze, *Zhur. eksp. teor Fiz. Pisma Red.*, 1969, **9**, 341 (*J.E.T.P. Letters*, 1969, **9**, 200).
47 N. G. Basov, L. V. Kulakov, E. P. Markin, A. I. Nikitin, and A. N. Oraevskii, *Zhur. eksp. teor. Fiz. Pisma Red.*, 1969, **9**, 613 (*J.E.T.P. Letters*, 1969, **9**, 375).
48 V. S. Burmasov, G. G. Dolgov-Savel'ev, V. A. Polyakov, and G. M. Chumak, *Zhur. eksp. teor. Fiz. Pisma Red.*, 1969, **10**, 42 (*J.E.T.P. Letters*, 1969, **10**, 28).
49 T. O. Poehler, J. C. Pirkle, jun., and R. E. Walker, *I.E.E.E.-J. Quantum Electron.*, 1973, **QE–9**, 83.
50 J. R. Airey, *J. Chem. Phys.*, 1970, **52**, 156.
51 J. H. Parker and G. C. Pimentel, *J. Chem. Phys.*, 1969, **51**, 91.

absorbing molecule. Recently, the developement of tunabl i.r. lasers has reached the point where they can be used as routine tools by the chemical physicist. They offer considerable promise and should ultimately supersede conventional fixed-frequency sources. The semiconductor laser is a potentially useful tunable device for selective pumping of molecular states. The output from the *p–n* junction laser can be tuned over broad regions of the i.r. spectrum. Tuning is accomplished by varying the band gap, which is a function of material composition, magnetic field, temperature, and pressure.[52, 53]

Most work on diode lasers has been concentrated on the technologically important GaAs laser.[53] Peak powers of 105 W in pulses 25 ns wide have been obtained.[54] The output power can be increased by fabricating laser arrays.[55] Some devices have also exhibited *Q*-switching effects to produce 200 ps pulses.[56] It is also possible to modulate the laser output at very high frequencies because of the short lifetime of the carriers.[56] A review of laser modulation techniques and results has been published.[57]

Tunable i.r. diode lasers are fabricated from compounds having a narrow band gap. Table 2 lists some of these diode lasers and their tunable range. The spectral region from 1370 to 320 cm^{-1} has been covered continuously with $Pb_{1-x}Sn_xSe$ diodes.[58] A review of the preparation and operating characteristics of $Pb_{1-x}Sn_xTe$ and Pb–Sn–Se lasers has been published by Melngailis.[59] Normally, these lasers are tuned by varying the current (temperature),

Table 2 *Infrared semiconductor laser diodes*

Compound	*Approx. wavelength* μm
Ga (As,Sb)	0.9—1.5
GaSb	1.5
(In,Ga)As; (In,Ga)P	0.9—3.1
InAs	3.1
In(As,Sb)	3.1—5.4
PbS	4.3
InSb	5.4
PbTe	6.5
PbSe	8.5
Pb(Sn,Te)	6.5—2.8
Pb(Sn,Se)	8.5

52 'Gallium Arsenide Laser', ed. C. H. Gooch, Wiley–Interscience, New York, 1969.
53 H. Kressel in 'Lasers', ed. A. K. Levine and A. J. DeMaria, Dekker, New York, 1971, vol. 3, p. 1.
54 H. Nelson and H. Kressel, 'High Pulse Power Injection Laser; Final Report, Contract No. DA28–043–AMC–02471 (E)', U.S. Army Electronics Command, Fort Monmouth, New Jersey, Oct. 1967.
55 W. E. Ahearn and J. W. Crowe, *I.E.E.E.-J. Quantum Electron.*, 1970, **6**, 377.
56 J. Nishizawa, *I.E.E.E.-J. Quantum Electron.*, 1968, **4**, 143.
57 J. E. Ripper and T. L. Paoli, *Proc. I.E.E.E.*, 1970, **58**, 1457.
58 T. C. Harman, A. R. Calawa, I. Melngailis, and J. O. Dimmock, *Appl. Phys. Letters*, 1969, **14**, 333; *Phys. Rev. Letters*, 1969, **23**, 7.
59 I. Melngailis, *J. Phys. Radium*, 1968, **29**, C4.

resulting in a limited wavelength coverage.[60] Recently, however, increased wavelength coverage has been obtained by a combination of magnetic field tuning and conventional bias-current tuning. The spectral region 4—7 μm was covered using $PbSn_{1-x}Se_x$ ($x = 0.2$ and 0.4) and PbTe lasers.[61] Recently, lead salt diode lasers have become available commercially. Diode injection lasers of the lead salt semiconductor system ($Pb_{1-x}Sn_xTe$, $PbS_{1-x}Se_x$, and related compounds) enable spectral coverage of the region 3—30 μm to be achieved. However, these lasers, although capable of producing very narrow bandwidths, suffer from a narrow tunable range for a single composition (typically less than 10 cm^{-1}), mode-hopping discontinuities in the tuning curves, and low power (a few tens of microwatts CW). They are therefore of limited utility as pumping sources at present, and parametric oscillators or the spin-flip Raman laser are much more promising tunable devices.

The parametric oscillator is a potentially very useful source of tunable i.r. radiation.[62–64] In its presently commercially available form it is limited to wavenumbers above 2800 cm^{-1} and is thus suitable for exciting hydrogen-stretching fundamentals as well as overtone and combination bands in suitable molecules. Using a *Q*-switched and frequency-doubled Nd–YAG laser, Wallace[63] was able to tune continuously from 0.548 to 3.65 μm. Peak powers of a few kilowatts and pulse widths of 130—700 ns were obtained. Shorter pulses of tunable radiation were recently obtained by Nath and Pauli.[65] They used a $LiIO_3$ crystal pumped with the second harmonic of a *Q*-switched ruby laser. Peak powers of 10 kW in 5 ns pulses were obtained in the frequency range 0.415—2.1 μm. Even shorter pulses were obtained using a barium sodium niobate crystal pumped by a chain of picosecond pulses from a mode-locked Nd^{3+} glass laser.[66] Stimulated paramagnetic pulses less than 10 ps in duration with peak powers of 300 W were produced by this technique. Only a relatively restricted frequency range, from 0.96 to 1.16 μm, can be covered using this method. The generation of difference frequencies, however, provides the opportunity of extending the tunable range to longer wavelengths. Dewey and Hocker[67] have produced intense pulses of tunable radiation in the 3—4 μm range by generating the difference frequencies between a tunable dye laser and a *Q*-switched ruby laser in a phase-matched $LiNbO_3$ crystal. This system is limited to the region below 4.5 μm by the transmission characteristics of $LiNbO_3$. Using proustite (Ag_3AsS_3) as the non-linear mixer, Decker and Tittel[68] demonstrated

[60] K. W. Nill, F. A. Blum, A. R. Calawa, and T. C. Harman, *Appl. Phys. Letters*, 1971, **19**, 79.
[61] K. W. Nill, F. A. Blum, A. R. Calawa, and T. C. Harman, *Appl. Phys. Letters*, 1972, **21**, 132.
[62] S. E. Harris, *Proc. I.E.E.E.*, 1969, **57**, 2096.
[63] R. W. Wallace, *Appl. Phys. Letters*, 1970, **17**, 497.
[64] L. S. Goldberg, *Appl. Phys. Letters*, 1970, **17**, 489.
[65] G. Nath and G. Pauli, *Appl. Phys. Letters*, 1973, **22**, 75.
[66] T. A. Robson, J. J. Ruiz, P. C. Shak, and F. K. Tittel, *Appl. Phys. Letters*, 1972, **21**, 129.
[67] C. F. Dewey and L. O. Hocker, *Appl. Phys. Letters*, 1971, **18**, 58.
[68] C. D. Decker and F. K. Tittel, *Appl. Phys. Letters*, 1973, **22**, 411.

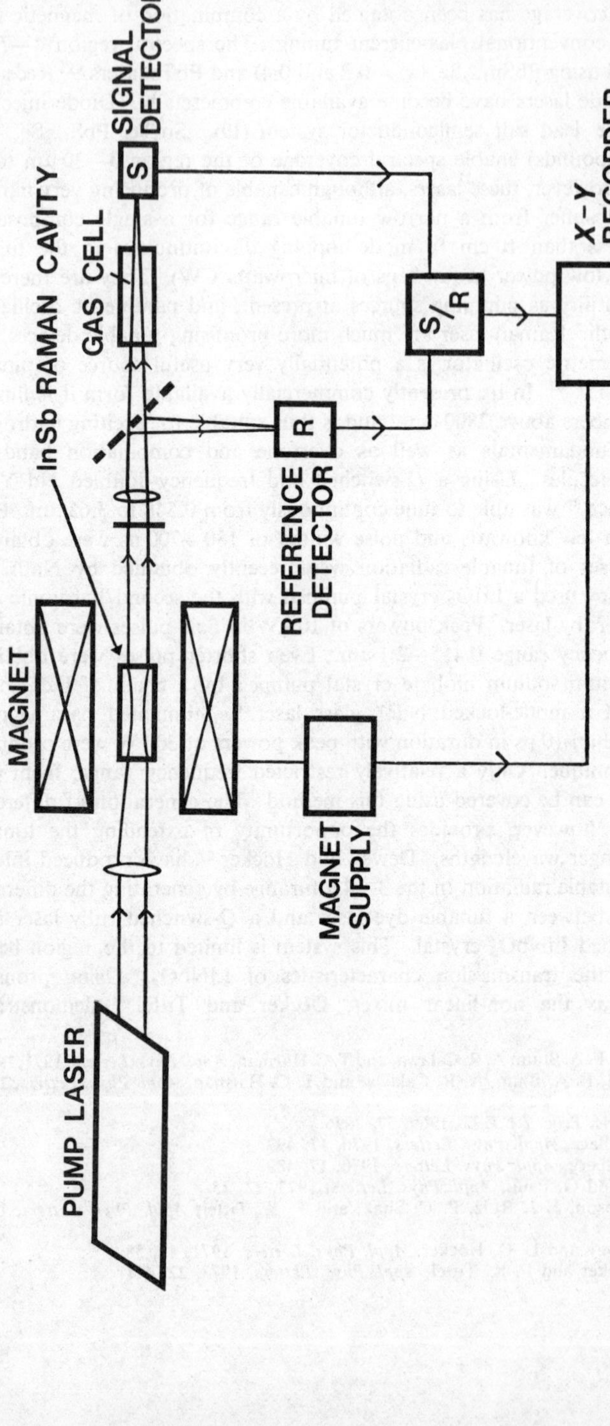

Figure 4 *Schematic diagram of commercial spin-flip Raman laser*
(Reproduced with permission of Edinburgh Instruments Ltd.)

tunability from 3.20 to 5.65 μm with peak power in the kilowatt range. By using appropriately cut proustite crystals, tunability from 5.65 to 13 μm should be attainable. Systems based on proustite should soon be available commercially. Tunable lasers have also been produced that emit in the far-i.r. region.[69–71] Pantell *et al.*[70] have used non-linear crystals and intense ruby radiation to yield simultaneous Raman-shifted and i.r. difference radiation. Continuous tunability over the frequency range 50—250 μm was achieved.

During the past few years, the work of two groups, headed by S. D. Smith at Heriot-Watt University and C. K. N. Patel at Bell Telephone Labs., has been principally responsible for the development of a continuously tunable source of i.r. radiation using an entirely different process.[72–78] This source, called the spin-flip Raman laser (SFRL), consists of a semiconducting crystal of *n*-type InSb mounted in a superconducting magnet. The system is pumped with radiation from a TEA CO_2 laser, the radiation being inelastically scattered from the conduction electrons in the InSb. The Raman process is associated with spin-reversal transitions, whose energy is proportional to the magnetic field in the crystal. Thus by varying the field of the superconducting magnet the frequency of the scattered radiation can be tuned. A SFRL system is now available commercially in modular form with a variety of laser pumps. This instrument is shown schematically in Figure 4. When pumped with a 40 MW TEA CO_2 laser the tuning range is from 1940 to 1560 cm^{-1} and from 1104 to 714 cm^{-1}. With the addition of a CW CO laser, additional lines are available in the range 1940—1560 cm^{-1}. Operated truly continuously, linewidths of about 0.1 cm^{-1} are obtained, while in the quasi-continuous mode a resolution of 10^{-6} cm^{-1} is obtainable. The limit on linewidth is known to be 3×10^{-8} cm^{-1}. Some characteristics of a commercially available system are shown in Table 3. The output power is about 1 kW in pulses about 30 ns wide in normal operation or about 3 ns in mode-locked operation. The tuning range may be extended by using different pump lasers and other semiconductors having narrow band gaps, such as $Pb_{1-x}Sn_xTe$ or $Ag_{1-x}Cd_xTe$, where the *g* value of the conduction electrons is larger than in InSb. The

[69] D. W. Farries, K. A. Gehring, P. L. Richards, and Y. R. Shen, *Phys. Rev.*, 1969, **180**, 363.
[70] J. M. Yarborough, S. S. Sussman, H. E. Puthoff, R. H. Pantell, and B. C. Johnson, *Appl. Phys. Letters*, 1969, **15**, 102.
[71] J. Gelbwacks, R. H. Pantell, H. E. Puthoff, and J. M. Yarborough, *Appl. Phys. Letters*, 1969, **14**, 258.
[72] C. K. N. Patel and E. D. Shaw, *Phys. Rev. Letters*, 1970, **24**, 451.
[73] R. L. Allwood, S. D. Devine, R. G. Mellish, S. D. Smith, and R. A. Wood, *J. Phys. (C)*, 1970, **3**, L186.
[74] R. L. Allwood, R. B. Dennis, S. D, Smith, B. S. Wherrett, and R. A. Wood, *J. Phys. (C)*, 1971, **4**, L63.
[75] R. L. Allwood, R. B. Dennis, R. G. Mellish, S. D. Smith, B. S. Wherrett, and R. A. Wood, *J. Phys. (C)*, 1971, **4**, L126.
[76] R. L. Allwood, R. B. Dennis, W. J. Firth, S. D. Smith, B. S. Wherrett, and R. A. Wood, *The Radio and Electronic Engineer*, 1972, **42**, 243.
[77] C. K. N. Patel, E. D. Shaw, and R. J. Kerl, *Phys. Rev. Letters*, 1970, **25**, 8.
[78] R. A. Wood, R. B. Dennis, and J. W. Smith, *Optics Comm.*, 1972, **4**, 6.

SFRL appears ideal for energy-transfer studies as well as very-high-resolution i.r. absorption spectroscopy,[79, 80] and should make possible entirely new lines of research not possible with present laser sources.

Table 3 *Characteristics of a tunable spin-flip Raman laser*

	Tuning range/cm^{-1}	*Peak power*/W	*Linewidth*/cm^{-1}
CO_2 *pump*			
Anti-Stokes	1104—990	>100	~0.04
Stokes	900—784	>1000	~0.04
II-Stokes	814—714	~100	~0.04
Doubled CO_2 *pump*			
Stokes	1940—1752	~1000	~0.04

Solid-state Lasers. Solid-state neodymium lasers are capable of producing nanosecond and picosecond pulses. The generation and detection of picosecond pulses has been reviewed by DeMaria *et al.*[81] By mode-locking, pulses as short as 0.3 ps have been produced. These lasers emit in the near-i.r. region at about 1 μm so that they cannot be used to excite fundamentals directly. However, they may be used to pump overtone and combination bands in suitable molecules. The very short pulse widths, combined with fast-risetime silicon avalanche photodiodes, should enable very fast energy-transfer processes to be studied. Bina and Jones [82] have demonstrated the direct optical excitation of the second overtone of HF ($v = 2$) using a 1.34 μm pulse from a temperature-tuned Nd–$YAlO_3$ laser.

Detectors.—The ideal detector for pulsed i.r. fluorescence work should be sensitive over a wide i.r. frequency range, have low noise, and a very short response time. Until recently, to obtain these characteristics it was necessary to cool to liquid-helium temperature, but recently developed sensitive detectors are now available which operate at liquid-nitrogen temperature. To obtain a wide frequency coverage, however, it is still necessary to use cryogenic temperatures. I.r. detectors can be broadly divided into two classes, thermal detectors and quantum detectors. Thermal detectors operate by sensing the change in temperature of an absorber. The output obtained may be in the form of a thermal e.m.f. (thermocouple detector), a change in resistance of a conductor (bolometer), or the movement of a diaphragm caused by the expansion of a gas (pneumatic detector), which may lead to the change in illumination of a subsidiary photocell (Golay cell). These detectors can be made to respond to radiation of a very wide range of wavelength but usually have a slow response (10^{-2}—10^{-1} s). Thermal detectors are therefore of no use for pulsed fluorescence measurements. They are,

[79] C. K. N. Patel, *Phys. Rev. Letters*, 1972, **28**, 649.
[80] L. B. Kreuzer and C. K. N. Patel, *Science*, 1971, **173**, 45.
[81] A. J. DeMaria, W. H. Glen, M. J. Brienza, and M. E. Mack, *Proc. I.E.E.E.*, 1969, **57**, 2.
[82] M. J. Bina and C. R. Jones, *Appl. Phys. Letters*, 1973, **22**, 44.

however, ideal for CW measurements, where a sensitive detector having a uniform response over a broad wavelength range is required. A recent article by Stanley [83] discusses the properties and performance of thermal detectors. The theory and practice of i.r. detection is covered comprehensively in a text by Smith, Jones, and Chasmar.[84]

A new class of thermal detector, the pyroelectric detector, offers the possibility of a thermal detector which is both sensitive and fast, covers a broad frequency range, and operates at room temperature.[85-87] The detectivity of the best pyroelectric detectors already approaches to within an order of magnitude of that of an ideal thermal detector at room temperature.

Pyroelectric detectors are made from materials which have a permanent electronic polarization which changes when the temperature is changed by the absorption of i.r. radiation. Single-crystal material gives the best performance but other forms, such as ceramics, are sometimes used. The detector is constructed from a thin slice of crystal with metal electrodes evaporated on to opposite sides to form a capacitor as shown in Figure 5.

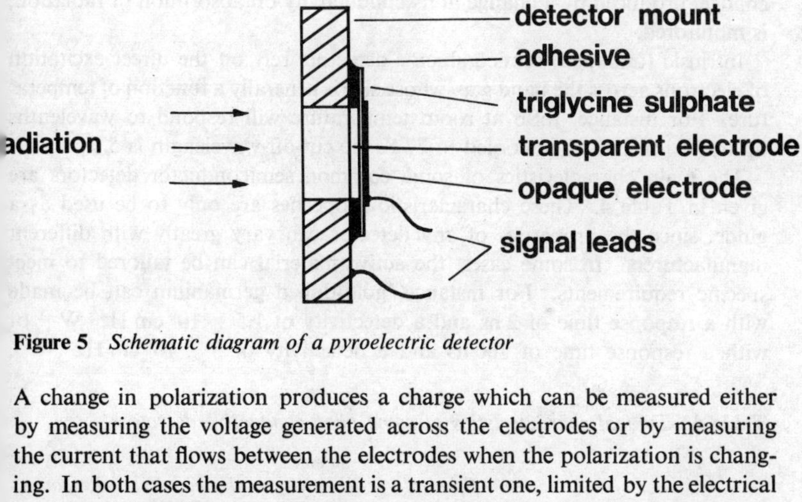

Figure 5 *Schematic diagram of a pyroelectric detector*

A change in polarization produces a charge which can be measured either by measuring the voltage generated across the electrodes or by measuring the current that flows between the electrodes when the polarization is changing. In both cases the measurement is a transient one, limited by the electrical or thermal time-constant. These are usually of the order of several seconds, so that these detectors can be used down to a frequency of 1 Hz or less. The upper frequency limit is determined by the load resistance across the detector. A commercial pyroelectric detector based on triglycine sulphate, which has a detectivity of $\sim 6 \times 10^8$ cm Hz$^{\frac{1}{2}}$ W^{-1}, is marketed by Mullard. With a

[83] C. R. Stanley, *Optics and Laser Technol.*, 1971, **3**, 144.
[84] R. A. Smith, F. E. Jones, and R. P. Chasmar, 'The Detection and Measurement of Infrared Radiation', Clarendon Press, Oxford, 1968.
[85] E. H. Putley, in 'Semiconductors and Semimetals', ed. R. K. Wellardson and A. C. Beer, Academic Press, New York, 1970, vol. 5, Ch. 6.
[86] H. P. Beerman, *Ferroelectrics*, 1971, **2**, 123.
[87] E. H. Putley, *Optics and Laser Technol.*, 1971, **3**, 150.

reduction in sensitivity, this detector may be used to detect pulsed or *Q*-switched laser outputs. Kimmit *et al.*[88] have used a lithium sulphate detector to monitor *Q*-switched laser pulses at 10.6 μm. Recent research has been directed towards the pyroelectric materials strontium barium niobate[89] and lithium tantalate.[90] Detectors with a response of less than 1 ns have been constructed using the former material and used to monitor the output of a mode-locked CO_2 laser. Considerable sacrifice in detectivity was necessary, however, in order to obtain such fast rise-times. Fast pyroelectric detectors are now available from a number of manufacturers. At present, pyroelectric detectors are not suitable for pulsed i.r. fluorescence work since they do not have sufficient sensitivity at high frequencies. They are, however, very useful for CW work and for monitoring laser pulses, where their low sensitivity is of no consequence.

Quantum Detectors. Quantum detectors for the i.r. may be operated in either the photoconductive or the photovoltaic mode. In the former, the sensitive element is biased under constant current, and the resultant voltage change, produced by a change in its conductivity on absorption of radiation, is monitored.

Intrinsic (undoped) semiconductor detectors rely on the direct excitation of electrons across the band gap, whose size is generally a function of temperature. For instance, InSb at room temperature will respond to wavelenths up to 7.5 μm but when cooled to 77 K the cut-off wavelength is 5.5 μm.

The main characteristics of some common semiconductor detectors are given in Table 4. These characteristic properties are only to be used as a guide, since the properties of any detector can vary greatly with different manufacturers. In some cases, the active material can be tailored to meet specific requirements. For instance, gold-doped germanium can be made with a response time of 2 ns and a detectivity of 1.5×10^9 cm Hz$^{\frac{1}{2}}$ W^{-1} or with a response time of 100 ns and a detectivity of 3×10^9 cm Hz$^{\frac{1}{2}}$ W^{-1}.

Table 4 *Typical characteristics of some photoconductive detectors*

Detector	*Temperature*	*Spectral range/μm*	*Peak sensitivity/ μm*	*Detectivity/ cm Hz$^{\frac{1}{2}}$ W^{-1}*	*Response time*
PbS	Room	vis.—3.0	2.2	4.0×10^{10}	~100 μs
InSb	Room	vis.—7.5	6.0	1.0×10^{8}	~50 ns
InSb	77 K	vis.—5.6	5.3	4.5×10^{10}	~5 μs
$Hg_{1-x}Cd_xTe$	77 K	3—15	8—15	5.0×10^{9}	400 ns
Au–Ge	77 K	1—10.6	4.5	3.0×10^{9}	<100 ns
Au–Ge	4 K	2—14.0	11.0	3.0×10^{10}	<100 ns
Cu–Ge	4 K	2—30.0	20.0	3.0×10^{10}	<100 ns

[88] M. F. Kimmitt, J. H. Ludlow, and E. H. Putley, *Proc. I.E.E.E.*, 1968, **56**, 1250.
[89] A. M. Glass, *Appl. Phys. Letters*, 1968, **13**, 147.
[90] A. M. Glass and R. L. Abrams, *J. Appl. Phys.*, 1970, **41**, 4455.

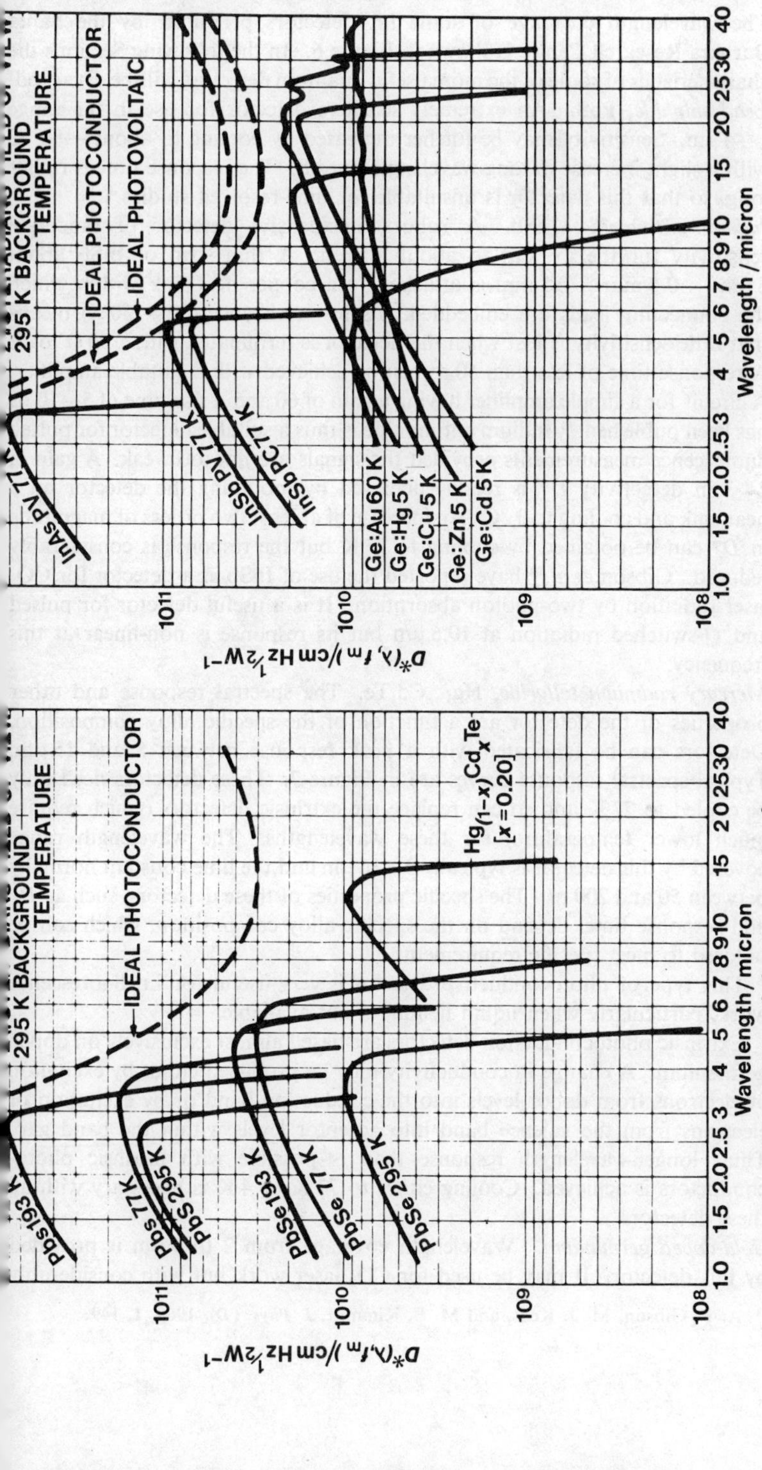

Figure 6 *Spectral detectivities of above-average i.r. detectors*
(Reproduced with permission of Santa Barbara Research Centre)

The wavelength coverage of some i.r. detectors produced by the Santa Barbara Research Centre is shown in Figure 6. In the following Sections the characteristics of some of the more useful quantum detectors will be discussed.

Lead sulphide, PbS. An extremely sensitive detector for use in the range 1—4 μm. Sensitivity may be further increased by cooling to about −80 °C with a slight increase in long-wavelength cut-off. The response time is rather long, so that this detector is unsuitable for time-resolved studies.

Indium antimonide, InSb. A robust, inexpensive detector of moderate sensitivity for the i.r. out to about 7.5 μm. A thin strip of InSb about 6 mm × 0.5 mm × 0.5 mm is attached to a copper heat-sink within which the connecting leads are embedded. Its low resistance (30—130 Ω) means that little sensitivity is lost when the detector is terminated with a 50 Ω load. A response time of less than 50 ns can be achieved with a suitable amplifier. A circuit for a simple amplifier having a gain of 60 and a rise time of 5—10 ns has been published.[83] Indium antimonide is thus a suitable detector for pulsed fluorescence measurements provided the signals are not too weak. A gain of 2—3 in detectivity D^* is readily achieved by mounting the detector on a heat-sink and cooling to 0 °C. An increase of at least two orders of magnitude in D^* can be obtained by cooling to 77 K but the response is considerably reduced. Gibson *et al.*[91] have reported the use of InSb as a detector for CO_2 laser radiation by two-photon absorption. It is a useful detector for pulsed and *Q*-switched radiation at 10.6 μm but its response is non-linear at this frequency.

Mercury cadmium telluride, $Hg_{1-x}Cd_xTe$. The spectral response and other properties of the detector are a function of the specific alloy composition. Detectors can be fabricated with a peak response between 8 and 15 μm. Typical spectral response curves are in Figure 7. These detectors need only be cooled to 77 K and so can replace the extrinsic detectors (which require much lower temperature) for these wavelengths. The wavelength range covered by this detector is typically 3—16 μm and the time constant normally between 50 and 200 ns. The specific properties of these detectors, such as D^* and response time, depend on the specific alloy composition, which can be tailored to meet specific requirements.

This type of photoconductive detector is very useful for i.r. fluorescence work, particularly when liquid helium is not available.

Extrinsic photoconduction detectors are based almost exclusively on doped germanium. A change in conductivity may be produced either by excitation of electrons from donor levels into the conduction band or by excitation of electrons from the valence band into acceptor levels within the band gap. Thus, longer-wavelength response than is possible with intrinsic photoconductors is achieved. Cooling either to 77 K or 4 K is necessary with all these detectors.

Gold-doped germanium. Wavelength coverage from 2 to 9 μm is provided by this detector. It may be used for CO_2 laser work but with considerably

[91] A. F. Gibson, M. J. Kent, and M. F. Kimmitt, *J. Phys.* (*D*), 1968, **1**, 149.

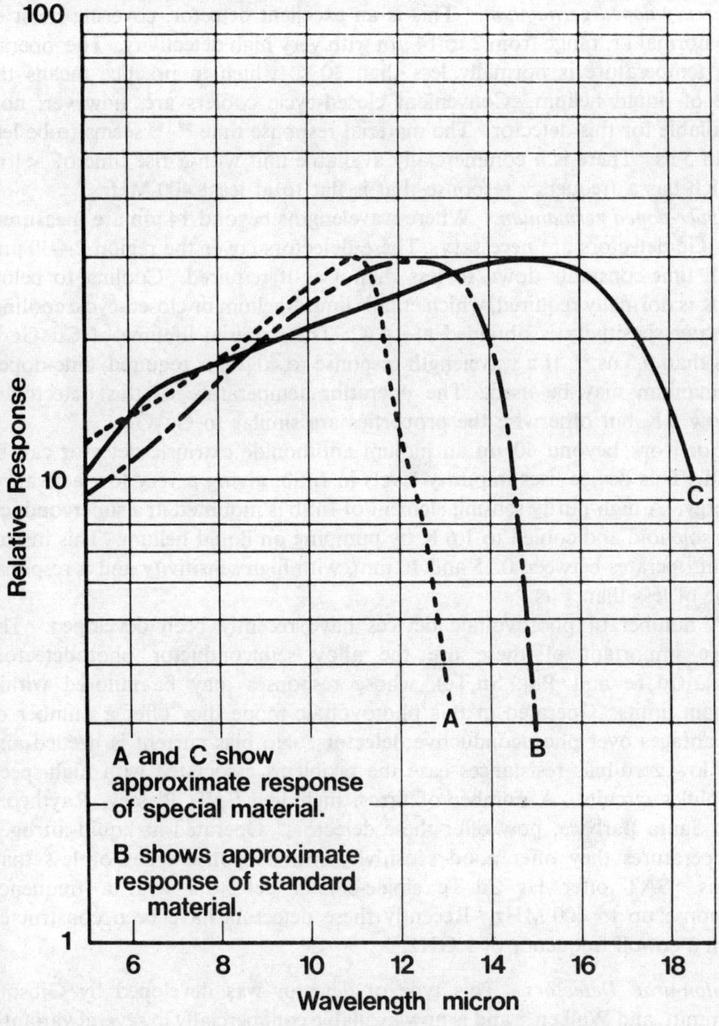

Figure 7 *Relative spectral response of some typical* HgCdTe *i.r. detectors*
(Reproduced with permission of Santa Barbara Research Centre)

reduced sensitivity. Optimum sensitivity is obtained at 60 K, but only a small loss in detectivity (2 to 3 times) occurs at liquid-nitrogen temperature. Using a pulse from a CO_2 laser, the material lifetime has been measured by Bridges *et al.*[92] and found to be 30 ns. Thus with 50 Ω termination this detector is useful for fluorescence work up to about 9 μm. It is more sensitive than InSb used at room temperature.

[92] T. J. Bridges, T. Y. Cheng, and P. K. Cheo, *Appl. Phys. Letters*, 1968, **12**, 297.

Mercury-doped germanium. This is an excellent detector, covering most of the normal i.r. range from 2 to 14 μm with very high detectivity. The operating temperature is normally less than 30 K, which in practice means the use of liquid helium. Convenient closed-cycle coolers are, however, now available for this detector. The material response time [92, 93] seems to be less than 5 ns. There is a commercially available unit with a rise time of <1 ns which has a frequency response that is flat to at least 400 MHz.

Copper-doped germanium. Where wavelengths beyond 14 μm are measured, Cu–Ge detectors are necessary. These detectors cover the region 2—30 μm, with time-constants down to less than 1 ns if required. Cooling to below 14 K is normally required, which entails liquid-helium or closed-cycle cooling. Greater sensitivity is obtained at 14 K. The material lifetime of Cu–Ge is less than 0.5 ns.[92] If a wavelength response to 40 μm is required, zinc-doped germanium may be used. The operating temperature of this detector is below 6 K but otherwise the properties are similar to Cu–Ge.

For work beyond 40 μm an indium antimonide extrinsic detector can be used. This device uses impurity levels in InSb, giving a very low activation energy. A high-purity sensing element of InSb is mounted in a superconducting solenoid and cooled to 1.6 K by pumping on liquid helium. This instrument operates between 0.15 and 10 mm, with high sensitivity and a response time of less than 1 μs.

A number of photovoltaic devices have recently been developed. The more important of these are the alloy semiconductor photodetectors ($Hg_{1-x}Cd_xTe$ and $Pb_{1-x}Sn_xTe$), whose responses may be tailored within certain limits. Operated in this photovoltaic mode they offer a number of advantages over photoconductive detectors. No bias current is needed and the low zero-bias resistances ease the problems associated with high-speed amplifier circuits. A number of firms, including SAT, Plessey, Raytheon, and Santa Barbara, now offer these detectors. Operated at liquid-nitrogen temperatures they offer good sensitivity with response times of less than 50 ns. SAT offer Hg–Cd–Te single-crystal detectors with a frequency response up to 400 MHz. Recently these detectors have been constructed with a cut-off frequency of 1 GHz.[94]

Photon-drag Detectors. This type of detector was developed by Gibson, Kimmitt, and Walker [95] and is now available commercially in several variants. The active element consists of a bar of doped germanium crystal through which the radiation is directed. Photons in the beam transfer their momentum to free carriers in the germanium. These can be holes in *p*-type or electrons in *n*-type germanium. The free carriers are physically driven down the bar by the energy loss from the photons. This creates a voltage gradient which is either amplified or fed directly to the recording equipment. The basic

[93] C. J. Buczek and G. S. Picus, *Appl. Phys. Letters*, 1967, **11**, 125.
[94] C. Verie and M. Sirieix, *I.E.E.E.-J. Quantum Electron.*, 1972, **QE–8**, 180.
[95] A. F. Gibson, M. F. Kimmitt, and A. C. Walker, *Appl. Phys. Letters*, 1970, **17**, 75.

detector has an inherent rise time of less than 1 ns. It is very insensitive to electrical interference and will withstand massive overloading without damage. Responsivity at 10.6 μm is typically 0.15 mV kW^{-1} into 50 Ω, so that the device is very insensitive. It is mainly used for monitoring the output from pulsed and *Q*-switched CO_2 lasers, for which purpose it is ideally suited. A straight-through version is also available which transmits about 75% of the radiation. This allows the CO_2 laser output to be monitored continuously.

Fluorescence Measurements.—*CW Measurements.* A typical experimental arrangement for a quasi-CW fluorescence experiment is shown in Figure 8.

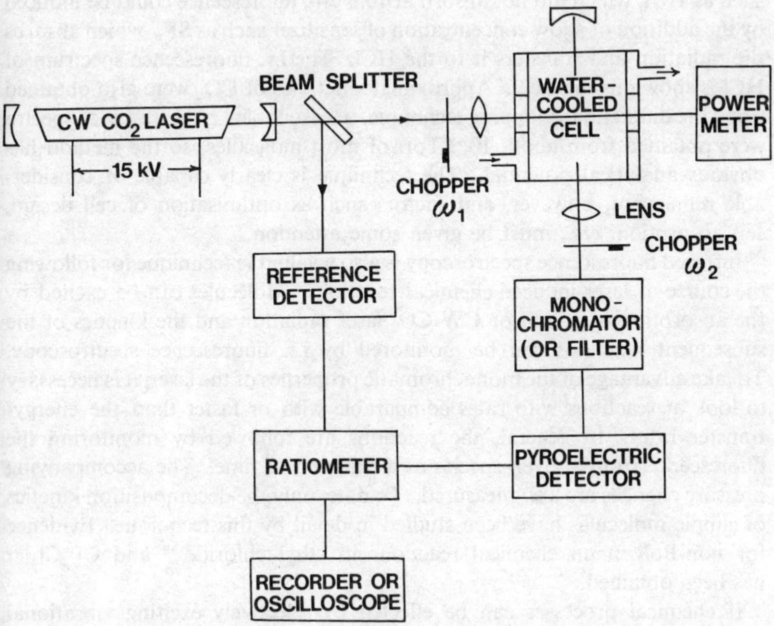

Figure 8 *Schematic arrangement for CW laser-excited i.r. fluorescence spectroscopy*

The output beam from a 200 W CO_2 laser is reduced to about 5 mm diameter with a lens system and passed vertically through the fluorescence cell. The fluorescence emission from the thin pencil of excited gas is imaged on to the slit of a monochromator with a lens and mirror system; the mirror which is placed behind the cell is used to increase the i.r. emission collected by the monochromator. The energy transmitted by the cell is monitored with a power meter. Both the incident laser beam and the fluorescent emission are modulated using a stepping-motor chopper system. In this way a known phase relationship between the incident and emitted radiation can be produced. The fluorescence cell consists of a stainless-steel cube with two 4 cm intersecting holes at right angles. A thermostatted water jacket is provided.

The cell windows were made from Irtran II. Pyroelectric detectors are very convenient, especially when modulation frequencies up to about 1 kHz are used. Pressure-sensitive transistors (Pitran) can be used to monitor the pressure profile in the laser-excited gas. These can be used at frequencies up to 100 kHz.

The quasi-CW technique has promising analytical applications. Robinson *et al.*[96, 97] first observed i.r. fluorescence from gases excited with a CW CO_2 laser. Subsequently, Bailey, Cruickshank, and Jones [98] studied the i.r. fluorescence spectra of a number of absorbing molecules. Fluorescence was observed from all the i.r. active modes in these molecules. For molecules such as HCl, which did not absorb at 10.6 μm, fluorescence could be induced by the addition of a low concentration of sensitizer such as SF_6, which absorbs the radiation and transfers it to the HCl. The i.r. fluorescence spectrum of HCl is shown in Figure 9. Approximate lifetimes of CO_2 were also obtained using the differential chopping technique. Reasonable i.r. fluorescence spectra were obtained from about 10^{-2} Torr of most molecules, so the method has obvious analytical potential. The technique is clearly capable of considerable refinement, however, and factors such as optimisation of cell design, self absorption, *etc.*, must be given some attention.

Infrared fluorescence spectroscopy is also a valuable technique for following the course of laser-induced chemical reactions. Molecules can be excited by the absorption of pulsed or CW CO_2 laser radiation and the kinetics of the subsequent reactions can be monitored by i.r. fluorescence spectroscopy. To take advantage of the monochromatic properties of the laser, it is necessary to look at reactions with rates comparable with or faster than the energy-transfer rates. In general, the reactions are followed by monitoring the fluorescence from a given species as a function of time. The accompanying pressure changes are also measured. To date, only the decomposition kinetics of simple molecules have been studied in detail by this technique. Evidence for non-Boltzmann chemical reactions in ethyl chloride [99] and CF_2Cl_2[100] has been obtained.

If chemical processes can be effected by selectively exciting vibrational levels in one of the reactants to produce non-equilibrium distributions, preferential reaction modes or accelerated reaction rates may be obtained. Possible applications of this technique include accelerated catalysis, more efficient fractionation of hydrocarbons, and separation of isotopes. Wilson [16] has recently discussed some laser-induced chemistry. Cohen *et al.*[101] have irradiated SF_6, ethylene, propene, allene, butadiene, NH_3, and PH_3 with

[96] J. W. Robinson, D. M. Hailey, and H. M. Barnes, *Talanta*, 1969, **16**, 1109.
[97] J. W. Robinson, C. Woodward, and H. M. Barnes, *Analyt. Chim. Acta*, 1968, **43**, 119.
[98] R. T. Bailey, F. R. Cruickshank, and T. R. Jones, *Nature*, 1971, **234**, 92.
[99] R. T. Bailey, F. R. Cruickshank, J. Farrell, D. S. Horne, A. M. North, P. B. Wilmot, and U. Tin Win, *J. Chem. Phys.*, in the press.
[100] R. T. Bailey, F. R. Cruickshank, and J. Gallagher, unpublished work.
[101] C. Cohen, C. Borde, and L. Henry, *Compt. rend.*, 1967, **265**, *B*, 267; 1966, **263**, *B*, 619; 1966, **262**, *B*, 1389.

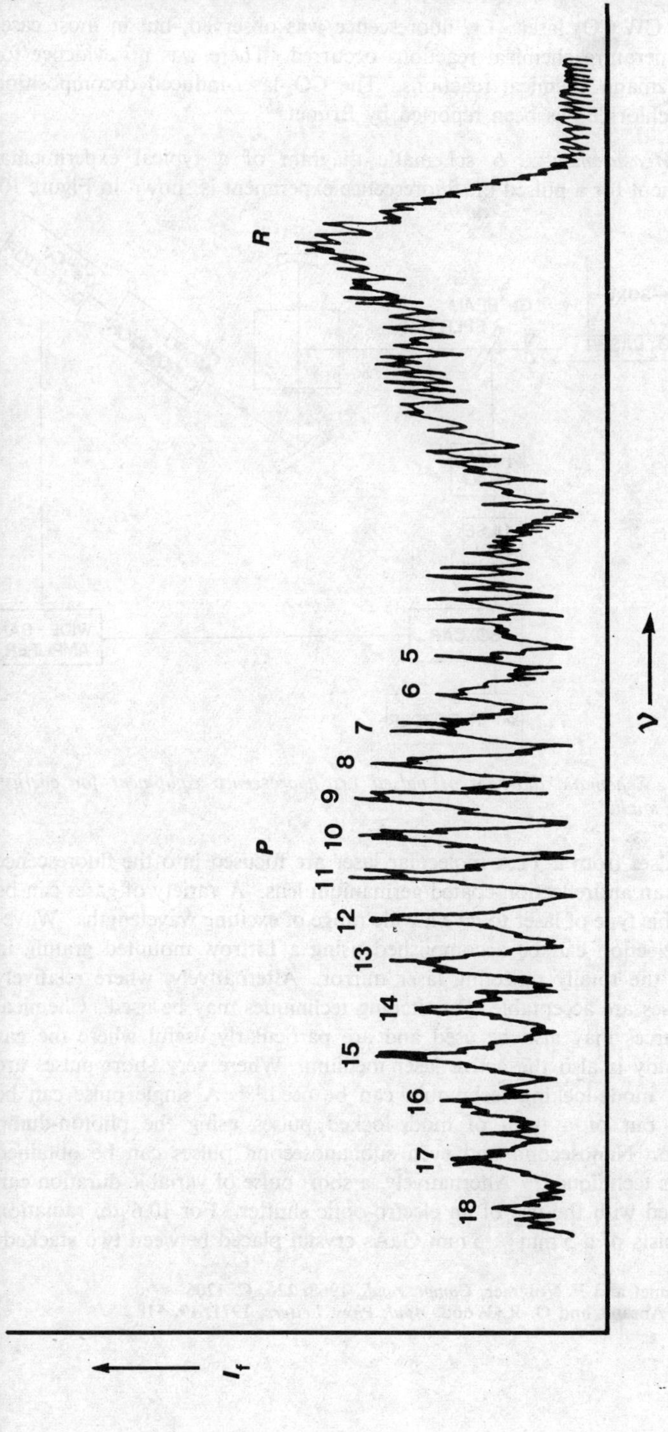

Figure 9 *Infrared fluorescence spectrum of* HCl
(Reproduced by permission from *Nature*, 1971, **234**, 92)

a 100 W CW CO_2 laser. I.r. fluorescence was observed, but in most cases high-temperature chemical reactions occurred. There was no evidence for non-Boltzmann chemical reactions. The CO_2-laser-induced decomposition of ethyl chloride has been reported by Brunet.[102]

Pulsed Measurements. A schematic diagram of a typical experimental arrangement for a pulsed i.r. fluorescence experiment is shown in Figure 10.

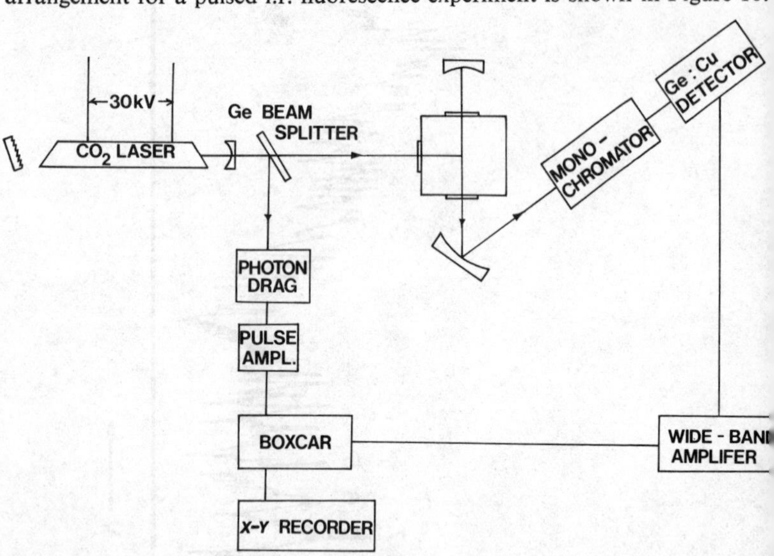

Figure 10 *Schematic diagram of pulsed i.r. fluorescence equipment for energy-transfer studies*

Short pulses from a TEA molecular laser are focused into the fluorescence cell with an antireflection-coated germanium lens. A variety of gases can be used in this type of laser to give a wide range of exciting wavelengths. Wavelength selection can be accomplished using a Littrow mounted grating in place of the totally reflecting laser mirror. Alternatively, where relatively long pulses are acceptable, *Q*-switching techniques may be used. Chemical laser sources may also be used and are particularly useful where the gas under study is also the active laser medium. Where very short pulses are required, mode-locking techniques can be used.[103] A single pulse can be switched out of a train of mode-locked pulses using the photon-dump technique. Nanosecond and even subnanosecond pulses can be obtained using this technique.[103] Alternatively, a short pulse of variable duration can be selected with the use of an electro-optic shutter. For 10.6 μm radiation this consists of a 5 mm × 5 mm GaAs crystal placed between two stacked-

[102] H. Brunet and F. Voignier, *Compt. rend.*, 1968, **226,** *C*, 1206.
[103] R. C. Abrams and O. R. Wood, *Appl. Phys. Letters*, 1971, **19,** 518.

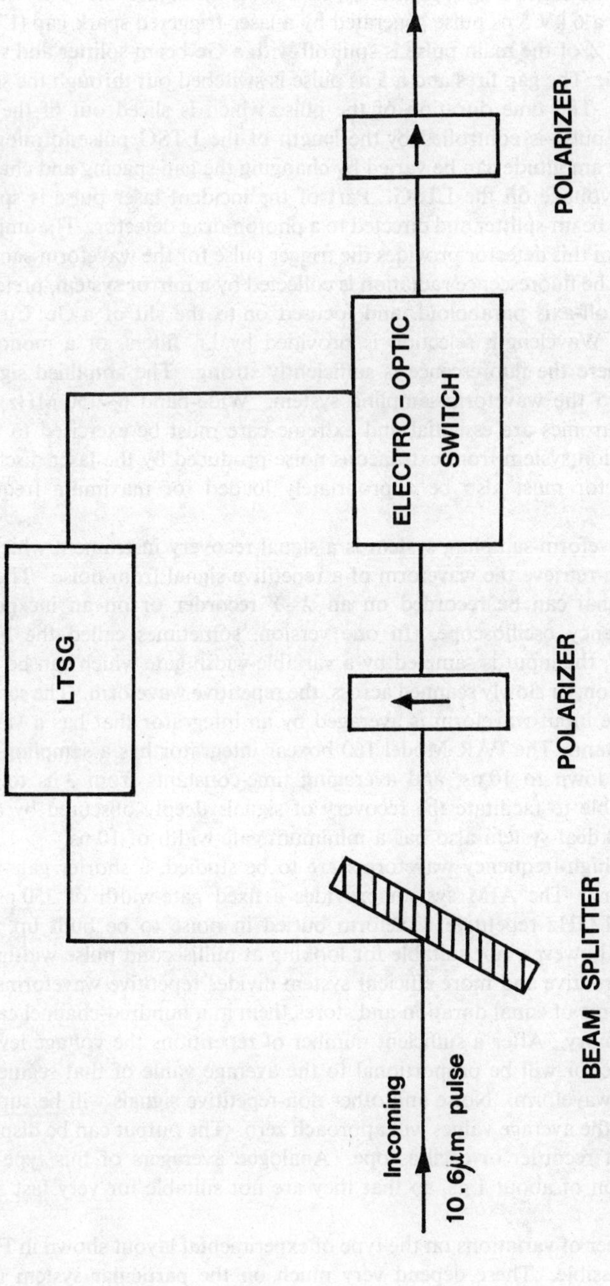

Figure 11 *Schematic diagram of Ga–As electro-optic switch*

plate Ge Brewster single polarizers, as shown in Figure 11. The GaAs is driven by a 6 kV 5 ns pulse generated by a laser-triggered spark gap (LTSG). About 15% of the main pulse is split off with a Ge beam-splitter and sent to the LTSG. The gap fires and a 5 ns pulse is switched out through the second polarizer. The time duration of the pulse which is sliced out of the main incoming pulse is controlled by the length of the LTSG pulse-forming line. The pulse amplitude can be varied by changing the gap spacing and changing the high voltage on the LTSG. Part of the incident laser pulse is split off with a Ge beam-splitter and directed to a photon-drag detector. The amplified signal from this detector provides the trigger pulse for the waveform-sampling system. The fluorescence radiation is collected by a mirror system, preferably using an off-axis paraboloid, and focused on to the slit of a Ge: Cu (4 K) detector. Wavelength selection is provided by i.r. filters, or a monochromator where the fluorescence is sufficiently strong. The amplified signal is directed to the waveform-sampling system. Wide-band (~150 MHz) low-noise electronics are essential and extreme care must be exercised to shield the detection system from extraneous noise produced by the laser discharge. The detector must also be appropriately loaded for maximum frequency response.

The waveform-sampling system is a signal-recovery instrument which can be used to retrieve the waveform of a repetitive signal from noise. The processed signal can be recorded on an *X–Y* recorder or on an inexpensive low-frequency oscilloscope. In one version, sometimes called the boxcar integrator, the input is sampled by a variable-width gate which can be set at any point on, or slowly scanned across, the repetitive waveform. The sampled part of the input waveform is averaged by an integrator that has a variable time-constant. The PAR Model 160 boxcar integrator has a sampling gate-width of down to 10 ns, and averaging time-constants from 3 ns to 100 s are available to facilitate the recovery of signals deeply obscured by noise. The Brookdeal system also has a minimum gate-width of 10 ns.

If very-high-frequency waveforms are to be studied, a shorter gate-width is necessary. The AIM system provides a fixed gate-width of 350 ps and allows a 1 GHz repetitive waveform buried in noise to be built up. This system is, however, not suitable for looking at millisecond pulse widths.

An alternative and more efficient system divides repetitive waveforms into 100 segments of equal duration and stores them in a hundred-channel capacitative memory. After a sufficient number of repetitions the voltage level on each capacitor will be proportional to the average value of that segment of the input waveform. Noise and other non-repetitive signals will be suppressed, since the average values will approach zero. The output can be displayed on a chart recorder or oscilloscope. Analogue averagers of this type offer a resolution of about 1 μs, so that they are not suitable for very fast signal recovery.

A number of variations on the type of experimental layout shown in Figure 10 are possible. These depend very much on the particular system under study. For instance, where a strong, moderately long fluorescence pulse is obtained it is often possible to record the pulse directly on a wide-bandwidth oscilloscope without any signal-averaging equipment. Relatively slow detection systems can be employed and the recorded pulse shape deconvoluted. Also, where the wavelength of the fluorescence signal is shorter than 8 μm the less expensive nitrogen-cooled Ge–Au or InSb detectors can be used. The choice of detector for any particular application often involves a compromise between such factors as cost, wavelength response range, sensitivity, rise time, and convenience of operation.

Pulsed tunable laser systems such as the SFRL or the parametric oscillator are fast beginning to make an impact on the energy-transfer scene. The spectral purity and high powers obtainable from these systems allow individual vibration–rotation states to be selectively populated. Alternatively, they can be used to probe the population of already excited individual vibrational–rotational states with great precision. Very recently, Leone and Moore [104] have used a tunable optical parametric oscillator to excite single vibration–rotation lines of the first and second vibrational levels of HCl directly. Any molecular transition having a wavelength less than 3.5 μm can be excited by this technique. The experimental apparatus used by these workers is shown in Figure 12. Typical oscillator pulses were 10^{-5} J in 10^{-7} s, 80 Hz repetition rate, and about 0.5 cm^{-1} spectral width. The spectral width

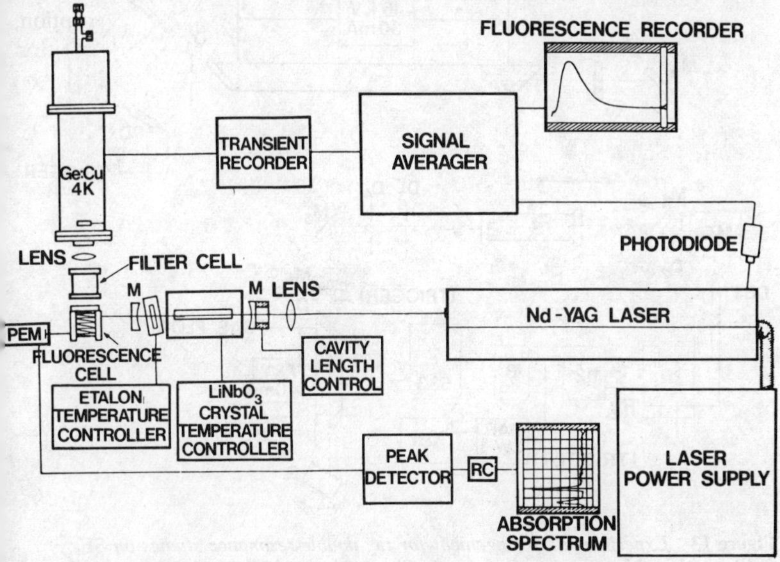

Figure 12 *Experimental arrangement for energy-transfer studies using a tunable parametric oscillator*
(Reproduced by permission from *Chem. Phys. Letters*, 1973, **19**, 340)

[104] S. R. Leone and C. B. Moore, *Chem. Phys. Letters*, 1973, **19**, 340.

was reduced dramatically by inserting a Fabry–Perot étalon into the cavity. The $LiNbO_3$ crystal was pumped with the 0.562 and 0.532 μm lines of a frequency-doubled Nd–YAG laser. Frequency selection is obtained by varying the temperature of the $LiNbO_3$ crystal. The laser radiation was multipassed through the gas and the fluorescence detected with a Ge–Cu detector at 4 K and processed with a boxcar integrator. A similar system was used by Sackett, Hordvik, and Schlossberg[105] to pump ground states of CO ($V = 0$) directly to the second vibrational level of CO ($V = 2$).

Double-resonance Techniques. Often the i.r. fluorescence signal is not sufficiently intense to permit analysis in a monochromator, even a high-aperture instrument. Thus, individual rotational energy levels cannot always be studied using this technique. A more sensitive method of probing the population of a particular rotational transition is to monitor the absorption, as a function of time, of laser radiation tuned to the molecular transition frequency. This technique is often called i.r. double resonance and can be a particularly sensitive probe.[106, 107] The basic experimental method involves the simultaneous exposure of the molecular transition to a stable monochromatic CW laser beam and to the i.r. excitation pulse. This can be

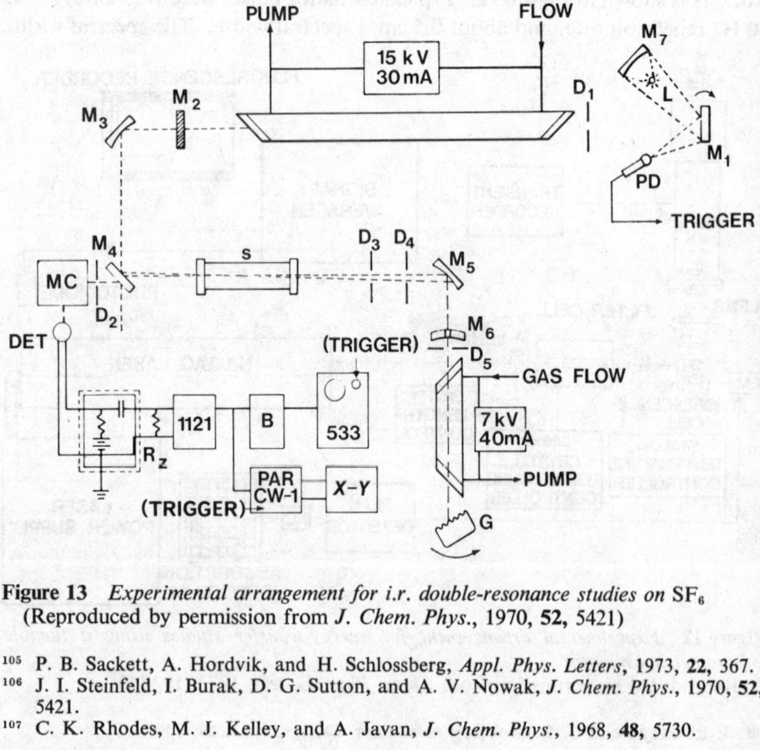

Figure 13 *Experimental arrangement for i.r. double-resonance studies on* SF_6 (Reproduced by permission from *J. Chem. Phys.*, 1970, **52**, 5421)

[105] P. B. Sackett, A. Hordvik, and H. Schlossberg, *Appl. Phys. Letters*, 1973, **22**, 367.
[106] J. I. Steinfeld, I. Burak, D. G. Sutton, and A. V. Nowak, *J. Chem. Phys.*, 1970, **52**, 5421.
[107] C. K. Rhodes, M. J. Kelley, and A. Javan, *J. Chem. Phys.*, 1968, **48**, 5730.

accomplished in the apparatus shown in Figure 13, which was used by Steinfeld *et al.*[106] A *Q*-switched laser produces a 10 kW pulse every 9 ms. The pulse width is controlled by the diaphragm D. When this is fully open a train of pulses 1—2 μs wide are produced. A trigger pulse is provided by a small photodiode which senses light reflected off the rotating-mirror *Q*-switch. The monitoring beam is produced by a CW CO_2 laser operating in the TEM_{00} mode. Wavelength selection is provided by the grating G, which allows the laser to be tuned over a large number of rotational lines. Low powers, <1 W, are necessary to avoid disturbing the molecular level populations. The two beams are superimposed in the sample cell S. The monitoring beam passes through the beam splitter and is directed on to the monochromator MC. A liquid-nitrogen cooled Ge–Au i.r. detector was used in this case which had a time constant of about 1 μs. Boxcar integration can be employed if the fluorescence is weak, or contains noise components. Obviously, the range and versatility of this technique can be enhanced considerably by the use of short pulses of tunable i.r. radiation and sensitive nanosecond detection systems. Individual vibration–rotation transitions can then be excited and the populations of these and other individual transitions accurately monitored. Changes in population produced by laser pumping can also be detected using microwave absorption techniques. Several such studies have been carried out using N_2O and CO_2 laser pumping frequencies.[108–112] Stark-tuned non-linear laser spectroscopy has been applied to a few transitions in CH_3F near 3.39 μm with a resolution of about 100 kHz.[113] Although not fluorescence spectroscopy, this work is of interest because dipole moments in excited vibrational states as well as ground states can be measured using this technique.

Stimulated-Raman Pumping. Conventional fluorescence measurements are limited to molecules which have suitable dipole transitions. Stimulated-Raman pumping is a potentially important method for producing a high density of vibrational excitation in transitions which are i.r.-inactive. When stimulated Stokes Raman scattering is produced in a gas, a high density of vibrationally excited molecules is simultaneously created. DeMartini and Ducuing [114] first demonstrated the feasibility of the technique in their work on hydrogen. Vibrational excitation was produced by focusing the output of a ruby laser into high-pressure hydrogen gas. Spontaneous anti-Stokes Raman scattering was used to follow the decay of the excited molecules. This method suffers from the disadvantage that rather high pressures and densities

108 T. Shimuzu and T. Oka, *Phys. Rev.*, 1970, **A2,** 1177.
109 J. Lemaire, J. Houriez, J. Thibault, and B. Maillard, *J. Phys.* (*Paris*), 1971, **32,** 35.
110 M. Faurrier, M. Redow, H. V. Lerberghe, and C. Borde, *Compt. rend.*, 1970, **270,** *B*, 537.
111 L. Frenkel, H. Maranty, and T. Sullivan, *Phys. Rev.*, 1971, **A3,** 1640.
112 J. Lemaire, J. Houriez, F. Herlemont, and J. Thibault, *Chem. Phys. Letters*, 1973, **19,** 373.
113 A. C. Luntz, J. D. Swalen, and R. C. Brewer, *Chem. Phys. Letters*, 1972, **14,** 512.
114 F. DeMartini and J. Ducuing, *Phys. Rev. Letters*, 1966, **17,** 117.

are required to produce stimulated-Raman scattering, resulting in very short and sometimes unmeasurable relaxation times. The flexibility and sensitivity of the method can be increased by producing the stimulated scattering in a separate cell optimized for this purpose and then combining the Stokes-shifted Raman radiation (ω') with the original ruby radiation (ω) and refocusing into a second cell, containing the gas under study.[115–117] In the second cell, the Raman-transition frequency $\omega_i \approx \omega - \omega'$ is stimulated by the laser and strongly coherent Stokes radiation. The density of excited molecules produced is proportional to the laser intensity, the Stokes intensity, and the pressure in the cell. A large fraction of the molecules in the focus of the two beams may be vibrationally excited. When vibrational relaxation occurs, the translational temperature of the gases in the focal zone increases and expansion takes place. This is monitored by observing the scattering of a helium–neon laser beam from the gas. This is a very sensitive technique for monitoring the energy-relaxation processes. The stimulated-Raman method, however, suffers from the disadvantage that it is difficult to produce stimulated Raman scattering from gases other than H_2, D_2, and CH_4 using conventional Q-switched pulse excitation. This is because stimulated Brillouin scattering (SBS) rather than SRS is often the dominant non-linear effect. The response times for the Raman and Brillouin processes are determined by the appropriate phonon lifetimes, either optical, in the case of Raman scattering, or acoustic, in the case of Brillouin scattering. Optical phonon lifetimes are typically 10—100 ps, while acoustic phonon lifetimes generally range from 1000 to 10 000 ps for the same pressure. Thus, by using picosecond pulse excitation, only the Raman process has sufficient time to establish an appreciable gain and the Brillouin competition that appears with nanosecond excitation is suppressed.[118] Kovacs and Mack [119] applied this technique of 'transient' stimulated-Raman scattering to study room-temperature vibrational relaxation in H_2, D_2, O_2, N_2, and CO. The pumping source used was a mode-locked ruby laser producing pulses 5—20 ps wide, having a peak power of 5 GW.

By using a suitably stabilized tunable dye laser together with a fixed-frequency laser, this Raman-pumping scheme could be more generally applicable to Raman-active transitions. Frey, Lukasik, and Ducuing [116] have used this technique to look at $V \rightarrow T$ transfer in pure N_2 and O_2 or in mixtures with H_2 and He. Their equipment is shown diagramatically in Figure 14. The dye cell is pumped by a 40 MW 30 ns ruby laser pulse (ω). The emitted dye radiation (10 MW, 30 ns, ω') together with the ruby radiation is focused inside the gas cell. Excitation occurs when the frequency of the Raman-active

[115] J. Ducuing, C. Joffrin, and J. P. Coffinet, *Opt. Comm.*, 1970, **2**, 6.
[116] R. Frey, J. Lukasik, and J. Ducuing, *Chem. Phys. Letters*, 1972, **14**, 514.
[117] M. M. Audibert and C. Joffrin, *Opt. Comm.*, 1972, **5**, 218; M. M. Audibert, C. Joffrin, and J. Ducuing, *Chem. Phys. Letters*, 1973, **19**, 26.
[118] M. E. Mack, R. L. Carman, J. Reintgis, and N. Bloembergen, *Appl. Phys. Letters*, 1970, **16**, 209.
[119] M. A. Kovacs and M. E. Mack, *Appl. Phys. Letters*, 1972, **20**, 487.

transition $\omega_i \approx \omega - \omega'$. The $V \rightarrow T$ relaxation is again followed by the refractive index change in the excited focal volume, which is monitored by the scattering of a He–Ne laser.

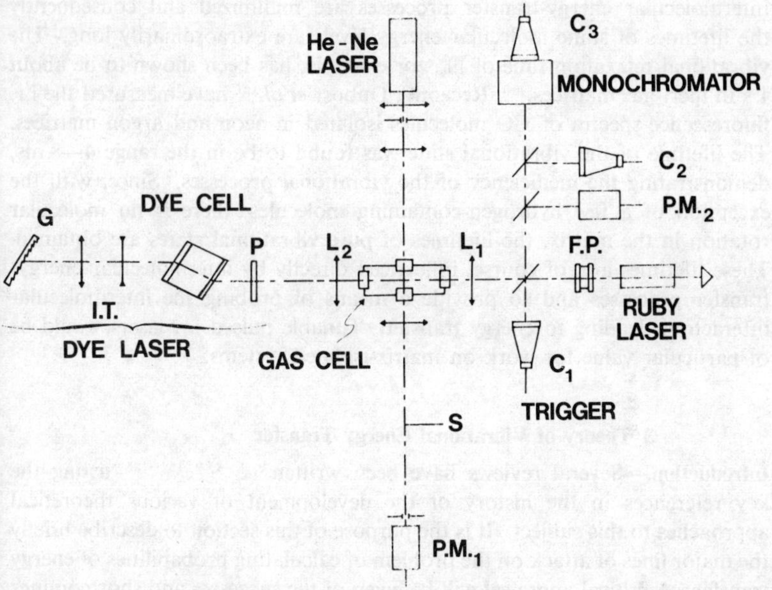

Figure 14 *Diagram of experimental arrangement for tunable stimulated-Raman studies of vibrational energy transfer*
(Reproduced by permission from *Chem. Phys. Letters*, 1972, **14**, 514)

This method has the advantage over the stimulated-Raman technique (even in the picosecond regime) that with presently available dye lasers the frequency ω' can be accurately chosen. Dye-laser linewidths as low as 10^{-2} cm^{-1} can be obtained and $\omega - \omega'$ can be adjusted to any desired value in the range 150—5 000 cm^{-1}.[116] This allows the intense and selective excitation of a broad range of Raman-active transitions.

Stimulated-Raman excitation methods have the advantage that extremely fast (picosecond) vibrational excitation can be obtained. Further more, spontaneous Raman scattering can provide an extremely fast method of monitoring the relaxation processes with the very fast response detectors available for these regions. Experiments could also be performed with condensed phases that are impracticable with i.r. laser excitation. Low-pressure gases are, however, not amenable to study with the Raman-pumping method.

Matrix-isolation Studies. The matrix-isolation technique provides a unique opportunity for the controlled study of intermolecular interaction and energy transfer. By using pulsed i.r. fluorescence, double resonance, or Raman excitation techniques the energy levels of relatively unperturbed ('isolated') molecules can be investigated. In noble-gas matrices at 4 K, the normal intermolecular energy-transfer processes are minimized and consequently the lifetimes of some molecular energy levels are extraordinarily long. The vibrational relaxation time of N_2, for example, has been shown to be about 1 s in inert-gas matrices.[120] Recently, Dubost *et al.*[121] have measured the i.r. fluorescence spectra of CO molecules isolated in neon and argon matrices. The lifetime of the vibrational state was found to be in the range 4—8 ms, demonstrating the inefficiency of the vibrational processes. Since, with the exception of a few hydrogen-containing molecules, there is no molecular rotation in the matrix, the lifetimes of pure vibrational states are obtained. These lifetimes are, of course, influenced directly by intermolecular energy-transfer processes and so provide a means of probing the intermolecular interactions leading to energy transfer. Tunable pulsed i.r. lasers would be of particular value for work on matrix-isolated systems.

3 Theory of Vibrational Energy Transfer

Introduction.—Several reviews have been written[3, 5, 6, 9, 122–126] listing the key references in the history of the development of various theoretical approaches to this subject. It is the purpose of this section to describe briefly the major lines of attack on the problem of calculating probabilities of energy transfer. A critical appraisal will be given of the successes and shortcomings of these various techniques, together with an indication of future trends. It is particularly appropriate that such a discussion should appear in this Report, since i.r. laser-initiated fluorescence has made available some of the most accurate data for the critical evaluation of these theories.

Theoretical Prediction of Transition Probability.—To calculate exact transition probabilities requires the knowledge of exact intermolecular potentials for all orientations of the two colliding species. These are rarely available and, even if they were, the solution of the resultant Schrödinger equation would be intractable in terms of computer time, if not for mathematical reasons. Approximations must, therefore, be relied upon and these fall essentially into three categories:

(*a*) A complete quantum-mechanical treatment.

[120] D. S. Tinti and G. W. Robinson, *J. Chem. Phys.*, 1968, **49**, 3229.
[121] H. Dubost, L. Abouaf-Marguin, and F. Legay, *Phys. Rev. Letters*, 1972, **29**, 145.
[122] K. Takayanagi, *Progr. Theor. Phys. Suppl. Kyoto*, 1963, **25**, 1.
[123] K. Takayanagi, *Adv. Atomic Mol. Phys.*, 1965, **1**, 149.
[124] K. Herzfeld and T. A. Litovitz, 'Absorption and Dispersion of Ultrasonic Waves', Academic Press, New York, 1959.
[125] K. F. Herzfeld in 'Thermodynamics and Physics of Matter', Princeton Univ. Press, Princeton, N.J., 1965.
[126] D. Rapp and T. Kassal, *Chem. Rev.*, 1969, **69**, 61.

(*b*) A classical treatment of trajectory of the incident particle with a quantum-mechanical treatment of the oscillator, coupling the time-dependent perturbation by the collision with the time-dependent Schrödinger equation to yield the final state of the oscillator and thence the transition probability.

(*c*) A totally classical treatment.

This last approach is best reserved for highly energized systems not usually found in i.r. laser fluorescence and thus will not be relevant to the systems discussed here. Each of the treatments (*a*), (*b*), and (*c*) can be subdivided into four classes according to the potential function approximation employed, *viz.*:

(i) A harmonic oscillator potential and exponential repulsive potential between colliding species.
(ii) A harmonic oscillator potential and anharmonic interaction potential.
(iii) Anharmonic oscillator potential and exponential repulsive interaction potential.
(iv) Anharmonic oscillator potential and anharmonic interaction potential.

Orientation corrections are applied in various ways and the Schrödinger equation is solved approximately also in various ways. Details of the various combinations so far studied may be found as follows:

(*a*) (i) refs. 3, 122, 124, 125, 127, 128, 129, 130;
(*b*) (i) refs. 128, 131, 132, 133;
(*c*) (i) ref. 133; (*a*) (ii) refs. 134, 135; (*b*) (ii) refs. 131, 135, 136;
(*c*) (iv) refs. 137, 138.

Clearly this list does not include all the possible combinations and the most realistic cases (*a*) (iii), (*a*) (iv), and (*b*) (iv) do not seem to have been formulated. In small molecules where the rotational level spacings are greater than kT,[137–143] rotational effects can be important.

[127] J. M. Jackson and N. F. Mott, *Proc. Roy. Soc.*, 1932, **A137**, 703.
[128] K. Takayanagi and T. Kishimoto, *Progr. Theor. Phys. Kyoto*, 1953, **9**, 578.
[129] K. Takayanagi, *J. Phys. Soc. Japan*, 1959, **14**, 75.
[130] R. N. Schwartz, Z. I. Slawsky, and K. F. Herzfeld, *J. Chem. Phys.*, 1952, **20**, 1591.
[131] C. Zener, *Proc. Cambridge Phil. Soc.*, 1932, **29**, 136.
[132] C. Zener, *Phys. Rev.*, 1931, **37**, 556.
[133] D. Rapp, *J. Chem. Phys.*, 1960, **32**, 735.
[134] A. F. Devonshire, *Proc. Roy. Soc.*, 1937, **A158**, 269.
[135] K. Takayanagi and Y. Mianoto, *Sci. Report Saitama Univ.*, 1959, **A111**, 101.
[136] T. L. Cottrell and N. Ream, *Trans. Faraday Soc.*, 1955, **51**, 1453.
[137] S. W. Benson and G. C. Berend, *J. Chem. Phys.*, 1964, **40**, 1289.
[138] E. B. Alterman and D. J. Wilson, *J. Chem. Phys.*, 1965, **42**, 1957.
[139] T. L. Cottrell, R. C. Dobbie, J. McLain, and A. W. Read, *Trans. Faraday Soc.*, 1964, **60**, 241.
[140] T. L. Cottrell and A. J. Matheson, *Trans. Faraday Soc.*, 1962, **58**, 2336.
[141] T. L. Cottrell and A. J. Matheson, *Trans. Faraday Soc.*, 1963, **59**, 824.
[142] S. W. Benson and G. C. Berend, *J. Chem. Phys.*, 1966, **44**, 4247.
[143] C. B. Moore, *J. Chem. Phys.*, 1965, **43**, 2979.

The most popular theory, against which current results are tested, is the Schwartz–Slawsky–Herzfeld (SSH) theory,[130] an *a priori* (*a*) (i) one-dimensional approach. Krauss and Mies [144] have calculated a nearly exact potential function for the He + H_2 system and the latter author has used this to discuss [145] the significance of the usual SSH approximations. Mies finds that potential functions generated by summing the exponential functions between each pair of atoms are grossly inaccurate. More sophisticated fitting to Lennard-Jones potentials derived from gas-kinetic experiments can also give very poor results if long-range interactions are important, since short-range interactions dominate the experimental Lennard-Jones potentials, rendering extrapolation inaccurate. Mies also found that the steric orientation averaging multiplier of $\frac{1}{3}$ is found to be a good approximation, although $\frac{1}{4}$ is better for He + H_2. Thus head-on collisions are *not* the most efficient in energy transfer. Perhaps the most significant of Mies' conclusions is that the energy-transfer probability can be several orders of magnitude in error if anharmonicity is not included in the calculation. The anharmonicity factor, *R*, is defined as ${}_1V_1/{}_0V_0$, where ${}_nV_n$ is the potential function for elastic scattering of He by H_2 in the vibrational state *n*. In fact *R* is about 1.1 and not the 1 of the harmonic approximation. Consequently the energy-transfer probability, depending on the square of the wavefunction, is dramatically reduced by two or three orders of magnitude.

If corrections for the above features are included, together with those for the shortcomings of the first-order distorted-wave approximation used and failure of the first-order approximation at high temperature, Rapp [126] deduces, for $N_2 + N_2$, a probability for first vibrational level quenching, $\langle P_{1\rightarrow 0}\rangle$, that is ten times less than the experimental result. However, at a given temperature this modified SSH theory correctly predicts the increase of $\langle P_{1\rightarrow 0}\rangle$ in the series N_2, CO, O_2, Cl_2, Br_2, and I_2. Long-range interactions were not included. This represents the current state of the art for a general theory.

In the original SSH derivation, series convergence in the averaging of $\langle P_{1\rightarrow 0}\rangle$ is possible only if the energy transferred in a collision, ΔE, is $\ll \frac{1}{2}\mu(v_0{}^*)^2$, where $v_0{}^*$ is the intermolecular velocity for which energy-transfer probability is maximized and μ is the reduced mass of the colliding pair. If the maximum velocity of the molecule is considered, $\Delta E \approx 200\ \text{cm}^{-1}$ for a molecule of molecular weight 10 at 300 K. Since one quantum must be transferred in the collision, serious errors will arise if ΔE exceeds 200 cm^{-1}, which it frequently does in laser studies.

In the simpler exponential potential treatments [146–148] the potential function is written as $\exp(-x/l)$ and the parameter *l* is usually predicted by matching

[144] M. Krauss and F. H. Mies, *J. Chem. Phys.*, 1965, **42**, 2703.
[145] F. H. Mies, *J. Chem. Phys.*, 1965, **42**, 2709.
[146] J. G. Parker, *Phys. Fluids*, 1959, **2**, 449.
[147] J. G. Parker, *J. Chem. Phys.*, 1964, **41**, 160.
[148] S. W. Benson and G. C. Berend, *J. Chem. Phys.*, 1966, **44**, 470.

experimental and theoretical $\langle P_{1 \to 0} \rangle$ values. 'Reasonable' values[124] of l typically result, *e.g.* as in Table 5.[146–148] In ref. 126 Rapp uses Herzfeld's procedure to fit the exponential to the Lennard-Jones function. Obviously all values are 'reasonable', but provide no test of the accuracy of the related theories. Thus the modified SSH procedure is to be preferred.

Table 5 *Values of the parameter l for collisions of various molecules that are quoted in literature*

Collision pair	*Ref.* 126	*Refs.* 146, 147	*Ref.* 148
N_2–N_2	0.21	0.25	0.27
CO–CO	0.21	0.25	0.25
O_2–O_2	0.28	0.26	0.26
Cl_2–Cl_2	0.23	0.28	0.30
Br_2–Br_2	0.24	0.32	0.32
I_2–I_2	0.29	0.38	0.37

As the transfer becomes more nearly resonant, *i.e.* the energy-transfer process becomes more thermoneutral, the SSH theory will become far less accurate, since long-range interactions are of prime importance, and these are poorly estimated by SSH procedures.

In a recent review[149] of energy-transfer processes important in the CO_2 laser it has been shown that, for the nearly resonant case:

$$CO_2^*(\nu_3) + N_2 \rightleftharpoons CO_2 + N_2^* + 18\ \text{cm}^{-1}\ \text{(to sink)}$$

it is now experimentally well established that the energy-transfer probability passes through a minimum at ~1000 K and thus at <1000 K the wrong temperature dependence is predicted by the SSH theory (Figure 15). No unified theory predicts this minimum.

Sharma and Brau[150, 151] expand the N_2–CO_2 interaction potential in terms of multipole moments. The first non-vanishing contribution is the interaction of the N_2 quadrupole moment Q with the CO_2 transition dipole moment. Consideration of rotational transitions is precluded by substitution of the spherical potential $V = \mu Q/2r^4$ corresponding to the r.m.s. values of the interaction averaged over all molecular orientations. Short-range forces are represented by a rigid sphere of collision diameter σ, independent of the internal states of the molecules. Procedure (*b*) above was followed in completing the calculation. The correct negative temperature dependence was predicted.

For exactly resonant transitions, *e.g.*

$$CO\,(V = 2) + CO\,(V = 0) \rightarrow 2CO\,(V = 1)$$

it has been shown[152] that dipole–dipole interactions dominate. Procedure (*b*) leads to a 1 to 10-fold greater cross-section than the gas-kinetic value.

149 R. L. Taylor and S. Bitterman, *Rev. Mod. Phys.*, 1969, **41**, 26.
150 R. D. Sharma and C. A. Brau, *Phys. Rev. Letters*, 1967, **19**, 1273.
151 R. D. Sharma, *Phys. Rev.*, 1969, **177**, 102.
152 B. H. Mahan, *J. Chem. Phys.*, 1967, **46**, 98.

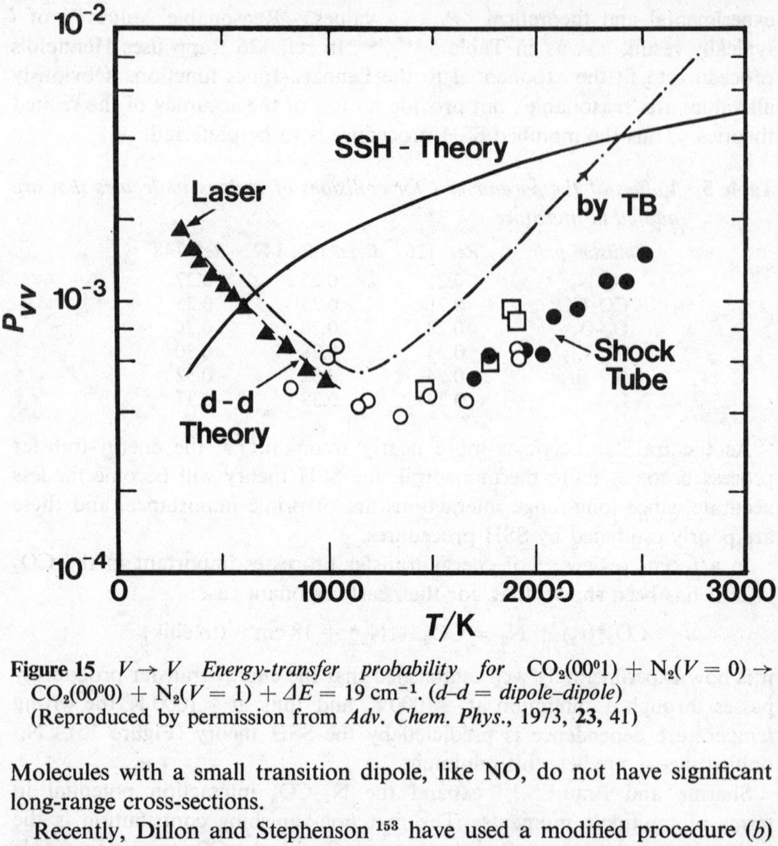

Figure 15 *$V \rightarrow V$ Energy-transfer probability for* $CO_2(00^01) + N_2(V = 0) \rightarrow CO_2(00^00) + N_2(V = 1) + \Delta E = 19\ cm^{-1}$. *(d–d = dipole–dipole)* (Reproduced by permission from *Adv. Chem. Phys.*, 1973, **23**, 41)

Molecules with a small transition dipole, like NO, do not have significant long-range cross-sections.

Recently, Dillon and Stephenson [153] have used a modified procedure (*b*) calculation with curved classical trajectories [154] to formulate the transition probability in terms of a scattering operator *S*. A diagonalization transformation is presented for its evaluation. They examine the dependence of transfer cross-sections on the molecular quantum numbers and on temperature. Their procedure eliminates the necessity for the Born approximation as used by Mahan or Sharma and Brau; indeed it is shown that if high vibrational states are involved this approximation becomes extremely poor. The exchange of multiple vibrational quanta in one collision is of significant importance: for example, the cross-section for the exchange of $n + 1$ quanta in such a collision is shown to be less than 10 times smaller than that for an n-quanta exchange. The Born approximation imposes the selection rules $\Delta V = \pm 1$, $\Delta J = 0$, which are thus invalid, particularly at higher quantum levels. The differences between the single-quantum exchange cross-sections of the Born approximation treatment and the results of Stephenson and Dillon arise

[153] T. A. Dillon and J. C. Stephenson, *Phys. Rev.*, 1972, **A6**, 1460.
[154] T. A. Dillon and J. C. Stephenson, *J. Chem. Phys.*, 1973, **58**, 2056.

from conservation of probability of energy transfer and attention to the weighting of rotational states. For example, a total vibrational excitation of 15 quanta yields cross-sections ten orders of magnitude greater than the Born approximation in the non-resonant limit and two orders of magnitude smaller in the resonant limit. In the former case the deviation from the Born result is particularly great at *low* molecular quantum number (n) of the energy-accepting partner in the collision. For nearly resonant cases the deviation is greatest at high (n). The calculation considers a harmonic oscillator, but it is claimed that, since an anharmonic oscillator wavefunction is a linear combination of harmonic oscillator wavefunctions, the results will not be seriously altered by anharmonicity. This treatment is able to give good *a priori* agreement with energy exchange between HF, DF, and HCl with $CO_2(\nu_3)$. Other current theories are unable to give such agreement.

From the above it seems that a rather full description of the system interactions is necessary for accurate prediction of $\langle P_{1 \to 0} \rangle$ values. Conversely, energy-transfer studies give highly detailed information on the nature of intermolecular potentials, although generally the calculations must be carried out for an assumed potential and the match of theoretical and experimental $\langle P_{1 \to 0} \rangle$ values must be tested. This problem does not arise with simpler exponential-type treatments, but these do not seem worth pursuing in view of the above results. Even in well-tested systems such as HCl and Ar, *etc.*[155] current theories do not seem to give the correct reduced-mass dependence.

Extraction of Rate Constants from Fluorescence-decay Curves.—The experimentally observed multiple decay curves are analysed in a way exemplified by the treatment[156] of the quaternary mixture DCl–CO–N_2–He. In this mixture a multiple $V \to V$ energy-transfer process occurs in the order DCl $\to$ CO $\to$ N_2 followed by $V \to T$ transfer, the DCl being excited by a DCl chemical laser. The He merely acts as a buffer gas. For CO the observed $V \to V$ energy-transfer fluorescence-decay curves occur in the early part of the total fluorescence-decay curve and may be expressed (and observed) as a double exponential of the form:

$$I_{\text{fluorescence}} = (1 - A) - \exp(-k'_{\text{exptl}}\, t) + A \exp(-k''_{\text{exptl}}\, t)$$

where $k'_{\text{exptl}} > 25\, k'_{\text{exptl}}$ and $(1 - A) =$ the fraction of energy in CO after $V \to V$ equilibration.

It may be shown by conventional kinetic analysis of the early part of the fluorescence-decay curves (where the double exponential is seen) that the relevant equations for the system:

$$N_2(V=1) + CO(V=0) \underset{k_{-1}}{\overset{k_1}{\rightleftharpoons}} N_2(V=0) + CO(V=1)$$

[155] D. J. Seery, *J. Chem. Phys.*, 1973, **58,** 1796.
[156] P. F. Zittel and C. B. Moore, *Appl. Phys. Letters*, 1972, **21,** 81.

$$CO(V=1) + DCl(V=0) \underset{k_{-2}}{\overset{k_2}{\rightleftharpoons}} CO(V=0) + DCl(V=1)$$

$$N_2(V=1) + DCl(V=0) \underset{k_{-3}}{\overset{k_3}{\rightleftharpoons}} N_2(V=0) + DCl(V=1)$$

are

$$k_{exptl}\,[DCl^*] = k_{-2}[CO][DCl^*] + k_{-3}[N_2][DCl^*] - k_2[CO^*][DCl] - k_3[N_2^*][DCl]$$

$$k_{exptl}\,[CO^*] = k_2[CO^*][DCl] + k_{-1}[N_2][CO^*] - k_1[N_2^*][CO] + k_{-2}[CO][DCl^*]$$

and

$$k_{exptl}\,[N_2^*] = k_3[N_2^*][DCl] + k_1[N_2^*][CO] - k_{-1}[N_2][CO^*] - k_{-3}[N_2][DCl^*]$$

where $[M^*] = [M(V=1)]$, *i.e.*

$$\begin{vmatrix} -\{k_{-2}[CO] + k_{-3}[N_2]\} + k_{exptl} & k_2[DCl] & k_3[DCl] \\ k_{-2}[CO] & -\{k_{-1}[N_2] + k_2[DCl]\} + k_{exptl} & k_1[CO] \\ k_{-3}[N_2] & k_{-1}[N_2] & -\{k_3[DCl] + k_1[CO]\} + k_{exptl} \end{vmatrix} \begin{Vmatrix} [DCl^*] \\ [CO^*] \\ [N_2^*] \end{Vmatrix} = 0 \qquad (1)$$

Values of k_2 and k_3 are measured in the appropriate ternary mixtures. k_{-2} is related to k_2 and k_{-3} to k_3 by the equation for the equilibrium constant, $k_n/k_{-n} = \exp(-\Delta E/RT)$. Thus substitution of k''_{exptl} into the above equation yields a value for k_1. k''_{exptl} is used since it is most sensitive to process 1,—1 and thus precision is increased. The third root of equation (1) is, of course, 0. Thus the typical evaluation of a relaxation rate-constant is achieved. Since this is a second-order rate constant, its reciprocal has dimensions of [concentration] [time] and this quantity, expressed as $p\tau$, will be used throughout this Report.

As an empirical correlation, the Lambert–Salter plot [157] may be used to obtain rough estimates of $p\tau$ values. This relation shows that, for polyatomic molecules containing hydrogen, $\log_{10} Z_{1,0}$ is proportional to the lowest frequency of the molecule. A similar relation holds for molecules not containing hydrogen. It has been suggested [158] that the higher rotational velocities of hydrogen-containing molecules should favour $V \rightarrow R$ transfer, particularly if the lowest frequency is associated with a bending mode.

This may account for the separation of the groups, but not for the nature of the correlation. Recent laser studies, particularly with the hydrogen halides, are beginning to uncover the significance of $V \rightarrow R$ transfer.

Extraction of Transition Probability from Experimental Rate Constants.—So far, only the problems of $\langle P_{1 \rightarrow 0} \rangle$ calculations have been discussed. Experimentally, of course, one measures the relaxation (Napier) time for a given

[157] J. D. Lambert and R. Salter, *Proc. Roy. Soc.*, 1959, **A253**, 277.
[158] T. L. Cottrell and A. J. Matheson, *Trans. Faraday Soc.*, 1962, **58**, 2336; 1963, **59**, 824.

observed vibrational level to reach 1/e of its original population. This has then to be related to a specific transition probability by a kinetic analysis of the system. Such analyses are in their early stages as yet, but are highly specific to each system studied. Generally, assumptions about relative rates are required in order to solve the kinetic equations, for example rotation–rotation, translation–translation, and rotation–translation processes are very fast. Other perhaps less obvious relations characteristic of the particular system may also be used. The frequency-specific nature of the laser fluorescence technique, particularly now that tunable spin-flip Raman laser systems of considerable power are available, will enable many more species to be followed in any one system, with a concomitant reduction in the necessary assumptions and an increase in precision. Relatively large molecules are now being tackled, *e.g.* HCl and DCl + CH_4 and CD_4.[159]

For pure compounds, if only a two-state system is considered, *e.g.* the $V = 0$ and $V = 1$ levels, with energies ε_0 and ε_1 and excitation and quenching constants $k'_{0,1}$ and $k'_{1,0}$, respectively, defined by equation (2), and if $N_j =$

$$k'_{0,1}/k'_{1,0} = \exp[-(\varepsilon_1 - \varepsilon_0)/kT] \quad (2)$$

population of level j after the laser pulse, equation (3) may be derived. For

$$dN_1/dt = k'_{0,1}(N_0 + N_1)(N_0) - k'_{1,0}(N_0 + N_1)(N_1) \quad (3)$$

most temperatures associated with laser studies, $(N_0 + N_1) \approx N_0$, and if we write $k'_{0,j}(N_0 + N_1) = k_{0,j}$, then $k_{0,j}$ is a pseudo-first-order rate constant, and we have equation (4). Now if we define N_j^* as the j-level population

$$dN_1/dt = k_{0,1}(N_0) - k_{1,0}(N_1) \quad (4)$$

prior to the laser pulse, $N_0 + N_1 = N_0^* + N_1^*$, and equations (5) may be derived.

$$dN_1/dt = k_{0,1}(N_1^* + N_0^* - N_1) - k_{1,0}(N_1) \quad (5a)$$

$$= k_{0,1}(N_1^* + N_0^*) - (k_{0,1} + k_{1,0})N_1 \quad (5b)$$

$$\approx k_{1,0}(N_1^*) - (k_{0,1} + k_{1,0})N_1 \quad (5c)$$

[from equation (2)]

Integration gives equation (6), where $N_j(t)$ = population of level j at time t

$$[k_{1,0}N_1^* - (k_{0,1} + k_{1,0})N_1(0)]/[k_{1,0}N_1^* - (k_{0,1} + k_{1,0})N_1(t)] = \exp(k_{0,1} + k_{1,0})t \quad (6)$$

after the laser pulse, and, if $k_{1,0} \gg k_{0,1}$, equation (7) may be derived, *i.e.*

$$[N_1^* - N_1(0)]/[N_1^* - N_1(t)] \approx \exp(k_{0,1} + k_{1,0})t \quad (7)$$

the Napier time, $\tau = (k_{0,1} + k_{1,0})^{-1} = N_0^{-1}(k'_{0,1} + k'_{1,0})^{-1}$. This implies that (pressure) $\times \tau$ = constant at a fixed temperature. Clearly this whole analysis

[159] P. F. Zittel and C. B. Moore, *J. Chem. Phys.*, 1973, **58**, 2004.

requires $\varepsilon_1 - \varepsilon_2 \gg kT$ and will fail for closely spaced energy levels and high temperatures. Again from equation (2) and the above result one may derive equations (8), where Z = rate of bimolecular collisions. Thus we obtain

$$k^{-1}_{1,0} = \tau\{1 + \exp[-(\varepsilon_1 - \varepsilon_0)/kT]\} \tag{8a}$$

$$k_{1,0} = k'_{1,0}N^*_0 = \langle P_{1\to 0}\rangle ZN^*_0 \tag{8b}$$

the popular expression for average transition probability [equation (9)],

$$\langle P_{1,0}\rangle = [ZN^*_0\{1 + \exp[-(\varepsilon_1 - \varepsilon_0)/kT\}]^{-1} \tag{9}$$

where $Z_{1,0}$, the collision number, is defined as $\langle P_{1\to 0}\rangle^{-1}$. This model is generally valid for most laser studies, since the frequencies concerned have so far been $\geqslant 1000\ \text{cm}^{-1}$ and temperatures have been $\leqslant 1000$ K. In any event $\langle P_{1\to 0}\rangle$ is theoretically predicted to within only a factor of ten of the experimental result because of inherent calculation approximations.

For higher temperatures, many states are appreciably excited and a multi-state model is used. An energy-relaxation equation is used in this case. Let E = energy in all states,

$$\therefore \quad E = h\nu \sum_j jN_j \tag{10}$$

and

$$\mathrm{d}E = h\nu \sum_j j\,\mathrm{d}N_j \tag{11}$$

Now

$$\mathrm{d}N_1/\mathrm{d}t = k_{0,1}N_0 - k_{1,0}N_1 - k_{1,2}N_1 + k_{2,1}N_2 \tag{12}$$

etc. and

$$\mathrm{d}N_j/\mathrm{d}t = k_{j-1,j}\,N_{j-1} - k_{j,j-1}N_j - k_{j,j+1}N_j + k_{j+1,j}N_{j+1} \tag{13}$$

$$\therefore \quad \partial E/\partial t = h\nu \sum_j j\,\mathrm{d}N_j/\mathrm{d}t \tag{14}$$

$$\begin{aligned} &= h\nu[-k_{1,0}N_1 + k_{0,1}N_0 - k_{j,j+1}N_j + k_{(j+1,j)}N_{j+1}] + \ldots \\ &\qquad + [-k_{j,j+1}N_j - k_{j,j-1}N_j + k_{j+1,j}N_{j+1} + k_{j-1,j}N_{j-1}] \\ &= h\nu[j[(j+1)k_{1,0}N_{j+1} - (j+1)k_{0,1}N_j] + \{k_{0,1}[N_0 + 2N_1 + \ldots + \\ &\qquad (j+1)N_j] - k_{1,0}[N_1 + 2N_2 + \ldots + jN_j]\}] \end{aligned} \tag{15}$$

from the Landau–Teller rules [160] which state that:

$$k'_{j+1,j} = (j+1)k'_{1,0} \tag{16a}$$

$$k'_{j,j+1} = (j+1)k'_{0,1} \tag{16b}$$

(These rules assume $N_0 \sim$ const. and $\Delta j = \pm 1$ only.) Thus

$$\mathrm{d}E/\mathrm{d}t \approx h\nu[k_{0,1}\sum_j (j+1)N_j - k_{1,0}\sum_j jN_j] \tag{17}$$

(*cf.* equation 5 in ref. 160)

i.e. equilibrium is assumed at the uppermost levels considered, which is approximately valid for large j.

[160] L. Landau and E. Teller, *Phys. Z. Sowjetunion*, 1936, **10**, 34.

$$\therefore \quad dE/dt \approx h\nu[k_{0,1}N_0^* - (k_{1,0} - k_{0,1})E/h\nu] \tag{18a}$$

$$= k_{0,1}E^* - (k_{1,0} - k_{0,1})E \tag{18b}$$

where E^* = energy in all states prior to the laser pulse. Integration with the same assumptions as for two states leads now to equation (19), where $\tau = (k_{1,0} - k_{0,1})^{-1}$

$$\langle P_{1\to 0}\rangle = [ZN_0^*\tau\{1 - \exp[-(\varepsilon_2 - \varepsilon_1)/kT\}]^{-1} \tag{19}$$

Note that the sign before the exponential term is the opposite of that in the two-state case.

To summarize, the kinetic treatment of a simple relaxation equation as above depends on $h\nu \gg kT$, which could readily be invalidated in some studies, particularly those concerned with temperature variation. The multi-state approach additionally requires that j be large.

For mixtures of gases or a pure anharmonic oscillator, the component with the lowest frequency, or the higher vibrational levels in the anharmonic case, will be preferentially populated at typical laser-pulse energies. Consider the equilibrium constant K_1 for the system:

$$AB(\nu : V) + C(\nu' : V') \rightarrow AB(\nu, V + 1) + CD(\nu', V - 1) + \Delta E$$

where $K_1 = \exp(\Delta E/kT)$ and $\Delta E = h(\nu' - \nu)$ is a positive number which is the energy released to translation or rotation, at temperature T. Clearly there is a large K_1, and therefore large concentrations of $AB(\nu + 1)$, when $\nu \ll \nu'$. At very high vibrational levels vibration → translation transfer limits the rise in population, although at the lower levels, of course, vibration → vibration equilibration is normally faster than vibration → translation transfer. Moore [9] has shown that inversion of population can even occur. The occurrence of such vibrational-level population disturbances is exemplified by the $CO–N_2$ laser which operates only on high V transitions and not on low.[161–163]

At high laser-pulse energies, corrections (of a few percent) must be applied to the final equilibrium fluorescence intensity, which is higher than the original pre-perturbation value.[159]

Summary.—Thus i.r. lasers have contributed considerably to the data for testing the *a priori* theories and as a result these have been greatly modified. As yet a unified theory, accurately (*i.e.* within ~50%) predicting, for example, temperature and mass dependence over a wide range, is still lacking, but is surely to be expected soon. Extraction of $\langle P_{1\to 0}\rangle$ from experimental data is improving in precision as more accurate rates are measured, and many experimental pitfalls have now been located and corrected.

Since τ is the Napier time for a given level population which decays by collision, *i.e.* a second-order process, we shall remove the pressure dependence

[161] F. Legay and N. Legay-Sommaire, *Canad. J. Phys.*, 1970, **48**, 1949.
[162] N. Legay-Sommaire and F. Legay, *Canad. J. Phys.*, 1970, **48**, 1966.
[163] K. P. Horn and P. E. Oettinger, *J. Chem. Phys.*, 1971, **54**, 3040.

of this quantity by quoting values of the customary $p\tau$ product, where p = pressure. This is slightly preferred to quoting $1/p\tau = k$, *i.e.* second-order rate constants, since the primary data are pseudo-first-order rate constants or τ values.

4 Experimental Measurements of the Hydrogen Halides

Hydrogen Fluoride.—*The Kinetic Model and $V \rightarrow T$ Transfer.* The interest in the vibrational relaxation of HF is twofold. First there is the current interest in HF chemical lasers and the consequent need to understand the processes governing their operations. Secondly, there is the phenomenon of extremely fast relaxations for hydrogen halides, HX, which no existing general theory predicts. It is supposed that intermolecular interactions, which are strong for HX, are of paramount importance and that their nature will be revealed by these studies.

For all HX the pure gas-kinetic model is the simplest possible. Under irradiation by the appropriate HX chemical laser (usually TEA) the HX is raised to the first vibrational level, whence it decays by gas-phase collisions. Thus only one exponential curve should be observed for the $V = 1 \rightarrow 0$ emission, there being only one vibrational mode. Again the $p\tau$ value should be constant (if heating effects are corrected for) and the first value, reported by Airey and Fried [164] is 10.7 μs Torr at 350 K, about 3 orders of magnitude greater than expected from the Lambert–Salter correlation. Now HF reacts with all oxides to form water, which is a highly efficient catalyst for energy-transfer processes. Its absence must be ensured and these workers seem to have done everything possible in this respect, keeping $[H_2O]$ below 300 p.p.m. Absorption of HF was countered by means of a flow system.

Low laser power (<100 W), essentially on the HF ($V = 1, J = 5 \rightarrow V = 0, J = 6$) transition, minimized heating effects, and only the R branch was monitored by the detector, thereby reducing pick-up of scattered laser P-branch radiation. HF partial pressure was 1 Torr and about 1% of the buffer gas pressure. At these pressures the concentration of dimers was kept below a significant level.

The terminology of P and R branches can be confusing in emission. In *absorption* the P branch is at lower frequency than the band centre and is associated with $\Delta J = -1$ transitions. In emission, in order to refer to the *same frequencies* by the label $P(n)$ the definition needs to be altered to associate $P(n)$ with the transition $P(n-1) \rightarrow P'(n)$, *i.e.* in fact with $\Delta J = +1$.

With Ar, N_2, and F_2 collision partners, rate limits obtained were, at 350 K, $p\tau > 3040$ μs Torr, >3040 μs Torr, and >760 μs Torr, respectively. This dramatic *increase* in decay time was attributed to a fast $V \rightarrow V$ transfer between HF and H_2 resulting in a slower, observable, $V \rightarrow T$ decay of both equilibrated oscillators. This phenomenon is to be expected for near-resonant processes such as:

$$\mathrm{HF}(V = 1, J') + \mathrm{H_2}(V = 0, J) \rightarrow \mathrm{HF}(V = 0, J' - 1) + \mathrm{H_2}(V = 1, J)$$

in which $\Delta E = 17$, 50, or 85 cm^{-1} for $J' = 5$, 4, and 3, respectively.[164] Indeed, this has already been observed for the $CO_2 + N_2$ system.[165] HF self relaxation has also been measured [166] as 11.5 μs Torr, the temperature being rather lower (294 K). The P_{10} (7) HF line was used in this study and the laser power was ~750 W (peak), rather higher than that of the first study (~90 W). A further value of 19 μs Torr at 350 K was obtained by Stevens and Cool,[167] using what must be one of the lowest laser powers ever used in laser fluorescence (~1 W)! At such low powers the use of buffer gas, to eliminate heating effects, is unnecessary, and applying the rule-of-thumb that the longest value is always the best, this last value should be the most accurate. It is also supported by Javan *et al.* (see ref. 173). It is, however, difficult to see why this longer value should be obtained, since heating effects should have been negligible in the first study.[164]

Although a simple two-level model is adequate for the above discussion, it is clear that the $V = 2$ level will be populated to an extent which can become significant. Such population is most probably achieved *via* collisions and has been observed in vibrational-band population.[98] The $V \rightarrow V$ relaxation will be very fast in comparison to a $V \rightarrow T$ process, of course.

Bott [168] has examined the temperature dependence of the $p\tau$ values for HF using a shock-heated sample and monitoring laser-induced HF fluorescence decay. The laser operated on P_{10} (4), P_{10} (3), and P_{10} (6) in order of decreasing line strength. Peak power was ~200 W. HF relaxation was monitored at temperatures from 460 to 1030 K and at 295 K. $p\tau$ at 295 K was found to be 17.8 μs Torr but a double exponential decay curve was observed. The faster decay was attributed to the $V \rightarrow V$ transfer process:

$$HF(V = 1) + HF(V = 1) \rightarrow HF(V = 0) + HF(V = 2)$$

Figure 16 shows a comparison of the temperature dependence of $p\tau$ together with the previously measured values and the predicted dependence according to Shin's theory,[169] modified for an attractive intermolecular potential of 3—3.5 kcal mol^{-1}.[170] The agreement is good.

V → V Transfer. At room temperature Bott [168] has obtained the value $p\tau = 0.77$ μs Torr for the process:

$$2HF(V = 1) \rightarrow HF(V = 0) + HF(V = 2)$$

by observing $V = 2$ level fluorescence using appropriate filters. Relaxation rates were measured [171] for the $V = 3 \rightarrow V = 5$ levels by monitoring the

[164] J. R. Airey and S. F. Fried, *Chem. Phys. Letters*, 1971, **8**, 23.
[165] C. B. Moore, R. E. Wood, B. L. Hu, and J. T. Yardley, *J. Chem. Phys.*, 1967, **46**, 4222.
[166] J. K. Hancock and W. H. Green, *J. Chem. Phys.*, 1972, **56**, 2474.
[167] R. R. Stevens and T. A. Cool, *J. Chem. Phys.*, 1972, **56**, 5863.
[168] J. F. Bott, *J. Chem. Phys.*, 1972, **57**, 96.
[169] H. K. Shin, *Chem. Phys. Letters*, 1970, **6**, 494; *J. Phys. Chem.*, 1971, **75**, 1079; *Chem. Phys. Letters*, 1971, **10**, 81.
[170] J. F. Bott and N. Cohen, *J. Chem. Phys.*, 1971, **55**, 3698.
[171] N. Cohen 'A Review of Rate Coefficients for Reactions in the $H_2 + F_2$ Laser System' TR–0172 (2779)–2, The Aerospace Corp., El Segundo, California, 3rd September 1971.

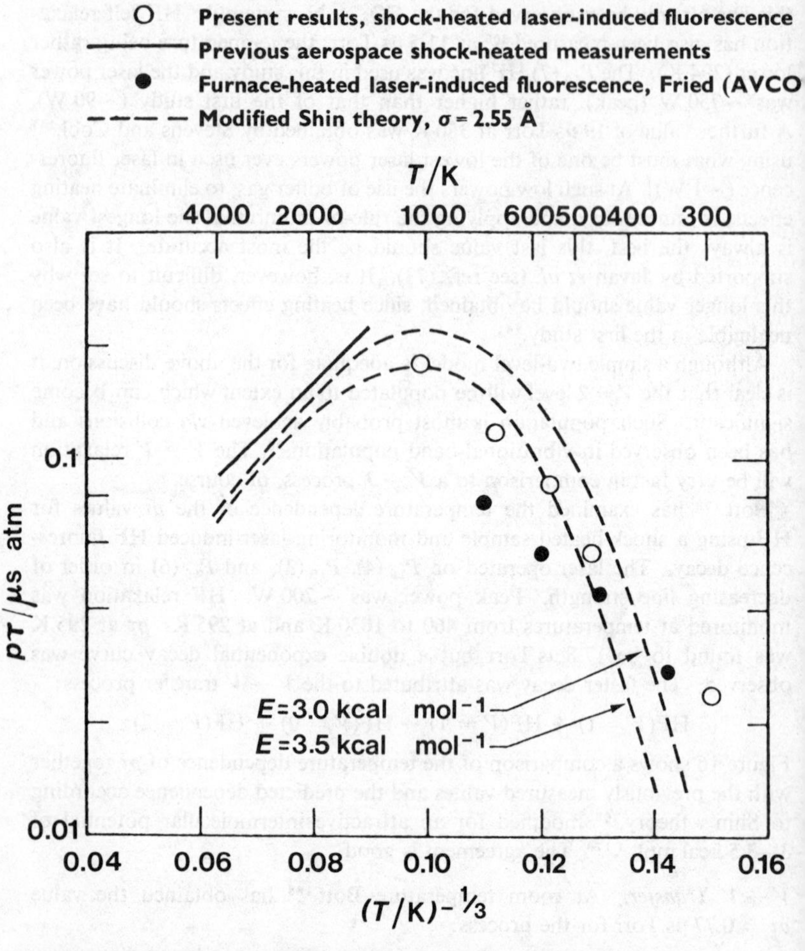

Figure 16 *Temperature dependence of pτ for vibrational relaxation of* HF (Reproduced by permission from *J. Chem. Phys.*, 1972, **57**, 96)

overtone emissions using various filters and photomultipliers. As required by the Landau–Teller rules, $Vp\tau$ should be, and is, constant at $\sim$18 μs Torr for such $V \rightarrow T,R$ processes, and indeed this value is very close to the required $p\tau$ for the $V = 1 \rightarrow T,R$ process. Bott comments that the same result would, however, arise if the $V \rightarrow V$ rates were large in comparison to the $V \rightarrow T,R$ rate. $p\tau_{1 \rightarrow 0}$ is virtually identical with Bott's value but this may be fortuitous since J was assumed equal to zero in both initial and final states and we have already seen the great differences arising from allowing ΔJ to assume a range of values (see Section 3, p. 299). The calculation was a semi-classical type after Rapp *et al.*[126, 172]

The $V \rightarrow V$ transfer rate for the process:

$$HF(V = 2) + HF(V = 0) \rightarrow 2HF(V = 1)$$

has also been evaluated from *spectral resolution* of the $V = 2$ and $V = 1$ fluorescence radiation by a monochromator. A helical pin TEA HF laser was used, delivering several kilowatts over ~0.5 μs.[173] The laser oscillated on the $P(4)$ line and the fluorescing medium was pure HF at 290 K. HF pressures of 0.05—0.5 Torr were studied. The $V \rightarrow T, p\tau$ value calculated was 19.2 μs Torr, in agreement with ref. 167, but of course the temperature is lower, thus indicating only a slight temperature dependence.

Heating effects seem to have been negligible both in view of the above and the observation that if argon was added (16 Torr) the result was unchanged. The $V \rightarrow V\, p\tau$ value obtained was 1.5 μs Torr. These workers[173] were also able to measure approximately the delay, (~5 μs) at 30 μmHg pressure, between excitation of the $P_{1,0}$ (1) line and fluorescence of the $P_{1,0}$ (6) line, thus observing *directly* rotation–rotation energy transfer in the $V = 1$ level. Such is the power of the laser fluorescence technique!

The same group measured the V–V transfer rate from the $V = 3$ level by simultaneously pumping the $V = 1$ and $V = 2$ levels[174] and observing the time dependence of the $V = 3$ level fluorescence in the 11 100 cm^{-1} overtone using a photomultiplier. The $p\tau$ value obtained was 0.63 μs Torr, so that for the work of this group $Vp\tau$ is 1.9 μs Torr for $V = 3$ and 3 μs Torr for $V = 2$.

Direct overtone excitation of the $V = 2$ level of HF has been achieved[175] by a pulsed Nd–YAlO$_3$ laser, temperature tuned. The $V = 0 \rightarrow V = 2$, $P(6)$ line was in fact used. The HF was present to the extent of 1 part in 20 in argon at a total pressure of ~20 Torr at 294 K. $p\tau$ values of 13.9 ± 1.6 μs Torr were measured for the $V = 1$ level. These values lie between the results of ref. 164 and subsequent work, for the $V \rightarrow T,R$ processes. A double exponential decay curve was observed but the electronics were not fast enough (⩾1 μs decay time) to observe the expected $V \rightarrow V$ relaxation. However, the $p\tau$ value for the presumed $V \rightarrow T,R$ process of the $V = 2$ level, extracted from the double exponential decay curve, was found to be 2.2 ± 0.4 times this value, in agreement with Bott[168] and Javan *et al.*,[173] although, as pointed out by Bott and as is clear from the above $V \rightarrow V\, p\tau$ values, the relaxation of upper levels is dominated by fast $V \rightarrow V$ exchange, and this makes it difficult to determine the V dependence of $p\tau$ for $V \rightarrow T,R$ processes.

[172] D. Rapp and P. Englander-Golden, *J. Chem. Phys.*, 1964, **40**, 573, 3120.
[173] R. M. Osgood, A. Javan, and P. B. Sackett, *Appl. Phys. Letters*, 1972, **20**, 469.
[174] R. M. Osgood, P. B. Sackett, and A. Javan, *Appl. Phys. Letters*, 1973, **22**, 254.
[175] M. J. Bina and C. R. Jones, *Appl. Phys. Letters*, 1973, **22**, 44.

$V \rightarrow V$ Transfer to Other Gases from HF *and* DF. These systems have been examined by Stephens and Cool,[167] and for HF($V = 1$) and DF($V = 1$) to $CO_2(00^01)$ the $p\tau$ values are 27 μs Torr and 5.7 μs Torr, respectively. For the $V = 0$ levels of HF and DF the $p\tau$ values for the analogous processes are 28 μs Torr and 53 μs Torr; also $p\tau$ = 50 μs Torr for the process:

$$\mathrm{DF}(V = 1) + \mathrm{DF}(V = 0) \rightarrow 2\mathrm{DF}(V = 0)$$

Unfortunately, in the case of HF and CO_2 the experiment cannot distinguish between:

$$\mathrm{CO_2}(00^01) \rightarrow \mathrm{CO_2}(00^00) + h\nu\ ;\ \Delta J = \pm 1$$

and the hot-band emissions:

$$\mathrm{CO_2}(10^01 \text{ or } 02^01) \rightarrow \mathrm{CO_2}(10^00 \text{ or } 02^00) + h\nu\ ;\ \Delta J = \pm 1$$

so that the mechanisms:

$$\mathrm{HF}(V = 1, J') + \mathrm{CO_2}(00^00) \rightarrow \mathrm{HF}(V = 0, J') + \mathrm{CO_2}(00^01) + \Delta E = 1609 \text{ cm}^{-1}$$

$$\mathrm{HF}(V = 1, J') + \mathrm{CO_2}(00^00) \rightarrow \mathrm{HF}(V = 0, J') + \mathrm{CO_2}(10^01) + \Delta E = 242 \text{ cm}^{-1}$$

or $$\mathrm{HF}(V = 1, J') + \mathrm{CO_2}(00^00) \rightarrow \mathrm{HF}(V = 0, J') + \mathrm{CO_2}(02^01) + \Delta E = 349 \text{ cm}^{-1}$$

and $$\mathrm{CO_2}(10^01) + \mathrm{CO_2}(00^00) \rightarrow \mathrm{CO_2}(10^00) + \mathrm{CO_2}(00^01) + \Delta E = -21 \text{ cm}^{-1}$$

or $$\mathrm{CO_2}(02^01) + \mathrm{CO_2}(00^00) \rightarrow \mathrm{CO_2}(02^00) + \mathrm{CO_2}(00^01) + \Delta E = -26 \text{ cm}^{-1}$$

cannot be distinguished from each other.

The high deactivation rate for DF($V = 1$) by CO_2 is well predicted by the Sharma and Brau theory (see Section 3, p. 297), but the others are not so well accounted for. Now the Brau approximation is unlikely to hold for these systems anyway, because the highly attractive nature of their intermolecular potentials can cause violation of the rotational selection rules. Indeed, Stephens and Cool deduce that a successful theoretical approach must allow the major portion of the energy defect, *i.e.* overall ΔE, to be assimilated by the rotational levels. Thus the Stephenson and Dillon concepts may well prove more successful in predicting these $p\tau$ values. The separation between successive absorption bands in the V–R spectrum of HF is given by equation (20).

$$\Delta G_{(V \rightarrow V+1)} = \omega_e - 2\omega_e x_e - 2\omega_e x_e V \tag{20}$$

For HF, ω_e = 4138.52 cm^{-1} and $\omega_e x_e$ = 90.07 cm^{-1}.

For the process:

$$\mathrm{HF}(V = 2) + \mathrm{HF}(V = 0) \rightarrow 2\mathrm{HF}(V = 1)$$

the value of $p\tau$ = 1.5 μs Torr whereas for its reverse [168] $p\tau$ = 0.77 μs Torr. Thus K_{eq} for the process as written is 0.77/1.5 = 0.513.

From the above anharmonicity data for HF,

$$-\log_{10}K = \Delta\varepsilon/2.303\,kT\,[\Delta\varepsilon = \mathrm{HF}(V=1) - \mathrm{HF}(V=0) + \mathrm{HF}(V=1) - \mathrm{HF}(V=2) \quad (21a)$$

$$= \omega_e - 2\omega_e x_e - \omega_e + 2\omega_e x_e + 2\omega_e x_e \quad (21b)$$

$$= 2\omega_e x_e = 180\ \mathrm{cm}^{-1}] \quad (21c)$$

$\therefore\ K_{eq} = 0.4$, in support of the relaxation work.

Hydrogen and Deuterium Chlorides.—*The Kinetic Model and $V \rightarrow (T,R)$ Energy Transfer*. The kinetic model is a simple two-state system as for HF. Again, as for HF, interest centres on the importance of rotation and the effects of strong dipole interactions. In HCl, $(V \rightarrow T,R)$ transfer takes place almost a factor of 16 faster than extrapolation of the shock-tube data between 2000 and 1000 K would predict on a log (cross section) *vs.* $(T)^{-\frac{1}{3}}$ plot.[176] These rates are compatible with $V \rightarrow R$ theory but not $V \rightarrow T$ processes.

Moore *et al.*[176] used an HI + Cl· chemical HCl laser oscillating on the $V = 1 \rightarrow 0$ transition (40%). The remaining 60% lay in the $V = 3 \rightarrow 2$ and $V = 2 \rightarrow 1$ bands. The strongest lines in the $V = 1 \rightarrow 0$ transition were the $P(9)$, $P(10)$, and $P(11)$. Peak power in the 10—20 μs pulses was ~1 kW. Precautions were again taken to exclude water, which was kept as low as 100 p.p.m. Temperature rise in the system was kept <40 K for pure HCl and was less with argon present. The calculation of this temperature rise assumed that the laser operated on high J transitions.

The value of $p\tau$ for pure HCl was found to be 1220 μs Torr ±15% over the pressure range 50—250 Torr. Slow diffusion to the walls and the long (30 ms) radiative lifetime make no significant contribution at these pressures.

With argon buffer gas present the $p\tau$ value observed was fitted to equation (22). The resultant value of $p\tau_{\mathrm{HCl-Ar}}$ is for HCl infinitely dilute in argon

$$(1/p\tau)_{\mathrm{obs}} = 1/(p\tau_{\mathrm{HCl-HCl}})X_{\mathrm{HCl}} + 1/(p\tau_{\mathrm{HCl-Ar}})X_{\mathrm{Ar}} \quad (22)$$

and was found to be 1—0.5 × 10^6 μs Torr, which must be regarded as a lower limit in view of the significant effects of impurities, of even low concentration, at this rate level; $p\tau_{\mathrm{HCl-HCl}}$ was found to be 1140 μs Torr.

The enhancement of the $1/p\tau$ value compared with that obtained by linear extrapolation of the 2000—1000 K data is predicted by the SSH theory expression [126] of the form shown in equation (23) when $h\nu > kT$ or when the

$$AT^{-\frac{1}{6}}\exp[(\tfrac{1}{2}h\nu + \varepsilon)/kT]\exp[\beta T^{-\frac{1}{3}}] \quad (23)$$

well depth of the attractive potential, ε, is $\geqslant kT$. As Moore [176] points out, the assumptions of the SSH theory are invalid for HCl at 296 K so that this formula cannot be used. Another study [177] of HCl fluorescence decay, with a Q-switched chemical laser source, obtained a $p\tau$ value of 1430 μs Torr. This is slightly larger than the result of Moore *et al.*[176]

176 H. L. Chen and C. B. Moore, *J. Chem. Phys.*, 1971, **54**, 4072.
177 M. Margottin-Maclou, L. Doyenette, and L. Henry, *Appl. Optics*, 1971, **10**, 1768.

The $p\tau$ value for DCl obtained was 4000 μs Torr, in agreement with the value obtained by Breshears *et al.*[178] who argued that, since $V \rightarrow T$ theory predicts $p\tau_{DCl-DCl} \ll p\tau_{HCl-HCl}$ ($\sim 10^{-3}\, p\tau_{HCl-HCl}$) and $V \rightarrow R$ theory predicts $p\tau_{DCl-DCl} \gg p\tau_{HCl-HCl}$ ($\sim 5\, p\tau_{HCl-HCl}$), $V \rightarrow R$ transfer predominates.

Provided that the probability for energy transfer per collision remains small, this predicted isotope effect will remain, despite the crude mathematical approximations used in the theoretical solutions. In view of the magnitude of the ratio $p\tau_{DCl-DCl}/p\tau_{HCl-HCl}$ it is concluded that virtually all the energy goes into the DCl–HCl rotational degrees of freedom.

$V \rightarrow V$ Transfer. As with all HX, this type of transfer must be from the upper V levels in the pure gas. In a study by Ronn *et al.*[179] a TEA chemical HCl laser was used, filtered to leave only the $V = 1 \rightarrow 0$ line at 1 kW over 0.5 μs. The $V = 2 \rightarrow 1$ fluorescence was monitored by means of a filter. It was first shown that the slow decay of the $V = 2$ level was twice the slow decay of the $V = 1$ level, *i.e.* the Landau–Teller rules were obeyed for HCl for the $V \rightarrow (T,R)$ process and a vibrational manifold equilibrium must have been established.

The $p\tau$ value for the $V \rightarrow V$ energy-transfer process was found to be constant at 11.1 μs Torr over a pressure range of 0.4—1.4 Torr HCl partial pressure in helium. The $p\tau$ value calculated from Sharma's theory [151] was 71 μs Torr, which is probably reasonable agreement. It is doubtful whether one can deduce that short-range repulsive forces may also play a significant role.

A further study [180] of the same system has been carried out using a pulsed HCl chemical laser. The fluorescence cell was partly coupled to the cavity by having a mirror reflect the unabsorbed radiation into the laser. Again the $V = 2 \rightarrow 1$ fluorescence was monitored and the $p\tau$ value for the $V = 2 \rightarrow 1$ process:

$$HCl(V = 2) + HCl(V = 0) \rightarrow 2HCl(V = 1) + \Delta E = 104\ cm^{-1}$$

was evaluated as 11.9 μs Torr ±14%, in agreement with Ronn *et al.* Again the Landau–Teller rules were verified, for several rare-gas buffers. An SSH-type calculation yielded $P_{2\rightarrow 1,\ 0\rightarrow 1} = 1.8 \times 10^{-2}$ whereas the experimental value is $1.3 \pm 0.2 \times 10^{-2}$. This excellent agreement is concluded to imply that short-range forces predominate. In view of the approximations inherent in these theories, we find it difficult to agree that, where two theories (*e.g.* SSH and Sharma) on different bases both yield predictions within a factor of 10 of the observed value, one can be deemed more correct than the other! A more recent study of the $V = 2 \rightarrow 1$ relaxation has now appeared [104] using a tunable parametric oscillator pumped by a frequency-doubled Nd^{3+}: YAG laser with pulse widths of ~100 ns at peak powers of 100 W. With

[178] W. D. Breshears and P. F. Bird, *J. Chem. Phys.*, 1969, **50**, 333.
[179] I. Burak, Y. Noter, A. M. Ronn, and A. Szoke, *Chem. Phys. Letters*, 1972, **17**, 345.
[180] B. M. Hopkins and H. L. Chen, *J. Chem. Phys.*, 1972, **57**, 3816.

such a system discrete vibration–rotation lines were excited for HCl ($V = 2$) and ($V = 1$) states. The two processes were:

$$HCl(V = 2) + HCl(V = 0) \rightarrow 2HCl(V = 1) \qquad (24)$$

and

$$H^{35}Cl(V = 1) + H^{37}Cl(V = 0) \rightarrow H^{35}Cl(V = 0) + H^{37}Cl(V = 1) \qquad (25)$$

This latter is our first example of isotopic $V \rightarrow V$ transfer processes and is the only possible intermolecular $V \rightarrow V$ process for the ($V = 1$) level of a pure gas. A Fabry–Perot étalon was placed in the cavity to narrow the laser line for these studies.

For HCl partial pressures in argon of 1—3 Torr and excitation from the $V = 0$ to $V = 2$ level, $p\tau$ was 10 μs Torr ±11% for process (24), in good agreement with the other two results. This close agreement is characteristic of laser fluorescence experiments and contrasts with other techniques for the study of energy transfer.

For process (25), either HCl species was excited from $V = 0$ to $V = 1$ [R(2)] and then decay or rise of fluorescence was monitored as appropriate. Both approaches gave $p\tau$ values of 1.61 μs Torr ±25% (error limits are the 95% confidence limits). Again, Sharma-type predictions put $P_{2\rightarrow1,\ 0\rightarrow1}$ $\sim$0.01, which can be raised somewhat by using 0.1 (Debye Å)^2 for the quadrupole moment of HCl.

Although it is logical to suppose that attractive forces should be significant in the intermolecular interactions in view of the dimerization in the gas phase which has been observed,[181, 182] it is clear that, as yet, neither theory (SSH or Sharma-type) is able to predict conclusively the nature of the interactions.

$V \rightarrow V$ Transfer to Foreign Gases. Two relaxation times were observed [176] in this case for the HCl. From the faster relaxation is deduced $p\tau = 300$ μs Torr ± 10%, for X_{HCl}* from 0.0093 to 0.083 and $X_{DCl} = 0.054$—0.079 in argon buffer, for the process:

$$HCl(V = 1) + DCl(V = 0) \rightarrow HCl(V = 0) + DCl(V = 1) + \Delta E = 795\ cm^{-1}$$

From the slow relaxation $p\tau = 1740$ μs Torr is deduced for the process:

$$DCl(V = 1) + HCl(V = 0) \rightarrow DCl(V = 0) + HCl(V = 0)$$

with DCl infinitely dilute in HCl.

Moore has deduced [176] qualitatively, from comparison of these values for isotopic mixtures with the values for pure isotopic species, that the rotational velocity of the vibrating (excited) molecule itself plays a dominating role, most of the vibrational energy of the excited species being transferred to its *own* rotation. The remaining energy is principally transferred to rotation of the collision partner, with translation playing little part.

* X_M = mole fraction of M.

[181] B. Katz, A. Ronn, and O. Schnepp, *J. Chem. Phys.*, 1967, **47,** 5303.
[182] D. H. Rank, P. Sitaram, W. A. Glickman, and T. A. Wiggins, *J. Chem. Phys.*, 1964, **39,** 2673.

Rates of energy transfer from HCl to other diatomic gases are much faster than for other diatomic–diatomic collisions, as expected. Moreover, these rates are much less dependent on the energy difference between donor and acceptor vibrational levels. This implies that the *chemical* nature of the collision partner is important.

Using the same apparatus as previously,[176] Moore *et al.*[183] have studied $HCl(V = 1)$ de-excitation by HBr, N_2, HI, CO, CH_4, and D_2. For these collision partners $p\tau$ values obtained were 27.8, 1150, 188, 3700, 11.9, and 18.6 μs Torr, respectively. The anomalously high efficiency of CH_4 is probably due to a complex mechanism of fast $V \rightarrow V$ transfer from HCl to CH_4 followed by fast $V \rightarrow (T,R)$ transfer from CH_4. Sample heating could thus be a problem, but the $V \rightarrow V$ process must undoubtedly be fast from the 2886 cm^{-1} HCl fundamental to the CH_4 levels at 2916, 3020, 3027, 2823, or 2600 cm^{-1}.

For N_2, HI, and CO collision partners, only one decay was observed, and in the case of CO the rate of rise of CO fluorescence equalled the rate of decay of HCl. Although the experiment actually measures $V \rightarrow V$ and $V \rightarrow (T,R)$ processes, in these cases the value is almost purely $V \rightarrow V$, unlike the CH_4 value.

In the case of HCl–HBr mixtures both HBr and HCl fluorescences were monitored and yielded the same results for HCl and HBr $p\tau$ values.

The fraction of collision partner molecules excited in the case of HBr, DCl, N_2, CO, and HI is such that $V = 2$ levels may be significantly populated. This was not experimentally observed in any of these cases, however, and, for HBr all $p\tau$ values over a range of HBr mole fractions from 0.0063 to 0.0162 were constant $\pm$ 20%.

For the process:

$$HCl(V = 1) + D_2(V = 0) \rightarrow HCl(V = 0) + D_2(V = 1) + \Delta E = -104 \text{ cm}^{-1}$$

the $p\tau$ values were 5830, 59 000, and 12 400 μs Torr for H_2, D_2, and HD, respectively.[176, 184] In the H_2 work both ortho- and para-hydrogen were used as collision partners and both gave the same result. This is consistent with the major role of rotation remaining with the excited molecules. More recently, Moore and Zittel,[185] using their normal HCl chemical laser, measured rate constants for $V \rightarrow (T,R)$ energy transfer from DCl to various species. The values for rare gases are quoted always as limits in view of the serious influences of a few p.p.m. of impurities at these long lifetimes. For He, Ne, Ar, DCl, H_2, p-H_2, HD, and D_2 they obtain $p\tau$ values of $465 \pm 10\%$, $>1.67 \times 10^6$, $>3.34 \times 10^6$, 4550, 1430, 1560, 3700, and 1.67×10^4 μs Torr, respectively.

Vibration–vibration energy-transfer rates were also measured between HCl and DCl and various gases. The $p\tau$ values are given in Table 6. From the $V \rightarrow (T,R)$ data it is concluded that the three-dimensional SSH calcula-

[183] H. L. Chen and C. B. Moore, *J. Chem. Phys.*, 1971, **54**, 4080.
[184] B. M. Hopkins and H. L. Chen, *J. Chem. Phys.*, 1972, **57**, 3161.
[185] P. F. Zittel and C. B. Moore, *J. Chem. Phys.*, 1973, **58**, 2922.

Table 6 *Values of $p\tau$ for energy transfer from gas X to gas Y*

X	Y	$p\tau/\mu s$ Torr [a]
^{12}CO	DCl	21.3
^{13}CO	DCl	14.9
DCl	NO	30.3
N_2	DCl	600
DCl	O_2	1 640
DCl	DBr	65.0
DCl	DI	565
DCl	$SF_6(\nu_3)$	2 900
HCl	NO	323
HCl	$^{18}O_2(V = 1, 2)$	5 720
HCl	$^{16}O_2(V = 1, 2)$	9 350
HCl	$SF_6(\nu_3)$	10 400

[a] $p\tau$ values are all $\pm 10\%$.

tion [186] inadequately describes the transfer probabilities for HX systems. For example, a factor of ten difference in probability is predicted between $\langle P_{1,0} \rangle(HCl \rightarrow He)$ and $\langle P_{1,0} \rangle(DCl \rightarrow He)$, the latter being the larger, whereas experimentally these probabilities are found to be equal. If vibrational anharmonicity is corrected for,[187] the difference is reduced to a factor of two. Rotation, it is concluded, must play a significant role in $V \rightarrow (T,R)$ transfer for HX.

DCl transfers vibrational energy to translation/rotation in D_2 25 times faster than in He. Although this can be explained by rotational participation, slight alterations in the slope of the repulsive potential could also account for an effect of this magnitude. This is particularly possible since the resonance ($\Delta E \approx 44$ cm^{-1}) between the DCl vibrational fundamental and the $J = 2 \rightarrow 6$ rotational transition of H_2 has little effect (Zittel and Moore) [185] on the relative energy-transfer rates for DCl to normal H_2 or p-H_2.

For near-resonant $V \rightarrow V$ transfer, Moore has compared the experimental probability with that calculated on the basis of dipole–dipole interactions and a 'resonant SSH approach'. Table 7 summarizes the experimental results. The calculations give results all within a factor of ten of the observed probability, the SSH treatment giving the closest agreement (within 50%). As these authors conclude, a temperature-dependence study is the best way to distinguish between dipole–dipole attractive potential cases and exponential repulsive potential cases. Unfortunately either theoretical approach can be made to fit, for any given narrow temperature range.

Table 7 *Experimental values of ΔE and P for near-resonant $V \rightarrow V$ transfers*

System		ΔE/cm^{-1}	*P*(exptl.)
^{13}CO	DCl	5	8.8×10^{-3}
CO	DCl	52	$6.3 \times 10^{-}$
CO	CD_4	$-116(\nu_3)$, $35(\nu_1)$	$1.0 \times 10^{-}$

186 S. L. Stretton, *Trans. Faraday Soc.*, 1965, **61**, 1053.
187 F. H. Mies, *J. Chem. Phys.*, 1964, **40**, 523.

Hydrogen Bromide.—*$V \rightarrow (T,R)$ Transfer and the Kinetic Model.* The kinetic model is the same as for HF. In the study by Chen,[188] an HBr chemical laser was used, giving pulses 4—8 μs wide with ~20% in $V = 2 \rightarrow 1$ transitions. In the $V = 1 \rightarrow 0$ transition the strongest lines were the $P(6)$—$P(9)$. Overall peak power was ~4 kW. In pure HBr only one exponential was observed, for which $p\tau = 1.75 \times 10^3$ μs Torr, in the pressure range 15—150 Torr. A value of $p\tau = 1.65 \pm 0.15 \times 10^3$ μs Torr was obtained when He, Ne, or Ar were used as buffer gases.

This value is in agreement with the simple $V \rightarrow R$ prediction that $p\tau_{HCl-HCl}$ be less than $p\tau_{HBr-HBr}$, although the values are rather close.

$V \rightarrow V$ Transfer. As for all HX, isotopic species or upper vibrational levels must be involved. In the work of Ronn *et al.*[189] a TEA chemical laser (1.5 μs pulse width), was used to pump the $V = 1$ level and the temporal development of the $V = 2 \rightarrow 1$ fluorescence was observed, *via* appropriate filters. The $p\tau$ value obtained for the $V = 2 \rightarrow 1$ transfer:

$$HBr(V = 2) + HBr(V = 0) \rightarrow 2HBr(V = 1)$$

was 7.14 μs Torr, rather less than the value for HCl but close to it. Examination of the predicted $p\tau$ value (330 μs Torr) based on dipole–dipole interactions reveals a considerable discrepancy, far greater than in the case of HF. It is possible to speculate on the increased significance of the repulsive potential, but this is probably not justified. Hopkins and Chen [190] have also measured $p\tau$ for the $V = 2 \rightarrow 1$ transfer using a chemical laser. They obtain $p\tau = 10.5$ μs Torr and conclude from comparison with the theory of Sharma and Brau and Stephenson *et al.*[191] that long-range multipolar forces are most significant.

$V \rightarrow V$ Transfer from HBr *to Foreign Gases.* Using the same apparatus as before, Chen [192] has shown that $p\tau$ values for $V \rightarrow V$ transfer from $HBr(V = 1)$ to D_2, N_2, CO, and O_2 are 2200, 312, 102, and 6350 μs Torr, all ±5—10%. Some doubt arises as to the nature of this last value, the $V \rightarrow (T,R)$ rate being indistinguishable from it, although presumably making only a minor contribution to it. Chen also gives $p\tau$ values [188] for $V \rightarrow (T,R)$ energy transfer between $HBr(V = 1)$ and HCl, H_2, He, Ne, or Ar. These are 760, 4800, 11.4×10^4, 19.8×10^4, and 8.4×10^5 μs Torr, respectively. No difference was observed for ortho- or para-hydrogen. For $V \rightarrow V$ energy transfer the $p\tau$ values for $HBr(V = 1)$ with HCl and HI are 137 and 46.4 μs Torr, respectively.

Again, from the results for ortho- and para-hydrogen, rotation of the collision partner does not seem important although ^{4}He is almost 4 times less

188 H. L. Chen, *J. Chem. Phys.*, 1971, **55**, 5551.
189 I. Burak, Y. Noter, A. M. Ronn, and A. Szoke, *Chem. Phys. Letters*, 1972, **16**, 306.
190 B. M. Hopkins and H. L. Chen, *Chem. Phys. Letters*, 1972, **17**, 500.
191 J. C. Stephenson, R. E. Wood, and C. B. Moore, *J. Chem. Phys.*, 1968, **48**, 4790
192 H. L. Chen, *J. Chem. Phys.*, 1971, **55**, 5557.

efficient than D_2 at quenching HBr. Of course, differences in the slope of the potential functions could also account for this result.

More recently[193] it has been shown that $p\tau$ values for DBr + DBr and DBr + HBr (excited species named first) are 5930 and 3880 μs Torr, respectively, for the $(V \rightarrow T,R)$ process. For HBr($V = 1$) transfer $(V \rightarrow V)$ to DBr, $p\tau = 493$ μs Torr.

For the $V \rightarrow (T,R)$ process we see that $p\tau_{\mathrm{HBr-HBr}} < p\tau_{\mathrm{DBr-HBr}} < p\tau_{\mathrm{DBr-DBr}}$, similar to the HCl results.[176] Since DBr transfers $V \rightarrow (T,R)$ more readily to HBr than to DBr the rotational velocity of the colliding partner would seem to be important in this case. Moreover, since $p\tau_{\mathrm{HBr-HBr}} < p\tau_{\mathrm{DBr-DBr}}$, translational degrees of freedom must play little or no part in energy transfer despite the 719 cm^{-1} difference in quantum size.

The $V \rightarrow V$ energy-transfer rates for HBr seem to be much less dependent on energy-level differences than other HX systems. This may be a result of weaker interactions in the HBr case.

Hydrogen Iodide.—$V \rightarrow (T,R)$ *Transfer*. For the usual kinetic model for HX using CO_2 + HI mixtures Moore *et al.*[43] fit the experimental data with $p\tau_{\mathrm{HI-HI}} = 2670 \pm 14\%$ μs Torr and $p\tau_{\mathrm{CO_2-HI}} = 445$ μs Torr. A *Q*-switched CO_2 laser was used, but the purification of the gases was not particularly rigorous so that these lifetimes may be too short and should be regarded as lower limits.

Summary.—For non-resonant $V \rightarrow V$ energy transfer, it is observed that energy-transfer probabilities involving diatomic molecules not containing hydrogen lie on a curve different from those for hydrogen-containing diatomics when plotted against energy discrepancy (Figure 17). From Figure 17 it is clear that the more hydrogen atoms that are in the system, the higher is the probability of energy transfer and the less dependent is it on ΔE. Neither the probability difference for systems with supposedly similar interaction potentials (HX–HX and DX–DX) nor the different dependencies on ΔE for isomers containing H and D is explicable by simple $V \rightarrow V$ theory.

Moore *et al.*[185] conclude that rotation may well play a significant role in $V \rightarrow V$ as in $V \rightarrow (T,R)$ transfer, decreasing the dependence on ΔE by absorbing some of the energy defect into the rotational levels. The para-hydrogen work does not seem to bear this out, but it may be exceptional. Moore also points out that since NO, CO, N_2, and O_2 have similar masses, vibrational matrix elements, and Lennard-Jones diameters, the considerable differences in the probabilities of $V \rightarrow V$ energy transfer with HX are probably due to differences in the potential either in the repulsive *or* attractive parts. Hydrogen-bonding has been suggested as an explanation of the high $\langle P_{1,0} \rangle$ and low ΔE dependence between NO and HX.[194] This type of

[193] M. Y. D. Chen and H. L. Chen, *J. Chem. Phys.*, 1972, **56**, 3315.
[194] A. B. Callear and G. J. Williams, *Trans. Faraday Soc.*, 1966, **62**, 2030.

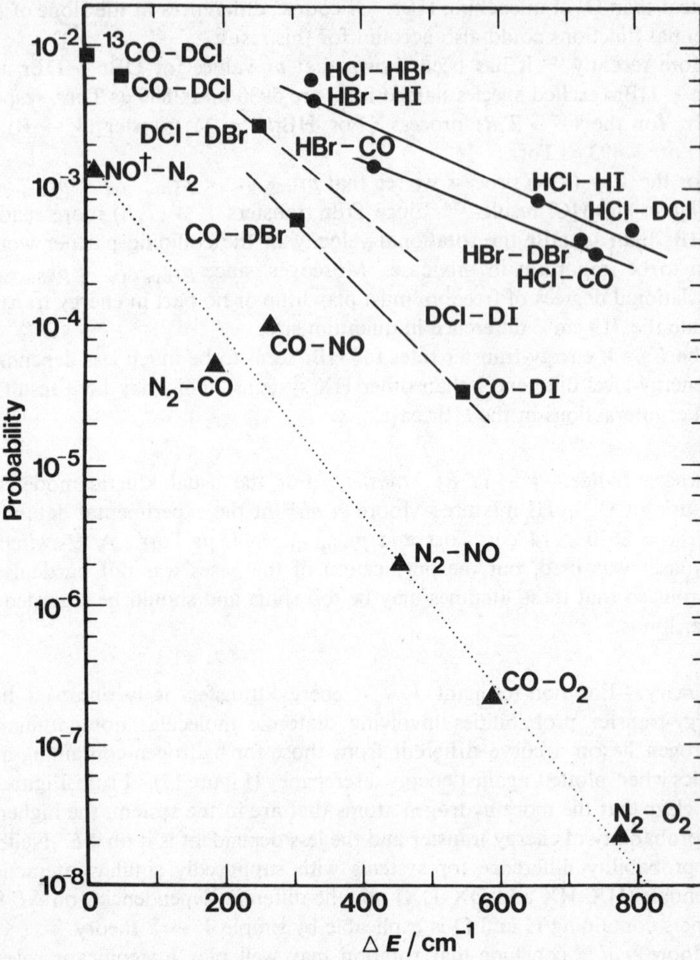

Figure 17 *V → V Energy-relaxation probability as a function of energy defect* ΔE (Reproduced by permission from *J. Chem. Phys.*, 1973, **58**, 2922)

comparative technique is a second method of extracting information on the potential function from energy-transfer probabilities.[195] Moore *et al.* also point out that for the process:

$$\mathrm{HCl}(V=1) + {}^{16}\mathrm{O}_2 \text{ or } {}^{18}\mathrm{O}_2(V=0) \rightarrow \mathrm{HCl}(V=0) + {}^{16}\mathrm{O}_2 \text{ or } {}^{18}\mathrm{O}_2(V=2) + \Delta E = -203, -30\ \mathrm{cm}^{-1}$$

[195] C. B. Moore, in 'Fluorescence', ed. G. G. Guilbault, Dekker, New York, 1967, p. 175.

near-resonant excitation of the O_2 overtone is possible. Current $V \rightarrow (T,R)$ and $V \rightarrow V$ theories would predict that $P_{HCl \rightarrow {}^{16}O_2}/P_{HCl \rightarrow {}^{18}O_2} \gg 1$. Experimentally this ratio is 0.6, which requires ~40% overtone excitation for $^{18}O_2$. Presumably this occurs because the 1500 cm^{-1} reduction in ΔE for the overtone excitation compensates for the hundredfold smaller matrix elements as estimated by SSH theory's harmonic oscillator expanded exponential potential.

5 Experimental Measurements of Methyl Halides

Methyl Fluoride.—*Kinetic Model and $V \rightarrow (T,R)$ Energy Transfer.* The full energy-level diagram is shown in Figure 18. For the purpose of $V \rightarrow (T,R)$ energy transfer these levels may be approximated to a three-level system as in Figure 19.

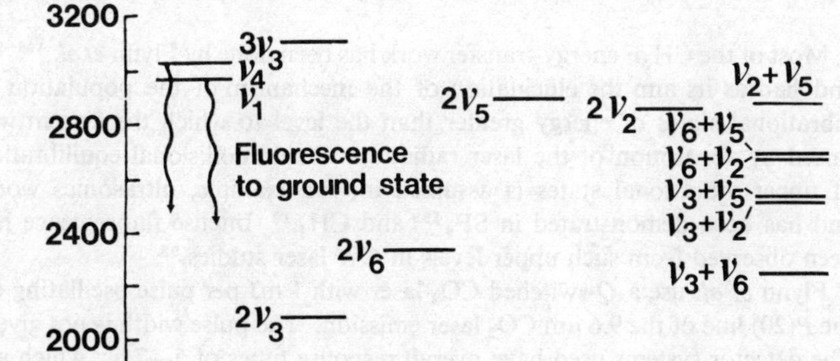

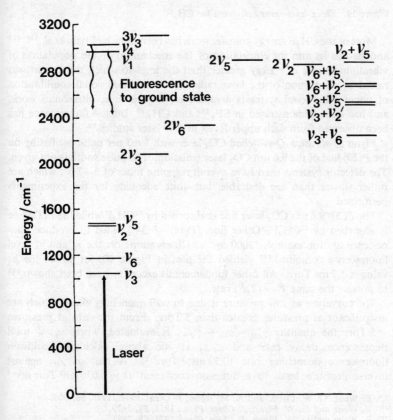

Figure 18 *Energy-level diagram for* CH_3F *up to* 3200 cm^{-1}
(Reproduced by permission from *J. Chem. Phys.*, 1972, **56,** 6060)

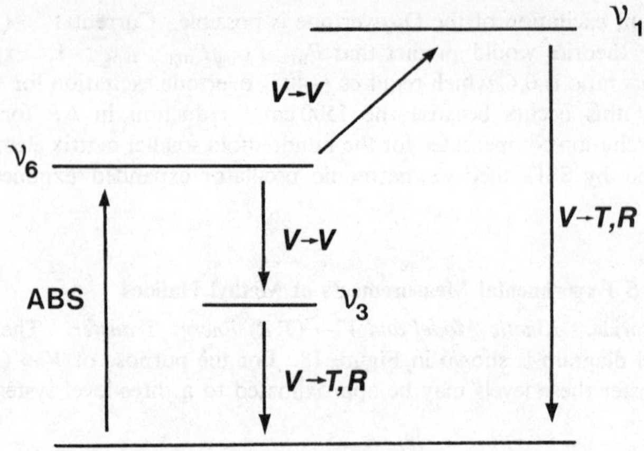

Figure 19 *Three-level energy diagram for* CH_3F

Most of the CH_3F energy-transfer work has been done by Flynn *et al.*,[196–198] and had as its aim the elucidation of the mechanism of the population of vibrational states of energy greater than the level to which the system was raised by absorption of the laser radiation. Rapid collisional equilibration of upper vibrational states is assumed in, for example, ultrasonics work, and has been demonstrated in SF_6[199] and CH_4.[18] Intense fluorescence has been observed from such upper levels in CW laser studies.[98]

Flynn *et al.* use a *Q*-switched CO_2 laser with 1 mJ per pulse oscillating on the *P*(20) line of the 9.6 μm CO_2 laser emission. The pulse width is not given. The detector systems used have overall response times of 5—7 μs, which are rather slower than are desirable, but quite adequate for the experiments performed.

The *P*(20) 9.6 μm CO_2 laser line is absorbed by $^{12}CH_3F$ while the *P*(32) line is absorbed by $^{13}CH_3F$. Other lines [*P*(14)—*P*(34)] failed to produce fluorescence at, for example, 3000 cm^{-1}. Observations on the ν_1 and ν_4 level fluorescence combined[196] yielded the plot in Figure 20, for which the $p\tau$ value is 1.7 ms Torr. All other fundamentals except ν_3 have been shown[198] to possess the same $V \rightarrow (T,R)$ rate.

The curvature at low pressure is due to wall-quenching effects which are insignificant at pressures greater than 5 Torr. From the data at pressures <5 Torr the quantity $\gamma_{diff} = \gamma_{tot} - \gamma_{V\rightarrow T}$ is evaluated, where γ_{tot} = total fluorescence decay rate and $\gamma_{V\rightarrow T}$ is the above calculated collision fluorescence quenching rate (0.59 ms^{-1} Torr^{-1}). A plot of γ_{diff} against inverse pressure leads to a diffusion coefficient $D = 0.076$ cm^2 Torr ms^{-1}

196 E. Weitz, G. W. Flynn, and A. M. Ronn, *J. Chem. Phys.*, 1972, **56**, 6060.
197 E. Weitz and G. W. Flynn, *J. Chem. Phys.*, 1973, **58**, 2679.
198 E. Weitz and G. W. Flynn, *J. Chem. Phys.*, 1973, **58**, 2781.
199 R. D. Bates, G. W. Flynn, J. T. Knudtson, and A. M. Ronn, *J. Chem. Phys.*, 1970, **53**, 3621.

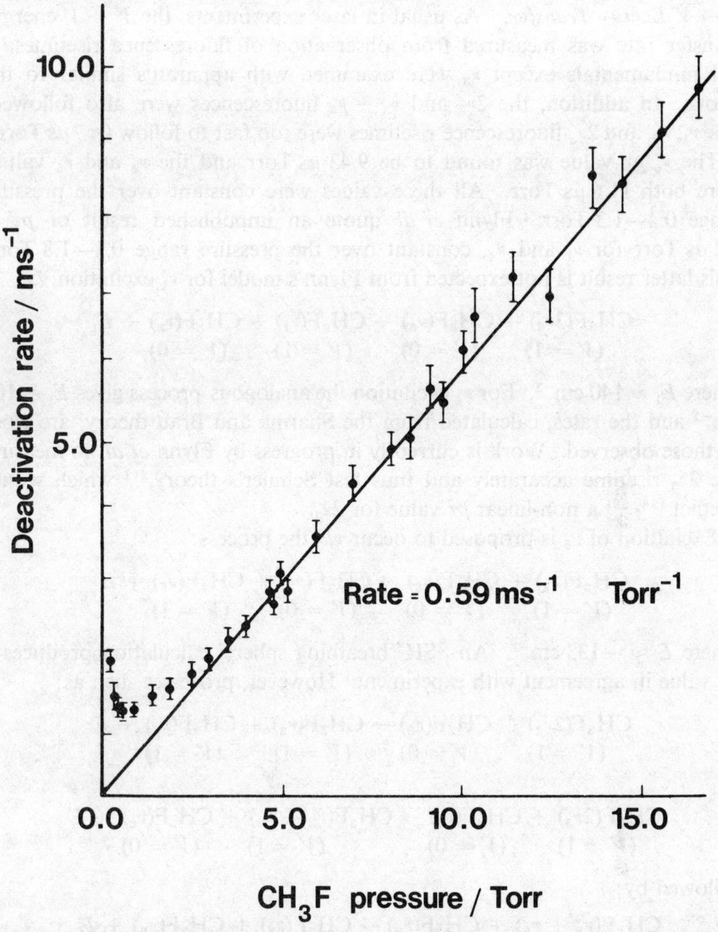

Figure 20 *Pressure dependence of* CH_3F *fluorescence. The least-squares line for collisional deactivation is shown*
(Reproduced by permission from *J. Chem. Phys.*, 1972, **56**, 6060)

$\pm 15\%$, in agreement with a classical [200] diffusion coefficient of $D = 0.082$ cm^2 Torr ms^{-1}.

Flynn *et al.* also examined the CW laser power dependence of the fluorescence intensity. Although the fluorescence intensity varies as (laser power)$^{1.9}$ it is very unlikely that a double-quantum absorption is responsible for population of the 3000 cm^{-1} levels. Indeed this result is readily explained on the basis of bulk heating, which does not, of course, occur in the pulsed experiments.

[200] G. A. Miller and R. D. Bernstein, *J. Phys. Chem.*, 1959, **63**, 710.

$V \rightarrow V$ Energy Transfer. As usual in laser experiments, the $V \rightarrow V$ energy-transfer rate was measured from observation of fluorescence risetimes.[198] All fundamentals except ν_3 were examined with apparatus similar to the above. In addition, the $2\nu_3$ and $\nu_3 + \nu_6$ fluorescences were also followed. The ν_1, ν_4, and $2\nu_3$ fluorescence risetimes were too fast to follow (<7 μs Torr).

The ν_6 $p\tau$ value was found to be 9.43 μs Torr and the ν_2 and ν_5 values were both 11.6 μs Torr. All these values were constant over the pressure range 0.4—1.3 Torr. Flynn *et al.* quote an unpublished result of $p\tau$ = 8.2 μs Torr for ν_1 and ν_4, constant over the pressure range 0.1—1.8 Torr. This latter result is not expected from Flynn's model for ν_1 excitation, *i.e.*

$$\underset{(V=1)}{CH_3F(3\nu_3)} + \underset{(V=0)}{CH_3F(\nu_3)} \rightarrow \underset{(V=1)}{CH_3F(\nu_1)} + \underset{(V=0)}{CH_3F(\nu_3)} + E_1$$

where $E_1 \approx 140$ cm^{-1}. For ν_4 excitation the analogous process gives $E_1' \approx 100$ cm^{-1} and the rates, calculated from the Sharma and Brau theory, are close to those observed. Work is currently in progress by Flynn *et al.* to measure the $2\nu_3$ risetime accurately and thus test Schuler's theory,[201] which would predict [198, 201] a non-linear $p\tau$ value for $2\nu_3$.

Excitation of ν_6 is proposed to occur *via* the process:

$$\underset{(V=1)}{CH_3F(\nu_3)} + \underset{(V=0)}{CH_3F(\nu_3)} \rightarrow \underset{(V=0)}{CH_3F(\nu_3)} + \underset{(V=1)}{CH_3F(\nu_6)} + E$$

where $E \approx -133$ cm^{-1}. An SSH 'breathing sphere' calculation produces a $p\tau$ value in agreement with experiment. However, processes such as:

$$\underset{(V=1)}{CH_3F(2\nu_3)} + \underset{(V=0)}{CH_3F(\nu_3)} \rightarrow \underset{(V=1)}{CH_3F(\nu_3)} + \underset{(V=1)}{CH_3F(\nu_6)} + E$$

or:

$$\underset{(V=1)}{CH_3F(2\nu_3)} + \underset{(V=0)}{CH_3F(\nu_3)} \rightarrow \underset{(V=1)}{CH_3F(\nu_3 + \nu_6)} + \underset{(V=0)}{CH_3F(\nu_3)} + E$$

followed by:

$$\underset{(V=1)}{CH_3F(\nu_3 + \nu_6)} + \underset{(V=0)}{CH_3F(\nu_3)} \rightarrow \underset{(V=1)}{CH_3F(\nu_3)} + \underset{(V=1)}{CH_3F(\nu_6)} + E$$

are possible.

Because of the large population of ν_6 these last two possibilities should show up as an anomalously fast $2\nu_3$ fluorescence decay time, and this is not found: $2\nu_3$ and $\nu_3 + \nu_6$ fluorescence decay at the same rate as the other fundamentals. ν_5 and ν_2 are excited most probably *via* ν_6 by the process:

$$\underset{(V=1)}{CH_3F(\nu_6)} + \underset{(V=0)}{CH_3F(\nu_3)} \rightarrow \underset{(V=1)}{CH_3F(\nu_2) \text{ or } (\nu_5)} + \underset{(V=0)}{CH_3F(\nu_3)} + E$$

where $E \approx -300$ cm^{-1}.

The 6.8 μm fluorescence intensity (ν_2 and ν_5) was found to vary as (CW

[201] K. E. Schuler and G. H. Weiss, *J. Chem. Phys.*, 1966, **45**, 1105.

laser power)$^{1.3 \pm 0.1}$. Again multiphoton absorption is shown to be unlikely in view of the close agreement between this power and that predicted by a bulk-heating model.

CH_3F *and Buffer Gases.* $p\tau$ values for collisional deactivation [197] of 3000 cm^{-1} fluorescence of CH_3F by various added gases are given in Table 8. The value for O_2 is corrected for energy sharing in the O_2 vibration. The ν_3 state gradually becomes more important in the energy transfer as the molecular weight of the collision partner increases, according to SSH theory. Although ^{3}He, ^{4}He, H_2, and D_2 $p\tau$ values are fitted reasonably well by acceptable range parameters for the SSH exponential repulsive potential, $V \rightarrow R$ theories need to be involved for heavier collision partners.

Table 8 *Collisional deactivation of* 3000 cm^{-1} *fluorescence of* CH_3F

Collision partner	$p\tau$/µs Torr
^{3}He	540
^{4}He	1 280
Ne	15 400
Ar	25 600
Kr	45 500
Xe	52 600
H_2	71.2
D_2	503
N_2	25 000
CH_3F	1 700
O_2	24 400

As a rough criterion for $V \rightarrow R$ significance the ratio (average rotational velocity)/(average relative translational velocity) may be examined. This ratio equals $(\mu d^2/I)^{\frac{1}{2}}$, where μ = reduced mass of the collision pair, d is the radius of the rotor, and I is the molecular moment of inertia. In the rare-gas series used, this ratio exceeds unity at Ne, and at this point $V \rightarrow R$ processes should become increasingly significant. It is concluded that both $V \rightarrow T$ and $V \rightarrow R$ processes operate, but current quantitative theories are over an order of magnitude in error in predicting the absolute probabilities. There is no unified theory of $V \rightarrow (T,R)$ energy transfer, and yet this is certain to be necessary as more data accrue on polyatomic species such as these.

Methyl Chloride.—$V \rightarrow (T,R)$ *Energy Transfer and the Kinetic Model.* The relevant energy levels are shown in Figure 21. The ν_3 (732 cm^{-1}) fluorescence was monitored,[202] that at 3000 cm^{-1} having such a poor signal/noise ratio that it could be used only to verify that the decay rate seemed similar to that at 732 cm^{-1}. Essentially the system may be reduced to a four-level scheme (Figure 22). Values of $p\tau$ for $V \rightarrow (T,R)$ energy transfer obtained from a study of ν_3 fluorescence are shown in Table 9 for various collision partners. The CH_4 case was corrected for ν_4 ($V = 1$) excitation of CH_4. As for CH_3F, CH_3Cl requires a combined $V \rightarrow (T,R)$ theory to predict properly these $p\tau$

[202] J. T. Knudtson and G. Flynn, *J. Chem. Phys.*, 1973, **58**, 2684.

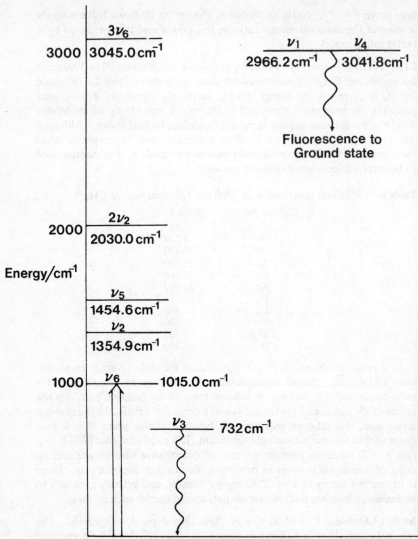

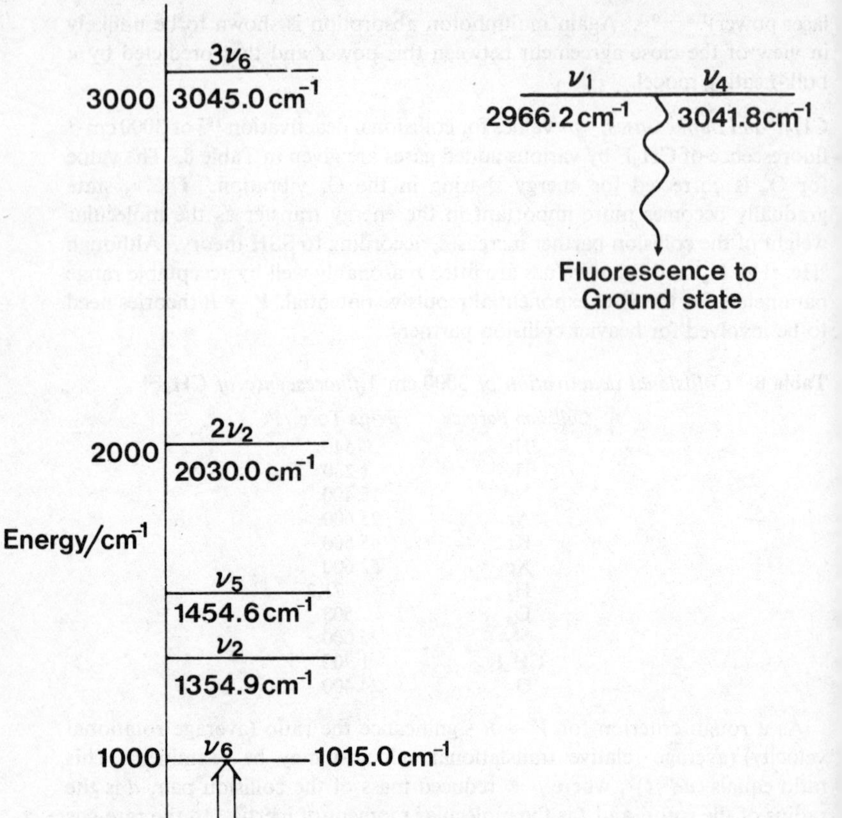

Figure 21 *Energy-level diagram for* CH_3Cl
(Reproduced by permission from *J. Chem. Phys.*, 1973, **58**, 2684)

data, and such a theory is not yet developed. Again the SSH theory was used to test for the significance of rotation and as for CH_3F it breaks down when the collision partner has a molecular weight greater than that of D_2. Also, as for CH_3F, only qualitative agreement is obtained with either $V \rightarrow T$ or $V \rightarrow R$ theory as appropriate.

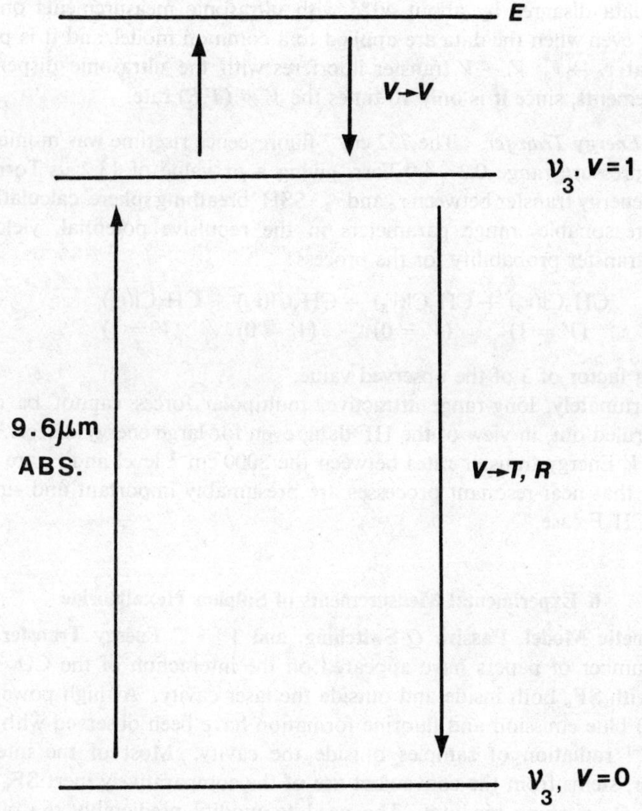

Figure 22 *Simplified* CH_3Cl *energy-level diagram*

Table 9 *Values of pτ obtained for collisions of* CH_3Cl *with various collision partners, as found from studies of the* ν_3 *fluorescence*

Collision partner	$p\tau/\mu s$ Torr
^3He	169
^4He	323
Ne	1 960
Ar	1 850
Kr	1 850
Xe	1 890
H_2	21.8
D_2	92.0
CH_4	116
CH_3Cl	134

The data disagree by about 40% with ultrasonic measurements on the system[3] even when the data are applied to a common model, and it is probable that $\nu_6 \rightarrow \nu_3$, $V \rightarrow V$ transfer interferes with the ultrasonic dispersion measurements, since it is only 10 times the $V \rightarrow (T,R)$ rate.

$V \rightarrow V$ Energy Transfer. The 732 cm^{-1} fluorescence risetime was monitored over a pressure range 0.3—4.0 Torr, giving a $p\tau$ value of 13.2 μs Torr for $V \rightarrow V$ energy transfer between ν_6 and ν_3. SSH 'breathing sphere' calculations, using 'reasonable' range parameters in the repulsive potential, yield an energy-transfer probability for the process:

$$\underset{(V=1)}{CH_3Cl(\nu_6)} + \underset{(V=0)}{CH_3Cl(\nu_3)} \rightarrow \underset{(V=0)}{CH_3Cl(\nu_3)} + \underset{(V=1)}{CH_3Cl(\nu_3)}$$

within a factor of 3 of the observed value.

Unfortunately, long-range attractive, multipolar forces cannot be completely ruled out, in view of the HF data, even for large energy defects.[203]

$V \rightarrow V$ Energy-transfer rates between the 3000 cm^{-1} level and ν_6 are very fast, so that near-resonant processes are presumably important and similar to the CH_3F case.

6 Experimental Measurements of Sulphur Hexafluoride

The Kinetic Model, Passive *Q*-Switching, and $V \rightarrow T$ Energy Transfer.—A large number of papers have appeared on the interaction of the CO_2 laser beam with SF_6 both inside and outside the laser cavity. At high powers[204] (150 W) blue emission and fluorine formation have been observed with CW 943 cm^{-1} radiation of samples outside the cavity. Most of the interest, however, stems from the convenient use of the comparatively inert SF_6 as a passive *Q*-switching medium. The need to modify predictably the output of such a *Q*-switching system has led to detailed studies of the interaction mechanism. Two recent papers summarize much of the earlier work.[205, 206] A four-state kinetic model was used to predict *Q*-switching behaviour and the results agreed well with the observed pulse width, repetition frequency, and peak power in the presence of added buffer gases mixed with the SF_6. The model may be represented as in Figure 23.

This model has also been used to explain saturation of SF_6 irradiated by the CO_2 laser[207] and has recently been simplified to a three-state model[208] where levels (1) and (4) are treated as one and form the ground state; level (2)

[203] J. R. Airey and I. W. M. Smith, *J. Chem. Phys.*, 1972, **57**, 1669.
[204] C. Borde, A. Henry, and L. Henry, *Compt. rend.*, 1966, **263**, 619.
[205] I. Burak, P. L. Houston, D. G. Sutton, and J. I. Steinfeld, *I.E.E.E.-J. Quantum Electron.*, 1971, **QE-7**, 73.
[206] D. G. Sutton, I. Burak, and J. I. Steinfeld, *I.E.E.E.-J. Quantum Electron.*, 1971, **QE-7**, 82.
[207] I. Burak, J. I. Steinfeld, and D. G. Sutton, *J. Quant. Spectroscopy Radiative Transfer*, 1969, **9**, 959.
[208] J. T. Knudtson and G. W. Flynn, *J. Chem. Phys.*, 1973, **58**, 1467.

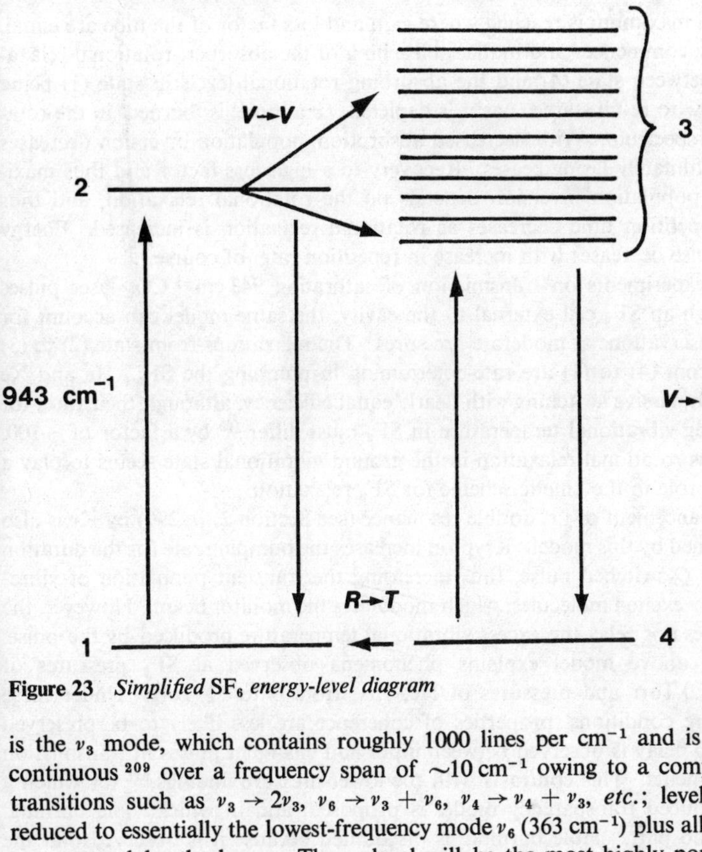

Figure 23 *Simplified* SF_6 *energy-level diagram*

is the ν_3 mode, which contains roughly 1000 lines per cm^{-1} and is nearly continuous [209] over a frequency span of ~10 cm^{-1} owing to a complex of transitions such as $\nu_3 \rightarrow 2\nu_3$, $\nu_6 \rightarrow \nu_3 + \nu_6$, $\nu_4 \rightarrow \nu_4 + \nu_3$, *etc.*; level (3) is reduced to essentially the lowest-frequency mode ν_6 (363 cm^{-1}) plus all others not pumped by the lasers. The ν_6 level will be the most highly populated of these and it is assumed that it is *via* this level that the previously equilibrated vibrational manifold will transfer energy to the translational degree of freedom. No absorptions to higher excited states are considered and this may be inaccurate if large numbers of molecules are removed from the ground state in high-intensity pulses or high-power CW systems. Under optimum *Q*-switching conditions the change in population of the ground state is almost negligible. The model predicts quenching of the *Q*-switching action at high buffer gas pressures, although the calculated pressure (5 Torr He) does not quite agree with the observed range (~19 Torr He). The importance of the rotational relaxation between states (4) and (1) (Figure 23) [205, 206] is not considered in the later work.[208]

The passive *Q*-switching action is explained as follows. After a pulse the bleached absorber population is restored to its equilibrium value, which increases the *Q* of the cavity. Population inversion begins and continues

[209] H. Brunet *I.E.E.E.-J. Quantum Electron.*, 1970, QE–6, 678.

until a maximum is reached where gain and loss factor of the tube are equal. Lasing commences and induces bleaching of the absorber, rotational relaxation between state (4) and the absorbing rotational levels in state (1) being too low to re-fill state (1) as it is depleted, *i.e* a 'hole' is 'burned' in the rotational spectrum. With decreased absorption, population inversion decreases and ultimately lasing ceases. Recovery to a high loss factor and thus maximum population inversion depends on the rotational relaxation, and thus the repetition time decreases as rotational relaxation is increased. Energy per pulse decreases with increase in repetition rate, of course.

In experiments on transmission of saturating 943 cm^{-1} CO_2 laser pulses through an SF_6 cell external to the cavity, the same model can account for the observations at moderate pressures. The relaxations from state (2) to (3) and from (4) to (1) are rate-determining in pumping the SF_6. He and Xe quench passive switching with nearly equal efficiency, although their rates for relaxing vibrational temperature in SF_6 must differ [210] by a factor of ~100. Thus rotational relaxation in the ground vibrational state seems to play a major role in the kinetic scheme for SF_6 relaxation.

Enhancement of i.r. double resonance (see Section 2, p. 290) by Kr is also explained by this model. Krypton increases the pumping rate for the duration of the *Q*-switched pulse, thus increasing the transient population of vibrationally excited molecules, which modulates the monitor beam. However, the Kr does not relax the excess vibrational temperature produced by the pulse.

The above model explains phenomena observed at SF_6 pressures of 0.4—1.0 Torr and pressures of He, *etc.*, from 0 to 11 Torr. Under these pressure conditions, properties of coherence are less likely to be preserved and no delay is observed between input and emergent pulses in transmission experiments. This contrasts with the lower-pressure findings,[211] for which a self-induced transparency model is proposed, and in which 'hole burning' plays no part. 'Hole burning' is discounted because it is observed that the output pulse is wider, more symmetrical than the input pulse, and is of greater intensity in its later part than the input pulse since it is delayed ~0.5 μs from the latter.

The mechanism is propagation of a '2π' pulse driving molecules from $V = 0 \rightarrow 1 \rightarrow 0$, *i.e.* absorption followed by stimulated emission. Such pulses propagate without attenuation and the medium appears transparent. This mechanism will operate only at low pressures where dephasing collisions that disrupt the coherence are few (0.01—0.05 Torr SF_6). Power levels of 10 W cm^{-2} were used. The collisional relaxation time represents an upper limit to the output pulse width, this being the condition that the pulse remains '2π'. The possibility of relaxation-time measurement by photon echo techniques (see Section 6, p. 331) is raised in Patel's paper.[211]

At pressures between those of these two studies (*i.e.* 0.1—1 Torr) it has

[210] J. I. Steinfeld, I. Burak, D. G. Sutton, and A. V. Novak, J. *Chem. Phys.*, 1970, **52**, 5421.

[211] C. K. N. Patel and R. E. Slusher, *Phys. Rev. Letters*, 1967, **19**, 1019.

been shown by Brunet [209] that the saturation intensity (I_{sat}) is proportional to the pressure. Indeed, (I_{sat}/pressure) = 6 ± 1.5 W cm^{-2} Torr^{-1}. The data agree with those of Steinfeld,[212] the model used being identical with that of Flynn *et al.* (three-state).[208] The vibration → translation relaxation of the lowest (ν_6) SF_6 level is associated with the laser intensity required for saturation. Unfortunately, this study could not measure this relaxation time with precision, because of uncertainty in the absorption cross-section. However, a $p\tau$ value of 95 μs Torr was deduced and is near the 140—180 μs Torr measured by ultrasonics [3] and 122 μs Torr of i.r. double resonance in SF_6.[212] Straightforward fluorescence experiments [213, 214] yield $p\tau$ = 150—160 μs Torr for the vibration → translation ($V \rightarrow T$) relaxation. In view of the proposed model [206] for the Q-switching action in an overlapping pressure regime it is possible that the $V \rightarrow T$ relaxation measured by fluorescence, *etc.*, considers a lower state which is not at rotational equilibrium.

This conclusion is at variance with the usual experience that rotational relaxation is very rapid when compared with the vibrational relaxation. If we suppose that the rotational relaxation of the SF_6 ground state is fast, the Steinfeld and Brunet experiments may be explained with a single three-state model, which is also well supported in recent work.[208] In Steinfeld's experiments few SF_6 molecules need be renewed from the ground state and, if a 'hole' is 'burned' in the spectrum, the fastest way of replenishing the population of the relevant rotational line is by rotational relaxation in the ground state. If the state (3) → (4) vibrational relaxation is much slower than this,[208] its kinetics will not enter the relevant expressions. The ground-state population could be re-equilibrated by this slow process in the quiescent period between Q-switch pulses.

In Brunet's experiments mechanically chopped CW radiation is used and, at saturation, large numbers of molecules would be removed from the ground state during the long irradiation period. Rotational relaxation will not then be an effective method of restoring the population and the slower vibrational relaxation will determine the kinetics.

Thus the (3) state model is probably consistent with the facts at reasonable pressures and wholly valid for discussion of the experiments of Flynn *et al.*[208, 215, 199] There seems little doubt now that above ~1 Torr bulk heating also occurs, giving extremely long fluorescence lifetimes and also contributing to transparency.[216–220]

[212] I. Burak, A. V. Novak, J. I. Steinfeld, and D. G. Sutton, *J. Chem. Phys.*, 1969, **51**, 2275.
[213] J. T. Yardley, *J. Chem. Phys.*, 1968, **49**, 2816.
[214] R. D. Bates, J. T. Knudtson, G. W. Flynn, and A. M. Ronn, *Chem. Phys. Letters*, 1971, **8**, 103.
[215] R. D. Bates, J. T. Knudtson, and G. W. Flynn, *J. Chem. Phys.*, 1972, **57**, 4174.
[216] H. Brunet, *Compt. rend.*, 1967, **264**, *B*, 1721.
[217] O. R. Wood, P. L. Gordon, and S. E. Schwarz, *I.E.E.E.-J. Quantum Electron.*, 1969, **QE-5**, 502.
[218] H. Brunet, *I.E.E.E.-J. Quantum Electron.*, 1968, **4**, 335.
[219] H. Brunet and M. Perez, *Compt. rend.*, 1968, **276**, *B*, 1084.
[220] O. R. Wood and S. E. Schwarz, *Appl. Phys. Letters*, 1970, **16**, 518.

Similar bulk heating has been observed in other gases,[98] and at pressures of 3 Torr and over in a cell 6.8 cm long by 2 cm diameter, acoustic shocks have been observed [220] as a result. These results have been investigated in two virtually simultaneous studies.[221, 222] In the first,[221] the rate constant for $V \rightarrow V$ energy transfer from the 943 cm^{-1} transition in SF_6 to the 615 cm^{-1} transition was determined as $p\tau < 20$ μs Torr from the risetime of the 615 cm^{-1} fluorescence. The decay curve of both 615 cm^{-1} and 943 cm^{-1} fluorescence exhibited peaks whose spacing depended on cell geometry, but not on SF_6 pressure, laser intensity, or fluorescing state. It was observed (Figure 24) that $x/y = a/(a + b)$, which is completely consistent with the passage of pulse-initiated shocks. The then current $V \rightarrow V$ and $V \rightarrow T$ transfer rates also supported this interpretation.

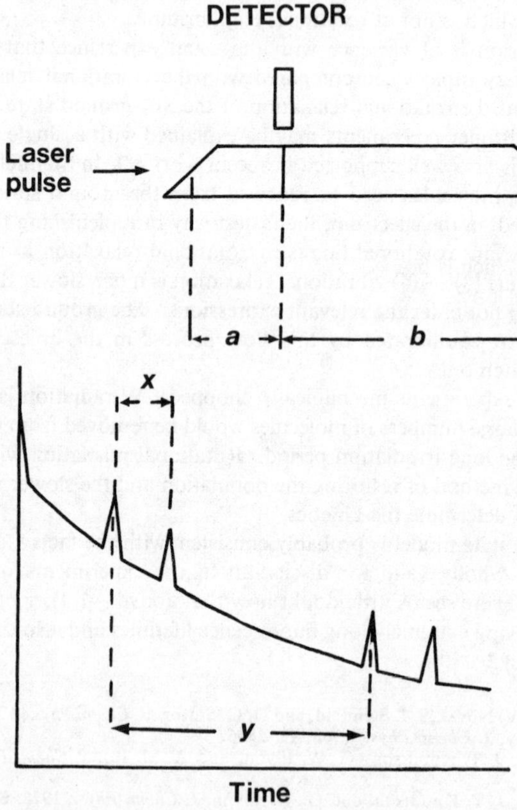

Figure 24 *The relationship between cell geometry and acoustic shocks as seen in* SF_6 *fluorescence*

[221] R. D. Bates, G. W. Flynn, and J. T. Knudtson, *J. Chem. Phys.*, 1970, **53**, 3621.
[222] I. Burak, P. Houston, D. G. Sutton, and J. I. Steinfeld, *J. Chem. Phys.*, 1970, **53**, 3632.

In the second study [222] the time dependence of the pressure wave was studied, and its initial time development indicated that $V \rightarrow V$ transfer in SF_6 occurred with $p\tau \leqslant 0.5$ μs Torr, in contrast to an earlier [220] result (45 ms Torr!) which would now appear to be associated with thermal relaxation.

Acoustic shocks have also been detected in CO_2.[223]

SF_6 has also been used as a passive mode-locking device,[224] giving 20 ns pulses at peak powers of 10^4—10^5 W, and pulse trains of over 100 μs have been observed at millisecond intervals. SF_6 can also be used in the laser cavity to tune a CO_2 laser over the various accessible *P*- and *R*-branch lines.[225] It has an advantage over prisms [226] and gratings [227] in that the *Q* of the cavity is not degraded to the same extent.

$V \rightarrow V$ Transfer and Lifetimes of the State of SF_6 Directly Excited by the CO_2 Laser.—Some values for $V \rightarrow V$ transfer have already been quoted above, but the most recent work seems to give the most accurate results. A *Q*-switched CO_2 laser was used at 400 Hz repetition rate with appropriate filters and a Cu–Ge detector (4 K). The detector was operated in the photoconductive mode with a 1.2 kΩ load, giving it a rather slow risetime. The output was boxcar-integrated at 1 μs gate-width and 10 ms time-constant using a 40 minute trace-time. Fluorescence risetimes, *Tr*, were determined from 10% to 90% of maximum, and $Tr = 2.2\tau$ for first-order kinetics. Rare gases were added to 0.6 Torr SF_6. For SF_6 $p\tau_{V \rightarrow T}$ for the 615 cm^{-1} level (presumed *via* the 363 cm^{-1}) as quoted already [219] is 150 μs Torr, which is so much (~150 times) slower than the risetime of 615 cm^{-1} fluorescence that two well-separated decay curves obtain. Even with rare gases added in quantity the $V \rightarrow T$ value was at least 20 times the 615 cm^{-1} fluorescence risetime. Deconvolution was necessary to correct for laser pulse width and electronics risetime.

$V \rightarrow V$ Energy-transfer pathways are traced, following the usual rules that high-probability processes will have low overall energy differences and few quanta will be transferred. The $V \rightarrow V$ paths deduced [208] are:

type (1) $SF_6(\nu_3) + SF_6 \rightarrow SF_6(\nu_4) + SF_6$; $\Delta E = 325\ cm^{-1}$

$$\Delta h\nu = 2$$

type (2) $SF_6(\nu_3) + SF_6 \rightarrow SF_6(\nu_4 + \nu_6) + SF_6$; $\Delta E = 38\ cm^{-1}$

$$\Delta h\nu = 3$$

type (3) $SF_6(\nu_3) + SF_6 \rightarrow SF_6(\nu_4) + SF_6(\nu_6)$; $\Delta E = 38\ cm^{-1}$

$$\Delta h\nu = 3$$

223 H. W. Brinkschulte, *J. Appl. Phys.*, 1970, **41**, 2298.
224 O. R. Wood and S. E. Schwarz, *Appl. Phys. Letters*, 1968, **12**, 263.
225 I. Burak, J. I. Steinfeld, and D. G. Sutton, *J. Appl. Phys.*, 1968, **39**, 4464.
226 P. Laures and X. Ziegler, *J. Chim. phys.*, 1967, **64**, 100.
227 G. Moeller and J. D. Rigden, *Appl. Phys. Letters*, 1966, **8**, 69.

The 615 cm^{-1} fluorescence will originate from all ν_4 $\Delta V = -1$ transitions and it is thought that mechanisms such as:

$$SF_6(2\nu_3) + SF_6 \rightarrow SF_6(3\nu_4) + SF_6;\ \Delta E = 35\ cm^{-1}$$
$$\Delta h\nu = 5$$

are involved. This is only one of many near-resonant paths arising from the high density of combination states near $2\nu_3$.

Transition probabilities for $SF_6 + SF_6$ (SSH theory) are theoretically about three times those expected for SF_6 and rare gases for the same reduced mass, and the latter transition probability changes little with reduced mass.

For transitions of type (1), *i.e.* large ΔE, the SSH theory predicts decreasing probability with increasing reduced mass, the *opposite* of the experimental result. For near-resonant paths where we expect the SSH theory to be worst, *e.g.* paths (2) and (4), SSH probability increases with increasing reduced mass, in agreement with experiment, but theory over-estimates the difference in dependence between He and Xe by a factor of at least ten. Correction for the greater number of energy-transfer paths in the case of the lighter gases improves this agreement but is difficult to quantify.

It would seem that the treatment of Dillon and Stephenson could well produce corrections in the desired direction, particularly if rotational degrees of freedom are important.

The effect of long-range interactions for SF_6 and rare gas (dipole–induced-dipole) using the Sharma and Brau approach [150] was tested. Although the probability varies in the correct sense with reduced mass, the dependence is grossly over-estimated. Thus such long-range interactions are deemed unimportant for this intramolecular $V \rightarrow V$ transfer in SF_6.

A weak dependence on reduced mass might indicate rotational transfer intervention, but it must be emphasized that the ratio rotational-velocity/transitional-velocity is 0.8, which is much lower than cases where rotation has been thought important previously.

The results obtained in the pressure range 0—5 Torr rare gas and 0.6 Torr SF_6 were,[208] for He, Ar, Kr, Xe, and SF_6 as collision partners, $p\tau = 1.9, 3.6, 4.0, 3.3$, and 1.1 μs Torr, respectively.

The observed $SF_6 + SF_6$ $V \rightarrow V$ transfer cross-section is only 2—3 times those of the SF_6–rare-gas cross-sections at the same reduced mass. Although this may seem surprising in view of the large difference in number of possible paths, the *absolute* difference is large and may be due to such energy-sharing paths in $SF_6 + SF_6$ cases.

These results [208] typify the elegant experimental results of recent laser experiments. Fluorescence risetime measurements can sometimes be coupled with a decay measurement of equal rate [18] conclusively assigning that rate to a given transition. This has not been done yet for SF_6, but is probably unnecessary in view of the nature of the absorption spectrum. The 615 cm^{-1} level is the first below the laser absorption region and a gap of 100 cm^{-1} exists between the two levels, giving a relatively slow $V \rightarrow V$ transfer step.

Provided that all other major separations do not exceed this by too much, this $V \rightarrow V$ step will be rate-determining in the equilibration of the vibrational manifold.

The Nature of the Absorbing State: Photon Echoes.—The technique of photon echoes[228–230] with polarized CO_2 laser irradiation of SF_6 indicates that the echoes are always polarized in the same direction as the second pulse but have intensity varying as $\cos^2 \varphi$, where φ is the angle between the polarization directions of the two pulses. The theory of echoes indicates that $j = 0 \leftrightarrow j = 1$ or $j = 1 \leftrightarrow j = 1$ transitions are most probably involved. Moreover, if this degree of discrimination is possible the individual vibration–rotation lines of the SF_6 spectrum must be separated and thus theoretically resolvable.

Homogeneous relaxation times for SF_6 + SF_6, He, Ne, and H_2 have been measured by this approach and were 22, 33, 46.4, and 16.4 ns Torr. This decay is the relaxation of the vibration–rotation levels of SF_6 excited by 10.5915 μm CO_2 laser radiation. The relaxation occurs by dephasing of the transition dipole by collision-induced change of rotational state or translational velocity and it seems strange therefore that these are referred to as homogeneous relaxations – a term usually reserved for Doppler broadening.

The cross-sections obtained are nearly pressure-independent in the range 0.002—0.015 Torr SF_6 and 0.001—0.02 Torr added gas. For all low j values ($\leqslant 6$) the theory shows that the polarization vector of the echo can identify the angular momenta of the states involved.

7 Experimental Measurements of Carbon Dioxide

Introduction.—The unparalleled interest in the energy-transfer processes occurring in gaseous CO_2 stems mainly from its importance as the active medium in the CO_2 laser. The general mechanism of the CO_2 laser process is basically understood but a detailed knowledge of the kinetics of the energy-transfer rates between the various normal modes is also necessary for optimizing the laser process.[20, 149, 231] While it is true that the $V \rightarrow T$ rate for the lower laser level is dominant in controlling the gain and power of the CO_2–N_2 laser, both $V \rightarrow V$, and $V \rightarrow T,R$ processes must be considered for a general understanding of the energy transfer within this system.

From the results below, it will be apparent that CO_2 is one of the most studied molecules. The variation in experimental techniques used is very wide and divergent results are frequently obtained. Clearly much remains to be done to resolve these difficulties, which are both experimental and theoretical in origin. One wonders how complex other, apparently simple, systems would appear if studied in such detail.

228 C. K. N. Patel and R. E. Slusher, *Phys. Rev. Letters*, 1968, **20**, 1087.
229 R. L. Abrams and A. Dienes, *Appl. Phys. Letters*, 1969, **14**, 237.
230 W. J. Tomlinson, J. P. Gordon, C. H. Wang, C. K. N. Patel, and R. E. Slusher, *Phys. Rev.*, 1969, **179**, 294.
231 C. K. N. Patel, *Phys. Rev. Letters*, 1967, **13**, 617.

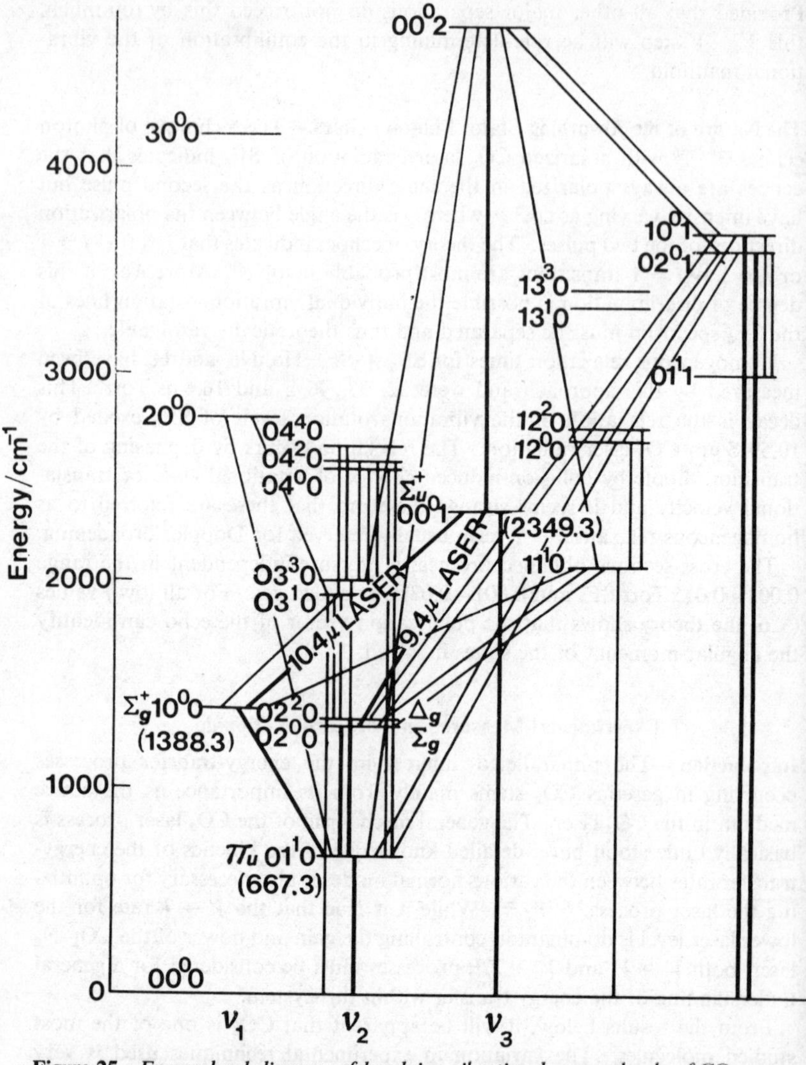

Figure 25 *Energy-level diagram of low-lying vibrational energy levels of* CO_2 (Reproduced by permission from 'Lasers', ed. A. K. Levine and A. J. DeMaria, Dekker, New York, 1971, vol. 3, p. 111)

An energy-level diagram of the low-lying vibrational levels of the CO_2 molecule is shown in Figure 25. The main laser transitions are indicated in Figure 26. The strongest transition near 10.4 μm occurs between the vibration–rotation transitions of the asymmetric stretching fundamental (00^01) at 2349 cm^{-1} and the symmetric stretching mode (10^00) at 1388 cm^{-1}. Laser action also arises near 9.4 μm from transitions between the (00^01) fundamental and the (02^00) level (*cf.* Figure 25). Laser transitions have been obtained from both the *P* branch ($\Delta J = -1$) and the *R* branch ($\Delta J = +1$) of each band. The lower lasing levels (10^00) and (02^00) are strongly mixed by Fermi-resonance interactions. Excess population of the upper lasing (00^01) state is generally achieved by resonant energy transfer from electrically excited $N_2(V = 1)$ molecules. Laser action then occurs to the (10^00) and (02^00) states. The relaxation of these lower laser levels plays an important part in the efficiency of the laser process.

In spite of the relatively large amount of data available for CO_2, a general theory which accounts for all the energy-transfer cross-sections is not available. The detailed nature of the intermolecular force field leading to energy transfer is still unknown. In some instances, satisfactory agreement between data and theory has been obtained with the SSH theory, while in other cases

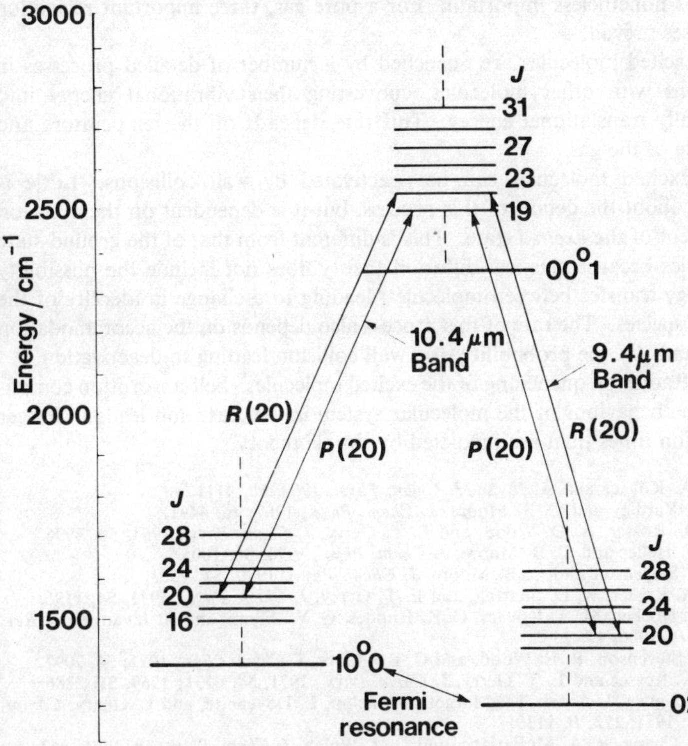

Figure 26 *Detailed laser-transition diagram for the* (00^01—10^00) *and* (00^01—02^00) *bands, including the rotational levels*
(Reproduced by permission from 'Lasers', ed. A. K. Levine and A. J. DeMaria, Dekker, New York, 1971, vol. 3, p. 111)

the observed energy-transfer cross-sections have been interpreted in terms of weak interactions proportional to the transitions involved. A universal theory is, however, still lacking.

Relaxation from the (00^01) **Level.**—A variety of laser-excited vibrational fluorescence studies of the energy relaxation of the CO_2 asymmetric stretching mode both in the pure gas and in mixtures with other gases has been made.[165, 232–246] Accounts can be found in reviews by Cheo,[20] Taylor and Bitterman,[149] and by Moore.[10] Basically, all the fluorescence experiments involve populating the (00^01) level either with a short pulse of CO_2 laser radiation or electrically, and then monitoring the decay of this level by i.r. fluorescence spectroscopy. The radiative relaxation of this level is monitored by recording the fluorescence at 10.4, 9.4, or 4.26 μm as shown in Figure 27. The mixing of the 1388 and 1285 cm^{-1} levels by Fermi resonance has only recently been taken into account in discussions of CO_2 relaxation processes,[247] but it is nonetheless important. For a pure gas, three important relaxation processes prevail.

(i) Excited molecules are quenched by a number of detailed processes in collisions with other molecules, converting their vibrational energy into essentially translational energy. This rate depends on the temperature and pressure of the gas.

(ii) Excited molecules can be deactivated by wall collisions. Little is known about the details of this process, but it is dependent on the diffusion coefficient of the *excited* state. This is different from that of the ground-state molecules because classical diffusion theory does not include the possibility of energy transfer between molecules, leading to a change in identity of the excited species. The rate of this process also depends on the accommodation coefficient, *i.e.* the probability of a wall collision leading to deactivation.

(iii) Radiation quenching of the excited molecules. Self-absorption complicates the behaviour of the molecular system in this case and leads to longer relaxation times than are predicted by simple theory.

[232] M. A. Kovacs and A. Javan, *J. Chem. Phys.*, 1969, **50,** 4111.
[233] J. T. Yardley and C. B. Moore, *J. Chem. Phys.*, 1967, **46,** 4491.
[234] W. A. Rosser, A. D. Wood, and E. T. Gerry, *J. Chem. Phys.*, 1964, **50,** 4996.
[235] D. F. Heller and C. B. Moore, *J. Chem. Phys.*, 1970, **52,** 1005.
[236] J. C. Stevenson and C. B. Moore, *J. Chem. Phys.*, 1970, **52,** 2333.
[237] W. A. Rosser, R. D. Sharma, and E. T. Gerry, *J. Chem. Phys.*, 1971, **54,** 1196.
[238] L. O. Hocker, M. A. Kovacs, C. K. Rhodes, G. W. Flynn, and A. Javan, *Phys. Rev. Letters*, 1966, **17,** 233.
[239] J. C. Stevenson, R. E. Wood, and C. B. Moore, *J. Chem. Phys.*, 1971, **54,** 3097.
[240] W. A. Rosser and E. T. Gerry, *J. Chem. Phys.*, 1971, **54,** 4131; 1969, **51,** 2286.
[241] H. Gueguen, I. Ardite, M. Margottin-Maclou, L. Doyenette, and L. Henry, *Compt. rend.*, 1971, **272,** *B*, 1139.
[242] R. S. Chang, R. A. McFarlane, and G. J. Wolga, *J. Chem. Phys.*, 1972, **56,** 667.
[243] J. C. Stevenson and C. B. Moore, *J. Chem. Phys.*, 1972, **56,** 1295.
[244] J. C. Stevenson, R. E. Wood, and C. B. Moore, *J. Chem. Phys.*, 1972, **56,** 4813.
[245] J. C. Stevenson, J. Finzi, and C. B. Moore, *J. Chem. Phys.*, 1972, **56,** 5214.
[246] Y. V. C. Rao, J. S. Rao, and D. R. Rao, *Chem. Phys. Letters*, 1972, **17,** 531.
[247] K. N. Seeber, *J. Chem. Phys.*, 1971, **55,** 5077.

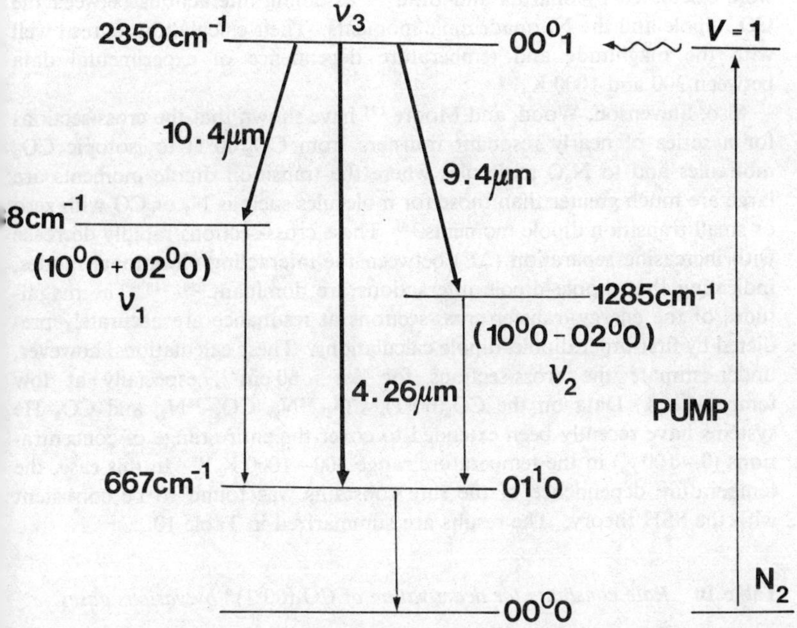

Figure 27 *A simplified energy-level diagram of the principal vibrational transitions (including laser transitions) for the* CO_2 *molecule*

An early study of CO_2 (00^01) relaxation used a *Q*-switched CO_2 laser for excitation, and the 4.26 μm fluorescence was monitored.[238] Departure from the expected first-order decay below 2 Torr was attributed to process (ii) becoming dominant. A $p\tau$ value of 2.60×10^3 μs Torr ($\sigma = 3.3 \pm 0.3 \times 10^{-19}$ cm^2) was obtained. Neglecting the accommodation coefficient and self absorption, *D*, the diffusion coefficient to the wall of CO_2 (00^01) at one atmosphere, was calculated to be 0.5 $cm^2\ s^{-1}$. In a similar study of 4.26 μm fluorescence a $p\tau$ value of 2.86×10^3 μs Torr was obtained.[233] When mixtures of CO_2 with ^3He, ^4He, Ar, Kr, and Xe were studied the mass dependence of deactivation efficiency was found to be slight.

Moore, Wood, Hu, and Yardley [165] analysed the kinetics of the CO_2 laser system using a three-level approximation involving only the (00^01), (01^10), and (00^00) levels. The probability of energy transfer was found to depend on the degree of mixing of the vibrational states due to anharmonicity and Coriolis coupling. When this mixing was accounted for, the usual exponential repulsion perturbation calculations were in order of magnitude agreement with the observed probabilities.

Energy-transfer cross-sections for the near-resonant process:

$$CO_2(00^01) + N_2(V = 0) = CO_2(00^00) + N_2(V = 1) + \Delta E = 19\ cm^{-1}$$

were calculated by Sharma and Brau [150] assuming interactions between the CO_2 dipole and the N_2 quadrupole moments. Their calculations agreed well with the magnitude and temperature dependence of experimental data between 300 and 1000 K.[234]

Also, Stevenson, Wood, and Moore [191] have shown that the cross-sections for a series of nearly resonant transfers from $CO_2(00^01)$ to isotopic CO_2 molecules and to N_2O molecules where the transition dipole moments are large are much greater than those for molecules such as N_2 or CO with zero or small transition dipole moments.[191] These cross-sections rapidly decrease with increasing separation (ΔE) between the interacting vibrational modes, indicating that dipole–dipole interactions are dominant.[236, 243] The magnitudes of the energy-transfer cross-sections at resonance are accurately predicted by first-order dipole–dipole calculations. These calculations, however, under-estimate the cross-sections for $\Delta E \geqslant 60\ \text{cm}^{-1}$, especially at low temperatures. Data on the $CO_2(00^01)$, CO_2–$^{14}N_2$, CO_2–$^{15}N_2$, and CO_2–He systems have recently been extended to cover the entire range of concentrations (0—100%) in the temperature range 300—1000 K.[246] In this case, the temperature dependence of the rate constants was found to be consistent with the SSH theory. The results are summarized in Table 10.

Table 10 *Rate constants for deactivation of* $CO_2(00^01)$[a] *by various gases*

CO_2 : *rate*/ μs Torr × 10^3	$^{14}N_2$: *rate*/ μs Torr × 10^3	$^{15}N_2$: *rate*/ μs Torr × 10^3	He : *rate*/ μs Torr × 10^3
2.60	9.43	10.6	11.8
2.73	9.09	10.3—8.33	12.2
2.86	8.70	6.85 (400 K)	20.0
2.99	9.43	4.67 (500 K)	11.9
3.03	4.65 (400 K)		12.2
3.20	2.94 (500 K)		16.7
2.95			
2.44 (400 K)			
1.20 (500 K)			

[a] All values are those obtained at room temperature unless otherwise indicated.

Rates associated with the deactivation of $CO_2(00^01)$ by collisions with CO have been measured by Rosser, Sharma, and Gerry [237] in the temperature range 300—900 K. These reactions were:

$$CO_2(00^01) + CO \underset{k_2}{\overset{k_1}{\rightleftharpoons}} CO_2 + CO(V = 1)$$

$$CO_2(00^01) + CO \xrightarrow{k_3} CO_2(\nu_1\nu_2) + CO$$

$$CO(V = 1) + CO_2 \xrightarrow{k_4} CO + CO_2(\nu_1\nu_2)$$

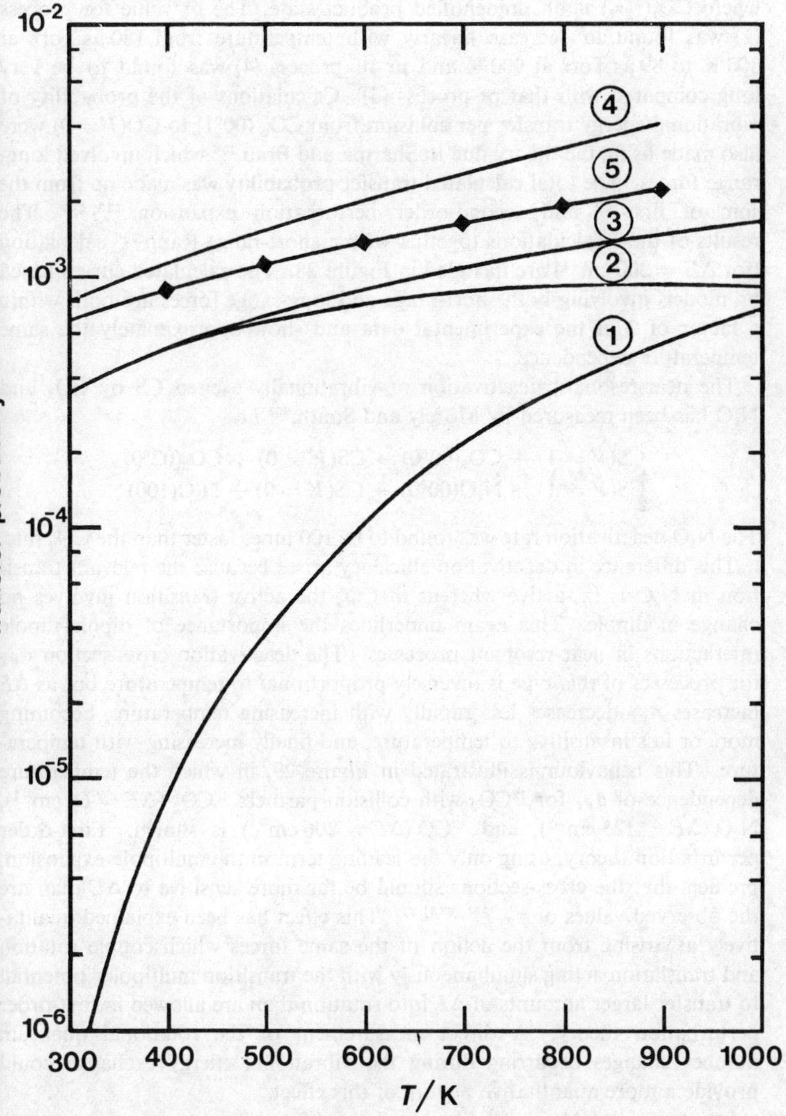

Figure 28 *Comparison of calculated and measured transition probabilities for* CO_2–CO *transfer.* (*i*) *First-order dipole–dipole probability;* (*ii*) *second-order probability;* (*iii*) *sum of* (*i*) *and* (*ii*); (*iv*) *short-range Rapp calculations for* $\Delta E = 200$ cm^{-1}; (*v*) *measured probabilities*
(Reproduced by permission from *J. Chem. Phys.*, 1971, **54**, 1196)

where $CO_2(\nu_1\nu_2)$ is an unidentified product state. The $p\tau$ value for process (1) was found to decrease linearly with temperature from 180 μs Torr at 300 K to 89 μs Torr at 900 K and $p\tau$ of process (4) was found to be very long compared with that of process (3). Calculations of the probability of vibrational energy transfer per collision from CO_2 (00^{0}1) to CO($V = 0$) were also made using the theory due to Sharma and Brau,[150] which involved long-range forces. The total calculated transfer probability was made up from the sum of first-[150] and second-order perturbation expansions.[248, 249] The results of these calculations together with a short-range Rapp [250] calculation (for $\Delta E = 200\ \text{cm}^{-1}$) are included in Figure 28. The calculated curves based on models involving both short-range and long-range forces are both within a factor of 2 of the experimental data and show approximately the same temperature dependence.

The near-resonant deactivation of vibrationally excited CS by CO_2 and N_2O has been measured by Morely and Smith,[251] *i.e.*

$$CS(V = 1) + CO_2(00^00) \rightarrow CS(V = 0) + CO_2(02^00)$$
$$CS(V = 1) + N_2O(00^00) \rightarrow CS(V = 0) + N_2O(100)$$

The N_2O deactivation rate was found to be 100 times faster than the CO_2 rate.

This difference in deactivation efficiency arises because the relevant transition in N_2O is i.r.-active whereas in CO_2 the active transition involves no change in dipole. This again underlines the importance of dipole–dipole interactions in near-resonant processes. The deactivation cross-section σ_{VV} for processes of this type is inversely proportional to temperature but as ΔE increases σ_{VV} decreases less rapidly with increasing temperature, becoming more or less insensitive to temperature, and finally increasing with temperature. This behaviour is illustrated in Figure 29, in which the temperature dependence of σ_{VV} for $^{12}CO_2$ with collision partners $^{13}CO_2$ ($\Delta E = 66\ \text{cm}^{-1}$), N_2O ($\Delta E = 125\ \text{cm}^{-1}$), and ^{12}CO ($\Delta E = 206\ \text{cm}^{-1}$) is shown. First-order perturbation theory, using only the leading term in the multipole expansion, predicts that the cross-sections should be far more sensitive to ΔE than are the observed values of σ_{VV}.[151, 243, 249] This effect has been explained qualitatively as arising from the action of the same forces which couple rotation and translation acting simultaneously with the transition multipolar potential to transfer larger amounts of ΔE into rotation than are allowed in first-order perturbation theory. A direct measurement of the rotational quantum number changes occurring during the vibrational energy exchange could provide a more quantitative picture of this effect.

Stevenson and Moore [236] have used the 4.3 μm fluorescence technique to study the near-resonant multiquantum deactivation of CO_2(00^{0}1) with a number of collision partners, CO_2, CH_4, C_2H_4, CH_3F, CH_3Cl, CH_3Br, CH_3I,

[248] R. D. Sharma and C. A. Brau, *J. Chem. Phys.*, 1969, **50**, 924.
[249] R. D. Sharma, *Phys. Rev.*, 1970, **A2**, 173.
[250] D. Rapp, *J. Chem. Phys.*, 1965, **43**, 316.
[251] I. W. M. Smith and C. Morley, *Trans. Faraday Soc.*, 1971, **67**, 2575.

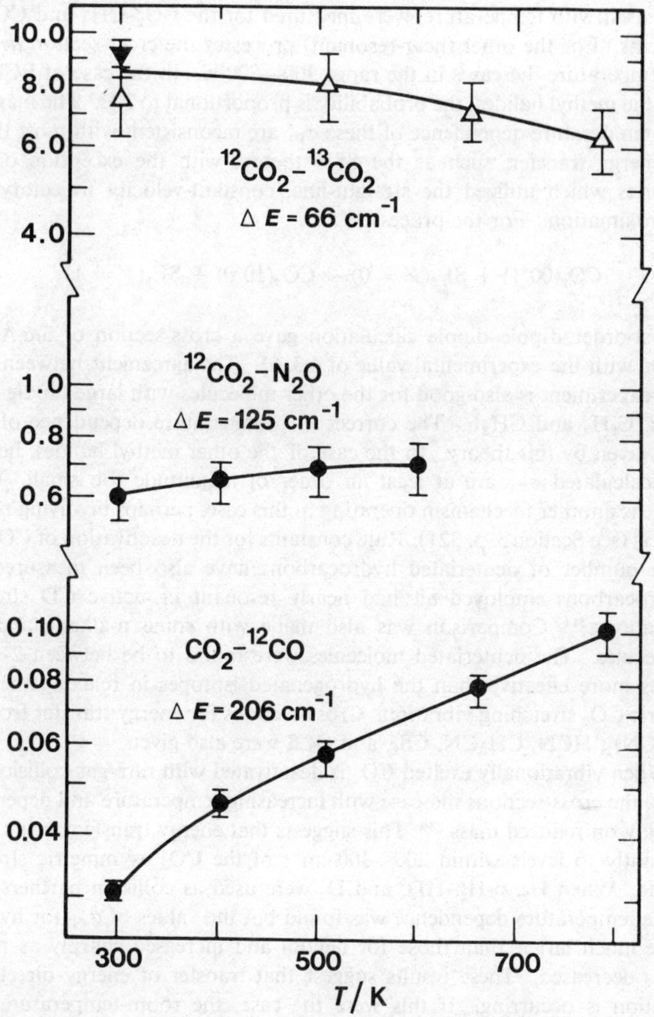

Figure 29 *Temperature dependence of cross-sections for nearly resonant $V \rightarrow V$ single-quantum transfers*
(Reproduced by permission from *Adv. Chem. Phys.*, 1973, **23**, 41)

BCl_3, and SF_6, in the temperature range 300—800 K. Large deactivation cross-sections were again found for systems in which near-resonant transfer was possible. For the last seven molecules, the room-temperature values of σ_{VV} were 3 or 4 orders of magnitude larger than in CO_2–rare-gas collisions[233] and 4—40 times larger than single-quantum vibrational energy transfer to $^{14}N_2$ or $^{15}N_2$,[191] a process with similar ΔE. Smaller cross-sections, which increased with temperature, were measured for the CO_2–CH_4 and CO_2–CO_2 systems. For the other (near-resonant) processes the cross-section *decreases* as temperature increases in the range 300—800 K. In the case of BCl_3, SF_6, and the methyl halides, the probability is proportional to T^{-1}. The magnitude and temperature dependence of these σ_{VV} are inconsistent with most theories of energy transfer, such as the SSH theory, with the exception of those theories which utilized the straight-line, constant-velocity trajectory, Born approximation. For the process:[236]

$$CO_2(00^01) + SF_6(V = 0) \rightarrow CO_2(10^00) + SF_6(V = 1)$$

a first-order dipole–dipole calculation gave a cross-section of 3.6 Å^2 compared with the experimental value of 4.3 Å^2. The agreement between theory and experiment is also good for the other molecules with large dipole derivatives, C_2H_4 and CH_3F. The correct T^{-1} temperature dependence of σ_{VV} is also given by this theory. In the case of the other methyl halides, however, the calculated σ_{VV} are at least an order of magnitude too small. Clearly there is another mechanism operating in this case, perhaps involving rotation effects (see Section 5, p. 321). Rate constants for the deactivation of $CO_2(00^01)$ by a number of deuteriated hydrocarbons have also been measured. The hydrocarbons employed all had nearly resonant i.r.-active CD stretching vibrations.[244] Comparison was also made with some n-alkane and other molecules. The deuteriated molecules were found to be between 2 and 20 times more effective than the hydrogenated isotopes in relaxing the asymmetric CO_2 stretching vibration. Cross-sections for energy transfer from CO_2 to $(CN)_2$, HCN, CH_3CN, CS_2, and OCS were also given.

When vibrationally excited CO_2 is deactivated with rare-gas collision partners, the cross-sections increase with increasing temperature and depend only weakly on reduced mass.[239] This suggests that energy transfer occurs predominantly to levels within 200—300 cm^{-1} of the CO_2 asymmetric stretching mode. When H_2, p-H_2, HD, and D_2 were used as collision partners [239] the same temperature dependence was found but the values of σ_{VV} for hydrogen were much larger than those for helium and increased sharply as reduced mass decreased. These results suggest that transfer of energy directly into rotation is occurring. If this were the case, the room-temperature cross-sections for normal and p-H_2 should be different whereas, in fact, the experimental values were found to be identical. Current SSH and Landau–Teller treatments could not adequately predict the observed cross-sections and their dependence on reduced mass and temperature. Coriolis and anharmonic coupling of normal vibrational modes probably play a part in energy-transfer processes where a multivibrational quantum-number change is involved. In hydrogen halide mixtures transfer of substantial amounts of vibrational energy into rotational degrees of freedom has been demonstrated. Long-range multipole moment calculations by Sharma have shown the cross-section for relaxation of the bending mode in $CO_2(01^10)$ by para-hydrogen to be about one order of magnitude smaller than that for ortho-hydrogen. Therefore o-H_2 should be about an order of magnitude more efficient than p-H_2 for deactivating the bending vibration of CO_2. More experimental as well as theoretical work is needed in this area to clarify the situation.

Heller and Moore [235] have also demonstrated the deactivation rates of $CO_2(00^01)$ by H_2O, D_2O, and HDO using the vibrational fluorescence technique. The experimental $V \rightarrow V$ transfer probabilities were large and temperature-insensitive. Qualitatively, this may be explained by the existence of a large attractive interaction between CO_2 and water, resulting in strong collisions in the repulsive region of the potential which are relatively insensitive to molecular vibrations. Recent fluorescence results are also available for the deactivation of $CO_2(00^01)$ with HCl, DCl, DBr, and DI.[245] This series of molecules allows a large systematic variation in molecular parameters such as ΔE, the rotational constant B_0, and permanent and transition multipole moments. Observed cross-sections were relatively large for $V \rightarrow V$ exchange between CO_2 and HX and decreased slightly with temperature in the range 298—510 K. This decrease was more rapid with increasing energy difference (ΔE) between the vibrations of DX than for HX. Transfer of energy into rotational degrees of freedom in the HX and DX molecules, giving a more resonant process, may be responsible for this difference. The theory of Sharma [151] which involves first-order dipole–dipole interactions could not explain the observed cross-sections for HX and DX. For HI the transition dipole is too small and for the other molecules the energy difference ΔE is too large for this interaction to be important. Dipole–quadrupole treatments developed by Sharma and Brau [150, 248] predict the observed HI, HBr, and DCl cross-sections when reasonable values of the transition quadrupole moment were assumed. This treatment was unsuccessful for HCl and DBr, however, owing to the large value of ΔE. Second-order perturbation treatments involving multipolar interactions were also unsuccessful for HCl. These treatments ignore the perturbation of the molecular trajectories caused by the attractive and repulsive forces between collision partners. Also, larger rotational quantum number changes, ΔJ, probably occur during collisions than are associated with first-order dipolar ($\Delta J = \pm 1$) and quadrupolar ($\Delta J = 0, \mp 2$) interactions. Thus theories based on long-range multipolar forces do not account for the observed energy-transfer data in these systems. Energy transfer among the CO_2 vibrations is probably aided by attractive forces and transfer of energy to rotational degrees of freedom.

Rotational relaxation in the $CO_2(00^01)$ upper laser level has recently been measured by Cheo and Abrams [252] using the i.r. double-resonance technique. A 10.6 μm, single-frequency, (P20), pulse 20 ns wide was used to cause selective depletion of the population of one rotational level ($J = 19$) of the (00^01) vibrational band in pure CO_2. Relaxation of this excited ($J = 19$) level by rotational transitions from all neighbouring levels was monitored by

[252] P. K. Cheo and R. L. Abrams, *Appl. Phys. Letters*, 1969, **14**, 47.

measuring the absorption of a CW single-frequency 9.6 μm ($P20$) CO_2 laser. Since the 10.6 and 9.6 μm laser transitions share a common upper level ($J = 19$), the recovery time of the gain at 9.6 μm was a direct measure of the rotational relaxation time of the upper 10.6 μm laser level. The lower 9.6 μm laser level (02^00) could be perturbed by resonant vibrational transfer (Fermi resonance) from the lower 10.6 μm laser level (10^00) but this effect should be small. The 10 kW, 20 ns CO_2 pulse was obtained by the cavity-dump method using a GaAs electro-optic switch. More than one relaxation time was found, with the fastest process occurring immediately following the pumping pulse. This relaxation was attributed to rotational relaxation in the (00^01) level ($J = 19$). The $p\tau$ value of this process was (0.09 ± 0.02) μs Torr. Other longer decay times may arise from resonant vibrational relaxation between the (00^01) and (02^00) levels or between the (02^00) and (01^10) levels.

In a subsequent publication Adams and Cheo [253] reported the effect on the (00^01) relaxation rate of adding N_2 and He. These gases were found to be slightly less effective than CO_2 in thermalizing the rotational levels. For helium the $p\tau$ value was 0.14 μs Torr, and for N_2 it was 0.10 μs Torr, compared with a $p\tau$ of 0.91 μs Torr for pure CO_2. The number of gas-kinetic collisions necessary to cause rotational thermalization in pure CO_2 is 0.6, implying that collisional-induced rotational mixing occurs more rapidly than simple hard-sphere collisions, *i.e.* long-range interactions are important.

Relaxation from CO_2 (10^00) and (02^00) Levels.—The relaxation of the CO_2 (10^00) and (02^00) levels is of fundamental importance in determining the efficiency of CO_2 lasers. The kinetics of energy transfer among the CO_2 symmetric stretching (10^00) and bending (02^00) modes were first investigated by Rhodes, Kelly, and Javan [107] using an i.r. double-resonance technique. The basic experimental technique has already been outlined in Section 2. One laser source was used to create population changes while the second was used to monitor the change in the energy-level population as a function of time. The basic energy-level scheme is illustrated in Figure 30. A short pulse of radiation from a Q-switched 9.6 μm CO_2 laser pumps molecules from the (02^00) level to the (00^01) level. Approximately equal population of these levels is achieved by this pulse. A second low-power CW CO_2 laser operating at 10.6 μm ($10^00 \rightarrow 00^01$) probes the population of the (10^00) and (00^01) levels. Since the absorption of the 10.6 μm laser radiation is proportional to the difference in population densities of the (00^01) and (10^00) states, the absorption coefficient of the gas exhibits a transient response due to the effect of the 9.6 μm pulse. The behaviour of the (00^01) state is known from other measurements so that the relaxation behaviour of the (10^00) state can be inferred. During the pump pulse, the transmission of the probe laser was found to increase initially, owing to the increased (00^01) population.

[253] R. L. Abrams and P. K. Cheo, *Appl. Phys. Letters*, 1969, **15**, 177.

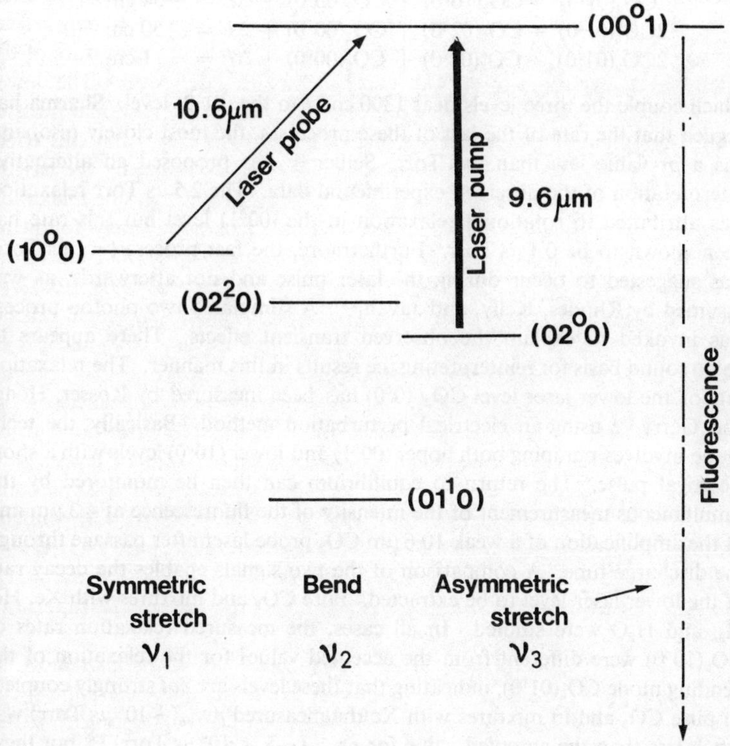

Figure 30 *Energy-level scheme for i.r. double-resonance experiment for the* CO_2 *molecule*

Immediately afterwards, a very rapid relaxation was observed which led to a further, very fast, increase in the transmission of the probe radiation. (The time resolution of the experiment was limited to about 10^{-6} s by the multichannel analyser.) This was followed by a decrease in the transmission of the probe with a relaxation rate given by $p\tau \approx 2.5 \pm 0.5$ μs Torr at pressures between 0.1 and 1.6 Torr. Finally, over a much longer time interval, known from fluorescence measurements, the probe transmission decreased further as the (00^01) state was depopulated. The first, very rapid, relaxation ($p\tau \approx 1.0$ μs Torr – upper limit) was attributed to a decrease in the population of the (10^00) level by collisional coupling with the (02^00) level, which is depleted by the pump laser.

$$CO_2(10^00) + CO_2(00^00) \rightarrow CO_2(02^00) + CO_2(00^00) + \Delta E = 103 \text{ cm}^{-1}$$

The slower exponential relaxation of the (10^00) level was attributed to near-resonant ($V \rightarrow V$) processes such as:

$$2CO_2(01^10) \rightarrow CO_2(10^00) + CO_2(00^00) + \Delta E = -54 \text{ cm}^{-1}$$
$$2CO_2(01^10) \rightarrow CO_2(02^00) + CO_2(00^00) + \Delta E = 50 \text{ cm}^{-1}$$
$$2CO_2(01^10) \rightarrow CO_2(02^20) + CO_2(00^00) + \Delta E = -1 \text{ cm}^{-1}$$

which couple the three levels near 1300 cm^{-1} to the (01^10) level. Sharma has argued that the rate of the last of these processes, the most closely resonant, has a $p\tau$ value less than 1 μs Torr. Seiber[247] has proposed an alternative interpretation of the observed experimental data. The 2.5 μs Torr relaxation was attributed to rotational relaxation in the (00^01) level but this rate has been shown to be 0.1 μs Torr. Furthermore, the fast process (<1 μs Torr) was suggested to occur during the laser pulse and not afterwards, as was assumed by Rhodes, Kelly, and Javan.[107] A stimulated two-photon process was invoked to explain the observed transient effects. There appears to be no sound basis for reinterpreting the results in this manner. The relaxation rate of the lower laser level $CO_2(10^00)$ has been measured by Rosser, Hoag, and Gerry[254] using an electrical perturbation method. Basically, the technique involves pumping both upper (00^01) and lower (10^00) levels with a short electrical pulse. The return to equilibrium can then be monitored by the simultaneous measurement of the intensity of the fluorescence at 4.3 μm and of the amplification of a weak 10.6 μm CO_2 probe laser after passage through the discharge tube. A comparison of the two signals enables the decay rate of the lower laser level to be extracted. Pure CO_2 and mixtures with Xe, He, N_2, and H_2O were studied. In all cases, the measured relaxation rates of $CO_2(10^00)$ were different from the accepted values for the relaxation of the bending mode $CO_2(01^10)$, indicating that these levels are *not* strongly coupled. In pure CO_2 and in mixtures with Xe the measured $p\tau_{100}$($\sim10^3$ μs Torr) was much less than the accepted value for $p\tau_{010}$ ($\sim5 \times 10^3$ μs Torr)[255] but these were both longer than the value obtained by Rhodes, Kelly, and Javan ($\sim$2.5 μs Torr). This discrepancy was attributed to the ambiguity present in the interpretation of the results of Rhodes *et al.* This was considered to arise from the single measurement of probe transmission (gain) which was dependent on two unknown laser-level populations and could not, therefore, be unambiguously interpreted in terms of a single decay rate. Moreover, since the relaxation of the laser levels was not studied to completion, the monitored process, which occurs early in the relaxation, could be followed by other slower processes similar to those observed by Rosser *et al.*[254] These latter workers would not have detected a fast ($\sim$10 μs Torr) process.

Bulthius and Ponsen[256–258] used the decay of the laser emission in the afterglow of a CO_2 laser discharge to determine the relaxation time of the (10^00) level in the presence of H_2O and D_2O. A $p\tau$ value of 71 μs Torr was evaluated for the rate of energy exchange between the (10^00) and (01^10) levels.

[254] W. A. Rosser, E. Hoag, and E. T. Gerry, *J. Chem. Phys.*, 1972, **57**, 4153.
[255] F. D. Shields, *J. Acoust. Soc. Amer.*, 1957, **29**, 450.
[256] K. Bulthius and G. J. Ponsen, *Chem. Phys. Letters*, 1972, **14**, 613.
[257] K. Bulthius and G. J. Ponsen, *I.E.E.E.-J. Quantum Electron.*, 1972, **8**, 597.
[258] K. Bulthius and G. J. Ponsen, *Phys. Letters*, 1971, **36A**, 123.

In another pulsed laser afterglow experiment, DeTemple, Sukre, and Coleman [259] found a relaxation rate for the (10^00) level which agreed with the data of Rhodes, Kelly, and Javan. The (10^00) level was found to be governed by two decay rates: a fast rate ($p\tau \sim 0.6$ μs Torr) and a slower rate ($p\tau \sim 6$ μs Torr). These decay rates were linearly dependent on inverse CO_2 pressure and were unaffected by the presence of He. They were, therefore, assigned to $V \rightarrow V$ equilibration processes within the ν_1 and ν_2 modes of CO_2. These data agree approximately with the results of Rhodes *et al.*, who observed both a fast ($p\tau < 1$ μs Torr) decay and a slower (2.5 ± 0.5 μs Torr) recovery rate. The decay rates of the 10^00 level measured by Bulthius and Ponsen [256–258] and by Rosser, Hoag, and Gerry [254] were at least one order of magnitude smaller than these values. The latter results are probably more characteristic of $V \rightarrow T$ processes under the influence of a perturbing radiation field while those of DeTemple, Sukre, and Coleman [259] are more representative of the intrinsic $V \rightarrow V$ relaxation processes in CO_2. Over the pressure range 0.5—2.6 Torr of CO_2 the decay of the 9.6 μm laser pulse was found to be a single exponential, consistent with a relaxation rate for the (02^00) level of about 0.8 μs Torr. This was supported by the recent work of Gower and Carswell,[260] who measured the $V \rightarrow T$ rates for the CO_2 (10^00) and (00^01) levels by observing the rise in pure CO_2 and in mixtures with N_2 and He. Rates of 2.8×10^3 μs Torr and 3.1×10^3 μs Torr were recorded for the relaxation of the (00^01) and (10^00) levels, respectively, for pure CO_2.

8 Experimental Measurements of Miscellaneous Compounds

Carbon Monoxide.—The $V \rightarrow T$ energy-transfer rate has been measured several times for CO, but not by laser-induced fluorescence.[261] The $V = 1$ evel is known to be exceptionally long-lived. Thus higher vibrational levels decay to this state *via* the process:

$$CO(V = 2) + CO(V = 0) \rightarrow 2CO(V = 1)$$

and the rate of this was first measured by laser (parametric oscillator) excited fluorescence.[103] The $V = 0 \rightarrow V = 2$ overtone was excited and the rapid fluorescence (2130 cm^{-1}) decay was followed. One of the great advantages of the parametric oscillator is its narrow pulse width, and in this study the pulse width was only 100 ns, with a peak of power 100 W. The pressure range studied was 0.5—5 Torr.

The $p\tau$ value obtained from the usual Stern–Volmer plot was 16.1 μs Torr $\pm 16\%$. Although this is from the sum of the $V \rightarrow V$ and $V \rightarrow (T,R)$ rate constants, it is reasonably assumed that $V \rightarrow (T,R)$ processes will be very slow compared with $V \rightarrow V$ processes and that the above $p\tau$ value is essentially that of the $V = 2 \rightarrow V = 1$ transfer.

259 T. A. DeTemple, D. R. Sukre, and P. D. Coleman, *Appl. Phys. Letters*, 1973, **22**, 349.
260 M. C. Gower and A. I. Carswell, *Appl. Phys. Letters*, 1973, **22**, 321.
261 R. C. Millikan, *J. Chem. Phys.*, 1963, **38**, 2855.

The tunability of the optical parametric oscillator allowed the excitation of different vibration–rotation lines but the $p\tau$ value remained constant within experimental error. Thus rotational equilibration is probably rapid when compared with $V \rightarrow V$ rates, since it is unlikely to be slow *and* of the same rate irrespective of the rotational levels involved.

Variation of the repetition rate of pulsing did not affect the $p\tau$ value, indicating that sample heating was insignificant. However, it is interesting to note that, since pulses appeared every 20 ns and the $V \rightarrow (T,R)$ and radiative lifetimes of the $V = 1$ state are so long, 'pumping' of the $V = 1$ level occurred. The resultant emission from the $V = 1$ state would have been troublesome were it not essentially trapped by the system.

The measured $p\tau$ value is longer than for the $V = 4$ level but shorter than for the $V = 5$ level.[262] This is interpreted as being due to the situation that the decrease in energy defect with V is offset by the decrease in dipole–dipole transition-matrix elements which predominate at low V and decrease as V increases.

Moore *et al.*[156] have recently measured the $V \rightarrow V$ energy-transfer $p\tau$ value for $CO \rightarrow N_2$. This was achieved in He–DCl–CO–N_2 quaternary mixtures using a DCl chemical laser providing 10 mJ pulses 10 μs wide, *i.e.* ~1 kW peak power.

The $p\tau$ values required for the kinetic analysis (see Section 3) were measured in the appropriate ternary mixtures and were:

$$p\tau_{CO \rightarrow DCl} = 21.0\ \mu s\ Torr$$
$$p\tau_{N_2 \rightarrow DCl} = 600\ \mu s\ Torr$$

For $CO \rightarrow N_2$ the $p\tau$ value is 5300 μs Torr or 2170 μs Torr in the exothermic direction $N_2 \rightarrow CO$.

The transition probability for this process agrees well with Callear's correlation[263] of reduced probability, P/mn, as a function of energy defect, ΔE, where P is the probability of the process:

$$N_2(V = n) + CO(V = m - 1) \rightarrow N_2(V = n - 1) + CO(V = m)$$

(see Figure 31).

For nearly resonant exchanges, the multipole contributions have been estimated as shown and, added to the repulsive values, give limiting values for P/mn. The temperature dependence was in good agreement with the SSH approach.

Methane.—$V \rightarrow (T,R)$ *Energy Transfer and the Kinetic Model.* Most of the work on CH_4 has been done by Moore *et al.*[18, 264, 265] A chopped He–Ne

[262] G. Hancock and I. W. M. Smith, *Chem. Phys. Letters*, 1971, **8**, 41; *Appl. Optics*, 1971, **10**, 1827.
[263] A. B. Callear, *Appl. Optics, Suppl.*, 1965, **2**, 145.
[264] J. T. Yardley, and C. B. Moore, *J. Chem. Phys.*, 1964, **45**, 1066.
[265] J. T. Yardley, M. N. Fertig, and C. B. Moore, *J. Chem. Phys.*, 1970, **52**, 1450.

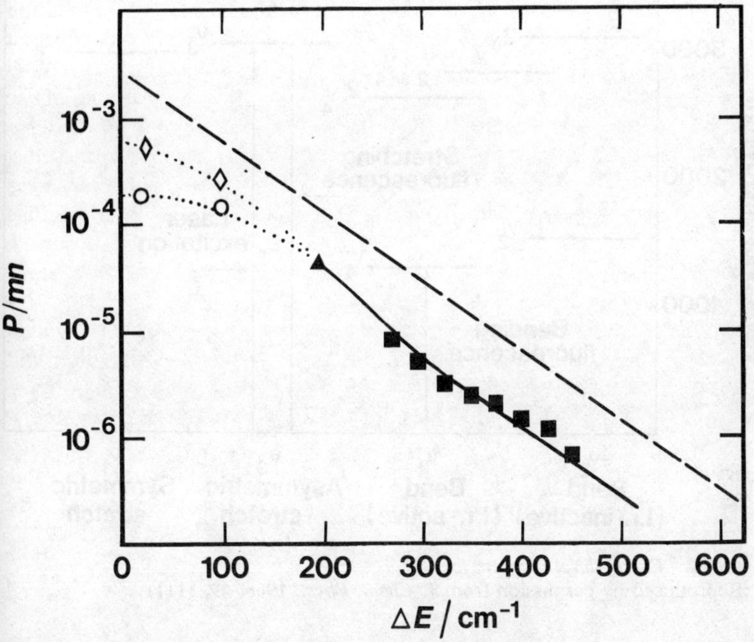

Figure 31 *The reduced probability P/mn for the process:* $N_2(V = n) + CO(V = m - 1) \rightarrow N_2(V = n - 1) + CO(V = m)$ *as a function of* ΔE *for* $11 \geqslant m \geqslant 4$ *and* $n = 1$. *The dotted line is Callear's correlation*
(Reproduced by permission from *Appl. Phys. Letters*, 1972, **21**, 81)

laser oscillating at 2947.9 cm^{-1} and a phase-shift technique were used. The relevant energy-level model is shown in Figure 32. Ge–Cu (20 K) detection was employed for the fluorescence with an RC time constant less than 800 ns for $V \rightarrow (T,R)$ measurement and less than 180 ns for $V \rightarrow V$ measurement.

$V \rightarrow (T,R)$ energy transfer will occur from the two lowest levels ν_2 and ν_4, as they are so close together, so that the pseudo-first-order rate constant observed for $V \rightarrow T,R$ processes, k_{obs}, is given by equation (26), where

$$k_{obs} = (1 + K)^{-1}(k_{\nu_4} + Kk_{\nu_2}) \qquad (26)$$

$K = g(\nu_2)/g(\nu_4)\exp[-h(\nu_2 - \nu_4)/kT]$, $g(\nu_n)$ being the degeneracy of level n. K is the equilibrium constant for $V \rightarrow V$ transfer between ν_2 and ν_4 and thus equation (26) is valid if $V \rightarrow V$ processes are faster than $V \rightarrow (T,R)$.

For the CH_4 model $K = 0.23$ and k_{ν_2}/k_{ν_4} is ~0.2 on the basis of $V \rightarrow T$ or $V \rightarrow R$ theory. Thus $k_{obs} \approx k_{\nu_4}/(1 + K)$ and $1/k_{obs} = p\tau = 1440 \pm 8\%$ μs Torr. Corrections have been applied for translational heating. At pressures below 10—20 Torr wall deactivation was significant. This $p\tau$ value is in agreement with ultrasonic data within experimental error.[140, 266]

[266] P. D. Edmonds and J. Lamb, *Proc. Roy. Soc.*, 1958, **72**, 940.

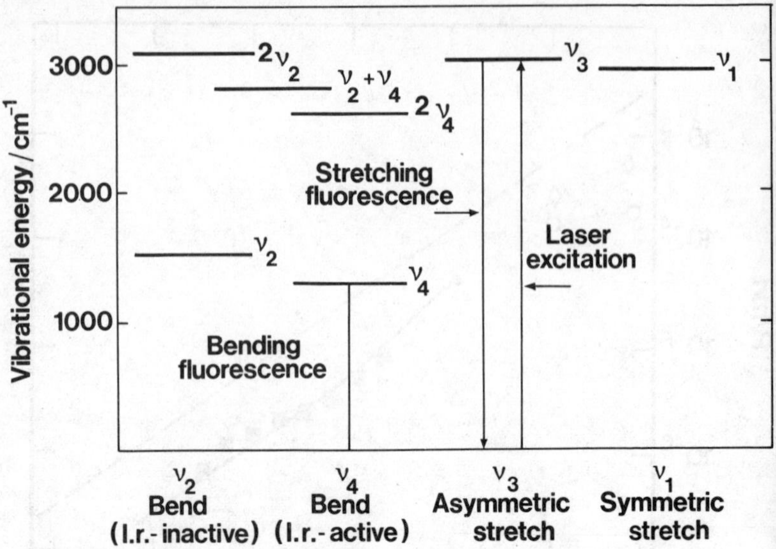

Figure 32 *Energy-level diagram for* CH_4
(Reproduced by permission from *J. Chem. Phys.*, 1968, **49**, 1111)

$V \rightarrow V$ Energy Transfer. $V \rightarrow V$ energy transfer from ν_3 to ν_4 was monitored both by the decay rate of ν_3 fluorescence and the appearance rate of ν_4 fluorescence. The former $p\tau$ value is 5.3 μs Torr and the latter 3.8 μs Torr, the difference being within experimental error ($\pm14\%$ for decay rates and $+60\%$, -40% for appearance rates).

Some experiments were carried out on ν_4 fluorescence in the presence of rare gases. With argon present, $p\tau_{V \rightarrow V}$ was 30 ± 15 μs Torr whereas with neon diluent $p\tau_{V \rightarrow V}$ was calculated as 25 ± 12 μs Torr. Thus the relaxation of ν_4 is 3—10 times slower in rare-gas buffer than in pure CH_4. Rare gases may affect this relaxation time through $\nu_3 \rightarrow \nu_4$ transfer or *via* $\nu_2 \rightarrow \nu_4$ transfer. Thus one should not conclude that the intramolecular $\nu_3 \rightarrow \nu_4$ process is slow compared to the intermolecular process.

From the above data, $V \rightarrow V$ energy transfer from ν_3 to ν_4 seems to be direct, although the experimental error is rather large. As expected, the $V \rightarrow V$ transfer is fast compared with $V \rightarrow T$ transfer. $V \rightarrow V$ transfer from $\nu_2 \rightarrow \nu_4$ requires less than a hundred CH_4–CH_4 collisions and the ν_4 $V \rightarrow T$ transfer needs 1.4×10^4 CH_4–CH_4 collisions.

The observed rates agree with SSH theory modified for $V \rightarrow V$ transfer and corrected for energy-sharing processes. However, this agreement could be fortuitous in view of the inherent approximations.

CH_4 and Collision Partners, $V \rightarrow (T,R)$ Relaxation. As for other relatively complex molecules, the lowest CH_4 frequency will relax by transfer of its

energy into its own rotation and/or translation together with transfer to the collision partner's rotation and/or vibration. Recent results [265] are presented in Table 11. The error in these values is ~ ±10% except for the last, where it is ~ ±15%. The $V \rightarrow T$ theories predict the wrong trend in

Table 11 *Values of $p\tau$ for collisions of methane with other molecules*

Collision partner	$p\tau$/μs Torr
CH_4	1 440
^3He	1 060
^4He	1 750
Ne	8 730
Ar	15 000
Kr	22 200
Xe	27 400
H_2	910
HD	1 830
D_2	685
CO	9 500
N_2	16 000
O_2	6 100
C_2H_6	4.87

range parameter from the above data, whereas elementary $V \rightarrow R$ theory predicts the correct trend but results in values which are too small. It is concluded that the real description lies between these extremes and some sensitive tests are available. Since, for example, $p\tau_{V\rightarrow(T,R)}$ for $CH_4 + {}^4He$ $> p\tau_{V\rightarrow(T,R)}$ for $CH_4 + {}^3He$, the transfer to translation of He is important. The question of distribution of energy in the case of the heavier rare gases, *etc.*, must await a more realistic theory.

Ethylene.—*$V \rightarrow (T,R)$ Energy Transfer and the Kinetic Model.* Ethylene has been used as a passive *Q*-switch for the CO_2 laser [267] and has been examined under powerful CW radiation (10.6 μm up to ~100 W) [101, 204, 268] whereupon it decomposes,[269] principally with the evolution of H_2, CH_4, and C_2H_2. Non-linear absorption by the ν_7 level has also been observed,[270] leading to a proposed 'negative' anharmonicity between the $0 \rightarrow 1$ and $1 \rightarrow 2$ levels of the ν_7 band, *i.e.* $h\nu_{0\rightarrow1} < h\nu_{1\rightarrow2}$. Most of the energy-transfer work on C_2H_4 has been done by Flynn *et al.*[271, 272] using a 1 kW peak-power, *Q*-switched CO_2 laser [*P*(26) only] with pulses 1 μs wide and having a 5 μs risetime, and an In–Sb detector. Non-linear absorption was observed at *P*(14) and *P*(16), so that it is unimportant in this work.

[267] S. M. Lee, L. A. Gamss, and A. M. Ronn, *Chem. Phys. Letters*, 1970, **7**, 463.
[268] R. Vanden-haute and X. de Hemptinne, *Ann. Soc. sci. Bruxelles*, 1971, **85**, 185.
[269] R. T. Bailey and F. R. Cruickshank, unpublished data.
[270] N. W. Ressler and R. W. Crain, *Appl. Phys. Letters*, 1972, **20**, 42.
[271] R. C. L. Yuan and G. W. Flynn, *J. Chem. Phys.*, 1972, **57**, 1316.
[272] R. C. L. Yuan and G. W. Flynn, *J. Chem. Phys.*, 1973, **58**, 649.

The relevant energy levels are shown in Figure 33. A pressure range of 2—30 Torr pure C_2H_4 was examined. The $p\tau$ value for $V \rightarrow (T,R)$ energy transfer was found to be 138 μs Torr from the decay rate of the 5 μm fluorescence, in excellent agreement with the ultrasonics measurement of 140 μs Torr.

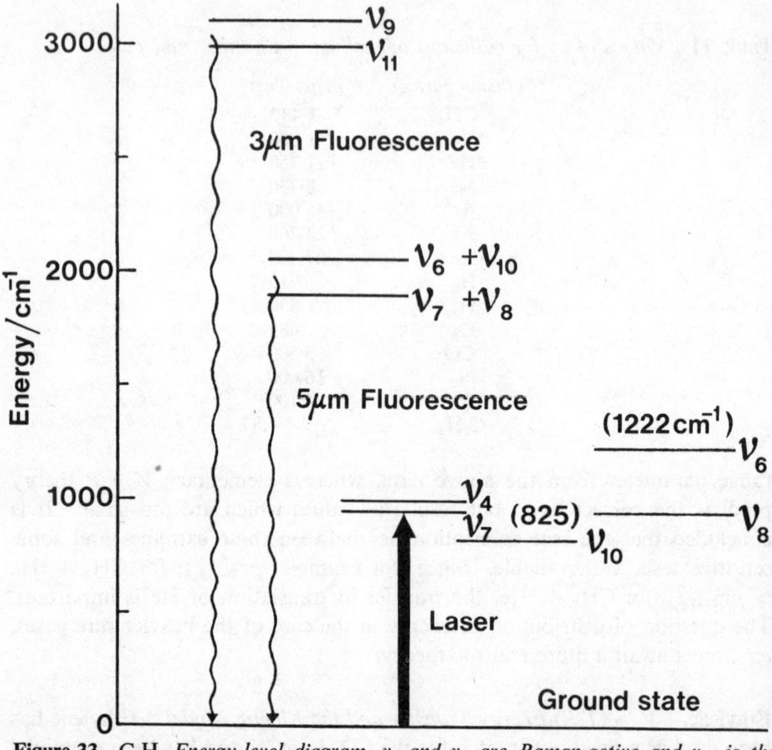

Figure 33 C_2H_4 *Energy-level diagram.* ν_6 *and* ν_8 *are Raman-active and* ν_{10} *is the lowest vibrational state*
(Reproduced by permission from *J. Chem. Phys.*, 1972, **57**, 1316)

As for CH_3F, the fluorescence intensity dependence on laser power absorbed was examined. For 5 μm and 3 μm fluorescence the slopes of ln(fluorescence intensity) *vs.* ln(power absorbed) plots were 2.2 and 3.1, respectively. These are larger than the calculated maximum slopes of 1.5 and 2.5, and Flynn *et al.* claim that the discrepancy is too large to be covered by uncertainties in the extinction coefficient *etc.* Thus they conclude that some element of multiphoton absorption must occur, especially in the population of the higher levels. Harmonics of the ν_7 absorption could also play a leading role, as in CH_3F, together with near-resonant collisions of the type:

$$\underset{(V=1)}{C_2H_4(\nu_7)} + \underset{(V=1)}{C_2H_4(\nu_7)} \rightarrow \underset{(V=0)}{C_2H_4(\nu_7)} + \underset{(V=1)}{C_2H_4(2\nu_7)} + \Delta E$$

where $\Delta E \approx -2\ \text{cm}^{-1}$.

V → V Energy Transfer. There seems only to be an as yet unpublished double-resonance experiment which indicates that $\nu_7 \rightarrow \nu_{10}$ $V \rightarrow V$ energy transfer occurs at ~10 μs Torr. This may be the fast relaxation which was observed in the ultrasonics work at ~69 μs Torr.

C_2H_4 *in Gas Mixtures.* Table 12 shows the $p\tau$ values for $V \rightarrow (T,R)$ transfer from C_2H_4 to various collision partners. Again by an analysis similar to the CH_3F case, $V \rightarrow T$ transfer seems to dominate in the light-partner collisions

Table 12 *Values of $p\tau$ for $V \rightarrow (T,R)$ energy transfer between* C_2H_4 *and various gases*

Collision partner	$p\tau$/μs Torr
^3He	90
^4He	184
Ne	1490
Ar	2130
Kr	2630
Xe	3220
H_2	15.2
D_2	56.7
O_2	700
N_2	1220
CH_4	58.5
C_2H_4	138

and $V \rightarrow R$ in the heavy-partner collisions. The poorer quantitative fit is again underlined and most probably stems from the difficulty of finding accurate potential functions. The potential functions used come essentially from gas viscosity experiments which measure quite different types of interactions and would not be expected to fit energy-transfer calculations without modification.

Nitrous Oxide.—Ronn *et al.*[273] have followed the decay of N_2O (001, 2224 cm^{-1}) fluorescence sensitized by a *Q*-switched N_2O–He–N_2 laser. Peak power was 0.5 kW with a pulse duration of 5 μs at 10.65 μm (001–100 transition). The decay of the 001 fluorescence was monitored at 2220 cm^{-1} with a Au–Ge (77 K) detector. The overall risetime of the detection electronics was < 10 μs, and the pressure range was 1—40 Torr. The $p\tau$ value obtained from the simple exponential decay is 1490 μs Torr and is attributed to intramolecular $V \rightarrow V$ transfer. This value agrees with two spectrophone measurements.[274, 275]

A separate study in similar apparatus [177] yielded the $p\tau$ value 1340 μs Torr ±5%, rather lower than Ronn *et al.*[273] but agreeing with the 1320 μs Torr of Yardley.[276] Table 13 lists $p\tau$ values obtained [276] for various collision

273 R. D. Bates, G. W. Flynn, and A. M. Ronn, *J. Chem. Phys.*, 1968, **49**, 1432.
274 P. V. Slobodskaya and N. F. Tkachenko, *Optika i Spektroskopiya*, 1967, **23**, 480 (*Optics and Spectroscopy*, 1967, **23**, 256).
275 T. L. Cottrell, I. M. MacFarlane, and A. W. Read, *Trans. Faraday Soc.*, 1967, **63**, 2093.
276 J. T. Yardley, *J. Chem. Phys.*, 1968, **49**, 2816.

Table 13 *Values of pτ for nitrous oxide in collision with other gases*

Collision partner	*pτ*/μs Torr
N_2O	1 320
He	3 840
Ne	10 300
Ar	9 720
Kr	18 500
H_2	179
D_2	920

partners, the errors being ±10%. Clearly, H_2 is unusually effective, but it is not possible to distinguish between potential effects, intervention of rotation ($V \rightarrow R$), or a simple mass ($V \rightarrow T$) effect.

As for the CO_2–N_2 system, the N_2O–N_2 transfer rates are in agreement with Sharma and Brau theory, but this may be fortuitous.[276]

Hydrogen.—Mixtures of hydrogen with rare gases are of especial importance in the $V \rightarrow T$ energy-transfer studies. The hydrogen–helium system is one of the few for which *ab initio* calculations of the potential-energy surface have been performed.[144, 277] This is due mainly to the complexity of the interaction potentials and their extreme sensitivity to energy-transfer cross-sections. In addition, recent accurate theoretical calculations for $V \rightarrow (T,R)$ energy transfer in H_2–He collisions [278, 279] have made possible the direct comparison between experiments and theory.

Shock-tube studies of hydrogen and deuterium in mixtures with rare gases have been performed [280–282] above 1400 K but these measurements could not be extended into the lower temperature regime. Since the hydrogen vibration is not i.r.-active, conventional direct laser excitation techniques cannot be used. Vibrational relaxation of pure hydrogen and deuterium has, however, been studied using the stimulated-Raman pumping technique.[117b, 119] Audibert, Joffrin, and Ducuing [117b] studied vibrational relaxation in H_2–rare-gas mixtures using the stimulated-Raman pumping technique with a *Q*-switched, single-mode, ruby laser. The relaxation process was followed by observing the scattering of He–Ne laser radiation. A differential technique which discriminates against side effects such as thermal diffusion or slow volume expansion was used to analyse the data. Pure hydrogen and mixtures with helium and argon were studied. The results are displayed in Table 14 where they are compared with values obtained by a Landau–Teller extrapolation of various high-temperature data. It appears that for H_2–He and H_2–Ar the room-temperature values do not exhibit the expected drop below

277 M. D. Gordon and D. Secrest, *J. Chem. Phys.*, 1970, **52,** 120.
278 D. Secrest and W. Eastes, *J. Chem. Phys.*, 1972, **56,** 2501.
279 W. H. Miller and T. F. George, *J. Chem. Phys.*, 1972, **56,** 5673.
280 J. H. Kiefer and R. W. Lutz, *J. Chem. Phys.*, 1966, **44,** 668.
281 P. F. Bird and W. D. Breshears, *Chem. Phys. Letters*, 1972, **13,** 529.
282 J. E. Dove, D. G. Jones, and H. Teitellbaum, unpublished data.

Table 14 *Vibrational relaxation rates for hydrogen*

$p\tau/\mu s$ Torr	H_2–H_2	H_2–He	H_2–Ar	D_2–D_2
(ref. 119)	0.38 ± 0.04	—	—	1.97 ± 0.40
(ref. 117*b*)	0.50	3.00	9.40	—

the Landau–Teller variation but uncertainties in the extrapolation do not allow definite conclusions to be drawn. The higher values [115, 119] of $p\tau_{H_2-He}$ found in this paper probably resulted from the elimination of extraneous effects. The relaxation rate calculated by Shin [169] for the H_2–He system was about 20 times larger than that measured experimentally. Shin [169] used the Krauss and Mies [144] potential and performed a zero-impact-parameter calculation, taking molecular orientation into account. The discrepancy between Shin's calculations and experiment has been attributed to $V \rightarrow R$ transfer but the three-dimensional distorted-wave calculations of Calvert [283] for pure $V \rightarrow T$ transfer correctly predict the experimental rates for H_2–H_2,[115] D_2–D_2, and D_2–H_2.[184] Thus the role of $V \rightarrow R$ transfer in these systems is uncertain. Kovacs and Mack [119] have also performed vibrational relaxation measurements on pure H_2 and D_2 using the stimulated-Raman pumping technique. In this case, picosecond pulse excitation was employed to limit competition from the stimulated Brillouin process. Their results are included in Table 14. Population changes produced by indirect excitation have been used by Hopkins and Chen [184] in their room-temperature relaxation studies on D_2 and HD mixtures with HCl, excited with an HCl laser to produce vibrationally excited HCl. The excess vibrational energy is rapidly redistributed between HCl and the hydrogen isotope in fast $V \rightarrow V$ processes such as:

$$\mathrm{HCl}(V=1) + \mathrm{D_2}(V=0) \rightarrow \mathrm{HCl}(V=0) + \mathrm{D_2}(V=1) + \Delta E = -104\ \mathrm{cm^{-1}}$$

Following the rapid vibrational energy exchange, both levels are deactivated simultaneously to the ground state by various $V \rightarrow (T,R)$ processes such as:

$$\mathrm{HCl}(V=1) + \mathrm{HCl}(V=0) \rightarrow 2\mathrm{HCl}(V=0) + \Delta E = 2886\ \mathrm{cm^{-1}}$$

$$\mathrm{HCl}(V=1) + \mathrm{D_2}(V=0) \rightarrow \mathrm{HCl}(V=0) + \mathrm{D_2}(V=0) + \Delta E = 2886\ \mathrm{cm^{-1}}$$

$$\mathrm{D_2}(V=1) + \mathrm{HCl}(V=0) \rightarrow \mathrm{D_2}(V=0) + \mathrm{HCl}(V=0) + \Delta E = 2990\ \mathrm{cm^{-1}}$$

which are followed by conventional fluorescence spectroscopy. Relaxation rates and cross-sections for $V \rightarrow (T,R)$ energy transfer were found for D_2–HCl, HD–HCl, D_2–H_2–HCl, and D_2–HD–HCl mixtures. To gain an insight into the mechanism of the energy-relaxation process, probabilities for hydrogen vibrational relaxation were calculated using the simple SSH ($V \rightarrow T$) and Moore ($V \rightarrow R$) [284] theories. These calculations indicated that, at room temperature, $V \rightarrow R$ energy transfer was not the dominant process involved in D_2–hydrogen collisions but did, however, contribute significantly to the

[283] J. B. Calvert, *J. Chem. Phys.*, 1972, **56**, 5071.
[284] C. B. Moore, *J. Chem. Phys.*, 1965, **43**, 2979.

observed probability. Furthermore, $V \rightarrow R$ energy transfer is probably the dominant process in HD self-relaxation since the interaction potential between HD molecules is slightly aspherical, allowing greater amounts of vibrational energy to be coupled into rotation. At room temperature the SSH calculations indicate the $V \rightarrow T$ process to be the dominant relaxation in the case of molecular H_2 and D_2 but not HD.

In a later study, Hopkins and Chen [285] measured the vibrational relaxation rates of hydrogen in ^3He and ^4He mixtures, again using the HCl laser technique. Several features of their results are worthy of note. First, the cross-sections did not depend strongly on collisional reduced mass or velocity as one goes from ^3He to ^4He as collision partners. Secondly, in spite of these differences in electronic structuıe or intermolecular potentials, the cross-sections remain about the same in changing from ^4He to D_2 as collision partner in deactivating vibrationally excited D_2. Finally, in comparing the cross-sections for D_2–^3He and D_2–HD energy transfer, values for HD were found to be six times greater than those for ^3He despite the similar translational reduced masses of both systems. The transfer of vibrational energy into both rotation and vibration in D_2–HD collisions probably accounts for this difference. The fact that only a slight increase in cross-section (about 30%) is observed for D_2–He systems in contrast to the 90—100% change expected from SSH ($V \rightarrow T$) theory further indicates that other processes in addition to $V \rightarrow T$ may be of importance for this system.

9 Summary of the Energy-transfer Work

Several points emerge from the above work. Clearly laser fluorescence is a powerful tool in the study of energy transfer. As in the case of CH_4 it is possible to observe directly $V \rightarrow V$ transfer both in the decay of population in one level and in the appearance of the energy in another level. Multiple relaxations are directly observable with the recent developments in electronics. Even rotational relaxation is becoming directly observable. Never before has such a powerful technique been applied to the energy-transfer problem.

As a result, the theoretical background, much of it based on ideas proposed long before the discovery of the laser, is found wanting. Unfortunately, we feel that there is insufficient consistency in the application of theories for a meaningful comparison to be made of any but the SSH and Sharma and Brau theories. Even in the simple HX systems both theories can be made to fit the experimental data within about an order of magnitude. This is most unsatisfactory and clearly there is too great a freedom of choice of parameters to enable statements to be made with confidence on the nature of the intermolecular potentials.

As detailed above, some elegant experiments have been performed to examine the mass dependence (isotopic substitution) and temperature

[285] B. M. Hopkins and H. L. Chen, *J. Chem. Phys.*, 1973, **58**, 1277.

dependence of the relaxation rates. Unfortunately such work is confined to very few systems and should be greatly extended, since it provides the best means of deciding which theoretical approach is most valid. A more rigorous comparison of the various theories is also desirable. For example, SSH and Sharma and Brau's theories using the *same* mathematical approximations should be applied to *all* available data and the quality of fit examined using different mathematical approximations and different corrections for anharmonicity *etc.*

It may well be that no single theory can cope with the many factors that apparently are necessary. These include anharmonic potential functions, multipolar effects, rotational velocity effects, and the inclusion of second- and higher-order terms to acknowledge the reduction of energy defect by allowing multiquantum rotational transitions. However, no attempt has been made so far to arrive at even the irreducible minimum number of theories.

10 Other Miscellaneous Applications of Laser-induced Fluorescence

Diffusion and Accommodation Coefficients.—The diffusion coefficient for an excited state need not equal that for the ground state since transfer of energy can be involved. Frequently, diffusion coefficients have been measured in the course of energy-transfer work and reports of these have already been included in the appropriate section (Sections 5 and 7).

Values have also been obtained [177] for HCl, N_2O, and CO (this last was not laser-excited). These were 0.125, 0.075, and 0.2 $cm^2\ s^{-1}$ atm for the diffusion coefficients and 0.45 and 0.025 for the accommodation coefficients, respectively. (The accommodation coefficient for CO was not reported.)

It is noted that the diffusion coefficient for N_2O is less than that deduced from gas viscosity (0.109 $cm^2\ s^{-1}$ atm) within the 6% experimental error.

Sensitized Decompositions.—Several workers have tried to excite molecules selectively with i.r. lasers in the hope of achieving non-thermal reactions and novel products. This idea, which resembles the WCM Lewis theory of unimolecular reactions,[286] could only succeed at high power densities. To date, no clear evidence of success in such a project is available, although several reports have appeared.[287]

Most recently, ethyl chloride has been examined,[99] and it has been shown that it is extremely difficult to determine whether a reaction is sensitized or not.

[286] A. Lamble and W. C. M. Lewis, *Trans. Chem. Soc.*, 1914, **105**, 2330; 1915, **107**, 233.

[287] K. L. Kompa, *Z. Naturforsch.*, 1972, **27b**, 89; K. Taki, P. H. Kim, and S. Namba, *Bull. Chem. Soc. Japan*, 1969, **42**, 823, 2377; *ibid.*, 1970, **43**, 1450; *ibid.*, 1970, **43**, 3278; M. F. Goodman, H. H. Seliger, and J. M. Minkowski, *Photochem. and Photobiol.*, 1970, **12**, 355; M. F. Goodman and E. Thiele, *Phys. Rev.*, 1972, **A5**, 1355; N. G. Basov, E. P. Markin, A. N. Oraevskii, A. V. Pankratov, and A. N. Schachkov, *J.E.T.P. Letters*, 1971, **14**, 165; J. L. Lyman and R. J. Jensen, *Chem. Phys. Letters*, 1972, **13**, 421; A. Yogev, R. M. J. Lowenstein, and D. Amar, *J. Amer. Chem. Soc.*, 1972, **94**, 1091; E. Quel and X. de Hemptinne, *Ann. Soc. sci. Bruxelles*, 1969, **83**, 262.

Ethyl chloride has the advantage over previous systems in that its thermal decomposition is well characterized, *viz.*

$$C_2H_5Cl \rightarrow C_2H_4 + HCl,$$

and unimolecular. However, the 10.6 μm CO_2 laser radiation is absorbed by the C—C stretch assigned [288] to the 972 cm^{-1} band. Thus selective excitation should cause C—C cleavage, CH_3 production, and thus CH_4 production. CH_4 is found, but the complicating absorption of laser radiation by C_2H_4 (also produced) and the long induction periods not only invalidate previous work but could allow a thermal mechanism to explain the results.

Further work is currently in progress on this system to resolve the question of a sensitized process, but the technique of using i.r. fluorescence to monitor reaction rates successfully is being expanded to include the examination of other high-temperature processes in an essentially wall-free environment.

Flynn *et al.*[198] have pointed out that the long $V \rightarrow (T,R)$ relaxation time of CH_3F and the low efficiency for deactivation by N_2O, *etc.*, mean that CH_3F can be vibrationally excited for long enough that non-thermal processes could occur. Perhaps the ethyl chloride system will also prove to be such a system.

[288] L. Y. Liang, L. W. Daasch, and J. R. Nielsen, *J. Chem. Phys.*, 1954, **22**, 1293.

5
Infrared Intensities

BY W. B. PERSON AND D. STEELE

1 Introduction

After a promising start in the 1950's and 1960's, the experimental study of i.r. intensities has slowed considerably over the past few years. There are, of course, several reasons for this, but the primary one seems to be the lack of adequate theoretical interpretation of the results. It requires, of course, a considerable effort to obtain an accurately measured experimental intensity. Techniques were developed in the 1950's to measure accurate values for i.r. intensities in the gas phase.[1] However, the difficulty in interpreting these results, together with the considerable effort required to obtain these values, discouraged any major exploitation of these techniques. Instead, applications have remained generally empirical; most of the studies of i.r. intensities during the past few years have been made on solutions or condensed phases. Many of these are empirical studies designed to obtain information about just what does happen to the intensity of a given absorption band when the molecules condense from gas to liquid and then to the solid phase, or when the molecules are dissolved in different solvents. Considerable effort has been devoted, however, to developing refined techniques for the very difficult problems of obtaining accurate i.r. absorption intensities in these condensed phases.

Having made these measurements, it was found that spectacular intensity changes (by factors of 20 or so) may occur for some absorption bands, but that others in the same molecule may not change intensity at all. Attempts to interpret such results are usually in terms of 'physical effects' (to explain the cases where little change in intensity occurs in going from gas to solution, for example) and specific 'chemical effects' (such as hydrogen bonding) when large changes occur. In the past few years there have been some advances in these areas, although such theories are necessarily not very satisfying.

Because of the ease of the experimental work, most intensity studies of the past few years have been made for solutions. As suggested in the preceding paragraph, it is not clear that results from such studies are characteristic only of the solute; intermolecular interaction between solute and solvent may often be important in determining the i.r. intensity. Furthermore, overlapping

[1] J. Overend, in 'Infrared Spectroscopy and Molecular Structure', ed. M. Davies, Elsevier, Amsterdam, 1963, p. 345.

absorption by even the most ideal solvent is likely to interfere with the absorption by other bands of the solute, so that studies in solution are usually made of only one band of the solute. Nevertheless, a great many such studies have been made. Because of the lack of any adequate theoretical interpretation for such results, the authors often grasp at any straw available, attempting correlations with various substituent constants (Hammett or Taft σ, *etc.*), usually ignoring any possible solvent effects on intensities. Again, these crude correlations do often work surprisingly well.

For classification purposes, then, it seems useful to consider the extent of experimental intensity information available for each molecule. We may classify the more complete and therefore potentially useful studies as follows: (i) studies of *all* the i.r.-active fundamental vibrations of the isolated molecule and possibly of its deuteriated derivatives (in the gas phase); (ii) studies in the gas phase that include also the overtones and combination bands, so that *all* the i.r. activity is measured; (iii) empirical studies of all the fundamental vibrations in condensed phases, for comparison with gas-phase values; (iv) studies similar to (iii) for species (such as ClO_4^- ions) that do not exist in the gas phase; (v) studies of the intensities of only one or two characteristic vibrations (for example, the C≡N or C=O stretching frequencies in a whole series of carbonyl or cyanide compounds). In (iii), (iv), and (v), one must usually also consider the changes in frequency of the vibration and any changes in half-intensity width (or other measure of line shape).

In attempting to interpret detailed gas-phase studies [(i) and (ii) above] there are several difficulties. First of all, the normal-co-ordinate transformation matrix (L or L^{-1}, where $\boldsymbol{R} = \boldsymbol{L}\,\boldsymbol{Q}$ defines the transformation from the column vector of internal co-ordinates to the column vector of the normal co-ordinates) is required, since the intensity of the ith vibration is interpreted in terms of the derivative of the dipole moment ($\boldsymbol{p}$) with respect to the ith normal co-ordinate,

$$\partial \boldsymbol{p}/\partial Q_i = \sum_j (\partial \boldsymbol{p}/\partial R_j)(\partial R_j/\partial Q_i) = \sum_j (\partial \boldsymbol{p}/\partial R_j)L_{ji} \tag{1a}$$

or

$$\partial \boldsymbol{p}/\partial R_j = \sum_i (\partial \boldsymbol{p}/\partial Q_i)(\partial Q_i/\partial R_j) = \sum_i (\partial \boldsymbol{p}/\partial Q_i)L^{-1}_{ij} \tag{1b}$$

The transformation from $\partial \boldsymbol{p}/\partial Q_i$ to $\partial \boldsymbol{p}/\partial R_j$ is necessary in order to obtain a well-defined quantity for further interpretation. However, it immediately introduces the necessity not only for the measurement of all vibrations [in order to evaluate the sum in equations (1)] but also for finding accurate normal co-ordinates. This problem is not considered to be part of our review, except to say that it has also proved to be a challenging and difficult one, and that considerable progress has been made during the past 20 years in obtaining accurate normal co-ordinates for small molecules.

The second problem encountered in the interpretation of the i.r. intensities is the question of the sign to be chosen for the experimental value of $\partial \boldsymbol{p}/\partial Q_i$. This difficulty arises because the measurable quantity is the integrated absorption coefficient (A or Γ; see later discussion) related (see ref. 1) to the *square* of transition moment $\sum_i |\langle i+1 | p | i \rangle|^2$

$$\Gamma_i \nu_i \simeq A_i = \frac{8\pi^2 N\nu_i}{3hc} \sum_i |\langle i+1 | p | i \rangle|^2 \tag{2}$$

Under the assumption that electrical anharmonicity is small [*i.e.* $(\partial^2 p/\partial Q_i^2)$ and higher derivatives are negligible] and that the vibration is harmonic, the transition-moment integral can be evaluated, to obtain:

$$A_i = \frac{N\pi g_i}{3c^2} |\partial p/\partial Q_i|^2 \tag{3}$$

Thus, the experimental measurement of A_i will determine only the *magnitude* of $\partial p/\partial Q_i$, and not its sign. As a result, when equation (1) is evaluated, we must consider all possible signs of the $\partial p/\partial Q_i$ values, so that there will be several possible values of each $\partial p/\partial R_j$, depending upon the sign choice. In principle, it is possible to determine $\partial p/\partial R_j$ for several isotopic derivatives of a molecule, as well as for the molecule. We expect that the value of $\partial p/\partial R_j$ will be isotopically invariant; the sign choices of $\partial p/\partial Q_i$ that result in such invariance are considered to be correct. Overend[1] has, however, emphasized the difficulty of such a procedure in practice.

Finally, even if both of these problems with the interpretation are solved, we come to the question of what the results do actually mean. Is there something that can be done with the results *besides* just plotting them against the Hammett σ values?

Because of all the problems in interpretation, we believe the major advance in the study of i.r. intensities during the past few years has been the moderate success in applying relatively simple quantum-mechanical theory to the interpretation of intensity data. In principle it is possible to calculate the transition moment from quantum mechanics; or the dipole moment can be calculated at different R_j values so that an approximate $\partial p/\partial R_j$ can be obtained, for example as follows:

$$\frac{\partial p}{\partial R_j} \simeq \frac{p(R) - p(R_e)}{R - R_e} = \left[\frac{\langle 0|p|0\rangle_R - \langle 0|p|0\rangle_{R_e}}{R - R_e}\right] \tag{4}$$

Here R is the value of R_j after a displacement from its equilibrium value (R_e) along that co-ordinate by, for example, 0.01 or 0.02 Å. In practice such calculations have been too expensive to make with standard *ab initio* SCF (self-consistent field) Hartree–Fock quantum-mechanical methods using large enough basis sets to give reasonably reliable calculations. However, approximate (CNDO) methods[2] have proved to be unexpectedly reliable[3]

[2] J. A. Pople and D. L. Beveridge, 'Approximate Molecular Orbital Theory', McGraw-Hill, New York, 1970.
[3] G. A. Segal and M. L. Klein, *J. Chem. Phys.*, 1967, **47**, 4236.

for predicting dipole-moment derivatives. These methods are so simple and inexpensive to use that a considerable body of results has accumulated to provide at least partial answers to the last two problems mentioned concerning the interpretation of data.

Before attempting to define the scope of this review, we need to say a little about units. First of all, we have chosen in this review to report the integrated molar absorption coefficient A_i, rather than the alternative measure, Γ_i:[1]

$$A_i = \left(\frac{1}{nl}\right)\int \ln(I_0/I)\mathrm{d}\nu \qquad (5)$$

$$\Gamma_i = \left(\frac{1}{nl}\right)\int \ln(I_0/I)\mathrm{d}(\ln \nu) \qquad (6)$$

Here n is the concentration in mol l^{-1}, or in SI units it is in mol dm^{-3}. When the pathlength, l, is in cm, then A_i has the units of dm^3 mol^{-1} cm^{-2}, which are the same as the 'dark' in ref. 1. When using SI units, the integrated absorbance [$\int \ln (I_0/I)\mathrm{d}\nu$] has units of m^{-1}, the concentration may be expressed in mol m^{-3}, and the pathlength in m, so that the SI units of A_i are m mol^{-1}. The conversion factor is 1 'dark' = 1 cm mmol^{-1} = 1 dm^3 mol^{-1} cm^{-2} = 10 m mol^{-1}. The values of A_i vary from about 10^2 to 10^5 m mol^{-1}, so that a more convenient unit of intensity is the km mol^{-1}, with 10^2 'dark' = 10^3 m mol^{-1} = 1 km mol^{-1}. We shall report all values of A_i in units of km mol^{-1}.

The reason we prefer A_i to Γ_i is that we shall usually interpret our results in terms of dipole-moment derivatives $(\partial \boldsymbol{p}/\partial Q_i)$ rather than transition moments $\langle i + 1|\boldsymbol{p}|i\rangle$, even when we recognize electrical and mechanical anharmonicity. Since A_i is just proportional [equation (3)] to $(\partial \boldsymbol{p}/\partial Q_i)^2$, it seems more natural to use this quantity. Furthermore, the value of A_i is related directly to the more usual plot of absorbance *vs.* wavenumber whereas Γ_i is not. Thus, if we know that A_1 is ten times as large as A_2, we know that the area under the plot of absorbance *vs.* wavenumber for A_1 is ten times larger than A_2, regardless of whether or not ν_2 is different from ν_1. For intensity as measured by Γ_i we must weight by the frequency (or wavenumber), according to equations (5) and (2), in order to obtain relative intensities.

The SI units of dipole moment are C m, with 1 e.s.u. cm = 1 D = 3.335 $\times$ 10^{-20} C m. Thus the familiar units of charge (D Å^{-1}) for $\partial \boldsymbol{p}/\partial R_i$ are expressed in C in SI units, with 1 D Å^{-1} = 3.335 $\times$ 10^{-20} C = 3.335 $\times$ 10^{-2} aC. However, it seems quite useful to report the values of $\partial \boldsymbol{p}/\partial R_i$ as multiples of the electronic charge e (e = 4.8 $\times$ 10^{-10} D Å^{-1} = 16.02 $\times$ 10^{-2} aC). In this review we shall report all dipole gradients $(\partial \boldsymbol{p}/\partial R_i)$ in terms of fractions of the electronic charge. If R_i refers to a bending co-ordinate ($R_i = \Delta\alpha$), so that $\partial \boldsymbol{p}/\partial R_i$ is reported in units of D rad^{-1}, we shall convert the dipole derivative by use of a weighting factor r_0 = 1 Å so that it has the dimensions of charge. Thus a reported value for $\partial\mu/\partial\gamma$ is converted, in effect, to $(\partial\mu/\partial\gamma)\cdot(1/r_0)$. The r_0 terms will not be explicitly written.

Three reviews on i.r. intensities appeared in the period 1961—4.[1, 4, 5] These were primarily concerned with measurement and interpretation of the absorption intensities of gases and vapours. That of Gribov and Smirnov[5] was subsequently developed into a small book[6] which is reviewed later in this article. Wexler published an excellent review of the liquid-state measurements up to 1966.[7] In this, emphasis is placed on the analytical applications of absolute intensities. The literature coverage on this subject was comprehensive. Only a very few more-recent papers have been mentioned in later literature reviews, none of which have been devoted specifically to intensities. In view of this long gap and of the very significant advances which have occurred in recent years, the present reviewers aim to cover the literature back to 1968 in reasonable depth, and in addition to discuss certain papers published prior to this date which seem to us to have been given inadequate attention.

2 Theory

One of the most exciting developments in molecular spectroscopy during the past few years has been the increasing sureness with which quantum-mechanical procedures have been used in calculating experimental observables. A general review of *ab initio* methods and some of the results has been given by Schaefer.[8] This survey collects enough data from various *ab initio* calculations and compares them with enough experimental results to indicate the quality of agreement that can be expected. No results are given for vibrational intensities; but the results for force constants and potential curves collected in this survey are good enough to indicate the need by the experimentalist to become acquainted with such calculations. Pople and his co-workers have been exploring the use of limited gaussian basis-set calculations in predicting all kinds of molecular properties.[9] These procedures are intermediate in expected predictive quality between the results from the very simple CNDO (Complete Neglect of Differential Overlap) procedures[10, 2] and the results from detailed SCF calculations, for example as reviewed by

[4] D. Steele, *Quart. Rev.*, 1964, **18**, 21.

[5] L. A. Gribov and V. N. Smirnov, *Soviet Physics Uspekhi*, 1962, **4**, 919.

[6] L. A. Gribov, 'Intensity Theory for Infrared Spectra of Polyatomic Molecules', trans. P. P. Sutton, Consultants Bureau, New York, 1964 (original, Academy of Sciences Press, Moscow, 1963).

[7] A. S. Wexler, *Appl. Spectroscopy Rev.*, 1967, **1**, 29.

[8] H. F. Schaefer, tert., 'The Electronic Structure of Atoms and Molecules' Addison-Wesley, Reading, Massachusetts, 1972.

[9] M. D. Newton, W. A. Lathan, W. J. Hehre, and J. A. Pople, *J. Chem. Phys.*, 1970, **52**, 4064; W. A. Lathan, W. J. Hehre, L. A. Curtiss, and J. A. Pople, *J. Amer. Chem. Soc.*, 1971, **93**, 6377; W. J. Hehre, R. F. Stewart, and J. A. Pople, *J. Chem. Phys.*, 1969, **51**, 2657; R. Ditchfield, W. J. Hehre, and J. A. Pople, *ibid.*, 1971, **54**, 724; see also W. A. Lathan, L. A. Curtiss, W. J. Hehre, J. B. Lisle, and J. A. Pople, *Progr. Phys. Org. Chem.*, to be published.

[10] J. A. Pople and G. A. Segal, *J. Chem. Phys.*, 1965, **43**, S136; J. A. Pople and G. A. Segal, *ibid.*, 1966, **44**, 3289 and others (see ref. 2).

Clementi.[11] While none of these papers has systematically examined the calculation of dipole-moment functions and i.r. intensities, it is quite clear that the number of high-quality calculations can be expected to increase rapidly.

While discussing the general problem of predicting vibrational properties of molecules from quantum-mechanical results, it is appropriate to mention the very impressive effort by Pulay and Meyer [12] who have calculated, in an *ab initio* quantum-mechanical calculation, a set of force constants for ethylene that is in excellent agreement with the latest experimental set.[13]

One of the major problems that discourages the calculation of dipole-moment derivatives for comparison with experimental values is the problem of communication between theoretician and experimentalist. Each must know, for example, what the other means by a value for a dipole derivative. A displacement co-ordinate, R_j, can be well defined by giving a drawing defining co-ordinates, including a definition of the sign of the co-ordinate. This will answer questions such as 'Is a C—H out-of-plane bend for ethylene that is up from the plane of the paper positive or negative?' For $3N-6$ internal co-ordinates there are 2^{3N-6} choices of sign. While most of these are 'obvious', it only takes one misinterpretation to give a different sign. It is *absolutely essential* for such definitions to be included in each paper, both experimental and theoretical. Furthermore, the 'experimental' values of $\partial \boldsymbol{p}/\partial R_j$ depend also on which normal-co-ordinate transformation was used, so that this transformation must also be given. Finally, the internal co-ordinates (R_j) of a molecule are 'molecule-fixed' (they rotate with the molecule as it vibrates), whereas the calculations are usually made in a 'space-fixed' co-ordinate system. As a result, a correction for rotation [14] must be applied to the results before a comparison can be made.

For these reasons it is of interest to examine two new formulations of the relation between band intensities and molecular parameters that appeared about ten years ago. Neither have yet been mentioned in other reviews; their full potential has not yet been explored; the original papers may not be readily available in most libraries; and the terminology used in the original papers is unusual. It seems desirable, therefore, to summarize and compare the formulations.

In the Gribov formulation [6] the total molecular dipole moment is expressed as a sum of dipoles, each of which is oriented along a bond. The author stresses that this expression does not imply the additivity of bond moments, but is just a formal resolution of all contributions to the total dipole moment $\boldsymbol{p}$ into components that are parallel to the bond vectors. Differentiation of the resulting expression (7)

$$\boldsymbol{p} = \sum_k \mu_k \boldsymbol{e}_k \tag{7}$$

[11] E. Clementi, *Proc. Nat. Acad. Sci. U.S.A.*, 1972, **69**, 2942.
[12] P. Pulay and W. Meyer, *J. Mol. Spectroscopy*, 1971, **40**, 59.
[13] J. L. Duncan, D. C. McKean, and P. D. Mallinson, *J. Mol. Spectroscopy*, 1973, **45**, 221.
[14] B. Crawford, jun., *J. Chem. Phys.*, 1952, **20**, 977.

leads to

$$\frac{\partial p}{\partial Q_i} = \left[\tilde{e}\left(\frac{\partial \mu}{\partial r} \,\Big\vert\, \frac{\partial \mu}{\partial \gamma}\right) + \tilde{\mu}\left(\frac{\partial e}{\partial r} \,\Big\vert\, \frac{\partial e}{\partial \gamma}\right)\right]\left(\frac{l_{ri}}{l_{\gamma i}}\right) \tag{8}$$

Here e and μ are column vectors formed from the unit vectors in the bond directions and from the effective bond dipoles [the μ_k of equation (7)]. [A tilde ($\sim$) denotes the matrix or vector transpose.] The array $\left(\frac{\partial \mu}{\partial r} \,\Big\vert\, \frac{\partial \mu}{\partial \gamma}\right)$ is a rectangular matrix of the derivatives of the scalar bond-dipole moments with respect to the set of displacement co-ordinates used in setting up the vibrational problem; r and γ denote the stretching and angle-bending co-ordinates, respectively, and $\left(\frac{l_{ri}}{l_{\gamma i}}\right) \equiv l_i$ is a column vector of the appropriate normal-co-ordinate terms defining Q in terms of the r and γ.

The bond-reorientation terms $\frac{\partial e}{\partial r}$, $\frac{\partial e}{\partial \gamma}$ are shown to be given by

$$\left(\frac{\partial e}{\partial r} \,\Big\vert\, \frac{\partial e}{\partial \gamma}\right) = R^{-1}\{\Delta M^{-1}\tilde{B}G^{-1} - [E \,\vert\, 0]\} \tag{9}$$

R^{-1} and M^{-1} are diagonal matrices of the reciprocal bond lengths and nuclear masses respectively. B is the transformation matrix between internal displacement co-ordinates and cartesian displacement co-ordinates and G is the inverse kinetic-energy matrix. Δ has the dimensions of the number of bonds by number of cartesian co-ordinates (or number of atoms if entries are considered as vectors). The row corresponding to a given bond shows the value of $+1$ in the columns for the terminal atom and -1 in the columns referring to the co-ordinates of the initial atom. E is a diagonal square matrix of bond direction vectors and 0 is the null matrix of dimensions of the number of bonds by the number of deformation co-ordinates less the number of bonds.

The second term on the right-hand side of equation (8) gives the contribution to the dipole-moment derivatives arising not only from angular re-orientations resulting from angular deformations but also from vibrational angular momentum. This clear formulation of the vibrational angular momentum correction is superior to the earlier Crawford 'hypothetical isotope' method.[14, 15] In fact, Barrow and Crawford had derived a formulation similar to Gribov's but unfortunately never published details.[16]

An equation equivalent to (9) is presented[6] for the dipole derivatives in terms of symmetry co-ordinates, and several examples of the application of the theory are followed through. Dipole-gradient expressions for H_2O, ethylene, and ethane are derived in detail, including explicit rotational correction terms. Intensity data are used to derive the electro-optical parameters for a considerable number of additional molecules. By electro-optical parameters we mean the bond dipole moments and their derivatives with

[15] A. D. Dickson, I. M. Mills, and B. Crawford, jun., *J. Chem. Phys.*, 1957, **27**, 445.
[16] G. M. Barrow and B. Crawford, jun., unpublished material.

respect to the deformation co-ordinates. Details for the vibrational angular momentum contribution are not specified in these cases. The systems treated in this way include the methyl halides, dimethylacetylene, propylene, acetonitrile, and hydrogen cyanide.

In Chapter IV of ref. 6 the conditions necessary for 'characteristic vibrational bands' to have characteristic intensities are examined and the theory is applied to the methyl chlorosilanes. Shigorin and his co-workers have used such an approach to explain the intensity behaviour in ethyl tin acetylenes.[17]

The final chapter presents a fascinating and, as far as we are aware, unused theory of i.r. combination bands. By differentiating the dipole-gradient equation (8), expressions are derived for $\partial^2 \boldsymbol{p}/\partial Q_i \partial Q_j$. The interesting feature is that in addition to second derivatives of the effective bond dipoles with respect to the internal co-ordinates, Gribov's expression contains terms depending on the first derivatives. Thus

$$\frac{\partial^2 \boldsymbol{p}}{\partial Q_i \partial Q_j} = \tilde{\boldsymbol{l}}_i \left(\frac{\partial^2 \mu}{\partial R_m \partial R_n} \right) \xi \, \boldsymbol{l}_j + \tilde{\boldsymbol{l}}_i \left[\widetilde{\left(\frac{\partial \mu}{\partial R} \right)} \left(\frac{\partial \boldsymbol{e}}{\partial R} \right) + \widetilde{\left(\frac{\partial \boldsymbol{e}}{\partial R} \right)} \left(\frac{\partial \mu}{\partial R} \right) \right] \boldsymbol{l}_j \tag{10}$$

where R designates an internal deformation co-ordinate and ξ is the matrix

$$\begin{bmatrix} \boldsymbol{e}_1 & & & & & \\ & \boldsymbol{e}_1 & & & & \\ & & \boldsymbol{e}_1 & & & \\ & & & \cdot & & \\ & & & & \cdot & \\ & & & & & \cdot \\ \boldsymbol{e}_2 & & & & & \\ & \boldsymbol{e}_2 & & & & \\ & & \boldsymbol{e}_2 & & & \\ & & & \cdot & & \\ & & & & \cdot & \\ & & & & & \cdot \\ \boldsymbol{e}_m & & & & & \\ & \boldsymbol{e}_m & & & & \\ & & \boldsymbol{e}_m & & & \\ & & & \cdot & & \\ & & & & \cdot & \\ & & & & & \cdot \end{bmatrix}$$

which has the dimensions of (number of bonds multiplied by number of internal co-ordinates) × number of internal co-ordinates. The $(\partial \mu/\partial R)$ and $(\partial \boldsymbol{e}/\partial R)$ represent matrices of bond dipole and bond vector derivatives.

[17] E. A. Gastilovich, D. N. Shigorin, K. V. Zhukova, and A. M. Sklyanova, *Optics and Spectroscopy*, 1970, **28**, 481.

It appears from this that in the approximation of the mechanically harmonic oscillator there is an intensity contribution to a combination band or overtone which depends directly on the first dipole derivatives. The derivation of equation (10) appears straightforward, but the resulting expression seems to violate certain requirements due to the different transformation properties of the first and second terms. For example, the first overtone of the a_{2u} vibration of benzene must belong to the A_{1g} species. As such it is i.r.-inactive. It is readily apparent that the first term of equation (10) will disappear as a consequence of the vectorial additions of the derivatives. Each out-of-plane mode, however, must have a de/dR which is perpendicular to the ring plane. We fail to see how this harmonic term can disappear as symmetry necessitates. Gribov himself claims on page 91 that 'the overtone and combination frequencies arising from non-planar vibrations of planar molecules are independent of the anharmonic term and the polarization is perpendicular to the plane of the molecule'. These statements are clearly incorrect.

A more detailed examination of the theory presented in this final chapter of ref. 6 appears to be necessary.

Biarge, Herranz, and Morcillo [18] formulated the results from i.r. intensity studies of a molecule in terms of a set of atomic 'polar tensors', instead of the usual dipole-moment derivatives with respect to internal co-ordinates. The 'polar tensor' $P_X^{(\alpha)}$ for the αth atom is defined as:

$$P_X^{(\alpha)} = \begin{pmatrix} \partial p_x/\partial x_\alpha & \partial p_x/\partial y_\alpha & \partial p_x/\partial z_\alpha \\ \partial p_y/\partial x_\alpha & \partial p_y/\partial y_\alpha & \partial p_y/\partial z_\alpha \\ \partial p_z/\partial x_\alpha & \partial p_z/\partial y_\alpha & \partial p_z/\partial z_\alpha \end{pmatrix} \tag{11}$$

Here p_x (*etc.*) is the component of the dipole moment along a space-fixed cartesian co-ordinate (x) of the molecule. The co-ordinates x_α, y_α, and z_α locate the αth atom in that co-ordinate system. The total dipole-moment change for small displacements of the nuclei is:

$$\Delta p = \sum_\alpha P_X^{(\alpha)} X_\alpha \tag{12}$$

Here X_α is the column vector of the three cartesian displacement co-ordinates of the αth atom.

This atomic polar tensor has the advantage over corresponding expressions in the internal co-ordinates in that its vectorial character is well defined. It is especially simple, at least in principle, to calculate the polar tensor for an atom from quantum mechanics. For a diatomic AB molecule with z along the AB axis, Sambe [19] has shown that the polar tensor for the A atom, for example, takes the simple form:

$$P_X^{(\mathrm{A})} = \begin{pmatrix} p^0/R & 0 & 0 \\ 0 & p^0/R & 0 \\ 0 & 0 & \partial p/\partial R \end{pmatrix} \tag{13}$$

[18] J. F. Biarge, J. Herranz, and J. Morcillo, *Anales de Quím.*, 1961, **A57**, 81.
[19] H. Sambe, *J. Chem. Phys.*, 1973, **58**, 4779.

Biarge, Herranz, and Morcillo [18] have shown that this is the form expected for the polar tensor of, for example, the H atom in a C—H bond, with the z-axis along the C—H bond, if the bond-moment hypothesis is satisfied. Polarization of an attached atom (or of the central C atom in a hydrocarbon) when the C—H bond is distorted by movement of the H atom causes off-diagonal elements to appear in the polar tensor, which in general is not symmetric.

The polar tensor can be derived from the experimental data as follows (see ref. 18; also see ref. 20, for details). The experimental value of A_i (or Γ_i) is measured for each i.r.-active fundamental. Using equation (2) the value of $\partial p/\partial Q_i$ is deduced. Designating the sign of $\partial p/\partial Q_i$ by σ_i ($\sigma_i = +1$ or -1, depending on the sign choice) we may then write an experimental $(3 \times 3N - 6)$ polar tensor, expressed in terms of the normal co-ordinates:

$$P_Q = \begin{pmatrix} \sigma_1 \partial p_x/\partial Q_1 & \sigma_2 \partial p_x/\partial Q_2 \ldots \sigma_{3N-6} \partial p_x/\partial Q_{3N-6} \\ \sigma_1 \partial p_y/\partial Q_1 & \sigma_2 \partial p_y/\partial Q_2 \ldots \sigma_{3N-6} \partial p_y/\partial Q_{3N-6} \\ \sigma_1 \partial p_z/\partial Q_1 & \sigma_2 \partial p_z/\partial Q_2 \ldots \sigma_{3N-6} \partial p_z/\partial Q_{3N-6} \end{pmatrix} \quad (14)$$

Here, since Q_i will result in a dipole change along one of the principal axes (say t) for molecules with a reasonable amount of symmetry, most of the entries will be zeros, with only the $\sigma_i \partial p_t/\partial Q_i$ values different from zero.

Utilizing the well-known normal-co-ordinate transformation $\boldsymbol{R} = \boldsymbol{LQ}$ (or $\boldsymbol{Q} = \boldsymbol{L}^{-1}\boldsymbol{R}$) we may derive the $(3 \times 3N - 6)$ polar tensor, P_R, expressed in internal co-ordinates:

$$P_R = P_Q L^{-1} \quad (15)$$

Further transformation from the $3N - 6$ molecule-fixed internal co-ordinates to the $3N$ space-fixed cartesian displacement co-ordinates is achieved by using the augmented matrix including the 6 translations and rotations (designated by ρ):

$$\begin{pmatrix} \boldsymbol{R} \\ \hdashline \rho \end{pmatrix} = \begin{pmatrix} B \\ \hdashline \beta \end{pmatrix} \boldsymbol{X} \quad \text{or} \quad \boldsymbol{X} = (A \,\vdots\, a) \begin{pmatrix} \boldsymbol{R} \\ \hdashline \rho \end{pmatrix} \quad (16)$$

Using equation (16), we find

$$P_X = P_R B + P_\rho \beta \quad (17)$$

Here the 3×3 block in the first term corresponding to the co-ordinates of atom α is called the 'vibrational polar tensor' in ref. 18 and ref. 20, and the 3×3 term corresponding to α in $P_\rho \beta$ in equation (17) is the 'rotational polar tensor'. The expression for P_ρ is given by Biarge, Herranz, and Morcillo: [18]

$$P_\rho = \begin{matrix} \\ p_x \\ p_y \\ p_z \end{matrix} \begin{matrix} \rho_{tx} & \rho_{ty} & \rho_{tz} & \rho_{Rx} & \rho_{Ry} & \rho_{Rz} \\ \end{matrix} \begin{pmatrix} 0 & 0 & 0 & 0 & p_z^0/I_y^{\frac{1}{2}} & -p_y^0/I_z^{\frac{1}{2}} \\ 0 & 0 & 0 & -p_z^0/I_x^{\frac{1}{2}} & 0 & p_x^0/I_z^{\frac{1}{2}} \\ 0 & 0 & 0 & p_y^0/I_x^{\frac{1}{2}} & -p_x^0/I_y^{\frac{1}{2}} & 0 \end{pmatrix} \quad (18)$$

[20] J. Morcillo, L. J. Zamorano, and J. M. V. Heredia, *Spectrochim. Acta*, 1966, **22**, 1969.

Here p_x^0 is the x component of the permanent dipole; I_x is the principal moment of inertia about the x-axis, *etc.* Thus, the rotation correction is seen to arise naturally in the transformation from molecule-fixed to space-fixed co-ordinates, and it is very simply evaluated as the matrix product $P_\rho \beta$. If the sign and magnitude of p_0 are known, for example from a CNDO calculation, then the rotation correction in this formulation is readily evaluated.

The resulting sets of atomic polar tensors, $P_X^{(\alpha)}$, have several very interesting properties. For example, neutrality of the molecule requires that the sum of the atomic polar tensors be the 3 × 3 null matrix:

$$\sum_\alpha P_X^{(\alpha)} = 0 \qquad (19)$$

Thus the polar tensor for C in CH_2F_2, for example, is determined from the sum of polar tensors for H and F.

If the cartesian axes for the atomic polar tensor are rotated so that z lies along the bond (for example along the C—H bond in ethylene) then the polar tensor of the end atom (H in this case) might be expected to have the diagonal form of equation (12) if the bond-moment hypothesis were obeyed.

The product of the polar tensor P_Q with its transpose $\tilde{P}_Q$ gives a product matrix $P_Q\tilde{P}_Q$ whose trace is:

$$\mathrm{Tr}[P_Q\tilde{P}_Q] = \sum_i \left[\left(\frac{\partial p_x}{\partial Q_i}\right)^2 + \left(\frac{\partial p_y}{\partial Q_i}\right)^2 + \left(\frac{\partial p_z}{\partial Q_i}\right)^2\right] = \sum_i \left|\frac{\partial \boldsymbol{p}}{\partial Q_i}\right|^2 \qquad (20)$$

Using equation (3), we see that the right-hand sum is just the intensity sum:

$$K\,\mathrm{Tr}[P_Q\tilde{P}_Q] = \sum_i A_i \qquad (21)$$

where $K = N\pi g_i/3c^2$. Applying equation (15), we find that

$$K\,\mathrm{Tr}[P_Q\tilde{P}_Q] = K\,\mathrm{Tr}[P_R L\tilde{L}\tilde{P}_R] = K\,\mathrm{Tr}[P_R G\tilde{P}_R] = \Sigma A_i \qquad (22)$$

Thus we have obtained the Crawford '*G*-sum' rule for intensities. [Note that the internal co-ordinates $\boldsymbol{R}$ may be chosen as symmetry co-ordinates, the G matrix may be for just one symmetry block, and the sum of intensities may be just over the vibrations of one symmetry species.]

In terms of cartesian co-ordinates, application of equation (17) yields:

$$K\,\mathrm{Tr}[P_X M^{-1}\tilde{P}_X] = \Sigma A_i + \Omega \qquad (23)$$

Here

$$\Omega = [(p_x^0)^2 + (p_z^0)^2]/I_y + [(p_z^0)^2 + (p_y^0)^2]/I_x + [(p_x^0)^2 + (p_y^0)^2]/I_z = \mathrm{Tr}[P_\rho\tilde{P}_\rho] \qquad (24)$$

Equations (23) and (24) are a particularly convenient form of the intensity-sum rule derived first in ref. 18, and independently by King, Mast, and Blanchette.[21]

[21] W. T. King, G. B. Mast, and P. P. Blanchette, *J. Chem. Phys.*, 1966, **56**, 4440.

The latter[21] defined an effective charge ξ_α for the αth atom by:

$$\xi_\alpha{}^2 = \mathrm{Tr}[P_X{}^{(\alpha)}\tilde{P}_X{}^{(\alpha)}] = \left|\frac{\partial p}{\partial x_\alpha}\right|^2 + \left|\frac{\partial p}{\partial y_\alpha}\right|^2 + \left|\frac{\partial p}{\partial z_\alpha}\right|^2 \qquad (25)$$

so that equation (17) is:

$$\sum_\alpha \left(\frac{1}{m_\alpha}\right) \xi_\alpha^2 = \sum_i A_i + \Omega$$

We shall say more about these quantities (ξ_α) later.

After developing the theory of the polar tensors, Morcillo and co-workers applied this concept to the interpretation of experimental gas-phase intensity studies of $CHCl_3$, $SiHCl_3$, and $CDCl_3$,[22] for CH_2Cl_2, CH_2F_2, and CH_2Br_2;[20] and for CCl_4 and $SiCl_4$.[23]

The difficulty with the polar tensors is that there are 2^{3N-5} choices for the signs (the σ_i): 2^{3N-6} for the signs of $\partial p/\partial Q_i$ and two choices for the sign of p^0, the permanent moment. In practice, the situation is not quite so bad; the number of distinct polar tensors is the same as the number of distinct sets of $\partial p/\partial S_i$ values (where S_i is a symmetry co-ordinate). (The $\partial p/\partial Q_i$ values for the symmetry species containing x determine the first row of the atomic polar tensors; those for the symmetry species containing y determine the second row; and those for the species containing z determine the third row.) However, the applications by Morcillo[20, 22, 23] do appear to be somewhat confusing because of the usual difficulties with sign choice.

The real value of the polar tensor formulation appears when it is combined with some theoretical treatment to limit the number of possible sign choices. Morcillo and co-workers do this by trying to use chemical intuition to predict the signs of some of the dipole derivatives. For example, the sign of the dipole moment of H_2CF_2 is assumed to be determined by locating the H atoms at the positive end; the sign of $\partial p/\partial Q_{C-F}$ is assumed to be positive (Q_{C-F} is the C—F stretching co-ordinate), *etc.* Such reasoning is probably usually correct; however, if the CNDO method can be used to calculate the signs of the derivative (see below), we may expect to be able to limit consideration to just one or two possible polar tensors. We may expect this approach to be more important in the future.

The beauty of the polar tensor is seen when the changes in dipole are plotted in a figure. In order to illustrate the results, we quote the polar tensor for $CHCl_3$ listed by Morcillo, Biarge, Heredia, and Medina[22] for the sign choice they believe most probable [$\sigma_0 = +1$ ($^+HCCl_3^-$); $\sigma_1 = +1$, $\sigma_2 = +1$, $\sigma_3 = -1$; $\sigma_4 = +1$, $\sigma_5 = -1$, $\sigma_6 = -1$]. (Another sign choice is listed as possible, but the corresponding atomic polar tensors are not very

[22] J. Morcillo, J. F. Biarge, J. M. V. Heredia, and A. Medina, *J. Mol. Structure*, 1969, **3**, 77.

[23] J. Morcillo, M. Lastra, and J. F. Biarge, *Anales de Quím.*, 1961, **A57**, 179; also A. Medina, Ph.D. thesis, University of Madrid, 1966, and J. M. V. Heredia, Ph.D. thesis, University of Madrid, 1965.

different.) The results are calculated from experimental measurements of the gas-phase intensities made by the authors (with one atmosphere N_2 for pressure broadening) and analysed using normal co-ordinates calculated from Decius' force field.[24] The atomic polar tensors for the H, Cl, and C atoms are shown in Table 1. The cartesian co-ordinate axes are chosen with the origin at the C atom, and with the *z*-axis lined up with the C—H bond for $P_X^{(H)}$ and $P_X^{(C)}$; for $P_{X_b}^{(Cl)}$, the cartesian axis is rotated so that the *z*-axis lies along the C—Cl bond.

Table 1 *The atomic polar tensors for atoms in* $CHCl_3$ *(in units e) for one of the 'most probable' choices of sign made by Morcillo, Biarge, Heredia, and Medina* [22]

$$\begin{array}{c} \\ p_x \\ p_y \\ p_z \end{array} \overset{\displaystyle P_X^{(H)}}{\begin{pmatrix} -0.115 & 0 & 0 \\ 0 & -0.115 & 0 \\ 0 & 0 & 0.037 \end{pmatrix}} \quad \overset{\displaystyle P_{X_b}^{(Cl)}}{\begin{pmatrix} -0.091 & 0 & -0.158 \\ 0 & -0.062 & 0 \\ -0.081 & 0 & -0.680 \end{pmatrix}} \quad \overset{\displaystyle P_X^{(C)}}{\begin{pmatrix} 1.24 & 0 & 0 \\ 0 & 1.24 & 0 \\ 0 & 0 & 0.208 \end{pmatrix}}$$

(columns x, y, z)

We see that $P_X^{(H)}$ is forced by symmetry to the simple form of equation (13), but that $P_{X_b}^{(Cl)}$ is somewhat more complicated, indicating that stretching the C—Cl bond (distorting by $+\Delta z_{Cl}$) causes the dipole moment to change

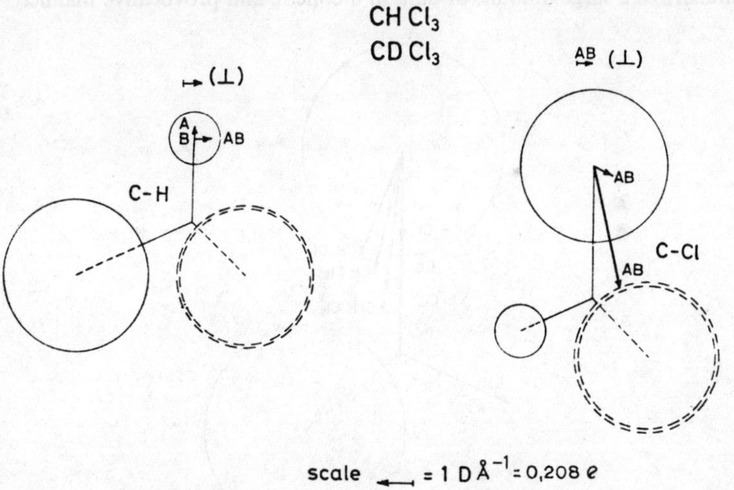

Figure 1 *Changes of the dipole moment in* $CHCl_3$ *and* $CDCl_3$ *due to C—Cl (right) and C—H (left) bond stretching and bending in the plane of the figure. The dipole-moment change due to a bond bending perpendicular to the plane of the figure is shown at top* (⊥)
(Reproduced by permission from *J. Mol. Structure*, 1969, **3,** 77)

[24] J. C. Decius, *J. Chem. Phys.*, 1948, **16,** 214.

both in the x direction (by —0.158 Δz_{Cl}) and in the z direction (by —0.68 Δz_{Cl}). The resultant was plotted in ref. 22 to give the result shown here as Figure 1. This Figure shows the change in dipole as an arrow (AB) when the C—H and C—Cl bonds are stretched or bent. As seen there, stretching the C—H bond ($+\Delta z_H$) results in a slight increase in the magnitude of the dipole moment of the molecule (directed along the $+z$ direction) and stretching the C—Cl bond results in a decrease along the z_b direction. Bending either bond results in a decrease in moment. Thus the Cl atom acts as though its effective charge were negative both for bending and stretching motions; the H atom has a negative effective charge in bending, but a positive effective charge for stretching, if this sign choice for the σ_i's is indeed correct.

For other molecules examined by Morcillo and co-workers (for example, see ref. 20) the sign choices are not easy to determine, and we may expect the application of CNDO techniques to be especially useful. However, this promising work has gone far enough to show its real power, especially for comparing polar tensors (for example for the Cl atom) from one molecule to another. Figure 2 gives an example taken from ref. 22, summarizing the data for the C—Cl stretch from all of the studies of XH_xCl_{4-x} (X = C or Si) compounds by Morcillo and co-workers, showing their best estimate of these values with 'preferred sign choices', *etc.* If these 'best estimates' are confirmed (for example, by CNDO calculations), then it is clear that Figure 2 summarizes a large amount of data in a concise and provocative manner.

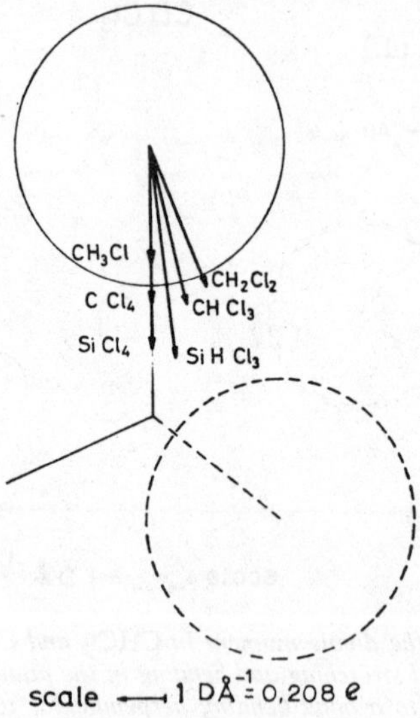

Figure 2 *Changes of the dipole moment due to a* C—Cl *bond stretching in different molecules*
(Reproduced by permission from *J. Mol. Structure*, 1969, **3**, 77)

Another variation on this theme has been given by Girijavallabhan, Joseph, and Venkateswarlu.[25] They define 'intensity constants' as elements of a $(3N - 6)$ square matrix τ that is defined in our notation as:

$$\tau = K\tilde{P}_R P_R = K \begin{pmatrix} (\partial p/\partial r_1)^2 & & & \text{symmetric} \\ \left(\frac{\partial p}{\partial r_1}\right)\left(\frac{\partial p}{\partial r_2}\right) & (\partial p/\partial r_2)^2 & & \\ \vdots & & & \\ \left(\frac{\partial p}{\partial r_1}\right)\cdot\left(\frac{\partial p}{\partial r_{3N-6}}\right) & \left(\frac{\partial p}{\partial r_2}\right)\cdot\left(\frac{\partial p}{\partial r_{3N-6}}\right) & \cdots & \left(\frac{\partial p}{\partial r_{3N-6}}\right)^2 \end{pmatrix} \quad (27)$$

It should be noted that P_R may be considered as a *row* vector with elements $\partial p/\partial R$.

Applying the transformation to symmetry co-ordinates, $S = UR$, they found the 'symmetrized intensity constant matrix', T:

$$T = U\tau\tilde{U} = K\,\tilde{P}_S P_S \quad (28)$$

Continuing, we write

$$K\tilde{P}_Q P_Q = \mathscr{A} = \tilde{L}\tilde{P}_S P_S L = \tilde{L}TL \quad (29)$$

Here $\mathscr{A}$ is the matrix whose diagonal elements are the $3N - 6$ values of the intensity, A_k, and whose off-diagonal terms are the cross-products $K\left(\frac{\partial p}{\partial Q_i}\right)\cdot\left(\frac{\partial p}{\partial_{Qj}}\right)$. Thus, T may be deduced from the intensity matrix $\mathscr{A}$:

$$T = (\tilde{L})^{-1}\mathscr{A}L^{-1}, \quad (30)$$

and then the values of τ are calculated from equation (28) (actually from $\tau = \tilde{U}TU$). The diagonal elements (τ_{ii}/K) of τ/K are the squares of the magnitudes of the vector $\partial p/\partial r_i$, and the off-diagonal elements (τ_{ij}/K) of τ/K are the dot products of the two vectors $\partial p/\partial r_i$ and $\partial p/\partial r_j$, so that

$$\tau_{ij}/K = (\sqrt{\tau_{ii}}\sqrt{\tau_{jj}}/K)\cos\theta_{ij} \quad (31)$$

where θ_{ij} is the angle between the two vectors. Thus, the τ matrix is analysed to give magnitudes of $\partial p/\partial r_i$ and θ_{ij}, the angles between these vectors.

The authors[25] state that the 'intensity-constant matrix' τ is independent of the signs of the $\partial p/\partial Q_i$. However, they have used for the $\mathscr{A}$ matrix just the diagonal elements (the intensity values A_i) and have neglected the cross terms. These latter terms are, of course, dependent upon the signs, and the combination of these terms through equation (30) gives elements of T and ultimately of τ whose magnitudes depend on the choice of signs for $\partial p/\partial Q_i$. Furthermore, the value of θ_{ij} depends upon the choice made for signs. Thus,

[25] C. P. Girijavallabhan, K. Babu Joseph, and K. Venkateswarlu, *Trans. Faraday Soc.*, 1969, **65**, 928.

there is a flaw in the application in ref. 25 to specific molecules because of failure to include these off-diagonal elements of $\mathscr{A}$ and to consider the different sign choices for $\sqrt{\tau_{ii}}, \sqrt{\tau_{jj}}$, but the theory,[25] as modified here, may be useful in manipulating the data from intensity studies.

While on the subject of the analysis of the data from experimental intensity studies we should like to discuss the concept of 'effective charges' as introduced by King, Mast, and Blanchette.[21] As had been emphasized by Biarge, Herranz, and Morcillo,[18] these sum laws [such as equation (26)] are independent of the normal co-ordinates. Thus, King, Mast, and Blanchette used equation (26) with experimental data for the intensities of hydrocarbons to write:

$$\left(\frac{n_H}{m_H}\right)\xi_H^2 + \left(\frac{n_C}{m_C}\right)\xi_C^2 = \Sigma A_i + \Omega$$

Here n_H (n_C) are the number of H atoms (or C atoms) in the hydrocarbon, and m_H (m_C) are the mass of the hydrogen (carbon) atom. If data are also available for the perdeuteriated molecule, then a corresponding equation can be written with n_D and m_D. Solving these two equations, King, Mast, and Blanchette were able to obtain a set of values for ξ_H for these hydrocarbons. The results are summarized in Table 2, where we note the remarkable constancy of ξ_H for these molecules. It seems likely that these values of ξ_H are expected to be somewhat different from one molecule to the next, so that the result for acetylene (for example) in Table 2 probably does not mean that the experimental results are badly in error. Instead, the H atom in acetylene may be appreciably different from those in methane, as measured by the value of ξ_H. It is interesting, for example, to compare the values in Table 2 with the values obtained for the H atom in $CHCl_3$ using the polar tensor in Table 1 with the definition in equation (26) to give ξ_H of 0.17 *e* for this H atom in $CHCl_3$. Clearly, this property is of considerable importance in determining the total i.r. intensity (ΣA_i) for the molecule. Since it can be fairly readily calculated from quantum mechanics, and since it does seem to be relatively constant, we believe it is of considerable interest for future work.

Table 2 *'Effective charges' for the* H *atoms in some hydrocarbons, taken from ref. 21*

Molecule	ξ_H/e [a]
CH_4	0.17 ± 0.01
C_2H_6	0.17
C_6H_6	0.17 ± 0.01
C_2H_4	0.18 ± 0.01
C_2H_2	0.32 ± 0.01
CH_3X (X = Cl, Br, or I)	0.13 ± 0.01

[a] Charges are converted from D Å^{-1} into multiples of the electronic charge *e* by using the factor 1 *e* = 0.208 D Å^{-1}.

The constancy of ξ_H values in Table 2, and the fairly smooth variation in polar tensors for the Cl atom indicated in Figure 2, suggests that it may not be unreasonably optimistic to expect to be able to predict i.r. intensities for unknown molecules from intensity parameters developed from measurements on a few related molecules. Wexler, in his review,[7] has indicated how such attempts might go by listing the data for the vibrations in hydrocarbons (for example) by 'integrated intensities of *structural units* in saturated hydrocarbons' (italics are ours).

The most impressive published attempt of this type was made by Snyder,[26] who attempted to predict the i.r. intensities of crystalline normal paraffins (which are not expected to be much different from the intensities of gaseous n-paraffins) from a set of 'group moment derivatives'. The latter were derived from experimental studies of a few molecules. These parameter values then could be expected to be transferrable to other n-paraffins, so that intensities could be predicted for the latter. Although this study was presented rather cautiously by the author,[26] and although its test against experimental results was hampered by the difficulty in obtaining enough reliable and accurate experimental data, it was remarkably successful. We are surprised that there do not yet appear to have been any other attempts to continue this very promising kind of study, and to extend it to other systems.

Probably the most important activity in the area of theory of i.r. intensities in the past few years has been the efforts (by many people) to calculate the dipole-moment functions (or at least to evaluate $\partial \boldsymbol{p}/\partial R_j$ at the equilibrium configuration) from quantum-mechanical methods. The potential importance of this possibility has been recognized for a long time. In fact the whole reason for measuring i.r. intensities has been to gain information about the change in electronic charge distribution when a molecule distorts in a vibrational mode. The early intensity work was interpreted in terms of the valence theory of the time; unfortunately, the results from the experimental studies were apparently not easily related to the simplified valence theory of the 1950's

One of the first attempts to calculate i.r. intensities was made by Coulson and co-workers;[27] these results were apparently so unsuccessful that Coulson later commented that it would be extremely difficult for quantum theory to predict accurate intensities.[28] Other attempts at that time and later, using simple, and sometimes more sophisticated, theory, led to similar conclusions.[29–31] Even more complicated calculations (for example the SCF calculation of HF by Nesbet[32] and the calculations for CH_4, NH_3, *etc.* by

[26] R. G. Snyder, *J. Chem. Phys.*, 1965, **42**, 1744.
[27] (*a*) N. V. Cohan and C. A. Coulson, *Trans. Faraday Soc.*, 1956, **53**, 1160; (*b*) C. A. Coulson and M. J. Stephen, *ibid.*, 1957, **53**, 272.
[28] C. A. Coulson, *Spectrochim. Acta*, 1959, **14**, 161.
[29] I. M. Mills, *Mol. Phys.*, 1958, **1**, 107.
[30] W. C. Hamilton, *J. Chem. Phys.*, 1957, **26**, 345.
[31] W. T. King, *J. Chem. Phys.*, 1963, **39**, 2141.
[32] R. K. Nesbet, *J. Chem. Phys.*, 1962, **36**, 1518.

Moccia and co-workers[33]) were not very encouraging. The apparent failure of these early attempts to calculate dipole derivatives, is not, of course, too surprising. The calculations were made using rather crude wavefunctions; the errors in the equilibrium dipole moments calculated using those wavefunctions were quite large, so that it is understandable that there should be errors in the calculated derivatives. Other SCF calculations of isolated molecules were also not very encouraging.

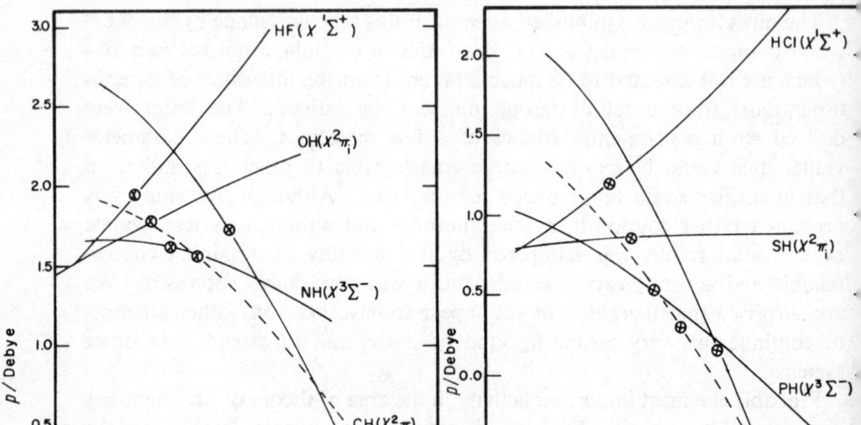

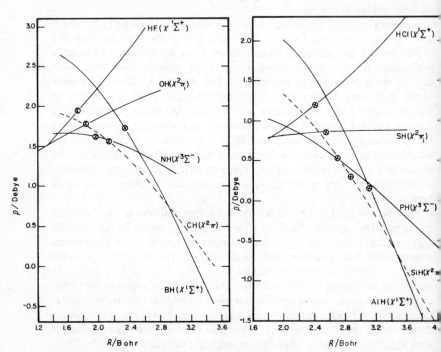

Figure 3 *Calculated dipole-moment functions for first- and second-row diatomic hydrides (from ref. 34). The calculated dipole moment at the calculated equilibrium distance is marked by* ⊗; *the slope of the calculated dipole-moment function at that point is the value of* $\partial p/\partial r$, *in units of* D Bohr^{-1}. *This value is converted to* D Å^{-1} *and to e by:* 1 D Bohr^{-1} = 1.89 D Å^{-1} = 0.394 *e*

The first real progress in this area came with the calculations by Cade and Huo[34] of properties of first- and second-row diatomic hydrides. Among the calculated properties (but not published) was a set of calculated curves of dipole moments as a function of X—H distance.[34c] These calculated curves are shown in Figure 3. Evaluating the dipole moment and the derivatives of

[33] E. Menna, R. Moccia, and L. Randaccio, *Theor. Chim. Acta*, 1966, **4**, 408.
[34] (*a*) P. E. Cade and W. M. Huo, *J. Chem. Phys.*, 1967, **47**, 614; (*b*) *ibid.*, p. 649; (*c*) personal communication from Dr. Cade.

these curves at the calculated X—H equilibrium distance gives the values shown in Table 3 for a few molecules for which comparisons can be made with experimental data. These results are quite encouraging; the calculated values of the dipole derivatives agree with the experimental values well within a factor of two for all the molecules, and the agreement is within about 20% for most, or almost within experimental error. These results renewed one's confidence in the ability of the theory to predict dipole-moment derivatives if one were able to make calculations with a large enough basis set.

Table 3 *Comparison of dipole moments* (p^0) *and dipole-moment derivatives* ($\partial p/\partial r$) *from near-Hartree–Fock (single configuration) wavefunctions with the experimental values for some diatomic* (AB) *molecules.*[a] [p^0 *values*/e Å; $(\partial p/\partial r)/e$.] (1 D Å^{-1} = 0.208 e)

AB	p^0_{calc}	p^0_{exp}	$(\partial p/\partial r)_{calc}$	$(\partial p/\partial r)_{exp}$	
LiH	1.25 [b]	1.22 [b]	0.49 [c]	±0.48 [d]	
HF	0.404 [b]	0.379 [b]	0.42 [e]	±0.33 [f],	0.38 [g]
HCl	0.243 [b]	0.226 [b]	0.30 [e]	±0.18 [h],	+ 0.193 [p]
CO	0.058 [i]	—0.024 [j]	1.00 [k]	±0.73 [j],	0.63 [l]
LiF	1.36 [i]	1.31 [m]	0.92 [i]	±0.91 [m]	
LiCl	1.50 [n]	1.48 [n]	0.91 [n]	±1.00 [n]	
NaF	1.73 [o]	1.69 [o]	0.86 [o]	±0.91 [o]	

[a] The signs in this table are positive if the dipole moment is directed $^+$A—B$^-$, and $\partial p/\partial r$ is positive if that moment increases as r_{AB} increases. Note: this table is not necessarily meant to be conclusive; we report these values to give an indication of the quality of agreement that can be expected for diatomic molecules. [b] Values are from Table I of P. E. Cade and W. M. Huo, *J. Chem. Phys.*, 1966, **45,** 1063. See also 'Compendium of *ab initio* Calculations of Molecular Energies and Properties', compiled by M. Krauss, Technical Note 438, National Bureau of Standards, 1967. [c] Values from I. G. Csizmadia, *J. Chem. Phys.*, 1966, **44,** 1849. [d] Value calculated from data for p^v as a function of vibrational quantum number given by L. Wharton, L. P. Gold, and W. Klemperer, *J. Chem. Phys.*, 1962, **37,** 2149. [e] Values from ref. 34. [f] From G. A. Kuipers, *J. Mol. Spectroscopy*, 1958, **2,** 75. [g] Quoted by E. Clementi, *J. Chem. Phys.*, 1962, **36,** 33. [h] W. S. Benedict, R. Herman, G. E. Moore, and S. Silverman, *J. Chem. Phys.*, 1957, **26,** 1671. [i] From A. D. McLean and M. Yoshimine, 'Tables of Molecular Wave Functions', International Business Machines Corp., San Jose, California. See also Krauss (footnote *b*). [j] B. Schurin and R. E. Ellis, *J. Chem. Phys.*, 1966, **45,** 2528. [k] W. M. Huo, *J. Chem. Phys.*, 1965, **43,** 624. [l] R. C. Herman and K. E. Schuler, *J. Chem. Phys.*, 1954, **22,** 481. [m] L. Wharton, W. Klemperer, L. P. Gold, R. Strauch, J. J. Gallagher, and V. E. Derr, *J. Chem. Phys.*, 1963, **38,** 1203. [n] R. L. Matcha, *J. Chem. Phys.*, 1967, **47,** 4595. [o] R. L. Matcha, *J. Chem. Phys.*, 1967, **47,** 5295. [p] E. W. Kaiser, *J. Chem. Phys.*, 1970, **53,** 1686.

The main reason why more calculated dipole derivatives of the same high quality as those by Cade and Huo [34] are not generally available is that such calculations are quite expensive, so that the early calculations were made at just one configuration of the nuclei—usually the experimental equilibrium configuration. During the past year or so, there have been a number of calculations made with large basis sets for several configurations of polyatomic molecules. Usually such calculations are made to determine the calculated equilibrium configuration, so that the dipole-moment function is still not reported. Very recently, however, detailed calculations of the entire dipole-moment functions for CO and OH were reported at a meeting,[35] and should soon be published.

Even for these complicated Hartree–Fock calculations, however, exact agreement with experimental results cannot be expected. The explanation is the same as that given for the fact that calculated equilibrium dipole moments are always too high; namely, the Hartree–Fock calculations for a single configuration tend to weight ionic structures too highly, and the calculations predict dissociation into ions instead of neutral atoms. As a result we may expect the calculated values of both the dipole moment and the dipole-moment derivatives to be too high, as illustrated in Figure 4 for the HF molecule. The experimental dipole-moment function is determined by the equilibrium dipole moment, by the gradient, and by the expectation that it goes smoothly to zero for this molecule at $R = 0$ and at $R = \infty$; it is indicated schematically by the dashed line. The calculated curve from Cade and

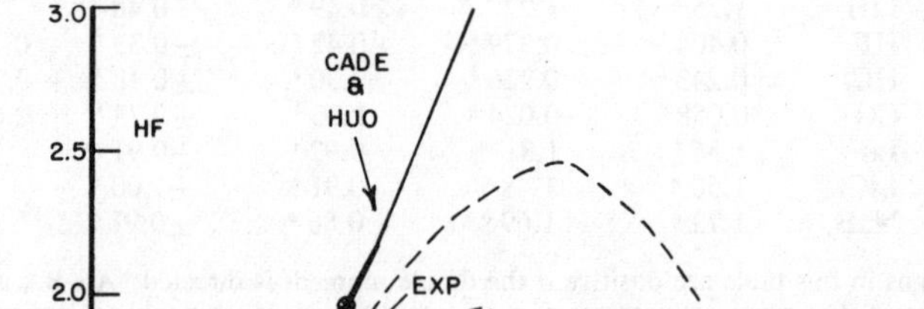

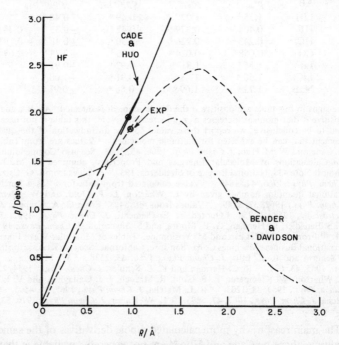

Figure 4 *Comparison of the calculated dipole-moment function (from ref. 27) with the experimental values for* HF. *The dashed line indicates schematically the expected behaviour, while —·—·— indicates the results calculated by Bender and Davidson* (*Phys. Rev.*, 1969, **183**, 23) *with configuration interaction*

[35] (*a*) M. Krauss and F. P. Billingsley, paper Q2; (*b*) M. Krauss, F. Mies, D. Newmann, W. Stevens, and A. C. Wahl, paper V2, Symposium on Molecular Structure and Spectroscopy, June 11—15, 1973, Columbus, Ohio.

Huo [34e] is shown as a full line, with our own extrapolation to zero. We have also shown a curve plotted from a CI SCF calculation by Bender and Davidson,[36] showing that the more complicated computation does predict a decreased dipole-moment derivative for HF (although the predicted decrease from this calculation is too large). Additional calculations including configuration interaction appear from time to time (for example, see ref. 35) and are apparently converging properly to the incompletely determined experimental dipole-moment functions.

One of the more interesting attempts to predict dipole-moment derivatives was made by Garrett and Mills.[37] They derived expressions for force constants and for dipole-moment derivatives of molecules in terms of the Hartree-Fock wavefunctions found at the equilibrium position of the nuclei, and the derivatives of that function. For example, they write for the dipole derivative:

$$\partial \boldsymbol{p}/\partial R_\alpha = Z_\alpha \hat{\boldsymbol{r}}_\alpha - 4 \sum_{i=1}^{n} \left\langle \frac{\partial \phi_i^\mu}{\partial R_\alpha} \middle| \boldsymbol{r}_\mu \middle| \phi_i^\mu \right\rangle \tag{32}$$

Here R_α is a generalized displacement co-ordinate of the αth atom in a direction given by the unit vector $\hat{\boldsymbol{r}}_\alpha$, Z_α is the charge on the αth nucleus, $\boldsymbol{r}_\mu$ is the position vector of the μth electron, and ϕ_i^μ is the ith Hartree–Fock spin orbital associated with the μth electron:

$$\phi_i^\mu(\boldsymbol{r}_\mu) = \sum_{t=1}^{m} \chi_t(\boldsymbol{r}_\mu) C_{ti} \tag{33}$$

Here $\chi_t(\boldsymbol{r}_\mu)$ is the tth basis function, and C_{ti} its coefficient. The derivative of this orbital is:

$$\partial \phi_i^\mu/\partial R_\alpha = \sum_{t=1}^{m} \left[\left(\frac{\partial \chi_t^\mu}{\partial R_\alpha} \right) C_{ti} + \chi_t^\mu \left(\frac{\partial C_{ti}}{\partial R_\alpha} \right) \right] \tag{34}$$

The authors go on to develop general equations for expressing the perturbed Hartree–Fock functions when the αth atom is displaced and to develop general expressions for the dipole derivative [equation (32)] in terms of the known Hartree–Fock wavefunctions evaluated at equilibrium. This subject was recently reviewed by Garrett,[37c] so that additional discussion of the theory will not be given here.

The application of this 'perturbed Hartree–Fock theory' was made [37b] to diatomic hydrides using a minimum-basis-set LCAO SCF MO wavefunction with somewhat disappointing results. Thus, the dipole derivative calculated for LiH was —3.9 *e* and for HF was —0.45 *e* (*cf.* Table 3; note the different sign predicted here [37b] for the dipole derivative of HF). However, this minimum-basis-set calculation was much better for force constants, and the calculation indicated the expected power of the method, given more accurate wavefunctions.

[36] C. F. Bender and E. R. Davidson, *Phys. Rev.*, 1969, **183**, 23.
[37] (*a*) J. Gerratt and I. M. Mills, *J. Chem. Phys.*, 1968, **49**, 1719; (*b*) *ibid.*, p. 1730; (*c*) J. Gerratt, *Ann. Reports*, 1969, **65**, *A*, 3.

This procedure [37] has recently been applied to calculate force constants and dipole derivatives of LiH and BH by Yde, Thomsen, and Swanstrom,[38] with quite satisfactory results. A related procedure has been used by Pulay and Meyer, who have recently published near-Hartree–Fock calculations of force constants and dipole derivative ($\partial p/\partial S_j$, where S_j is a symmetry co-ordinate) for CH_4[39a] and NH_3,[39b] using several different fairly large gaussian-type lobe-function basis sets. They found convergence of calculated values, both for force constants and for dipole derivatives, with a reasonable size for the basis set. Their results are compared with experimental values in Table 4, and with some unpublished results, obtained by Person from an analysis of the near-Hartree–Fock wavefunctions from a calculation by Body, McClure, and Clementi.[40] Another calculation of dipole-moment derivatives and electro-optical parameters has been reported by Lazarev and Kovalev, using wavefunctions published by Duncan.[41] They report even better agreement with experiment than that shown here in Table 4 for these more recent computations. All these results are considerably improved over the earlier one-centre wavefunction calculations.[33]

In general the results in Table 4 are quite encouraging. The results calculated by Meyer and Pulay [39a] for CH_4 appear quite definitely to lead to a preferred choice of signs for $\partial p/\partial Q_3$ and $\partial p/\partial Q_4$ that is (— +). This conclusion is in contrast with that drawn from the CNDO results for CH_4. However, the magnitude of $\partial p/\partial S_4$ is so small that it may not be surprising that the calculated results are in disagreement. Certainly, the near-Hartree–Fock wavefunctions used by Meyer and Pulay are expected to give nearly correct results, as indicated by the agreement with the experimental values shown in Table 4. For NH_3, the calculated results for $\partial p/\partial S_1$ are all lower than the experimental value by about a factor of 10. A possible explanation has been offered by Segal and Klein, who have explored the dipole moment of NH_3 as a function of the S_1 symmetry co-ordinate. They find that the calculated dipole moment goes through a maximum near the equilibrium NH bond length, so that the calculated value is very sensitive to the value assumed for the equilibrium NH bond length. While this sensitivity of the dipole-moment function may not be found for the better calculations,[39, 41] it remains an attractive possible explanation.

For the other vibrations, the agreement between the Pulay–Meyer calculation [39b] and the experimental results is quite satisfactory.

Perhaps one other attempt to use near-Hartree–Fock wavefunctions to calculate dipole derivatives should be mentioned. Bruns and Person used

[38] P. B. Yde, K. Thomsen, and P. Swanstrom, *Mol. Phys.*, 1972, **23**, 691.

[39] (*a*) W. Meyer and P. Pulay, *J. Chem. Phys.*, 1972, **56**, 2109; (*b*) P. Pulay and W. Meyer, *ibid.*, 1972, **57**, 3337.

[40] R. G. Body, D. S. McClure, and E. Clementi, *J. Chem. Phys.*, 1968, **49**, 4916. Dipole-moment data from this calculation were supplied to W. B. Person who obtained the dipole-moment derivatives listed in Table 4.

[41] A. G. Lazarev and I. F. Kovalev, *Optics and Spectroscopy*, 1971, **31**, 367. Their wavefunctions were from A. B. F. Duncan, *J. Chem. Phys.*, 1957, **27**, 423.

Table 4 *Comparison of dipole-moment derivatives calculated from near-Hartree–Fock wavefunctions for* CH_4 *and* NH_3. *(Units are e)*

		CH_4					NH_3			
				exp [c]						
Symmetry co-ordinate derivative		*calc* [a]	*calc* [b]	(− +)	(− −)		*calc* [d]	*calc* [e]	*calc* [b]	*exp* [f]
$\partial p/\partial S_1$ (XH stretch)	(a_1)	0	0	0	0	(a_1)	−0.003	−0.004	−0.019	∓0.083
$\partial p/\partial S_2$ (XH bend)	(e)	0	0	0	0	(a_1)	−0.388	−0.50	−0.24	∓0.32
$\partial p/\partial S_3$ (XH stretch)	(f)	−0.202	−0.13	**∓0.16**	∓0.14	(e)	+0.043	—	—	±0.037
$\partial p/\partial S_4$ (XH bend)	(f)	−0.068	+0.050	**∓0.067**	±0.073	(e)	+0.091	—	—	±0.072

[a] From ref. 39*a*. [b] Calculation by CNDO methods, from ref. 3. [c] From E. T. Ruf, M.S. thesis, University of Minnesota 1959; see also J. Heicklen, *Spectrochim. Acta*, 1961, **17**, 201. Results obtained using normal co-ordinates from ref. 29. Preferred sign choice (from experimental studies) shown in bold face. [d] From ref. 39*b*. [e] From an unpublished analysis of the a_1 vibrations by Person, using wavefunctions by Body, McClure, and Clementi (ref. 40). [f] From D. C. McKean and P. N. Schatz, *J. Chem. Phys.*, 1956, **24**, 316.

published near-Hartree–Fock wavefunctions to calculate dipole-moment derivatives for N_2O and for HCN.[42] Comparison of the Hartree–Fock and CNDO results calculated for these molecules with the experimental values is similar to that shown for CH_4 and NH_3 in Table 4. The results for N_2O from both calculations agreed in determining that the signs of $\partial p/\partial Q_1$ and $\partial p/\partial Q_3$ are both negative. The two sets of calculated values for $\partial p/\partial r_{NN}$ and $\partial p/\partial r_{NO}$ bracketed the experimental values, with one calculated value about 50% higher and the other lower than the experimental value by about the same amount. For HCN the results were more discouraging, with an excellent agreement between the calculated (Hartree–Fock) and experimental *magnitudes* of $\partial p/\partial r_{CH}$ and $\partial p/\partial r_{CN}$, but the calculated derivatives both had the same sign, while the experimental values were found (by studies of the intensities of HCN and DCN by Hyde and Hornig [43]) to have opposite signs. The magnitude of $\partial p/\partial r_{CH}$ calculated by the CNDO method was markedly smaller than the experimental values. The conclusion was reached [42] that the Hartree–Fock wavefunctions were still not necessarily accurate enough to predict derivatives with absolute confidence. However, a possible reason for the difficulty with HCN (namely, an unbalanced basis set) suggests that near-Hartree–Fock calculations designed specifically to predict dipole derivatives could probably be expected to be more reliable.

Although progress during the past few years in the *ab initio* calculation of dipole-moment functions with 'near-Hartree–Fock' wavefunctions has been spectacular and promises to be even better in the future, the most useful theoretical work at present appears to be the calculations based on the CNDO methods. These techniques were developed by Pople and co-workers.[2, 10] Their success in predicting equilibrium dipole moments [2] suggested their application to calculate dipole-moment derivatives, first done by Segal and Klein.[3] The latter reported a survey of about 21 calculated dipole derivatives from 11 different small polyatomic and diatomic molecules, compared with the corresponding experimental values. The calculated values of $\partial p/\partial S_j$ for NO, CO, the symmetric stretch and the bend of CO_2, the bending modes of H_2O and HCN, the a_1 modes of NH_3 and CH_3F, all i.r.-active modes of CH_4 and C_2H_4, the symmetric bend of BF_3, and for S_3 and S_5 of acetylene, were in surprisingly good agreement with experimental values. The computer program for carrying out CNDO calculations is readily available (see refs. 2 and 44) and a calculation for a molecule as large as benzene (or even hexafluorobenzene) is relatively inexpensive. Thus one can afford to carry out the full calculation of the wavefunctions and dipole moments at several different nuclear configurations in order to obtain the $\Delta p/\Delta S_j$ values of equation (4). Thus, the experimentalist now has avail-

[42] R. E. Bruns and W. B. Person, *J. Chem. Phys.*, 1970, **53**, 1413.
[43] G. E. Hyde and D. F. Hornig, *J. Chem. Phys.*, 1952, **20**, 647.
[44] The CNDO program is available from the Quantum Chemistry Program Exchange, University of Indiana, Bloomington, Indiana, in the original version as QCPE 91 CNDO/2 by G. A. Segal, or in a revised and expanded version as QCPE 141 CNINDO: CNDO and INDO Molecular Orbital Program (FORTRAN IV), by P. A. Dobosh.

able a theoretical tool for helping to understand intensities that is just as easy to use as were the older arguments about resonance and rehybridization, and considerably more reliable.

The first application of this technique to aid in the interpretation of experimental i.r. intensity studies was made independently and simultaneously by Segal and by Bruns and Person (and published [45] jointly), attempting to determine the correct relative signs of $\partial p/\partial Q_i$ values found experimentally by Hopper, Russell, and Overend [46] in a gas-phase i.r. intensity study of F_2CO. The calculated values of $\partial p/\partial S_j$ for all i.r.-active modes apparently agreed with the experimental values for only one of the possible choices for the signs of $\partial p/\partial Q_i$, presenting strong theoretical support for preferring that sign choice.

This apparently straightforward conclusion was questioned by McKean, resulting in a detailed re-examination of the F_2CO problem.[47] Several inconsistencies were found in the analysis of experimental data: differences in the definitions of symmetry co-ordinates, in relations between $\partial p/\partial S_j$ and $\partial p/\partial R_k$'s (the derivatives in internal co-ordinates), and in a number of other places. The many possible inconsistent sign choices that can occur in the analysis between the experimental measurement and the calculated $\partial p/\partial S_j$ values became painfully obvious. These difficulties had been realized before, but one of the major contributions of the application of CNDO methods to predict dipole derivatives was to provide values that could be checked against experimental results and so reveal any inconsistencies. The 'correction' of the F_2CO calculation [47] was three times as long as the original note,[45] but it also presented $\partial p/\partial R_k$ values and attempted to compare $\partial p/\partial R_k$ values in a useful way. The only error that was *still* uncorrected in ref. 47 was the one made, originally apparently as a typographical error, in Hopper's thesis, changing (again) the assignment of the overlapping pair, ν_3 and ν_5, from that currently accepted [48] ($\nu_3 = 580\ \text{cm}^{-1}$; $\nu_5 = 620\ \text{cm}^{-1}$) back to the original assignment.[49] The only effect of this change was to interchange the intensities A_3 and A_6 and thus the values of $\partial p/\partial Q_3$ and $\partial p/\partial Q_5$, since the original analyses [45–47] had used the normal-co-ordinate transformation that corresponded to the accepted assignment.[48] The values for $\partial p/\partial S_j$ of F_2CO now believed to be correct (corresponding to the assignment in ref. 48), based on the analysis of experimental data as reported previously [46, 47] but with corrected values of A_3 and A_5, are listed in Table 5 and compared with the values from the CNDO calculation; [47] the earlier experimental values [47] are given in parentheses. Since the changes are rather small, perhaps no further 'corrections' are needed.

[45] G. A. Segal, R. Bruns, and W. B. Person, *J. Chem. Phys.*, 1969, **50**, 3811.
[46] M. J. Hopper, J. W. Russell, and J. Overend, *J. Chem. Phys.*, 1968, **48**, 3765.
[47] D. C. McKean, R. E. Bruns, W. B. Person, and G. A. Segal, *J. Chem. Phys.*, 1971, **55**, 2890.
[48] J. Overend and J. R. Scherer, *J. Chem. Phys.*, 1960, **32**, 1296; see also V. W. Laurie and D. T. Pence, *J. Mol. Spectroscopy*, 1963, **10**, 155.
[49] A. H. Nielsen, T. G. Burke, P. J. H. Woltz, and E. A. Jones, *J. Chem. Phys.*, 1952, **20**, 596.

Table 5 *Corrected*[a] *experimental values of* $\partial p/\partial S_j$ *for* F_2CO *compared with previously published results.*[b] *(Units have been converted from* D Å^{-1} *to e, shown here)*

		Exp.[c]	*Calc.*[b]
A_1			
(signs of $\partial p/\partial Q_i$	$\partial p/\partial S_1$ (CO str.)	—0.77 (—0.86)	—1.13
are — + —)	$\partial p/\partial S_2$ (CF str.)	+0.92 (0.81)	+0.96
	$\partial p/\partial S_3$ (CF bend)	—0.36 (—0.38)	—0.40
B_1			
(signs are — —)	$\partial p/\partial S_4$ (CF str.)	1.07 (1.09)	1.06
	$\partial p/\partial S_5$ (CF bend)	—0.29 (—0.26)	—0.49
B_2			
(—)	$\partial p/\partial S_6$ (CF bend)	—0.37 (—0.37)	—0.64

[a] Corrections made by J. H. Newton and W. B. Person; unpublished results reported as paper H3, Symposium on Molecular Structure and Spectroscopy, June 12—16, 1972, Columbus, Ohio. [b] Ref. 47. [c] The corrected values obtained as described in the text are compared with the previous values in parentheses.

One additional word of warning may be helpful. The calculations of '$\partial p/\partial S_j$' have not always been made using proper definitions of symmetry co-ordinates involving displacements of all atoms involved in a bending co-ordinate, for example. Thus, $\partial p/\partial S_3$ is calculated for a displacement of each F atom perpendicular to the CF bond to give an increase in the FCF angle of 2°. These imprecise definitions are not expected to lead to very serious errors in the calculated values of $\partial p/\partial S_j$, especially for non-polar molecules, or for cases where proper rotational corrections are made to these calculated 'S_j' co-ordinates, but a reader does need to be aware of this difficulty with most of the published CNDO calculations.

Although the history of this application of CNDO theory to the dipole derivatives of F_2CO has been plagued by this series of almost comic errors, it does illustrate very well the kind of difficulty that is typical of this work. It is difficult to overstate the necessity of very careful work, preferably by one man to avoid inconsistencies, all the way from the original experimental data (the A_i values) through the normal-co-ordinate analysis to the $\partial p/\partial S_j$ values, and on to the $\partial p/\partial R_k$ values, which are then compared with the values calculated from CNDO (or *ab initio*) methods. For this reason it is not sufficient just to calculate and publish a general survey of results such as that given by Segal and Klein,[3] but instead one must very carefully and critically re-examine the entire problem. Again we emphasize that one of the major contributions from the CNDO calculation is to provide an easy prediction that causes the analysis of experimental data to be re-examined.

The agreement shown in Table 5 between the final experimental results for F_2CO and those from the CNDO calculation is perhaps misleadingly good. We have already seen in Table 4 for CH_4 that the CNDO calculation apparently predicts the wrong sign for $\partial p/\partial S_4$, the derivative with respect to the CH_4 bending motion. A number of tests have been made, comparing the CNDO-calculated results with experimental values where relative signs of the $\partial p/\partial Q_i$'s are known from studies of the intensities of the molecule and its deuteriated derivatives (see ref. 1 for a discussion of this procedure). Besides the previously discussed results for CH_4 and HCN, the comparison for CH_4, ethylene, ethane, and acetylene has been given by Lewis and Levin;[50] tests of the ability of the theory to predict the correct sign choice have also been made for ethane by Kondo and Saëki[51] and for benzene.[52] The comparison between the CNDO calculation for CH_4 and the experimental results is not conclusive, because the experimental error in the determination of relative signs from the study of the intensities of CH_4 and its deuteriated derivatives quoted in Table 4 was too large to permit a clear-cut choice. Nevertheless, the comparison with the Meyer and Pulay near-Hartree–Fock calculation[39a] suggests that the signs indicated by the CNDO calculation for CH_4 are incorrect. However, the CNDO calculation successfully predicts the correct sign choice for the a_1 vibrations of NH_3 (shown in Table 4).

Table 6 presents the comparison between the values of $\partial p/\partial S_j$ calculated by the CNDO method with the four different sets of values from experiment for benzene.[52] The experimental study of the intensities of C_6H_6 and of C_6D_6 leads to the conclusion that the values in bold-face type in Table 6 (or the set obtained from these by multiplying each term by —1) are correct (see ref 1). We see that this set is also in the best agreement with the CNDO calculation. Comparisons such as those in Table 6 (including also some additional unpublished results by Bruns and Person) and other comparisons[50, 51] support the conclusion that the relative signs of $\partial p/\partial S_j$ values calculated by CNDO methods are expected to agree with the experimental values in the great majority of cases.

Table 6 *Comparison of experimental values of $\partial p/\partial S_j$ and $\partial p/\partial R_k$ for benzene with values calculated by CNDO methods.*[a] *(Units are e.)*

Signs of $\partial p/\partial Q_i$	e_{1u}: $\partial p/\partial S_{18a}$ (CH bend)	e_{1u}: $\partial p/\partial S_{19b}$ (ring stretch)	e_{1u}: $\partial p/\partial S_{20a}$ (CH stretch)	a_{2u}: $\partial p/\partial S_{11}$ (CH bend, oop)
+ + +	**+0.110**	**+0.077**	**+0.179**	—
+ + −	+0.108	+0.112	−0.179	±0.298
+ − +	−0.021	+0.269	+0.200	—
− + +	0.023	−0.304	+0.161	—
CNDO	+0.104	+0.088	+0.359	+0.167
	$\partial p/\partial \gamma$ (CH bend, oop)	$\partial p/\partial \beta$ (CH bend, ip)	$\partial p/\partial R$ (C=C stretch)	$\partial p/\partial r$ (CH stretch)
+ + +; +	+0.127	+0.067	+0.004	−0.104
CNDO	+0.071	+0.062	0.010	−0.206

[a] See ref. 52 for definitions. Preferred values (see ref. 1) are in bold-face type.

[50] T. P. Lewis and I. W. Levin, *Theor. Chim. Acta*, 1970, **19**, 55.
[51] S. Kondo and S. Saëki, *Spectrochim. Acta*, 1973, **29A**, 735.
[52] R. E. Bruns and W. B. Person, *J. Chem. Phys.*, 1972, **57**, 324.

Segal and Klein [3] have explained that a dipole-moment gradient calculated by CNDO methods is given as a sum of three terms: (i) a contribution $\partial p_1/\partial r$ ($= q_0$) due to the equilibrium charge, q_0, on the atoms involved in the co-ordinate r; (ii) a contribution $\partial p_2/\partial r$ ($= \partial p/\partial r$) due to the *change* in charge on those atoms when r changes; and (iii) a contribution $\partial p_3/\partial r$ due to the change in atomic *sp*-type polarization when r changes. When the CNDO calculation indicates that all three of these contributions have the same sign (or when one of these terms is much larger than the others), then we may expect the calculated values to have correct signs. Such is the case for the $\partial p/\partial S_j$ values for benzene listed in Table 6. Thus, it helps us to evaluate the reliability of the CNDO calculated values if these various contributions are listed.

The values of $\partial p/\partial S_j$ can be converted to dipole gradients with respect to the internal co-ordinates, $\partial p/\partial R_k$. The values for benzene are listed also in Table 6. We note there that $\partial p/\partial r_{CH}$ is negative. This prediction that the H atom will become less positive (or that the dipole in the sense $^+$C—H$^-$ will increase) as the C—H bond distance increases is contrary to some kinds of chemical intuition, and is one of the more surprising results from the CNDO calculation. However, it is typical of the C—H bond in hydrocarbons. The negative sign comes from the relatively large negative value of $\partial p_2/\partial r_{CH}$, which overwhelms the usually quite small values of $\partial p_1/\partial r_{CH}$ and $\partial p_3/\partial r_{CH}$. For fluorocarbons, however, the value of $\partial p_1/\partial r_{CF}$ is dominant, and the CNDO theory predicts that the CF dipole increases in the sense $^+$C—F$^-$ as the CF bond stretches, in better agreement with chemical intuition.

For the bending co-ordinates [β, the in-plane (ip)C—H bend, and γ, the out-of-plane (oop)C—H bend] the CNDO calculation indicates major contributions to the dipole gradients from $\partial p_2/\partial\beta$ (—0.065 *e*) and from $\partial p_3/\partial\beta$ (+0.131 *e*) and $\partial p_3/\partial\gamma$ (+0.073 *e*).[52] The magnitude of $\partial p_3/\partial\gamma$ is of some special interest, since this quantity corresponds to a 'rehybridization moment' that is believed (by chemical intuition) to be responsible for the observed high intensity of out-of-plane C—H bending vibrations of benzene [53] and related compounds. From the comparison of experimental values (see Table 6) of $\partial p/\partial\beta$ and $\partial p/\partial\gamma$, Steele and Wheatley [53] suggested that the magnitude of the rehybridization moment was +0.060 *e*. Jalsovszky and Orville-Thomas [54] attempted to evaluate this rehybridization moment for benzene and HCN by examining the CNDO calculated value of $\partial p_3/\partial\gamma$; they were pleased that the calculated value (see above) agrees with the magnitude expected from experiment.[53] However, an even larger contribution from this term is calculated for the *in-plane* CH bending motion ($\partial p_3/\partial\beta$) (see above) so that the CNDO calculation does not really support the existence of a special 'rehybridization moment' for the out-of-plane bending motion. It seems quite possible that the CNDO calculation may give the correct order

[53] D. Steele and W. Wheatley, *J. Mol. Spectroscopy*, 1969, **32**, 265.
[54] G. Jalsovszky and W. J. Orville-Thomas, *Trans. Faraday Soc.*, 1971, **67**, 1894.

of magnitude for the total dipole derivatives, but that the finer breakdown into the different contributions is less meaningful.

An attempt was made also by Galabov and Orville-Thomas [55] to compute the rehybridization moment for ethylene. Again this term ($\partial p_3/\partial \alpha$) was found to be quite large for all three i.r.-active C—H bending motions in ethylene.

Most of the CNDO results discussed so far have been obtained with the parameters determined by Pople and Segal.[56, 2] Lewis and Levin [50] and also Kondo and Saëki [51] have examined the effect of varying these parameters to attempt to obtain better fits to the experimental derivative for hydrocarbons. The re-parametrization was slightly different in each case, but it may be significant that the new values for the β_{AB} parameters chosen as better by Kondo and Saëki [51] are nearly the same as the ones determined independently by del Bene and Jaffé [57] to fit the u.v. spectra of benzene, pyridine, and diazines.

After having surveyed the comparison between $\partial p/\partial S_j$ values calculated from CNDO methods and the experimental values, Lewis and Levin then attempted, reasonably successfully, to predict the relative intensities in the i.r. spectrum of cyclopropane.[50] Kondo and Saëki,[51] having calibrated the CNDO calculation by testing predictions for ethane, then apply the CNDO method to calculate values of $\partial p/\partial S_j$ for *propane* to select the correct set of signs from the 256 different possibilities for the a_1 class of propane (and from the 128 possibilities for b_1 and the 64 possibilities for the b_2 class). This impressive demonstration of the potential power of the CNDO method reduces the possible sets of $\partial p/\partial S_j$ values to only 8 in the a_1 class, 5 for b_1, and 6 for b_2 that need be considered. Kondo and Saëki were able to pick only one set in each symmetry class with some confidence, based on minimizing the residuals (between experimental and calculated values of $\partial p/\partial S_j$) but the variation in the possible values of $\partial p/\partial S_j$ was not excessive, even if other possible sign choices were included.

Lewis and Levin [50] also investigated the use of INDO methods [2] to calculate the dipole derivatives of the hydrocarbons. The CNDO calculations appear to be in significantly better agreement with experiment. This conclusion agrees with other observations (for example, unpublished work at the University of Florida by Bruns). Apparently this is because the parameters for the CNDO calculation have been fortuitously chosen.

Levin and Lewis [58] have used the CNDO and INDO calculated values of dipole derivatives to decide between the possible sets of $\partial p/\partial S_j$ values for S_3 and S_4 from a reinvestigation of the experimental intensities of CF_4. They calculated $\partial p/\partial S_{3a} = -0.69\,e$ (CNDO) or $-0.83\,e$ (INDO) *vs.* $\pm 0.99\,e$ (exp.); and $\partial p/\partial S_{4a} = -0.51\,e$ (CNDO) or $-0.39\,e$ (INDO) *vs.* $\pm 0.29\,e$

[55] B. Galabov and W. J. Orville-Thomas, *J. C. S. Faraday II*, 1972, **68**, 1778.
[56] J. A. Pople and G. A. Segal, *J. Chem. Phys.*, 1966, **44**, 3289.
[57] J. del Bene and H. H. Jaffé, *J. Chem. Phys.*, 1968, **48**, 1807.
[58] I. W. Levin and T. P. Lewis, *J. Chem. Phys.*, 1970, **52**, 1608.

(exp.). Here, the INDO calculation does agree somewhat better with experiment than does the CNDO result. Perhaps more interesting is the comparison between the value of $\partial p/\partial r_{CF}$ derived for CF_4 (—0.99 *e*) with the values from F_2CO[47] (—0.98 ± 0.01 *e*) and from C_6F_6[56] (—1.12 *e*) Such agreement is probably mostly coincidental, but it revives hope that the confident assignment of signs by the CNDO calculation will lead to experimental values of dipole derivatives that do make chemical sense, after all. Comparisons of this type are just now being made, and it is too early to tell whether or not they will continue to be encouraging.

Levin [59] calculated the dipole derivatives for NF_3 and compared them to different sets of experimental values, arising from different estimates of the force constants, consistent with the Coriolis coupling constants. He used the agreement as a basis for choosing the correct force field for the *e*-type symmetry species. Carrying this theme even further, Levin [60] calculated the relative intensities expected from the CNDO calculated values of $\partial p/\partial S_j$ values for the symmetry co-ordinates of SF_4, and compared them with the experimentally observed spectrum to reach a vibrational assignment. A CNDO calculation of the dipole derivatives $\partial p/\partial S_{3a}$ and $\partial p/\partial S_{4a}$ was made first for SF_6. The agreement between the calculated values and one of the possible sets of experimental values was quite good indeed, adding confidence that the predicted spectrum of SF_4 (from the CNDO results) could be expected to be in good agreement with the experimental spectrum, as indeed it was. A similar, but more extensive, application of CNDO calculations of $\partial p/\partial S_j$ values has been made by Levin and Adams to interpret their recent measurement of the absolute i.r. intensities of PF_3.[61] Even though one may wish in the future for more thorough testing of these kinds of applications against molecules whose normal co-ordinates and other properties are better known, these imaginative applications [58–61] illustrate the potential importance of CNDO-calculated values of $\partial p/\partial S_j$ in all areas of vibrational spectroscopy.

Another example of the use of this kind of theory to aid in the understanding of the i.r. spectrum of a molecule is provided by the *ab initio* calculation of the spectrum of CO_3 made by Sabin and Kim,[62] this time with a gaussian basis set using the IBMOL program. It is hoped that such calculations will become a routine tool for the spectroscopist interested in vibrational assignments of unstable species.

To continue the catalogue of CNDO calculations of dipole derivatives we mention the calculations for BF_3[63] and for H_2CO[64a] by Bruns and Person. In the former, a comparison is made with the results for the related F_2CO molecule. The calculated value for $\partial p/\partial \gamma$, the BF out-of-phase bending

[59] I. W. Levin, *J. Chem. Phys.*, 1970, **52**, 2783.
[60] I. W. Levin, *J. Chem. Phys.*, 1971, **55**, 5393.
[61] I. W. Levin and O. W. Adams, *J. Mol. Spectroscopy*, 1971, **39**, 380.
[62] J. R. Sabin and H. Kim, *Chem. Phys. Letters*, 1971, **11**, 593.
[63] R. E. Bruns and W. B. Person, *J. Chem. Phys.*, 1971, **55**, 5401.
[64] (*a*) R. E. Bruns and W. B. Person, *J. Chem. Phys.*, 1973, **58**, 2585; (*b*) I. C. Hisatsune and D. F. Eggers, *J. Chem. Phys.*, 1955, **23**, 487.

motion, is —0.57 *e*, compared to an experimental value of —0.36 *e*. The over-estimation of this derivative is believed [63] characteristic of attempts to calculate out-of-plane derivatives by CNDO methods, resulting in the scepticism previously mentioned (see ref. 52) concerning CNDO estimates of $\partial p_3/\partial \gamma$ (the *sp*-polarization term, or 'rehybridization moment'). The experimental results for CH_2O are less certain, since the measurements are old [64b] (as are, of course, most of the measurements for the molecules discussed here) and did not adequately resolve overlapping bands into the component intensities, and excluded isotopic (CD_2O) molecules. The comparison [64b] between experimental and calculated results, however, does suggest that it would be worthwhile repeating the experimental study. The results are analysed to show the magnitudes of the different contributions ($\partial p_1/\partial R_k$, $\partial p_2/\partial R_k$, and $\partial p_3/\partial R_k$) to $\partial p/\partial R_k$.

Most of the CNDO calculations mentioned above were for compounds formed from atoms in the first row of the Periodic Table (H through F). The reason is that the parameters that go into the CNDO theory have been quite well evaluated for these atoms.[2] Less reliable parameters have been suggested for second-row atoms (Na through Cl), but calculated dipole derivatives are not expected to agree as well with experiment as are those for first-row atoms. Nevertheless, we have seen above that the calculated values appear to be in surprisingly good agreement with experiment.[60, 61] This conclusion has been further tested by Bruns and Nair [65] for Cl_2CO; the calculated results are compared with the experimental values [46] and then with the values for F_2CO,[47] and the comparison is then used to interpret the results for Br_2CO. A further test has been made by Bruns [66] of the ability of the CNDO technique to calculate dipole derivatives for F_2CS and Cl_2CS, to be compared with the recently reported experimental values from gas-phase intensity measurements.[67] Again, the results are in surprisingly close agreement, although there are occasional discrepancies by a factor of two or so.

Having been able to determine with some confidence the correct set of experimental derivatives for F_2CS and Cl_2CS (and for the other molecules in the series), Bruns [66] was then able to examine these values for trends from one molecule to another, for example, from F_2CO to F_2CS or from Cl_2CO to Cl_2CS. He found that the bond-stretching derivatives do obey a remarkable empirical relationship, expressed as:

$$(\partial p/\partial r_{\mathrm{CCl}})_{\mathrm{Cl_2CS}} - (\partial p/\partial r_{\mathrm{CF}})_{\mathrm{F_2CS}} = (\partial p/\partial r_{\mathrm{CCl}})_{\mathrm{Cl_2CO}} - (\partial p/\partial r_{\mathrm{CF}})_{\mathrm{F_2CO}} \quad (35)$$

This simple relationship exists between $\partial p/\partial r_{\mathrm{CS}}$ and $\partial p/\partial r_{\mathrm{CO}}$, also. Bruns [66] tested the relationship by using it to predict successfully the dipole derivatives for Cl_2CS from the experimental values for Cl_2CO, F_2CO, and F_2CS. This kind of additivity relationship is of obvious importance in any hopes of

65 R. E. Bruns and R. K. Nair, *J. Chem. Phys.*, 1973, **58**, 1849.
66 R. E. Bruns, *J. Chem. Phys.*, 1973, **58**, 1855.
67 M. J. Hopper, J. W. Russell, and J. Overend, *Spectrochim. Acta*, 1972, **28A**, 1215.

predicting quantitatively the i.r. intensities of molecules without resorting to a quantum-mechanical calculation (even an easy one such as CNDO).

As an example of the kind of trend that can be examined, we show in Table 7 (taken from ref. 64*a*) the 'substituent effect' of an X atom on the value of $\partial p/\partial r_{CO}$ (or $\partial p/\partial r_{CS}$) for the series of X_2CO and X_2CS molecules, using the CNDO-calculated values of dipole derivatives to determine the correct experimental sets of derivatives, and then examining these sets for trends. This trend is qualitatively what might be expected from chemical intuition; its quantitative meaning (if any) has yet to be explored. Certainly such trends have been expected before, but the uncertainty in the analysis of data has always made them difficult to observe.

Table 7 *Comparison of experimental and CNDO-calculated values for* $\partial p/\partial r_{CO}$ *and* $\partial p/\partial r_{CS}$ *for* X_2CO *and* X_2CS *molecules.* (*Units are e*)

	H_2CY [a]	F_2CY	Cl_2CY [c]	Br_2CY [c]
$(\partial p/\partial r_{CO})_{exp}$	−0.73	−0.77 [b]	−0.98	−1.10
$(\partial p/\partial r_{CO})_{CNDO}$	−0.73	−1.13 [b]	−1.47	—
$(\partial p/\partial r_{CS})_{exp}$	—	−0.64 [c]	−0.77	—
$(\partial p/\partial r_{CS})_{CNDO}$	—	−0.96 [c]	−1.03	—

[a] Values from ref. 63. Y is O or S. [b] Values from Table 5; corrected from ref. 47.
[c] Values from refs. 65 and 66.

In summary, it is clear that the CNDO calculations of $\partial p/\partial S_j$ are relatively easy to make and the results are in reasonably good agreement with experiment. These calculated values are expected to be reliable enough to permit the choice of the correct set of $\partial p/\partial S_j$ values for many molecules. The predicted values of $\partial p/\partial r_k$, where r_k is a stretching co-ordinate, are expected to be quite good for a polar bond, such as CF or CO, and reasonably reliable also for CH. The calculated values of $\partial p/\partial \alpha$, where α is a bending co-ordinate, are expected to be somewhat less reliable, especially when the charge-transfer contribution ($\partial p_2/\partial \alpha$) opposes the *sp*-polarization contribution ($\partial p_3/\partial \alpha$), as often happens. Nevertheless, the enormous help in analysing the data that are provided by these calculations fully justifies the effort of making them. Whether or not the CNDO-calculated contributions to the derivatives are reliable enough to provide any useful insight into the nature of charge redistribution during vibration remains to be seen. At any rate, *ab initio* calculations for polyatomic molecules are appearing increasingly often, so that the necessary theoretical guidance appears likely to be available soon.

The relation between atomic polarization, P_A, and i.r. band intensities is well known.[68, 69] Atomic polarizations of benzene, hexafluorobenzene, and mixtures of these two compounds have been determined from polarization and refraction data.[70] The values of 1.3 and 3.8 $\times$ 10^{-6} m^3 so obtained for

[68] D. H. Whiffen, *Trans. Faraday Soc.*, 1958, **54**, 327.
[69] R. J. W. LeFevre and D.A.A.S.N. Rao, *Austral. J. Chem.*, 1955, **8**, 39.
[70] M. E. Baur, D. A. Horsma, C. M. Knobler, and P. Perez, *J. Phys. Chem.*, 1969, **73**, 641.

benzene and for hexafluorobenzene are to be compared with values of 1.07 and 4.7×10^{-6} m^3 derived from the vapour-phase band-intensity measurements. The polarization and refraction data refer to the liquid state, of course, but inclusion of a contribution to P_A from the far-i.r. absorption of the liquid must increase the spectroscopic estimate. It seems very unlikely that this contribution will be very large for either system. The discrepancies between the two sets of experimental values of the atomic polarizations are probably due to difficulties in the dielectric polarization measurements and in determining the correct value of the extrapolated refractive index, n_∞. The dielectric data and refractive index show a nonlinear dependence of the atomic polarization on the fractional composition of the benzene–hexafluorobenzene mixtures. Such behaviour implies that the i.r. absorption intensities will depend on the C_6H_6/C_6F_6 ratio. The intensity of the a_{2u} mode of C_6F_6 has been observed to depend on the concentration of benzene in the three-component system C_6F_6–C_6H_6–cyclohexane. This behaviour was interpreted in terms of a weak interaction complex.[71] Yarwood and Orville-Thomas obtained an approximate value of 0.89×10^{-6} m^3 for the atomic polarization of ethylene using their refractive index data on the liquid.[72] This value is based on a low-frequency refractive index measured at 16.6 μm. As the refractive index at this wavelength is still being affected by the intense absorption near 950 cm^{-1}, the derived value for the atomic polarization in the frequency interval investigated is probably too high. The value of 0.89×10^{-6} m^3 so obtained is about twice that derived from the measured intensities (0.44×10^{-6} m^3). Rao, however, has quoted a value of 0.39×10^{-6} m^3 for P_A derived from dielectric and refraction studies. Rao has made a number of comparisons of values obtained from i.r. studies with those obtained by dielectric constant and molar refraction studies.[73] Agreement is generally very satisfactory. It is noteworthy that the dielectric measurements from which P_A are computed by Rao are for the gas phase and were reported in 1936.[74]

3 Experimental Techniques

General.—In addition to the intensity problems associated with molecular force fields and with the signs of the $\partial p/\partial Q$, which were mentioned in the introduction, there are serious problems associated with intensity measurements. In the gas or solid phase these difficulties arise mainly from inadequate instrumental resolution and band overlapping, and in the liquid state they are also due to internal field effects.

In view of these problems, new methods of making absolute intensity measurements are always welcome in that they present new ways of checking

[71] W. Wheatley, Ph.D. thesis, London University, 1968.
[72] J. Yarwood and W. J. Orville-Thomas, *Trans. Faraday Soc.*, 1966, **62,** 3011.
[73] D.A.A.S.N. Rao, *Trans. Faraday Soc.*, 1963, **59,** 43.
[74] H. E. Watson, K. L. Ramaswamy, and G. P. Kane, *Proc. Roy. Soc.*, 1936, **A156,** 130, 144.

data. Several interesting techniques have been proposed in the period covered by this review.

A direct measurement of the energy incident on a gaseous sample and the amount absorbed has been made by one investigator.[75] The pressure rise arising from sample heating is reproduced by electrical heating and the absolute energy absorption is thereby deduced. In view of the small amounts of energy involved, a broad-frequency-pass filter is used to limit the frequency range studied at any one time. This means that all absorptions in the selected frequency range are measured as a whole and all energy-distribution information is lost. Nevertheless the technique was useful in that it confirmed values of earlier measurements on methane, nitrous oxide, and acetylene in the 3—5 μm range.

Computer simulation of absorption bands is a process that is now commonly used. It is frequently used in the formulation of the transition energies in terms of molecular parameters, but in the past only relative intensities of the transitions have been considered. In fact, for a symmetric-top molecule the intensity distribution is given by:

$$\alpha(\nu) = C' A_{J,K} \nu_{J,K} \exp[-F_{J,K}\frac{hc}{kT}] \,|\, \langle \psi_{v'} | \mu | \psi_{v''} \rangle \,|^2 S(\nu, \nu_{JK}) \qquad (36)$$

$A_{J,K}$ is a known function of the rotational quantum numbers; $F_{J,K}$ is the rotational term value of the initial state; $S(\nu, \nu_{JK})$ is a line shape function satisfying the normalization condition that $\int_0^\infty S(\nu, \nu_{JK})\, d\nu = 1$, C' is a constant given, as shown in ref. 76, by:

$$C' = \frac{8\pi^3 N}{3hcg}\left[\left(\frac{hc}{kT}\right)^3 \frac{B^2}{C\pi}\right]^{\frac{1}{2}} \qquad (37)$$

g is a parameter equal to one for the nondegenerate species and two for the E species, B and C are the rotational constants, and the other symbols have their usual meanings. All the parameters in equations (36) and (37) with the exception of the transition-moment term are generally known or can be estimated with adequate accuracy. Comparison of experimental absorption coefficients with those computed from (36) and (37), but excluding the vibrational transition-moment factor, allows the evaluation of the latter. The results obtained for benzene, C_6D_6, and C_6F_6 were in good accord with conventional band-integration techniques. The method ought to be of considerable value for light symmetric-top systems whose absorption extends over a large frequency range and for which the molecular energy parameters (vibrational and rotational constants) have been well determined.

In such computer simulation and comparison procedures, compatibility of instrumental data output with computer inputs is obviously very advantageous. Indeed for a considerable number of corrections or operations on the

[75] T. F. Hunter, *J. Chem. Soc.* (*A*), 1967, 374.
[76] D. Steele and W. Wheatley, *J. Mol. Spectroscopy*, 1969, **32**, 260, 265.

data, computer processing is essential. Such processes include corrections due to non-linearity of the instrument's response, slit-width distortions, and refractive-index effects, as well as separation of overlapping bands, translation of transmissions into absorption coefficients, *etc*. Commercial instrumentation is now being produced to ease the difficulties in interfacing with computers. Thus the Perkin Elmer 180 has digitized output facilities, and encoders are readily attached to the wavelength and transmission drives of the 25 series. One should also note here the considerable potential of the Fourier-transform system marketed by Digilab. A high photometric accuracy is claimed from very low to very high transmissions; resolution can be as high as 0.1 cm^{-1}; its fast recording time and reproducibility permits averaging of spectra and each run can produce data for the entire spectral range of the system. The wavenumber range 10 000—10 cm^{-1} is covered using three different beam splitters.

A digital recording and computer-processing system for intensities has been described.[77] Using this system, measurements have been made on the ν_8 band of methylene chloride and on ethane, propane, and various deuteriated derivatives.

Overlapping bands continue to pose a difficult problem. For gaseous systems it may be possible in some cases to separate the intensity contributions of the various bands by computer simulation, but this is inevitably an expensive method and necessitates a detailed knowledge of the vibrational–rotational constants. For liquid systems it is not possible to give a general analytical form to the band contour and it is therefore necessary to resort to analytical band-fitting procedures. The most detailed study of analytical methods of separation is that of Pitha and Jones,[78] who examined various mathematical iteration techniques for producing an optimum fit of overlapping band systems as sums of bands of given parametric form. They found the method of Meiron [79] much superior to the other techniques examined. A variant of this is used in Fortran IV programs which are available from the National Research Council of Canada.[80]

A very different approach stems from the work of Clifford and Crawford.[81] It appears from this study that bands are reasonably well described by the Van Vleck–Weisskopf–Lorentz absorption model. It is shown how this leads to an analytical method of determining intensities which are corrected for internal field effects. It seems that the method might allow the separate contributions of the different bands to be deduced. This ability to separate contributions from different bands stems from the Kramers–Krönig relations between refractive indices and absorption coefficients. These can, of course, be used directly to deduce the band intensities without field corrections (see

77 K. Tanabe, S. Saëki, and M. Mizuno, *Bunseki Kagaku*, 1969, **18**, 1347.
78 J. Pitha and R. N. Jones, *Canad. J. Chem.*, 1966, **44**, 3031.
79 J. Meiron, *J. Opt. Soc. Amer.*, 1965, **55**, 1105.
80 R. N. Jones, *et al.*, N.R.C. Bulletins, No. 11 (1968), No. 12 (1969), and No. 13 (1969).
81 A. A. Clifford and B. Crawford, jun., *J. Phys. Chem.*, 1966, **70**, 1536.

for example ref. 82) employing the well-known relation

$$n_\nu = n_e + \frac{1}{2\pi^2} \sum_j A_j/(\nu_j^2 - \nu^2) \tag{38}$$

The disadvantage of equation (38) is that information is lost about the band contour. There is a great deal of information in this contour. This aspect of spectroscopy is discussed elsewhere in this volume by R. T. Bailey (Chapter 3).

High pressures and preferably various values of path lengths are required in the more conventional Wilson–Wells extrapolation studies of absorption. A high-pressure cuvette capable of withstanding up to 70 atm and of variable path length has been developed.[83] The length can be varied stepwise by insertion of packing washers. Band intensities of methane and acetylene are reported to have been measured.[83] A high-pressure cell with continuously variable path length has also been described.[84] It was designed specifically for use in the far-i.r, but it would appear to be applicable for use in other spectral regions with appropriate changes in the window material.

The measurement of strong absorption bands presents considerable difficulties. These arise from reflection effects at the cell–sample interfaces and from the necessity of using very short path lengths in transmission measurements. A number of studies have been made of these problems as experienced in transmission measurements. Also there has been a continued interest in alternative techniques based on reflection measurements.

Conventionally in double-beam spectroscopy an empty cell, a cell window, or a cell filled with non-absorbing liquid is placed in a reference beam to correct for cell-transmission losses. Fujiyama, Herrin, and Crawford have shown by computational and experimental techniques that the use of an alkali-metal halide plate or of an empty cell in the reference beam does not give an accurate base line.[85] Using optical data on the 1526 cm^{-1} band of hexafluorobenzene, the measured transmittance in the band wings lies between that of an empty cell and of an alkali-metal halide plate of a thickness equal to the sum of the thicknesses of the cell windows. The most disturbing feature is the displacement between the relative transmissions in the high- and low-frequency wings. Distortion of the band as a result of the refractive index variations with wavelength was shown to be considerable. The effects are a shift in band centre, an asymmetry in the contour, and a very marked change in the band width. In the examples given the apparent width is about 20% less than the true one. It is important to realize that these are errors due to the wavelength dependence of reflection at the sample–window interfaces and are independent of the field effects discussed in refs. 81 and 86.

[82] J. A. Ramsey, J. A. Ladd, and W. J. Orville-Thomas, *J.C.S. Faraday II*, 1972, **68**, 193.
[83] L. A. Gribov and V. P. Kruglov, *Isvest. Timiryazev. Sel'skokhoz. Akad.*, 1970, **70**, 194.
[84] V. P. Kruglov and L. A. Gribov, *Zhur. priklad. Spektroskopii*, 1971, **14**, 161.
[85] T. Fujiyama, J. Herrin, and B. Crawford, jun., *Appl. Spectroscopy*, 1970, **24**, 9.
[86] N. G. Bakhshiev, O. P. Girin, and V. S. Libov, *Optics and Spectroscopy*, 1963, **14**, 395.

It is usually assumed that the reflection errors are negligible for weak bands or for dilute solutions in spectral regions remote from strong solvent bands. It appears from the results reported on the weak e_{1u} band of hexafluorobenzene at 315 cm^{-1} and on the a_{2u} band of benzene as measured in carbon tetrachloride solution that this might be wishful thinking.[85] The errors are worse when an empty cell is used in the reference beam. For weak absorbers the lesson appears to be to use a single plate in the reference beam and to acknowledge a 2% uncertainty arising from the background variations. Alternatively, the reflection corrections can be computed as shown in ref. 85. In practice most spectroscopists would use a non-absorbing or weakly absorbing material in the reference cell and correct for the resulting apparent fluctuations in the value of 100% sample transmission. One presumes that the errors discussed in ref. 85 would be significantly less owing to the decrease in interface effects where this technique is practical. A very useful set of equations relating the transmissions of multilayer systems to the complex refractive indices and the angles of incidence of the incident beam are presented. It is said that the basic equations from which they are derived come from Born and Wolf.[87] As several groups have found, at the expense of much wasted time, the reflection and transmission formulae (section 13.4), even in the fourth edition of Born and Wolf's volume, are incorrect. The sum of the fractions of transmitted, reflected, and absorbed radiation is not equal to unity. The formulae given by Fujiyama, Herrin, and Crawford [85] are free of this defect.

Young and Jones have described how to make an ultra-thin i.r. transmission cell (path length as low as 0.1 μm).[88] The interference fringes for such cells are so extended as to give no significant fluctuation of the baseline intensity. However, interference must still exist, and for accurate intensity work either an area of zero absorption, if such exists, must be matched with the transmission of the reference cell, or preferably the reflectance should be calculated as described in ref. 85 and a suitable correction applied. It has been shown that the reflection distortion can be derived from transmission measurements alone using the Kramers–Krönig transform relation. Results for the C_6F_6 e_{1u} band [88] at 1526 cm^{-1} are in excellent agreement with those reported in ref. 89.

A novel type of i.r. transmission cell uses convex internal faces. These faces touch in the centre and the optical beam is focused in the plane of the cell. Movement of the cell perpendicular to the beam axis results in a thickness variation proportional to the square of the displacement. Knowing the cell curvature, the effective cell length at any point can be deduced.[90]

Using the methods of attenuated total reflection (ATR), both the real and imaginary parts of the complex refractive index can be directly derived.

[87] M. Born and E. Wolf, 'Principles of Optics', 4th edn., Pergamon Press, Oxford, 1970.
[88] R. P. Young and R. N. Jones, *Canad. J. Chem.*, 1969, **47**, 3463.
[89] T. Fujiyama and B. Crawford, jun., *J. Phys. Chem.*, 1968, **72**, 2174.
[90] P. K. Khripunov, L. I. Alperovich, and V. M. Zolotarev, *Optics and Spectroscopy*, 1970, **29**, 576.

Knowledge of the angles of incidence and reflection to 1′ is desirable, but even at this high accuracy it has been shown that careful selection of the reflection angles is necessary in order to deduce meaningful results.[91] An ATR spectrometer for intensity studies has been described.[92] The results obtained for C_6F_6 were found to satisfy the Kramers–Krönig relation. This indicated the reliability of the measured real and imaginary parts of the complex refractive indices.[89] It would appear that this test should be more widely applied than it has been. Libov and Bakhshiev studied the optical constants of chloroform and of carbon tetrachloride by normal-incidence reflection studies.[93] The real and imaginary components of the refractive indices were deduced by making reflection measurements from two different window materials. They also found their data to be consistent with the Kramers–Krönig relation.

A review on band shapes by Young and Jones [94] contains a summary of much pertinent information on simple reflection and ATR measurements.

Finite spectral slit widths are well known to affect measured band intensities. Important reviews of, *inter alia*, methods of correcting for slit-width distortions have been published.[95, 96] Iogansen has proposed some methods of deducing the integrated intensity based on extrapolations of linear functions of the measured transmittances.[97] The aim here is both to overcome the finite slit-width effect and to utilize directly the instrument measurable—namely the transmittance ratio.

A different source of error in liquid-phase measurements is the uncertainty in the proportion of intensity in the wings. Evrard and Duculot [98] have proposed a modification to the method of Cabana and Sandorfy [99] which, at least for nitriles, appears to yield improved values. The optimum parameters as determined for 4-halogenobenzonitriles are applied to other aromatic nitriles and the results are shown to be in good accord ($<5\%$ error) with the values obtained by direct integration.[100]

An i.r. intensity standard has been sought for many years. The out-of-plane C—H band of 1,2,4,5-tetrachlorobenzene has been proposed as a strong isolated and narrow band suitable for this purpose.[101] Unfortunately, at values of $\ln (I_0/I)_{max}$ greater than about one many subsidiary bands appear in the wings, so that it appears that a good standard still eludes us.

Gas-phase Studies.—*Diatomics.* The detailed studies now being reported on the intensities of the fundamental and its overtones of diatomic molecules

91 A. C. Gilby, J. Burr, W. Krueger, and B. Crawford, jun., *J. Phys. Chem.*, 1966, **70**, 1525.
92 A. C. Gilby, J. Burr, and B. Crawford, jun., *J. Phys. Chem.*, 1966, **70**, 1520.
93 V. S. Libov and N. G. Bakhshiev, *Optics and Spectroscopy*, 1964, **16**, 122.
94 R. P. Young and R. N. Jones, *Chem. Rev.*, 1971, **71**, 219.
95 S. G. Rautian, *Soviet Physics Uspekhi*, 1958, **1**, 245.
96 K. S. Seshadri and R. N. Jones, *Spectrochim. Acta*, 1963, **19**, 1013.
97 A. V. Iogansen, *Optics and Spectroscopy*, 1964, **16**, 442.
98 J. Evrard and C. Duculot, *Compt. rend.*, 1970, **271**, *B*, 503.
99 A. Cabana and C. Sandorfy, *Spectrochim. Acta*, 1960, **16**, 335.
100 J. Evrard, C. Duculot, and C. Dagnelie, *Compt. rend.*, 1970, **271**, *B*, 781.
101 J. T. Shimozawa and M. K. Wilson, *Spectrochim. Acta*, 1966, **22**, 1591.

pose a very exacting challenge to the theoretician. With polyatomic systems the experimental and interpretational difficulties are such that approximate agreement between theory and experimentally determined transition moments for fundamental transitions must generally be considered satisfactory. With diatomic systems the dipole-moment expansions are being determined in some cases to the third power of the displacement co-ordinate. Naturally, as more detail is sought from experimental data, so the theory employed must be more critically assessed. It is not surprising then to find developments in the formulation of transition moments as a function of molecular parameters accompanying the more ambitious experimental studies.

Bunker has examined the effects of relaxing the Born–Oppenheimer approximation on the dipole-moment and nuclear-quadrupole expansions in the vibrational quantum number for heteronuclear diatomics in a $^1\Sigma$ state.[102] The effects on the ratio of the vibrationally independent part of the dipole moment for HCl to that of DCl are found to be much larger than the effects due to electrical and mechanical anharmonicity. The non-adiabatic effects, allowing for the mixing of rovibronic states, dominate the corrections. However, within the approximations made, there are no effects on the dipole derivatives with respect to bond stretching. This theoretical conclusion is given experimental support by the analysis of accurate molecular-beam dipole moments in excited states for ^{1}HCl and ^{2}HCl.[103] This is discussed more fully below.

Several theoretical studies have been presented of the distribution of intensity within a vibrational–rotational band of a heteronuclear diatomic molecule. Toth, Hunt, and Plyler have used perturbation theory up to third order to derive expressions for transition probabilities.[104] Integrated intensities were measured for the fundamental, and its first and second overtones, of HCl using a spectral resolution of 0.02—0.04 cm^{-1}. An uncertainty of 3% is claimed, and within this value the intensity of the fundamental (1.35 mm^{-2} atm^{-1} at 300 K, ~33.2 km mol^{-1}) agrees well with that given in ref. 105 (1.30 ± 0.05 mm^{-2} atm^{-1} at 300 K, ~32.0 ± 1.2 km mol^{-1}) and, though lower than the value of Penner and Weber[106] (1.60 ± 0.3 atm^{-1} at 300 K, ~39.4 ± 7.4 km mol^{-1}), still agrees with this latter result within the quoted uncertainty limits. For the overtone the value of 0.0373 ± 0.0013 mm^{-2} atm^{-1} is in excellent accord with values of 0.0368 and 0.0364 given in two earlier studies[106, 107] but is significantly higher than the value of ref. 108. Now the transition integral $|R_v^{v'}(J' \leftarrow J)|$ for the transition indicated in paren-

102 P. R. Bunker, *J. Mol. Spectroscopy*, 1973, **45**, 151.
103 E. W. Kaiser, *J. Chem. Phys.*, 1970, **53**, 1686.
104 R. A. Toth, R. H. Hunt, and E. K. Plyler, *J. Mol. Spectroscopy*, 1969, **32**, 74; 1970, **35**, 110.
105 W. S. Benedict, R. Herman, G. E. Moore, and S. Silverman, *Canad. J. Phys.*, 1956, **34**, 850.
106 S. S. Penner and D. Weber, *J. Chem. Phys.*, 1953, **21**, 649.
107 J. H. Jaffe, S. Kimel, and M. A. Hirschfeld, *Canad. J. Phys.*, 1962, **40**, 113.
108 W. S. Benedict, R. Herman, G. E. Moore, and S. Silverman, *J. Chem. Phys.*, 1957, **26**, 1671.

theses can be written as the product of two terms, one of which is independent of the rotational transition involved [designated $R_v^{v'}(0)$], and the second of which [$F_v^{v'}(J' \leftarrow J)$] can be expressed as a power-series expansion in the variable $m = 0.5[J'(J'+1) - J(J+1)]$. Thus the transition probability can be written as

$$|R_v^{v'}(0)|^2 |F_v^{v'}(J' \leftarrow J)|$$

where $F_v^{v'}(J' \leftarrow J) = 1 + C_v m + D_v m^2$.

As is well known, the sign of the Herman–Wallis coefficient C_v is determined by the relative signs of the molecular dipole and its first derivative. From the observed line intensity dependence on the rotational quantum numbers, the known molecular dipole moment, and the theoretical expressions for $R_v^{v'}(J' \leftarrow J)$, the signs of the $d^n p/dQ^n$ were determined for the fundamental and its first and second overtones (the value of the intensity of the second overtone was taken from ref. 109). This leads to the dipole expansion:

$$p = 3.653 + 3.02(\pm 0.06)(r - r_e) - 0.22(\pm 0.08)(r - r_e)^2 - 2.43(\pm 0.23)(r - r_e)^3$$

The absolute sign of $p°$ is taken as positive in the sense $^+$H—Cl$^-$ from the quantum-mechanical calculation by Cade and Huo, a result in accord with chemical intuition. In the above and in the following dipole expressions p is in units of 10^{-30} C m and the displacements $(r - r_e)$ are in Ångstroms. Thus to convert $\partial p/\partial r$ into units of e, it is necessary to divide by 3.3356×4.803. (3.02×10^{-30} C m Å$^{-1} \equiv 0.188\ e$). As there is rather a large scatter in the data for the second overtone, an alternative solution corresponding to the alternative sign for C_3 cannot be positively discounted. Kaiser's results for the dipole expansion [103] based on i.r. intensities from ref. 108 and on molecular-beam values for the dipole moments in different vibrational states are in good agreement with the above for the leading two terms, but differ in higher terms. Thus for $^1H^{35}Cl$ he obtains

$$p = 3.6469(\pm 0.0017) + 3.085(\pm 0.07)(r - r_e) + (0.53 \pm 0.37)(r - r_e)^2 - (12.8 \pm 3.0)(r - r_e)^3 - (31 \pm 15)(r - r_e)^4$$

and for $^2H^{35}Cl$

$$p = 3.6432(\pm 0.0017) + (3.12 \pm 0.08)(r - r_e) + (0.47 \pm 0.43)(r - r_e)^2 - (12.7 \pm 3.7)(r - r_e)^3 - (25 \pm 16)(r - r_e)^4.$$

Intuitively, one would expect $\partial^2 p/\partial r^2$ to be negative, as dissociation is expected to lead to neutral ions and the 'combined atom' would have zero dipole. Kaiser's value for this derivative is therefore surprising and his results imply that it is the fourth derivative, which he computes to be large and negative, that dominates the dipole-moment changes for large displacements. It would be interesting to compare a plot of Kaiser's expansion and of Toth, Hunt, and Plyler's expansion of the dipole moment with the theoretical expansion of Figure 4.

[109] R. A. Toth, R. H. Hunt, and E. K. Plyler, *J. Mol. Spectroscopy*, 1969, **32**, 85.

The results from a detailed study of carbon monoxide using absorption techniques have also been published.[109]

A different approach to deriving the Herman–Wallis factors for diatomics has been used by Herman, Tipping, and Short.[110] They used an iterative method for determining analytic wavefunctions for vibration–rotation levels. The resulting expansion in $2B/\tilde{\nu}_e$ agrees with the results of Toth, Hunt, and Plyler except in the inclusion of small terms in the fourth derivatives of the dipoles, which were neglected by the latter authors. The results are applied to literature values of the absorption intensities for HI.[110]

A series of papers on the hydrogen halides in gas and solution phases [111–113] is described in more detail in the liquid-state study section (p. 420). The primary aim here was to establish the way in which the dipole expansion is modified by liquid-state interactions, so one should not perhaps be too critical in comparing the dipole expansions in refs. 113, 108, and 103.

It has been shown that electric resonance studies in molecular beams can lead to precise values for the dipole moments in ground and excited states.[114] As has been discussed by several authors (see refs. 115, 113, and earlier references therein), the dependence of the dipole on the vibrational quantum number is insufficient in itself to determine the dipole expansion. One also needs information on the molecular wavefunctions. Kaiser combined the resonance data on $^{1}H^{35}Cl$ and $^{2}H^{35}Cl$ with intensity data, as mentioned earlier, to derive the dipole expansions. In addition, values for atomic quadrupoles and spin–rotation interaction constants in the different states were derived. As the data for $^{1}H^{35}Cl$ and $^{2}H^{35}Cl$ are quite independent it was possible to obtain information on the validity of the Born–Oppenheimer approximation. As the dipole *versus* quantum-number curves of the two species were parallel, it is to be concluded that the vibrational transition moments are independent of the Born–Oppenheimer separation to a very high approximation. This is in accord with the predictions of Bunker.[102] There is, however, a displacement in the two curves of about 0.003 C m, indicating that the absolute value of the dipole is sensitive to this approximation.

Alexander has considered the problem of deducing the dipole-moment function by combining the molecular-beam data with values of the dipole function computed from *ab initio* studies.[115] The wave-mechanical results are used to define constraints on the dipole parameters. To interpret the intensity results the Rydberg–Klein–Rees potential curves are employed to determine the 'experimental wavefunctions', which are then used in the equations relating the observed transition moments to the dipole function.

[110] R. M. Herman, R. H. Tipping, and S. Short, *J. Chem. Phys.*, 1970, **53**, 595.
[111] M. Scrocco, R. Giuliani, and C. Costarelli, *Spectrochim. Acta*, 1972, **28A**, 761.
[112] M. Scrocco, *Spectrochim. Acta*, 1972, **28A**, 771.
[113] M. Scrocco, *Spectrochim. Acta*, 1972, **28A**, 777.
[114] M. Kaufman, L. Wharton, and W. Klemperer, *J. Chem. Phys.*, 1965, **43**, 943.
[115] M. H. Alexander, *J. Chem. Phys.*, 1972, **56**, 3030.

By using the dipole data of Wharton, Gold, and Klemperer [116] and of Rothstein,[117] the spectroscopic literature data, and the theoretical dipole function of Brown and Shull,[118] values of the transition moments were deduced for $^7Li^1H$ and $^7Li^2H$. A similar approach gave results for KF.

Polyatomics. Surprisingly few papers on vapour-phase intensity studies have appeared during the past few years. Presumably there has been too much despondence over the difficulties, both of experimental and theoretical natures, which we have discussed earlier. Those investigations which have been reported have shown a fairly traditional approach, and therefore we shall be quite brief in our reviewing.

In seeking to derive the polar properties of bonds (electro-optical parameters) from measured intensities, errors can arise from several sources. These include the molecular force field, the geometry, and the harmonic frequencies, in addition to the intensity measurements. Expressions have been derived [119] for the statistical dispersions in the dipole parameters as a function of the uncertainties, or dispersions, in these various error sources. The authors point out that the full expressions may be not very practical in many cases since the number of degrees of freedom in the determining equations may be too small for a proper statistical analysis to be appropriate and, perhaps more important, the number of terms in the variance–covariance matrices is very large. A much simpler set of equations is given from which an approximate uncertainty in the eigenvector matrices, L, can be derived from a knowledge of the uncertainties in the force constants and in the geometry. Thus, for example,

$$\Delta(L_{ij}) = \sum_{k,l} | [\partial L_{ij}/\partial F_{kl}]\Delta F_{kl}| + \sum_{k,l} | [\partial L_{ij}/\partial G_{kl}]\Delta G_{kl}|$$

The partial derivatives are readily derived from expressions given by the authors; the dispersions in the force constants will usually be known from a normal-co-ordinate calculation; and ΔG is easily derived by adjusting the geometrical parameters by amounts equal to their uncertainties. The principal approximation, of course, is in assuming that the contributing uncertainties to G and F are uncorrelated. Both the rigorous and the approximate treatment were used on the best available methyl halide data taken from the literature. The following conclusions were drawn.

(i) The statistical dispersions of the eigenvectors were of the same magnitude when computed from the approximate treatment as when they were derived by the rigorous approach.

(ii) The errors arising from frequencies and geometry were quite negligible in all the cases examined.

(iii) The experimental uncertainties in the intensities were the primary source of the dispersions in the dipole parameters, though the uncertainties

[116] L. Wharton, L. P. Gold, and W. Klemperer, *J. Chem. Phys.*, 1962, **37**, 2149.
[117] E. Rothstein, *J. Chem. Phys.*, 1969, **50**, 1899; 1970, **52**, 2804.
[118] R. E. Brown and H. Shull, *Internat. J. Quantum Chem.*, 1968, **2**, 663.
[119] J. W. Russell, C. D. Needham, and J. Overend, *J. Chem. Phys.*, 1966, **45**, 3383.

in the force field were very significant in many cases. It has been realized for a long time that when the force fields are ill-defined the force-field uncertainties could be major sources of error. The present treatment allows an actual estimate to be made for the contributions to the dispersion of an electro-optical parameter arising from force-field dispersions. It is worth noting in this context that a seriously constrained force field might reproduce closely the experimental frequencies and yet lead to an inaccurate picture of the eigenvectors. This may show up in a failure to reproduce centrifugal distortion constants or Coriolis constants. It follows that the dispersion on force constants as derived from least-squares fitting of frequencies to force constants must usually be taken as a lower estimate.

The intensities of all six fundamentals of both thiocarbonyl fluoride and thiocarbonyl chloride have been measured using absorption techniques on samples mixed with argon at 1000 p.s.i.[67] Previous intensity measurements exist only for three of the $CSCl_2$ bands.[120] Agreement for the strongest band (ν_4) is good, but for the other two fundamentals the earlier values were low. This has been explained as due to inadequate pressure broadening in the earlier study. Sets of effective bond moments and dipole derivatives are derived from which the 'optimum' one was chosen by analogy with the corresponding parameters of F_2CO [45–47] (see p. 381).

An exemplary study of gas-phase intensities is that of Kondo and Saëki, who have used isotopically substituted systems, least-squares analysis, and CNDO calculations to deduce the $\partial p/\partial S_i$ of ethane and propane.[51] For ethane and its deuteriated derivatives eight sets of sign combinations appeared feasible on the basis of the CNDO/2 studies. A least-squares analysis of the $\partial p/\partial S$ using the 8 different sign sets led to only 4 different solutions. Two of these had much lower residuals than the other two and are therefore preferred. A similar analysis of intensities of the propane bands led to a set of dipole derivatives which were in very good agreement with their ethane and methane counterparts (see Table 8). The agreement between comparable derivatives is generally satisfactory. Certain dipole derivatives show a definite trend in the series CH_3X (X = Et, Me, H). The effective atomic charges as defined by King, Mast, and Blanchette,[24] derived from the intensity data, are $\xi_H = 0.183\,e$ and $\xi_C = 0.225\,e$ for ethane and $\xi_H = 0.181\,e$ and $\xi_C = 0.217\,e$ for propane. The ξ_H values should be compared with the values of this parameter given for a variety of molecules in Table 2.

Another very thorough study of gas-phase intensities is that by Tanabe and Saëki [121] on all active bands of chloroform, [^{2}H]chloroform, and carbon tetrachloride. The data were combined with earlier data on methane, methyl chloride (data sources not given), and methylene chloride [122] and effects were made to find a common set of electro-optical parameters which would reproduce the intensities of all bands. It was found necessary to allow the

[120] R. J. Lovell and E. A. Jones, *J. Mol. Spectroscopy*, 1960, **4**, 173.
[121] K. Tanabe and S. Saëki, *Spectrochim. Acta*, 1970, **26A**, 1469.
[122] J. Morcillo, J. Herranz, and J. Fernandez Biarge, *Spectrochim. Acta*, 1959, **15**, 110.

Table 8 *The derivatives of the total dipole moment with respect to the local symmetry co-ordinates of the methyl group for propane, ethane, and methane (in units of e).* (From *Spectrochim. Acta*, 1973, **29A,** 735)

S_j	$\partial p/\partial S_j$		
	Propane [a]	*Ethane*	*Methane*
$(r_1 + r_2 + r_3)/\sqrt{3}$	−0.155 (−0.016)	−0.153	−0.087
$(\phi_1 + \phi_2 + \phi_3 - \theta_1 - \theta_2 - \theta_3)/\sqrt{6}$	−0.009 (0.038)	−0.074	−0.085
$(2r_1 - r_2 - r_3)/\sqrt{6}$	−0.182 (−0.003)	−0.177	−0.173
$(2\phi_1 - \phi_2 - \phi_3)/\sqrt{6}$	0.039 (−0.014)	0.035	0.060
$(2\theta_1 - \theta_2 - \theta_3)/\sqrt{6}$	−0.037 (−0.024)	−0.027	−0.060
$(r_2 - r_3)/\sqrt{2}$	−0.160	−0.177	−0.173
$(\phi_2 - \phi_3)/\sqrt{2}$	0.045	0.035	0.060
$(\theta_1 - \theta_3)/\sqrt{2}$	−0.015	−0.027	−0.060

[a] The values in parentheses are those from components of $\partial p/\partial S_j$ which are perpendicular to the main component and result from the collapse of axial symmetry in the propane molecule.

effective bond moment μ_{CH} and the diagonal terms of the P_R matrix $\partial\mu_{CH}/\partial r_{CH}$ to vary from molecule to molecule. Several solution sets were obtained, of which four with the lowest residuals are listed. It appears that the C—Cl parameters are more transferable. This is in accord with deductions based on studies of the liquid methane derivatives, which will be discussed in the next section. The conclusions of the authors are rendered more questionable by the necessity to freeze some parameters to zero and due to the conclusion reached at one stage that μ_{CCl} and $\partial\mu_{CCl}/\partial r_{CCl}$ were also not transferable (see Table 8 of ref. 121). The necessity of freezing certain parameters is probably of little importance as the terms are off-diagonal ones and the indeterminacy of these is almost certainly due to redundancies in the angle-deformation co-ordinates.

The CCl_4 intensity values quoted are in good accord with previous literature values. For $CHCl_3$ the values seem to be about 25% less than those of Morcillo, Herranz, and Biarge [122] and for $CDCl_3$ no earlier data are quoted. There appears to be a need for inter-laboratories collaboration to obtain reliable values for the intensities of a number of bands. Such bands could then be used for testing procedures and calibrations.

It would be interesting to compare the atomic polar tensors computed from Tanabe and Saëki's intensities for chloroform with those derived from the Morcillo, Biarge, Heredia, and Medina results [21] (see Table 1). The necessary computations are being made.

The intensity of the a_2 band of BBr_3 has been measured and from this the effective bond moment has been deduced from zero-order bond-moment theory.[123] Using earlier values for the corresponding B—F and B—Cl bonds

[123] M. C. Pomposiello and O. Brieux de Mandirola, *J. Mol. Structure*, 1972, **11**, 191.

in BF_3 and BCl_3, a very good linear relation is shown between these effective moments and the halogen electronegativities. Such a relation is probably fortuitous.

Secroun, Barbe, and Jouve have used the method of contact transformations to obtain high-order expressions for intensities as a function of dipole parameters for diatomic and bent XY_2 molecules.[124] Their diatomic expressions are in accord with earlier studies, but their XY_2 results differ from expressions given earlier by two of the authors [125] and derived using first-order perturbation methods. The results are applied to measured SO_2 intensities [126] and the dipole expansion is derived.

Other papers on gas-phase intensities give results on:

(*a*) CO_2 bands 1.43—1.65 nm range.[127] The resolution employed was in the range 0.035—0.06 cm^{-1}.

(*b*) N_2O in 2.9 nm region.[128] Again very high resolution was used.

(*c*) C_6F_6 (lowest frequency e_{1u} and a_{2u} bands) and also C_6H_6 (a_{2u}).[53] A computer simulation method was used.

(*d*) PF_3.[61] All fundamentals were studied.

(*e*) Tetrahydrofuran.[129] The intensities of all fundamentals above 400 cm^{-1} were measured.

(*f*) CF_4.[58] The intensities of the active fundamentals were measured with 40 atm pressure of nitrogen to broaden the lines. The values obtained are compared with earlier dispersion and absorption values.

Liquid-phase and Solution Studies.—Considerable progress has been made in this field during the past decade, primarily as a result of the realisation that one must treat the susceptibility C as a complex quantity. Clifford and Crawford have found the Lorentz–Lorenz equation perfectly adequate in their formulation of the overall problem.[81] As they have shown, the real and imaginary parts of the susceptibility are given by:

$$C' = \frac{3}{4\pi}\left[1 - \frac{3(\varepsilon' + 2)}{(\varepsilon' + 2)^2 + \varepsilon''^2}\right] \tag{39}$$

and

$$C'' = \frac{9\varepsilon''}{4\pi[(\varepsilon' + 2)^2 + \varepsilon''^2]} \tag{40}$$

(In these definitions they have opted to extract M/N, the molecular mass.) Since the permeability is unity in this spectral range, Maxwell's equations

[124] C. Secroun, A. Barbe, and P. Jouve, *J. Mol. Spectroscopy*, 1973, **45**, 1.
[125] C. Secroun and P. Jouve, *J. Phys. (Paris)*, 1971, **32**, 871.
[126] C. Secroun and P. Jouve, *Compt. rend.*, 1970, **270**, *B*, 1610.
[127] R. A. Toth, R. H. Hunt, and E. K. Plyler, *J. Mol. Spectroscopy*, 1971, **38**, 107.
[128] R. A. Toth, *J. Mol. Spectroscopy*, 1971, **40**, 588.
[129] L. A. Evseeva, A. G. Finkel, L. M. Sverdlov, and L. V. Pronina, *J. Appl. Spectroscopy*, 1970, **12**, 232.

give $\hat{\varepsilon} = \hat{n}^2$, where $\hat{\varepsilon}$ is the complex dielectric constant and $\hat{n}$ is the complex refractive index. The novel approach of Clifford and Crawford was to examine the consequences of applying at this stage the Van Vleck and Weisskopf model to the absorption, from which they deduce

$$C' = K + 2S\left[\frac{\gamma^4 + \gamma^2(2\nu_0^2 + \nu^2) + \nu_0^2(\nu_0^2 - \nu^2)}{\gamma^4 + 2\gamma^2(\nu_0^2 + \nu^2) + (\nu_0^2 - \nu^2)^2}\right] \tag{41}$$

and

$$C'' = 2S\left[\frac{\nu\gamma(\gamma^2 + \nu_0^2 + \nu^2)}{\gamma^4 + 2\gamma^2(\nu_0^2 + \nu^2) + (\nu_0^2 - \nu^2)^2}\right] \tag{42}$$

From equations (41) and (42)

$$C' = K + 2S + \left(\frac{1}{\gamma}\right) C''\,\nu\,\frac{(\nu_0^2 - \nu^2)}{(\nu_0^2 + \nu^2)} \tag{43}$$

$$= K + 2S + \left(\frac{1}{\gamma}\right) C^0 \tag{44}$$

ν_0 is the frequency of the band centre and γ is a damping constant related to the relaxation time τ by $\tau = 1/2\pi c\gamma$; K is the contribution to the susceptibility from bands other than that under consideration and S is the strength factor. The strength factor is related to the absolute intensity through

$$\Gamma = 8\pi^3\nu_0 MS/\rho$$

where M and ρ are the molecular weight and density, respectively. The procedure in determining Γ from measurements of $\hat{n}$ is as follows. Knowing $\hat{n}$, and hence $\hat{\varepsilon}$, then from equations (39) and (40) C' and C'' can be deduced. The maximum of C'' occurs at $\nu = \sqrt{\nu_0^2 + \gamma^2} \approx \nu_0$, which allows ν_0 to be derived. From equation (44) a plot of C' *versus* C^0 leads to γ, and finally S is derived from equation (42). The following interesting points emerged from this study:

(*a*) the C'' *versus* frequency graph is much more symmetrical than the graph of the imaginary part of the refractive index *versus* frequency or the apparent absorption curve.

(*b*) the plots of C' and C'' against the real and imaginary parts of the shape factor [equations (41) and (42)] gave good linear plots except in the region of overlap of bands. This indicates that the assumptions were justified.

(*c*) the intensities as derived above are corrected for field effects in the medium and compare well with those obtained using the Polo–Wilson equation. The problem of integration limits, overlapping bands, *etc.* should make the latter method less reliable. The results are also in satisfactory accord with gas-phase results in cases where intermolecular interactions can be expected to be weak (see Table 9).

Table 9 *A comparison of frequencies and intensities in the pure liquid phase, before and after correction for the internal field dielectric effects, with vapour-phase parameters*

Compound	*Apparent* ν_0 *as max. of* n''/cm^{-1}	*Max. of* C''/cm^{-1}	A^{liq}/km mol^{-1}	$A^{\text{P.W.}}$/km mol^{-1}	A^{c}/km mol^{-1}	A^{gas}/km mol^{-1}	*ref.*
Benzene	674	678	142	107	104 [a]	85.4	*c*
Benzene	1035	1035	11.4	8.8	10.2 [a]	8.5	*c*
Chloroform	762	769	286	231	231 [a]	277	*c*
Chloroform	760	769	297	—	260	193	*d*
CCl_4	786	792 }	452	350	238 [a] }	386	*c*
	762	766 }			113 [a] }		
CCl_4	787	793	470	—	380	323	*d*
CH_3I	521.6 [b]		3.75	2.80	2.8	—	*e*
					2.6 [a]		
CH_3I	884.2 [b]		17.4	13.0	13.0	—	*e*
					12.3 [a]		
CD_3I	492.6 [b]		1.37	1.01	1.01	—	*e*
					1.05 [a]		
	656.1 [b]		5.28	3.92	3.94		*e*
					3.98 [a]		
C_6F_6	1527	1534	695	—	564	546	*f, g*
					565 [a]		
C_6F_6	1019	1021	488	—	399	408	*f, g*
	993	996					
C_6F_6	315	315	3.56	—	2.9	2.5	*f, g*
					3.3 [a]		

A^{c} is the integrated intensity after applying field corrections.
A^{liq} is the integrated apparent intensity (no field corrections).
$A^{\text{P.W.}}$ is the intensity computed from A^{liq} using the Polo–Wilson correction factor.
[a] Indicates A^{c} obtained from the Clifford–Crawford method based on the Van Vleck–Weisskopf plot. [b] shift not quoted. [c] A. A. Clifford and B. Crawford, jun., *J. Phys. Chem.*, 1966, **70**, 1536. [d] V. S. Libov, N. G. Bakhshiev, and O. P. Girin, *Optics and Spectroscopy*, 1964, **16**, 549. [e] T. Fujiyama and B. Crawford, jun., *J. Phys. Chem.*, 1969, **73**, 4040. [f] T. Fujiyama and B. Crawford, jun., *J. Phys. Chem.*, 1968, **72**, 2174. [g] D. Steele and W. Wheatley, *J. Mol. Spectroscopy*, 1969, **32**, 265.

It should be noted that the energy dissipated per unit time in a system is proportional to the imaginary part of the susceptibility.[130] This is in accord with observation (*a*). To a first approximation C'' is proportional to $n'n''$, as is evident from equation (40). The correlation function of the transition moment is related to the Fourier transform of this imaginary part of the susceptibility. In the formulations of this function by Fulton [131] and by Gordon [see equation (11) of ref. 132] C'' appears as proportional to the product $n'n''$.

Bakshiev, Libov, and collaborators have used an approach [133] similar to that of Clifford and Crawford.[81] However, instead of directly employing the Van Vleck–Weiskopf line-shape function, the molecular polarizability is left as a parameter to be determined by experiment. Thus they use the Onsager–Bötcher model [134] giving the effective field E_{eff} acting on the molecule as a function of the average applied field E_{av}

$$E_{\text{eff}} = \frac{3n^2}{(2n^2+1)(1-f\alpha)} E_{\text{av}}$$

$$f = \frac{1}{r^3}\left[\frac{2(n^2-1)}{2n^2+1}\right]$$

and r is the effective Onsager radius of the molecule. To determine f and α the following relation [135] is assumed to hold in regions remote from any absorption (generally from data in the 300—1000 nm region):

$$\frac{12\pi n^2 N}{(n^2-1)(2n^2+1)} = \frac{1}{\alpha} - \frac{1}{r^3}\left[\frac{2(n^2-1)}{2n^2+1}\right] \tag{45}$$

A plot of the left-hand side against $\dfrac{2(n^2-1)}{2n^2+1}$ yields good linear graphs, from which the required parameters are derived. Since the probability of a transition at a frequency ν is proportional to the real part of $n(\nu)[E_{\text{av}}(\nu)/E_{\text{eff}}(\nu)]^2$, the true absorption curve can be derived.

The conclusions of the Russian group [133] are similar to those given in ref. 81, in that considerable frequency shifts and intensity changes are found to result from the dispersion corrections, especially when resonance doublets are being considered. When strong intermolecular interactions are absent, the corrected frequencies and intensities are much closer to the gas-phase results than are the apparent absorptions.

130 B. J. Berne in 'Physical Chemistry – An Advanced Treatise', Volume VIIIB on the Liquid State, ed. H. Eyring, D. Henderson, and W. Jost, Academic Press, New York and London, 1970, p. 544.
131 R. L. Fulton, *J. Chem. Phys.*, 1971, **55**, 1386.
132 R. G. Gordon, *Adv. Magn. Resonance*, 1968, **3**, 1.
133 V. S. Libov, N. G. Bakhshiev, and O. P. Girin, *Optics and Spectroscopy*, 1964, **16**, 549.
134 C. Bötcher 'Theory of Electric Polarization', Elsevier, Amsterdam, 1952.
135 Ref. 134, p. 239.

In addition to using the Onsager–Bötcher relation,[134] use of the simple Lorentz equation

$$E_{eff} = \frac{n^2 + 2}{3} E_{av}$$

was explored. This amounts to assuming that the molecule may be treated as a uniformly polarizable sphere of radius r.[136] It may be surprising that the results are significantly different from those obtained using the Onsager–Bötcher model and are in less satisfactory agreement with the vapour data (see Figure 5). In equation (45) a and r are clearly effective polarizabilities and molecular radii. It is interesting to note that equation (45) yields effective values for these parameters that apparently reproduce correctly the reaction-field correction.

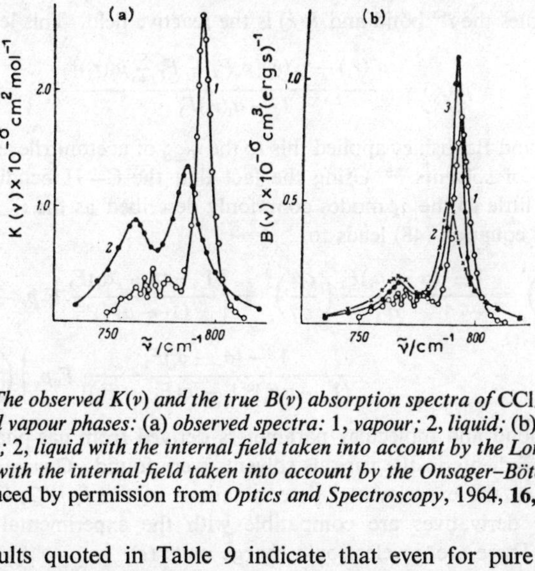

Figure 5 *The observed K(ν) and the true B(ν) absorption spectra of* CCl_4 *molecules in liquid and vapour phases:* (a) *observed spectra:* 1, *vapour;* 2, *liquid;* (b) *true spectra:* 1, *vapour;* 2, *liquid with the internal field taken into account by the Lorentz method;* 3, *liquid with the internal field taken into account by the Onsager–Bötcher method.* (Reproduced by permission from *Optics and Spectroscopy*, 1964, **16**, 549)

The results quoted in Table 9 indicate that even for pure liquids the intensity changes from the gas phase are, in these cases, primarily due to reaction-field and dielectric effects. Nevertheless deviations do occur, and these are particularly apparent when intensities of a given solute are compared in different solvents.

The interactions responsible can be classified into two groups. In the first, the interactions are polar in nature, but do not involve changes in the intramolecular bonding. In the second, specific bonds, such as hydrogen bonds, are formed. An interesting approach to this problem has been formulated

[136] J. O. Hirschfelder, C. F. Curtiss, and R. B. Bird 'Molecular Theory of Gases and Liquids', John Wiley and Sons, New York, and Chapman and Hall, London, 1954, p. 852.

by Girin and Bakhshiev.[137] The intermolecular potential, $V(r)$, is given by

$$V(r) = V_0' - p(r)[\tfrac{1}{2}F_1 p(r) + F_2 p_e] - \alpha(r)F_3 \qquad (46)$$

V_0' represents that part of the intermolecular potential which is independent of the vibrational co-ordinate r. p_e is the permanent dipole moment and $p(r)$ is the instantaneous value of the dipole. The second, third, and fourth terms on the right-hand side of equation (46) represent the induction, orientation, and dispersion interactions, respectively. F_1, F_2, and F_3 are given by dielectric theory as functions of the refractive indices, dielectric constants, and ionization potentials of the solvent and solute. The specific interactions, by which the formulation diverges from the earlier theory,[133] appear through allowing the bond dipoles $\mu_i(r)$ to vary as a result of the reaction field. Thus

$$\mu_i(r) = \mu_i^0(r) + \alpha_i(r_i)R(r) \qquad (47)$$

where i denotes the ith bond and $R(r)$ is the reactive field. This leads to

$$\mu_i(r_i) = \frac{\mu_i^0(r_i) + \alpha_i(r_i)[p_e F_2 + F_1 \sum_{j \neq i} \mu_j(r_j)]}{1 - \alpha_i(r_i)F_1} \qquad (48)$$

Akopyan and Bakhshiev applied this to the $\nu_{C\equiv N}$ of acetonitrile as measured in a variety of solvents.[138] Using the fact that the C—H bending motion contributes little to the a_1 modes commonly described as the ν_{C-C} and $\nu_{C\equiv N}$ modes, then equation (48) leads to

$$\left(\frac{\partial\mu}{\partial q}\right)_i^s = \frac{1-(\alpha-\alpha_i)F_1}{1-\alpha F_1}\left(\frac{\partial\mu}{\partial q}\right)_i^0 + \left\{\frac{1-(\alpha-\alpha_i)F_1}{(1-\alpha F_1)^2} F_2 p_e + \frac{1-(\alpha-\alpha_i)F_1}{(1-\alpha F_1)^2[1-\alpha(F_1+F_2)]} F_2 p_e\right\}\left(\frac{\partial\alpha}{\partial q}\right)_i$$

Using the bond and molecular parameters derived from gas-phase studies, the dipole gradients in the various solvents were found. Owing to the sign ambiguities in the experimental determinations of the $\partial\mu/\partial Q$, four sets of bond-dipole derivatives are compatible with the experimental gas-phase intensities. These are, in electronic charge units (e)

Set no.	$\left(\frac{\partial\mu}{\partial q}\right)_{CN}$	$\left(\frac{\partial\mu}{\partial q}\right)_{CC}$
I	−0.158	−0.087
II	0.158	0.087
III	−0.025	0.121
IV	0.025	0.121

which lead to the predicted sets quoted in Table 10. It is clear from these results that sets I and III are unacceptable but that the theory, along with sets II and IV, gives a pleasing reproduction of the experimental trends. The

[137] O. P. Girin and N. G. Bakhshiev, *Optics and Spectroscopy*, 1968, **24**, 488.
[138] S. Kh. Akopyan and N. G. Bakhshiev, *Optics and Spectroscopy*, 1969, **26**, 203.

greatest discrepancies are for benzene, chloroform, and nitromethane. It is with these solvents that chemical interactions might be most expected.

Table 10 *Integral intensities*/km mol^{-1} *of the* ν_{CN} *mode of acetonitrile in solution.* (From *Optics and Spectroscopy*, 1969, **26**, 203)

Solvent	*A*	A_{calc}			
		I	II	III	IV
C_6H_{14}	3.2	0.8	3.0	0.7	2.9
C_7H_{16}	3.4	0.7	3.1	0.7	2.9
CCl_4	4.4	0.6	3.3	0.6	3.2
C_6H_5Cl	6.0	0.1	5.1	0.1	4.9
C_6H_6	6.2	0.6	3.5	0.5	3.3
CH_2Cl_2	7.4	0.1	5.6	0.0	5.4
$C_6H_5NO_2$	8.4	0.0	6.5	0.0	6.2
$CHCl_3$	8.6	0.2	4.8	0.1	4.6
$C_2H_5NO_2$	9.6	0.0	6.3	0.0	6.1
CH_3NO_2	10.2	0.0	6.4	0.0	6.2
Gas	1.4	—	—	—	—

Akopyan and Bakhshiev have extended the studies of solvent effects on i.r. intensities to the ν_{C-C} mode of acetonitrile.[139] The same procedure that they applied to the $\nu_{C\equiv N}$ mode led them to the conclusion that the parameter set II with the larger $\left(\frac{\partial\mu}{\partial q}\right)_{CN}$ is correct (see Table 11).

Table 11 *Experimental and calculated values for the integrated intensity of* ν_{CC} *mode of acetonitrile.* (Data extracted from *Optics and Spectroscopy*, 1969, **27**, 375)

Solvent	A_{CC}*(exp.)*/km mol^{-1}	A_{CC}*(calc.)*/km mol^{-1}	
		II	IV
Gas	1.1	—	—
CCl_4	1.6	1.7	0.7
$CHBr_3$	2.0	2.1	0.5

Trichloroacetonitrile,[138] hydrogen cyanide, and deuterium cyanide[140, 141] have also been studied in this way. For Cl_3CCN the solution data indicate

$$(a) \text{ that } \left(\frac{\partial u}{\partial q}\right)_{CN} \simeq -0.129\,e, \quad \left(\frac{\partial u}{\partial q}\right)_{CC} \simeq +0.158\,e$$

[139] S. Kh. Akopyan and N. G. Bakhshiev, *Optics and Spectroscopy*, 1969, **27**, 375.
[140] O. P. Girin, S. Kh. Akopyan, and N. G. Bakhshiev, *Optics and Spectroscopy*, 1968, **25**, 111.
[141] S. Kh. Akopyan and N. G. Bakhshiev, *Optics and Spectroscopy*, 1969, **26**, 297.

and (*b*) that the bond dipole of the CN bond is not opposed to the dipoles of both the CC and CCl bonds. The data do not lead to a rejection of the other two possibilities for the relative signs of the dipoles of (CN, CC) or (CN, CCl) bonds.

With HCN and DCN there is no ambiguity in the signs and values of the dipole derivatives. Application of the known values of the parameters to the calculation of the intensity changes for the stretching modes produced reasonable results.[140] In the case of the deformation vibration the intensity decreases to about half of its value in going from gas to solution state. The model fails to reproduce this. The authors make the remarkable suggestion that the degeneracy is removed and hence the intensity is halved.[141]

The intensity changes accompanying dissolution of HCl and HBr were not found to be remarkable and were correctly predicted.[142] Of greater note is the large frequency shift on going from the gas phase. The model also allows the calculation of this frequency shift [137] and it is correctly predicted.[142] The shift is about 50 cm^{-1} for the $0 \rightarrow 1$ transition of HCl in CCl_4. It would appear then that one must be very cautious in taking such shifts as evidence for chemical interaction. Further evidence for the lack of such interactions comes from the ability of the theory to explain the observed temperature variation of the HCl intensity in CCl_4[143] (discussed in more detail on p. 437).

While the results of the Girin–Bakhshiev theory look impressive, a closer look at the form of the F_1, F_2, and F_3 parameters [137] is likely to dampen one's enthusiasm. These parameters are given by

$$F_1 = \frac{2}{a^3}\left(\frac{\varepsilon_v - 1}{2\varepsilon_v + 1}\right)$$

$$F_2 = \frac{2}{a^3}\left[\left(\frac{\varepsilon - 1}{2\varepsilon + 1}\right) - \left(\frac{\varepsilon_v - 1}{2\varepsilon_v + 1}\right)\right]$$

$$F_3 = \frac{3}{2a^3}\left(\frac{J_1 J_2}{J_1 + J_2}\right)\left(\frac{n^2 - 1}{n^2 + 2}\right)$$

where ε is the static dielectric constant of the medium, ε_v is the dielectric constant in the frequency region investigated, n is the refractive index of the medium, a is the Van der Waals radius of the molecule, and J_1 and J_2 are the ionization potentials of the interacting molecules.

These imply isotropic interactions at some average distance '*a*' irrespective of the site locations of the interacting dipoles. The inverse-cube dependence on '*a*' of all the terms makes it indeed remarkable that quantitative agreement with experiment is obtained.

A more traditional empirical approach to the effects of solvents on nitrile band intensities was employed by Figeys and Mahieu.[143] The ν_{CN} band intensities were measured from the absorption maxima and half-band-widths

[142] O. I. Arkhangelskaya, O. P. Girin, and N. G. Bakhshiev, *Optics and Spectroscopy*, 1969, **27**, 424.
[143] H. P. Figeys and V. Mahieu, *Spectrochim. Acta*, 1968, **24A**, 1553.

assuming a Lorentzian band shape. A plot of the square root of the intensity (proportional to $\partial p/\partial Q$) of various CN bands in a given solvent against the square root of the intensities in a reference solvent produced good linear graphs. This is an extension of the approach of Bayliss, Cole, and Little.[144] There was a general increase in gradient with increasing dipole moment of the solvent, but where there is a tendency to complex the slopes were larger than would otherwise be expected. Thus, chloroform and methylene chloride produced relatively large gradients. The gradients correlated well with Reichardt's E_T values [145] with the exception that the gradient for chloroform was much too high. E_T represents the u.v. transition energy of the solvatochromic intramolecular charge-transfer band of diphenylpyridinium *N*-phenolbetaine. Increased polarity of the solvent stabilizes the ground state. Aromatic nitriles were found to be more sensitive to dipolar interactions than aliphatic nitriles.

Solvents may have very different effects on the intensities of different bands of the compound. The studies of Bakhshiev and colleagues have shown one way in which such variations may arise, but chemical interactions are not covered by this theory. The difference in behaviour between the symmetric and antisymmetric CH_2 stretching modes of methylene chloride was explained [146] by Evans and Lo as the combined effect of two different processes. In methylene chloride the ν_{CH} mode of the b_1 species (ν_6) has virtually zero intensity. ν_1 of the a_1 species is weak but nevertheless considerably stronger. The relative strengths of the bands are attributed to a very small $\dfrac{\partial \mu_{CH}}{\partial r_{CH}}$ and a larger $\dfrac{\partial \mu_{CCl}}{\partial r_{CH}}$ of opposite sign. In the a_1 mode the intensity is largely governed by the latter derivative whereas this $\partial \mu_{CCl}/\partial r_{CH}$ increases in value, probably due to a charge-transfer state of the type A^-HB^+, caused through hydrogen-bonding. The observed intensity trends with solvents of different proton-acceptor strengths, and the observed independence of the Raman intensities of these bands, are quoted as supporting this hypothesis.

In methylene bromide the gas-phase intensity of ν_6 is about half of that of ν_1 whilst in methylene iodide ν_6 is stronger than ν_1. Solvent effects similar to those mentioned above are observed and are in accord with the proposed mechanism for the intensity changes in CH_2Cl_2; there is an increasing value for $\partial \mu_{CH}/\partial r_{CH}$ in the series CH_2Cl_2, CH_2Br_2, and CH_2I_2.

In an effort to gain a more quantitative knowledge of this sort of behaviour the solvent dependence of the a_1 modes of methyl iodide and [$^2H_3$]methyl iodide has been measured and the results have been interpreted using Gribov theory formulation.[147] It was found that intensities of the ν_{CH}, ν_{CD}, and ν_{CI}

144 N. S. Bayliss, A. R. H. Cole, and L. H. Little, *Spectrochim. Acta*, 1959, **15,** 12.
145 C. Reichardt, *Angew. Chem. Internat. Edn.*, 1965, **4,** 29.
146 J. C. Evans and G. Y. S. Lo, *Spectrochim. Acta*, 1965, **21,** 33.
147 S. Higuchi, S. Tanaka, and H. Kamada, *Spectrochim. Acta*, 1968, **24A,** 1929.

modes were linearly related to the reaction-field constant $(\varepsilon - 1)/(2\varepsilon + 1)$, but that the CH (CD) gradient was of opposite sign to the CI gradient. By contrast, the symmetric bending modes show an intensity maximum when plotted against $(\varepsilon - 1)/(2\varepsilon + 1)$ with the maximum roughly separating polar and non-polar solvents. A plot against $(n_D^2 - 1)/(2n_D^2 + 1)$† effectively removes the dielectric contribution of the dipole. Such a plot is linear and this implies that those factors which determine intensities and depend on the permanent dipoles cancel for this mode. Using Gribov formulation, the intensity changes are related to functions of the electro-optical parameters. The relatively simple and orderly fashion of the intensity changes is taken to indicate that only one or two terms contribute to the changes in each combination. The authors' reasoning (equivalent to accepting the bond-moment hypothesis) leads them to suggest that $\partial\mu_{CH}/\partial r_{CH}$ becomes more positive with increasing polarity and at the same time $\partial\mu_{CH}/\partial r_{CI}$ becomes more negative. An explanation of the behaviour of the bending terms is more complex. It appears that as much as can be said at the present is that the dipole gradient is the sum of two competing terms—probably $\partial\mu/\partial\alpha$ and $\partial\mu/\partial\beta$.

† n_D is the refractive index obtained using the sodium D lines.

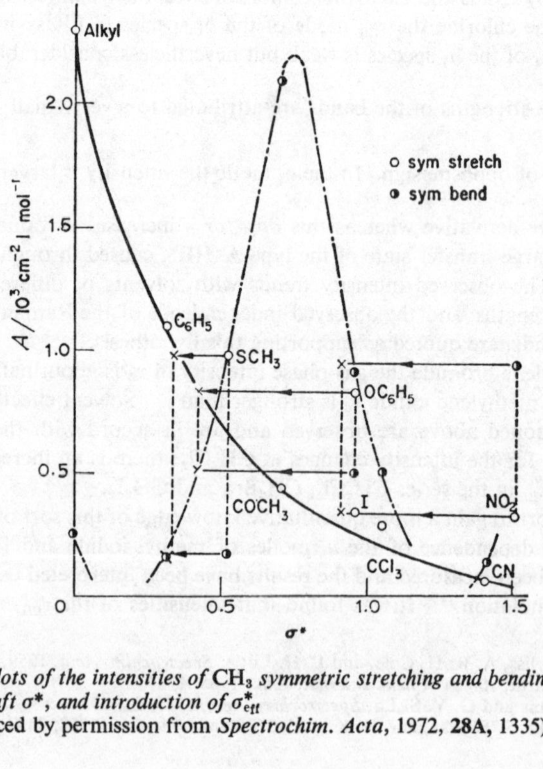

Figure 6 *Plots of the intensities of* CH_3 *symmetric stretching and bending vibrations against Taft* σ^**, and introduction of* σ^*_{eff}
(Reproduced by permission from *Spectrochim. Acta*, 1972, **28A**, 1335)

A subsequent paper on the symmetric stretching and bending vibrations of the methyl group [148] examines the effect of different groupings attached to the methyl group. All measurements were made in carbon tetrachloride. The general trend was for the intensity of the symmetric stretching vibration to increase with increasing polarity of the substituent. The bending mode, as in the solvent-dependence studies, gave an intensity maximum for a substituent of intermediate electronegativity. Plots of intensity against Taft σ^* values yield smooth curves for those compounds in which the methyl group was attached directly to a carbon atom (see Figure 6), but for those molecules in which the group was attached to atoms having lone pairs of electrons, the points were displaced to higher σ^* values. A relatively smooth curve containing all points was obtained when the intensities of the stretching and bending modes were plotted against one another. This latter graph implied that subtraction of a constant from the σ^* value simultaneously pulled the intensities of both stretching and bending modes on to the intensity $- \sigma^*$ curves. These corrections were taken to represent the partial cancellation of the polarization $CH_3^+—R^-$ by delocalization of the *n*-electrons. Since σ^* reflects the action of a group through a saturated carbon atom, this procedure appears to be justified. A clear example of inversion of the normal order is given by the methyl halides, where the intensity of the symmetric stretching mode decreases as we proceed through the series Cl, Br, to I (see Table 12).

Table 12 *Gas-phase intensity data of symmetric* CH_3 *stretching vibration of the methyl halides.* (From L. A. Gribov, 'Intensity Theory for Infrared Spectra of Polyatomic Molecules', translated by P. P. Sutton, Consultants Bureau, New York, 1964, p. 61)

Compound CH_3X	$\sigma^*(X)$	$\tilde{\nu}/cm^{-1}$	$A/km\ mol^{-1}$
CH_3Cl	1.15	3062	10.00
CH_3Br	1.10	3061	9.00
CH_3I	0.95	3060	6.20

MO calculations are reported as demonstrating that the degree of delocalization of *n*-electrons increases in an orderly fashion, from iodide, through bromide, to chloride.[149] No new light has been shed on the problem of the intensities of the bending modes by this study. The range of intensities of the methyl modes is quite remarkable, the symmetric stretch and the bend varying by factors of 35 and 13, respectively. Facts like these emphasize the importance of gaining a deeper understanding of the factors involved.

Methylene chloride has been the subject of a number of investigations in addition to that of Evans and Lo. The measured intensities in the different investigations have been compiled in Table 13. Evans and Lo do not quote their actual measured intensities but say that they agree within experimental

[148] S. Higuchi, E. Kuno, S. Tanaka, and H. Kamada, *Spectrochim. Acta*, 1972, **28A**, 1335.
[149] Y. Nikei, Doctoral Thesis, Dependent of Industrial Chemistry, Faculty of Engineering, University of Tokyo, 1969.

Table 13 *Reported absorption intensities*/km mol^{-1} *of the vibrational bands of methylene chloride in vapour* (A_v), *solution in* CCl_4(A_s), *and in the pure liquid* (A_L) *states.* *Wavenumbers*/cm^{-1}

Symmetry	$\tilde{\nu}$	A_v[a]	A_v[b]	A_v[c]	A_s[a]	A_s[d] (*conc.*/mol l^{-1})	A_s[e] (*conc.*/mol l^{-1})	A_L[e]	A_L[f]	A_L[g]
a_1	3137	6.90	7.26	7.2	4.60	5.50 (0.059)	5.60 (0.2)	3.70	1.90	—
						5.20 (0.275)	4.70 (1.1)	—	—	—
						4.90 (0.5)	—	—	—	—
	1430	0.60	2.28	—	2.10	—	—	—	3.80	2.90
	714	8.00	7.92	—	9.00	—	—	—	19.90	7.3
	283	0.60	—	—	0.60	—	—	—	—	—
b_1	3195	0.0	—	—	2.60	2.90 (0.059)	1.30 (0.2)	5.20	4.50	—
						3.40 (0.275)	2.10 (1.1)	—	—	—
						4.10 (0.5)	—	—	—	—
	896	1.20	1.17	1.0	3.00	—	—	—	3.80	1.6
b_2	1268	26.6	32.8	—	27.0	—	—	—	28.8	27.2
	737	95	119	106	115	—	—	—	123	133

[a] From S. Saëki and K. Tanabe, *Spectrochim. Acta*, 1969, **25A**, 1325. A_s values are for dilute solution in CCl_4 (concentration not specified). [b] J. Morcillo, L. J. Zamorano, and J. M. V. Heredia, *Spectrochim. Acta*, 1966, **22**, 1969. The quoted A^G for $\tilde{\nu} = 3137$ cm^{-1} is the sum of those listed by these authors for ν_1 and ν_6. For $\tilde{\nu} = 1430$ cm^{-1} the intensity contains a large contribution from a band not considered as fundamental by ref. 152. [c] J. W. Straley, *J. Chem. Phys.*, 1955, **23**, 2183. [d] E. S. Jayadevappa and S. K. Haibatti, *Z. phys. Chem.* (*Leipzig*), 1969, **242**, 67; (in CCl_4). [e] M. T. Forel, A. Lafaix, P. Saumagne, and M. L. Josien, *J. Chim. phys.*, 1963, **60**, 50. Data confirmed by Evans and Lo. [f] W. H. Prichard and W. J. Orville-Thomas, *Trans. Faraday Soc.*, 1963, **59**, 2218. [g] J. A. Ramsey, J. A. Ladd, and W. J. Orville-Thomas, *J.C.S. Faraday II*, 1972, **68**, 193.

error with those of Forel *et al.*[150] In comparing the results it should be noted that, owing to assignment changes which now appear to be accepted, the gas-phase intensity of Morcillo *et al.*[151] for the 1430 cm^{-1} a_1 band (ν_2) is dominated by what Saëki and Tanabe believed to be a combination band.[152] Saëki and Tanabe therefore excluded this from their measurement. In addition, the intensity quoted for the ν_1 a_1 band is equally divided between the 3137 and 3195 cm^{-1} (a_1 and b_2) assignments in ref. 151. Saëki and Tanabe measured the vapour and solution intensities for CH_2Cl_2 $CDHCl_2$, and CD_2Cl_2.[152] Crawford's sum rules are regarded to be well satisfied by the results for the different isotopic species in both phases. It can be seen from Table 13 that the vapour results agree well with those of ref. 151. At first sight the solution results of refs. 150, 152, and 153 are in poor accord. The difficulty lies in separation of the overlapping symmetric and antisymmetric stretches. The sums of the intensities are in far superior agreement. Use of the Kramers–Krönig relations may be of some use here in determining the relative strengths. The measurements of Prichard and Orville-Thomas [154] on the pure liquid are from dispersion studies and appear to give a rather low total C—H stretching intensity. At the opposite end of the frequency scale their value for the $a_1\nu_{C-Cl}$ intensity appears to be very high when compared with the vapour and solution results. However, these results were obtained from the dispersion data using an incomplete Schatz dispersion equation and have been significantly modified by applying the full dispersion formula,[155] as shown in the last column of Table 13.

Saëki and Tanabe sought to interpret the vapour and solution results using a complete Gribov formulation.[152] It turns out that there are eleven combinations of bond parameters which may be derived from the intensities of the three isotopic related systems. Of course, there are not $3(3N - 6) = 27$ independent values, as a first reading of the paper would imply, but only about half of this number (the reviewers make it 13; 8 for CH_2Cl_2 and another 5 for CD_2Cl_2 allowing for the three product ratios, and the $CHDCl_2$ probably gives no further constants as a result of higher-order intensity rules.) It is not surprising then that infinite uncertainties are quoted for 7 of the 11 parameters. Three different force fields were developed in order to determine the eigenvectors, but this treatment suffers from the same problem as the subsequent intensity-parameter evaluation—the majority of them are just not determinable as a result of correlations between the parameters. It seems unfortunate that no attempts were made to constrain some of the probably least important parameters to see if probable ranges of force constants and intensity parameters could be deduced. Nevertheless the work served its main purpose well: those terms involving dipole gradients with respect to

[150] M. T. Forel, A. Lafaix, P. Saumagne, and M. L. Jodien, *J. Chim. phys.*, 1963, **60**, 50.
[151] J. Morcillo, L. J. Zamorano, and J. M. V. Heredia, *Spectrochim.* Acta, 1966, **22**, 1969.
[152] S. Saëki and K. Tanabe, *Spectrochim. Acta*, 1969, **25**, 1325.
[153] E. S. Jayadevappa and S. K. Haibatti *Z. phys. Chem.* (*Leipzig*), 1969, **242**, 67.
[154] W. H. Prichard and W. J. Orville-Thomas, *Trans. Faraday Soc.*, 1963, **59**, 2218.
[155] J. A. Ramsey, J. A. Ladd, and W. J. Orville-Thomas, *J.C.S. Faraday II*, 1972, **68**, 193.

bond stretches were the four well-determined intensity parameters for a given assumed field, the values did not vary much between force fields, and the changes in the terms on going from vapour to liquid form a clear pattern. Eight different sets of the 11 parameters were found which adequately reproduced the intensities. Four sets were considered to be unlikely in that they predicted a higher intensity for the CD_2 rocking mode (masked by ν_7 band) than for the CH_2 rocking mode. The remaining sets of parameters for the vapour and the definitions of the parameters are given in Table 14. The authors then follow a suggestion of Gribov [6] that if an interaction force constant linking two internal co-ordinates, j and k, is small then the corresponding dipole terms $\partial\mu_j/\partial r_k$ and $\partial\mu_k/\partial r_j$ are likely to be small, and from this they deduce the bond gradients shown in Table 14. The principle behind this

Table 14 *The definitions of those symmetry co-ordinates of methylene chloride which involve stretching motions; the preferred values of the dipole derivatives with respect to these symmetry co-ordinates for the vapour; and the derived bond-dipole-stretching gradients for both vapour and solution phases.* (From S. Saëki and K. Tanabe, *Spectrochim. Acta*, 1969, **25A**, 1325)

$$P_3 = \sqrt{2}\cos\left(\frac{\alpha}{2}\right)\left(\frac{\partial\mu_{CH}}{\partial r_{CH}} + \frac{\partial\mu_{CH}}{\partial r'_{CH}}\right) - \sqrt{8}\cos\left(\frac{\beta}{2}\right)\left(\frac{\partial\mu_{CCl}}{\partial r_{CH}}\right)$$

$$P_4 = \sqrt{8}\cos\left(\frac{\alpha}{2}\right)\left(\frac{\partial\mu_{CH}}{\partial r_{CCl}}\right) = \sqrt{2}\cos\left(\frac{\beta}{2}\right)\left(\frac{\partial\mu_{CCl}}{\partial r_{CCl}} + \frac{\partial\mu_{CCl}}{\partial r'_{CCl}}\right)$$

$$P_8 = \sqrt{2}\sin\left(\frac{\alpha}{2}\right)\left(\frac{\partial\mu_{CH}}{\partial r_{CH}} - \frac{\partial\mu_{CH}}{\partial r'_{CH}}\right)$$

$$P_{10} = \sqrt{2}\sin\left(\frac{\beta}{2}\right)\left(\frac{\partial\mu_{CCl}}{\partial r_{CCl}} - \frac{\partial\mu_{CCl}}{\partial r'_{CCl}}\right)$$

	(1)	(2)	(3)	(4)	
P_3	0.066	−0.066	0.066	−0.066	
P_4	−0.339	0.339	−0.339	0.339	(in units of *e*—for vapour and
P_8	0.018	0.018	0.018	0.018	general force field quadratic)
P_{10}	0.881	0.881	−0.739	−0.739	

	Vapour				*Liquid*			
	(1)	(2)	(3)	(4)	(1)	(2)	(3)	(4)
$\frac{\partial\mu_{CH}}{\partial r_{CH}}$	0.017	0.017	0.017	0.017	−0.029	−0.029	−0.029	−0.029
$\frac{\partial\mu_{CH}}{\partial r_{CCl}}$	0.162	0.591	−0.531	−0.102	0.181	0.610	−0.552	−0.123
$\frac{\partial\mu_{CCl}}{\partial r_{CCl}}$	0.752	0.752	−0.631	−0.631	0.789	0.789	−0.675	−0.675
$\frac{\partial\mu_{CCl}}{\partial r_{CH}}$	0.035	0.050	0.035	0.050	−0.044	0.017	−0.044	0.017

assumption appears to be of rather doubtful validity and it would be interesting to check this by, for example, CNDO calculations. The result that $\partial\mu_{CH}/\partial r_{CH}$ changes sign in going from vapour to liquid is unlikely to be affected by the assumption. Also the gradient $\partial\mu_{CCl}/\partial r_{CH}$ changes by -0.06 to -0.08 *e* in the corresponding phase transition. These changes are sufficient to explain the observed behaviour of the C—H stretching bands and thus it seems that this study has furnished an alternative and quantitative picture of the effects of the condensed-state interactions.

The absorption intensities of a number of aliphatic nitriles XCN, where X = Me, Et, Pr^i, Bu^t, Br, or I, and also CH_2=CHCN and *cis*- and *trans*-CH_2=CHCH=CHCN have been measured in solution and as pure liquids.[156] The correlation of intensities as measured in solution with Taft σ^* factors was satisfactory. Intensities in chloroform were approximately twice as great as those measured in CCl_4 or C_2Cl_4. This parallels earlier observations on benzonitriles [157, 158] and is strong evidence for hydrogen-bonding between the nitrogen atom of the nitrile and the chloroform proton. The intensity trends in RCN systems with varying R are explained through inductive effects on the resonance structures R—C≡N and R—$\overset{+}{C}$=$\overset{-}{N}$. In the pure liquid phase, the $\nu_{C\equiv N}$ band intensities in the series $(CH_3)_nCH_{3-n}$ CN do not exhibit smooth changes with variations in σ^*, and this is explained as due to association of the molecules in antiparallel pairs. As with the methyl systems, the enormous range of intensities of the characteristic vibrations should be noted. For example, in CCl_4 solution $ClCH_2CN$ has an A_{CN} of 0.39 km mol^{-1} whereas BrCN has a corresponding intensity of 12.4 km mol^{-1}.

Dispersion measurements on XCH_2CN (X = F, Cl, or I) have been made [155] and the data interpreted using the dispersion equation

$$n(\nu) = n_\infty + \frac{1}{2\pi^2}\sum_j \frac{A_j}{\nu_j{}^2 - \nu^2} \tag{49}$$

Results are compared with those obtained by plots of refractive indices against $(\nu_j^2 - \nu^2)^{-1}$ using data near the centre of the band. The results reported are significantly different, in some cases the results differing by a factor of two. It is clear that one must exercise caution in using the partial form

$$n(\nu) = n + \frac{A_j}{\nu_j^2 - \nu^2} \tag{50}$$

It must be realized that equation (49) is valid only outside the main region of the absorption band. Once within the frequency range of intense absorption it is necessary to resort to the fuller form of Kramers–Krönig relation,

$$n(\nu)^2 - 1 = \frac{2}{\pi}\mathscr{P}\int \frac{2(n_{\nu'}\kappa_{\nu'})\nu'}{\nu'^2 - \nu^2}\,d\nu \tag{51}$$

[156] B. H. Thomas and W. J. Orville-Thomas, *J. Mol. Structure*, 1971, **7**, 123.
[157] T. L. Brown, *J. Amer. Chem. Soc.*, 1958, **80**, 794.
[158] P. J. Krueger and H. W. Thompson, *Proc. Roy. Soc.*, 1958, **A250**, 22.

where $n(\nu)$ and κ are the real and imaginary parts of the complex refractive index and $\mathscr{P}$ indicates that the principal part of the integral is to be taken, that is, the singularity at $\nu' = \nu$ is omitted from the integration. In addition to new data on the XCH_2CN systems a re-analysis of previous dispersion data on CH_3CN, CH_3I, CH_2Cl_2, CH_2Br_2, and CH_2I_2 is reported. Table 15 compares the recent liquid and solution intensities for CH_3I.

Table 15 *Comparison of the reported intensities*/km mol^{-1} *for the vibrational bands of* CH_3I *in liquid and solution states*

	$\tilde{\nu}$/cm^{-1}	$A_L{}^a$	$A_L{}^b$	$A_s{}^{c,d}$	$A_s{}^{c,e}$
a_1	2950	—	9.9	—	8.0
	1242	—	31.8	21.4	21.2
	522	3.75	3.8	2.3	2.2
e	1429	—	22.6	—	—
	887	17.4	20.4	—	—

[a] C. E. Favelukes, A. A. Clifford, and B. Crawford, jun., *J. Phys. Chem.*, 1968, **72**, 962.
[b] J. A. Ramsey, J. A. Ladd, and W. J. Orville-Thomas, *J.C.S. Faraday II*, 1972, **68**, 193.
[c] S. Higuchi, S. Tanaka, and H. Kamada, *Spectrochim. Acta*, 1968, **24A**, 1929. [d] solvent n-pentane. [e] solvent carbon tetrachloride. Intensities in other solvents are also quoted.

Chloroform, bromoform, and their deuteriated forms have been studied by the same procedure.[159] The values obtained for the bromoform intensities are in moderate agreement with earlier values for ν_4 and ν_5 (10% and 2%, respectively). The results for chloroform are compared with earlier literature values in Table 16. Values for ν_5 are reasonably consistent. We can exclude here the absorption value of Lisitsa and Tsyashchenko as bands of such high intensity present formidable problems in absorption studies. The previous literature data on other bands are very sparse. As the weaker bands are best studied by absorption methods the case is strong for further investigation using modern medium-resolution spectrometers.

Zelano and King used equation (49) to interpret their interferometric data on liquid benzene and [$^2H_6$]benzene.[160] Results are compared with earlier measurements. Agreement is not at all satisfactory—especially with the absorption values. Thus for the 1037 cm^{-1} band of C_6H_6 values of 11.5,[161] 11.4 and 12.9,[162] 18.0,[163] and 19.3 km mol^{-1} [160] have been reported by absorption, ATR, dispersion, and the interferometric method, respectively (*cf.* Table 9). It is extremely disappointing to find such large discrepancies still occurring between recently measured intensities—especially between such accurate studies as reported in refs. 81 and 160. However, the authors of the interferometric study have recently reported measurements of the intensities of the benzene bands and the [$^2H_6$]benzene bands in various

159 H. Ratajczak, T. A. Ford, and W. J. Orville-Thomas, *J. Mol. Structure*, 1972, **14**, 281.
160 A. J. Zelano and W. T. King, *J. Chem. Phys.*, 1970, **53**, 4444.
161 I. C. Hisatsune and E. S. Jayadevappa, *J. Chem. Phys.*, 1960, **32**, 565.
162 A. C. Gilby, J. Burr, jun., and B. Crawford, jun., *J. Phys. Chem.*, 1966, **70**, 1525.
163 S. Maeda and P. N. Schatz, *J. Chem. Phys.*, 1962, **36**, 571.

Table 16 *The integrated absorption intensities*/km mol^{-1} *of the vibrational bands of chloroform*

	$\tilde{\nu}$/cm^{-1}	A_L^a	A_L^b	A_L^c	A_L^d	A_L^e	A_L^f	A_L^g	A_L^h
A_1	3040	5.7	7.1	6.5	—	—	—	—	—
	667	6.8^i	—	2.9	—	—	—	—	5.6
	370	0.13	—	—	—	—	—	—	—
E	1214	38.2	—	21.6	—	—	—	—	—
	755	307.0	—	178	342	281	286	297	329
	261	1.2	—	—	—	—	—	—	—
Measurement		D	A	A	R	D	D	R	D

[a] H. Ratajczak, T. A. Ford, and W. J. Orville-Thomas, *J. Mol. Structure*, 1972, **14**, 281. [b] G. J. Boobyer, *Spectrochim. Acta*, 1967, **23A**, 335. [c] M. P. Lisitsa and Yu. P. Tsyashchenko, *Optics and Spectroscopy*, 1959, **6**, 396. [d] P. N. Schatz, S. Maeda, J. L. Hollenberg, and D. A. Dows, *J. Chem. Phys.*, 1961, **34**, 175. [e] L. P. Lindsay, S. Maeda, and P. N. Schatz, see ref. *d*. [f] A. C. Gilby, J. Burr, jun., W. Krueger, and B. Crawford, jun., *J. Phys. Chem.*, 1966, **70**, 1525. [g] V. S. Libov, N. G. Bakhshiev, and O. P. Girin, *Optics and Spectroscopy*, 1964, **16**, 549. [h] M. Cameo and J. Vincent Geisse, *Compt. rend.*, 1961, **252**, 1579. [i] Absorption value taken from R. E. Kagarise, NRL Report No. 5906, Washington D.C., 1963.

D = Dispersion; A = Absorption; R = Reflection.

solvents.[164] The values quoted in this paper for the pure liquid intensities are different. For the 1037 cm^{-1} band it is given as 14.8 km mol^{-1} in pure benzene. Broun and Iogansen have measured the CD stretching vibration intensities of [$^2H_6$]benzene and C_6D_{12} in a variety of solvents of greatly varying polarity.[165] Variations from solvent to solvent were small (less than 20%). For C_6D_6 the reported solution value of A_{CD} is in good agreement with Zelano and King's recent liquid and solution values.

The factors which govern the intensity variations in carbonyls continue to attract attention. Wexler reviewed this work thoroughly up to 1967.[7] A further contribution to the study of the effects of ring substitution on the carbonyl band of aromatic aldehydes shows that *A* increases from 230 to 290 km mol^{-1} with *para*-substitution of the *NN*-dimethylamino-group on benzaldehyde.[166] Neither chloro- nor fluoro-alkyl groups on the nitrogen affect the intensity significantly. Methyl groups *ortho* to the aldehyde group do not significantly alter it either, indicating that there is already a strong steric hindrance to conjugation of the C=O with the ring.

Heess and Kriegsmann have measured the intensities of a wide variety of carbonyl groups in solvents of varying donor–acceptor ability.[167] The intensity values in CCl_4 are in the ranges expected on the basis of Wexler's tables. Plots of the intensities of the $\nu_{C=O}$ of a reference compound in various solvents against intensities of other selected $\nu_{C=O}$ bands in corresponding solvents give good straight lines. The reverse problem of the dependence of

[164] A. J. Zelano and W. T. King, *J. Chem. Phys.*, 1972, **57**, 4370.
[165] E. V. Broun and A. V. Iogansen, *Optics and Spectroscopy*, 1968, **24**, 207.
[166] M. I. Zavadskaya and I. I. Chizhevskaya, *J. Appl. Spectroscopy*, 1970, **12**, 134.
[167] R. Heess and H. Kriegsmann, *Spectrochim. Acta*, 1968, **24A**, 2121.

the intensity of bands of the solvents on the nature of the carbonyl group was examined. Bands such as ν_{CD} of $CDCl_3$ and $\nu_{C\equiv N}$ of CH_3CN show a considerable dependence on the nature of the carbonyl group (for example, see Table 17).

Table 17 *The intensities*/km mol^{-1} *of some stretching vibrations in a variety of solvents.* (From *Spectrochim. Acta*, 1968, **24A**, 2121)

Vibration	C_6H_{12}	MeCOMe	$HCONMe_2$
ν_{CD} $CDCl_3$	1.9	30.2	51.6
$\nu_{C\equiv N}$ CH_3CN	2.5	8.1	11.1
ν^{E}_{CCl} CCl_4	147	162	158

The intensity variations of the carbonyl bands between C_6H_{12}, CCl_4, and CH_3CN are within 20%. This is substantiated by a comprehensive study of carbonyl bands in acetonitrile by Wood and Buckingham.[168] The results agreed satisfactorily with literature values in non-polar solvents. The results are used to supplement n.m.r., u.v., and chemical data in elucidating the structures of some toxic fungal metabolites, rubratoxins A and B.

The strengths of metal carbonyl bands have been used in a number of instances to monitor metal–ligand interactions. Thus in a series of cobalt carbonyl nitrosyl complexes $Co(CO_2)(NO)L$ [where L = PPh_3, $AsPh_3$, $P(OMe)_3$, $SbPh_3$, $P(n\text{-}C_4H_9)_3$, CO, and $P(OPh)_3$] it was found that the absorption strength per CO group increased over that of the parent $Co(CO)_3(NO)$ on substitution while the strength of the $\nu_{N=O}$ decreased.[169] The mechanism of the interaction was discussed. From the relative strengths of the symmetric and the antisymmetric modes the changes in the CCoN angles were estimated by using the simple bond-moment hypothesis. The ability to compute the interbond angles in the metal carbonyls from intensities has been refuted by Darensbourg.[170] This latter author has analysed the intensities of the $\nu_{C\equiv O}$ modes of trigonal-bipyramidal cobalt and iron tetracarbonyl compounds (C_{3v} symmetry) in some detail. The intensities of the A_1 modes are expressed in terms of the dipole derivatives with respect to C≡O symmetry co-ordinates (μ'_{MCO}) plus a contribution from electronic migration along the three-fold axis during the stretching of the equatorial carbonyl groups. The latter of course is a term which is extracted from the definitions of the μ'_{MCO}. Since arguments based on π electronic charge migration during the CO stretches suggest that the dipole derivative with respect to the degenerate symmetry co-ordinate should be greater than those of the symmetric stretches, and this result is contrary to observation, it is concluded that there is a transfer of π electronic character from the silicon or phosphorus ligands through the metal to the carbonyl group as the CO groups

168 A. B. Wood and B. Buckingham, *Spectrochim. Acta*, 1970, **26A**, 465.
169 A. Poletti, A. Foffani, and R. Cataliotti, *Spectrochim. Acta*, 1970, **26A**, 1063.
170 D. J. Darensbourg, *Inorg. Chim. Acta*, 1970, **4**, 597.

stretch. References to many papers published before 1970 on this subject are to be found in refs. 169 and 170.

Frequency shifts have been demonstrated to give anomalous measures of bond strengths in metal–ligand systems. For instance, the frequency shift of $\nu_{N\equiv N}$ in $IrN_2(PPh_3)_2Cl$ is 221 cm^{-1} whereas for $\Delta\nu_{C\equiv O}$ in $IrCO(PPh_3)_2Cl$ it is 178 cm^{-1}. The nitrogen is considerably more unstable than the carbon monoxide complex. Attempts have been made to rationalize these results, but clearly the frequency shift remains an unsatisfactory measure of metal to ligand bonding—at least when comparing different ligands. Darensbourg has measured the intensities of the group vibrations near 2000 cm^{-1} and noted that the effective group dipole of N_2 is less than that of CO.[170, 171] From this he deduces that the intensities are a better measure of bonding power. Regrettably, it seems too naïve to take the relative intensities of a bonded homonuclear and of a heteronuclear diatomic molecule as measures of the metal–ligand bond strengths. The dipole gradients of the metal-to-ligand stretching vibration would have been a much more convincing measure. Unfortunately, the complexity of the far-i.r. spectra foiled an attempt to measure these gradients.[172]

The integrated intensities of the phosphoryl absorption bands of 16 compounds have been correlated with inductive and resonance parameters.[173] The results indicate that the effects of the substituents are additive and that inductive and resonance effects are both operative. Measured intensities are a factor 2—3 times less than those reported by other investigators [174, 175] (could this be $\log_e 10$?). The Russian group have examined the intensity problem of $POCl_3$ in detail [174] using a complete electro-optical parameter formulation and have shown that the PO intensity is determined by $\partial\mu_{PO}/\partial r_{PO}$ and $\partial\mu_{PX}/\partial r_{PO}$. Intensity changes with change of X will arise from both terms.

Both the carbonyl and phosphoryl stretching intensities have been shown to correlate well with the hydrogen-bonding ability of the solvent as measured by the $\Delta\nu_{OH}$ of phenol in the solvent.[176]

Intensities have been listed for a band shown by several thioketones in the range 1270—1244 cm^{-1}.[177] The number of compounds studied was only four. For Me_2CS the intensity is reported as 75 km mol^{-1} whereas for the other compounds with more-complex alkyl groups the intensities are 20—26 km mol^{-1}. This may reflect the differing contributions of alkyl distortions in the vibrations.

171 D. J. Darensbourg and C. L. Hyde, *Inorg. Chem.*, 1971, **10**, 431.
172 D. J. Darensbourg, *Inorg. Chem.*, 1971, **10**, 2399.
173 H. Goldwhite and J. Previdi, *Spectrochim. Acta*, 1970, **26A**, 1403.
174 O. G. Strukov, V. A. Shiyapochnikov, and S. S. Dubov, *J. Appl. Spectroscopy*, 1967, **7**, 150.
175 R. Mathis-Noel, M. T. Boisdon, J. P. Vives, and F. Mathis, *Compt. rend.*, 1963, **257**, 402.
176 T. Gramstad and H. J. Storesund, *Spectrochim. Acta*, 1970, **26A**, 426.
177 C. Andrieu and Y. Mollier, *Spectrochim. Acta*, 1972, **28A**, 785.

The ν_3 band of CS_2 has been measured in chloroform and carbon tetrachloride using a rather large spectral slit width of 7 cm^{-1}.[178] It is not surprising that the values reported are about 30% low as compared with those measured using a grating spectrometer.[179] David and Hallam examined the applicability of the Buckingham equation for solution intensities[180] to solutions of HCl, DCl, SO_2, CS_2, and $SOCl_2$ in CCl_4, CCl_2CCl_2, C_6H_6, and MeCN. Plots of intensities against $(\varepsilon - 1)/(2\varepsilon + 1)$ were reasonably satisfactory for SO_2. SO_2 in benzene gave points well off the line, indicating strong intermolecular interactions. Thionyl chloride and CS_2 results do not correlate at all well. The authors explored the effects of an ellipsoidal solvent cavity on the predicted intensity changes going from vapour to solution.[181] These can be quite large.

An almost ideal application of Taft constants is given by correlation of the integral absorptions of OH, CCl, CBr, CI, and NO_2 stretching vibrations in a homologous series of unbranched hydrocarbons with the Taft constants of the alkyl groups.[182] It is suggested that the slopes are a good measure of the electronegativities of the functional groups.

Solvent and concentration effects on the integrated intensities of the symmetric and antisymmetric NH stretching intensities of aliphatic and aromatic primary amines have been studied.[183] Earlier work had showed that the frequency separation increased when only one hydrogen-bond was formed, but remained unaltered when two formed.[184] Values of $\partial\mu_{NH}/\partial q_{NH}$ and $\partial\mu_{NH}/\partial q'_{NH}$ are tabulated for a wide variety of amines and solvents but not discussed.

The spectral intensities of the hydrogen halides in solution have been the subject of two series of papers. Scrocco and collaborators[111] measured the intensities of the $1 \leftarrow 0$ and $2 \leftarrow 0$ bands of HCl, HBr, and HI in polar solvents. The gas-phase data of other authors were first analysed using a dipole function obtained with the aid of a Morse potential function to derive the dipole-moment expansion up to the second power in the distortion. This was then compared with values obtained from solution data. Two sets of solutions exist, corresponding to the sign ambiguities in $\sqrt{A_{2 \leftarrow 0}}$. Results should be compared with those of the more detailed analyses given in refs. 103 and 104. A model is described which leads to the derivation of bond and lone-pair moment contributions to the dipole moment from the measured molecular static dipole and the intensities of the stretching fundamental and its overtone.[112] Application of the model to the hydrogen halides in CCl_4[113] suggests that bond and lone-pair dipoles increase in solution for all the

[178] S. Badilescu, *Rev. Roumaine Chim.*, 1971, **16**, 9.
[179] J. G. David and H. E. Hallam, *Trans. Faraday Soc.*, 1969, **65**, 2838.
[180] A. D. Buckingham, *Proc. Roy. Soc.*, 1958, **A248**, 169; 1960, **A255**, 32.
[181] J. G. David and H. E. Hallam, *Trans. Faraday Soc.*, 1969, **65**, 2843.
[182] G. Geiseler, J. Fruwert, L. Schmidt, and S. Schwalbe, *Spectrochim. Acta*, 1968, **24A**, 1007.
[183] E. L. Zhukova and I. I. Shmanko, *Optics and Spectroscopy*, 1969, **26**, 295.
[184] E. L. Zhukova and I. I. Shmanko, *Optics and Spectroscopy*, 1968, **25**, 279.

halides. It is not clear to what extent the dielectric effects of the medium have been taken into account.

Girin, Bakhshiev, and Maksimova [185] employed a different model from that due to Girin and Bakhshiev [137] to interpret the intensity changes observed for HCl in going from gas to solution in CCl_4. Both frequency shifts and intensity changes were adequately given by the theory. This does not of course imply an invariant dipole derivative, but rather that the electrostatic interactions arising from induced dipoles are adequate to explain the changes. It does not appear necessary to involve specific bonding interactions. The application of the method to the evaluation of the dipole and polarizability expansions in solution has been discussed.[186]

A large number of papers have now appeared on the dependence of aromatic ring vibrations on substituents. An expansion of the original work on monosubstituted benzenes [187] has led to a correlation of the intensities of the ν_{16} bands near 1600 cm^{-1} of 110 monosubstituted benzenes with the σ_R^0 of the substituents.[188] The correlation is good. Certain discrepancies were rationalized. Thus the σ_R^0 values derived from ^{19}F n.m.r. for Cl, Br, SH, and SMe appear to be low, and this is explained as due to *d*-orbital participation in the corresponding *para*-substituted fluorobenzenes used for the n.m.r. determinations. In substituted durenes substituents with non-cylindrical symmetry, such as OMe and NMe_2, have low apparent σ_R^0 values, implying loss of conjugation due to steric effects.

For *para*-substituted systems the sum of the intensities of ν_{16a} and ν_{16b} does not correlate well directly with the sums of the σ_R^0. Allowance has to be made for interaction between the substituents.[189] Various types of interactions (*e.g.* *d*-orbital interactions of Cl, Br, and I with strong donors) are discussed. In a similar manner the intensities of the ν_{13} band near 1500 cm^{-1} [mainly β(CH) vibration] of both mono- and *para*-di-substituted benzenes correlate satisfactorily with the resonance parameters.[190] Donor–acceptor 'through conjugation effects' are much less important in ν_{13a} than in ν_{16}. As in all this series of papers, such observations are rationalized in terms of the forms of the normal mode using MO arguments. Further studies in the series include the ν_{16} band of *ortho*- and *meta*-disubstituted benzenes;[191] of pyridine, pyridine 1-oxide, and their monosubstituted derivatives;[192] of aryl derivatives

185 O. P. Girin, N. G. Bakhshiev, and O. I. Maksimova, *Optics and Spectroscopy*, 1968, **25**, 22.
186 O. I. Arkhangelskaya and N. G. Bakhshiev. *Optics and Spectroscopy*, 1971, **31**, 26.
187 R. T. C. Brownlee, A. R. Katritzky, and R. D. Topsom, *J. Amer. Chem. Soc.*, 1965, **87**, 3260.
188 R. T. C. Brownlee, R. E. J. Hutchinson, A. R. Katritzky, T. T. Tidwell, and R. D. Topsom, *J. Amer. Chem. Soc.*, 1968, **90**, 1757.
189 P. J. Q. English, A. R. Katritzky, T. T. Tidwell, and R. D. Topsom, *J. Amer. Chem. Soc.*, 1968, **90**, 1767.
190 R. T. C. Brownlee, P. J. Q. English, A. R. Katritzky, and R. D. Topsom, *J. Phys. Chem.*, 1960, **73**, 557.
191 A. R. Katritzky, M. V. Sinnott, T. T. Tidwell, and R. D. Topsom, *J. Amer. Chem. Soc.*, 1969, **91**, 628.
192 A. R. Katritzky, C. R. Palmer, P. J. Swinbourne, T. T. Tidwell, and R. D. Topsom, *J. Amer. Chem. Soc.*, 1969, **91**, 636.

of metalloids;[193] of alkylbenzenes;[194] of isocyanates, isothiocyanates, and azides;[195] of furans and thiophens;[196] and of the ν_{CN} mode of substituted benzonitriles.[197]

For the 4-substituted pyridines and pyridine oxides the ν_{16} intensities furnish direct evidence for interaction of the substituent and the hetero-group. For the pyridines only strong electron donors deviated from the expected intensity dependence on σ_R^0. For the pyridine oxides both strong donors and acceptors can deviate, indicating the dual character of the N=O group. An attempt was made to correlate the intensity of the ν_{CN} band with σ, with σ_R, and with σ^+ (σ value appropriate for electron-demanding situations).[196] The correlation of $A^{\frac{1}{2}}$ with σ^+ was far superior to the other correlations. A dual parameter treatment fitting against inductive and resonance parameters shows that the resonance term σ_R^+ is much more important for the *para*- than for *meta*-substitution.

The intensities of the ν_{16} modes of a variety of mono- and *meta*- and *para*-di-substituted benzenes have been calculated using both simple HMO and CNDO/2 methods.[198] The results of the CNDO/2 are significantly better and correlate extremely well with the observed values. Agreement on A is within 40% except where the intensity is very weak, in which cases the predicted intensity is also weak.

In a recent publication [199] new measurements on the intensities of the 1600 cm^{-1} (now listed as ν_8) bands and of the 1380 (ν_{19}) bands of a variety of monosubstituted benzenes are reported. In this new work the bands as measured in the liquid were separate from one another by fitting to Gaussian–Lorentzian functions (product or sum?). The resulting intensities were slightly higher than earlier values as a result of the extra wing contributions obtained in this way. Several of the bands were also studied in the vapour phase with the purpose of showing that the intensity trends were not affected by the effects peculiar to the condensed state. The only peculiarity in the results was that the gas-phase intensities were consistently a few per cent greater than the measured values in carbon tetrachloride solution, whereas dielectric effects would be expected to produce the reverse situation. Intensities were computed by the CNDO/2 method and the results show a very

[193] J. M. Angelelli, R. T. C. Brownlee, A. R. Katritzky, R. D. Topsom, and L. Yakhontov, *J. Amer. Chem. Soc.*, 1969, **91**, 4500.
[194] T. J. Broxton, L. W. Deady, A. R. Katritzky, A. Liu, and R. D. Topsom, *J. Amer. Chem. Soc.*, 1970, **92**, 6845.
[195] A. R. Katritzky, H. J. Keogh, S. Ohlenrott, and R. D. Topsom, *J. Amer. Chem. Soc.*, 1970, **92**, 6855.
[196] L. Deady, A. R. Katritzky, R. A. Shanks, and R. D. Topsom, *Spectrochim. Acta*, 1973, **29A**, 115.
[197] J. M. Angelelli, A. R. Katritzky, R. F. Pinzelli, and R. D. Topsom, *Tetrahedron*, 1972, **28**, 2037.
[198] R. T. C. Brownlee, A. R. Katritzky, M. V. Sinnott, M. Szafran, R. D. Topsom, and L. Yakhontov, *Tetrahedron Letters*, 1968, **55**, 5773; *J. Amer. Chem. Soc.*, 1970, **92**, 6850.
[199] R. T. C. Brownlee, D. G. Cameron, R. D. Topsom, A. R. Katritzky, and A. J. Sparrow, *J. Mol. Structure*, 1973, **16**, 365.

good agreement with measured intensities for all bands, especially for the ν_8 vibrations. It is tentatively suggested that the poorer agreement for the ν_{19} modes is due to the contribution of the motion of the hydrogen atoms to the intensity, the implication being that the CNDO technique is not good in reproducing the hydrogen contributions.

In this review we have referred to many papers in which the intensities of bands arising from 'characteristic' vibrations have been correlated with substituent Hammett or Taft factors. The purpose of such correlations may be to study the factors which control the transition moments, or alternatively to determine a scale of constants which represent the electronic influences of the substituent on the grouping whose vibrational properties are being studied. This latter aim is not just a variation on the main theme, which is the dependence of rate constants for given type reactions on the substituents. There is an inherent problem in determining 'substituent' constants from ionization constants and reaction constants, which, though discussed [200] as long ago as 1934, is often still blandly ignored. Ionization constants, for instance, are very dependent on temperature, showing a maximum at some temperature, θ, which may be below or above room temperature. Furthermore for a wide range of θ, even in a given solvent, the pK–temperature graphs cross and the substituent constants change order and may even change sign (electron-attracting to/from electron-donating). It seems remarkable that Hammett and Taft σ constants are in any way transferable as a means of correlating molecular properties with donor or acceptor power of the substituents. Clearly a spectroscopic scale of parameters, provided it has a firm, clear conceptual basis, ought to be far more satisfactory. Many of the difficulties in reactivity-constant scales, as given by thermodynamic data, are well discussed in an article by Kubler, Lui, Newton, and Patterson.[201]

Schmid has examined the C—H stretching intensities in a series of substituted benzenes [202] and fitted the intensities to a relation

$$A_{\mathrm{CH}} = F(a_j, \sigma_{\mathrm{I}}) \tag{52}$$

where a_j is an empirical parameter and σ_{I} is the Taft substituent constant. It is found that A_{CH} goes through a minimum, from which it is deduced that the dipole derivative has gone through zero and changed sign. It is also deduced that in benzene itself the bond polarity is $\overset{+}{\mathrm{C}}$—$\overset{-}{\mathrm{H}}$, but that in certain *para*-substituted benzenes with highly electronegative substituents (*e.g.* NO_2, CN) the polarity is reversed. Similar conclusions were reached in a study of *meta*-substituted benzenes.[203] This result is contrary to the conclusions of Bell, Thompson, and Vago [204] and of Brownlee *et al.*[190] These

[200] H. S. Harned and N. D. Embree, *J. Amer. Chem. Soc.*, 1934, **56,** 1050.
[201] D. G. Kubler, P. J. Lui, J. H. Newton, and C. S. Patterson, *Furman University Bulletin*, June, 1970, XVII, 37.
[202] E. D. Schmid, *Spectrochim. Acta*, 1966, **22,** 1659.
[203] E. D. Schmid and H. H. Seydewitz, *Ber. Bunsengesellschaft phys. Chem.*, 1971, **75,** 141.
[204] R. P. Bell, H. W. Thompson, and E. E. Vago, *Proc. Roy. Soc.*, 1948, **A192,** 498.

latter studies are based on effective bending moments. This apparent paradox must be due to the assumption that in the stretching or bending motions the equilibrium charge distribution moving with the nuclei dominates the dipole gradient. It would seem that either this is not so for the stretching motion or it is not so both for the out-of-plane and the in-plane bending motions. As has already been mentioned (p. 384), CNDO calculations [52] support the idea that the sign of the C—H stretching gradient is opposite to that which would be expected on the basis of the predicted bond dipole.

The intensities of the acetylenic vibration in acetylene derivatives have been the subject of several papers. The intensity variations in the $\nu_{C=C}$ and $\nu_{C\equiv C}$ bands of the series YCH=CH—C≡CX (Y = C_4H_9S, C_4H_9O, or Et_2NO; X = H or SiMe$_3$) have been shown to be due mainly to internal co-ordinate contributions from the ≡C—S, ≡C—O, ≡C—N, and ≡C—Si bonds.[205] An analysis of the absorption band intensities of the series Et_{4-n}-Sn(C≡CH)$_n$[206] led to the interesting result that the electro-optical parameters $\partial\mu/\partial q_{C\equiv C}$ and $\partial\mu/\partial q_{Sn-C}$ are unchanged by increased substitution. Non-additivity of the intensities is apparently due to a non-zero $\partial\mu_{Sn-C}/\partial q'_{Sn-C}$, where the prime on ∂q denotes a different bond from that to which $\partial\mu$ refers. The effect of hydrogen-bonding between the acetylenic hydrogen and solvents on the C—H and C≡C band intensities has been investigated.[207] Values of the bond-dipole derivative of the hydrogen were derived. Interaction decreases for the molecules studied in the order

$$
\begin{array}{ccc}
 & C\equiv CH & & C\equiv CH \\
 & / & & / \\
Me_3Si\,C\equiv CH > Me_2Si & & \approx Me_2Si \\
 & \backslash & & \backslash \\
 & C\equiv CSi\,Me_3 & & Ph
\end{array}
$$

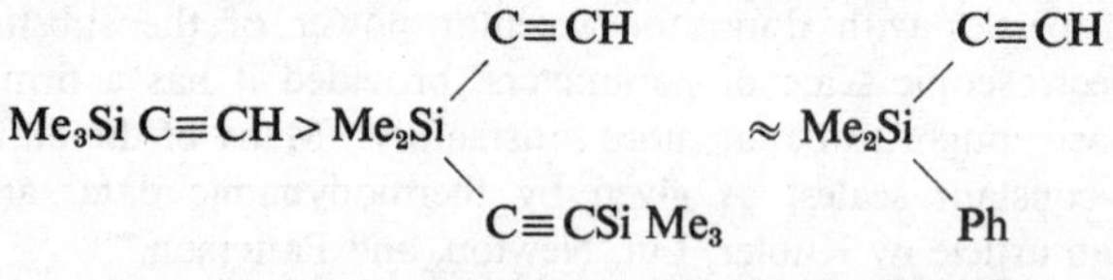

$$> Me_2Si(C\equiv CH)_2 > Et_3SnC\equiv CH > Me_3CC\equiv CH.$$

The intensities of spectral bands of the molecules Me$_3$XC≡CY (X = C or Si; Y = H or D) have also been discussed in ref. 208. Another series of conjugated acetylenes similar to those examined in ref. 205 have been studied by Shumetov.[209]

The band intensities of the methyl chlorosilanes Me$_n$SiCl$_{4-n}$ have been measured as a function of n. The results are explained in terms of mixing of Si—C and Si—Cl motions and by inductive effects on the C—H dipoles.[210]

[205] E. A. Gastilovich, D. N. Shigorin, K. V. Zhukova, T. D. Burnashova, and N. V. Komarova, *Optics and Spectroscopy*, 1968, **25**, 429.

[206] E. A. Gastilovich, D. N. Shigorin, K. V. Zhukova, and A. M. Sklyanova, *Optics and Spectroscopy*, 1970, **28**, 481.

[207] E. A. Gastilovich, K. V. Zhukova, D. N. Shigorin, O. G. Yarosh, and I. S. Aknurina, *Optics and Spectroscopy*, 1970, **29**, 22.

[208] V. Hoffmann, G. Stehlick, and W. Zeil, *Z. Naturforsch.*, 1970, **25a**, 572.

[209] V. G. Shumetov, *Zhur. priklad. Spektroskopii*, 1970, **13**, 174.

[210] E. N. Tikhomirova, I. F. Kovalev, M. G. Voronkov, and E. Ya. Lukevits, *Optics and Spectroscopy*, 1969, **27**, 334.

Intensity measurements of fundamental vibrational transitions in the low-frequency spectral region are few.[211, 212] It is encouraging then to note that Yarwood has measured the intensities of the ν_{NI} and ν_{ICl} bands of the pyridine–ICl complex at 292 and 140 cm^{-1} on three different instruments,[211] including a Beckman IR 11 and an RIIC FS 720 interferometer. The results agreed reasonably well, the standard errors from 18 and 17 measurements on the two bands being about $1\frac{1}{2}\%$ and 5%, respectively.

It was found that the intensity of the I—Cl stretching vibrational band in complexes of ICl with methyl-substituted pyridines was insensitive to the number and position of methyl groups. By contrast, the intensity of the N—I stretching vibration decreased with methyl substitution in the 2-position. The intensity decrease appears to correlate reasonably well with the frequency decrease.

Further work on pyridine with ICl, IBr, and I_2 and on [$^2H_5$]pyridine with IBr led to a set of estimates for the bond-dipole stretching gradients.[212] The analysis of the intensities requires a knowledge of the vibrational modes. This was achieved by treating the complexes pyridine–IBr and [$^2H_5$]pyridine–IBr as triatomics. The force-constant ellipsoids indicated an interaction constant of about 1 or 6 N m^{-1}. The actual values of the bond-dipole gradients derived are sensitive to the value of the interaction constant – and presumably therefore to any mixing which might occur of the N—I—X modes with the ring vibrations. The values of the bond gradients are given as a function of the interaction force constant, and trends in the gradients in the series X = Cl, Br, and I are discussed. Unfortunately, even the order of the gradients $\partial\mu/\partial R_{NI}$ and $\partial\mu/\partial R_{IX}$ in the series is altered in changing the value of the interaction constant from 1 to 6 N m^{-1}. As the value of this force constant is bound to change on changing X it would seem that conclusions regarding these gradients must be very tentative at the present time.

The values of the dipole gradients were calculated using the simple model of a dipole interacting with a polarizable iodine molecule. The computed intensities are lower than observed—though not greatly so. The range of values corresponding to the possible effective intermolecular interaction distances overlaps with the feasible range of $\partial\mu/\partial R_{II}$ derived from the observed intensities. The author favours a low value for the interaction constant, however, and for this situation the model predicts less than half of the observed gradient.

Ratajczak and Orville-Thomas have shown [213] that Mulliken's charge-transfer theory [214] when applied to weak halogen complexes leads to an almost linear relationship between the enthalpy of formation of the complex and change in vibrational transition moment of the halogen. Application of the equations to results from iodine complexes with benzene, dioxan, and

211 J. Yarwood, *Spectrochim. Acta*, 1970, **26A**, 2099.
212 G. W. Brownson and J. Yarwood, *J. Mol. Structure*, 1971, **10**, 147.
213 H. Ratajczak and W. J. Orville-Thomas, *J. Mol. Structure*, 1972, **14**, 155.
214 R. S. Mulliken, *J. Amer. Chem. Soc.*, 1952, **74**, 811.

various heteronuclear aromatics indicates that such a relation does indeed exist within the rather large uncertainty limits of the available data.

A much more detailed discussion of the effect of complexing on i.r. intensities is presented in a recent multi-author text.[215]

Solid-phase Studies.—It is not our purpose in this review to provide an *extensive* analysis of studies of i.r. intensities of molecular solids and their comparison with gas-phase intensity studies, since this subject was rather thoroughly surveyed in preliminary studies published in the early 1960s, with activity peaking about 1964, so it preceded the period covered by our review. However, a few studies have been published since then, and the entire subject has not been properly covered in any review (of which we are aware) so that it may be useful to give a brief survey of these studies. [However, there is a review by Yamada [216] for readers familiar with Japanese.]

The quality of experimental studies of i.r. intensities in the solid phase ranges all the way from very careful and accurate reflection studies of painstakingly prepared single crystals of MgO [217] to very crude and inaccurate attempts to study relative intensities of moderately complicated molecular crystals such as solid benzene.[218] The experimental problems hindering the attainment of accurate results are extremely great. Almost all the problems that occur for the measurements of intensities in the pure liquid phase (see above) are encountered in the solid phase. Thus, one must worry about the accuracy of determination of the sample pathlengths (of the order of 1 μm) and of the density and hence concentration, as well as about the photometric problems of defining and measuring an appropriate background I_0, correcting for reflection and for interference phenomena from these thin films, and correcting for the effect on the spectrum from the relatively large spectral slit widths as compared to the relatively narrow natural half-band-widths. The work during the 1960s developed techniques for solving all these experimental problems; some techniques are more satisfactory than others.

However, the astonishing thing about these preliminary studies of molecular crystals is that the changes observed in the i.r. intensities of some of the vibrations of the molecule, compared to the gas phase, were much larger than even the admittedly large experimental errors. It is for this reason that we think it worth bringing these preliminary results to the attention of others. One reason for the faltering advance of experimental work since the mid-1960s has been the lack of a decent theoretical explanation for the observed results, in spite of the challenge by Steele in his intensity review.[4] For this reason, the emphasis of the recent studies has not been on the comparison of gas-phase and solid-phase intensities but instead on other aspects, such as the intensities of lattice vibrations (which can be explained fairly well by a relatively simple theory).

[215] 'Spectroscopy and Structure of Molecular Complexes', ed. J. Yarwood, Plenum Press, London, 1973.
[216] H. Yamada, *Infrared Spectra*, 1965, **17**, 29 (in Japanese).
[217] G. Andermann and E. Duesler, *J. Opt. Soc. Amer.*, 1970, **60**, 53.
[218] C. A. Swenson and W. B. Person, *J. Chem. Phys.*, 1960, **33**, 56.

The molecules whose absorption intensities have been studied in both gas and solid phases are listed in Table 18 together with the experimental values of A_i in both phases. No 'field corrections' (see previous sections) have been given.

The experimental data in Table 18 have been obtained, for the most part, using the techniques developed by Hollenberg and Dows [219] and described in detail by Friedrich and Person [220] and by Yamada and Person.[221–223] Briefly, this technique consists of depositing, by sublimation on to a cold window, a thin film of what is believed to be a compact mass of randomly oriented microcrystals of the sample. The path length is measured by interference fringes observed by measuring the transmitted intensity, at a given wavelength, of the window plus film as a function of time of deposit. If the film is uniform, a series of fringes of constant peak-to-peak intensity are observed; non-uniformity of the path length of the film as it deposits over the area of the light beam results in a decrease in peak-to-peak intensities with time. From the total number of fringes (N_f) observed for a given film, the path length is calculated using:[224]

$$l = N_f/2n\tilde{\nu} \quad (53)$$

Here n is the refractive index of the film at $\tilde{\nu}$, and $\tilde{\nu}$ is the monitoring wave-number/cm^{-1}. The refractive index at $\tilde{\nu}$ is usually not known, of course, but it can be measured from the observed peak-to-peak intensities of fringes formed on different substrate windows with different known refractive indices.[221, 225] The concentration is calculated using densities for the pure solids, often obtainable at temperatures near those of the measurement from *X*-ray crystallographic studies.

The limiting error in these studies is the determination of the integrated absorbance [$\int \ln (I_0/I)\, d\nu$]. The previous section outlined some of the advances made on solving this problem for measurement of intensities in the liquid phase. For the solid phase one usually makes measurements at only one angle of incidence and analyses the measured transmittance using the Kramers–Krönig theory as described by Maeda, Thyagarajan, and Schatz [226a] and by Kozima, Suëtaka, and Schatz.[226b] This procedure yields n and κ and thus the value of the integrated intensity (A_i or Γ_i) from the measured apparent absorbance, with corrections for reflectance, interference, and all 'thin-film effects', except for scattering. However, for relatively crude intensity measurements accurate to ± 10 or 20%, Yamada and Person [221]

[219] J. L. Hollenberg and D. A. Dows, *J. Chem. Phys.*, 1961, **34**, 1061; *ibid.*, 1962, **37**, 1300.
[220] H. B. Friedrich and W. B. Person, *J. Chem. Phys.*, 1963, **39**, 811.
[221] H. Yamada and W. B. Person, *J. Chem. Phys.*, 1964, **40**, 309.
[222] H. Yamada and W. B. Person, *J. Chem. Phys.*, 1964, **41**, 2478.
[223] H. Yamada and W. B. Person, *J. Chem. Phys.*, 1965, **43**, 2519.
[224] O. S. Heavens, 'Optical Properties of Thin Solid Films', Butterworths, London, 1955.
[225] A. Bandy, Ph.D. thesis, University of Florida, 1968.
[226] (*a*) S. Maeda, G. Thyagarajan, and P. N. Schatz, *J. Chem. Phys.*, 1963, **39**, 3474; (*b*) K. Kozima, W. Suëtaka, and P. N. Schatz, *J. Opt. Soc. Amer.*, 1966, **56**, 181.

Table 18 *Infrared absorption intensities in gas (A_g) and solid (A_s) phases for several molecules*

		ν_g/cm^{-1}	ν_s/cm^{-1}	A_g/km mol^{-1}	A_s/km mol^{-1}
HCl [a]		2886	2730	39.0	240
HBr [a]		2560	2422	13.4	176
DBr [b]		—	—	(6.7)*	106
CO_2 [c]	ν_2(bend)	668	658	54	77
	ν_3(CO, *a.s.*)	2349	2342	635	459
N_2O [c]	ν_1	1285	1293	59.0	44.0
	ν_2(bend)	589	589	8.2	11.5
	ν_3	2224	2235	366	238
CS_2 [d]	ν_2(bend)	397	394	57	8.5
	ν_3(CS)	1523	1504	567	800
COS [e]	ν_1	859	856	8.0	7.4
	ν_2(bend)	521	518	2.9	7.6
	ν_3	2062	2040	590	680
ClCN [f]	ν_1(CCl) + $2\nu_2$	{714, 784}	{734(?), 830(?)}	9.08 + 1.80	0 + 0.27
	ν_2(bend)	380	398	4.55	6.52
	ν_3(CN)	2219	2209	18.20	36.36
BrCN [f]	ν_1(CBr) + $2\nu_2$	{575, 691}	{573, 740, 762}	1.06 + 2.32	6.75 + 0
	ν_2(bend)	342	363	3.30	5.22
	ν_3(CN)	2200	2193	7.70	35.14
ICN [f]	ν_1(CI)	485	451	5.30†	37.60
	ν_2(bend)	304	327	2.60†	1.72
	ν_3(CN)	2188	2176	5.30†	30.23
CCl_4 [g]	ν_{3a}(CCl) + (ν_1 + ν_4)	—	{761, 781}	390	450
	ν_4(bend)	—	316	0.20	0.80
SF_6 [h]	ν_5(SF)	947	906	1065	1040
	ν_6(bend)	615	608	69	122
$CHCl_3$ [i]	a_1 {ν_1(CH)	3034	3007	0.44	12
	a_1 ν_2(CCl)}	681	673	6	16
	e {ν_4(CH bend)	1221	1214	42	27
	e ν_5(CCl)}	769	759	190	270
$CDCl_3$ [i]	a_1 {ν_1(CD)	2262	2250	—	3
	a_1 ν_2(CCl)}	651	651	—	11
	e {ν_4(CD bend)	915	902	—	89
	e ν_5(CCl)}	747	730	—	200
C_2H_4 [j]	b_{1u} ν_7(oop bend)	949	942	79.8	95.0
	b_{2u} {ν_9(CH)	3105	3089	24.9	7.5
	b_{2u} ν_{10}(bend)}	810	823	(0.53)	0.47
	b_{3u} {ν_{11}(CH)	2989	2974	13.5	7.0
	b_{3u} ν_{12}(bend)}	1443	1437	9.76	16.4

	ν_g/cm^{-1}	ν_s/cm^{-1}	A_g/km mol^{-1}	A_s/km mol^{-1}
C_2D_4[j] b_{1u} ν_7(oop bend)	720	722	41.7	57.4
b_{2u} ν_9(CD)	2345	2328	12.2	4.6
b_{2u} ν_{10}(bend)	(589)	591	(0.043)	0.31
b_{3u} ν_{11}(CD)	2200	2189	7.6	3.6
b_{3u} ν_{12}(bend)	1078	1072	5.2	9.7
C_6H_6[k] a_{2u} ν_{11}(oop bend)	673	697	87.8	92
e_u $\nu_{18}(\nu_8 + \nu_{19}$, CH)	3080	{3084, 3029}	59.8	22
e_u ν_{19}(ring str.)	1486	1477	13	35
e_u ν_{20}(CH bend)	1038	1036	8.8	17.5
C_6D_6[k] a_{2u} ν_{11}(oop bend)	—	512	49.8	53.7
e_u ν_{18}(CD)	—	{2280, 2266}	35.3	14.9
e_u ν_{19}(ring str.)	—	1329	2.9	11.2
e_u ν_{20}(CD bend)	—	812	8.0	14
C_6F_6[l] a_{2u} ν_{11}(oop bend)	210	—	4.5	9
e_u ν_{18}(CF)	1000	—	410	460
e_u ν_{19}(ring str.)	1530	—	550	470
e_u ν_{20}(CF bend)	315	—	3.0	3.7

[a] Ref. 220; see also J. M. P. J. Verstegen, H. Goldring, S. Kimel, and B. Katz, *J. Chem. Phys.*, 1966, **44**, 3216. [b] H. B. Friedrich, H. Yamada, and W. B. Person, *J. Chem. Phys.*, 1965, **43**, 4180. [c] Ref. 222. [d] Ref. 221. [e] Ref. 223. [f] Ref. 227. [g] C. F. Cook, W. B. Person, and L. C. Hall, *Spectrochim. Acta*, 1967, **23A**, 1425. [h] Ref. 232. [i] A. Kimoto and H. Yamada, *Bull. Chem. Soc. Japan*, 1967, **40**, 243; *ibid.*, 1968, **41**, 1096. [j] Ref. 233. [k] Ref. 219; see also H. Yamada and W. B. Person, *J. Chem. Phys.*, 1963, **38**, 1253, and ref. 236. [l] A. R. Bandy, W. B. Person, and M. F. (Axford) Herkes, unpublished results (University of Florida).

* Estimated from A_{HBr}. † Measured in solution.

suggested that the values of the integrated intensity obtained by evaluating numerically the area between the measured absorption curve and a base line 'drawn-in' between the shoulders of the absorption band were much easier to measure, and were automatically corrected approximately for most of the 'thin-film effects'. Bandy [225, 227] later verified that the values of A_i from a 'drawn-in' base line agreed with the values obtained from a Kramers–Krönig analysis [226] within ±10% for bands in the cyanogen halides. Thus, the experimental intensities from even the relatively crude measurements with 'drawn-in base lines' are expected to be accurate to about ±20%, for reasonable estimates of all the errors involved, including also the resolution error, from the fact that the half-intensity widths may be smaller than the spectral slit widths for some of the bands in solids.

The interpretation of the results in Table 18 is another matter. It is clear from the results listed there that the integrated intensities per mole (A_i) are

[227] A. R. Bandy, H. B. Friedrich, and W. B. Person, *J. Chem. Phys.*, 1970, **53**, 674.

quite different in the solid phase than they are in the gas phase. These differences are far greater than any reasonable estimate of the experimental errors. In general, the intensities in the solid phase are greater than those in the gas phase. It is expected that the intensities will be larger because of the change in the 'effective field' from the light wave acting on the molecule. This correction is similar to the correction made by the Polo–Wilson equation in liquids (see above), except for the anisotropy in the solid phase. Yamada and Person [221–223] tried to account for this effect using a model with the absorbing molecule in an ellipsoidal cavity (instead of in a spherical cavity as for the Polo–Wilson equation) following a suggestion by Kalman and Decius.[228] Since then, Decius [229] has reformulated a procedure suggested by Mandel and Mazur [230] to give the correction in the form:

$$\partial p/\partial Q_i = (\partial/\partial Q_i)[(U + \boldsymbol{\alpha} S)^{-1} \boldsymbol{p}_s] \quad (54)$$

Here $\partial p/\partial Q_i$ is the dipole-moment derivative per molecule to be compared with the gas phase, $\boldsymbol{p}_s$ is the intrinsic molecular dipole moment in the solid, U is a unit tensor, $\boldsymbol{\alpha}$ is the polarizability tensor, and S is the field propagation tensor. The latter is evaluated using idealized dipole lattice summations.[231] In this procedure the value of $\partial p/\partial Q_i$ obtained depends quite strongly upon the assumed location of the ideal dipole, so that some uncertainty still exists in the estimates of the 'field corrections'.

Nevertheless the field corrections (calculated for 'reasonable' ellipsoidal cavities) apparently account quite well for the observed changes in intensities shown in Table 18 for studies of systems whose intermolecular interactions are expected to be weak ('physical' interactions) and where frequency shifts, for example, are small. Such systems include CO_2, N_2O, CS_2, and COS; the agreement between calculated and experimental intensity ratios (A_s/A_g, where A_s is the intensity measured in the solid phase and A_g is the value measured in the gas phase; $A_s = fA_g$, if only the field effect, f, causes the intensity change) was illustrated for those molecules in Figure 4 of ref. 223, reproduced here as Figure 7. Other molecules in Table 18 for which the 'field effect' apparently explains most of the observed intensity changes include CCl_4, C_6F_6, and probably SF_6.

There are at least three factors that can cause intensity changes from the gas to solid (or liquid) phase. These are: (i) the 'field effect'; (ii) changes in normal co-ordinates for the vibration, and (iii) intermolecular 'charge-transfer' effects that cause the effective charge of a vibrating atom on one molecule to be different from the gas-phase value due to the presence in the crystal of the neighbouring molecule.

[228] O. F. Kalman and J. C. Decius, *J. Chem. Phys.*, 1961, **35**, 1919.
[229] J. C. Decius, *J. Chem. Phys.*, 1968, **49**, 1387; see also H. Mueller, *Phys. Rev.*, 1953, **47**, 947.
[230] M. Mandel and P. Mazur, *Physica*, 1958, **24**, 116.
[231] For example see D. B. Dickmann, M. S. thesis, U.S. Naval Postgraduate School, 1966; and F. W. de Witte and G. E. Schacher, *Phys. Rev.*, 1965, **137**, A78.

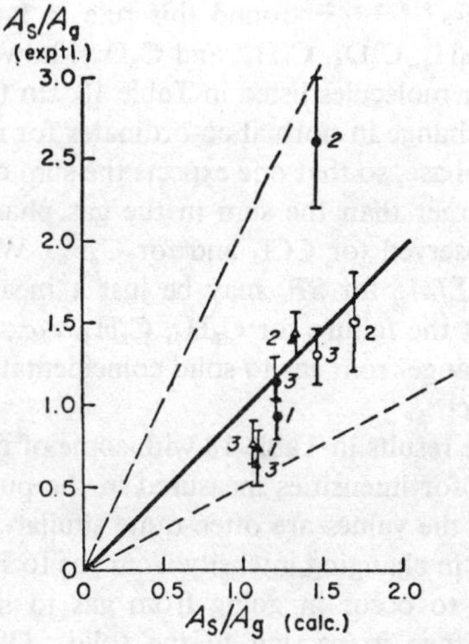

Figure 7 *Summary, showing a plot of the experimental values of the intensity ratio* A_s/A_g vs. *the calculated ratio for the i.r.-active fundamentals of linear triatomic molecules. The calculated ratios are determined for the 'field effect' only. The symbol designates the molecule; the number refers to the vibration* (i.e., ▲1 *means the point is for* ν_1 *of* N_2O.) *The solid line has unit slope; if the calculated values agreed with the experimental ratios, all points would fall on this line. The vertical lines give estimates of the maximum experimental error.* ○, CS_2; ●, COS; △, CO_2; ▲, N_2O
(Reproduced by permission from *J. Chem. Phys.*, 1965, **43**, 2519)

We have seen that the 'field effect' is expected to result in fairly small changes (f may range from about 1 to 2) from gas to solid; the 'ellipsoidal cavity' effect can result in fairly large differences in the value of f (1.87 for the a_{2u} mode of C_6F_6, *cf.* 1.21 for the e_{1u} modes) for vibrations of different symmetry. However, vibrations of the same symmetry are expected to have approximately the same field correction.

Dows and Wieder [232] noticed that the intensity and frequency of ν_5 decreased and those of ν_6 increased on changing SF_6 from gas to solid. They calculated the change in the normal co-ordinates corresponding to the frequency changes and found that those changes explained the intensity changes very well. They ignored the correction for the effective field in this calculation, but it is expected to be small ($f = 1.2$) and the same for both vibrations, so its effect may tend to cancel.

If this second effect (change in normal co-ordinates with change in phase) were the only factor affecting the change in intensity, then one should expect an intensity sum rule to hold.[219, 232, 233] Specifically, the Crawford *G*-sum rule may be expected to hold, so that the sum of all the intensities (for a non-polar molecule) may be expected to be a constant, independent of phase.

[232] D. A. Dows and G. M. Wieder, *Spectrochim. Acta*, 1962, **18**, 1567.
[233] G. M. Wieder and D. A. Dows, *J. Chem. Phys.*, 1962, **37**, 2990.

Dows and co-workers [219, 232, 233] found this rule to hold for the molecules they studied (SF_6, C_2H_4, C_2D_4, C_6H_6, and C_6D_6); however, it does not hold for most of the other molecules listed in Table 18. In fact, one expects both a field effect and a change in normal co-ordinates for molecules going from the gas to the solid phase, so that one expects the sum of all intensities in the solid phase to be larger than the sum in the gas phase by the field effect, approximately as observed for CCl_4 and for C_6F_6. We believe that failure to find that $\Sigma A_{si} = \Sigma f_i A_{gi}$ for SF_6 may be just a measure of experimental uncertainty, but that the failure for C_6H_6, C_2H_4, *etc.*, is an indication that the $\partial p/\partial Q_i$ values change from gas to solid coincidentally to cancel (approximately) the field effect.

Comparison of the results in Table 18 with some of the results reported in the previous section for intensities measured in the pure liquid phase or in solution reveals that the values are often quite similar. If only the first two effects are important in changing intensity from gas to solid, we might expect most of the change to occur in going from gas to liquid, with relatively minor additional change in passing to the solid. Of course the average forces between molecules in the solid are stronger than in the liquid, so that effects due to changing normal co-ordinates are expected to be somewhat larger. Effects due to intermolecular charge transfer are also expected to be stronger in the solid than in the liquid phase. One does expect to observe a regular progression in intensity from gas to liquid to solid; when experimental data suggest a maximum or minimum for the liquid-phase intensity, one must question the data.

The third effect causing the intensity to change from gas to liquid ('charge-transfer' effects changing the effective charge on the vibrating atom) have been called 'chemical effects', 'specific effects', 'vibronic effects', and 'delocalization effects', to mention a few of the names. A detailed understanding of this effect may be expected only when quantum-mechanical calculations have been made for clusters of molecules in models for solids or liquids. A review of the effect on vibrational spectra of electron donor–acceptor interaction between D, A pairs forming complexes has recently been published.[215] Some of these ideas can be applied to the interaction between pairs of like molecules in the solid and liquid phases. For example, the studies of complexes suggest that the C—H stretching vibration will be susceptible to the greatest change in intensity from gas to solid for unsaturated hydrocarbons, in qualitative agreement with the results in Table 18. A decrease in intensity suggests that the intrinsic moment $(\partial p/\partial Q_{CH})_g$ opposes the 'delocalization moment' due to the intermolecular charge transfer. The latter is expected to have a positive sign, consistent with the negative sign for $\partial p/\partial Q_{CH}$ suggested from the CNDO calculations (see Theory section). Thus these ideas concerning the intensity changes from gas to solid shown in Table 18 appear to be qualitatively correct. Similar concepts appear to account, at least qualitatively, for the larger changes noted in Table 18. (See refs. 220 and 227, for example.)

In addition to the results in Table 18, aimed primarily at the study of the differences in intensities between gas and solid phases, several other studies have appeared in the years covered by our review. In particular, one should mention again the experimental aspects of the study by Snyder [26] of the intensities of n-paraffins that has previously been mentioned for its theoretical value (p. 373). He measured the intensities, in the solid phase, of the i.r.-active fundamentals of the normal paraffins from n-C_4H_{10} through n-C_8H_{18}. He began to observe formation of oriented (instead of random) films with the longer hydrocarbons, which might introduce sizeable errors if not detected and if the data are analysed as though they are from a random film.

Tsuji and Yamada have studied the i.r. intensities of several i.r.-active bands of naphthalene.[234] They measured intensities by study of the ATR spectrum of a single crystal. Optical contact between the crystal and the KRS-5 cylinder is a problem. The technique and corrections to be applied to the analysis of data due to incompleteness of contact between the KRS-5 element and the crystal are described in other papers.[235] The results from the ATR study of the single crystal were compared with results from an ATR study of a polycrystalline disk and from an absorption study of an oriented thin film. The results are comfortingly comparable. Application to iodoform has also been made.[235 a, b]

Szczepaniak and Person have studied the spectrum of benzene in mixed crystals ('matrices') with HCl in 13:1 excess.[236] The intensities (measured roughly) are compared with results for the solid, and 'glassy', deposits of pure benzene, and for benzene in a Br_2 'matrix'.[237] The results are crude by comparison to the measurements in pure solid benzene, but they indicate the potential interest of studies of 'solvent effects' on the intensities in solid phase. Such studies are similar to studies in liquid solutions, but the bands are sharper in the solid and there are some advantages over liquid solution studies.

Glover and Hollenberg have raised some questions concerning the accuracy of i.r. intensities measured in solids.[238] They have used combinations of transmission and reflection measurements of thin films with the Kramers–Krönig relations in an attempt to obtain (with considerable difficulty) unique solutions for the optical constants, n and κ. An improved Kramers–Krönig analysis method has been reported by Neufeld and Andermann [239] for treating i.r. transmittance data from studies of thin solid films to get the optical constants. They have applied it to results for the far-i.r. spectra of a single crystal of $NaClO_3$ (whose near-i.r. reflectance spectrum had previously

[234] K. Tsuji and H. Yamada, *J. Phys. Chem.*, 1972, **76**, 260.

[235] (*a*) H. Yamada, K. Suzuki, and I. Nitta, *Spectrochim. Acta*, 1967, **23A**, 1735; (*b*) K. Tsuji, H. Yamada, and K. Suzuki, *ibid.*, 1970, **26A**, 475; (*c*) K. Tsuji, and H. Yamada, *Bull. Chem. Soc. Japan*, 1968, **41**, 1975.

[236] K. Szczepaniak and W. B. Person, *Spectrochim. Acta*, 1972, **28A**, 15.

[237] W. B. Person, C. F. Cook, and H. B. Friedrich, *J. Chem. Phys.*, 1976, **46**, 2521.

[238] D. E. Glover and J. L. Hollenberg, *J. Phys. Chem.*, 1969, **73**, 889.

[239] C. J. D. Neufeld and G. Andermann, *J. Opt. Soc. Amer.*, 1972, **62**, 1156.

been quantitatively studied by Andermann and Dows).[240] These techniques are expected to be of considerable value in obtaining more accurate measurements of optical constants of thin solid films in the future.

There have been a number of other studies reported recently[241] of this kind, designed to help obtain more accurate values of n and κ for thin solid films. Since we are now interested in the comparison of results for molecular solids between phases, as in Table 18, we shall not pursue this matter further here, except to point out the obvious applications to studies of vibrational intensities of molecular ions (NO_3^-, ClO_3^-, *etc.*) in their ionic crystals[242]—properties that cannot be studied for molecular ions isolated in the gas phase.

Measurements of absolute intensities of librational lattice modes provide data that can be explained very nicely by a simple dipolar coupling model. Schnepp has given one form of this theory;[243] a test using experimental i.r. intensities of lattice modes in N_2, CO_2, and CO is quite satisfying.[243b] Applications have also been made to the interpretation of the far-i.r. intensities of solid ethylene[244] and of solid cyanogen.[245] However, in a quantitative test of the theory against more accurately measured intensities of the lattice modes of solid N_2, St. Louis and Schnepp[246] find that the experimental intensities are smaller by a factor of five than the theoretical prediction,[243a] leading them to conclude that the theoretical model may be quantitatively inadequate.

Friedrich[247] has measured the intensities of lattice motions in crystals of polar molecules, explaining them quantitatively in terms of the permanent dipole moment using the dipolar formalism of Mandel and Mazur[230] and Decius.[229] Carlson and Friedrich report results for the lattice modes of HCl and DCl[247a] and Friedrich[247b] for ClCN and BrCN.

We should like to close this section by mentioning the imaginative and useful application by Rochkind of solid-phase i.r. intensity techniques to the problem of analysis of gaseous mixtures.[248] This technique was called 'pseudo-matrix isolation'; it consists of depositing the gaseous mixtures to be analysed on the cold sample window of a commercial cryogenic cell, and measuring the absorbance of a band belonging to the described component. From a previous calibration (Beer's law plot) the nl (concentration $\times$ path-

[240] G. Andermann and D. A. Dows, *J. Phys. and Chem. Solids*, 1967, **28**, 1307.

[241] For example, see J. Bock and G. J. Su, *J. Chem. Phys.*, 1972, **57**, 1464; N. J. Harrick, *Appl. Optics*, 1971, **10**, 2344; B. A. Seiber, A. M. Smith, B. E. Wood, and P. R. Müller, *ibid.*, 1971, **10**, 2086; G. R. Field and E. Murphey, *ibid.*, p. 1402; and others.

[242] For example, see V. Schettino and I. C. Hisatsune, *J. Chem. Phys*, 1970, **52**, 9; and also R. Frech and J. C. Decius, *ibid.*, 1971, **54**, 2374.

[243] (*a*) O. Schnepp, *J. Chem. Phys.*, 1967, **46**, 3983; (*b*) A. Ron and O. Schnepp, *ibid.*, p. 3991.

[244] M. Brit and A. Ron, to be published.

[245] P. M. Richardson and E. R. Nixon, *J. Chem. Phys.*, 1968, **49** 4276.

[246] R. V. St. Louis and O. Schnepp, *J. Chem. Phys.*, 1969, **50**, 5177.

[247] (*a*) R. E. Carlson and H. B. Friedrich, *J. Chem. Phys.*, 1971 **54**, 2794; (*b*) H. B. Friedrich, *ibid.*, 1970, **52**, 3005.

[248] (*a*) M. M. Rochkind, *Analyt. Chem.*, 1967, **39**, 567; (*b*) M. M. Rochkind, *Environ. Sci. Technol.*, 1967, **1**, 434; (*c*) M. M. Rochkind, *Science*, 1968, **160**, 196; (*d*) M. M. Rochkind, *Analyt. Chem.*, 1968, **40**, 762.

length) product is determined. The latter is obtained from differential pressure measurements as the sample deposits, so that the total number of micromoles of sample in the solid film is known. The technique is claimed to be quite reproducible and would appear to be a useful tool for this difficult analytical problem.

Combination Bands and Overtones.—Since combination bands and overtones are normally weak and densely distributed in a given frequency range it is not surprising that the intensities of such bands have rarely been studied. The C—H out-of-plane combination bands of benzenes and ethylenes are noteworthy exceptions. Kakiuti gave expressions for the intensities of such bands.[249] Dunstan and Whiffen measured the intensities of these bands for gaseous benzene and its deuteriated species.[250] By use of the relation

$$\frac{\partial^2 \mathbf{p}}{\partial Q_i \partial Q_j} = \sum_{mn} \frac{\partial^2 \mathbf{p}}{\partial S_m \partial S_n} l_{im} l_{jn} \tag{55}$$

they showed that the intensities were adequately accounted for using only two intensity parameters—namely $\dfrac{\partial^2 \mu}{\partial \gamma_i^2} = \pm 0.23\ e\ \text{rad}^{-2}$ and $\dfrac{\partial^2 \mu}{\partial \gamma_i \partial \gamma_{i+1}} = 0.03\ e$ rad^{-2}. Note that we have a factor of $1/r_0$ in our definitions of angular dipole derivatives as indicated in the introduction. Using liquid-phase data recorded in CS_2, Kakiuti, Suzuki, and Onda found that the corresponding bands of methyl- and chloro-substituted benzenes could be expressed using a limited set of similar electro-optical parameters.[251] These authors refined the expression for the dipole expansion by inclusion of a dipole term proportional to the static bond dipole and to the square of the angular displacement. Such a term arises from the in-plane expansion of $\boldsymbol{\mu}_i \sin \gamma$ in powers of γ. This means that the dipole parameters quoted are not directly comparable with those of Dunstan and Whiffen. The resulting dependence of the other parameters on the relative signs of $\boldsymbol{\mu}_i$ and $\dfrac{\partial^2 \mu}{\partial \gamma_m \partial \gamma_n}$ leads to pairs of optimum solutions. Unfortunately the authors assume that the bond-dipole moment is that derived from application of the zeroth-order model to the out-of-plane a_{2u} fundamental of benzene (2.17×10^{-30} C m). As this is probably twice as great as the static bond dipole, the zero-order correction is over-estimated. Even so, the zero-order intensity is computed to be only 10% of the observed. Two sets of parameters are given, depending on the definition of the out-of-plane bending co-ordinate. It was found, as in the case of benzene,[250] that a single parameter of the type $\partial^2\mu/\partial\gamma_i^2$ was adequate to give a reasonable picture of the intensity distribution. Addition of *ortho-* and *meta*-terms, $\partial^2\mu/\partial\gamma_i\partial\gamma_{i+1}$ and $\partial^2\mu/\partial\gamma_i\partial\gamma_{i+2}$, naturally improved the agreement. In view

[249] Y. Kakiuti, *J. Chem. Phys.*, 1956, **25**, 777.
[250] F. E. Dunstan and D. H. Whiffen, *J. Chem. Soc.*, 1960, 5221.
[251] Y. Kakiuti, Y. Suzuki, and M. Onda, *J. Mol. Spectroscopy*, 1968, **27**, 402.

of the Fermi-resonance problems and overlapping bands, the significance of this improvement is obscured. No statistical uncertainty limits were computed for the parameters. The authors do not make it clear as to whether or not they make a dielectric field correction.

The fundamental γ_{CH} modes and their combination bands in thiophen and its deuteriated and methyl derivatives have also been measured in CS_2 by Akiyama.[252] Using the same formulation as in ref. 121, $\partial^2\mu/\partial\gamma_i^2$ was determined as —0.17 or +0.41 e rad^{-2}, assuming a value for μ_{CH} of 3.3×10^{-30} C m as determined from the out-of-plane fundamentals. The former value is preferred and is in good accord with the value of —0.18 e rad^{-2} derived for substituted benzenes [251] assuming $\partial^2\mu/\partial\gamma_i\partial\gamma_{i+1}$ and $\partial^2\mu/\partial\gamma_i\partial\gamma_{i+2} \neq 0$.

By measuring the intensities of the C—H stretching fundamentals and their first and second overtones for a number of molecules containing one C—H bond, Boobyer [253] has sought to investigate the dependence of the C—H bond dipole on bond length. Using a diatomic model for the C—H bond and Dunham's expressions for the intensities of an anharmonic oscillator,[254] values for $\partial\mu/\partial r$ and $\partial^2\mu/\partial r^2$ were deduced. The $\partial\mu/\partial r$ values vary smoothly with C—H frequency and bond lengths, though at least three turning points in the function are indicated. The values of $\partial\mu/\partial r$ deduced agree closely with those deduced simply from the fundamental using the harmonic model. This is so even when the overtone is relatively strong, as in liquid chloroform, and when the fundamental intensity is strongly perturbed, as in chloroform or dichloroacetonitrile. There is an ambiguity in the values derived for $\partial^2\mu/\partial r^2$ arising from the problem of the relative signs of the first and second derivatives. This makes it difficult to make any generalizations about the values. No indications are given of attempts to correlate the values with the dipole–bond-length curve. It is very interesting that whereas the intensity of the fundamental of chloral increases by a factor of ten on going from a CCl_4 solution to pure liquid, the first overtone is practically unaffected. This is true also for dichloroacetonitrile, which is the only other case listed for which both solution and liquid-state data are given. This has obvious analytical applications.

Temperature Dependence.—The temperature dependence of the absorption bands of several materials in the liquid phase has been measured over a wide temperature range (up to 300 °C).[255–259] In accord with earlier observations, the intensities are generally observed to decrease with increase in temperature,

[252] M. Akiyama, *J. Mol. Spectroscopy*, 1972, **43,** 226.
[253] G. J. Boobyer, *Spectrochim. Acta*, 1967, **23A,** 335.
[254] J. L. Dunham. *Phys. Rev.*, 1929, **34,** 438; 1930, **35,** 1347.
[255] Yu. E. Zabiyakin and N. G. Bakhshiev, *Optics and Spectroscopy*, 1968, **25,** 29.
[256] Yu. E. Zabiyakin and N. G. Bakhshiev, *Optics and Spectroscopy*, 1969, **25,** 525.
[257] Yu. E. Zabiyakin and N. G. Bakhshiev, *Optics and Spectroscopy*, 1969, **26,** 38.
[258] V. S. Smirnov, Yu. E. Zabiyakin, O. P. Girin, and N. G. Bakhshiev, *Optics and Spectroscopy*, 1969, **27,** 138.
[259] T. M. Barakat, *Optics and Spectroscopy*, 1971, **30,** 627.

there being a linear dependence on temperature in many, but not all, cases.[255] In one or two cases an increase in temperature was observed to lead to an increase in intensity. Such cases included the 3040 cm^{-1} band of benzene, the 898 and 1268 cm^{-1} bands of CH_2Cl_2, and the 902 cm^{-1} band of acetone. The changes in the optical constants (n) with temperature were found to be very considerable for the 780 cm^{-1} doublet of CCl_4.[256] This indicated at least one of the factors responsible. Correction of the intensities for modification of the effective field of the light wave by the dielectric properties of the medium reduced the temperature dependence of the intensity in all cases studied.[260]

Table 19 *Showing the effect of field correction on the ratio of band intensities measured at two different temperatures.* (Compiled from *Optics and Spectroscopy*, 1959, **7**, 123)

			Ratio of A	
Substance	$\tilde{\nu}/cm^{-1}$	$T/°C$	*No field corr.*	*With field corr.*
CCl_4	792	0, 160	1.40	1.33
Me_2CO	1720	—50, 110	1.28	1.20
	1224	—20, 50	1.05	1.00
$CHBr_3$	1143	10, 140	1.08	1.00
$CHCl_3$	1215	—50, 50	1.08	1.00
C_6H_6	1486	10, 180	1.45	1.33
	1032	10, 200	1.10	1.06

As seen from Table 19, for 3 of the 7 bands the dependence was entirely accounted for. The behaviour of solutions is found to be similar to those of pure liquids when specific strong interactions are absent.[257] By contrast, the behaviour of gases appears to be different. It has been reported that intensities increase markedly with temperature for CCl_4, CS_2, $CHCl_3$, and Me_2CO for instance.[261, 262] These results are surprising and contrary to the experience of others.[263, 264] The temperature dependence of HCl in CCl_4 has been computed to be in very satisfactory agreement with experiment,[258] using the theory of Girin and Bakhshiev. This supports the idea that it is unnecessary to invoke bonding between HCl and CCl_4 to explain the spectroscopic properties of this system.

One of the problems in studying spectra at temperatures well away from room temperature is that emissions from the surfaces at different temperatures lead to apparent sample transmissions which differ from the true transmissions. A simple technique of correcting for this has been given by Neszmélyi and Imre.[265] It involves placing an iris between the sample and

260 Yu. E. Zabiyakin and N. G. Bakhshiev, *Optics and Spectroscopy*, 1969, **26**, 102.
261 P. A. Bazhulin and V. N. Smirnov, *Optics and Spectroscopy*, 1959, **7**, 123.
262 N. A. Borisevich and E. A. Zalesskaya, *Optics and Spectroscopy*, 1964, **16**, 420.
263 T. N. Adams, D. M. Weston, and R. A. Matula, *J. Chem. Phys.*, 1971, **55**, 5674.
264 W. Wheatley, Ph.D. thesis, University of London, 1968.
265 A. Neszmélyi and L. Imre, *Spectrochim. Acta*, 1968, **24A**, 297.

source. Since partial closing of the iris cannot affect that part of the emission from the cell which reaches the detector, but has a predictable effect on the apparent transmission, then the true transmission can be deduced. Using this technique, the intensity of the 524 cm^{-1} band of cyclohexene was found to be independent of temperature up to 70 °C.

6
Raman Intensities

BY R. E. HESTER

1 Introduction

Considerable progress has been made in recent years in the development of theoretical understanding of the factors governing Raman intensities. It is the aim of this chapter to review the work of the main contributors to this progress, and to relate the theory to experimental results.

The simple classical theory due to Placzek [1] has formed the basis for most of the early work in this field, relating the intensity of Raman bands to molecular polarizability derivatives with respect to nuclear co-ordinates. This classical theory is limited in its applicability to situations where the electromagnetic radiation used to excite the Raman effect is of a frequency far from those frequencies at which the molecules absorb. However, within this zone of applicability, which may be designated the 'off-resonance' zone, the Placzek theory and developments of it have been extremely useful. Recent examples of this are presented in the following sections of this chapter.

When the incident radiation in a Raman scattering experiment is coincident in frequency with that of an absorption band, the resonance Raman effect (RRE) is observed, wherein the intensity relationships appropriate to the off-resonance zone break down. Typically, in this on-resonance zone, the scattering cross-sections of a small selection of the molecular vibrational modes active in the normal (off-resonance) Raman effect are greatly enhanced. A great deal of attention currently is being devoted to this intensity effect, which has important applications in biological chemistry and in the elucidation of electronic structure of molecules. The total break-down of classical polarizability theory in the resonance condition has necessitated the development of a quantum mechanical approach to account for the intensity behaviour of Raman bands, and much progress has been made in this respect. As will be seen from the following development, the newer quantum mechanical expressions govern the full frequency regime, from normal (off-resonance) scattering to resonance Raman scattering, and including the intermediate 'near-resonance' zone, where deviations in intensity patterns from those predicted by classical considerations first begin to manifest themselves.

[1] G. Placzek in 'Handbuch der Radiologie', ed. E. Marx, Akademische Verlagsgesellschaft, Leipzig, 1934, vol. 6, p. 205; English translation by Ann Werbin, U.C.R.L. Trans. —526(L), from the U.S. Dept. of Commerce, clearing house for Federal Scientific and Technical Information.

2 Ground-state Polarizability Theory

The ground-state polarizability theory of Placzek [1] has been reviewed in some detail by Woodward [2] and by Hester.[3] This theory depends on special simplifications, admissible only in the off-resonance condition, of the general theory of light scattering by an individual particle as developed by Kramers and Heisenberg.[4] Together with closely related derivations based on radiation field theory by Dirac,[5] and the subsequent utilization of these expressions by Van Vleck [6] in determining selection rules for the activity of vibrational modes in Raman scattering, these constitute the basis for essentially all further developments.

Labelling the initial vibrational state wavefunction as m and the final state n, both of these being time-independent functions within the electronic ground state of the molecule, the induced transition moment of a molecule subjected to a radiation field may be expressed in the form

$$\boldsymbol{P}_{mn} = \langle n|\boldsymbol{P}|m\rangle \tag{1}$$

where $\boldsymbol{P}$ is the induced molecular dipole moment. The classical treatment of the induced dipole involves the relationship

$$\boldsymbol{P} = \boldsymbol{\alpha E} \tag{2}$$

with the radiation electric field strength, $\boldsymbol{E}$, and the molecular polarizability, $\boldsymbol{\alpha}$. In the simplest case of a non-rotating molecule referred to a space-fixed co-ordinate system, the vector components of the induced dipole moment may be expressed as

$$\begin{aligned} P_x &= \alpha_{xx}E_x + \alpha_{xy}E_y + \alpha_{xz}E_z \\ P_y &= \alpha_{yx}E_x + \alpha_{yy}E_y + \alpha_{yz}E_z \\ P_z &= \alpha_{zx}E_x + \alpha_{zy}E_y + \alpha_{zz}E_z \end{aligned} \tag{3}$$

These relationships demonstrate the tensor character of the molecular polarizability.

Until very recently, all cases of molecular vibrational Raman scattering were found to be governed by symmetric polarizability tensors, with the equalities

$$\alpha_{xy} = \alpha_{yx},\ \alpha_{yz} = \alpha_{zy},\ \alpha_{xz} = \alpha_{zx} \tag{4}$$

bringing about a reduction of the number of distinct components of the tensor from nine to six. The new experimental situations wherein an asymmetric scattering tensor is indicated will be examined later in this chapter,

[2] L. A. Woodward, in 'Raman Spectroscopy', ed. H. A. Szymanski, Plenum Press, New York, 1967, ch. 1.
[3] R. E. Hester, in 'Raman Spectroscopy', ed. H. A. Szymanski, Plenum Press, New York, 1967, ch. 4.
[4] H. A. Kramers and W. Heisenberg, *Z. Physik*, 1925, **31**, 681.
[5] P. A. M. Dirac, *Proc. Roy. Soc.*, 1927, **114**, 710.
[6] J. H. Van Vleck, *Proc. Nat. Acad. Sci. U.S.A.*, 1929, **15**, 754.

but it may be useful here to see how the concept of a molecular polarizability ellipsoid arises from the foregoing theory. In a conceptual determination of the Raman-allowed or -forbidden character of a vibrational mode, this polarizability ellipsoid performs a function analogous to the dipole moment in infrared absorption. Rotation of the co-ordinate system relative to the molecule produces a transformation of the polarizability tensor of the same form as that given by the ellipsoidal equation:

$$\alpha_{xx}x^2 + \alpha_{yy}y^2 + \alpha_{zz}z^2 + 2\alpha_{xy}xy + 2\alpha_{yz}yz + 2\alpha_{zx}zx = 1 \quad (5)$$

Further specifying a coincidence of the co-ordinate axes and principal axes X, Y, Z of the ellipsoid causes the off-diagonal tensor components to vanish, leaving the reduced form of the equation

$$AX^2 + BY^2 + CZ^2 = 1 \quad (6)$$

with A, B, and C representing the diagonal tensor components α_{XX}, α_{YY}, and α_{ZZ}, respectively.

Returning to the quantum mechanical treatment, the induced transition moment may be expressed in the following form:

$$P_{mn} = (\alpha)_{mn}E \quad (7)$$

The basic Kramers–Heisenberg–Dirac dispersion theory has been reviewed by Behringer [7, 8] and by Tang and Albrecht,[9] and expressions have been derived for $(\alpha)_{mn}$. The most general form for a $\rho\sigma$-th component of this tensor is given by the following sum over all vibronic states of the molecule:

$$(\alpha_{\rho\sigma})_{mn} = \sum_e \left[\frac{\langle m|M_\sigma|e\rangle\langle e|M_\rho|n\rangle}{E_e - E_m - E_0 + i\gamma_e} + \frac{\langle m|M_\rho|e\rangle\langle e|M_\sigma|n\rangle}{E_e - E_n + E_0 + i\gamma_e}\right] \quad (8)$$

Here $|e\rangle$ represents an intermediate state in the Raman transition, E_0 is equal to $h\nu_0$, the energy of the incident light, γ_e is a damping constant, and the many-electron operators M_σ and M_ρ are defined by

$$M_\sigma = R_\sigma - (g|R_\sigma|g) \quad (9)$$

R_σ and R_ρ are the electronic dipole moment operators, and $(g|R_\sigma|g)$ is an integral over electronic co-ordinates which vanishes in the case of molecules with no dipole moment.[9] The damping factor in equation (8) commonly is neglected, and the energy denominators are written as frequency differences, *viz.*

$$E_e - E_m - E_0 = h(\nu_{em} - \nu_0)$$

and (10)

$$E_e - E_n + E_0 = h(\nu_{en} + \nu_0)$$

[7] J. Behringer, *Z. Elektrochem.*, 1958, **62**, 906.
[8] J. Behringer, in 'Raman Spectroscopy', ed. H. A. Szymanski, Plenum Press, New York, 1967, ch. 6.
[9] J. Tang and A. C. Albrecht, in 'Raman Spectroscopy', ed. H. A. Szymanski, Plenum Press, New York, 1970, vol. 2, ch. 2.

Since the vibrational state energy separation $E_n - E_m$ usually is much smaller than the separation of either of these states from the intermediate state(s) E_e, the approximation $\nu_{em} \approx \nu_{en} = \nu$ may be used to further simplify equation (8). It should be noted that the operators in equation (8) are concerned only with electronic scattering, the polarizability being considered as a function purely of electronic structure in the frequency region (visible and ultraviolet) normally of interest in Raman spectroscopy. The nuclear motion involved in molecular vibrations may thus be active in the Raman effect only through its influence on the molecular (electronic) polarizability.

The Raman intensity of the line associated with the induced transition $m \rightarrow n$ is proportional to the fourth power of the scattered radiation frequency. Thus the total intensity (over the solid angle 4π) as given by Behringer [8] and by Tang and Albrecht [9] is

$$I_{mn} = \frac{2^7\pi^5}{3^2c^4} I_0(\nu_0 \pm \nu_{mn})^4 \sum_{\rho,\sigma} |(\alpha_{\rho\sigma})_{mn}|^2 \tag{11}$$

where I_0 is the intensity of the incident light.

As stated earlier, the polarizability tensor (α_{mn}) may be treated as a function of the nuclear co-ordinates, thus giving rise to a vibrational Raman effect. Placzek [1] performed a Taylor-series expansion of the polarizability in terms of the vibrational normal co-ordinate Q, truncating the expansion after the linear term, *viz.*

$$\alpha = \alpha_0 + \left(\frac{\partial \alpha}{\partial Q}\right)_0 Q \tag{12}$$

This simplification leads rather directly [2] to the prediction that vibrational mode overtones and combination tones are forbidden in the Raman effect and that the intensity of Raman scattering at right angles to the direction of the incident light may be expressed in terms of invariants of the derived polarizability tensor:

$$I_n = \text{const.} \tfrac{1}{45} \{45(\bar{\alpha}')^2 + 13(\gamma')^2\} \tag{13}$$

where $\bar{\alpha}'$ is the derivative $(\partial\bar{\alpha}/\partial Q)_0$ of the mean molecular polarizability:

$$\bar{\alpha} = \tfrac{1}{3}(\alpha_{xx} + \alpha_{yy} + \alpha_{zz}) \tag{14}$$

and γ' is the derivative of the anisotropy of the polarizability:

$$\gamma^2 = \tfrac{1}{2}[(\alpha_{xx} - \alpha_{yy})^2 + (\alpha_{yy} - \alpha_{zz})^2 + (\alpha_{zz} - \alpha_{xx})^2 + 6(\alpha_{xy}^2 + \alpha_{yz} + \alpha_{zx}^2)] \tag{15}$$

Equation (13) is valid only for right-angle scattering of natural (unpolarized) incident light and takes on the following form for the case of plane-polarized incident light:

$$I_p = \text{const.} \tfrac{1}{45}\, 45(\bar{\alpha}')^2 + 7(\gamma')^2 \tag{16}$$

This equation is appropriate to the usual experimental configuration used with laser sources, *i.e.* light incident along the *x*-direction, polarized in the *z*-direction, and scattered light collected in the *y*-direction. For this arrangement it may further be seen [2] that an analysis of the polarization properties of the scattered light should yield a degree of depolarization:

$$\rho_n = \frac{I_y}{I_z} = \frac{6(\gamma')^2}{45(\bar{\alpha}')^2 + 7(\gamma')^2} \tag{17}$$

for unpolarized incident light, and

$$\rho_p = \frac{I_y}{I_z} = \frac{3(\gamma')^2}{45(\bar{\alpha}')^2 + 4(\gamma')^2} \tag{18}$$

for incident light plane-polarized in the *z*-direction. The former expression, equation (17), results in a depolarization ratio of 6/7 for situations where $\bar{\alpha}' = 0$ but $\gamma' \neq 0$, *i.e.* for antisymmetric vibrational modes. Equation (18) yields $\rho_p = \frac{3}{4}$ in this case. When $\bar{\alpha}' \neq 0$, as will be the case for symmetric vibrational modes, the depolarization ratios ρ_n and ρ_p will be less than $\frac{6}{7}$ and $\frac{3}{4}$, respectively, and both will be exactly zero for $\gamma' = 0$, $\bar{\alpha}' \neq 0$. Of course, when $\gamma' = 0$ and $\bar{\alpha}' = 0$ the associated vibrational mode gives rise to no Raman activity, *i.e.* is forbidden.

Further development of the ground-state polarizability theory, due largely to Woodward and Long,[10] has been summarized by Hester [3] and the following intensity expression derived:

$$I = \frac{KM(\nu_0 - \nu_{mn})^4}{\nu_{mn}\{1 - \exp(-h\nu_{mn}/kT)\}}(45)(\bar{\alpha}')^2 \frac{6}{6 - 7\rho} \tag{19}$$

In this equation, *K* is an experimental constant, *M* is the molar concentration of scattering molecules, and ρ is the depolarization ratio determined by taking the ratio of the total intensity of light scattered in the *y*-direction due to light incident along the *x*-direction and polarized first in the *y*-direction and then in the *z*-direction. This equation has been extensively used to derive molecular polarizability derivatives, $\bar{\alpha}'$, for solution species.[3]

Values of $\bar{\alpha}'$ derived from measured Raman band intensities may be reduced to the more useful form of bond polarizability derivatives if the normal co-ordinates are known. This reduction was first proposed by Wolkenstein [11] and Eliashevich,[12] was given a general mathematical formulation by Long,[13] and subsequently modified by Long *et al.*[14] An assumption inherent in this procedure [2, 3] is that a bond polarizability derivative is independent of all molecular motions other than the stretching of that particular bond. The procedure, which involves first a normal-mode analysis of the molecular

[10] L. A. Woodward and D. A. Long, *Trans. Faraday Soc.*, 1949, **45**, 1131.
[11] M. W. Wolkenstein, *Compt. rend. Acad. Sci. U.S.S.R.*, 1941, **32**, 185.
[12] M. Eliashevich and M. W. Wolkenstein, *J. Phys. (Moscow)*, 1945, **9**, 101, 326.
[13] D. A. Long, *Proc. Roy. Soc.*, 1953, **A217**, 203.
[14] D. A. Long, A. H. S. Matterson, and L. A. Woodward, *Proc. Roy. Soc.*, 1954, **A224**, 33.

vibrations by the Wilson FG matrix method,[15, 16] has been reviewed by Hester.[3] The prime significance of this concept of the bond polarizability derivative lies in its demonstrated relationship to bond type. Examples will be presented later in this chapter which suggest that the nature of chemical bonding may be inferred rather directly from a knowledge of bond polarizability derivatives as determined from Raman band intensities.

3 The Delta-function Model

A variation of the ground-state polarizability theory discussed above, providing an alternative approach to the determination of bond polarizability derivatives, is embodied in the delta-function potential model as used by Lippincott and Nagarajan [17] and by Long and Plane.[18] This model involves the assumption that, instead of a Coulomb potential, at each nucleus in a polyatomic molecule there exists an infinite potential, while the potential is zero everywhere else. A 'delta function strength' is defined as the (finite) integral of the potential over all space, this being analogous to the Slater 'effective nuclear charge'. An essential feature of the model is its one-dimensional quality: each chemical bond is considered to be a separate unidirectional entity. The linear combination of atomic wavefunctions results in a particularly simple bond potential, this being non-zero at only two points along a bond.

Lippincott and Nagarajan's version involves single-electron delta-function potentials of the form:

$$V = -\left[A_1 g\delta\left(x - \frac{a}{2}\right) + A_2 g\delta\left(x + \frac{a}{2}\right)\right] \tag{20}$$

for the simple case of a diatomic molecule, with x being the co-ordinate of motion along the internuclear axis, and a being the delta-function spacing; g is the unit delta-function strength (the value for the hydrogen atom), and $\delta\left(x \pm \frac{a}{2}\right)$ is a normalized delta-function of x, implying that the potential is everywhere zero except at the delta-function positions, *i.e.* $x = a/2$ and $x = -a/2$. A_1 and A_2 are the delta-function strengths for nuclei 1 and 2, respectively, these being related to atom electronegativities, and being calculated according to the empirical relationship [19]

$$A = [\chi/(2.6N - 1.7P - 0.8D + 3.0F)]^{\frac{1}{2}} \tag{21}$$

where χ is the atom electronegativity, N is the principal quantum number,

[15] E. B. Wilson, *J. Chem. Phys.*, 1939, **7**, 1047; 1941, **9**, 76.
[16] E. B. Wilson, J. C. Decius, and P. C. Cross, 'Molecular Vibrations', McGraw-Hill, New York, 1955.
[17] E. R. Lippincott and G. Nagarajan, *Bull. Soc. chim. belges*, 1965, **74**, 551.
[18] T. V. Long, jun., and R. A. Plane, *J. Chem. Phys.*, 1965, **43**, 457.
[19] E. R. Lippincott and M. O. Dayhoff, *Spectrochim. Acta*, 1960, **16**, 807.

$P = 1$ for atoms with p electrons in the valence shell, and $P = 0$ for atoms with no p-electrons in the valence shell, D is the total number of completed p and d shells in the atom, and F is the number of completed f shells in the atom.

The Long and Plane treatment employs the following form of one-dimensional perturbing potential for a homonuclear diatomic molecule:

$$H_1 = \varepsilon_x(g/Z)^{\frac{1}{2}}\left(-\sum_{i=1}^{j} x_i + \sum_{\alpha=1}^{2} Zx_\alpha\right) \tag{22}$$

Here ε_x is the electric field along the internuclear axis; $(g/Z)^{\frac{1}{2}}$ is identified with the charge on an electron; x_i is the position of any outer-shell electron, with the summation extending over all outer-shell electrons; Z is the magnitude of the unshielded nuclear charge; and x_α is the position of the αth nucleus. The delta-function potential strength, g, is given the value of unity for the hydrogen atom, and arbitrarily the square root of the valence state electronegativity on the Pauling scale, $g = \chi^{\frac{1}{2}}$, for all other atoms. This choice of values is said to be justified by the success of calculations [19] of bond dissociation energies, force constants, anharmonicities, and equilibrium internuclear distances using the Pauling electronegativity relation. Moreover, this surprisingly simple relationship with the square root of electronegativity does appear to give directly a proper magnitude to the potential sought, and it has the required one-dimensionality.

The parallel component of a homonuclear bond polarizability is given by Long and Plane in terms of the delta-function potential theory as

$$\boldsymbol{\alpha}_{\|b} = 8(g/Z)(1/a_0)(\tfrac{1}{2}n)(\langle x^2\rangle)^2 \tag{23}$$

where a_0 is the radius of the first Bohr orbit of the hydrogen atom, $\langle x^2\rangle$ is the expectation value of the square of the position of a bonding electron along the internuclear axis, and $\frac{1}{2}n$ is the number of bonds (the bond order, or simply half the number of bonding electron pairs). The following assumptions are implicit in the foregoing model: (i) $\langle x\rangle = 0$ for a homonuclear diatomic molecule because of the symmetric placement of the delta-function potentials; (ii) $(x_1 - \langle x\rangle)(x_2 - \langle x\rangle) = 0$ since the model allows no electron correlation; and (iii) $x = x_1 = x_2 = x_3 = \ldots$, since all bonding electrons are considered to be equivalent. These may be seen as rather severe simplifications.

An expression for $\langle x^2\rangle$ in terms of delta-function wavefunctions has been derived by Lippincott and Stutman [20] as follows:

$$\langle x^2\rangle = \tfrac{1}{4}r^2 + \frac{1}{2c^2} - \frac{r^3ce^{-cr}}{6\{1 + (rc + 1)e^{-cr}\}} \tag{24}$$

Here r is the equilibrium internuclear distance, and $c = (-2E)^{\frac{1}{2}}$, where E is the energy of a separated atom. The third term in equation (24) always is

[20] E. R. Lippincott and J. M. Stutman, *J. Phys. Chem.*, 1964, **68**, 2926.

negligibly small compared with the first two terms, so that the analytical expression for the parallel component of the bond polarizability becomes

$$\boldsymbol{\alpha}_{\parallel b} = (8g\sigma/Za_0)(\tfrac{1}{2}n)\left(\frac{1}{4}r^2 + \frac{1}{2c^2}\right)^2 \tag{25}$$

In this expression the parameter σ is the Pauling covalent bond character, defined by $\sigma = \exp\{-\tfrac{1}{4}(\chi_1 - \chi_2)^2\}$ for a bond between two atoms of electronegativities χ_1 and χ_2, respectively. This is introduced as a weighting factor on $\boldsymbol{\alpha}_{\parallel b}$ to correct for the polar character of a heteronuclear bond. For such heteronuclear bonds the value of $(g/Z)^{\frac{1}{2}}$ is taken as the geometric mean of the two atomic values. In differentiating equation (25) with respect to change in bond length it is further assumed [17, 18] that the second term in equation (24) makes only a negligible contribution, resulting in the following form for the derivative of the parallel component of bond polarizability:

$$(\partial\boldsymbol{\alpha}_{\parallel b}/\partial r) = (2g\sigma/Za_0)(\tfrac{1}{2}n)(r^3) \tag{26}$$

The mean polarizability of a diatomic molecule may be written as a sum of contributions from purely parallel and perpendicular components:

$$\bar{\boldsymbol{\alpha}} = \tfrac{1}{3}(\boldsymbol{\alpha}_{\parallel} + 2\boldsymbol{\alpha}_{\perp}) \tag{27}$$

in which $\boldsymbol{\alpha}_{\parallel}$ is the polarizability component along the internuclear axis, and $\boldsymbol{\alpha}_{\perp}$ is the component perpendicular to this axis. A consequence of the use of the one-dimensional delta-function model is that the derivative $(\partial\boldsymbol{\alpha}_{\perp}/\partial r)$ is exactly zero, so that we may write $(\partial\bar{\boldsymbol{\alpha}}/\partial r) = \tfrac{1}{3}(\partial\boldsymbol{\alpha}_{\parallel}/\partial r)$. Moreover, the implicit neglect of contributions to polarizability from non-bonding electrons leads to the further equality $(\partial\boldsymbol{\alpha}_{\parallel}/\partial r) = (\partial\boldsymbol{\alpha}_{\parallel b}/\partial r)$. On the basis of these many simplifications, Long and Plane [18] have derived finally an expression for the bond polarizability derivative:

$$(\partial\bar{\boldsymbol{\alpha}}/\partial r) = \tfrac{2}{3}(g\sigma/Za_0)(\tfrac{1}{2}n)(r^3) \tag{28}$$

The corresponding expression derived by Lippincott and Nagarajan [17] is

$$(\partial\bar{\boldsymbol{\alpha}}/\partial r) = NA_{12}\sigma(\tfrac{1}{3}a_0)(r^3) \tag{29}$$

this stemming from their use of the root-mean-square delta-function strength of the two nuclei, A_{12}, each atom value being defined by equation (21). In this equation, N is the bond order $(=\tfrac{1}{2}n)$. Thus, the Lippincott and Nagarajan treatment parallels that of Long and Plane but takes as its starting point the following alternative [to equation (23)] equation for the parallel component of a bond polarizability:

$$\boldsymbol{\alpha}_{\parallel b} = 4NA_{12}(1/a_0)(\langle x^2\rangle)^2 \tag{30}$$

Both treatments arrive at expressions for total bond polarizability derivatives which, owing to the absence of interaction between neighbouring bonds implied by the nature of the delta-function potential, are applicable to polyatomic molecules. Both are seen to involve a series of approximations and assumptions and, as will be seen from data presented later in this chapter, their calculated values for bond polarizability derivatives are not in particularly good agreement. Nonetheless, it must be emphasized that in many cases the predictions of the delta-function model do correlate with measured Raman band intensities through the proportionality $I \propto (\partial\bar{\alpha}/\partial r)^2$ [see equation (19)], and appear further to relate quite well to the nature of chemical bonds (degree of covalent character, and bond order). Data in support of this statement will be presented later in this chapter.

4 Vibronic Expansion Theories

An extensive review by Behringer of the theory of resonance Raman scattering is given in Chapter 2 of this volume. The treatment given in this section and in Section 7 relates only to particular aspects of Raman intensities.

The foregoing theories were based on the properties of the electronic ground state only, the explicit dependence of the Raman scattering process on excited molecular states implied by equation (8) being eliminated by various approximations. As has been pointed out in a recent review by Tang and Albrecht,[9] an alternative approach is possible wherein each term in the dispersion equation represents a contribution from the transition between the initial (or final) state of the molecule and an excited virtual state. The relationship between Raman scattering and absorption spectroscopy evidently is an important one, and much progress may be made in understanding the properties of excited electronic states through study of vibrational Raman intensities associated primarily with the ground electronic state. A significant contribution was made in this area by Albrecht,[21] who developed an explicit relationship between the Raman intensities of the ground-state vibrations and the vibronic intensities of allowed transitions of a molecule. This relationship has subsequently, however, been further developed in a more general form by Tang and Albrecht.[9]

In deriving their general vibronic representation of the Raman intensity theory, Tang and Albrecht take the basic dipole version of the Kramers–Heisenberg–Dirac dispersion expression without damping, assume the adiabatic approximation for the wavefunctions, and expand the electronic parts of the wavefunctions about the ground-state equilibrium configurations of the nuclei in a Herzberg–Teller [22] series of the first order. This expansion in terms of a Taylor series in the nuclear displacement co-ordinates from an equilibrium position in the electronic ground state involves the identification of the coefficients of the linear terms with a vibronic coupling operator through the perturbation method. The final result of the expansion is an

[21] A. C. Albrecht, *J. Chem. Phys.*, 1961, **34**, 1476.
[22] G. Herzberg and E. Teller, *Z. phys. Chem. (Leipzig)*, 1933, **B21**, 558.

expression for a general polarizability tensor component as a simple sum of three terms, *viz.*

$$(\alpha_{\rho\sigma})_{gi,\,gj} = A + B + C \tag{31}$$

The notation here relates back to that used in equations (1), (7), (8), and (9), modified by the use of the following notation for the adiabatic (Born–Oppenheimer) approximation, *i.e.* the assumption that the wavefunctions may be written as products of electronic and vibrational parts:

$$\begin{aligned} |m\rangle &= |g)||i\rangle \quad \text{or} \quad \Psi_m(r, Q) = \theta_g(r, Q)\phi_i^g(Q) \\ |n\rangle &= |g)||j\rangle \quad \text{or} \quad \Psi_n(r, Q) = \theta_g(r, Q)\phi_j^g(Q) \\ |e\rangle &= |e)||v\rangle \quad \text{or} \quad \Psi_e(r, Q) = \theta_g(r, Q)\phi_v^e(Q) \end{aligned} \tag{32}$$

Here gi, gj, and ev represent the vibronic states, with $|\,)$ signifying a function of electron co-ordinates and $|\,\rangle$ signifying a function of nuclear co-ordinates. The three terms in the sum expressed in equation (31) have the following forms:

$$\sum_{e\neq g}{}' \sum_{v} \left[\frac{(g^0|R_\sigma|e^0)(e^0|R_\rho|g^0)}{E_{ev} - E_{gi} - E_0} + \frac{(g^0|R_\rho|e^0)(e^0|R_\sigma|g^0)}{E_{ev} - E_{gj} + E_0}\right]\langle i||v\rangle\langle v||j\rangle \tag{33}$$

$$\begin{aligned} B = {} & \sum_{e\neq g}{}' \sum_{v} \sum_{s\neq e}{}' \sum_{a} \\ & \times \left[\frac{(g^0|R_\sigma|e^0)(e^0|h_a|s^0)(s^0|R_\rho|g^0)}{E_{ev} - E_{gi} - E_0} + \frac{(g^0|R_\rho|e^0)(e^0|h_a|s^0)(s^0|R_\sigma|g^0)}{E_{ev} - E_{gj} + E_0}\right] \\ & \times \langle i||v\rangle\langle v||Q_a||j\rangle/(E_e^0 - E_s^0) \\ & + \left[\frac{(g^0|R_\sigma|s^0)(s^0|h_a|e^0)(e^0|R_\rho|g^0)}{E_{ev} - E_{gi} - E_0} + \frac{(g^0|R_\rho|s^0)(s^0|h_a|e^0)(e^0|R_\sigma|g^0)}{E_{ev} - E_{gj} + E_0}\right] \\ & \times \langle i||Q_a||v\rangle\langle v||j\rangle/(E_e^0 - E_s^0) \end{aligned} \tag{34}$$

$$\begin{aligned} C = {} & \sum_{e\neq g}{}' \sum_{t\neq g}{}' \sum_{v} \sum_{a} \\ & \times \left[\left\{\frac{(g^0|h_a|t^0)(t^0|R_\sigma|e^0)(e^0|R_\rho|g^0)}{E_{ev} - E_{gi} - E_0} + \frac{(g^0|h_a|t^0)(t^0|R_\rho|e^0)(e^0|R_\sigma|g^0)}{E_{ev} - E_{gj} + E_0}\right\}\right. \\ & \times \frac{\langle i||v\rangle\langle v||Q_a||j\rangle}{E_g^0 - E_t^0} \\ & + \left\{\frac{(g^0|R_\sigma|e^0)(e^0|R_\rho|t^0)(t^0|h_a|g^0)}{E_{ev} - E_{gi} - E_0} + \frac{(g^0|R_\rho|e^0)(e^0|R_\sigma|t^0)(t^0|h_a|g^0)}{E_{ev} - E_{gj} + E_0}\right\} \\ & \left.\times \frac{\langle i||Q_a||v\rangle\langle v||j\rangle}{E_g^0 - E_t^0}\right] \end{aligned} \tag{35}$$

The vibronic coupling operator h_a^0 is simply $(\partial H/\partial Q_a)^0$ for a normal mode Q_a, with H being the electronic Hamiltonian, and the functions $|e)$ and $|g)$ are expressed in the following expanded form:

$$|e) = |e^0) + \sum_a \sum_{s \neq e}{}' \frac{(h_a)^0_{es} \cdot \Delta Q_a}{E^0_e - E^0_s} |s^0) \tag{36}$$

$$|g) = |g^0) + \sum_a \sum_{t \neq g}{}' \frac{(h_a)^0_{gt} \cdot \Delta Q_a}{E^0_g - E^0_t} |t^0) \tag{37}$$

This general vibronic representation of the Raman intensity theory governs the full frequency region and includes all three zones which have been recognized earlier – from the normal non-resonance to the rigorous resonance effect.

A useful approximate form of this general treatment was presented earlier by Albrecht,[21] who showed that with certain assumptions about the energy denominators in equation (31) the A term may be seen to contribute only to Rayleigh scattering. Further, neglecting the second term in equation (37), *i.e.* assuming $|g) = |g^0)$, the C term is made to vanish. The following simplified expression for a polarizability tensor component is then derived,[9, 21] writing $(E_{ev} - E_{gi}) \approx (E_{ev} - E_{gj}) \approx (E^0_e - E^0_g)$ when far away from resonance:

$$(\alpha_{\rho\sigma})_{gi,\,gj} = B \quad \text{for} \quad i \neq j \tag{38}$$

where

$$B = -\sum_e \sum_{s \neq e} \sum_a \Bigg[[(g^0|R_\rho|e^0)(e^0|h_a|s^0)(s^0|R_\sigma|g^0) + (g^0|R_\sigma|e^0)(e^0|h_a|s^0)(s^0|R_\rho|g^0) \times \frac{[(E^0_e - E^0_g)(E^0_s - E^0_g) + E^2_0]\langle i||Q_a||j\rangle}{[(E^0_e - E^0_g)^2 - E^2_0]\,[(E^0_s - E^0_g)^2 - E^2_0]}\Bigg]$$

From this approximate form it is predicted that, as the incident-light frequency in a Raman scattering experiment approaches a given allowed electronic transition, those normal modes which are vibronically active in the electronic transition should exhibit particularly striking enhancement of their Raman intensities.

Alternative treatments of the vibronic method have been presented by a number of authors, and most of these have been reviewed in the context of the foregoing theory by Tang and Albrecht.[9] One of the most successful early attempts at formulating this theory is the so-called semi-classical treatment due to Shorygin.[23, 24] In this, only one excited state e and one totally symmetric normal vibrational mode are considered. Development of this theory by Krushinskii and Shorygin[25] has yielded expressions for the

23 P. P. Shorygin, *Doklady Akad. Nauk S.S.S.R.*, 1952, **87**, 201.
24 P. P. Shorygin, *Izvest. Akad. Nauk S.S.S.R., Ser. fiz.*, 1953, **17**, 581.
25 L. L. Krushinskii and P. P. Shorygin, *Optika i Spektroskopiya*, 1961, **11**, 24, 151; 1965, **19**, 562 (*Optics and Spectroscopy*, 1961, **11**, 12, 80; 1965, **19**, 312).

polarizability tensor component which have been found useful in predicting effective absorption frequencies for comparison with measured absorption spectra. The recent very thorough treatment of resonance Raman scattering by Behringer [8] has reviewed this theory in some detail. Savin [26] and Verlan [27] also have performed general vibronic expansions of the basic Kramers–Heisenberg–Dirac dispersion equation in developing expressions for the polarizability. Both authors ultimately arrive at expressions closely analogous to equation (31). A comparison in detail of these parallel developments has been given by Tang and Albrecht, who also have summarized the conclusions of Verlan's theory. One important result is that totally symmetric vibrational modes derive most of their intensity from the term *A*, while non-totally symmetric modes derive intensity only from terms *B* and *C*. Since term *C* usually will be small compared with *A* and *B*, this represents a convenient separation.

Formulae for the transition probabilities, and hence the intensities of molecular vibrational spectra in both infrared absorption and Raman scattering, have recently been derived through a unified quantum-field treatment by Peticolas and co-workers.[28] This treatment assumes a mechanism of time-ordered steps involving the destruction of an incident radiation quantum, the creation or destruction of a vibrational quantum, and the creation of the emitted radiation quantum. These three-quanta processes are treated by third-order time-dependent perturbation theory in the interaction representation, and result in an expression similar to those obtained by Albrecht and Savin, Yet another approach has been taken by the French workers in this fleld,[29, 30] who have used a resolvent operator formalism in developing expressions for the Raman scattering tensor. Interactions between scattering molecules have been considered specifically in this treatment, in addition to the fundamental radiation–molecule interaction, thus generating terms relevant to band-profile analysis as well as straightforward intensity determination. Two cases are treated by the resolvent formalism: the pre-resonance Raman effect together with resonance at excitation in the absorption continuum, and resonance at excitation into a discrete level. The further theoretical treatment of the rigorous resonance effect is currently in an active state of development. Some recent experimental results will be discussed later in this Report and related to aspects of the existing theories.

26 F. A. Savin, *Optika i Spektroskopiya*, 1965, **19**, 555, 743; 1966, **20**, 989 (*Optics and Spectroscopy*, 1965, **19**, 308, 412; 1966, **20**, 549).
27 E. M. Verlan, *Optika i Spektroskopiya*, 1966, **20**, 605, 802 (*Optics and Spectroscopy*, 1966, **20**, 341, 447).
28 W. L. Peticolas, L. Nafie, P. Stein, and B. Fanconi, *J. Chem. Phys.*, 1970, **52**, 1576.
29 M. Jacon, R. Germinet, M. Berjot, and L. Bernard, *J. Phys.* (*Paris*), 1971, **32**, 517.
30 M. Jacon, M. Berjot, and L. Bernard, *Compt. rend.*, 1971, **272**, *B*, 595; 1971, **273**, *B*, 585, 956; 1972, **274**, *B*, 344; *Optics Comm.*, 1971, **4**, 3, 117, 246; 1972, **5**, 94; *Canad. Spectroscopy*, 1972, **17**, 60.

5 Measurement of Raman intensities

The experimental difficulties inherent in the determination of absolute intensities of Raman bands have been discussed previously by Hester,[3] and Shorygin [31] recently has reviewed the literature dealing with such measurements. The most common approach used has been the measurement of integrated intensity relative to a band of some standard compound. The band of interest may be measured from the spectrum of a pure compound, the spectrum being generated under identical conditions to those used for the reference material: this is the so-called *external* standard method. More commonly, the reference material is intimately mixed with the compound of interest and the intensity measurements made directly on the mixture. This *internal* standard method has the advantage that the results are automatically corrected for refractive index effects, which govern the effective optical path within the sample, and for absorption and scatter losses, temperature variations, and stability of the radiation source. The difficulty of defining an exact scattering angle, with the associated convergence errors, is also effectively compensated by the use of an internal standard. However, as has been pointed out by Woodward,[2] with liquid samples, wherein intermolecular forces may have a substantial influence, there may be so-called *internal field effects* which are not exactly compensated. These arise from the fact that the intrinsic scattering power of a molecule in a liquid, and also the effective field strength of the exciting radiation at the molecule, may be affected by the immediate molecular environment, and this environment may be different for the reference material and the compound of interest.

Raman band intensities of solutes commonly have been measured relative to the solvent. CCl_4 has most often been used for non-polar solutes, this being chosen for its inertness and simplicity; cyclohexane also has been used. Care must be taken in interpretation of these relative intensity measurements, particularly where solvent effects known to affect even the frequencies of bands have been established.[32] For electrolyte solutions, the perchlorate ion commonly has been incorporated as an internal intensity standard owing to its inert, non-complexing character.

Equation (19) incorporates the prediction that the intensity of a Raman band should be directly proportional to the molar concentration of the active compound. This prediction has been verified for mixtures of aliphatic hydrocarbons, and a variety of other systems, where the molar intensity coefficient has been found experimentally to be independent of concentration.[33, 34] However, a significant concentration dependence has been found in many cases where a concentration-dependent association reaction is involved.

[31] P. P. Shorygin, *Russ. Chem. Rev.*, 1971, **40**, 367.
[32] P. P. Shorygin and B. V. Lopatin, *Zhur. fiz. Khim.*, 1965, **39**, 2868 (*Russ. J. Phys. Chem.*, 1965, **39**, 1533).
[33] H. W. Schrotter, *Z. Elektrochem.*, 1960, **64**, 853.
[34] H. W. Schrotter and J. Brandmuller, 'Advances in Molecular Spectroscopy', Pergamon, Oxford, 1962, p. 1128.

The need for an internal standard to compensate for refractive index changes with concentration also is apparent in this. It has been demonstrated that variations in refractive index of the solvent medium over the range 1.25—1.60 may result in variations of $\pm 40\%$ in the molar intensity coefficient for non-resonance excitation.[35] These solvent effects are still more marked under near-resonance conditions, as has been demonstrated by Shorygin.[31]

Proposals have been made for Raman band-intensity constants which are less dependent on measurement conditions.[36] An absolute intensity, or 'activity coefficient' for scattering, may be derived from equation (19) by correcting the measured intensity for polarization, and eliminating the factor $(\nu_0 - \nu_{mn})^4/\nu_{mn}\{1 - \exp(h\nu_{mn}/kT)\}$. In his recent very thorough review of Raman intensities, however, Shorygin [31] has tabulated extensive data for a wide range of compounds, both organic and inorganic, based on different intensity coefficients I^s. These are molar integrated band intensities referred to the intensity of the 313 cm^{-1} band of liquid carbon tetrachloride, this being taken as 100 units when excited by non-polarized 435.8 nm light and observed at a scattering angle of about 90°. The sheer range of compounds covered by Shorygin argues strongly in favour of the use of his I^s intensity coefficient.

6 Bond Order Determinations

From a purely chemical point of view, one of the most interesting aspects of the study of Raman band intensities is the possibility of using such intensity data for the determination of the nature of chemical bonds. Various aspects of this link between intensities and electron-density distributions in bonds have previously been discussed in some detail by Hester.[3] Key experiments in the forging of this link were those of Woodward and Long [10] and of Yoshino and Bernstein.[37] The former authors found that $\bar{\alpha}'$ values determined from Raman band intensities of the totally symmetric modes of Group IV element tetrahalides were proportional to the percent covalent character of the chemical bonds. Yoshino and Bernstein, from a series of intensity measurements on gaseous hydrocarbons, concluded that the associated C—C bond polarizability derivatives were directly proportional to C—C bond order. Although Chantry and Plane,[38] on the basis of Raman intensity measurements with polyatomic inorganic ions containing multiple bonds, have found that σ and π bonds in general do not have equal polarizability derivatives, it is established firmly that a meaningful connection between bond polarizability derivatives and the nature of chemical bonds does exist.

35 G. Fini, P. Mirone, and P. Patella, *J. Mol. Spectroscopy*, 1968, **28**, 144.
36 W. F. Murphy, W. Holzer, and H. J. Bernstein, *Appl. Spectroscopy Rev.*, 1969, **23**, 211.
37 T. Yoshino and H. J. Bernstein, *J. Mol. Spectroscopy*, 1958, **2**, 213; *Spectrochim. Acta*, 1959, **14**, 127; 'Petroleum Conference on Molecular Spectroscopy', Pergamon, New York, 1960.
38 G. W. Chantry and R. A. Plane, *J. Chem. Phys.*, 1960, **32**, 319.

In the formulation of the theoretical basis for the delta-function potential model [see Section 3 of this chapter, equations (20)—(30)], it was demonstrated that the approximations inherent in the method are severe, and the empirical nature of some of the parameters used was emphasized. In spite of the unpromising appearance of this model, however, it has been extensively used to obtain chemical bond orders from measured Raman intensities. The results, on the whole, are in surprisingly fair agreement with experiment, justifying our consideration of them in some detail.

Since two variations of the delta-function potential model have been presented, it is informative first to examine how well the two agree with one another in their calculated bond polarizability derivatives, as well as seeing how the calculated values compare with those determined experimentally through Raman band intensity measurements. In Table 1 are set out the

Table 1 *Bond polarizability derivatives (in* 10^{-2} nm²) *based on the delta-function potential model and compared with experimental values from symmetric stretching modes*

Molecule/Ion	*Assumed bond order*	$\partial\bar{\alpha}/\partial r$ *experimental*[17]	$\partial\bar{\alpha}/\partial r$ *from equation* (28)[18]	*equation* (29)[17]
CH_4	1	1.04	1.20	0.68
C_2H_6 (C—C)	1	1.37	1.81	1.73
(C—H)	1	1.08	1.20	0.72
C_2H_4 (C—C)	2	1.89	2.46	2.24
(C—H)	1	1.04	1.11	0.65
C_2H_2 (C—C)	3	2.99[37] or 3.36[39]	2.94	3.27
(C—H)	1	1.02	0.87	0.63
CCl_4	1	2.08	2.02	2.43
$CHCl_3$ (C—Cl)	1	1.93	1.90	2.46
CH_2Cl_2 (C—Cl)	1	1.73	1.75	2.46
$SiCl_4$	1	1.96	2.06	2.30
$GeCl_4$	1	2.66	2.28	2.47
$SnCl_4$	1	3.37	3.08	3.18
CBr_4	1	3.33	2.76	3.02
$SnBr_4$	1	6.73	4.02	3.89
NO_3^-	4/3	1.71	0.83	1.47
CO_3^{2-}	4/3	1.08	1.05	1.37
PO_4^{3-}	1	0.95	0.85	1.12

results of $(\partial\bar{\alpha}/\partial r)$ calculations for molecules and ions which have been treated both by Long and Plane [18] and by Lippincott and Nagarajan.[17] In obtaining these results the bond orders, $n/2$, have been assumed as obvious, and σ and π bonds have been assumed to contribute equally to $(\partial\bar{\alpha}/\partial r)$ values (see, however, ref. 38). It may be seen that the two sets of calculated values of $(\partial\bar{\alpha}/\partial r)$, though exhibiting a general rough agreement with the experimental values, do not agree particularly well even with one another. The level of

[39] H. W. Schrotter and H. J. Bernstein, *J. Mol. Spectroscopy*, 1964, **12**, 1.

Table 2 *Bond orders for some polyatomic ions calculated from the delta-function potential model (bond-polarizability derivatives in* 10^{-2} nm^2)

Ion	$\partial\bar{\alpha}/\partial r$ *experimental*	*Bond order from equation* (28)[18]	*equation* (29)[17]
SO_4^{2-}	1.37	1.47	1.08
ClO_4^-	1.73	1.76	1.26
IO_4^-	2.74	1.80	1.29
$ZnCl_4^{2-}$	0.87	0.25	0.35
$CdCl_4^{2-}$	1.04	0.21	0.33
$HgCl_4^{2-}$	2.10	0.35	0.56
$GaCl_4^-$	1.12	0.43	0.46
$ZnBr_4^{2-}$	1.80	0.38	0.61
$CdBr_4^{2-}$	2.46	0.40	0.67
$HgBr_4^{2-}$	5.11	0.66	1.14
$GaBr_4^-$	3.08	0.91	1.07

agreement apparent between the values listed in Table 1 must be taken as an indication of the validity of the delta-function potential method for determining bond orders. With this cautionary note, a series of calculated bond orders for some polyatomic ions are presented in Table 2, together with the experimental (from Raman band intensities) bond-polarizability derivatives from which they were derived. Again, comparison of bond-order values as calculated by the two variations of the method summarized by equations (28) and (29) shows the limitations of the method. In the face of these discrepancies, deductions such as those made by Lippincott and Nagarajan [17] about the relative contributions made by σ and π bonds to the $(\partial\bar{\alpha}/\partial r)$ values must be treated with some caution. Similarly, some of Long and Plane's conclusions, such as those relating the halogenomethane calculated polarizability derivatives to orbital hydridization variations, must be considered somewhat speculative.

In spite of the difficulties referred to above, the delta-function potential model has continued to find application in the interpretation of Raman intensities. An interesting series of papers by Spiro and co-workers,[40–43] treating a variety of inorganic and organometallic systems, will serve to illustrate the utility of the model. A common theme of these investigations is the determination of the characteristics of metal–metal bonds. The vibrational frequencies for hexamethyldi-silicon, -germanium, -tin, and -lead, for example, have been used [42] in approximate normal-co-ordinate analyses to obtain a consistent set of skeletal force constants for the molecules. This type of analysis is a necessary prerequisite to the calculation of $(\partial\bar{\alpha}/\partial r)$ values from the Raman band intensities, since a knowledge of the eigenvectors is needed to perform the transformations from $(\partial\bar{\alpha}/\partial Q)$ to $(\partial\bar{\alpha}/\partial r)$.[3] The

[40] P. A. Bullier, V. A. Maroni, and T. G. Spiro, *Inorg. Chem.*, 1970, **9**, 1887.
[41] W. M. Scovell, B. Y. Kimura, and T. G. Spiro, *J. Coordination Chem.*, 1971, **1**, 107.
[42] B. Fontal and T. G. Spiro, *Inorg. Chem.*, 1971, **10**, 9.
[43] P. A. Bùlliner, C. O. Quicksall, and T. G. Spiro, *Inorg. Chem.*, 1971, **10**, 13.

approximations inherent in the usual normal-co-ordinate analysis procedure are well known. These add further uncertainty to the bond orders ultimately evaluated through this method. From measurements of the A_1 Raman band intensities and depolarization ratios relative to the A_1 bond of CCl_4 as an internal standard, $\partial\bar{\alpha}/\partial Q = \bar{\alpha}_Q'$ values were found and converted to $\partial\alpha/\partial U = \bar{\alpha}_U'$ values, *i.e.* polarizability derivatives with respect to the internal co-ordinates (metal–carbon stretching, metal–metal stretching, *etc.*). This conversion was accomplished through the transformation [3]

$$\bar{\alpha}'_{Q_i} = \sum_U \sqrt{N_U} l_{Ui} \bar{\alpha}'_U \qquad (40)$$

where N_U is the number of internal co-ordinates in the set, and l_{Ui} is the eigenvector element connecting the normal mode with the symmetry co-ordinate constructed from the internal-co-ordinate set.[16] This transformation introduces still more problems of interpretation owing to uncertainty about the sign of $\bar{\alpha}'_{Qi}$ [it is the *square* of this quantity which is obtained from the Raman intensity data – see equation (19)] and a redundancy condition in the transformation of co-ordinates, so that, finally, solutions for $\bar{\alpha}'_{CM}$ and $\bar{\alpha}'_{MM}$ have to be selected on the grounds that they have 'reasonable magnitudes'. Fontal and Spiro's [42] data are listed in Table 3. These authors are careful

Table 3 *Frequencies, force constants, polarizability derivatives, and bond orders for hexamethyldi-silicon, -germanium, and -tin*[42]

	A_1 *mode frequencies** /cm^{-1}	*Force constants* /10^{-16} N m^{-1}	$\bar{\alpha}_u'/10^{-20}$ m^2	*Bond orders*
$(CH_3)_6Si_2$	184 (CMC)	0.20	−0.05	—
	404 (MM)	1.70	1.69	0.31
	628 (MC)	2.49	1.33	0.48
$(CH_3)_6Ge_2$	165 (CMC)	0.17	−0.03	—
	273 (MM)	1.54	2.20	0.37
	569 (MC)	2.47	1.81	0.58
$(CH_3)_6Sn_2$	135 (CMC)	0.09	−0.21	—
	192 (MM)	1.39	3.89	0.42
	509 (MC)	2.08	2.93	0.70

* The frequencies listed are, of course, associated with the normal modes. The internal co-ordinate designations are in each case the major components of the normal modes.

to point out that the bond orders given in Table 3 may not be meaningful in any absolute sense, but claim that they may be useful in comparing chemically related systems. Within the context of this modest claim, it is interesting that although the metal–metal bond polarizability derivatives for the $(CH_3)_6M_2$ molecules more than double on going from M=Si to M=Sn, the calculated bond orders vary by only 30%. Although these clearly are M—M single bonds, the absolute values of the bond-order parameter (0.32—0.42) are only about one-third as large as might be expected. However, it is claimed that the approximate constancy of the calculated bond orders means that through this series of organometallic compounds the number of bonding electrons in the M—M bonds remains roughly fixed, although the M—M bond *strengths*, as reflected in the calculated force constants (Table 3), vary considerably. Similar work with the compounds hexaphenyl-ditin and -dilead has lead to the conclusion [43] that the Sn—Sn bond in the first of these has a bond order in the 'normal range', 0.3—0.4, for metal-metal bonds, but the calculated Pb—Pb bond order is anomolously high at 0.78.

Bond-order calculations based on Raman intensities and the delta-function potential model have been used to infer the extent of metal–metal bonding in a variety of metal 'cluster' compounds.[40, 41] These are of interest for comparison with other methods used for probing the nature of metal–metal bonds in such compounds. In Table 4 are listed data which may be used in

Table 4 *Cluster frequencies, metal–metal force constants, bond lengths, and bond orders for some bridged polynuclear complexes* [40, 41]

Species	M—M *frequency* /cm^{-1}	*Force constant* /10^{-16} N m^{-1}	*Bond length*/pm	*Bond order*
$\{(CH_3)_3PtOH\}_4$	137	0.79	342	0.12
$\{(CH_3)_3PtCl\}_4$	99	0.50	373	0.06
$\{(CH_3)_3PtI\}_4$	88	0.49	400	~0
$(t\text{-}C_4H_9Li)_4$	566	0.27	252	0.01
$[Bi_6(OH)_{12}]^{6+}$	177	0.97	371	0.06
$[Pb_4(OH)_4]^{4+}$	130	0.54	385	0.02
$Tl_4(OC_2H_5)_4$	102	0.27	383	0.02

this comparative manner. These are frequencies assigned primarily to metal-metal stretching modes, force constants determined from the frequencies, bond lengths, and the delta-function bond orders. The data all refer to bridged polynuclear complexes, the formulae of which are given in Table 4. It may be seen that although the magnitude of the metal–metal force constants appears to be inconsistent with the observed intensities, as reflected in the bond-order values, the variations in the two quantities do seem to be reasonably consistent within the series of compounds. In all these cases, the strength of the metal–metal bond interaction appears to be quite low, as judged by the magnitudes of the calculated bond orders.

7 Frequency Dependence of Raman Intensities in the Pre-resonance Region

The widespread availability of a variety of laser light sources, providing a wide range of excitation frequencies in well-collimated beams, has led to the accumulation in recent years of many new data on the frequency dependence of Raman intensities. The current development of high-power lasers which are tuneable over the full visible spectrum and into the ultraviolet is certain to have a major impact on this intriguing aspect of Raman scattering. The data available in the literature up to 1965 have been most excellently reviewed by Behringer,[8] who also has given a detailed account of the theoretical considerations governing the manner in which Raman intensities vary with changing frequency of excitation – particularly as an electronic transition in the scattering molecule is approached.

Albrecht and Hutley [44] recently have modified the earlier expressions of the vibronic expansion theory [see equations (31)—(38)] to derive an explicit form for the frequency dependence of Raman scattering:

$$(\alpha_{\rho\sigma})_{gi,\,gj} = \sum_{e}\sum_{s} [(\nu_e\nu_s + \nu_0^2)/h^2(\nu_e^2 - \nu_0^2)(\nu_s^2 - \nu_0^2)] \times \langle i||Q_a||j\rangle[(g^0|R_\rho|e^0)(e^0|h_a|s^0)(s^0|R_\sigma|g^0) + (g^0|R_\sigma|e^0)(e^0|h_a|s^0)(s^0|R_\rho|g^0)] \quad (41)$$

This polarizability tensor component relates to the Raman intensity as expressed by equation (11) given earlier. Subscripts e and s here refer to virtual electronic states with eigenfrequencies at ν_e and ν_s, respectively, and the superscript zero indicates the nuclear co-ordinate parameters at the ground-state equilibrium configuration. Other symbols are as defined earlier, in Section 4 of this chapter. The diagonal terms ($e = s$) in equation (41) have been designated as 'A' terms, and the off-diagonal terms ($e \neq s$) as 'B' terms,[9] as already seen in Section 4. In order to simplify the summation over virtual states expressed in equation (41), Albrecht and Hutley have considered the special case where the incident frequency, ν_0, gradually approaches a molecular absorption involving an excited electronic state which is serving as an active virtual state for the vibrational mode responsible for the scattering. Assuming that only one component, $\alpha_{\rho\rho}$, of the polarizability tensor is active, and that e is the major active electronic state in the near-resonance condition, the following approximation form is obtained:[44]

$$(\alpha_{\rho\rho})_{ij} \approx A_{\rho\rho} + B_{\rho\rho}, \quad (42)$$

where

$$A_{\rho\rho} = 2[(\nu_e^2 + \nu_0^2)/(\nu_e^2 - \nu_0^2)^2]\,[(e^0|R_\rho|g^0)^2/h^2] \times (e^0|h_a|e^0)\langle i||Q_a||j\rangle \quad (43)$$

and

$$B_{\rho\rho} = 4\sum_{s}[h^{-2}(\nu_e\nu_s + \nu_0^2)/(\nu_e^2 - \nu_0^2)(\nu_s^2 - \nu_0^2)] \times (g^0|R_\rho|e^0)(s^0|R_\rho|g^0)(e^0|h_a|s^0)\langle i||Q_a||j\rangle \quad (44)$$

Then, assuming that the active virtual states, s, lie at energies sufficiently high to permit use of some average ν_s in term $B_{\rho\rho}$, the A and B terms may be rewritten:

$$A_{\rho\rho} = (2e^2/h^2)[(\nu_e^2 + \nu_0^2)/(\nu_e^2 - \nu_0^2)^2]\sigma_e\varepsilon_{ee}(a_{ij}), \quad (45)$$

$$B_{\rho\rho} \approx (2e^2/h^2)[2(\nu_e\nu_s + \nu_0^2)/(\nu_e^2 - \nu_0^2)(\nu_s^2 - \nu_0^2)] \times \Sigma R_e R_s \varepsilon_{es}(a_{ij}) \quad (46)$$

[44] A. C. Albrecht and M. C. Hutley, *J. Chem. Phys.*, 1971, **55**, 4438.

where a new notation has been introduced for the cross-section for electronic transitions:

$$\sigma_k = |R_k|^2 \quad \text{with} \quad R_k = e^{-1}(k^0|R_\rho|g^0) \tag{47}$$

and the vibronic energy, $\varepsilon_{kl}(a_{ij})$:

$$\varepsilon_{ki}(a_{ji}) = (k^0|h_a|l^0)\langle i||Q_a||j\rangle \tag{48}$$

Defining the Raman scattering cross-section, σ_{RAM}, as

$$\sigma_{\mathrm{RAM}}(m, n) = \left(\frac{2^7\pi^5}{3^2}\right)\left(\frac{\nu^4}{c^4}\right)(\alpha_{mn})^2 \tag{49}$$

in order to re-express equation (11) in the simplified form:

$$I_{mn} = \sigma_{\mathrm{RAM}}(m, n)\, I_0, \tag{50}$$

Albrecht and Hutley [44] finally obtain the following simple expression for σ_{RAM} in units of barns (per orientally averaged scattering centre):

$$\sigma_{\mathrm{RAM}} \approx 0.88778[F_A\sigma_e{}^*\varepsilon_{ee}{}^*(a_{ij}) + F_B\sum_s R_e{}^*R_s{}^*\varepsilon_{es}{}^*(a_{ij})]^2 \tag{51}$$

where the asterisk denotes use of the dimensionless atomic (hartree) units (energy in 2 Ry, distance in the Bohr radius), and F_A and F_B are the *dimensionless* frequency factors:

$$F_A = \nu^2(\nu_e^2 + \nu_0^2)/(\nu_e^2 - \nu_0^2)^2, \tag{52}$$

$$F_B = 2\nu^2(\nu_e\nu_s + \nu_0^2)/(\nu_e^2 - \nu_0^2)(\nu_s^2 - \nu_0^2). \tag{53}$$

These new functions F_A and F_B carry the full frequency dependence (note that $\nu = \nu_0 - \nu_{mn}$) of the intensity and are easily evaluated for real examples.

Raman intensity measurements with the 1334 cm^{-1} band of *p*-nitroaniline (PNA) have been used by Albrecht and Hutley to test the foregoing theory. This molecule has a strong near-ultraviolet absorption band centred at $\nu_e = 28\,570$ cm^{-1} ($\lambda = 350$ nm). A value of $\nu_s = 55\,560$ cm^{-1} ($\lambda = 180$ nm) was assumed as a representative far-ultraviolet absorption, the calculations being quite insensitive to the choice of ν_s. The experimental intensity dependence of the 1334 cm^{-1} band of PNA is shown in Table 5, normalized to the value determined with 488.0 nm excitation. The numbers listed in Table 5 for $(F_A^2)_{\mathrm{rel}}$, $(F_B^2)_{\mathrm{rel}}$, and $(F_A + F_B)^2_{\mathrm{rel}}$ are determined from the calculated *F*-factors divided by the experimental intensity values and scaled to 488.0 nm. Thus, an exact fit between the observed and calculated frequency dependence of the Raman band intensity would result in a full column of 1.00 entries for the correct *F*-factor. The $(F_A + F_B)^2$ term is evaluated for the special case of $\nu_e = \nu_s = 55\,560$ cm^{-1}. It is readily apparent from Table 5 that the F_B^2 term alone is quite successful in representing the observed frequency dependence. Figure 1 shows these data in an alternative form. Within the limitations of the simplifications made in their treatment, Albrecht and

Table 5 *Experimental and calculated frequency dependence of the* p-*nitroaniline* 1334 cm^{-1} *Raman band intensity* [44]

Exciting line λ_0/nm	$\nu_0/10^3$ cm^{-1}	I_{rel} *experimental*	$(F_A^2)_{rel}$	$(F_B^2)_{rel}$	$(F_A + F_B)^2_{rel}$
632.8	15.80	0.106	0.55	1.11	2.17
514.5	19.43	0.572	0.87	1.06	1.28
488.0	20.49	(1.000)	(1.000)	(1.000)	(1.000)
476.5	20.98	1.265	1.12	1.01	0.91
457.9	21.83	2.084	1.28	0.95	0.71

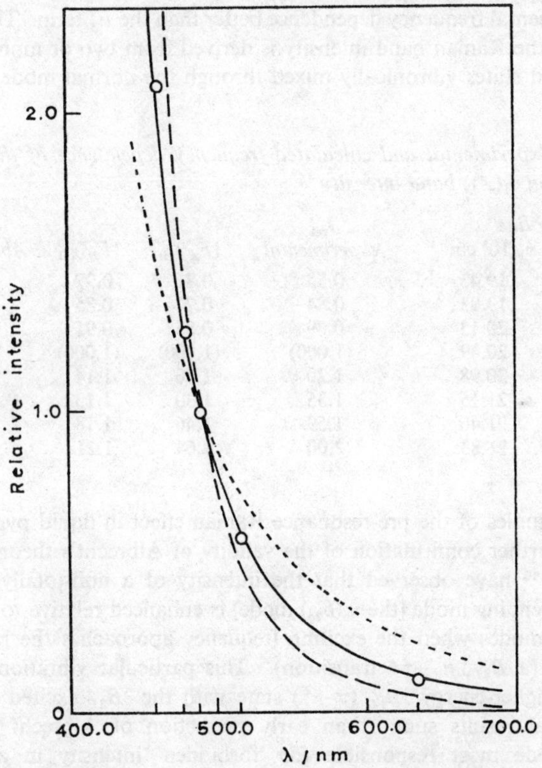

Figure 1 *Frequency dependence of a resonance-enhanced type B band in the spectrum of* p-*nitroaniline (PNA). Data scaled to unity at* 488.0 nm: (- - -), $F_A^2 = F_B^2$ ($\nu_e = \nu_s$ = 180 nm); (—), F_B^2(ν_e = 350 nm, ν_s = 180 nm); (— — —), F_A^2($\nu_e = \nu_s$ = 350 nm). *Experimental points are circled and laser wavelengths are indicated* (Reproduced by permission from *J. Chem. Phys.*, 1971, **55**, 4438)

Hutley[44] conclude from this that the source of intensity for the 1334 cm^{-1} Raman band of PNA is a vibronic mixing through this normal mode of the first excited singlet state of the molecule with excited states of similar symmetry (polarization) lying in the deep ultraviolet. The poor fit with the F_A^2 term indicates that the intensity is not derived from the low-lying electronic transition alone.

The foregoing type of analysis also has been used by Kiefer and Bernstein[45] in interpreting their data on the frequency dependence of Raman band intensities excited from an aqueous solution of the chromate ion. Two absorption bands are observed for CrO_4^{2-} in water. These peak at about 26 900 (ν_e) and 36 600 cm^{-1} (ν_s), both corresponding to electronic transitions of the same symmetry type, $^1A_1 \rightarrow {}^1T_2$. Using these values for ν_e and ν_s, F_A^2 and F_B^2 terms have been calculated and compared with the measured intensities. The data listed in Table 6 show that, as with the PNA results, the F_B^2 term represents the experimental frequency dependence better than the F_A^2 term. This suggests again that the Raman band intensity is derived from two or more electronically excited states vibronically mixed through the normal mode.

Table 6 *Experimental and calculated frequency dependence of the chromate ion* $\nu_1(A_1)$ *band intensity*[45]

Exciting line		I_{rel}			*Molar*
λ_0/nm	$\nu_0/10^3$ cm^{-1}	*experimental*	$(F_A^2)_{rel}$	$(F_B^2)_{rel}$	*Absorbance*, ε_0
514.5	19.43	0.55	0.81	0.97	0.4
501.7	19.93	0.84	0.77	0.85	1.8
496.5	20.13	0.89	0.85	0.91	3.2
488.0	20.49	(1.000)	(1.000)	(1.000)	7.8
476.5	20.98	1.20	1.26	1.14	24
472.7	21.15	1.35	1.30	1.13	33
465.8	21.46	1.59	1.46	1.18	56
457.9	21.83	2.00	1.64	1.21	97

Recent studies of the pre-resonance Raman effect in liquid pyrazine have provided further confirmation of the validity of Albrecht's theory. Ito and co-workers[46] have observed that the intensity of a non-totally-symmetric hydrogen bending mode [the $\nu_5(b_{2g})$ mode] is enhanced relative to the totally symmetric modes when the exciting frequency approaches the lowest-lying absorption (a B_{3U}, $n \rightarrow \pi^*$ transition). This particular vibration is known to mix a higher-energy $^1B_{1U}$ (π–π^*) state with the $^1B_{3U}$ excited state. The experimental results support an early prediction of Albrecht[21] that the normal mode most responsible for 'forbidden' intensity in an allowed electronic transition should show particularly striking Raman intensity

[45] W. Kiefer and H. J. Bernstein, *Mol. Phys.*, 1972, **23**, 835.
[46] M. Ito, I. Suzuka, Y. Udagawa, N. Mikami, and K. Kaya, *Chem. Phys. Letters*, 1972, **16**, 211.

enhancement as the exciting frequency approaches that of the given allowed electronic transition. This conclusion is supported by the work of Kalantar and co-workers,[47] also with liquid pyrazine, which involved a comparison of the $\nu_5(b_{2g})$ and the $\nu_4(b_{2g})$ modes. It is known that $\nu_4(b_{2g})$ does *not* cause vibronic mixing of the B_{1U} and B_{3U} excited states and, as expected from the theory, this mode does not show the Raman intensity enhancement displayed by $\nu_5(b_{2g})$.

8 Dependence of Raman Intensity on Absorbance

From their results of studies with the species S_2^- and S_3^- doped in alkali-metal halide crystals, Holzer and co-workers [48] reported a linear dependence of Raman intensity on λ_0 in the region of an allowed electronic transition. For S_3^-, λ_0 and ε_0 are linearly related, so that a linear dependence of intensity on absorbance, ε_0, is indicated. This also is the conclusion arrived at empirically by Kiefer and Bernstein from their study of the resonance Raman spectrum of the permanganate ion.[45] Earlier experimental results reported by Mortensen [49] on the I_2 solution spectrum excited over a range of frequencies within the visible absorption band also were found by Kiefer and Bernstein to have a linear intensity dependence on ε_0, though Mortensen himself had fitted these data to an ε_0^2 relationship. This squared dependence relates directly to a theoretical result derived by Mortensen,[49] expressing the Raman intensity in the following form:

$$I_{mn} \propto I_0 \varepsilon_{me}^{\alpha} \varepsilon_{ne}^{\beta} \tag{54}$$

where α and β refer to incident and scattered beams, respectively, and e is the virtual state intermediate (in time) between states m and n. Thus $\varepsilon_{me}(=\varepsilon_0)$ is the absorbance for the exciting line, and ε_{ne} is the absorbance for the Raman line. For small Raman frequency shifts, *i.e.* ν_{me} small, the approximation $\varepsilon_{me} \approx \varepsilon_{ne}$ results in the rough proportionality: $I \propto \varepsilon_0^2$.

The apparent contradictions implied by the foregoing may be resolved by further consideration of the A and B terms derived by Tang and Albrecht from the general vibronic expansion theory [see equations (31)—(34)]. It may be concluded that for those transitions where the A term is dominant the Raman intensities of totally symmetric vibrational modes will be particularly enhanced as a resonance condition is approached. In this case the Franck–Condon allowed transition is expected to be selectively activated. From the form of the transition moments (TM) it may be seen that we may crudely write $A \propto (\mathrm{TM})^2$, and since $I \propto \alpha^2$, and $(\mathrm{TM})^2 \propto \varepsilon$, then $I \propto \varepsilon^2$ for transitions governed by the A term. This condition will apply strictly

[47] A. H. Kalantar, E. S. Franzosa, and K. K. Innes, *Chem. Phys. Letters*, 1972, **17**, 335.
[48] W. Holzer, W. F. Murphy, and H. J. Bernstein, *J. Mol. Spectroscopy*, 1970, **32**, 13; *Chem. Phys. Letters*, 1970, **4**, 641.
[49] O. S. Mortensen, *J. Mol. Spectroscopy*, 1971, **39**, 48; *Mol. Phys.*, 1971, **22**, 179.

only to an on-resonance situation where the sum over virtual states is dominated by a single term. Other aspects of this situation are that it implies a symmetric scattering tensor, overtones and combination tones become allowed through an expansion of the energy denominator, and the frequency dependence of the Raman intensity is governed by the factor F_A^2, as discussed previously [equation (52)].

When the B term is dominant in equation (31), and the Raman spectrum is excited by frequencies within an absorption band of the molecule, the condition $B \propto$ (TM) holds, since only the transition moments $(g|R|e)$ match the energy of excitation, while the moments $(s|R|g)$ involve virtual states away from resonance. In this case we see that $I \propto \varepsilon$, *i.e.* a linear relationship between intensity and absorbance. The B term involves vibronic terms in the numerator and governs the off-diagonal tensor elements. As has already been seen [equation (53)], the frequency dependence of the B term is quite different from that of the A term, and it is predicted that when $I \propto \varepsilon$, the frequency dependence of I is given by F_B^2. Correspondingly, when $I \propto \varepsilon^2$, the frequency dependence of I is given by F_A^2.

9 Resonance Raman Intensities

As has been clearly explained by Behringer, the competing processes of scattering and absorption make for experimental difficulties in measuring resonance enhancement quantitatively. Other problems are fluorescence emission and photolysis. The difficulties may be relieved to some extent by the versatility of laser techniques, using such tricks as grazing incidence illumination and the rotating sample cell,[50] though these are relatively ineffective against efficient fluorescence and photolysis processes.

The mode selectivity of resonance Raman scattering [8] has particularly advantageous consequences when the technique is applied to the study of large, complex, biologically active molecules in dilute aqueous solution. Examples of such studies are those with haemoglobin and related compounds,[51–53] and with vitamin B_{12} and its derivatives.[54] The extensive studies of haem proteins already have yielded results of considerable theoretical, as well as structural and biological, interest. Vibrational modes associated specifically with the chromophore, the iron complex of protoporphyrin IX, are selectively enhanced by resonance. Some modes have been identified as type B according to the criteria established in the previous section of this chapter, and others of type A.

The absorption spectra of all low-spin metalloporphyrins are similar,

[50] W. Kiefer and H. J. Bernstein, *Appl. Spectroscopy*, 1971, **25**, 500.
[51] T. C. Strekas and T. G. Spiro, *Biochim. Biophys. Acta*, 1972, **263**, 830; 1972, **278**, 188.
[52] H. Brunner, A. Meyer, and H. Sussner, *J. Mol. Biol.*, 1972, **70**, 153.
[53] H. Brunner, *Biochem. Biophys. Res. Comm.*, 1973, **51**, 888.
[54] E. Mayer, D. J. Gardiner, and R. E. Hester, *J. C. S. Faraday II*, 1973, **69**, 1350.

containing an intense feature called the Soret,[55] or γ-band, at ~400 nm, and a weaker pair of visible (500—600 nm) features called Q-bands, or α and β, in order of decreasing wavelength.[56] These features arise from two electronic π–π* transitions, both terminating in the lowest empty orbital which is degenerate and of e_g symmetry under an assumed D_{4h} molecular symmetry. The α-band is associated with a transition from the highest-filled orbital, which has a_{2U} symmetry. The Soret band originates in a transition from the next highest filled orbital, which has a_{1U} symmetry. Thus the two transitions both are symmetry-allowed (type E_U), but they are strongly mixed by configuration interaction, with the Soret band gaining most of the intensity.[55] The β-band is assigned as a vibronic 0–1 component of the α-band. From the direct product $E_u \times E_u = A_{1g} + B_{1g} + B_{2g} + A_{2g}$ we may determine the vibrational modes active in mixing the two electronic transitions and thereby capable of 'borrowing' intensity from the Soret band. It has been established that the A_{1g} modes are ineffective in the vibronic mixing,[57] and correspondingly these modes, which give rise to polarized Raman bands, are either weak or absent from the spectra of metalloporphyrins in resonance with the αβ-bands.[56] Resonance active modes of type B are therefore of B_{1g}, B_{2g}, and A_{2g} symmetry.

The expected polarizations of the resulting Raman bands have been determined by reference to tables of the forms of the scattering tensors.[58, 59] For B_{1g} modes, $\boldsymbol{\alpha}_{xx} = -\boldsymbol{\alpha}_{yy}$, and for B_{2g} modes, $\boldsymbol{\alpha}_{xy} = \boldsymbol{\alpha}_{yx}$, with all other tensor elements being zero. From consideration of a form of Placzek's polarizability theory,[1] Spiro and Strekas have shown that this leads to a depolarization ratio for polarized incident light [see equation (18)] of 3/4 for the B_{1g} and B_{2g} modes, in agreement with experiment.[56] For the A_{2g} modes, which are forbidden in non-resonance Raman scattering, the scattering tensor is asymmetric, *viz.* $\boldsymbol{\alpha}_{xy} = -\boldsymbol{\alpha}_{yx}$. This leads to the prediction of an infinite depolarization ratio. The observation of bands in the resonance Raman spectrum of haemoglobin and cytochrome *c*, which appear strongly in perpendicular polarization (relative to the polarization of the incident light) but only weakly or not at all in parallel polarization, provided the first experimental confirmation of this prediction.[56] These bands are assigned to the normally forbidden A_{2g} modes and are said to be *inverse* polarized.

A well-characterized example of a resonance-enhanced A_{1g} mode of the porphyrin chromophore is the 1373 cm^{-1} polarized band studied by Spiro and Strekas.[60] The frequency dependence of this band intensity has been shown to follow the F_A^2 factor of equation (52) when the excitation frequencies approach the frequency of the Soret band. The experimental data and theore-

[55] M. Gouterman, *J. Chem. Phys.*, 1959, **30**, 1139; *J. Mol. Spectroscopy*, 1961, **6**, 138.
[56] T. G. Spiro and T. C. Strekas, *Proc. Nat. Acad. Sci. U.S.A.*, 1972, **69**, 2622.
[57] M. W. Perrin, M. Gouterman, and C. L. Perrin, *J. Chem. Phys.*, 1969, **50**, 4137.
[58] L. N. Ovander, *Optics and Spectroscopy*, 1960, **9**, 302.
[59] W. M. McClain, *J. Chem. Phys.*, 1971, **55**, 2789.
[60] T. G. Spiro and T. C. Strekas, *J. Raman Spectroscopy*, 1973, **1**, 22.

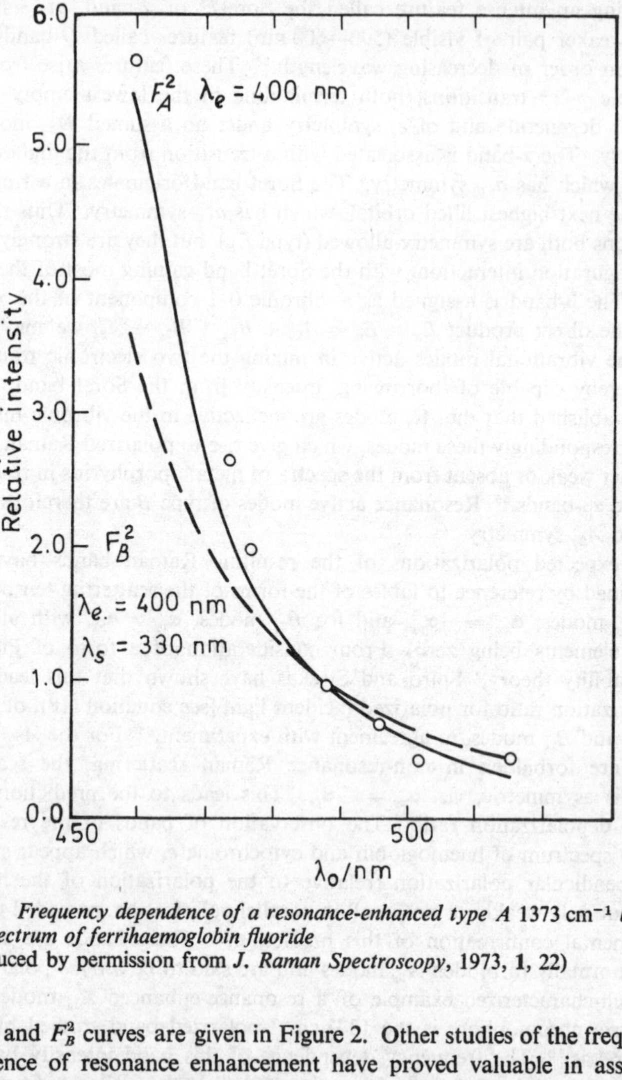

Figure 2 *Frequency dependence of a resonance-enhanced type A* 1373 cm^{-1} *band in the spectrum of ferrihaemoglobin fluoride*
(Reproduced by permission from *J. Raman Spectroscopy*, 1973, **1**, 22)

tical F_A^2 and F_B^2 curves are given in Figure 2. Other studies of the frequency dependence of resonance enhancement have proved valuable in assigning features in the absorption spectra of both porphyrin and corrin derivatives.[60–62]

One of the more interesting of the many other studies of resonance Raman spectra which have been published during the past few years is that of Holzer, Murphy, and Bernstein,[63] made with the halogen gases. By appropriate choice

[61] L. Rimai, M. E. Heyde, H. C. Heller, and D. Gill, *Chem. Phys. Letters*, 1971, **10**, 207.
[62] E. Mayer, D. J. Gardiner, and R. E. Hester, *Mol. Phys.*, in the press.
[63] W. Holzer, W. F. Murphy, and H. J. Bernstein, *J. Chem. Phys.*, 1970, **52**, 399.

of an exciting line from the many possibilities offered by an argon laser, these workers demonstrated how either a resonance Raman effect (RRE) or resonance fluorescence (RF) could be induced. For I_2 gas they determined overtone sequences up to the 14th harmonic, and measured Raman frequencies, depolarization ratios, and relative scattering cross-sections for the fundamentals and overtones of Cl_2, Br_2, I_2, BrCl, ICl, and IBr. Criteria for distinguishing between RRE and RF were formulated in terms of the band envelope, the overtone pattern, depolarization ratio, behaviour with increasing gas pressure, and behaviour with foreign gases added. For the RRE, observed with exciting light in the dissociation continuum, continuous broadening of the *Q*-branch and steadily decreasing peak intensity with the progression to higher overtones was found. Excitation in the discrete band region produced sharp RF doublets ($\Delta J = \pm 1$) displaying irregular overtone sequences with no broadening at higher orders, illustrating the different selection rules and relevance of the Franck–Condon factors for the two effects. The differences in behaviour of the two effects may in general be related to the fact that the RRE occurs very much faster than RF, so that the latter is more sensitive to intermolecular collisions and diffusion effects. However, theoretical consideration[64] of the details of RF has led to the conclusion that the lifetime of a resonance fluorescence process may be as short as 10^{-14} s, which would invalidate conclusions made above. Further uncertainty has been introduced by reported observations of RRE in I_2 gas upon excitation below the dissociation limit.[65]

In conclusion, it may be stated with confidence that the quantitative study of Raman intensity effects, particularly in the resonance or near-resonance condition, is likely to constitute an area of intense activity during the next few years. Results already available show great promise for the determination of the nature of chemical bonding, providing structural and conformational data on complex molecules of biological significance, and for probing the electronic structure of molecules.

The Reporter is grateful to A. C. Albrecht and G. M. Korenowski, both of Cornell University, for helpful discussions on Raman intensity theory.

[64] A. Nitzan and J. Jortner, *J. Chem. Phys.*, 1972, **57**, 2870.
[65] D. G. Fouche and R. K. Chang, *Phys. Rev. Letters*, 1972, **29**, 536.

7
Diatomic Predissociation Linewidths

BY M. S. CHILD

1 Introduction

The past five years have seen a revival of theoretical interest in predissociation phenomena, coupled with increasing sophistication in experimental technique [1] and a continuing search for related resonance effects in the molecular scattering field.[2–5] Given the established characterizations of predissociation types [6, 7] and selection rules,[8, 9] and the accepted value of predissociation measurements in locating the dissociation limit,[6, 10] the object of recent theoretical work has been to underline further information available from individual linewidth (or lifetime) measurements. This information relates primarily to the forms of the relevant potential-energy curves.

The purpose of this Report is to summarize our present understanding of predissociation by rotation (type III) [6] and by curve crossing (type I) [6] in diatomic molecules. It is also convenient to sub-divide predissociations of type I into 'near diabatic' and 'near adiabatic' categories as illustrated in Figure 1. The material therefore falls into two parts, each of which includes a formal quantum mechanical introduction, which may also serve to emphasize the connection between spectroscopic measurements and molecular scattering data. Subsequent sections cover the available computational and analytical techniques, followed by a detailed summary of conclusions. An attempt has been made to include all linewidth or lifetime calculations for specific systems published before May 1973. Useful general references in the molecular scattering field have been given by Levine [11, 12] and Micha.[13]

1 See R. A. Sutherland and R. A. Anderson, *J. Chem. Phys.*, 1973, **58**, 1226; also see refs. 71, 102, and 103.
2 R. B. Bernstein, *Phys. Rev. Letters*, 1966, **16**, 385.
3 W. C. Stwalley, A. Niehaus, and D. R. Herschbach, in 'Fifth International Conference on the Physics of Electronic and Atomic Collisions', Leningrad, 1967, Abstracts of Papers, ed. I. P. Flaks and E. S. Solovoyov, Nauka, Leningrad, 1967, p. 639.
4 A. Schutte, D. Bassi, F. Tommasini, and G. Scoles, *Phys. Rev. Letters*, 1972, **29**, 979.
5 J. M. Peek, *Physica*, 1973, **64**, 93.
6 G. Herzberg, 'Spectra of Diatomic Molecules', Van Nostrand, New York, 1950.
7 R. S. Mulliken, *J. Chem. Phys.*, 1960, **33**, 247.
8 R. de L. Kronig, *Z. Phys.*, 1932, **75**, 468; J. H. Van Vleck, *Phys. Rev.*, 1932, **40**, 544.
9 I. Kovács, 'Rotational Structure of the Spectra of Diatomic Molecules', Adam Hilger, New York, 1969.
10 A. G. Gaydon, 'Dissociation Energies and Spectra of Diatomic Molecules', 3rd edn., Chapman and Hall, London, 1968.
11 R. D. Levine, *Accounts. Chem. Res.*, 1970, **3**, 273.
12 R. D. Levine, in 'Theoretical Chemistry', ed. W. Byers Brown, MTP International Review of Science, Physical Chemistry Series One, Butterworths, London, 1972, Vol. 1.
13 D. A. Micha, *Accounts. Chem. Res.*, 1973, **6**, 138.

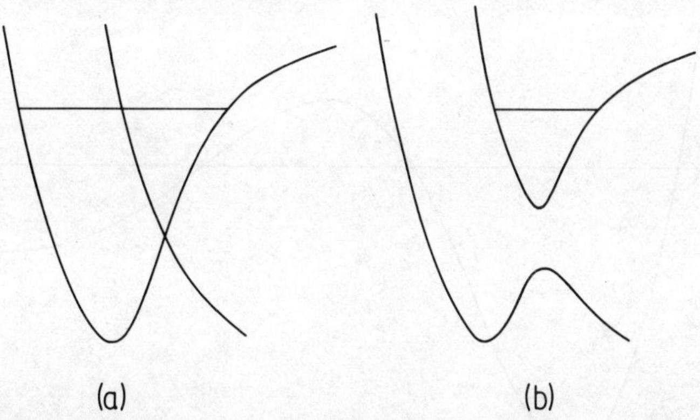

Figure 1 *Potential curves for type I predissociations of the* (a) *near diabatic and* (b) *near adiabatic categories*

The first general conclusion is that the pattern of predissociation linewidths provides a highly sensitive yardstick for the determination of unknown potential curves. Secondly, the computation of such a pattern for given potential curves is now a matter of routine, unless the predissociation is of the near adiabatic type (see Figure 1). As an alternative to exact numerical computation one may expect to apply available, relatively rapid, analytical (semi-classical) estimates within an accuracy of 10%. One general conclusion is that the linewidth arising from rotational predissociation (type III) is predicted to increase monotonically with increasing J, whereas type I predissociation is generally expected to cause an oscillatory linewidth pattern. These analytical formulae also provide physical insight into the details of the predissociation pattern, to the extent that a direct inversion procedure has been developed for determination of the repulsive potential curves responsible for type I predissociations.

2 Predissociation by Rotation

The relation between predissociation by rotation [6] and the shape or orbiting resonances of collision theory [2–5, 14] is by now well established. Both are due to the existence of a quasi-bound state separated from the continuum by a potential barrier, as depicted in Figure 2, but the two types of experiment are, in fact, complementary since the scattering resonances most readily detected experimentally correspond to lines which would appear indiscernibly diffuse in the spectrum.

[14] See also M. E. Gersch, and R. B. Bernstein, *Chem. Phys. Letters*, 1969, **4**, 221 for a computed total cross-section for the scattering of two H(1*S*) atoms.

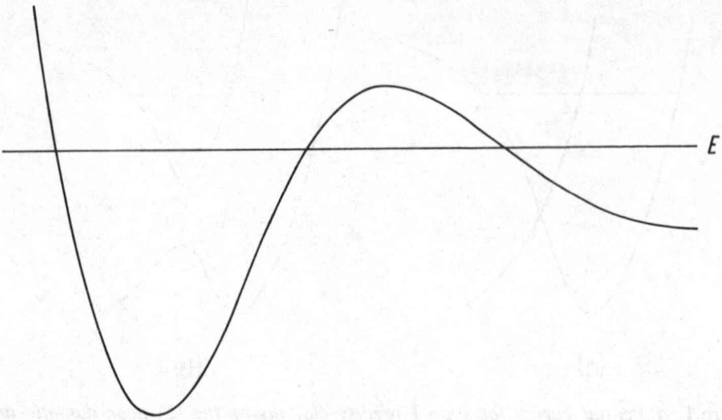

Figure 2 *The condition for possible predissociation by rotation*

Much of the relevant theory is to be found either in the scattering literature, or at least couched in the language of scattering theory. Advances have been made by both computational[15–22] and semi-classical methods.[23–38] Some computational workers have used the resonant jump in the scattering phase shift,[15–18] or the closely associated peak in the collision delay time (see below),[21, 39–41] to determine the positions and energy widths of the quasi-

[15] R. A. Buckingham and J. W. Fox, *Proc. Roy. Soc.*, 1962, **A267,** 102.
[16] R. A. Buckingham, J. W. Fox, and E. Gal, *Proc. Roy. Soc.*, 1965, **A284,** 237.
[17] R. B. Bernstein, C. F. Curtiss, S. Imam Rahajoe, and W. T. Wood, *J. Chem. Phys.*, 1966, **44,** 4072.
[18] T. G. Waech and R. B. Bernstein, *J. Chem. Phys.*, 1967, **46,** 4905.
[19] A. C. Allison, *Chem. Phys. Letters*, 1969, **3,** 371.
[20] J. L. Jackson and R. E. Wyatt, *Chem Phys. Letters*, 1970, **4,** 643.
[21] R. J. LeRoy and R. B. Bernstein, *J. Chem. Phys.*, 1971, **54,** 5114.
[22] R. A. Bain and J. N. Bardsley, *J. Chem. Phys.*, 1971, **55,** 4535.
[23] K. W. Ford, D. L. Hill, M. Wakano, and J. A. Wheeler, *Ann. Phys.*, 1959, **7,** 239.
[24] R. A. Buckingham and A. Dalgarno, *Proc. Roy. Soc.*, 1952, **A213,** 506.
[25] M. V. Berry, *Proc. Phys. Soc.*, 1966, **88,** 285.
[26] G. V. Dubrovskii, *Optika i Spektroskopiya*, 1964, **17,** 771; *Optics and Spectroscopy*, 1964, **17,** 416.
[27] P. M. Livingston, *J. Chem. Phys.*, 1966, **45,** 601.
[28] R. R. Herm, *J. Chem. Phys.*, 1967, **47,** 4290.
[29] W. H. Miller, *J. Chem. Phys.*, 1968, **48,** 1651.
[30] J. N. L. Connor, *Mol. Phys.*, 1968, **15,** 621; *ibid.*, 1969, **16,** 525.
[31] A. S. Dickinson, *Mol. Phys.*, 1970, **18,** 441.
[32] A. S. Dickinson, personal communication, 1973.
[33] N. Fröman and P. O. Fröman, *Nuclear Phys.*, 1970, **A147,** 606.
[34] J. N. L. Connor, *Mol. Phys.*, 1972, **23,** 717.
[35] J. N. L. Connor, *Mol. Phys.*, 1973, in the press.
[36] G. D. Mahan and M. Lapp, *Phys. Rev.*, 1969, **179,** 19.
[37] A. S. Dickinson and R. B. Bernstein, *Mol. Phys.*, 1970, **18,** 305.
[38] W. E. Baylis, *Phys. Rev.*, 1970, **1,** 990.
[39] L. Eisbud, dissertation, Princeton University, 1948.
[40] E. P. Wigner, *Phys. Rev.*, 1955, **98,** 145.
[41] F. T. Smith, *Phys. Rev.*, 1960, **118,** 349; erratum, 1960, **119,** 2098.

bound states. Others have pointed to the spectroscopically important peak in the relative amplitude of the wavefunction inside and outside the barrier region;[15, 19–21] however, the intimate connection [42] between this characteristic and the behaviour of the phase shift has received relatively little emphasis in the molecular literature. Again, in the semi-classical field attention has concentrated on the behaviour of the phase shift,[26–29] or equivalently (see below) on the location of complex energy poles in the scattering matrix,[30–32] but the behaviour of the amplitude ratio has been largely ignored.

Our first purpose will therefore be to demonstrate the necessary connection between resonant characteristics of the phase shift, $\eta(E)$, the collision delay time, $\tau_d(E)$, and the amplitude ratio, $a(E)$. This general discussion will involve the introduction of complex energy levels

$$E = E_n - i\Gamma_n/2, \Gamma_n > 0 \tag{1}$$

and lead to Breit–Wigner parameterizations of $\eta(E)$, $\tau_d(E)$, and $a(E)$. We shall then specialize to a semi-classical (JWKB) description in order to relate the resonance positions, E_n, and widths, Γ_n, to the form of the potential function. Finally, the practical applications of the theory will be discussed.

Resonance Characteristics.—Given that the quasi-bound state with angular momentum J, supported by a potential $V(r)$, lies in the continuum, it is convenient to take the energy zero at the dissociation limit, $V(\infty) = 0$, and to define

$$k^2 = 2\mu E/\hbar^2$$
$$U(r) = 2\mu V(r)/\hbar^2 \tag{2}$$

The wavefunction must then satisfy

$$\left[\frac{d^2}{dr^2} + k^2 - U(r) - \frac{J(J+1)}{r^2}\right] \chi_{EJ}(r) = 0 \tag{3}$$

subject to the boundary condition $\chi_{EJ}(0) = 0$. The most convenient normalization is to a delta-function of energy [43]

$$\int_0^\infty \chi_{EJ}(r)\chi_{E'J}(r)dr = \delta(E - E') \tag{4}$$

This fixes the asymptotic amplitude of the wavefunction as

$$\chi_{EJ}(r) \overset{r\to\infty}{\simeq} \left[\frac{2\mu}{\pi\hbar^2 k}\right]^{\frac{1}{2}} \sin\left[kr - J\pi/2 + \eta_J(E)\right] \tag{5}$$

where $\eta_J(E)$ is the phase shift, the term $-J\pi/2$ being included in equation (5) so that $\eta_J(E) = 0$ in the absence of any distortion potential $[V(r) = 0]$.

[42] See V. de. Alfaro and T. Regge, 'Potential Scattering' North Holland, Amsterdam, 1965.

[43] P. A. M. Dirac, 'The Principles of Quantum Mechanics', 4th edn., Oxford U.P., 1958.

Closely associated with $\eta_J(E)$ is the collision delay-time function:[39–41]

$$\tau_d(E, J) = 2\hbar \frac{\partial \eta_J}{\partial E} \tag{6}$$

the resonance behaviour of which is more conveniently characterized than that of the phase shift.[21]

The significance of the energy normalization given by equation (4) is that it ensures unit density of states.[44] Hence the differential oscillator strength for a spectroscopic transition from a bound level $\psi_b(r)$ of some other electronic state varies as

$$\nu \left| \int_0^\infty \psi_b(r) M(r) \chi_{EJ}(r) \mathrm{d}r \right|^2 \tag{7}$$

where $M(r)$ denotes the electronic transition moment. Thus, with the external amplitude fixed by equation (5), a peak in the internal-to-external amplitude ratio leads to strongly enhanced overlap with $\psi_b(r)$, and hence to a peak in the spectrum.

The necessary connection between the behaviour of this amplitude ratio and that of the phase shift is, however, most readily demonstrated by the introduction of a generalized solution $\chi_{kJ}(r)$ of equation (3), renormalized at the origin such that [42]

$$\lim_{r \to 0} r^{-J-1} \chi_{kJ}(r) = 1 \tag{8}$$

Note that for a non-singular potential, $\chi_{EJ}(r)$, given by equation (5), varies as Cr^{J+1} at the origin. This solution is extended to complex values of k:

$$k = k' + \mathrm{i}k'' \tag{9}$$

and hence, according to equation (2), to the complex energy levels given by equation (1), with

$$E_n = (k_n'^2 - k_n''^2)\hbar^2/2\mu$$
$$\Gamma_n = -2k_n' k_n'' \hbar^2/\mu \tag{10}$$

It may be shown [42] that this generalized solution may be expressed in the asymptotic region, $r \to \infty$, in terms of two Jost functions, $f_J(\pm k)$:

$$\chi_{kJ}(r) \overset{r \to \infty}{\simeq} \frac{1}{2\mathrm{i}} [f_J(k) \mathrm{e}^{\mathrm{i}kr} - f_J(-k) \mathrm{e}^{-\mathrm{i}kr}] \tag{11}$$

where for complex k

$$f_J(-k) = [f_J(k^*)]^* \tag{12}$$

[44] See A. Messiah, 'Quantum Mechanics', North Holland, Amsterdam, 1962, p. 734.

and $f_J(k)$ is analytic in k for Im $k < 0$. It follows that for real k (and hence real E):

$$\chi_{kJ}(r) \overset{r\to\infty}{\simeq} |f_J(k)| \sin [kr - J\pi/2 + \eta_J(E)] \quad (13)$$

where

$$\eta_J(E) = \arg f_J(k) + J\pi/2 \quad (14)$$

in agreement, apart from a change in normalization, with equation (5). We therefore have a wavefunction with a standard amplitude close to the origin, set by equation (8) and the short-range form of the potential. Equations (13) and (14) show that the external amplitude (and hence the amplitude ratio) and the phase shift depend on the behaviour of the function $f_J(k)$.

Particular attention attaches to complex zeros of the $f_J(\pm k)$, of which two types may be recognized. The simplest are zeros of $f_J(k)$ on the negative imaginary axis, $k = -i\gamma_{vJ}$ say, so that according to equation (11)

$$\chi_{kJ}(r) \overset{r\to\infty}{\simeq} Ae^{-\gamma_{vJ}r} \quad (15)$$

These therefore correspond to the bound states with real negative energies, $E_{vJ} = -\gamma_{vJ}^2\hbar^2/2\mu$. The quasi-bound states, on the other hand, are associated with zeros of $f_J(-k)$, given by equation (9) with $k' > 0$, $k'' < 0$. Hence Γ_n given by equation (10) is positive. For roots of this type, $\chi_{kJ}(r)$ has purely outgoing characteristics:

$$\chi_{kJ}(r) \overset{r\to\infty}{\simeq} \frac{1}{2i} f_J(k)e^{ikr} \quad (16)$$

and, as might be expected for a quasi-bound state, the imaginary component in the energy given by equation (1) ensures an exponential decrease in the time evolution factor:

$$|\exp(-iEt/\hbar)|^2 = \exp(-\Gamma_n t/\hbar) \quad (17)$$

with time constant

$$\tau_n = \hbar/\Gamma_n \quad (18)$$

Complex energy (and hence non-physical) solutions of this type are termed Siegert states,[45] while the energies themselves are termed poles in the S matrix, because (with the elements of S defined by the ratios of the outgoing to the incoming amplitudes of the wavefunction):

$$S_J = f_J(k)/f_J(-k)e^{iJ\pi} \quad (19)$$

in the present single-channel problem. Hence, S_J has a pole at every zero of $f_J(-k)$.

A linear expansion in the neighbourhood of such a pole,

$$f_J(k) = [f_J(-k)]^* = F(E)(E - E_n - i\Gamma_n/2) \quad (20)$$

[45] A. F. J. Siegert, *Phys. Rev.*, 1939, **56**, 750.

[where $F(E)$ may vary slowly with E] may now be used to demonstrate the characteristic resonance behaviour in the physical (real energy) solutions of equation (3). Since the internal amplitude is standardized by equation (8), and the external amplitude is given by equation (13), the amplitude ratio must vary as

$$|a(E, J)|^2 = |f_J(k)|^2 = |F(E)|^{-2}[(E - E_n)^2 + \Gamma_n^2/4]^{-1} \tag{21}$$

while

$$\eta_J(E) = \eta_J^0(E) + \arctan\,[\Gamma_n/2(E_n - E)] \tag{22}$$

where

$$\eta_J^0(E) = \arg F(E) + J\pi/2 \tag{23}$$

Hence the collision delay time defined by equation (6) becomes

$$\tau_d(E, J) = 2h\frac{\partial \eta_J^0}{\partial E} + \frac{\Gamma_n \hbar}{(E - E_n)^2 + \Gamma_n^2/4} \tag{24}$$

These are the standard Breit–Wigner parameterizations of $a(E, J)$, $\eta_J(E)$, and $\tau_d(E, J)$. Note that $a(E, J)$, being defined in terms of a standard behaviour at the origin, may not be the most convenient quantity in spectroscopic applications. It is therefore replaced in the following section by a quantity $A(E, J)$, which differs from $a(E, J)$ in magnitude but shows the same resonance characteristics; $A(E, J)$ is defined to depend on the ratio of the actual internal amplitude of the energy-normalized wavefunction to the amplitude of a bound wavefunction at a similar energy, normalized to unity over the potential well shown in Figure 2.

Clearly, if Γ_n is sufficiently small that variations in $F(E)$ may be ignored over the resonance region, both the spectroscopic lineshape and the collision delay-time functions follow the same Lorentzian form. The linewidth parameter Γ_n may therefore be taken as the full width at half height of either function. Γ_n may also be deduced from the maximum resonance increment, $4\hbar/\Gamma_n$, in the time-delay function, or in a purely analytical treatment (see below) from the imaginary part of the pole in the S matrix.

Possible distortions in the lineshape and changes in the peak position as the linewidth increases have recently been noted.[21] Since such distortions to $A(E, J)$ and $\tau_d(E, J)$ depend on $|F(E)|$ and $\arg F(E)$, respectively, they may be expected to differ in kind. Hence in any comparison with experiment, $a(E, J)$ must be employed in spectroscopic, and $\tau_d(E, J)$ [or $\eta_J(E)$] in scattering, applications.

These general considerations underline the intimate connection between the resonance behaviour of $\eta_J(E)$, $\tau_d(E, J)$, and $a(E, J)$, and indicate the origin of the Breit–Wigner parameterizations given by equations (19)—(22) in the sharp resonance limit. The next step is to relate the resonance positions E_n and widths Γ_n to the form of the potential function $V(r)$.

Semi-classical Theory.—The semi-classical (JWKB) method of solution of equation (3) relies on establishing connections between solutions of the form [46–48]

$$\chi_{EJ}(r) = [k_J(r)]^{-\frac{1}{2}}\left[X'e^{i\int_x^r k_J(r)\mathrm{d}r} + X''e^{-i\int_x^r k_J(r)\mathrm{d}r}\right] \tag{25}$$

where

$$k_J^2(r) = k^2 - U(r) - J(J+1)/r^2 \tag{26}$$

which are valid in the regions $a \ll r \ll b$, $r \gg c$ in Figure 3, such that

$$\left|\frac{\mathrm{d}}{\mathrm{d}r}\left[\frac{1}{k_J(r)}\right]\right| \ll 1 \tag{27}$$

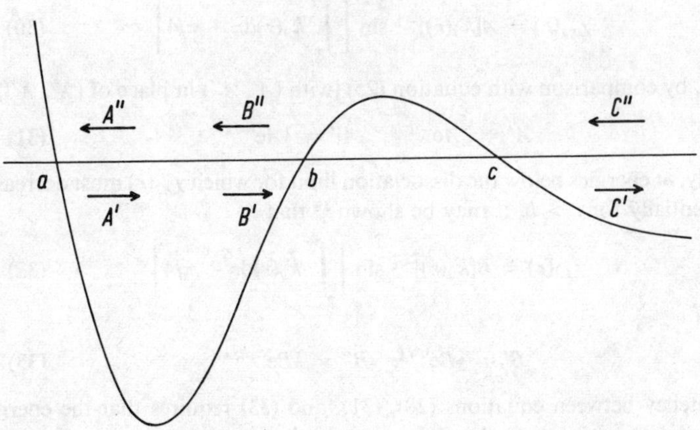

Figure 3 *Notation for the semi-classical theory. a, b, and c denote the classical turning-points. Capital letters denote the amplitudes of travelling waves, with primes and double primes used to indicate outgoing and incoming motion, respectively*

These connections are based on more accurate model solutions of equation (3) in the non-classical regions around the turning points a, b, and c, where $k_J(r) = 0$. A good recent introduction to this general technique has been given by Berry and Mount.[47]

Changes in $\chi_{EJ}(r)$ will be followed by setting (X', X'') in equation (25) equal to (A', A''), (B', B''), or (C', C'') according to the choice of phase reference point x as a, b, or c; single and double primes are used to denote outgoing and incoming motion, respectively, as illustrated in Figure 3. This means that since the same function may be referred either to a or to b in the region $a \ll r \ll b$, then

$$\begin{pmatrix} B' \\ B'' \end{pmatrix} = \begin{pmatrix} e^{i\alpha_J^0} & 0 \\ 0 & e^{-i\alpha_J^0} \end{pmatrix}\begin{pmatrix} A' \\ A'' \end{pmatrix} \tag{28}$$

[46] See B. S. Jeffreys, in 'Quantum Theory', ed. D. R. Bates, Academic Press, London and New York, 1961, p. 229.
[47] M. V. Berry and K. E. Mount, *Reports Progr. Phys.*, 1972, **35**, 315.
[48] L. D. Landau and E. M. Liftshitz, 'Quantum Mechanics', Adison-Wesley, 1965.

where

$$\alpha_J^0 = \int_a^b k_J(r)\mathrm{d}r \tag{29}$$

The significance of the superscript zero will become apparent later.

The correct behaviour of the wavefunction at an isolated left-hand turning point, a, now requires that $\chi_{EJ}(r)$ should behave as the asymptotic form of an Airy function,[49] which is the exact solution of equation (3) for a linear approximation to the potential, decreasing exponentially into the non-classical region $r < a$. The appropriate form is [46, 48]

$$\chi_{EJ}(r) = A[k_J(r)]^{-\frac{1}{2}} \sin\left[\int_a^r k_J(r)\mathrm{d}r + \pi/4\right] \tag{30}$$

Hence, by comparison with equation (25) [with (A', A'') in place of (X', X'')],

$$A' = \tfrac{1}{2}A\mathrm{e}^{-\mathrm{i}\pi/4}, \quad A'' = \tfrac{1}{2}A\mathrm{e}^{\mathrm{i}\pi/4} \tag{31}$$

Equally, at energies below the dissociation limit for which $\chi_{EJ}(r)$ must decrease exponentially for $r > b$, it may be shown [48] that

$$\chi_{EJ}(r) = B[k_J(r)]^{-\frac{1}{2}} \sin\left[\int_r^b k_J(r)\mathrm{d}r + \pi/4\right] \tag{32}$$

so that

$$B' = \tfrac{1}{2}B\mathrm{e}^{\mathrm{i}\pi/4}, \quad B'' = \tfrac{1}{2}B\mathrm{e}^{-\mathrm{i}\pi/4} \tag{33}$$

Consistency between equations (28), (31), and (33) requires that the energy shall be chosen to satisfy the Bohr quantization condition

$$\alpha_J^0 = \textstyle\int_a^b k_J(r)\mathrm{d}r = (n + \tfrac{1}{2})\pi, \quad n = 0, 1, 2\ldots \tag{34}$$

for a bound state. A semi-classical normalization [48] of the wavefunction then yields values for A and B in equations (30) and (32):

$$A = (-1)^n B = \left[\frac{2\mu\hbar\bar{\omega}}{\pi\hbar^2}\right]^{\frac{1}{2}} \tag{35}$$

where $\hbar\bar{\omega}$ is the local energy-level spacing, given by the quadrature

$$\hbar\bar{\omega} = \frac{\partial E}{\partial n} = \frac{\pi\hbar^2}{\mu}\left[\int_a^b k_J^{-1}(r)\mathrm{d}r\right]^{-1} \tag{36}$$

Analysis of the quasi-bound states requires a proper treatment of the barrier region, in place of a simple boundary condition at b. Different models for the barrier yield different relations between (B', B'') and (C', C'') in Figure 3,[24–33] but any connection formula is subject to the equations of flux conservation

[49] M. Abramowitz and I. A. Stegun, 'Handbook of Mathematical Functions', Dover, 1965.

$$|C'|^2 - |C''|^2 = |B'|^2 - |B''|^2 \tag{37}$$

and of symmetry under time reversal, so that the substitutions

$$(C', C'') \rightarrow (C''^*, C'^*) \quad \text{imply} \quad (B', B'') \rightarrow (B''^*, B'^*) \tag{38}$$

Hence we may express the general connection in the form

$$\begin{pmatrix} C' \\ C'' \end{pmatrix} = \begin{pmatrix} \cosh\tilde{\gamma}\,\mathrm{e}^{-\mathrm{i}\tilde{\varphi}}, & -\mathrm{i}\sinh\tilde{\gamma}\,\mathrm{e}^{\mathrm{i}\tilde{\theta}} \\ \mathrm{i}\sinh\tilde{\gamma}\,\mathrm{e}^{-\mathrm{i}\tilde{\theta}}, & \cosh\tilde{\gamma}\,\mathrm{e}^{\mathrm{i}\tilde{\varphi}} \end{pmatrix} \begin{pmatrix} B' \\ B'' \end{pmatrix} \tag{39}$$

and give specific expressions for the parameters $\tilde{\gamma}$, $\tilde{\theta}$, and $\tilde{\phi}$ for different barrier models in equations (59)—(76) below. Solution of equation (39) when $B' = 1$, $C'' = 0$ gives the following expressions for the barrier transmission and reflection probabilities:

$$\begin{aligned} T &= |C'|^2 = \operatorname{sech}^2\tilde{\gamma} \\ R &= |B''|^2 = \tanh^2\tilde{\gamma} \end{aligned} \tag{40}$$

It now remains merely to combine equations (28), (31), and (39) in order both to determine the phase shift and to express the external amplitudes (C', C'') in terms of the internal coefficient A:

$$\begin{aligned} \begin{pmatrix} C' \\ C'' \end{pmatrix} &= \begin{pmatrix} \cosh\tilde{\gamma}\,\mathrm{e}^{-\mathrm{i}\tilde{\varphi}}, & -\mathrm{i}\sinh\tilde{\gamma}\,\mathrm{e}^{\mathrm{i}\tilde{\theta}} \\ \mathrm{i}\sinh\tilde{\gamma}\,\mathrm{e}^{-\mathrm{i}\tilde{\theta}}, & \cosh\tilde{\gamma}\,\mathrm{e}^{\mathrm{i}\tilde{\varphi}} \end{pmatrix} \begin{pmatrix} \mathrm{e}^{\mathrm{i}\alpha_J^0}, & 0 \\ 0\ , & \mathrm{e}^{-\mathrm{i}\alpha_J^0} \end{pmatrix} \begin{pmatrix} \frac{1}{2}A\mathrm{e}^{-\mathrm{i}\pi/4} \\ \frac{1}{2}A\mathrm{e}^{\mathrm{i}\pi/4} \end{pmatrix} \\ &= \tfrac{1}{2}A \begin{pmatrix} [\mathrm{e}^{\tilde{\gamma}}\cos\alpha_J + \mathrm{i}\mathrm{e}^{-\tilde{\gamma}}\sin\alpha_J]\,\mathrm{e}^{(\mathrm{i}/2)(\tilde{\theta} - \tilde{\varphi} - \pi/2)} \\ [\mathrm{e}^{\tilde{\gamma}}\cos\alpha_J - \mathrm{i}\mathrm{e}^{-\tilde{\gamma}}\sin\alpha_J]\,\mathrm{e}^{-(\mathrm{i}/2)(\tilde{\theta} - \tilde{\varphi} - \pi/2)} \end{pmatrix} \\ &= \tfrac{1}{2}A\,(\cosh 2\tilde{\gamma} + \sinh 2\tilde{\gamma}\cos 2\alpha_J)^{\frac{1}{2}} \begin{pmatrix} \mathrm{e}^{\mathrm{i}\delta} \\ \mathrm{e}^{-\mathrm{i}\delta} \end{pmatrix} \end{aligned} \tag{41}$$

where

$$\begin{aligned} \alpha_J &= \alpha_J^0 - \tfrac{1}{2}(\tilde{\phi} + \tilde{\theta}) \\ \delta &= \arctan(\mathrm{e}^{-2\tilde{\gamma}}\tan\alpha_J) + \tfrac{1}{2}(\tilde{\theta} - \tilde{\phi}) - \pi/4 \end{aligned} \tag{42}$$

This means that, with $|C'|$ chosen to satisfy the asymptotic normalization given by equation (5)

$$\chi_{EJ}(r) \overset{r \to \infty}{\simeq} \left[\frac{2\mu}{\pi\hbar^2 k}\right]^{\frac{1}{2}} \sin\left[kr - J\pi/2 + \eta_J^0(E) + \eta_J^{(\mathrm{res})}(E)\right] \tag{43}$$

where

$$\eta_J^0(E) = \lim_{r \to \infty}\left[\int_c^r k(r)\mathrm{d}r - kr + (J + \tfrac{1}{2})\pi/2 + \tfrac{1}{2}(\tilde{\theta} - \tilde{\phi})\right] \tag{44}$$

$$\eta_J^{(\mathrm{res})}(E) = \arctan(\mathrm{e}^{-2\tilde{\gamma}}\tan\alpha_J) \tag{45}$$

while in the internal region, $a \ll r \ll b$,

$$\chi_{EJ}(r) = [\cosh 2\tilde{\gamma} + \sinh 2\tilde{\gamma} \cos 2\alpha_J]^{-\frac{1}{2}} \left[\frac{2\mu}{\pi\hbar^2 k(r)}\right]^{\frac{1}{2}} \times \sin\left[\int_a^r k(r)\mathrm{d}r + \pi/4\right] \quad (46)$$

This differs from the bound state solution given by equations (30) and (35) by the factor

$$A(E, J) = [\hbar\bar{\omega}(\cosh 2\tilde{\gamma} + \sinh 2\tilde{\gamma} \cos 2\alpha_J)]^{-\frac{1}{2}} \quad (47)$$

Hence $|A(E, J)|^2$ governs the spectroscopic lineshape.

Clearly, at least for large $\tilde{\gamma}$ when according to equation (40) the barrier transmission probability is low, resonances in $\eta_J(E)$ and $A(E, J)$ occur at energies E_{nJ} such that

$$\alpha_J = \textstyle\int_a^b k_J(r)\mathrm{d}r - \frac{1}{2}(\tilde{\theta} + \tilde{\phi}) = (n + \frac{1}{2})\pi \quad (48)$$

This may be regarded as the quantization condition of equation (34) for a bound level subject to a small (see below) level shift due to the phase terms $\frac{1}{2}(\tilde{\theta} + \tilde{\phi})$; as a rough estimate based on a linear expansion for α_J^0 and neglect of any energy variation in $(\tilde{\theta} + \tilde{\phi})$,

$$\Delta E_{nJ} = E_{nJ} - E_{nJ}^0 \approx (\tilde{\theta}_{nJ} + \tilde{\phi}_{nJ})\hbar\bar{\omega}_{nJ}/2\pi \quad (49)$$

where E_{nJ}^0 is the unperturbed level given by equation (34) and $\hbar\bar{\omega}_{nJ}$ is defined by equation (36). The subscripts imply that $\hbar\bar{\omega}$, $\tilde{\theta}$, and $\tilde{\phi}$ are evaluated at the resonance point; E_{nJ} may also be obtained directly (and more accurately) from equation (48).

It may be seen with the help of equations (45) and (47), to the extent that energy variations in $\tilde{\gamma}$ are unimportant over the resonance region, first that the resonance part of collision delay time $\tau_d(E)$ and the lineshape $|A(E, J)|^2$ have the same functionality:

$$\begin{aligned}\tau_d^{\mathrm{res}}(E, J) &= 2\hbar\frac{\partial\eta^{(\mathrm{res})}}{\partial E} = 2\pi\hbar|A(E, J)|^2 \\ &= (2\pi/\bar{\omega}_{nJ})[\cosh 2\tilde{\gamma}_{nJ} + \sinh 2\tilde{\gamma}_{nJ} \cos 2\alpha_J]^{-1} \end{aligned} \quad (50)$$

and secondly that the total integrated intensity from a single line is equal to unity:

$$\begin{aligned}&\int_{E_{nJ} - \frac{1}{2}\hbar\bar{\omega}}^{E_{nJ} + \frac{1}{2}\hbar\bar{\omega}} |A(E, J)|^2\mathrm{d}E \\ &\approx \frac{1}{\hbar\bar{\omega}_{nJ}}\left(\frac{\partial E}{\partial\alpha_J}\right)\int_{n\pi}^{(n+1)\pi} \frac{\mathrm{d}\alpha_J}{\cosh 2\tilde{\gamma}_{nJ} + \sinh 2\tilde{\gamma}_{nJ} \cos 2\alpha_J} = 1 \end{aligned} \quad (51)$$

This analysis therefore adds a final link to the theory of intensities in discrete and continuous spectra [50, 51] by including the quasi-bound states.

[50] A. C. Allison and A. Dalgarno, *J. Chem. Phys.*, 1971, **55**, 4342.
[51] A. L. Smith, *J. Chem. Phys.*, 1971, **55**, 4344.

The above general expressions are readily reduced to Breit–Wigner form over the sharp resonance region ($e^{2\tilde{\gamma}_{nJ}} \gg 1$), by the linear expansion

$$\alpha_J \approx (n + \tfrac{1}{2})\pi + \left(\frac{\partial \alpha_J}{\partial E}\right)(E - E_{nJ})$$

$$\approx (n + \tfrac{1}{2})\pi + \frac{\pi}{\hbar\bar{\omega}_{nJ}}(E - E_{nJ}) \tag{52}$$

with the results

$$\eta_J^{(\mathrm{res})}(E) = \arctan\,[\Gamma_{nJ}/2(E_{nJ} - E)] \tag{53}$$

$$\tau_{\mathrm{d}}^{(\mathrm{res})}(E, J) = 2\pi\hbar|A(E, J)|^2 = \frac{\Gamma_{nJ}\hbar}{\frac{1}{4}\Gamma_{nJ}^2 + (E - E_{nJ})^2} \tag{54}$$

where

$$\Gamma_{nJ} = (2\hbar\bar{\omega}_{nJ}/\pi)e^{-2\tilde{\gamma}_{nJ}} \tag{55}$$

Connor [30] and Dickinson [31] would obtain the same result directly from equation (41) by applying the outgoing boundary condition

$$C'' = e^{\tilde{\gamma}} \cos \alpha_J - ie^{-\tilde{\gamma}} \sin \alpha_J = 0 \tag{56}$$

and the linear expansion of equation (52), to obtain the complex energy of the Siegert state:

$$E = E_{nJ} - \frac{i}{2}\left(\frac{\hbar\bar{\omega}_{nJ}}{\pi}\right) \ln(\coth \tilde{\gamma}_{nJ})$$

$$\approx E_{nJ} - \frac{i}{2}\left(\frac{\hbar\bar{\omega}_{nJ}}{\pi}\right) \ln(1 + 2e^{-2\tilde{\gamma}})$$

$$\approx E_{nJ} - \frac{i}{2}\left(\frac{2\hbar\bar{\omega}_{nJ}}{\pi}\right) e^{-2\tilde{\gamma}_{nJ}} \tag{57}$$

Limitations of this approach have been discussed by Connor.[35] Note that in this sharp resonance limit ($e^{-2\tilde{\gamma}_{nJ}} \ll 1$), Γ_{nJ} is directly related to the barrier transmission probability given by equation (40) in the semi-classical form:[21]

$$\Gamma_{nJ} = T\hbar/t_{\mathrm{vib}} \tag{58}$$

where $t_{\mathrm{vib}} = 2\pi/\bar{\omega}_{nJ}$.

It remains to determine the parameters $\tilde{\gamma}$, $\tilde{\theta}$, and $\tilde{\phi}$ in equation (39) for different barrier models. The simplest theory employs independent semi-classical connection formulae at the turning points b and c with the results [24, 25]

$$\tilde{\gamma}_{nJ} = \textstyle\int_b^c |k_{nJ}(r)|\mathrm{d}r + \ln 2 \tag{59}$$

$$\tilde{\theta}_{nJ} = \tilde{\phi}_{nJ} = 0 \tag{60}$$

where the subscript on $k_{nJ}(r)$ implies that the energy is chosen to satisfy equation (48). This may be termed the simple semi-classical estimate. The first improvements [26, 28, 30] employ a quadratic approximation at the barrier maximum:

$$V_J(r) = V_J^{\max} - \tfrac{1}{2}\varkappa_J(r - r'_{\max})^2 \qquad (61)$$

to obtain

$$\tilde{\gamma}_{nJ} = \cosh^{-1}[1 + e^{-2\pi\varepsilon}]^{\frac{1}{2}} \qquad (62)$$

$$\tilde{\phi}_{nJ} = \phi(\varepsilon) = \arg\Gamma(\tfrac{1}{2} + i\varepsilon) - \varepsilon \ln|\varepsilon| + \varepsilon$$

$$\tilde{\theta}_{nJ} = 0 \qquad (63)$$

where

$$\varepsilon = (E_{nJ} - V_J^{\max})/\hbar\omega_J^*$$

$$\omega_J^* = (\varkappa_J/\mu)^{\frac{1}{2}} \qquad (64)$$

Methods employed by other workers give the same form for $\tilde{\gamma}_{nJ}$ but either a different phase correction,[23] or an undetermined phase connection.[27, 29, 33] Values of the function $\phi(\varepsilon)$ which determines the level shift in equation (49) and some limiting properties [23, 28, 52] are given in Table 1. A function of this form also plays an important part in scattering theory by removing a classical singularity in the deflection function.[23] Equations (62)—(64) define what we

Table 1 *Values of the level shift function* $\phi(\varepsilon)$ *in the associated quadratic approximation* [*equation* (63)]: $\phi(-\varepsilon) = -\phi(\varepsilon)$; $\phi(\varepsilon) \approx \varepsilon^{-1}/24 + 7\varepsilon^{-3}/2880$ *for* $\varepsilon > 1$

ε	$\phi(\varepsilon)$	ε	$\phi(\varepsilon)$
0	0	0.6	0.080
0.1	0.137	0.7	0.068
0.2	0.150	0.8	0.058
0.3	0.135	0.9	0.051
0.4	0.115	1.0	0.045
0.5	0.096	—	—

shall term the JWKB quadratic approximation. Miller and Good [53] have shown how this quadratic approximation may be replaced by a more general association between the physical barrier and the quadratic form (61), the association being defined by the identity

$$\varepsilon = -\frac{1}{\pi}\int_b^c |k_{nJ}(r)|\,dr \qquad (65)$$

which takes place of equation (64). This will be termed the uniform quadratic approximation. First suggested without proof by Connor,[52] it reduces to

[52] J. N. L. Connor, *Mol. Phys.*, 1968, **15**, 37.
[53] S. C. Miller and R. H. Good, *Phys. Rev.*, 1953, **91**, 174; see also C. E. Hecht and J. E. Mayer, *Phys. Rev.*, 1957, **106**, 1156, and R. B. Dingle, *Appl. Sci. Res.*, 1956, **B5**, 345.

equation (64) in the quadratic approximation and causes the expression (62) for $\tilde{\gamma}_{nJ}$ to go over to the simple semi-classical form (59) when $e^{-2\pi\varepsilon} \gg 1$ (recall that $\varepsilon < 0$ for energies below the barrier maximum). Since, however,[23, 29, 52]

$$\phi(\varepsilon) \sim \frac{1}{24}\,\varepsilon^{-1} + \frac{7}{2880}\,\varepsilon^{-3} \quad \text{for } \varepsilon > 1 \tag{66}$$

the level shifts predicted by equations (49) and (63) (but not covered by the simple semi-classical theory) depend inversely, rather than exponentially, on ε. Hence the level shift may be orders of magnitude larger than the line-width in the sharp resonance region.

Dickinson [31] and Soop [54] have taken the association method one stage further in order to obtain a two-parameter description. Dickinson [31] relies on an association between $k^2_{nJ}(r)$ and the inverted Morse form

$$t^2\{q^2 - [2e^{-s} - e^{-2s}]\} \tag{67}$$

where q^2 is the ratio of the available energy to the barrier height (measured from the dissociation limit) and t is defined by the equation

$$\pi t(1-q) = \int_b^c |k_{nJ}(r)|\,\mathrm{d}r \tag{68}$$

where, as in equation (65), $k_{nJ}(r)$ contains the true potential function. The expressions (62)—(65) are then replaced in the present notation by

$$\tilde{\gamma}_{nJ} = \sinh^{-1}[e^{-\pi\varepsilon_-}\cosh \pi\varepsilon_+/\sinh 2\pi qt] \tag{69}$$

$$\tilde{\phi}_{nJ} = \chi(2qt) - \phi(\varepsilon_-) \tag{70}$$

$$\tilde{\theta}_{nJ} = \chi(2qt) - \phi(\varepsilon_+) \tag{71}$$

where

$$\varepsilon_{\pm} = t(1 \pm q) \tag{72}$$

$$\chi(y) = \arg\Gamma(iy) - y\ln|y| + y + \pi/4 = \chi(2y) - \phi(y) \tag{73}$$

and $\phi(\varepsilon)$ is defined by equation (63). This gives a uniform Morse approximation. Note that ε_- is identical with $-\varepsilon$ defined by the quadratic association formula (65). Furthermore, t is typically large unless the barrier is very low (t lies in the range 4—15 for J levels of the H_2 ground state considered in Table 1), and $\chi(y) \simeq -1/12y$ for $y > 2$. Hence it may be verified that equations (69)—(72) reduce to expressions (62) and (63) with t given by equation (65), except for very low barriers (low J values) or for (very sharp) levels close to the base of the barrier.

A final somewhat similar (uniform sech²) form may be obtained from the results of Soop [54] based on association between $k^2_{nJ}(r)$ and

$$t^2\{q^2 - \operatorname{sech}^2 s\} \tag{74}$$

[54] M. Soop, *Arkiv Fysik*, 1965, **30**, 217.

The parameters q and t have the same definitions as those of Dickinson, but now

$$\tilde{\gamma}_{nJ} = \sinh^{-1}\left\{\frac{\cosh[\pi(t^2 - \frac{1}{4})]}{\sinh \pi q t}\right\}$$

$$\tilde{\phi}_{nJ} = 2\chi(qt) - \phi(\varepsilon_-') + \phi(\varepsilon_+') + \delta$$

$$\tilde{\theta}_{nJ} = 0 \qquad (75)$$

where

$$\varepsilon_\pm' = (t^2 - \tfrac{1}{4})^{\frac{1}{2}} \pm qt$$

$$\delta = t \ln\left|\frac{1+q}{1-q}\right| - (t^2 - \tfrac{1}{4})^{\frac{1}{2}} \ln(\varepsilon_+'/\varepsilon_-') - qt \ln\left[\frac{\varepsilon_+'\varepsilon_-'}{(1-q^2)t^2}\right] \qquad (76)$$

Quasi-bound Energy Levels.—Any attempt to calculate the width of a quasi-bound state requires prior knowledge of its position. Calculated level positions have been reported for the ground states of the H_2,[15, 16, 18, 20, 21, 24, 25] HD and D_2,[21] OH,[20] $^4HeH^+$, $^3HeH^+$, $^4HeD^+$, and $^3HeH^+$,[5] and BeH and HgH [55] molecules, and for a standard model based on the Lennard-Jones potential.[17, 22, 31]

The first calculations [15, 16, 18] employed the points of inflexion in the phase shift, $\eta_J(E)$, derived by numerical integration of equation (3), but this approach has been superseded by use of the time delay function, $\tau_d(E, J)$,[21] after recognition that $\tau_d(E, J)$ may be derived from the same scattering solution, normalized in the form

$$\tilde{\chi}_{EJ}(r) \overset{r\to\infty}{\simeq} (4\mu/\hbar k)^{\frac{1}{2}} \sin[kr - J\pi/2 + \eta_J(E)] \qquad (78)$$

by the quadrature,[41]

$$\tau_d(E, J) = \int_0^\infty [|\tilde{\chi}_{EJ}(r)|^2 - |\tilde{\chi}_{EJ}(\infty)|^2]dr + (\mu/\hbar k^2)\sin[2\eta_J(E) - J\pi] \qquad (79)$$

Hence the resonance points, and particularly the level widths, may be obtained more conveniently from the maxima in $\tau_d(E, J)$. The positions of peaks in the amplitude ratio, $A(E, J)$, first demonstrated by Buckingham and Fox,[14] have also been employed.[19, 21] Results derived from $A(E, J)$ for the ground state of H_2 are found to agree within 0.1 cm^{-1} with those derived [21] from $\tau_d(E, J)$ for the sharp resonances ($\Gamma < 5$ cm^{-1}), but discrepancies, due to variation of the ambient phase shift $\eta_J^0(E)$ in equation (22), of the order of 1—6 cm^{-1} are reported for the broad levels (see Table 2).[19, 21] Resonance positions discussed below refer to the maxima in $A(E, J)$ as having the greatest spectroscopic significance.

A major problem encountered in these calculations is that in seeking the scattering solution [such that $\chi_{EJ}(0) = 0$] there is no efficient algorithm

[55] L. Gottdiener and J. N. Murrell, *Mol. Phys.*, 1973, **25**, 1041.

Table 2 *Comparative resonance positions and widths for the* H_2 *ground state.*[21] *Values under BC(Airy) refer to the Airy-function boundary condition at the outermost turning point*

		E_{nJ}/cm^{-1}			Γ_{nJ}/cm^{-1}	
v	J	τ_d(max)	A(max)	$BC(Airy)$	$A(E, J)$	$JWKB$
0	38	7510.0	7514.0	7508.7	98.1	87.0
0	37	6513.3	6513.5	6513.3	5.98	5.60
1	35	5549.8	5550.0	5549.7	14.3	14.9
2	33	4688.4	4689.0	4688.2	20.8	22.5
3	31	3925.0	3925.4	3924.9	25.1	26.7
4	29	3254.7	3255.4	3254.8	25.4	28.4
5	27	2673.0	2673.8	2673.4	25.8	29.3
6	25	2175.0	2176.0	2175.7	27.4	31.4
7	23	1755.3	1756.7	1756.4	31.7	36.9
8	21	1407.0	1409.4	1409.4	42.1	48.3
9	19	1121.6	1127.2	[a]	66.2	[a]
9	18	725.9	725.9	726.0	0.53	0.55
10	16	586.0	586.0	586.1	2.93	3.22
11	14	480.1	481.0	481.7	18.5	22.3
11	13	199.4	199.4	199.4	0.005	0.0053
12	12	385.0	398.6	[a]	11.6	[a]
12	11	215.5	215.5	215.6	2.62	3.09
13	9	195.0	205.6	[a]	89.6	[a]
13	8	89.9	90.0	90.1	1.89	2.38
14	6	81.9	121.0	[a]	79.0	[a]
14	5	45.7	49.2	[a]	26.4	[a]
14	4	3.76	3.76	3.76	0.007	0.0085

[a] The boundary condition places this level above the barrier maximum.

for converging on a resonance, a problem which becomes acute for very sharp resonances;[21] the method proposed by Johnson, Balint-Kurti, and Levine[56] suffers from the same disadvantage. Bain and Bardsley[22] have therefore devised a different exact procedure specifically for the treatment of sharp resonances. Two solutions of equation (3) are introduced, one of which, $f(r)$, is defined to have zero derivative at the origin and to behave as

$$f(r) \overset{r\to\infty}{\simeq} \sin(kr + \delta) \tag{80}$$

at infinity. The second solution, $g(r)$, is chosen such that

$$g(r) \overset{r\to\infty}{\simeq} \cos(kr + \delta) \tag{81}$$

and integrated back to the origin. The resonances are then shown to occur at energies for which $g(0) = 0$, and the level widths for rotationless states ($J = 0$) to be given by an expression analogous to equation (79),

$$\Gamma_{nJ} = \frac{\hbar^2 k}{\mu} \lim_{R\to\infty} \left\{ \left[\int_0^R \{g^2(r) - f^2(r)\}\mathrm{d}r - \frac{1}{2k}\sin(2kR + 2\delta) \right]^{-1} \right\} \tag{82}$$

[56] B. R. Johnson, G. G. Balint-Kurti, and R. D. Levine, *Chem. Phys. Letters*, 1970, **1**, 268.

Table 3 *Comparative resonance positions and widths for the Lennard-Jones model, with the potential parameters employed by Bernstein* et al.[17] ε_0 *denotes the well depth; energies and widths are expressed as a fraction of the well depth*

	E_{nJ}			Γ_{nJ}		
l	*Phase shift*[a]	*BB*[b]	*JWKB*[c]	*Phase shift*[a]	*BB*[b]	*JWKB*[d]
118	0.310 05	0.310 04	0.310 03	—	1.2×10^{-17}	1.21×10^{-17}
197	0.310 05	0.310 08	0.310 07	—	9.6×10^{-28}	9.36×10^{-28}
25	0.348 55	0.348 62	0.348 46	6.2×10^{-4}	6.0×10^{-4}	6.00×10^{-4}
77	0.349 35	0.349 35	0.349 38	—	6.9×10^{-8}	6.89×10^{-8}
128	0.350 05	0.350 07	0.350 07	—	2.0×10^{-11}	2.02×10^{-11}
123	0.389 67	0.389 61	0.389 62	1.4×10^{-4}	8.4×10^{-5}	8.64×10^{-5}
87	0.390 06	0.390 09	0.390 08	4.1×10^{-4}	4.2×10^{-4}	4.22×10^{-4}
137	>0.401 2	0.402 33	0.402 5	—	3.4×10^{-3}	3.20×10^{-3}

[a] Bernstein *et al.*[17] [b] Bain and Bardsley.[22] [c] Dickinson.[31] [d] Dickinson.[32] These values are more accurate than those in the original paper.[31]

A similar more complicated expression is given for $J \neq 0$. A comparison between the level positions obtained for the Lennard-Jones model and those derived from the phase shift is given in Table 3.

Other more approximate methods have also been suggested. One approach is to convert the problem into a bound-state form by imposing a boundary condition either at the barrier maximum,[20, 21] or at the outermost turning point.[5, 21] Studies on the H_2 ground state [21] (see Table 2) favoured the requirement that the wavefunction should behave as an Airy function of the second type (increasing exponentially into the barrier region) at the outermost turning point; a similar suggestion [5] is that $\chi_{EJ}(r)$ should have the same logarithmic derivative at this point as the irregular Bessel function $Y_{J+\frac{1}{2}}(kr)$.[49] This approach still requires numerical solution of the Schrödinger equation. The semi-classical theory of the previous section shows that this may be replaced merely by two quadratures (over the potential well $a < r < b$, and the barrier $b < r < c$) at each point on a suitable energy grid. Dickinson [31] has applied the modified quantization condition, equation (48), with $(\tilde{\theta} + \tilde{\phi})$ given by equations (63) and (65), to obtain level positions for the Lennard-Jones model in good agreement with the exact values (see Table 3). It is interesting that almost equally good agreement may be achieved with a different phase correction by taking the JWKB theory to second order.[17]

Finally, methods for estimating the number rather than the positions of quasi-bound states may also be referenced at this point.[36–38]

Level Widths.—Given the resonance position, computation of the level width presents little difficulty. Values have been obtained from the slope of $\eta_J(E)$,[15, 16, 18] from the width and resonance value of $\tau_d(E, J)$,[21] from the width [19, 21] and the peak value [5] of the amplitude ratio $A(E, J)$, from the special formula of Bain and Bardsley,[22] and from the semi-classical equations

(55) and (59)—(76).[21, 28, 31–33] Comparisons are given for the Lennard-Jones system [17, 22, 31, 32] and for the ground state of H_2[21] in Tables 3 and 2, respectively. The most striking feature is the accuracy of the very rapid JWKB estimates. The values obtained by Dickinson [32] for the Lennard-Jones system in the uniform quadratic approximation [equations (55), (62), (63), and (65)] are seen to be in excellent agreement with the exact values of Bain and Bardsley,[22] and with the less accurate values derived from the slope of the phase shift by Bernstein *et al.*[17] Table 2 also shows close agreement between the exact linewidths and the simplest semi-classical estimates [equation (59)];[21] and this agreement may be improved for the broader levels ($\Gamma > 20$ cm^{-1}), with no more effort, by use of the second of equations (57), together with the more refined uniform quadratic equations (62) and (65) in place of (59). Thus the simple semi-classical estimate $\Gamma^{(0)}_{nJ}$ may be improved to a value $\Gamma^{(1)}_{nJ}$ by the algorithm

$$\Gamma^{(1)}_{nJ} = \left(\frac{\hbar\bar{\omega}_{nJ}}{2\pi}\right) \ln\left(1 + \frac{2\pi\Gamma^{(0)}_{nJ}}{\hbar\bar{\omega}_{nJ}}\right) \tag{83}$$

Table 4 *Simple, $\Gamma^{(0)}$, and improved, $\Gamma^{(1)}$, semi-classical level widths for the H_2 ground state; q and t in the final columns are the parameters for the associated Morse approximation; note that $\pi qt \gg 1$*

v	J	$\hbar\bar{\omega}$ [a]	$\Gamma^{(0)}$ [b]	$\Gamma^{(1)}$ [c]	Γ_{exact} [b]	q [d]	t
0	38	703.4	87.0	64.4	98.1	0.9983	62.5
1	35	1050.2	14.9	14.3	14.3	0.9753	15.2
2	33	979.7	22.5	21.0	20.8	0.9775	13.5
3	31	920.3	26.7	24.5	25.1	0.9778	12.1
4	29	864.3	28.4	25.8	25.4	0.9773	11.1
5	27	806.4	29.3	26.4	25.8	0.9765	9.97
6	25	737.6	31.4	27.8	27.4	0.9767	9.13
7	23	651.7	36.9	31.6	31.7	0.9799	8.44
8	21	529.1	48.3	38.2	42.1	0.9875	8.66
11	14	413.5	22.3	19.2	18.5	0.9608	4.27

[a] R. J. LeRoy (personal communication). [b] Γ_{exact} taken from the amplitude ratio.[21] [c] Equations (83). [d] From energy levels in Table 2 and $V_J^{\max}$ given by LeRoy.[57]

The comparison between $\Gamma^{(0)}_{nJ}$, $\Gamma^{(1)}_{nJ}$, and the exact level widths derived by LeRoy and Bernstein [21] given in Table 4 shows that the correction given by equation (83) may be quite significant for the broadest levels. The apparent exception for the $(v, J) = (0,38)$ level probably arises from severe distortion of the lineshape since it lies within 4 cm^{-1} of the barrier maximum.[57] If so, the Breit–Wigner parameterizations (52)—(55) would no longer be valid, but the lineshape could still be computed from equation (47). The final columns of Table 4 also contain values of Dickinson's parameters q and t defined by

[57] R. J. LeRoy, University of Wisconsin, Theoretical Chemistry Institute, Report WIS-TCI-387, 1971.

equation (68), in order to demonstrate, by the magnitude of t ($\pi t \gg 1$), the essential equivalence between the uniform quadratic and uniform Morse approximations for the levels in question. Hence the uniform quadratic approximation which requires the two quadratures defined in equations (36) and (65) may be taken to be the optimum semi-classical form for most purposes.

From an experimental viewpoint the most important theoretical conclusion is the very strong (exponential) energy dependence of the level width. Thus, as confirmed by experiment,[58] a typical rotational progression for a hydride system will terminate in absorption in the form sharp, broad, very diffuse, followed by a complete loss of structure which may obscure the existence of one or possibly two quasi-bound ($E_{nJ} < V_J^{max}$) upper-state levels. This must result in a discrepancy between the curve of limiting predissociation and the locus of the barrier maximum, V_J^{max}, which is estimated by LeRoy and Bernstein [21] to range from zero at $J = 0$ to as much as 1000 cm^{-1} at $J = 38$ for the ground state of H_2. Hence, although extrapolation of the limiting curve should yield the correct dissociation limit,[6, 10] deductions about the long-range behaviour of the potential [59–61] must be treated with caution. Any attempt to extrapolate from observed data to the true locus of V_J^{max} requires an understanding of both the measurably sharp and the very broad levels discussed above.

Summary and Conclusions.—The properties of non-physical Siegert states with complex energy levels, exponentially decreasing time evolution, and purely outgoing characteristics have been used on pp. 469—472 to unify the resonance behaviour of the phase shift, the collision delay time, and the spectroscopically observed internal to external amplitude ratio. The existence of such states also explains the Breit–Wigner parameterizations of $\eta_J(E)$, $\tau_d(E, J)$, and $a(E, J)$ [equations (21)—(24)].

A general semi-classical theory leads first on pp. 472—480 to a quantization condition [equation (48)] and to expressions for the level shift, (49), and for the linewidth (55) in terms of three parameters $\tilde{\gamma}_{nJ}$, $\tilde{\theta}_{nJ}$, and $\tilde{\phi}_{nJ}$. Formulae for these parameters are then given for four different treatments of the barrier region, of which the associated quadratic model, equations (62), (63), and (65), appears to be the most generally applicable.

Different techniques for the location of quasi-bound states are discussed on pp. 480—484. The most significant practical conclusion is that the (associated quadratic) semi-classical quantization condition appears to be remarkably accurate.

Recent semi-classical formulae for the level width are also shown to

[58] See, for example, L. Farkas and S. Levy, *Z. Phys.*, 1933, **84**, 195, (AlH), and T. L. Porter, *J. Opt. Soc. Amer.*, 1962, **52**, 1201 (HgH).
[59] R. B. Bernstein, *Phys. Rev. Letters*, 1966, **16**, 385.
[60] M. A. Byrne, W. G. Richards, and J. A. Horsley, *Mol. Phys.*, 1967, **12**, 273.
[61] J. A. Horsley and W. G. Richards, *J. Chim. phys.*, 1969, **66**, 41.

achieve high accuracy even for very broad levels. Thus there would appear to be little necessity for exact numerical computations in this field. The existence of reliable analytical formulae may also be expected to lead to more accurate location of barrier maxima, in cases where the highest quasi-bound levels are indiscernibly diffuse.

3 Predissociation by Internal Excitation

Just as predissociation by rotation may be seen as another manifestation of a shape or orbiting resonance, so is predissociation by internal (electronic) excitation (Herzberg's type I)[6] associated with the Fano[62] or Feshbach[63] resonance phenomenon in scattering theory. The situation is complicated by the necessity to consider at least two electronic states. Physical mechanisms and selection rules governing the necessary mixing of electronic states have been well characterized[8, 9] for many years. Our present interest is in the origin of the detailed dependence of the linewidth or lifetime pattern as a function of vibrational and rotational state. Recent work[64–76] has confirmed an early suggestion by Rice[77] that fluctuations in the linewidth, owing to interference between the nuclear wavefunctions for the interacting state, provide a sensitive measure of forms of the relevant potential-energy curves.

The subsequent theory is set in context by a short resumé of the Fano[62] theory of interaction between discrete levels and the continuum; alternative approaches have been suggested by Feshbach[63] and more recently by Van Santen.[78] We also include for the sake of completeness a brief account of possible electronic origins of the interaction term.

The Fano Lineshape.—Fano[62] first considers the coupling between a single discrete level ϕ_n and a continuum of states ψ_E. Second-order interactions between ϕ_n and other discrete states ϕ_n' are therefore neglected at this stage. The normalizations are chosen such that

[62] U. Fano, *Phys. Rev.*, 1961, **124**, 1866.
[63] H. Feshbach, *Ann. Phys.*, 1958, **5**, 357.
[64] J. N. Murrell and J. M. Taylor, *Mol. Phys.*, 1969, **16**, 609.
[65] I. M. Riess and Y. Ben Aryeh, *J. Quant. Spectroscopy Radiative Transfer*, 1969, **9**, 1463.
[66] A. Chutjian, *J. Chem. Phys.*, 1969, **51**, 5414.
[67] F. Fiquet-Fayard and O. Gallais, *Mol. Phys.*, 1970, **20**, 527; *Chem. Phys. Letters*, 1972, **16**, 18.
[68] R. D. Levine, B. R. Johnson, and R. B. Bernstein, *J. Chem. Phys.*, 1969, **50**, 1694.
[69] J. Czarny, P. Felenbok, and H. Lefebvre-Brion, *J. Phys. (B)*, 1971, **4**, 124.
[70] M. Leoni and K. Dressler, *J. Appl. Math. and Phys.*, 1971, **22**, 794.
[71] G. D. Chapman and P. R. Bunker, *J. Chem. Phys.*, 1972, **57**, 2951.
[72] O. Atabek and R. Lefebvre, *Chem. Phys. Letters*, 1972, **17**, 167.
[73] M. S. Child, *J. Mol. Spectroscopy*, 1970, **33**, 487.
[74] A. D. Bandrauk and M. S. Child, *Mol. Phys.*, 1970, **19**, 95.
[75] D. S. Ramsay and M. S. Child, *Mol. Phys.*, 1971, **22**, 263.
[76] M. S. Child, *J. Mol. Spectroscopy*, 1973, **45**, 293.
[77] O. K. Rice, *J. Chem. Phys.*, 1933, **1**, 375.
[78] R. A. Van Santen, *Physica*, 1971, **62**, 51.

$$(\phi_n|H|\phi_n) = E_n$$
$$(\psi_{E'}|H|\psi_{E''}) = E'\delta(E'' - E')$$
$$(\psi_{E'}|H|\phi_n) = (\phi_n|H|\psi_{E'})^* = V_{E'n} \tag{84}$$

This implies, according to equation (5), that

$$\psi_{E'} \overset{r\to\infty}{\simeq} \tilde{\psi}_{E'} \left[\frac{2\mu}{\pi\hbar^2 k'}\right]^{\frac{1}{2}} \sin\,[k'r + \delta(E')] \tag{85}$$

where $\tilde{\psi}_{E'}$ depends on the other (electronic and angular) co-ordinates of the system. Our purpose is to find an (energy-dependent) eigenfunction

$$\Psi_E = A(E)\phi_n + \int B(E, E')\psi_{E'}\mathrm{d}E' \tag{86}$$

subject to the same asymptotic normalization. The functions $a(E)$ and $b(E, E')$ will then determine the lineshape.

It follows from equations (84) and (86) that, at the energy eigenvalue E,

$$E_n A(E) + \int V_{nE'} B(E, E')\mathrm{d}E' = EA(E) \tag{87}$$
$$V_{E'n}A(E) + E'B(E, E') = EB(E, E') \tag{88}$$

$B(E, E')$ is now eliminated in favour of an admixture ratio $Z(E)$, which becomes the central quantity in the theory by a rearrangement of equation (88):[62]

$$B(E, E') = \left[\frac{1}{E - E'} + Z(E)\delta(E - E')\right] V_{E'n}A(E) \tag{89}$$

At the same time the presence of the resulting pole due to the term $(E - E')^{-1}$ makes it necessary to specify how the integrations in equations (86) and (87) are to be carried out, otherwise $Z(E)$ is not uniquely defined. The most convenient choice is the Cauchy principal value,[79] denoted by P, below.

Hence on substitution for $B(E, E')$ in equation (87) and removal of the common factor $A(E)$

$$Z(E) = \frac{E - E_n - F(E)}{|V_{nE}|^2} \tag{90}$$

where

$$F(E) = P\int_{-\infty}^{\infty} \frac{|V_{nE'}|^2\mathrm{d}E'}{E - E'} \tag{91}$$

It remains to determine $A(E)$ by applying the asymptotic normalization condition of equation (85) to Ψ_E in equation (86). Equations (85), (86), and (89), together with the identity

[79] See E. G. Phillips, 'Functions of a Complex Variable', Oliver and Boyd, 1957.

$$P\int_{-\infty}^{\infty}\frac{\sin(k'r+\delta)}{(E-E')}\,\mathrm{d}E' = P\int_{-\infty}^{\infty}\left(\frac{2k'}{k+k'}\right)\frac{\sin[(k'-k)r+kr+\delta]}{(k-k')}\mathrm{d}k'$$

$$= P\int_{-\infty}^{\infty}\left(\frac{2k'}{k+k'}\right)\cos(kr+\delta)\,\frac{\sin(k'-k)r}{(k-k')}\mathrm{d}k'$$

$$= -\pi\cos(kr+\delta) \qquad (92)$$

imply that

$$\Psi_E \overset{r\to\infty}{\simeq} \tilde{\psi}_E\left[\frac{2\mu}{\pi\hbar^2 k}\right]^{\frac{1}{2}} A(E)V_{En}[-\pi\cos(kr+\delta)+Z(E)\sin(kr+\delta)]$$

$$\simeq \tilde{\psi}_E\left[\frac{2\mu}{\pi\hbar^2 k}\right]^{\frac{1}{2}} A(E)V_{En}[\pi^2+Z^2(E)]^{\frac{1}{2}}\sin[kr+\delta+\delta^{(r)}] \qquad (93)$$

where

$$\tan\delta^{(r)} = -[\pi/Z(E)] \qquad (94)$$

because $\phi_n \overset{r\to\infty}{\simeq} 0$. Hence for the correct asymptotic behaviour

$$A(E) = V_{nE}^{-1}[\pi^2+Z^2(E)]^{-\frac{1}{2}} \qquad (95)$$

This function, which represents the amplitude of the discrete component ϕ_n in the energy-normalized solution (86), is the direct analogue of the amplitude ratio $A(E)$ of the previous section.

With $Z(E)$ given by expression (90), equations (94) and (95) show that $\delta^{(r)}$ and $A(E)$ have the characteristic Breit–Wigner resonance behaviour, given by equations (21) and (22):

$$\delta^{(r)} = \arctan[\Gamma_n/2(E_n'-E)] \qquad (96)$$

$$|A(E)|^2 = \frac{1}{2\pi}\left[\frac{\Gamma_n}{(E-E_n')^2+\frac{1}{4}\Gamma_n^2}\right] \qquad (97)$$

where

$$E_n' = E_n + F(E) \qquad (98)$$

$$\Gamma_n = 2\pi|V_{nE}|^2 \qquad (99)$$

Furthermore, by analogy with equation (51):

$$\int_{E_n'-\Gamma_n/2}^{E_n'+\Gamma_n/2}|A(E)|^2\mathrm{d}E = 1 \qquad (100)$$

Notice that the level width Γ_n depends, *via* V_{nE}, only on the resonance continuum state ψ_E, but that the level shift function $F(E)$ appears in equation (92) as an integral over the full continuum. Both Γ_n and $F(E)$ are strictly energy dependent but they may be evaluated with sufficient accuracy at the resonance point E_n if the linewidth is small compared with the level spacing. Experimental implications of the level-shift formula have been discussed by Bardsley.[80]

Finally, Fano[62] obtains a compact expression for the spectroscopic transition moment by collecting the above results in the form

$$\Psi_E = \frac{1}{\pi V_{nE}} \Phi_n \sin \delta^{(r)} - \psi_E \cos \delta^{(r)} \tag{101}$$

where Φ_n represents the original discrete state ϕ_n modified by an admixture of continuum states:

$$\Phi_n = \phi_n + P \int \frac{V_{nE'} \psi_{E'}}{E - E'} \mathrm{d}E' \tag{102}$$

Hence the transition moment between Ψ_E and another state ϕ_0 becomes

$$(\Psi_E|T|\phi_0) = \frac{1}{\pi V^*_{nE}} (\Phi_n|T|\phi_0) \sin \delta^{(r)} - (\psi_E|T|\phi_0) \cos \delta^{(r)} \tag{103}$$

The spectroscopic absorption lineshape therefore differs according to the relative values of $(\Phi_n|T|\phi_0)$ and $(\psi_E|T|\phi_0)$. It takes the Lorentzian form

$$|(\Psi_E|T|\phi_0)|^2 = \frac{\Gamma_n}{2\pi[(E - E_n')^2 + \Gamma_n^2/4]} |(\phi_n|T|\phi_0)|^2 \tag{104}$$

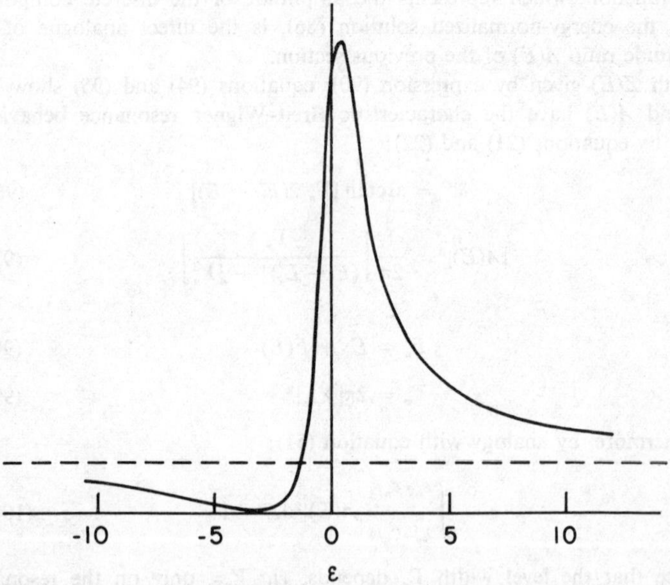

Figure 4 *The Beutler–Fano lineshape for q* = 3 [see equation (106)]. *The corresponding form for q* = −3 *is obtained by reflection in the origin*

[80] J. N. Bardsley, *J. Phys. (B)*, 1968, **1**, 349.

when transitions between ϕ_0 and the continuum ψ_E are forbidden, but appears as a Lorentzian emission line, against the continuum absorption, if $(\Phi_n|T|\phi_0) = 0$:

$$|(\Psi_E|T|\phi_0)|^2 = \left\{1 - \frac{\Gamma_n^2/4}{[(E - E_n')^2 + \Gamma_n^2/4]}\right\}|(\psi_E|T|\phi_0)^2 \quad (105)$$

Interference between the two terms in equation (103) results in a distorted lineshape (see Figure 4) characterized by Fano[62] in terms of the ratio

$$q = (\Phi_n|T|\phi_0)/\pi V_{nE}^*(\psi_E|T|\phi_0) \quad (106)$$

This characteristic Fano or Beutler[81] predissociation lineshape, familiar in atomic spectra, has recently been observed in the spectrum of H_2.[82-84]

Fano[62] also extends the argument to cover several discrete states ϕ_n interacting with one continuum ψ_E, and several continua $\psi_E^{(i)}$ interacting with one discrete state. The former naturally always applies in diatomic systems, but corrections to the one-level analysis (above) become significant only when the linewidth given by equation (99) becomes comparable with the level spacing. The latter situation may arise in certain diatomic molecules, but it is of more general significance in the polyatomic case because each (vibrational-rotation) channel has an associated continuum of states.[85]

The major correction to the theory in the many-level, one-continuum case is that the perturbed levels appear as the eigenvalues, E_ν, of a level-shift matrix[62] with elements $E_n\delta_{nm} + F_{nm}(E)$, where

$$F_{nm}(E) = P\int \frac{V_{nE'}V_{E'm}\mathrm{d}E'}{E - E'} \quad (107)$$

This leads to the introduction of transformed discrete state combinations

$$\phi_\nu = \sum_n c_{\nu n}\phi_n \quad (108)$$

where the $c_{\nu n}$ form the eigenvectors of the above matrix. The energy-normalized solution Ψ_E finally appears in the form

$$\Psi_E = \cos\delta^{(r)}\left\{\sum_\nu \frac{V_{E\nu}}{E - E_\nu}\Phi_\nu - \psi_E\right\} \quad (109)$$

where

$$\tan\delta^{(r)} = \sum_\nu \frac{\pi|V_{E\nu}|^2}{E - E_\nu} \quad (110)$$

$$\Phi_\nu = \phi_\nu + P\int \frac{V_{\nu E'}\psi_{E'}\mathrm{d}E'}{E - E'} \quad (111)$$

[81] H. Beutler, *Z. Phys.*, 1935, **93**, 177.
[82] F. J. Comes and U. Wenning, *Z. Naturforsch.*, 1969, **24a**, 587.
[83] F. J. Comes and G. Schumpe, *Z. Naturforsch.*, 1971, **26a**, 538.
[84] G. Herzberg, 'Topics in Modern Physics—A Tribute to Edward U. Condon', Colorado Associated Univ. Press, 1971, p. 191.
[85] See C. E. Caplan and M. S. Child, *Mol. Phys.*, 1972, **23**, 249.

The conditions for the validity of equations (98)—(101) are therefore first that $|F_{nm}(E)| \ll E_n - E_m$, so that

$$E_\nu \approx E_n + F_{nn}(E) \tag{112}$$

and secondly that $|V_{En}| \ll E_\nu - E_{\nu'}$, in which case the sums in equations (110) and (111) may be approximated by a single term at each resonance, and the term $[\cos \delta^{(r)} V_{E\nu}/(E - E_\nu)]$ reduces to $[\sin \delta^{(r)}_\nu/\pi V_{nE}]$ as in equation (101).

The results for the one discrete, several-continua case may be summarized in the form:[62]

$$\Psi_E = \frac{1}{\pi V_{nE}} \Phi_n \sin \delta^{(r)} - \sum_i \frac{V_{nE}^{(i)}}{V_{nE}} \psi_E^{(i)} \cos \delta^{(r)} \tag{113}$$

where

$$\Phi_n = \phi_n + \sum_i P \int \frac{V_{nE'}^{(i)} \psi_{E'}^{(i)} \mathrm{d}E'}{E - E'}$$

$$V_{nE}^2 = \sum_i |V_{nE}^{(i)}|^2 \tag{114}$$

and $\delta^{(r)}$ is given by equation (96). Both the level shift, $F(E)$, and the line-width, Γ_n, therefore appear, *via* V_{nE} in equation (108), as a simple sum of the previous single-continuum formulae taken over the continua in question.

The Interaction Term.—The theory in the previous section is quite general. When specialized to a diatomic system the interaction term may be written

$$V_{nE} = \int_0^\infty \chi_{2n}(r) H_{21}(r) \chi_{1E}(r) \mathrm{d}r \tag{115}$$

where

$$H_{21}(r) = \langle P_2 | H(q, r) | P_1 \rangle \tag{116}$$

Here n denotes a particular vibrational, total angular momentum level, $|P_1\rangle$ and $|P_2\rangle$ are used to designate appropriate electronic and rotational angular momentum states, and the integral in equation (116) is taken over the electronic co-ordinates, $\boldsymbol{q}$, and the angular co-ordinates of the nuclear position variable, $\boldsymbol{r}$; $\chi_{2n}(r)$ and $\chi_{1E}(r)$ are the corresponding vibrational and continuum wavefunctions, normalized to unity and to a delta function of energy, respectively.

Predissociation may be induced by both internal (intramolecular) and external interactions. The important internal cases are covered by the selection rules and expressions for $H_{12}(r)$ given by Kronig.[8] These require in every case

$$\Delta J = 0, \quad + \leftrightarrow +, \quad g \leftrightarrow u \tag{117}$$

but interactions governed by

$$\Delta\Lambda = 0, \quad \pm 1 \quad \text{in cases } (a) \text{ and } (b)$$

$$\Delta\Omega = 0, \quad \pm 1 \quad \text{in case } (c) \tag{118}$$

are allowed. There is also a weak selection rule

$$\Delta S = 0 \qquad \text{in cases } (a) \text{ and } (b)$$

which may be broken by spin–orbit coupling. An important observational distinction is made between homogeneous and heterogeneous interactions [86] for which $\Delta\Lambda = 0$ [or $\Delta\Omega = 0$ in case (*c*)] and $\Delta\Lambda = \pm 1$ (or $\Delta\Omega = \pm 1$), respectively, in that the latter (which arise from rotational–electronic coupling [87]) yield an expression for $H_{12}(r)$ proportional to $[J(J+1) - \Lambda(\Lambda \pm 1)]^{\frac{1}{2}}$. A short table of expressions for $H_{12}(r)$ evaluated in the approximation of pure precession has been given by Czarny, Felenbok, and Lefebvre-Brion.[69]

Observed predissociations have also been attributed to the external interactions induced by magnetic fields [9, 88] and molecular collisions.[89–91] A condensed form of Van Vleck's theory [9] of magnetically induced predissociation has recently been given by Chapman and Bunker.[71] Qualitative aspects of the theory of pressure-induced predissociation have been considered,[90, 92] but a comprehensive theory remains to be developed.

The Franck–Condon Approximation.—The existence of an intersection between the repulsive and bound state curves, $V_1(r)$ and $V_2(r)$, respectively, may be used to justify a significant simplification in equation (115), because the effective integration region is known [93] to be localized at the crossing point, R; an analytical estimate obtained on pp. 498—499 suggests an effective range between ± 0.05 Å and ± 0.20 Å in typical systems. Under these conditions it may be permissible to replace $H_{12}(r)$ by its value, H_{12}^0, at the crossing point, so that

$$\Gamma_n = 2\pi |V_{nE}|^2 = 2\pi |H_{12}^0|^2 \,|\textstyle\int_0^\infty \chi_{2n}(r)\chi_{1E}(r)\mathrm{d}r|^2 \qquad (119)$$

The validity of this Franck–Condon approximation to the interaction element is supported by exact numerical calculations,[64, 69] and by analytical arguments on pp. 496—499, below. The use of a similar Franck–Condon approximation in the computation of the level shift by means of equation (91) is, however, open to greater risk because the significant integration range in the evaluation of $V_{nE'}$ varies, as E' spans the continuum.

A second limitation on the validity of equation (119) is that the linewidth obtained should be small compared with the level spacing [see equation (84)]. A significant case not covered by (119) therefore arises when $H_{12}(r)$ is suffi-

[86] R. S. Mulliken, *J. Phys. Chem.*, 1937, **41**, 5.
[87] O. Gallais, *Mol. Phys.*, 1973, **25**, 949.
[88] L. A. Turner, *Z. Phys.*, 1930, **65**, 464; *Phys. Rev.*, 1931, **38**, 574.
[89] L. A. Turner, *Phys. Rev.*, 1933, **41**, 627.
[90] J. I. Steinfeld and W. Klemperer, *J. Chem. Phys.*, 1965, **42**, 3475.
[91] J. I. Steinfeld, *J. Chem. Phys.*, 1966, **44**, 2740.
[92] C. Zener, *Proc. Roy. Soc.*, 1933, **A140**, 660.
[93] See L. D. Landau and E. M. Lifshitz, ref. 48, p. 178.

ciently large that the predissociation occurs from the upper adiabatic potential curve, $V_+(r)$, where

$$V_{\pm}(r) = \tfrac{1}{2}[V_1(r) + V_2(r)] \pm \tfrac{1}{2}\{[V_1(r) - V_2(r)]^2 + 4H_{12}^2(r)\}^{\frac{1}{2}} \qquad (120)$$

rather than from the diabatic curve $V_2(r)$ (see Figure 1). The observed predissociation of ICl [94] and IBr [95, 96] are assumed to be of this type. In these circumstances equation (119) remains valid provided that $\chi_{2n}(r)$ and $\chi_{1E}(r)$ in equation (115) are replaced by the bound, $\chi_{+n}(r)$, and continuum, $\chi_{-E}(r)$, wavefunctions determined by $V_+(r)$ and $V_-(r)$, respectively, and the interaction term is written as [68, 97, 98]

$$H_{+-}(r) = -H_{-+}(r) = \frac{\mathrm{d}^2\theta}{\mathrm{d}r} + \frac{\mathrm{d}\theta}{\mathrm{d}r}\frac{\mathrm{d}}{\mathrm{d}r} \qquad (121)$$

where $\theta(r)$ is the adiabatic mixing parameter

$$\theta(r) = \arctan\left[\frac{2H_{12}(r)}{V_1(r) - V_2(r)}\right] \qquad (122)$$

However, the presence of a peak in $(\mathrm{d}\theta/\mathrm{d}r)$, and the disappearance of a crossing point, precludes the use of a Franck–Condon approximation in the evaluation of $V_{nE'}$ given by equation (115).

Intermediate cases between these near diabatic and near adiabatic limits may be covered by diagonalization of the Fano level-shift matrix defined by equation (107), in order to determine the appropriate admixture [equation (108)] of bound states. A simpler alternative may be to seek a direct solution of the appropriate coupled differential equations, either by numerical [68] or by analytical [74] techniques. It is interesting to find that an exact model computation of this type, covering the full range of possible interaction strengths,[68] shows very few resonance positions which depart significantly from those obtained by either a strict diabatic (distortion) or adiabatic approximation. This study gave, however, only a graphical indication of the level widths.

Finally, attention may be drawn to the possibility of predissociation in the absence of a potential curve crossing [98] (type a^0, b^0, or c^0 in Mulliken's classification.[7] Such cases may fall into either the 'non-diabatic' [predissociation from $V_2(r)$] or 'non-adiabatic' [predissociation from $V_+(r)$] category, and no decision can be made without examination of possible changes in the electronic state. Even in the non-diabatic case, however, the Franck–Condon approximation (119) would require careful justification in viewof the absence of a crossing point.

[94] W. G. Brown and G. E. Gibson, *Phys. Rev.*, 1932, **40**, 529.
[95] W. G. Brown, *Phys. Rev.*, 1932, **42**, 355.
[96] L. E. Selin, *Arkiv Fysik*, 1962, **21**, 529.
[97] R. S. Berry, *J. Chem. Phys.*, 1957, **27**, 1288.
[98] F. Fiquet-Fayard, *J. Chim. phys.*, (*Special Issue*), 1970, 57.

Exact Numerical Results in the Franck–Condon Approximation.—Within the validity of the Franck–Condon approximation (119), a computation of the linewidth rests on knowledge of the exact numerical wavefunctions $\chi_{1E}(r)$ and $\chi_{2n}(r)$. Henceforth n will be taken to include a pair (v, J) of vibrational rotational quantum numbers. There is no problem such as that encountered for shape resonances [see equation (3)] in locating the level position because $\chi_{2n}(r)$ may be taken as a vibrational-rotational eigenfunction of the experimental RKR bound-state curve $V_2(r)$. For any given repulsive curve $V_1(r)$, the continuum state $\chi_{1E}(r)$ is also readily determined; the only problem concerns normalization. The use of a true continuum state normalized as in equation (85) by comparison with a spherical Bessel function in the asymptotic region [99] is strictly preferable on grounds both of computational speed and flexibility, but a pseudo-continuum approximation is frequently employed,[66, 67] according to which

$$\chi_{1E}(r) \approx \left(\frac{\partial N}{\partial E_N}\right)^{\frac{1}{2}} \chi_{1N}(r) \tag{123}$$

where $\chi_{1N}(r)$ is the N-th normalized bound wavefunction for the potential $V_1(r)$ modified by an infinite wall at some large distance. Given such wavefunctions, it is possible to vary the assumed repulsive curve until the computed linewidth pattern agrees with that obtained by experiment. Computations of this type have been performed for predissociations from the states of $O_2\,(B\,^3\Sigma_u^-)$,[64, 65] $N_2\,(b\,^1\Pi_u)$,[70] H_2 and $D_2\,(D\,^1\Pi_u)$,[67] $OD\,(A\,^2\Sigma^+)$,[69] and $I_2\,[B\,^3\Pi(0_u^+)]$.[66, 71] A similar calculation for the level shift has been reported for the $B(0_u^+)$ state of Se_2.[72]

The first general conclusion is that the results are indeed sensitive to the form of the repulsive curve. This is dramatically illustrated by the calculations of Reiss and Ben Aryah [65] and Murrell and Taylor.[64] Secondly, the linewidth pattern is found, as first predicted by Rice,[77] to fluctuate with v at a frequency which is high or low according to whether the intersection occurs on the attractive or repulsive branch of $V_2(r)$ (types c^+ or c^- in Mulliken's notation).[7] This behaviour may be attributed to a variation in the relative phases of the wavefunctions $\chi_{1E}(r)$ and $\chi_{2n}(r)$ at the crossing point. Hence the oscillation frequency depends on the rate of divergence of $V_1(r)$ and $V_2(r)$ above the crossing point. Typical patterns obtained by Murrell and Taylor [64] are shown in Figure 5. The irregular nature of the fluctuations in case c^+ means that a single repulsive curve may give rise to two or more regions of strong predissociation separated by intermediate v levels for which the predissociation is weak. Theory [64, 71, 76] suggests that the observed predissociations of $O_2\,(^3\Sigma_u^-)$ [100–102] and $I_2\,[^3\Pi(0_u^+)]$ [66, 71, 103] may be understood in this way.

[99] See R. B. Bernstein, *J. Chem. Phys.*, 1960, **33**, 795.
[100] P. G. Wilkinson and R. S. Mulliken, *Astrophys. J.*, 1957, **125**, 594.
[101] P. K. Carroll, *Astrophys. J.*, 1959, **129**, 794.
[102] M. Ackerman and F. Biaume, *J. Mol. Spectroscopy*, 1970, **35**, 73.
[103] E. O. Degenkolb, J. I. Steinfeld, E. Wassermann, and W. Klemperer, *J. Chem. Phys.*, 1969, **51**, 615.

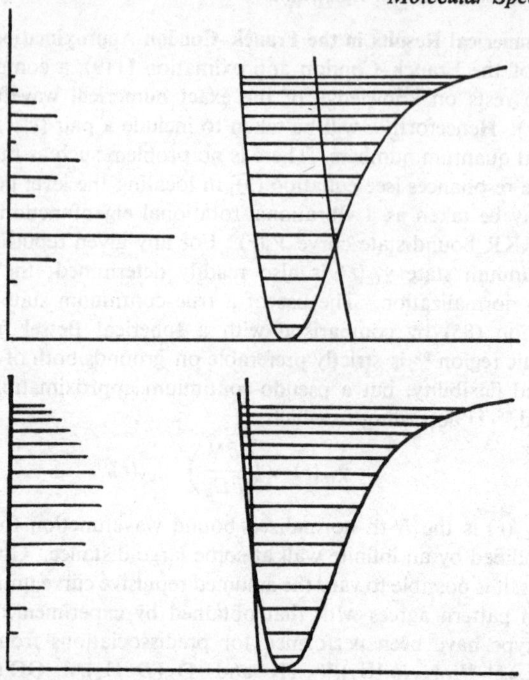

Figure 5 *Computed linewidth variations*[64] *with v for cases* c^+ *(upper diagram) and* c^- *(lower diagram). Note the contrast between the irregular fluctuations in case* c^+ *with the smooth variation in case* c^-*; compare also Figure 7*

The situation with respect to O_2 ($^3\Sigma_u^-$) is, however, still obscure because the most recent experimental linewidth pattern[102] is only in qualitative agreement with the numerical results of Murrell and Taylor;[64] furthermore, Schaefer and Miller[104] conclude that the potential curves for three possible interacting states ($^1\Pi_u$, $^5\Sigma_u^-$, and $^5\Pi_u$) must cut the attractive branch of the $^3\Sigma_u^-$ curve. The case of I_2 [$^3\Pi(0_u^+)$] shows an interesting peculiarity in that there are two types of predissociation with very similar lifetime patterns; the allowed heterogeneous predissociation into the $^1\Pi_u$ state,[66, 105] and a magnetically induced predissociation probably into the $^3\Pi(0_u^-)$ state.[71, 103] This suggests, as confirmed by analysis of the magnetic fluorescence quenching data,[71, 76] and of the continuum absorption $^1\Pi_u \leftarrow X\,^1\Sigma_g^+$,[106] that the two repulsive potential curves must be very close together.

A second consequence of the oscillatory nature of the linewidth pattern is that even when the predissociation becomes so strong that all levels appear indiscernibly broad, one may expect to find a few scattered sharp lines; the visible spectrum of IBr[96] is a clear example of this. The observation of such

[104] H. F. Schaefer and W. H. Miller, *J. Chem. Phys.*, 1971, **55**, 4107.
[105] J. Tellinghuisen, *J. Chem. Phys.*, 1972, **57**, 2397.
[106] J. Tellinghuisen, *J. Chem. Phys.*, 1973, **58**, 2821.

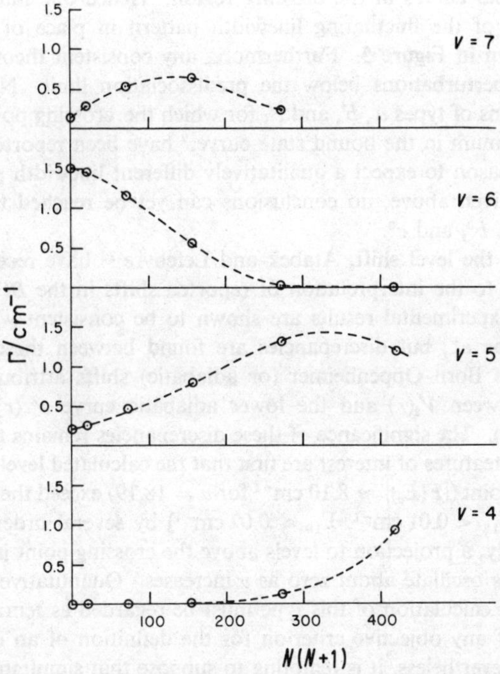

Figure 6 *Computed linewidth variations*[69] *with v and J for homogeneous predissociation from the A* ($^2\Sigma^+$) *state of* OH

'vestigial remains'[77] would itself provide important information about the repulsive curve.

Similar fluctuations in the theoretical linewidth as a function of the rotational quantum number (in this case N) have been reported for the A ($^2\Sigma^+$) state of OD.[69] The variation with $N(N + 1)$ is now, however, sufficiently slow that a more regular oscillatory pattern may be recognized (see Figure 6). Again, within the accuracy of the measured linewidths, the experimental pattern is found to be consistent with a single repulsive curve of type c^+, attributable on the evidence of the magnitude of linewidths to the state $\sigma\pi\sigma^*$ ($^4\Pi$). A similar calculation for the heterogeneous predissociation from the D ($^1\Pi_u$) to the B' ($^1\Sigma_u^+$) of H_2 shows,[67] over the small J range considered ($0 < J < 10$), no Franck–Condon contribution to the linewidth variation with J; the entire variation obtained may be attributed to the J dependence of H_{12}^0 in equation (119).

The above conclusions are derived in the main for predissociation of types c^+. For an extension to cases $a^\pm$ and $b^\pm$ for which the asymptotes lie at and above the curve crossing point, respectively,[7] one may note that the predissociation pattern depends by the above arguments only on the local forms of the potential curves in the crossing region. Hence one must expect an abrupt onset of the fluctuating linewidth pattern in place of the smooth increase shown in Figure 5. Furthermore, any consistent theory must also account for perturbations below the predissociation limit. No calculated predissociations of types a^i, b^i, and c^i, for which the crossing point coincides with the minimum in the bound state curve,[7] have been reported, but there appears no reason to expect a qualitatively different linewidth pattern. For the reasons given above, no conclusions can yet be reached for the cases designated a^0, b^0, and c^0.

Turning to the level shift, Atabek and Lefebvre [72] have recently applied equation (91) to the interpretation of reported shifts in the $B(^3\Sigma_u^-)$ state of Se_2.[107] The experimental results are shown to be consistent with an intersection of type c^+, but discrepancies are found between these Fano level shifts and the Born–Oppenheimer (or adiabatic) shifts attributable to the difference between $V_2(r)$ and the lower adiabatic curve $V_-(r)$ defined by equation (120). The significance of these discrepancies remains to be investigated. Other features of interest are first that the calculated level shifts below the crossing point ($|F(E_v)| \approx 8,10$ cm^{-1} for $v = 18,19$) exceed the corresponding widths ($\Gamma_{18} < 0.01$ cm^{-1}, $\Gamma_{19} \approx 0.02$ cm^{-1}) by several orders of magnitude. Secondly, a projection to levels above the crossing point indicates that the level shifts oscillate about zero as v increases. Quantitative conclusions drawn from a calculation of this type must be regarded as tentative in view of the lack of any objective criterion for the definition of an experimental level shift. Nevertheless, it is tempting to suppose that simultaneous predictions about the behaviour of the linewidth and the level shift above the crossing point might lead to experimental detection of the fragmentary remains of the spectrum which could be used to refine the details of the model, even when most lines are indiscernibly diffuse.

Analytical approximations which shed some light on the above numerical conclusions are developed in the following section.

Analytical Approximations.—Analytical approximations to the linewidths and level shifts are conveniently developed in the notation indicated in Figure 7. Here $V_i^J(r)$ denote the centrifugally corrected potential curves

$$V_i^J(r) = V_i(r) + J(J+1)\hbar^2/2\mu r^2 \qquad (124)$$

E_{xJ} is the energy at the crossing point, R,

$$E_{xJ} = V_1(R) + J(J+1)\hbar^2/2\mu R^2 = V_2(R) + (J+\tfrac{1}{2})^2\hbar^2/2\mu R^2 \qquad (125)$$

and ΔF is the difference between the potential derivatives,

$$\Delta F = F_{1J} - F_{2J} \qquad (126)$$

where

$$F_{iJ} = -(\partial V_i^J/\mathrm{d}r)_{r=R} \qquad (127)$$

[107] R. F. Barrow, G. G. Chandler, and C. B. Meyer, *Phil. Trans. Roy. Soc.*, 1960, **260**, 395.

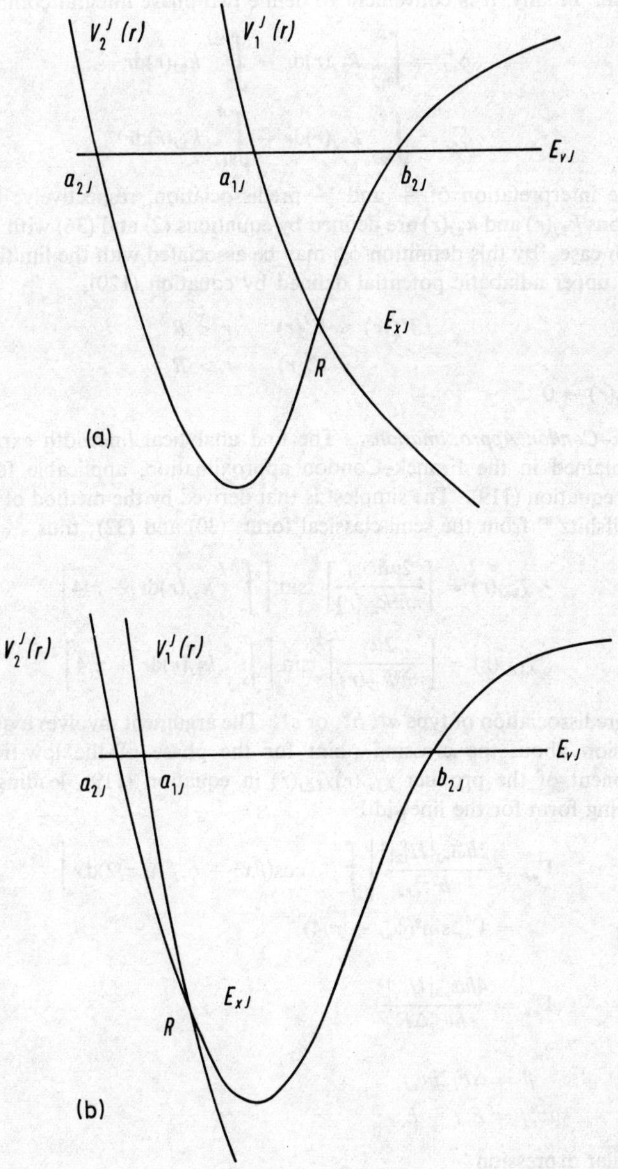

Figure 7 *Notation for the semi-classical theory* (a) *in case* c^+, *and* (b) *in case* c^-

ΔF is in fact independent of J, and (in the notation of Figure 7) necessarily positive. Finally, it is convenient to define two phase integral combinations

$$\phi_{vJ}^{+} = \int_{a_{1J}}^{R} k_{1J}(r)\mathrm{d}r + \int_{R}^{b_{2J}} k_{2J}(r)\mathrm{d}r$$

$$\phi_{vJ}^{-} = \int_{a_{2J}}^{R} k_{2J}(r)\mathrm{d}r - \int_{a_{1J}}^{R} k_{1J}(r)\mathrm{d}r \tag{128}$$

for the interpretation of + and − predissociation, respectively; here the functions $k_{1J}(r)$ and $k_{2J}(r)$ are defined by equations (2) and (36) with $E = E_{vJ}$ in each case. By this definition ϕ_{vJ}^{+} may be associated with the limiting form of the upper adiabatic potential defined by equation (120),

$$\begin{aligned} V_{+}^{J}(r) &\approx V_{1}^{J}(r) \qquad r < R \\ &\approx V_{2}^{J}(r) \qquad r > R \end{aligned} \tag{129}$$

as $H_{12}(r) \to 0$.

Franck–Condon Approximations. The first analytical linewidth expressions are obtained in the Franck–Condon approximation, applicable for small H_{12}^{0} in equation (119). The simplest is that derived by the method of Landau and Lifshitz [93] from the semi-classical forms (30) and (32); thus

$$\chi_{2vJ}(r) = \left[\frac{2\mu\hbar\bar{\omega}_{vJ}}{\pi\hbar^{2}k_{2J}(r)}\right]^{\frac{1}{2}} \sin\left[\int_{r}^{b_{2J}} k_{2J}(r)\mathrm{d}r + \pi/4\right]$$

$$\chi_{1EJ}(r) = \left[\frac{2\mu}{\pi\hbar^{2}k_{1J}(r)}\right]^{\frac{1}{2}} \sin\left[\int_{a_{1J}}^{r} k_{1J}(r)\mathrm{d}r + \pi/4\right] \tag{130}$$

for a predissociation of type a^{+}, b^{+}, or c^{+}. The argument involves a quadratic expansion about the crossing point for the phase of the low-frequency component of the product $\chi_{2vJ}(r)\chi_{1EJ}(r)$ in equation (119), leading to the following form for the linewidth

$$\begin{aligned} \Gamma_{vJ} &= \frac{2\hbar\bar{\omega}_{vJ}|H_{12}^{0}|^{2}}{\pi\hbar^{2}v_{vJ}^{2}}\left|\int_{-\infty}^{\infty} \cos(\beta x^{2} + \phi_{vJ}^{+} + \pi/2)\mathrm{d}x\right|^{2} \\ &= \Gamma_{vJ}^{0} \sin^{2}(\phi_{vJ}^{+} + \pi/4) \end{aligned} \tag{131}$$

where

$$\Gamma_{vJ}^{0} = \frac{4\hbar\bar{\omega}_{vJ}|H_{12}^{0}|^{2}}{\hbar v_{vJ}\Delta F} \tag{132}$$

and

$$\begin{aligned} \beta &= \Delta F/2\hbar v_{vJ} \\ \tfrac{1}{2}\mu v_{vJ}^{2} &= E_{vJ} - E_{xJ} \end{aligned} \tag{133}$$

A similar expression

$$\Gamma_{vJ} = \Gamma_{vJ}^{0} \sin^{2}(\phi_{vJ}^{-} + \pi/4) \tag{134}$$

is obtained in cases a^{-}, b^{-}, or c^{-}.

The relation between the Fresnel integral in equation (131) and the error function [49] indicates that 95% of the integral arises from the range $|x| < 2(2/\beta)^{\frac{1}{2}}$. Hence equations (131) and (134) depend on the validity of the semi-classical approximations (130), and on the assumptions of linearity in $V_i^J(r)$ and constancy in $H_{12}(r)$ over the range

$$\begin{aligned} r &= R \pm 2[\hbar v_{vJ}/\Delta F]^{\frac{1}{2}} \\ &\approx R \pm 5[(E_{vJ} - E_{xJ})/\mu\Delta F^2]^{\frac{1}{4}} \end{aligned} \tag{135}$$

where in the second line of (135), the units are taken to be r/Å, E/cm^{-1}, μ/(atomic mass units), and F/cm^{-1} Å^{-1}. The significance of this transition zone, the width of which increases with increasing energy, was first recognised by Bates.[108] Parameters applicable to the predissociations of $O_2\,(^3\Sigma_u^-)$, $I_2\,[B\,^3\Pi(0_u^+)]$ and OH $(A\,^2\Sigma^+)$ are listed in Table 5. Note that since ΔF may become quite small for inner (a^-, b^-, or c^-) curve crossings, the Franck-Condon approximation may require special justification in these cases, particularly for systems of low reduced mass.

Table 5 *Parameters for some observed predissociations*

State	*Type*	$E - E_x$ /cm^{-1}	ΔF /cm^{-1} Å^{-1}	*R* /Å	*Range* /Å	*Ref.*
$O_2\ B\,^3\Sigma_u^-$	c^+	5000	50 000	1.89	±0.10	64,73
$I_2\ B\,^3\Pi(0_u^+)$	c^-	3000	9 000	2.87	±0.15	71,76
OH $A\,^2\Sigma^+$	c^+	4000	24 000	1.65	±0.20	69,75

A second analytical approximation is available to cover energies near and below the crossing point, for which overlap between the transition zone and the classical turning-point regions invalidates the semi-classical approximation (130). One is necessarily now interested only in predissociation of types c^+ and c^-. In the latter case, for example, the following Airy function [108a] approximations are applicable:

$$\begin{aligned} \chi_{2vJ}(r) &\approx \left[\frac{2\mu\hbar\bar{\omega}_{vJ}}{\alpha_{2J}\hbar^2}\right]^{\frac{1}{2}} \mathrm{Ai}[-\alpha_{2J}(r - a_{2J})] \\ \chi_{1EJ}(r) &\approx \left[\frac{2\mu}{\alpha_{1J}\hbar^2}\right]^{\frac{1}{2}} \mathrm{Ai}[-\alpha_{1J}(r - a_{1J})] \end{aligned} \tag{136}$$

where

$$\alpha_{iJ} = (2\mu F_{iJ}/\hbar^2)^{\frac{1}{3}} \tag{137}$$

and the normalizations are chosen to be consistent with those in equation (130). Equations (136) define exact wavefunctions for the potential curves

$$V_{iJ}(r) = E_{xJ} - F_{iJ}(r - R) \tag{138}$$

[108] D. R. Bates, *Proc. Roy. Soc.*, 1960, **A257**, 22.
[108a] See L. D. Landau and E. M. Lifshitz, ref. 49, p. 161.

with $F_{iJ} > 0$. Hence the turning points at energy E_{vJ} are given by

$$a_{iJ} = (E_{vJ} - E_{xJ})/F_{iJ} \tag{139}$$

With the necessary overlap integral derived in the Appendix, the final line-width expression may be shown to be

$$\Gamma_{vJ} = \pi\Gamma^0_{vJ}(v_{vJ}/v^*_J)\mathrm{Ai}^2[-(E_{vJ} - E_{xJ})/E^*_J] \tag{140}$$

where

$$E^*_J = \tfrac{1}{2}\mu v^{*2}_J = (\hbar^2 F^2_{1J} F^2_{2J}/2\mu\Delta F^2)^{\frac{1}{3}} \tag{141}$$

This formula was first given by Rice,[77] and rediscovered independently by Degenkolb, Steinfeld, Wasserman, and Klemperer [103] and the Reporter.[73] It may be shown to go over to equation (134) when $(E_{vJ} - E_{xJ}) \gg E^*_J$ because [49]

$$\mathrm{Ai}(-z) \overset{z \gg 1}{\simeq} \pi^{-\frac{1}{2}} z^{-\frac{1}{4}} \sin(\tfrac{2}{3}z^{\frac{3}{2}} + \pi/4) \tag{142}$$

and within the validity of equation (138)

$$\begin{aligned}\phi^-_{vJ} &= \frac{(2\mu)^{\frac{1}{2}}}{\hbar}\left\{\int_{a_{2J}}^{R} [E_{vJ} - E_{xJ} + F_{2J}(r - R)]^{\frac{1}{2}}\,\mathrm{d}r \right.\\ &\quad \left. - \int_{a_{1J}}^{R} [E_{vJ} - E_{xJ} + F_{1J}(r - R)]^{\frac{1}{2}}\,\mathrm{d}r\right\} \\ &= \tfrac{2}{3}[(E_{vJ} - E_{xJ})/E^*_J]^{\frac{3}{2}}\end{aligned} \tag{143}$$

Equations (140), (141), and (143) (with ϕ^+_{vJ} in place of ϕ^-_{vJ}) may also be shown to be valid in case c^+; the argument requires merely a reversal of sign in the argument of $\chi_{2vJ}(r)$ and the introduction of $|F_{2J}|$ in the definition of α_{2J}. A similar more complicated uniform version of equation (140), designed to allow for some deviation from linearity in the potential curves, has been suggested by Miller:[109]

$$\Gamma_{vJ} = \pi\Gamma^0_{vJ} t^{\frac{1}{2}} \mathrm{Ai}^2(-t), \quad t = (\tfrac{3}{2}\,\phi^{\pm}_{vJ})^{\frac{2}{3}} \tag{144}$$

General techniques for the evaluation of WKB matrix elements have also been discussed by Pack and Dahler,[110] and Smith and Pack.[111]

As a measure of the relative validities of equations (131) or (134) and (140) we may note that equation (142) reproduces the Airy function within ± 0.025 at the first maximum, $\mathrm{Ai}(-1.012) = 0.1536$, and within ± 0.005 for $Z < -3.0$. Hence as a general rule the semi-classical forms (131) or (134), which require a linear approximation to the potential only over the transition zone [see equation (135)], may be taken to be superior for $E_{vJ} - E_{xJ} > 1.5\,E^*_J$.

Before examining the physical significance of these results, we may note

109 W. H. Miller, *J. Chem. Phys.*, 1968, **48**, 464.
110 R. T. Pack and J. S. Dahler, *J. Chem. Phys.*, 1969, **50**, 2397.
111 R. D. Smith and R. T. Pack, *J. Chem. Phys.*, 1970, **52**, 1381.

that an extension of the above argument to include an exponential interaction term

$$H_{12}(r) = H_{12}^0 \exp[-\alpha(r - R)] \quad (145)$$

may be shown to yield [112]

$$\Gamma_{vJ} = \pi\Gamma_{vJ}^0 \left(\frac{v_{vJ}}{v_{vJ}^*}\right) \exp\left[\frac{\hbar^2\alpha^3(F_{1J} + F_{2J})}{3\mu\,\Delta F^3}\right] \times \mathrm{Ai}^2\left[-\left(\frac{E_{vJ} - E_{xJ} - E_J^{\ddagger}}{E_J^*}\right)\right]^2 \quad (146)$$

where

$$E_J^{\ddagger} = (\hbar^2\alpha^2|F_{1J}F_{2J}|2\mu\Delta F^2) \quad (147)$$

The asymptotic form of equation (146),

$$\Gamma_{vJ} \approx \Gamma_{vJ}^0 \exp\left[\frac{\hbar^2\alpha^3(F_{1J} + F_{2J})}{3\mu\Delta F^3}\right] \sin^2\left[\frac{2}{3}\left(\frac{E_{vJ} - E_{xJ}}{E_J^*}\right)^{\frac{3}{2}} - \frac{\alpha^2\hbar v_J}{\Delta F} \cdots + \frac{\pi}{4}\right] \quad (148)$$

differs from that derived from equation (140) by the presence of the exponential factor and the additional phase term, $\alpha^2\hbar v_{vJ}/\Delta F$. The former is probably unimportant in most cases since $[\hbar^2\alpha^3(F_{1J} + F_{2J})/3\mu\Delta F^3] \approx 10^{-3}$ for $F_{1J} = 1600\ \mathrm{cm^{-1}\ \AA^{-1}}$, $F_{2J} = 8000\ \mathrm{cm^{-1}\ \AA^{-1}}$, $\alpha = 1\ \mathrm{\AA^{-1}}$, and $\mu = 10$ a.m.u., for

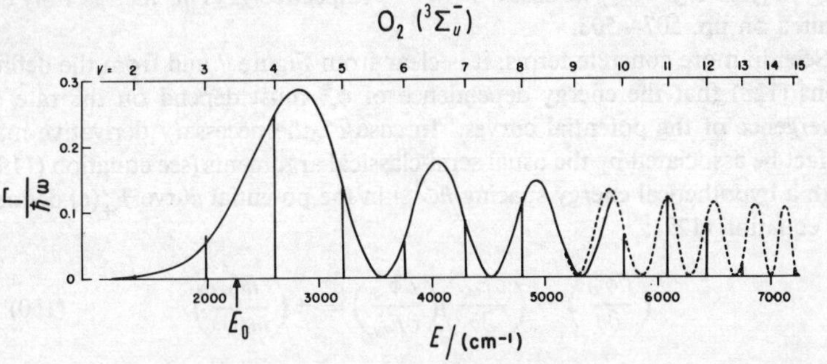

Figure 8 *A comparison between the computed linewidth pattern [64] for predissociation of $O_2\ (^3\Sigma_u^-)$ and the analytical expressions* (131) *(dashed line) and* (140) *(solid line). Note that a satisfactory Airy function representation (solid line) as far as the third minimum should be regarded as atypical. Curvature in the potential-energy functions may normally be expected to cause deviations between the semi-classical (dashed line) and the Airy-function behaviour at a point between the first maximum and the first minimum (see text)*

[112] M. S. Child, *Mol. Phys.*, 1972, **23**, 269.

example. The form of the phase correction again underlines the significance of equation (135); it will be small provided the range, α^{-1}, of $H_{12}(r)$ is large compared with the width of the transition zone.

We turn now to consider the physical factors affecting the linewidth variation with v at given J. Figure 8 gives a comparison between the predictions of equations (131) and (140) and the computed values of Murrell and Taylor.[64] The properties of the Airy function [49] indicate that the linewidth envelope should reach a maximum at $E_{xJ} + 1.019\,E_J^*$, with a point of inflexion at the crossing point and an exponential decrease

$$\Gamma_J \approx \tfrac{1}{4}|\Gamma^0_{vJ}| \exp\left[-\frac{4}{3}\left(\frac{E_{xJ} - E_{vJ}}{E_J^*}\right)^{\frac{3}{2}}\right] \tag{149}$$

for $E - E_{vJ} < -E_J^*$. The much slower decline at higher energies is governed according to equation (132) by the ratio $(\bar{\omega}_{vJ}/v_{vJ})$; note that since the local energy spacing, $\hbar\bar{\omega}_{vJ}$, must vanish at the dissociation limit, even when the predissociation appears very strong, a discrete spectrum must be regained as the energy increases.

The frequency of oscillations within this envelope depends on the energy variation of the appropriate phase integral ϕ^+_{vJ} or ϕ^-_{vJ}. Seen in this light the oscillations may be attributed to semi-classical interference [113] between the two possible escape trajectories associated with transitions at the crossing-point during outward or inward motion, respectively, the latter being followed by reflection at the classical turning point on the repulsive curve. Hence the linewidth pattern is in effect an interferogram for the nett path difference $b_{2J} - a_{1J}$ or $a_{1J} - a_{2J}$ in cases $+$ or $-$, respectively. This idea is fully exploited on pp. 507—508.

Seen in more concrete terms, it is clear from Figure 7 and from the definitions (128) that the energy dependence of $\phi^{\pm}_{vJ}$ must depend on the rate of divergence of the potential curves. In case c^+ the necessary derivative may in fact be associated by the usual semi-classical arguments [see equation (119)] with a hypothetical energy spacing $\hbar\bar{\omega}^{(+)}$ in the potential curve $V^J_+(r)$ defined by equation (129):

$$\left(\frac{\partial\phi^+_{vJ}}{\partial v}\right) = \left(\frac{\partial E_{vJ}}{\partial v}\right)\left(\frac{\partial\phi^+_{vJ}}{\partial E_{vJ}}\right) = \pi\left(\frac{\hbar\bar{\omega}_{vJ}}{\hbar\bar{\omega}^{(+)}}\right) \tag{150}$$

This means that close to the dissociation limit, where both $\hbar\bar{\omega}_{vJ}$ and $\hbar\bar{\omega}^{(+)}$ must depend only on the long-range form of $V^J_2(r)$,[114]

$$\frac{\partial\phi^+_{vJ}}{\partial v} = \pi \tag{151}$$

The factors affecting the linewidth variation with J may also be understood

[113] W. H. Miller, *J. Chem. Phys.*, 1970, **53**, 1949.
[114] R. J. LeRoy and R. B. Bernstein, *J. Chem. Phys.*, 1970, **52**, 3869.

in the light of equations (131), (134), and (140). The overriding consideration is that the energy difference in (140) varies as

$$E_{vJ} - E_{xJ} = E_v - E_{x0} + [hcB_v - \hbar^2/2\mu R^2]J(J+1) \qquad (152)$$

Hence for vibrational levels near the crossing point the linewidth must increase or decrease with increasing J in cases c^+ or c^-, respectively. Similarly, the phase terms ϕ_{vJ}^+ and ϕ_{vJ}^- may be seen to be increasing and decreasing functions, respectively, with $(\partial\phi_{vJ}^+/\partial J)$ approximately given by [75]

$$\left(\frac{\partial\phi_{vJ}^+}{\partial J}\right) = (2J+1)\pi\left(\frac{B_v - B^{(+)}(E_v)}{\hbar\bar{\omega}^{(+)}}\right) \qquad (153)$$

where $\hbar\bar{\omega}^{(+)}$ and $B^+(E_v)$ are, respectively, the hypothetical vibrational spacing and rotational constant (at the measured energy E_v) for the potential curve $V_+(r)$.

Finally, the effects of isotopic substitution may be considered. Comparisons are most conveniently made at equivalent mass reduced J values [115] [having a common value of $J(J+1)/\mu$] and at energies measured from a common zero. Equations (131) and (134) then show that the reduced linewidths $(\Gamma_{vJ}/\mu^{\frac{1}{2}}\hbar\bar{\omega}_{vJ})$ should oscillate over the high-energy sinusoidal region within a common envelope bounded by a curve of the form $C(E - E_{XJ})^{-\frac{1}{2}}$. Since the phases $\phi_{vJ}^{\pm}$ defined by equation (128) carry a factor $\mu^{\frac{1}{2}}$, the oscillation frequency is higher for the heavier isotopic substituent. Turning to the lower energy, Airy function, region, we see that the mass dependence of E_J^*, which varies as $\mu^{-\frac{1}{3}}$, leads again to a higher oscillation frequency and to a sharper onset of the predissociation for the higher mass value. Furthermore, the presence of the term $(v_J^*)^{-1}$ in equation (140) implies that the maximum

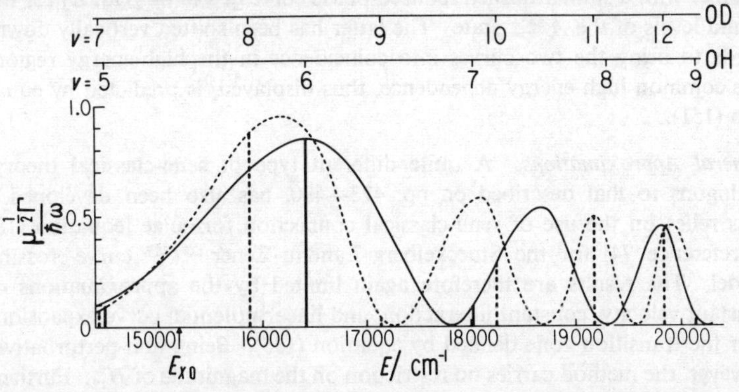

Figure 9 *Linewidth isotope effects in the $A\,^2\Sigma^+$ states of* OH *and* OD. *Solid and dashed lines refer to* OH *and* OD, *respectively. The verticals are computed values given by Czarny* et al.[69] *The curves follow equations* (131) *and* (140)

[115] W. C. Stwalley, to be published.

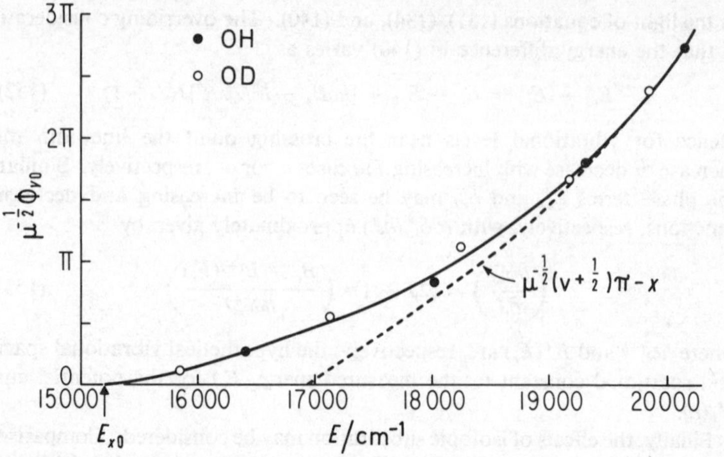

Figure 10 *Reduced phases for the $A\ ^2\Sigma^+$ states of* OH *and* OD. *The solid line passes through the points derived from computed linewidths of Czarny* et al.[69] *by the methods described on pp.* 507—508. *The dashed line, which follows the energy variation of $\mu^{-\frac{1}{2}}(v+\frac{1}{2})\pi - x$, is predicted by equation* (151) *to show the same energy variation as $\mu^{-\frac{1}{2}}\phi_{v0}^{+}$ near the dissociation limit*

in the reduced linewidth envelope should vary as $\mu^{\frac{1}{2}}$. These general features are illustrated in Figure 9, which has been constructed by the methods described on pp. 507—508 from computed ($N = 0$) linewidths for the $A\ ^2\Sigma^+$ states of OH and OD given by Czarny *et al.*[69] Figure 10 shows the smooth reduced-phase curve ($\mu^{-\frac{1}{2}}\phi_{v0}^{+}$ *vs.* E) used in the construction of Figure 9, together with a similar dashed reduced-phase curve [$\mu^{-\frac{1}{2}}(v+\frac{1}{2})$ *vs.* E] for the bound levels of the $A\ ^2\Sigma^+$ state. The latter has been shifted vertically downwards to bring the two curves into coincidence in the high-energy region. The common high-energy dependence, thus displayed, is predicted by equation (151).

General Approximations. A quite different type of semi-classical theory, analogous to that described on pp. 473—480, has also been developed.[74] This relies on the use of semi-classical connection formulae [equation (18) of reference 74] for the Stueckelberg–Landau–Zener [116–118] curve-crossing model. The results are therefore again limited by the approximations of constant velocity, constant interaction, and linear potential curve expansions over the transition zone defined by equation (135). Being non-perturbative, however, the method carries no restriction on the magnitude of H_{12}^{0}. Furthermore, it yields analytical expressions both for the linewidth and for the level shift.

[116] E. C. G. Stueckelberg, *Helv. Phys. Acta*, 1932, **5**, 369.
[117] L. D. Landau, *Z. Phys. Sow.*, 1932, **1**, 46.
[118] C. Zener, *Proc. Roy. Soc.*, 1932, **A137**, 696.

The results may be expressed in terms of one or other of the Landau–Zener transition probabilities [116–118]

$$P_{vJ} = 1 - \exp[-2\pi H_{12}^{0\,2}/\hbar v_{vJ}\Delta F]$$

$$\tilde{P}_{vJ} = 1 - P_{vJ} = \exp[-2\pi H_{12}^{0\,2}/\hbar v_{vJ}\Delta F] \tag{154}$$

P_{vJ} gives the probability of a transition from $V_2^J(r)$ to $V_1^J(r)$ on a single passage through the crossing zone, while $\tilde{P}_{vJ}$ refers to a transition from one adiabatic curve $V_+^J(r)$ or $V_-^J(r)$ to the other. Predissociations of measurable linewidth are predicted when either

$$P_{vJ} \ll 1 \quad \text{or} \quad \tilde{P}_{vJ} \ll 1 \tag{155}$$

The former correspond to what we have termed the non-diabatic predissociations discussed in the first part of this section (pp. 491—492), and formulae for the linewidths are identical with those given by equations (131) and (134). The associated level shifts may be written

$$\Delta E_{vJ} = \tfrac{1}{2}\Gamma_{vJ}^0 \sin(\phi_{vJ}^+ + \pi/4)\cos(\phi_{vJ}^+ + \pi/4) \tag{156}$$

in cases a^+, b^+, or c^+, and

$$\Delta E_{vJ} = -\tfrac{1}{2}\Gamma_{vJ}^0 \sin(\phi_{vJ}^- + \pi/4)\cos(\phi_{vJ}^- + \pi/4) \tag{157}$$

in cases a^-, b^-, or c^-.

The second inequality in equation (155) refers to predissociation out of the upper adiabatic potential curve $V_+^J(r)$ defined by equation (120), such as that observed in the visible spectra of ICl [94] and IBr.[95, 96] It is convenient to coin the notations $\tilde{a}^\pm$, $\tilde{b}^\pm$, or $\tilde{c}^\pm$ to distinguish predissociations of this type from the more common non-adiabatic phenomena. The expressions for the linewidth and level shift now become

$$\Gamma_{vJ} = \tilde{\Gamma}_{vJ}^0 \cos^2 \phi_{vJ}^{(2)} \tag{158}$$

$$\Delta E_{vJ} = \tfrac{1}{2}\tilde{\Gamma}_{vJ}^0 \sin \phi_{vJ}^{(2)} \cos \phi_{vJ}^{(2)} \tag{159}$$

where

$$\tilde{\Gamma}_{vJ}^0 = 2\left(\frac{\hbar\bar{\omega}_{vJ}}{\pi}\right) \exp[-2\pi H_{12}^{0\,2}/\hbar v_{vJ}\Delta F] \tag{160}$$

and

$$\phi_{vJ}^{(2)} = \int_{a_{2J}}^{b_{2J}} k_{2J}(r)\mathrm{d}r = \frac{(2\mu)^{\frac{1}{2}}}{\hbar}\int_{a_{2J}}^{b_{2J}} [E_{vJ} - V_2^J(r)]^{\frac{1}{2}}\mathrm{d}r \tag{161}$$

Note that $\hbar\bar{\omega}_{vJ}$ now refers to the level spacing in $V_+^J(r)$. The ratio of the linewidth to the local energy spacing, $\Gamma_{vJ}^0/\hbar\bar{\omega}_{vJ}$, is therefore predicted to oscillate within an increasing envelope as indicated schematically in Figure 11. The disappearance of $\hbar\bar{\omega}_{vJ}$ at the dissociation limit, however, again implies that the lines must eventually become sharp.

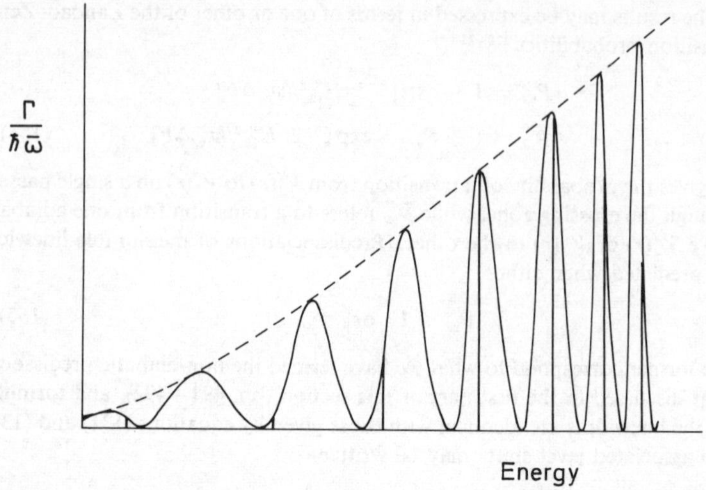

Figure 11 *Schematic linewidth variation with v for a $\tilde{c}^+$ case of non-adiabatic predissociation from the potential curve $V_+(r)$. Vertical lines indicate the linewidths*

These equations raise several points of interest. First, the inequalities (152) give quantitative tests for the validity of a non-diabatic or a non-adiabatic description. Secondly, equations (158) and (161) predict a sharp line in the spectrum whenever a bound level of $V_+^J(r)$ coincides with a bound level of $V_2^J(r)$ because, by the Bohr quantization condition (34), $\phi_{vJ}^{(2)} = (n + \frac{1}{2})\pi$ under these conditions. This behaviour is strikingly demonstrated by the visible spectrum of IBr.[95, 96] Finally, the intimate connection between the linewidth and the level shift, which implies zero level shift for the sharp resonances under all conditions, may have some general significance. Although this behaviour is not confirmed by the numerical results of Atabek and Lefebvre,[72] possibly owing to strong curvature in the repulsive curve $V_1(r)$, it means within the validity of the present model that the measured energy variation with J contains a systematic error in the rotational constant. Thus at the sharp resonance points

$$\left[\frac{\partial E_{vJ}}{\partial J(J+1)}\right] = hcB_v \pm \tfrac{1}{2}\Gamma_{vJ}^0\left(\frac{\partial \phi_{vJ}^{\pm}}{\partial J(J+1)}\right) \qquad (162)$$

for non-diabatic predissociations, while

$$\left[\frac{\partial E_{vJ}}{\partial J(J+1)}\right] = hcB_v - \tfrac{1}{2}\tilde{\Gamma}_{vJ}^0\left(\frac{\partial \phi_{vJ}^{(2)}}{\partial J(J+1)}\right) \qquad (163)$$

in the non-adiabatic case. The behaviour of $[\partial\phi_{vJ}^{\pm}/\partial J(J+1)]$ has been discussed above [see equation (153)]; the application of similar arguments shows that

$$\frac{\partial \phi_{vJ}^{(2)}}{\partial J(J+1)} = \pi \left(\frac{B_v - B^{(2)}(E_v)}{\hbar \bar{\omega}^{(2)}} \right) \tag{164}$$

This term will be negligible in cases a^-, b^-, and c^- because the minima in $V_+^J(r)$ and $V_2^J(r)$ must coincide, while $[\partial\phi^{(2)}/\partial J(J+1)]$ will clearly be negative in cases $\tilde{a}^+$, $\tilde{b}^+$, and $\tilde{c}^+$. Hence equations (153) and (162)—(164) carry a general prediction that anomalies of this type must lead to an increase in the apparent B value. The magnitude of this increase clearly depends on Γ_{vJ}^0 or $\tilde{\Gamma}_{vJ}^0$. It must be negligible for predissociations which are weak to the extent that all rotational lines are observed, but it may become significant in the interpretation of fragmentary systems, such as that shown by IBr,[95, 96] which contains only a few measurably sharp lines.

Direct Inversion of the Linewidth Pattern.—A direct method for determination of the repulsive curve $V_1(r)$ and the interaction strength H_{12}^0 has been developed [73, 75, 76] from the analytical theory of the previous section. Closely akin to the RKR method [119, 120] of bound-state spectroscopy, it is based on recognition that the phase integrals $\phi_{vJ}^{\pm}$ in equations (128), (131), and (134) contain the same information about the upper branches of $V_1(r)$ and $V_2(r)$ in Figure 7 as that supplied by the Bohr quantization condition $[\phi = (n + \frac{1}{2})\pi]$ for a normal bound-state curve. The only difference is that the energy dependence of $\phi_{vJ}^{\pm}$ must be extracted from the linewidth pattern. Once this function is known, however, the separation between the turning on $V_1(r)$ and $V_2(r)$ at energy U is given by [73, 76]

$$b_2(U) - a_1(U) = \frac{\hbar}{\pi}\left(\frac{2}{\mu}\right)^{\frac{1}{2}} \int_{E_{x0}}^{U} \left(\frac{\partial \phi_{v0}^{+}}{\partial E}\right) \frac{\mathrm{d}E}{(U-E)^{\frac{1}{2}}} \tag{165}$$

in case c^+ or [76]

$$a_1(U) - a_2(U) = \frac{\hbar}{\pi}\left(\frac{2}{\mu}\right)^{\frac{1}{2}} \int_{E_{x0}}^{U} \left(\frac{\partial \phi_{v0}}{\partial E}\right) \frac{\mathrm{d}E}{(U-E)^{\frac{1}{2}}} \tag{166}$$

in case c^-. Since the turning-point $a_2(U)$ or $b_2(U)$ is normally known from an experimental RKR curve $V_2(r)$, equations (165) or (166) are sufficient to determine $V_1(r)$. Similarly, the J dependence of $\phi_{vJ}^{\pm}$ may be used [75] to obtain an analogue of the second RKR equation, but this is unlikely to be of practical value except for confirmatory purposes.

The determination of $\phi_{v0}^{\pm}$ rests primarily on location of the crossing point E_{x0}. As a first step it is convenient to rewrite equations (131), (134), and (140) in the form

[119] J. R. Rydberg, *Z. Phys.*, 1931, **73**, 376.
[120] O. Klein, *Z. Phys.*, 1932, **76**, 226.

$$\gamma_{v0} = (\Gamma_{v0}/\hbar\bar{\omega}_{v0})^{\frac{1}{2}} = \pi^{\frac{1}{2}}KE_0^{*-\frac{1}{4}}\,\mathrm{Ai}\left[-\left(\frac{E_{v0}-E_{x0}}{E_0^*}\right)\right]$$

$$\text{for } E_{v0} \approx E_{x0} \qquad (167)$$

$$= K(E_{v0}-E_{x0})^{-\frac{1}{4}}\sin(\phi_{v0}+\pi/4+n_v\pi)$$

$$\text{for } E_{v0} \gg E_{x0} \qquad (168)$$

where

$$\bar{\omega}_{v0} = \frac{\partial G_v}{\partial v} \qquad (169)$$

γ_{v0} and E_{v0} are therefore experimental quantities; the unknowns are K, E_0^*, E_{x0}, $\phi_{v0}^{\pm}$, and the integers n_v. Two equations for K, E_0^*, and E_{x0} may be taken from the position and magnitude of the first Airy-function maximum estimated from three neighbouring linewidths (or from the greatest measured width in the absence of adequate data);[76] the equations are

$$\gamma^{\max} = \pi^{\frac{1}{2}}\,\mathrm{Ai}(-1.01879)K\,E_0^{*-\frac{1}{4}} = 0.94943K\,E_0^{*-\frac{1}{4}} \qquad (170)$$

$$E^{\max} - E_{x0} = 1.01879E_0^* \qquad (171)$$

For a third equation it may be assumed that the maximum product $\gamma_{v0}(E_{v0} - R_{x0})^{\frac{1}{4}}$ in the higher-energy region corresponds with a sinusoidal maximum (a crude estimate of E_{x0} will suffice for this purpose). Hence

$$K = [(E_{v0} - E_{x0})^{\frac{1}{4}}\gamma_{v0}]_{\max} \qquad (172)$$

As an alternative, depending on the quality of the data one might employ four measured linewidths in the Airy-function region to determine the first maximum and a point of inflexion in the linewidth envelope, the latter being taken to determine the crossing point. Preliminary estimates of K, E_{xJ}, and E_J^* obtained in this way may be improved by comparison between computed and measured linewidth patterns in the Airy region. The phases in this region are now given by equation (143):

$$\phi_{v0}^{\pm} = \tfrac{2}{3}[(E_{v0} - E_{x0})/E_0^*]^{\frac{3}{2}} \qquad (173)$$

It remains to eliminate ambiguities of $n_v\pi$ in the phases obtained by inversion of equation (168). A plot of

$$\Delta\phi_{v0} = \phi_{v0}^{+} - \int_R^{b_2} k_2(r)\mathrm{d}r \qquad \text{in case } c^+$$

$$= \int_{a_2}^{R} k_2(r)\mathrm{d}r - \phi_{v0}^{-} \qquad \text{in case } c^- \qquad (174)$$

against $(E_{v0} - E_{x0})^{\frac{3}{2}}$ is reeommended for this purpose. Here $k_2(r)$ is derived from the known RKR curve $V_2(r)$ by means of equations (2) and (6). This graph is suggested by a linear expansion for the potential $V_1(r)$ [see equation (143)]; the present model requires that it should pass through, and behave linearly at, the origin. Not only will this graph remove the necessary ambiguities of $n_v\pi$, it may also be used to smooth out experimental uncertainties in the linewidth, to refine the location of the crossing point in cases offering insufficient data in the Airy region, and to combine data for different isotopic substituents (see Figure 10).

Finally, one may note an internal check on the consistency of the method, because the value of E_0^* used to fit the data in the Airy-function region may be compared with that deduced from equation (141) in terms of the slopes F_{10} and F_{20} derived from the calculated curves. A valuable general property of the method is its stability. Refinement of a preliminary analysis [76] of the magnetic fluorescence quenching data for $I_2\,[B\,^3\Pi(0_u^+)]$,[71, 103] by means of the $\Delta\phi$ plot (Figure 12), is found to change the crossing-region parameters (E_{x0}, E_0^*, and K) from 306 cm^{-1}, 468 cm^{-1}, and 3.217 to 475 cm^{-1}, 340 cm^{-1}, and 3.162, but to alter the turning-points on $V_1(r)$ by -0.002 Å for $v = 6$ or 11, ± 0.001 Å for $v = 14, 16, 18, 21, 24, 28$, or 32, and -0.004 Å for $v = 38$. There are insufficient experimental data near the predicted crossing point for a meaningful consistency check on the value of E_0^*. A comparison between the refined potential curve $V_1(r)$ and the form deduced by Chapman and Bunker [71] from a numerical Franck–Condon calculation is given in Figure 12.

This direct inversion procedure has been successfully applied to the com-

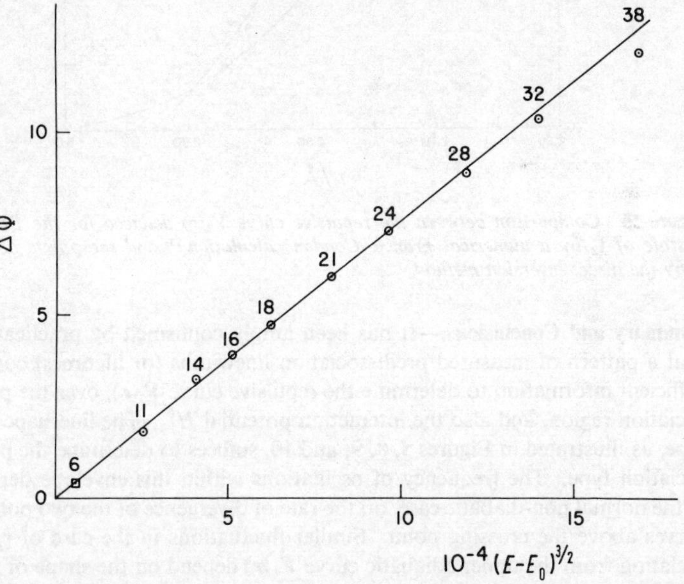

Figure 12 *Refinement of the magnetic fluorescence quenching data for the* $B[^3\Pi(0_u^+)]$ *state of* I_2. *The point* $v = 6$ *is derived* a postiori *from equation* (143) *with* E^* *given in terms of the slope of the inverted curve by equation* (141)

puted predissociation data [73] for O_2 ($^3\Sigma_u^-$) and to experimental data for OD ($A\ ^2\Sigma^+$) [75] and I_2 [$B\ ^3\Pi(0_u^+)$].[76] The resulting repulsive potential curves are closely comparable to those deduced by numerical Franck–Condon calculations (see Figure 13). The advantages of the direct approach are speed and flexibility, since the only prior assumption is of linearity in the potential curves over the transition zone. This suggests its general future use as a preliminary to the potentially more exact numerical Franck–Condon procedure.

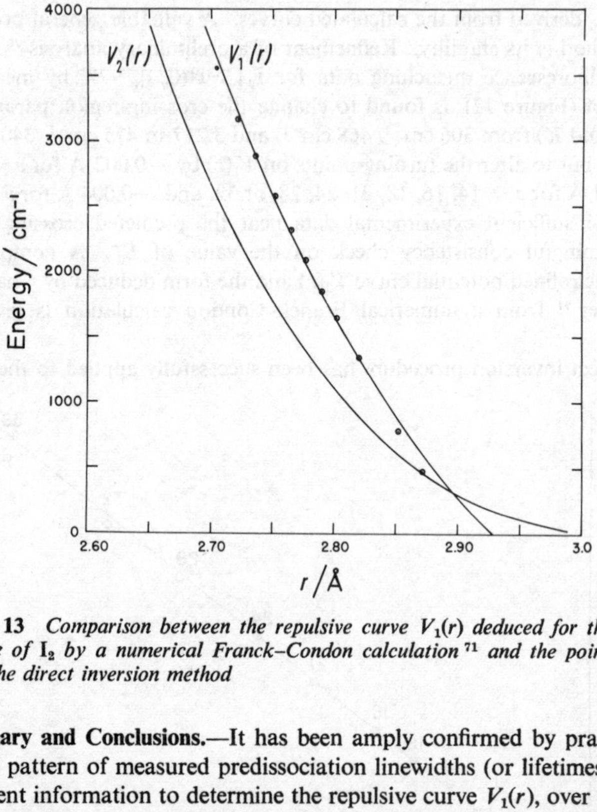

Figure 13 *Comparison between the repulsive curve $V_1(r)$ deduced for the $B\ ^3\Pi(0_u^+)$ state of I_2 by a numerical Franck–Condon calculation [71] and the points obtained by the direct inversion method*

Summary and Conclusions.—It has been amply confirmed by practical tests that a pattern of measured predissociation linewidths (or lifetimes) contains sufficient information to determine the repulsive curve $V_1(r)$, over the predissociation region, and also the interaction potential H_{12}^0. The lineshape envelope, as illustrated in Figures 5, 8, 9, and 10, suffices to determine the predissociation type. The frequency of oscillations within this envelope depends, in the normal non-diabatic case, on the rate of divergence of the two potential curves above the crossing point. Similar fluctuations in the case of predissociation from the upper adiabatic curve $V_+(r)$ depend on the shape of $V_2(r)$. Subsidiary fluctuations with J are also predicted, at a frequency roughly governed according to equations (153) and (164) by the displacement between the crossing point and the bound-state potential minimum. An analytical characterization of these fluctuations has led to development of the direct inversion procedure described on pp. 508—509. The interpretation of a well-developed predissociation pattern in cases $c^{\pm}$ is therefore now a routine matter.

Preliminary attention has also been directed towards the concurrent level shift. Specifically, at energies below the crossing point one may expect to observe shifts which are several magnitudes larger than the linewidth. However, it is not yet clear whether these should be interpreted as Fano shifts in the sense of equation (91), or simply as adiabatic shifts due to an avoided curve crossing. The sign of such shifts is negative for crossings of the + type, and may be expected by the adiabatic interpretation to be positive in the opposite case. At energies above the crossing point the level shift is predicted to oscillate about a mean value zero in such a way that within the validity of a Landau–Zener model a small positive additional term is predicted in the rotational constant.

Turning to the future, one may expect to see the direct connection between perturbations and predissociations of types *a* and *b* more fully exploited, particularly when the whole spectrum is substantially discrete. A more challenging situation for the theoretician arises when the predissociation is strong in the spectroscopic sense that all discrete structure appears lost, but weak to the extent that the linewidth is small compared with the vibrational spacing. Here one might hope that analysis of low-energy level shifts or possible perturbations might lead, *via* the appropriate repulsive curve, to the prediction and detection of fragmentary remains of the spectrum at the oscillation minima in Figures 8, 9, and 10.

Another possibility is that spectroscopic predictions of strong predissociation into states which correlate with ground-state atoms might stimulate the observation of such resonances by scattering techniques, as a complement to the achievements of Bernstein,[2] Stwalley *et al.*,[3] and Schutte *et al.*[4] in the somewhat simpler shape resonance field. A possible example might be the IBr system,[95, 96] but there may be others which are preferable on experimental grounds.

4 Appendix: The Airy Function Overlap Integral

The integral

$$I = \int_{-\infty}^{\infty} \mathrm{Ai}[\gamma(r - c)]\, \mathrm{Ai}[\gamma'(r - c')]\mathrm{d}r \qquad (175)$$

with $\gamma > \gamma'$, may be evaluated by use of the integral representation [40]

$$\mathrm{Ai}(z) = \frac{1}{\pi}\int_{0}^{\infty} \cos\left(\frac{1}{3}u^3 + uz\right)\mathrm{d}u = \frac{1}{2\pi}\int_{-\infty}^{\infty} \exp\left[\frac{\mathrm{i}u^3}{3} + \mathrm{i}uz\right]\mathrm{d}u \qquad (176)$$

Thus

$$I = \frac{1}{(2\pi)^2}\int_{-\infty}^{\infty}\int_{-\infty}^{\infty}\int_{-\infty}^{\infty}\exp\left[\frac{i}{3}(u^3+v^3)+i(\gamma u+\gamma' v)r - i(u\gamma c + v\gamma' c')\right]drdudv$$

$$= \frac{1}{2\pi}\int_{-\infty}^{\infty}\int_{-\infty}^{\infty}\exp\left[\frac{i}{3}(u^3+v^3) - i(u\gamma c + v\gamma' c')\right]\delta(\gamma u + \gamma' v)dudv$$

$$= \frac{1}{2\pi\gamma}\int_{-\infty}^{\infty}\exp\left\{\frac{i}{3}[1-(\gamma'/\gamma)^3]v^3 - i(c'-c)\gamma' v\right\}dv$$

$$= (\gamma^3 - \gamma'^3)^{-\frac{1}{3}}\,\mathrm{Ai}\left[\frac{\gamma\gamma'(c-c')}{(\gamma^3-\gamma'^3)^{\frac{1}{3}}}\right] \tag{177}$$

Here the identity [35]

$$\int_{-\infty}^{\infty}\exp ikr\,dr = 2\pi\,\delta(k) \tag{178}$$

has been used to obtain the second line of equation (177).

The Reporter is grateful for the hospitality of the Theoretical Chemistry Institute, University of Wisconsin, during the preparation of this Report. He wishes particularly to acknowledge the interest and advice of Professor R. B. Bernstein. Helpful correspondence with Professors F. Fiquet-Fayard, H. Lefebvre-Brion, R. J. LeRoy, and W. C. Stwalley, and Drs. A. S. Dickinson and J. N. L. Connor, is also gratefully acknowledged.

8

Rotational Structure in the Rydberg Series of Diatomic Molecules

BY J. W. C. JOHNS

1 Introduction

Absorption to Rydberg states of molecules generally lies in the far-u.v. – in the region beyond about 2000 Å. In recent years the resolution attained by large vacuum spectrographs has improved much and it is now possible to resolve spectral lines as close as 1 cm^{-1} at 800 Å. Although this may not be regarded as high resolution by spectroscopists used to working at longer wavelengths, it does nevertheless allow the study of rotational fine structure of relatively simple molecules in the far-u.v. As a consequence, there has been an increase in interest in the detailed study of Rydberg series of molecules such as H_2, N_2, NO, and even some simple polyatomic molecules.

For the purpose of the present article we will confine our attention to diatomic molecules and, moreover, to those where fine structure analyses have been made for relatively extensive Rydberg series. The aim will be to present the theory of the rotational structure of transitions involving these series and then to describe how this theory has been used in recent years to gain information on the properties of the molecular and ionic species involved.

We will define a Rydberg state as one in which an electron has been promoted to an orbital with principal quantum number, *n*, at least one greater than that of the outermost valence orbitals. In many respects this is not a very useful definition since *n* may be assigned different values according as one starts from a united-atom approximation or from a separated-atom approximation. Since the united-atom approximation would seem to be more appropriate for highly excited Rydberg states we will use it here despite the fact that it may not be realistic at low values of *n*. As is well known, the energy of such a Rydberg state can be expressed quite well by the simple formula:

$$E = \text{I.P.} - \frac{R}{(n - \delta)^2} \qquad (1)$$

where I.P. is the ionization potential of the molecule, *R* is the Rydberg constant, and δ is the quantum defect. The Rydberg orbital will be characterized not only by *n* but also by the orbital angular momentum quantum number *l*. Just as in atoms, the quantum defect becomes smaller as *l* increases and Walsh [1] has pointed out that, for hydrides of elements in the first row of the

[1] A. D. Walsh, *J. Phys. Radium*, 1954, **15**, 501.

Periodic Table, δ has values of ≈ 1.0, ≈ 0.6, and ≈ 0.1 for s ($l = 0$), p ($l = 1$), and d ($l = 2$) orbitals, respectively. For second-row hydrides these values are increased by about one and become still larger for subsequent rows. Rydberg series observed at low resolution have often been assigned on the basis of observed values of the quantum defects. This procedure can, however, be dangerous as it is clearly possible to change δ by unity simply by changing the value of n. It is much better to rely on a fine-structure analysis, which often makes it possible to determine l unambiguously.

In each Rydberg state the molecule can be regarded as being made up of a core and an outer electron in a Rydberg orbital. The core will be in an electronic state, not necessarily the lowest, of the ion and the Rydberg orbital can be described by values of n and l. For an atom this is sufficient, but for a molecule the degeneracy associated with values of l greater than zero will be removed by the non-spherical symmetry of the core, resulting in $l + 1$ distinct electronic states.

The orbital angular momentum of the Rydberg electron will have a component, λ, along the internuclear axis which can take on the values 0, ± 1, $\pm 2, \ldots \pm l$. This angular momentum must then be combined with the angular momentum of the core in order to obtain the total component of the orbital angular momentum along the internuclear axis. This total is denoted by Λ and according as $|\Lambda| = 0, 1, 2, 3, \ldots$ we have $\Sigma, \Pi, \Delta, \Phi, \ldots$ electronic states. In the second column of Table 1 we list some possible values of l and λ. States which arise from common values of n and l will usually lie close together and they can be said to form Rydberg 'complexes'. In column 3 of the Table we give the resulting molecular states in the trivial situation where the core has no orbital angular momentum, and in column 4 the states expected when the core has one unit of orbital angular momentum, *i.e.* it has Π symmetry. In general, however, the situation is more complicated

Table 1 *Rydberg series of diatomic molecules*

l	*Rydberg electron* $nl\lambda$	*Core*[a] $= \Sigma_g^+$	*Core*[a] $= \Pi_g$
0	$ns\sigma_g$	Σ_g^+	Π_g
1	$np\sigma_u$	Σ_u^+	Π_u
	$np\pi_u$	Π_u	$\Sigma_u^+, \Sigma_u^-, \Delta_u$
2	$nd\sigma_g$	Σ_g^+	Π_g
	$nd\pi_g$	Π_g	$\Sigma_g^+, \Sigma_g^-, \Delta_g$
	$nd\delta_g$	Δ_g	Π_g, Φ_g
3	$nf\sigma_u$	Σ_u^+	Π_u
	$nf\pi_u$	Π_u	$\Sigma_u^+, \Sigma_u^-, \Delta_u$
	$nf\delta_u$	Δ_u	Π_u, Φ_u
	$nf\phi_u$	Φ_u	Δ_u, Γ_u

[a] For heteronuclear molecules the g and u labels should be disregarded. For homonuclear molecules the core has been assumed to be of g symmetry even though either g or u is possible.

than shown here because we have ignored electron spin. With a singlet core, in some sense the simplest case, there will be doublet Rydberg states and with a doublet core there will be singlet and triplet Rydberg states, thus doubling the number of states listed in Table 1. It is fortunate that singlet and triplet states are not often observed at one and the same time. It is much more usual for only one set, those with the same multiplicity as the combining state, to be observed (see, however, discussion of HCl in Section 8).

When the various states of a complex come relatively close together compared to BJ, the orbital angular momentum becomes uncoupled from the internuclear axis, Λ ceases to be a good quantum number, and we have a transition from Hund's case (*a*) or case (*b*) to case (*d*). An alternative way to look at this phenomenon is to regard it as a simple perturbation – the various components of the complex become mixed (perturb each other) as a result of rotation–electronic interaction. It is this mixing which gives the rotational structure of many Rydberg transitions an unfamiliar appearance and has in the past been responsible for erroneous conclusions being made as to the geometrical structure of the molecule in the Rydberg states involved.

Up to the present time, detailed studies of the fine structure of Rydberg series have been limited to situations in which the ion core has had no orbital angular momentum. To be sure, Rydberg series have been observed in which the core does have angular momentum but, for various reasons, detailed analyses have not yet been completely successful. This article is thus concerned almost exclusively with situations which can be described by column 3 of Table 1. Situations described by column 4 of that Table will undoubtedly be studied in greater detail in the near future.

Section 2 describes the theory of the interaction which occurs amongst the components of a Rydberg complex using a singlet *p*-complex as an example. The remaining sections describe examples of observed spectra taken from the literature.

2 Theory of *l*-Uncoupling

The rotational Hamiltonian for a diatomic molecule can be written according to Van Vleck,[2] as:

$$H_{\text{rot}} = B(\boldsymbol{J} - \boldsymbol{L} - \boldsymbol{S})^2 \tag{2}$$

where B is the rotational constant, and $\boldsymbol{J}$, $\boldsymbol{L}$, and $\boldsymbol{S}$ are respectively the total, orbital, and spin angular momentum operators. Equation (2) can be expanded to give

$$H_{\text{rot}} = B[(J^2 - J_z^2) + (L - L_z^2) + (S - S_z^2) - (J^+L^- + J^-L^+) - (J^+S^- + J^-S^+) + (L^+S^- + L^-S^+)] \tag{3}$$

where

$$J^\pm = (J_x \pm \mathrm{i}J_y),\ L^\pm = (L_x \pm \mathrm{i}L_y),\ \text{and}\ S^\pm = (S_x \pm \mathrm{i}S_y).$$

[2] J. H. Van Vleck, *Rev. Mod. Phys.*, 1951, **23**, 213.

The *p*-Complex.—It will be instructive to examine in detail the consequences of equation (3) in a simple case. We shall consider the situation where there is no core angular momentum, where $l = 1$ and where the spin can be ignored – *i.e.* a singlet *p*-complex. The matrix of H_{rot} can be set up using $|l, \lambda; J, M\rangle$ as base functions. Since we are not interested in the effects of electric or magnetic fields, M can be ignored. Noting further that λ can take on the values 0 and ± 1 we find:

$$H_{\text{rot}} = B\begin{bmatrix} J(J+1)+l(l+1)-2 & -\sqrt{l(l+1)}\sqrt{J(J+1)} & 0 \\ -\sqrt{l(l+1)}\sqrt{J(J+1)} & J(J+1)+l(l+1) & -\sqrt{l(l+1)}\sqrt{J(J+1)} \\ 0 & -\sqrt{l(l+1)}\sqrt{J(J+1)} & J(J+1)+l(l+1)-2 \end{bmatrix} \quad (4)$$

where the rows and columns have been labelled with the possible values of $\lambda \equiv \Lambda$, namely $-1, 0$, and $+1$. Contributions from electronic and vibrational terms must be added to this rotational Hamiltonian. We will, following Hill and Van Vleck,[3] take them to be:*

$$H_{\text{ev}} = T + C\Lambda^2 \quad (5)$$

Thus

$$H = H_{\text{ev}} + H_{\text{rot}} = \begin{bmatrix} T+C+BJ(J+1) & -B\sqrt{2}\sqrt{J(J+1)} & 0 \\ -B\sqrt{2}\sqrt{J(J+1)} & T+B[J(J+1)+2] & -B\sqrt{2}\sqrt{J(J+1)} \\ 0 & -B\sqrt{2}\sqrt{J(J+1)} & T+C+BJ(J+1) \end{bmatrix} \quad (6)$$

This matrix can be factored by taking suitable sum and difference wavefunctions, when one obtains:

$$H = \begin{bmatrix} T+C+BJ(J+1) & 0 & 0 \\ 0 & T+B[J(J+1)+2] & -2B\sqrt{J(J+1)} \\ 0 & -2B\sqrt{J(J+1)} & T+C+BJ(J+1) \end{bmatrix} \quad (7)$$

The wavefunctions here may be labelled 1^-, 0^+, and 1^+ (or Π^-, Σ^+, and Π^+) where the + refers to series of levels which have + parity for even J and the

* The quadratic dependence of the energy on Λ cannot of course be verified for a *p*-complex for which $\Lambda = 0$ or 1 only. It has, however, been observed in several higher Rydberg complexes, as will be described later.

[3] E. L. Hill and J. H. Van Vleck, *Phys. Rev.*, 1928, **32**, 250.

– refers to levels which have – parity for even J. The energies associated with the Π^- levels are given simply by:

$$F(J) = T + C + BJ(J + 1) \tag{8}$$

and these levels are unaffected by the l-uncoupling interaction. The energies associated with the other component, Π^+, of the Π state and with the Σ state depend on the separation, C, between the Σ and Π states, and are given by the eigenvalues of the 2×2 block of equation (7). These energies are given by:

$$T(J) = T + C/2 + B + BJ(J + 1) \pm \sqrt{(C - 2B)^2/4 + 4B^2J(J + 1)} \tag{9}$$

Several comments may be made regarding equations (4)—(9).

The expressions are identical to those which describe the energies of a $^3\Sigma$ electronic state. This is hardly surprising since it matters little from the point of view of the mathematical formulation whether there is one unit of orbital angular momentum or one unit of spin angular momentum. Indeed, as we shall see later, a transition from a $^1\Sigma^+$ state to a p-complex can be identical in appearance to a transition from a $^1\Sigma^+$ state to a $^3\Sigma^-$ state. The reader may find it instructive to compare the expressions (4)—(9) given above with those given by Hougen [4] (pages 13 and 14).

The simple expressions (8) and (9) will rarely fit experimental data as well as they can be measured. Various extensions have to be made to the theory, just as has to be done for the $^3\Sigma$ state, so that experimental data can be fitted accurately. The most obvious extension is to allow for the effects of centrifugal distortion. Since this is well understood we shall, for the moment, simply assume that the model has accounted for it in a proper manner. Nevertheless, modifications are still needed and these modifications sometimes give interesting insights into perturbations which may be affecting the complex being studied. The simplest approach has been to allow the various components of a complex to have different effective rotational constants, B_{eff} (see for example Ginter [5]). The off-diagonal matrix elements may then be computed either by using averages of the appropriate diagonal B_{eff}'s or by using yet other independent B_{eff}'s. For d and higher complexes such a procedure can lead to an embarrassing number of parameters to which little physical significance can be attached. An alternative approach is to use only one B_{eff} for each symmetry class [*i.e.* one for each block of equation (7)] and to allow for small variations in the off-diagonal matrix elements by multiplying them by a factor $(1 + \gamma)$.* The parameter γ can sometimes

* The introduction of γ follows quite closely the treatment often given for the $^3\Sigma$ state. It turns out that the modification of off-diagonal matrix elements by a factor $(1 - \gamma/2B)$ has exactly the same effect on the energy levels as the introduction of the spin rotation term $\gamma\, N{\cdot}S$ into the Hamiltonian (see Hougen [4]).

[4] J. T. Hougen, 'The Calculation of Rotational Energy Levels and Rotational Line Intensities in Diatomic Molecules', NBS Monograph 115, U.S. Dept. of Commerce, Washington, 1970.

[5] M. L. Ginter, *J. Chem. Phys.*, 1966, **45**, 248.

be interpreted as a result of a perturbation by a state or complex with another value of l. As a consequence l ceases to be a good quantum number and may either increase from the expected integral value if the perturbing state has a higher value of l or decrease if the perturbing state has a lower value of l.

The derivation given above assumes, as does that of the equations for a $^3\Sigma$ state, that the separation of the two components of the complex is small compared with vibrational spacings and to separations from other electronic states. For light molecules this is certainly true in the $^3\Sigma$ situation but it may not be true for low values of n in a Rydberg series, when the separation C may be large and where other, perhaps non-Rydberg, states may be close by.

At low values of n, the splitting C between Σ and Π components is often large, and the eigenvalues of equation (7) may be obtained by perturbation theory. The result is to modify the diagonal energies of the Σ^+ and Π^+ levels by an amount $\Delta\nu$, as defined in equation (10). In other words the two compo-

$$\Delta\nu = \pm\frac{2B^2l(l+1)J(J+1)}{|C|} \qquad (10)$$

nents Π^+ and Π^- of the Π state can be regarded as having two effective B values which differ by $2B^2l(l+1)/C$. This is, of course, just the Λ-doubling of the Π state and the expression is that for Van Vleck's 'case of pure precession'. As Mulliken and Christy [6] have pointed out, equation (10) often comes surprisingly close to the truth even for low-lying non-Rydberg states.

At high values of n, C becomes small and the mixing between Σ^+ and Π^+ becomes more complete. In the limit when $C = 0$ we have Hund's case (d). The effect this has on the energy levels can be seen by rewriting the matrix of equation (7) in a case (d) basis, which may be obtained with the aid of the transformation matrix:

$$1/\sqrt{2J+1}\begin{vmatrix} \sqrt{2J+1} & 0 & 0 \\ 0 & \sqrt{J} & \sqrt{J+1} \\ 0 & \sqrt{J+1} & -\sqrt{J} \end{vmatrix} \qquad (11)$$

which, when $C = 0$, leads to the diagonal matrix:

$$H = \begin{vmatrix} T + BJ(J+1) & 0 & 0 \\ 0 & T + B\dfrac{(2J^3 - J^2 - J)}{(2J+1)} & 0 \\ 0 & 0 & T + B\dfrac{(2J^3 + 7J^2 + 7J + 2)}{(2J+1)} \end{vmatrix} \qquad (12)$$

If the substitutions $R = J$, $R = J + 1$, and $R = J - 1$ are made to the three

[6] R. S. Mulliken and A. Christy, *Phys. Rev.*, 1931, **38**, 87.

diagonal elements respectively, it can readily be seen that each element is given by:

$$E = T + BR(R + 1) \tag{13}$$

which is the expected case (*d*) energy expression and is just the energy of the rotating ion core.

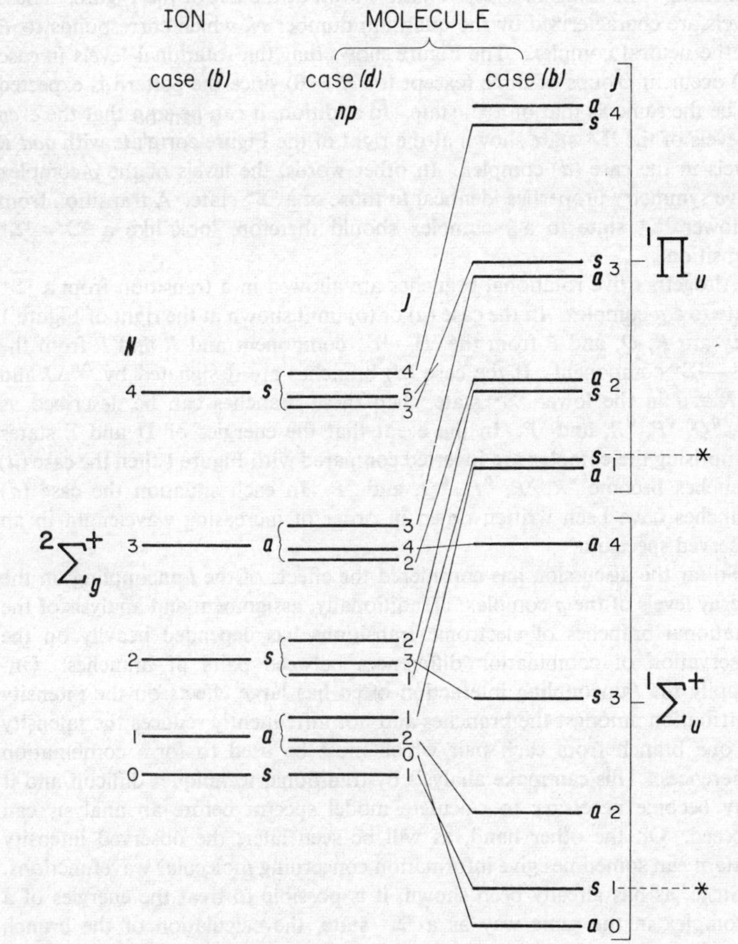

Figure 1 *Correlation of rotational levels of an ion in a* $^2\Sigma^+$ *state (at the left) with those of the* $^1\Sigma^+$ *and* $^1\Pi$ *states of a p-complex (at the right). Pure case (d) levels are shown in the centre. The two levels marked with asterisks are accessible from* $J = 0$ *level of a* $^1\Sigma^+$ *state as discussed in Section 4*

(Adapted with permission from *J. Mol. Spectroscopy*, 1972, **41**, 425)

Figure 1 illustrates schematically the transition from case (*a*) or case (*b*), represented by the matrix of equation (7), to case (*d*), represented by the matrix of equation (12). On the right-hand side of the Figure are shown the non-degenerate levels of a $^1\Sigma_u^+$ electronic state and the nearly degenerate levels of a $^1\Pi_u$ electronic state. The levels of the pure case (*d*) complex are shown in the centre of the Figure. The correlation between these two sets of levels is indicated. The levels of a $^2\Sigma_g^+$ ion are shown at the left of the Figure. These levels are characterized by the quantum number N, which corresponds to R in the neutral complex. The Figure shows that the rotational levels in case (*d*) occur in groups of three (except for $R = 0$) since the pattern is expected to be the same as that of a $^3\Sigma$ state. In addition, it can be seen that the *even* J levels of the $^1\Sigma^+$ state shown at the right of the Figure correlate with *odd* R levels in the case (*d*) complex. In other words, the levels of the *p*-complex have symmetry properties identical to those of a $^3\Sigma^-$ state. A transition from a lower $^1\Sigma^+$ state to a *p*-complex should therefore look like a $^3\Sigma^-$—$^1\Sigma^+$ transition.

Altogether five rotational branches are allowed in a transition from a $^1\Sigma^+$ state to a *p*-complex. In the case (*a*) or (*b*) limit shown at the right of Figure 1 they are R, Q, and P from the $^1\Pi$—$^1\Sigma^+$ component and R and P from the $^1\Sigma^+$—$^1\Sigma^+$ component. If the case (*d*) branches are designated by $^{\Delta R}\Delta J$ and if $R \equiv J$ in the lower $^1\Sigma^+$ state, then these branches can be described as ^SR, ^QQ, ^QP, ^QR, and ^OP. In the event that the energies of Π and Σ states comprising the complex are inverted compared with Figure 1 then the case (*d*) branches become ^SR, ^QP, ^QR, ^QQ, and ^OP. In each situation the case (*d*) branches have been written down in order of increasing wavelength in an observed spectrum.

So far the discussion has considered the effects of the *l*-uncoupling on the energy levels of the *p*-complex. Traditionally, assignment and analysis of the rotational branches of electronic transitions has depended heavily on the observation of combination differences between pairs of branches. Unhappily the *l*-uncoupling interaction often has large effects on the intensity distribution amongst the branches and not infrequently reduces the intensity of one branch from each pair which must be used to form combination differences. This can make analysis by traditional techniques difficult and it may become necessary to calculate model spectra before an analysis can succeed. On the other hand, as will be seen later, the observed intensity pattern can sometimes give information concerning molecular wavefunctions.

Since, as has already been shown, it is possible to treat the energies of a *p*-complex in the same way as a $^3\Sigma^-$ state, the calculation of the branch intensities of a p—$^1\Sigma^+$ transition can follow in the same way as for a $^3\Sigma^-$—$^1\Sigma^+$ transition. This calculation has been set out in some detail by Hougen [4] (pages 34—36, see also Watson [7]) and will not be repeated here. Hougen's results are:

[7] J. K. G. Watson, *Canad. J. Phys.*, 1968, **46**, 1637.

$$I(^{S}R) \propto [+\mu_{\parallel} + \mu_{\perp}]^2 (J + 1)(J + 2)/3(2J + 3) \quad (14a)$$

$$I(^{Q}Q) \propto [+\mu_{\perp}]^2 (2J + 1)/3 \quad (14b)$$

$$I(^{Q}P) \propto [+\mu_{\parallel}(J) - \mu_{\perp}(J - 1)]^2/3(2J - 1) \quad (14c)$$

$$I(^{Q}R) \propto [+\mu_{\parallel}(J + 1) - \mu_{\perp}(J + 2)]^2/3(2J + 3) \quad (14d)$$

$$I(^{O}P) \propto [+\mu_{\parallel} + \mu_{\perp}]^2 J(J - 1)/3(2J - 1) \quad (14e)$$

and are valid for pure case (*d*), *i.e.* $C = 0$. The quantities $\mu_{\parallel}$ and $\mu_{\perp}$ are matrix elements of the molecule-fixed components of the dipole-moment operator; they are defined as:

$$\mu_{\parallel} = \langle l', 0|\mu_z|l'', 0\rangle \quad (15a)$$

$$\mu_{\perp} = \langle l', 1|\mu_x + i\mu_y|l'', 0\rangle \quad (15b)$$

Since we are considering a transition to a *p*-complex, l' can be safely taken as 1. The value of l'' on the other hand may not be well defined. Let us consider a transition which involves the promotion of an electron from a σ-orbital to the *p*-orbital. The lower σ-orbital can be considered as a linear combination of *s, p, d, f*, . . . atom-like orbitals, of which we need only be concerned with the *s, p*, and *d* components. Values of I using $l'' = 0$, 1, and 2 (*i.e. s, p*, and *d*) have been calculated from equations (14) and (15) and are presented in Table 2. From this Table it can be seen that if the transition is to be regarded as np—$s\sigma$ then the ^{S}R and ^{O}P branches will vanish. In this situation no useful combination differences can be formed from the remaining three branches which, if $B' \approx B''$, will all be bunched together near the band origin, giving the transition a line-like appearance at low resolution. If lines are resolved one will find that the strong lines at the high-frequency end of the band are *P* lines and the strong lines at the low-frequency end of the band are *R* lines!

Table 2 *Intensity factors at* $J = 10$ *for an* $np \leftarrow {}^1\Sigma^+$ *transition*

	^{S}R	^{Q}Q	^{Q}P	^{Q}R	^{O}P
$l'' = 0$	0	7	6.3	7.7	0
$l'' = 1$	1.9	7	1.4	2.1	1.6
$l'' = 2$	17.2	7	2.1	1.5	14.2

Of course, the results of Table 2 are only valid for pure case (*d*) whereas most real examples will be intermediate between case (*a*) and case (*d*). Computed band profiles for such an intermediate case are presented in Figure 2. Each spectrum in the Figure was computed with identical molecular constants and the separation of the Σ and Π components was set at about $10B$.

The preceding discussion and Figure 2 have shown that the intensities of the rotational branches are strongly affected by the *l*-uncoupling interaction. Just which branches are enhanced in intensity and which are reduced depends strongly on the effective *l* values of the participating states and such information should, therefore, be available from studies of line intensities in Rydberg transitions.

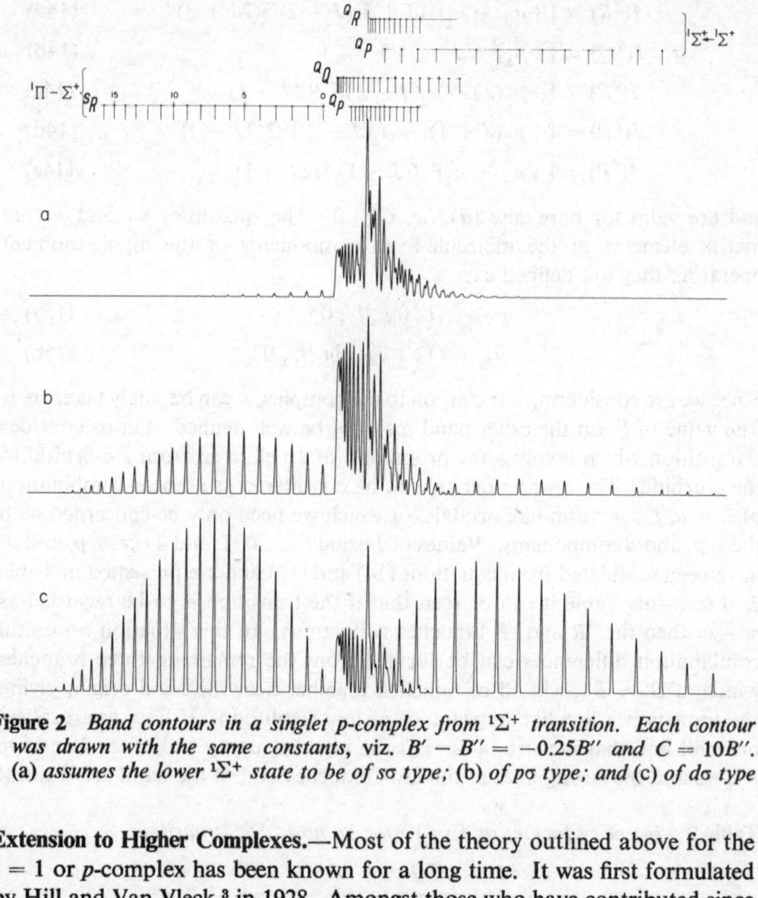

Figure 2 *Band contours in a singlet p-complex from* $^{1}\Sigma^{+}$ *transition. Each contour was drawn with the same constants,* viz. $B' - B'' = -0.25B''$ *and* $C = 10B''$. (a) *assumes the lower* $^{1}\Sigma^{+}$ *state to be of* $s\sigma$ *type;* (b) *of* $p\sigma$ *type; and* (c) *of* $d\sigma$ *type*

Extension to Higher Complexes.—Most of the theory outlined above for the $l = 1$ or *p*-complex has been known for a long time. It was first formulated by Hill and Van Vleck [3] in 1928. Amongst those who have contributed since that time are Chiu [8] and Kovács,[9] who has also given intensity formulae valid for intermediate coupling but only for the $l'' = 0$ example discussed above. MacDonald,[10] Chiu, Kovács and Ginter [5] are amongst those who have treated the *d*-complex. These workers have restricted attention to singlet systems. More recently Johns and Lepard [11] have written a computer program which can construct the relevant Wang-type matrices [*cf.* equation (7)] for any value of *l* and for any multiplicity. In addition they expanded the effective nuclear potential, composed of the vibrational and rotational Hamiltonian, as a Taylor series about the equilibrium internuclear separation,

[8] Y. N. Chiu, *J. Chem. Phys.*, 1964, **41**, 3235.
[9] I. Kovács, 'Rotational Structure in the Spectra of Diatomic Molecules', Elsevier, New York, 1969.
[10] J. K. L. MacDonald, *Proc. Roy. Soc.*, 1932, **A138**, 183.
[11] J. W. C. Johns and D. W. Lepard, to be published.

and obtained an effective rotational Hamiltonian for a given vibrational state which is valid to third order (*i.e.* terms in the sixth power of the angular-momentum operators are included). In other words, centrifugal distortion effects are included in the formalism in a consistent manner. Actually there are two programs: one calculates line positions and intensities from an assumed set of parameters and can be made to plot a band profile, and the other accepts assigned line positions and executes a non-linear least-squares fit to improve a set of initial molecular parameters. The profiles of Figure 2 were obtained from the first of these programs, and results obtained from both programs will be used from time to time in the later parts of this article.

3 Rydberg Series of N_2

The *np* Series.—In 1938 Worley and Jenkins [12] first reported a Rydberg series in N_2 which converged to a limit just below 800 Å. It was evident that the limit of this series represented the first ionization potential of N_2, that is the formation of the N_2^+ ion in its ground electronic state, $X\,^2\Sigma_g^+$. With the benefit of higher resolution, Ogawa and Tanaka [13] suggested that an observed 'doublet' structure of the series could be explained if it was assumed that two series of states* were involved, *viz.*

$$\ldots(\pi_u)^4(3\sigma_g)\ldots np\sigma_u\ {}^1\Sigma_u^+ \tag{16}$$

$$\ldots(\pi_u)^4(3\sigma_g)\ldots np\pi_u\ {}^1\Pi_u \tag{17}$$

Since the ground state of N_2 is $X\,^1\Sigma_g^+$, transitions to the two series are allowed and should, therefore, be strong. Other Rydberg series can be formed by excitation to *s* and *d* orbitals but these will give rise to *g* molecular states so that, in absorption from the ground state, the two series indicated in equations (16) and (17) are the only allowed series converging on the ground state of N_2^+.

Much more recently, with the benefit of even higher resolution, Carroll and Yoshino [14] were able to confirm the tentative conclusion given by Ogawa and Tanaka that there were indeed two series; the one at slightly longer wavelengths consisted of bands with *P*, *Q*, and *R* branches whereas the one at shorter wavelengths consisted of bands with only *P* and *R* branches. In the last year or so detailed studies have been made of the rotational fine structure of the bands by Carroll and Yoshino [15] and by Carroll [16] and simultaneously by Johns and Lepard.[11]

* While Ogawa and Tanaka [13] concluded correctly that there are indeed two series, they actually observed only the *R* and *Q* heads of the Π—Σ transitions. The nearly headless Σ—Σ transitions are almost invisible at low resolution.

[12] R. E. Worley and F. A. Jenkins, *Phys. Rev.*, 1938, **54**, 305.
[13] M. Ogawa and Y. Tanaka, *Canad. J. Phys.*, 1962, **40**, 1593.
[14] P. K. Carroll and K. Yoshino, *J. Chem. Phys.*, 1967, **47**, 3073.
[15] P. K. Carroll and K. Yoshino, *J. Phys.* (*B*), 1972, **5**, 1614.
[16] P. K. Carroll, *J. Chem. Phys.*, 1973, **58**, 3597.

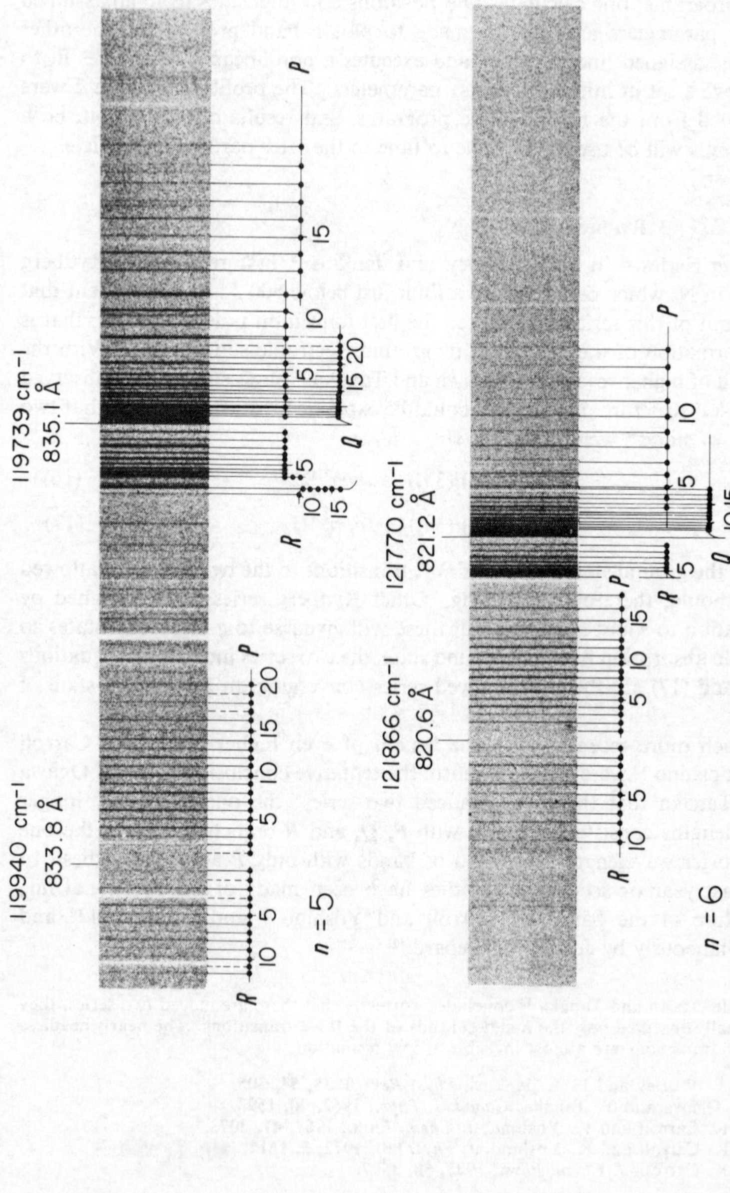

Figure 3 *Spectrograms of the $n = 5$ (top) and $n = 6$ (bottom) p-complexes in the Worley–Jenkins series of* N_2

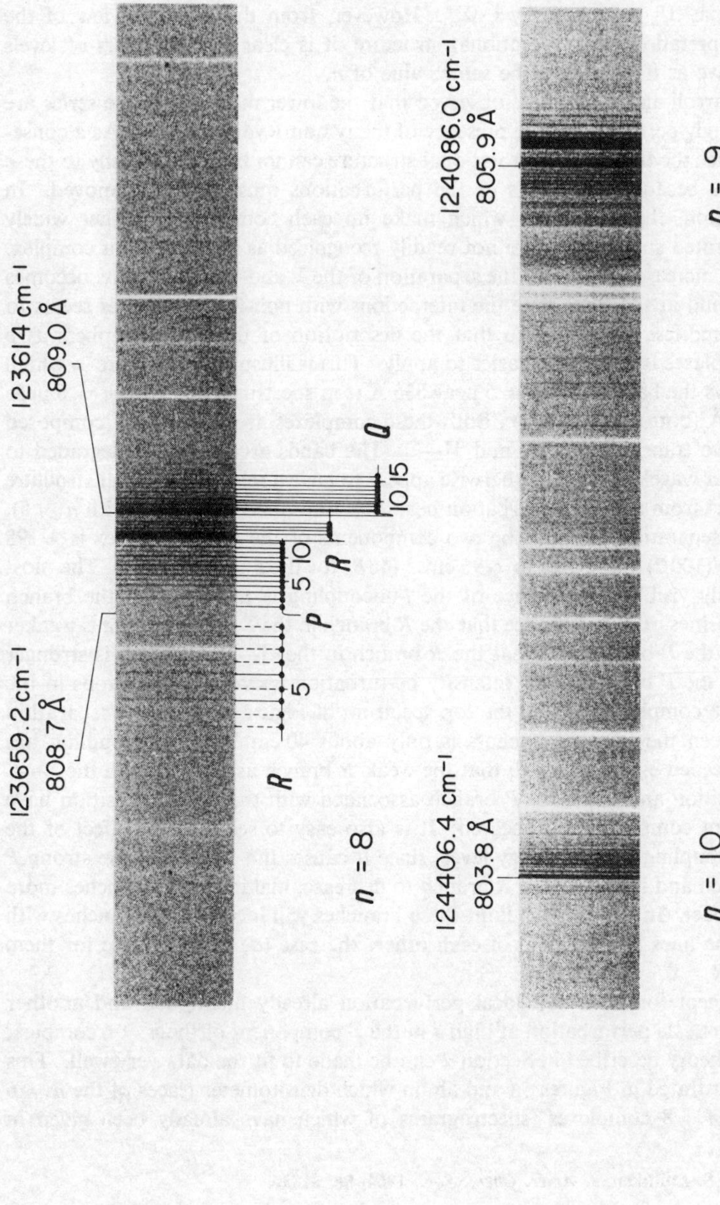

Figure 4 *Spectrograms of the n = 8 (top) and several higher (bottom) p-complexes in the Worley–Jenkins series of* N_2

For convenience in the following description the pairs of bands (Σ—Σ and Π—Σ) which go together to form each complex have been assigned the same value of n despite the fact that, as Mulliken [17] has pointed out, the nodal behaviour of the wavefunctions requires that each pair of bands be described as $np\pi_u\ {}^1\Pi_u$ and $(n+1)p\sigma_u\ {}^1\Sigma_u^+$. However, from the point of view of the interpretation of the rotational structure, it is clear that the pairs of levels behave as if they have the same value of n.

Carroll and Yoshino [15] observed that the lower members of the series are strongly perturbed by the presence of many non-Rydberg states. As a consequence the theory of the rotational structure cannot be applied easily to these states because the effects of the perturbations must first be removed. In addition, the two levels which make up each complex are rather widely separated so that they are not readily recognized as members of a complex. As n increases, however, the separation of the Σ and Π components becomes less and at the same time the interactions with non-Rydberg states seems to become less important so that the description of the series members as p complexes is easier and easier to apply. This is illustrated in Figure 3, which shows the bands with $n = 5$ near 835 Å (top spectrum) and with $n = 6$ near 821 Å (bottom spectrum). Both these complexes are very clearly composed of two transitions, Σ—Σ and Π—Σ. The bands are somewhat degraded to longer wavelengths but otherwise appear to have a relatively normal structure (apart from a local perturbation near $J = 10$ in the Π—Σ band with $n = 5$). The separation between the two components of the $n = 5$ complex is ≈ 198 cm^{-1} ($100B$) and drops to ≈ 96 cm^{-1} ($48B$) for the $n = 6$ complex. The most readily visible consequence of the l-uncoupling is the effect on the branch intensities. It is easy to see that the R branch in the Σ—Σ transition is weaker than the P branch and that the R branch in the Π—Σ transition is stronger than the P branch. This intensity perturbation is even more obvious in the $n = 8$ complex shown in the top spectrum of Figure 4. Here the separation between the two components is only about 40 cm^{-1} and l-uncoupling has proceeded even further so that the weak R branch associated with the Σ—Σ transition and the weak P branch associated with the Π—Σ transition have all but completely disappeared. It is also easy to see here the effect of the l-uncoupling on the energy levels since it causes the spacing in the strong P branch and in the strong R branch to decrease, making these branches more compact. In the case (d) limit these branches will look like Q branches with all the lines piled on top of each other; the case (d) selection rule for them is $\Delta R = 0$.

Except for the small local perturbation already mentioned and another larger-scale perturbation at high J in the Σ component of the $n = 6$ complex, the theory described in Section 2 can be made to fit the data very well. This is illustrated in Figures 5a and 5b, in which densitometer traces of the $n = 6$ and $n = 8$ complexes (spectrograms of which have already been given in

[17] R. S. Mulliken, *J. Amer. Chem. Soc.*, 1964, **86**, 3183.

Figures 3 and 4) are compared with computer-simulated band profiles generated by the program described by Johns and Lepard.[11] It is easy to see that the agreement between the experimental and theoretical profiles is excellent. The results of the fine-structure analyses are presented in Table 3. Except for $n = 7$, taken from Carroll and Yoshino,[15] and for $n = \infty$, taken from Herzberg,[18] the results are from Johns and Lepard.[11] A few points are worthy of comment.

Table 3 *Constants obtained from analysis of the Worley–Jenkins Rydberg series of* N_2 [a]

n	Λ	ν_0 /cm^{-1}	B^+ [b] /cm^{-1}	B^- [b] /cm^{-1}	$D \times 10^5$ /cm^{-1}	γ	$\Delta\nu$/cm^{-1} (obs.)	$\Delta\nu$/cm^{-1} (calc.)
5	1	119 739.42(3)	1.9245(5)	1.9272(4)	2.1 (1)	−0.038 (3)	197.52	168
	0	119 936.94(4)						
6	1	121 770.69(3)	1.9153(3)	1.9247(4)	1.89(1)	−0.071 (1)	92.02	97
	0	121 862.71(3)						
7[c]	1	122 897.3	1.922	1.922	—	—	64.3	61
	0	122 961.5						
8	1	123 614.74(3)	1.9190(3)	1.9248(5)	1.91(1)	−0.0052(7)	40.38	41
	0	123 655.12(3)						
9	1	124 079.1 (4)	1.921 (2)	1.921 (2)	1.8[d]	≈0.0	25.1	29
	0	124 104.2 (1)						
∞[e]		125 666	1.922	1.922	0.6	—	—	—

[a] Standard deviations, obtained from the least-squares analyses, are indicated in parentheses and refer to the least significant digit quoted; [b] B^+ and B^-, the two effective B values, one for each symmetry class as described in Section 2; [c] Constants for $n = 7$ were obtained from ref. 15; [d] Assumed value; [e] Obtained from Herzberg.[18] The rotational constants are those of the ground state of N_2^+.

First, as expected for Rydberg series, the molecular constants become very much like those of the ion as n increases. This is hardly surprising since an electron in a highly excited Rydberg orbital can have little influence on the molecular binding. For $n = 9$ the observed B value is indistinguishable from that of the ion. On the other hand, it is a little surprising that the centrifugal distortion constants D seem to be significantly larger than that found in the ion.

Second, the quantity γ [the off-diagonal matrix elements are multiplied by an empirical factor $(1 + \gamma)$ in the theoretical approach described by Johns and Lepard [11]] is moderately large for low n but decreases as n increases, becoming too small to be measured at $n = 9$. The size of γ can be regarded as a measure of the degree of purity of the complex, and with $\gamma = 0$, as found for $n = 9$, one can say that l is exactly equal to unity and that there are no measurable interactions with other molecular states.*

* It should be pointed out that the zero value found for γ may be due to the lower quality of the data for $n = 9$. This is partly because of decreasing intensity and partly because the l-uncoupling results in the strong lines all being bunched near the centre of the complex, thus requiring higher resolution.

[18] G. Herzberg, 'Spectra of Diatomic Molecules', Van Nostrand, New York, 1950.

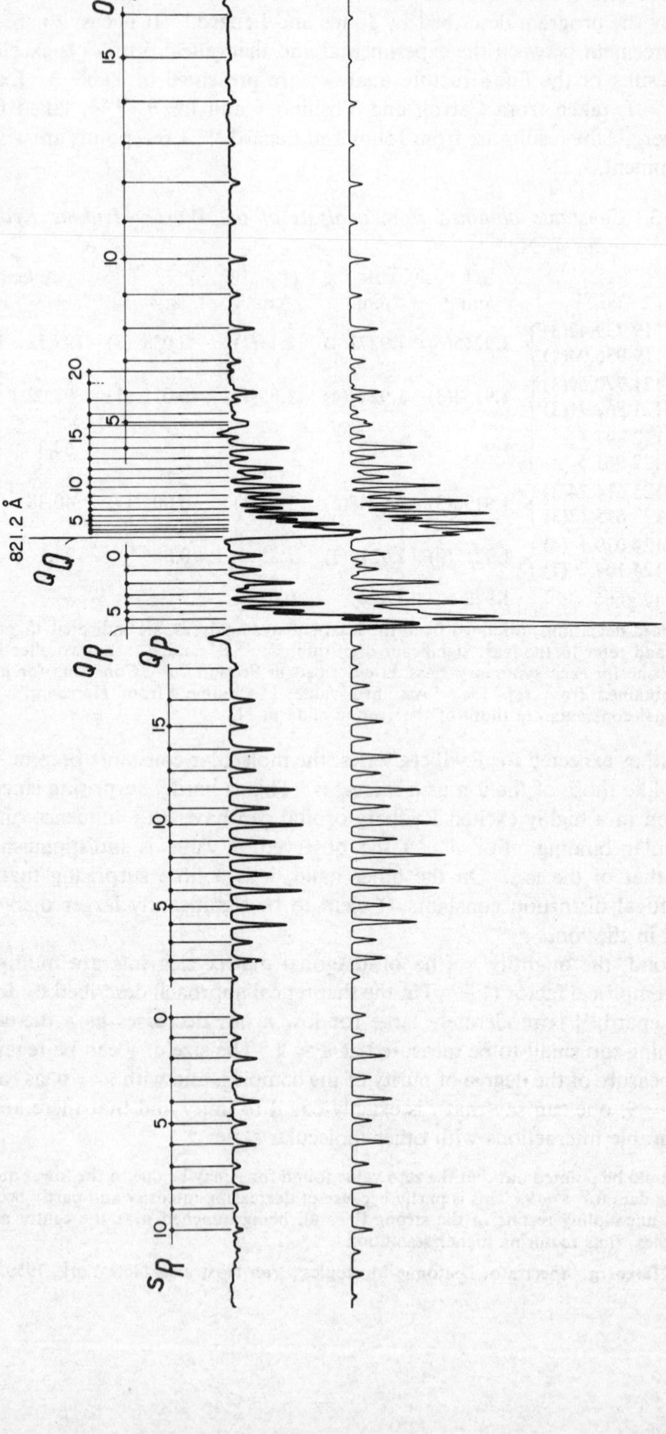

Figure 5a

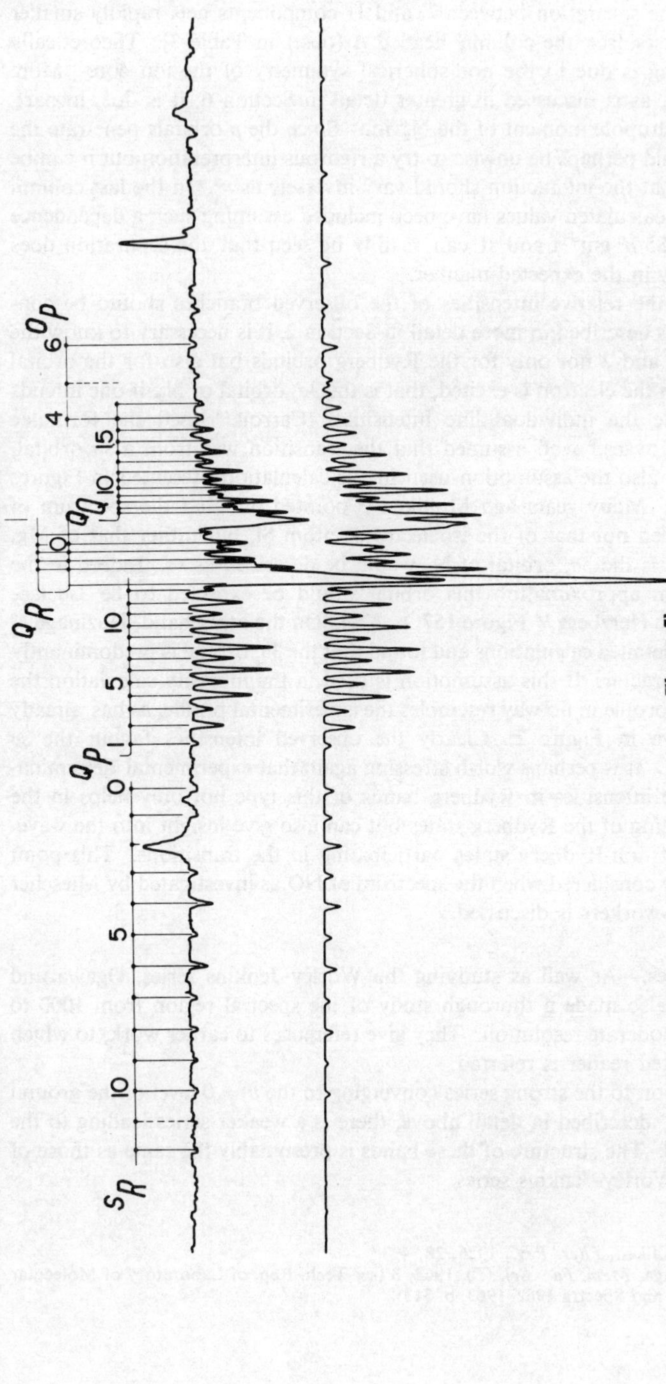

Figure 5b

Figures 5(a), (b) *Densitometer traces of the n = 6 (Figure 5a) and n = 8 (Figure 5b) complexes of N_2. In each case the lower of the two traces is a computer-generated plot obtained by means of the program described by Johns and Lepard*[11] *and using the parameters listed in Table 3*

Third, the separation between Σ and Π components gets rapidly smaller as n increases [see the column headed Δν(obs.) in Table 3]. Theoretically this splitting is due to the non-spherical symmetry of the ion core. More specifically, as is discussed in greater detail in Section 6, it is due, in part, to the quadrupole moment of the N_2^+ ion. Since the p-orbitals penetrate the core it would perhaps be unwise to try a rigorous interpretation but it can be shown [43] that the interaction should vary inversely as n^3. In the last column of Table 3 calculated values have been included assuming such a dependence ($\Delta\nu = 20\,885/n^3\ \text{cm}^{-1}$) and it can readily be seen that the separation does indeed vary in the expected manner.

Finally, the relative intensities of the observed branches should be considered. As described in more detail in Section 2, it is necessary to know the values of l and λ not only for the Rydberg orbitals but also for the orbital from which the electron is excited, that is the $3\sigma_g$ orbital of N_2, if one intends to calculate the individual line intensities. Carroll [16] used the formulae given by Kovács,[9] who assumed that the transition was from a $s\sigma$ orbital, and this is also the assumption used in the calculation presented in Figures 5a and 5b. Many years ago Mulliken [19] pointed out that the spectrum of N_2 resembled not that of the isoelectronic atom Si, but rather that of Mg. On this basis the $3\sigma_g$ orbital of N_2 would be described as $s\sigma$. Indeed in the united-atom approximation this orbital would be expected to be $3s\sigma$ (see for example Herzberg,[18] Figure 157, p. 329). On the other hand, Huzinaga [20] has made detailed calculations and found that the $3\sigma_g$ orbital is predominantly $3d\sigma_g$ in character. If this assumption is used in the intensity calculation the calculated profile in no way resembles the experimental profile, as has already been shown in Figure 2. Clearly the observed intensities favour the $s\sigma$ assignment. It is perhaps worth stressing again that experimental determination of the intensities in Rydberg bands of this type not only helps in the understanding of the Rydberg states but can also give insight into the wave-function of non-Rydberg states participating in the transitions. This point will also be considered when the spectrum of NO as investigated by Miescher and his co-workers is discussed.

Other Series.—As well as studying the Worley–Jenkins series, Ogawa and Tanaka [13] also made a thorough study of the spectral region from 1000 to 600 Å at moderate resolution. They give references to earlier work, to which the interested reader is referred.

In addition to the strong series converging to the $v = 0$ level of the ground state of N_2^+ described in detail above, there is a weaker series leading to the $v = 1$ level. The structure of these bands is presumably the same as those of the main Worley–Jenkins series.

[19] R. S. Mulliken, *Phys. Rev.*, 1926, **28**, 493.
[20] S. Huzinaga, *Mem. Fac. Sci.* (*B*), 1962, **3** (see Tech. Rep. of Laboratory of Molecular Structure and Spectra 1962–1963, p. 313).

Rydberg series have also been identified which lead to the $A\,^2\Pi_u$ ($v = 0$, 1, 2, 3) state of N_2^+ (Worley's series) and to the $B\,^2\Sigma_u^+$ ($v = 0$, 1) state (Hopfield's series). The situation is summarized in Figure 6, which has been taken from Ogawa and Tanaka [13] and which shows some of the observed states of N_2 together with those of N_2^+.

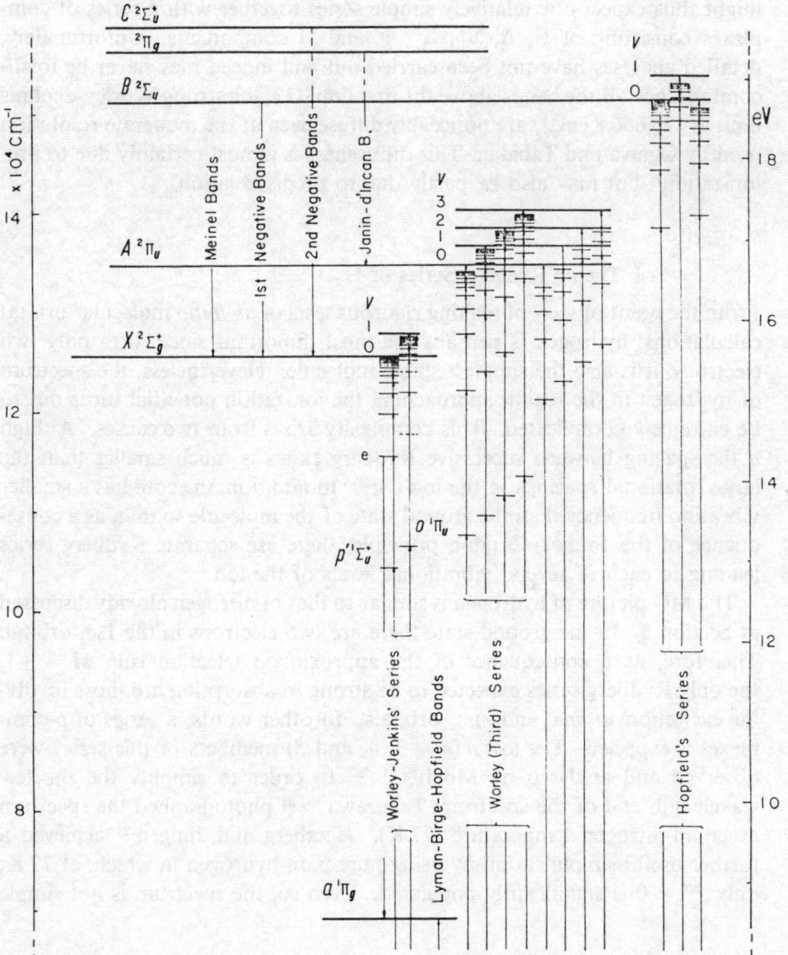

Figure 6 *Energy-level scheme for* N_2 *and* N_2^+
(Adapted with permission from *Canad. J. Phys.*, 1962, **40**, 1593)

The electron configuration of the series leading to $A\ ^2\Pi_u$ can be represented by:

$$(1\pi_u)^3\,(3\sigma_g)^2\,nl\lambda \tag{18}$$

and we have a $(\pi_u)^3$ or Π_i core and a situation corresponding to the configurations listed in column 4 of Table 1. In order to form states which can combine with the ground state, $(nl\lambda)$ can only be $(ns\sigma_g)$ or $(nd\sigma_g, \pi_g, \delta_g)$. We might thus expect one relatively simple series together with a series of complexes consisting of Π, Δ, Σ^+, Σ^-, Φ, and Π components. Unfortunately, detailed analyses have not been carried out and indeed may never be forthcoming since all the bands above the first limit (*i.e.* the strong Worley–Jenkins limit at 125 666.8 cm^{-1}) are noticeably diffuse even at the moderate resolution used by Ogawa and Tanaka. This diffuseness is almost certainly due to pre-ionization* but may also be partly due to predissociation.

4 The *np* Rydberg Series of H_2

From the point of view of making rigorous tests of *ab initio* molecular orbital calculations, hydrogen is perhaps the most important since, with only two electrons, it is also the simplest stable molecule. Nevertheless, the spectrum of hydrogen in the region approaching the ionization potential turns out to be extremely complicated. This complexity arises from two causes. At high n the spacing between successive Rydberg states is much smaller than the large rotational spacings in the ion core. In addition, the core has a smaller vibration frequency than the ground state of the molecule so that, as a consequence of the Franck–Condon principle, there are separate Rydberg series leading to each of several vibrational levels of the ion.

The MO picture of hydrogen is similar to that of nitrogen already discussed in Section 3. In the ground state there are two electrons in the $1s\sigma_g$ orbital. Therefore, as a consequence of the approximate selection rule $\Delta l = \pm 1$, the only Rydberg series expected to be strong in absorption are those involving excitation to $np\sigma_u$ and $np\pi_u$ orbitals. In other words, a series of *p*-complexes is expected. The low n ($n = 3$, 4, and 5) members of this series were observed and analysed by Monfils.[21, 22] In order to simplify the shorter-wavelength end of the spectrum, Takezawa [23, 24] photographed the spectrum at liquid-nitrogen temperature (77 K). Herzberg and Jungen [25] achieved a further useful simplification by using pure para-hydrogen in which, at 77 K, only $J'' = 0$ is significantly populated. Even so, the spectrum is not simple

* Following Herzberg [18] we use the term pre-ionization rather than auto-ionization.

[21] A. Monfils, *J. Mol. Spectroscopy*, 1965, **15**, 265.
[22] A. Mofils, *J. Mol. Spectroscopy*, 1968, **25**, 513.
[23] S. Takezawa, *J. Chem. Phys.*, 1970, **52**, 2575.
[24] S. Takezawa, *J. Chem. Phys.*, 1970, **52**, 5793.
[25] G. Herzberg and Ch. Jungen, *J. Mol. Spectroscopy*, 1972, **41**, 425.

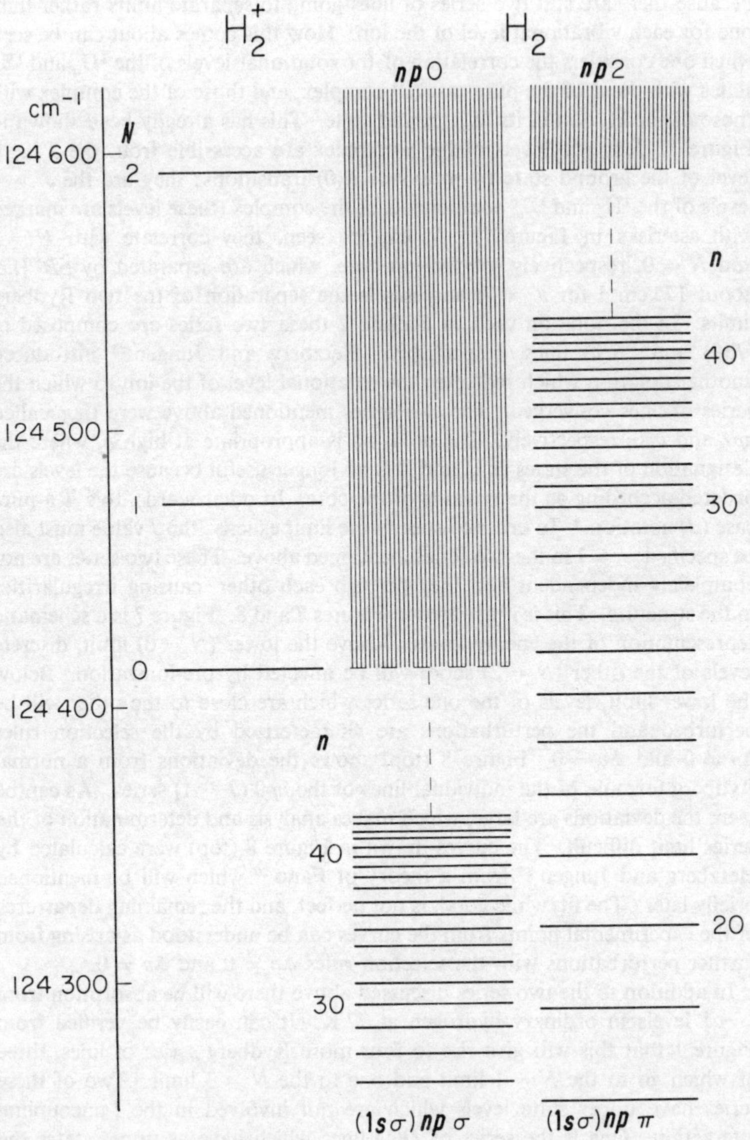

Figure 7 *Schematic representation of the energy levels of np0 and np2 Rydberg series in* H_2. *Note that each level of* H_2 *illustrated here has* $J = 1$
(Reproduced by permission from *J. Mol. Spectroscopy*, 1972, **41**, 425)

because there are still two series of lines going to separate limits rather than one for each vibrational level of the ion. How this comes about can be seen when one considers the correlation of the rotational levels of the $^1\Pi_u$ and $^1\Sigma_u^+$ states with those of the pure case (*d*) complex, and those of the complex with those of the H_2^+ ion in its $^2\Sigma_u^+$ ground state. This has already been shown in Figure 1. Just two levels of the *p*-complex are accessible from the $J'' = 0$ level of the ground state by means of $R(0)$ transitions: they are the $J' = 1$ levels of the $^1\Pi_u$ and $^1\Sigma_u^+$ components of the complex (these levels are marked with asterisks in Figure 1). As can be seen, they correlate with $N = 2$ and $N = 0$, respectively, of the ion core, which are separated by $6B'$ (*i.e.* about 175 cm^{-1} for $v' = 0$) and this is the separation of the two Rydberg limits. In the notation used in Section 2 these two series are composed of $^SR(0)$ and $^QR(0)$ lines, respectively. Herzberg and Jungen [25] introduced another notation which indicated the rotational level of the ion to which the series of lines converged. The two series mentioned above were thus called *np*2 and *np*0, respectively. The notation is appropriate at high *n*, where the designation of the states as Π and Σ is no longer useful because the levels are ordered according as the *N* value of the core. In other words, this is a pure case (*d*) notation.* In order to specify the limit exactly, the *J* value must also be specified: $J = 1$ in the two series mentioned above. These two series are not completely independent and they perturb each other, causing irregularities in the structure. This is illustrated in Figures 7 and 8. Figure 7 is a schematic representation of the energy levels. Above the lower ($N = 0$) limit, discrete levels of the other ($N = 2$) series will be affected by pre-ionization. Below the lower limit, levels of the one series which are close to the other will be perturbed and the perturbations are characterized by the selection rules $\Delta n \neq 0$ and $\Delta v = 0$. Figure 8 (top) shows the deviations from a normal Rydberg formula of the individual lines of the *np*0 ($J = 1$) series. As can be seen, the deviations are large, which makes analysis and determination of the series limit difficult. The curves drawn in Figure 8 (top) were calculated by Herzberg and Jungen [25] from a theory of Fano [26] which will be mentioned briefly later. The fit, while good, is not perfect, and the remaining departures of the experimental points from the curves can be understood as arising from further perturbations with the selection rules $\Delta n \neq 0$ and $\Delta v \neq 0$.

In addition to the two series discussed above there will be absorption from $J = 1$ levels in ordinary hydrogen at 77 K. It can easily be verified from Figure 1 that this will give rise to four more Rydberg series of lines, three of which go to the $N = 1$ limit and one to the $N = 3$ limit. Two of these series have upper state levels which are not involved in the *l*-uncoupling interaction. One is the series of $Q(1)$ lines which have as upper states the

* The Worley–Jenkins series of N_2 furnished good examples of *p*-complexes intermediate between Hund's case (*a*) and case (*d*). That part of the spectrum of H_2 described here consists of essentially *pure* case (*d*) *p*-complexes with $C \approx 0.1B$ (see Section 2) at $n = 30$.

[26] U. Fano, *Phys. Rev.* (*A*), 1970, **2**, 353.

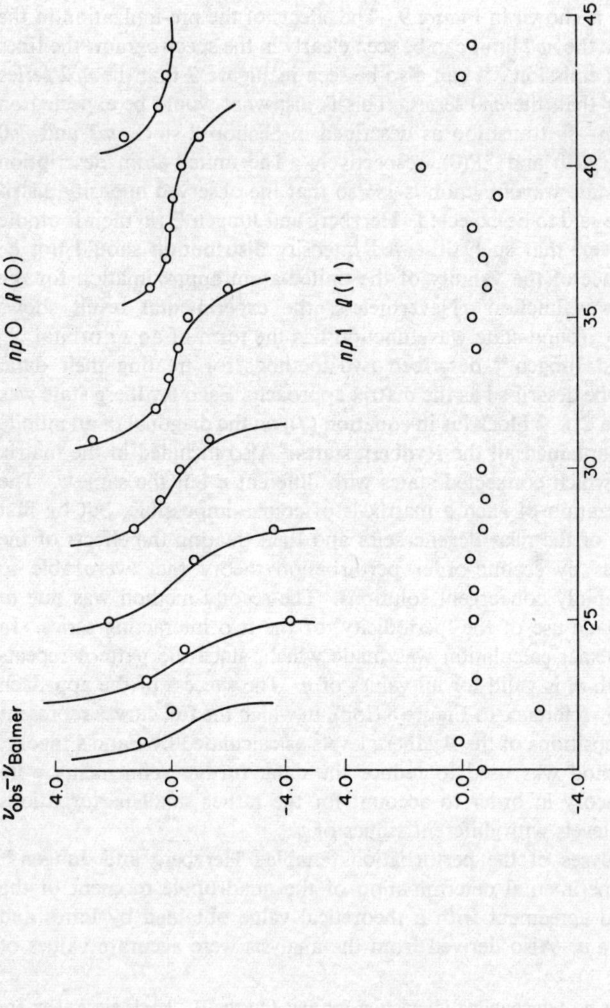

Figure 8 *Deviations of the $v' = 1$, $J' = 1$ R(0) (top) and Q(1) (bottom) Rydberg series in H_2 from simple Rydberg formulae as a function of n. The curves in the top part of the Figure represent the calculated deviations according to the theory of Fano*[26] (Adapted with permission from *J. Mol. Spectroscopy*, 1972, **41**, 425)

$J = 1$ level of Π^-.* The other is the series of $P(1)$ lines which go to the unique $J = 0$ level of the complex. The result of this is that there are no $\Delta n \neq 0$, $\Delta v = 0$ perturbations which cause so much trouble in the *np*0 and *np*2 series although the much weaker $\Delta n \neq 0$, $\Delta v \neq 0$ perturbations are still present. A plot, similar to Figure 8 (top), showing the deviation from a simple Rydberg formula for the $Q(1)$ *np*1 series is shown in Figure 8 (bottom), where it is apparent that there are much smaller deviations from the calculations. A spectrogram of the six series discussed above leading to the $v = 1$ vibrational state of the ion is shown in Figure 9. The effects of the pre-ionization in the *np*2 series above the *np*0 limit can be seen clearly in the spectrogram: the lines appear to be in emission. It can also be seen in Figure 9 that the *np*2 series is much weaker than the *np*0 series. This is just what would be expected on the basis of a p—$s\sigma$ transition as described in Section 2 since *np*2 and *np*0 correspond to ${}^{S}R(0)$ and ${}^{Q}R(0)$, respectively. The united-atom description of the ground-state wavefunction is $1s\sigma$ so that the observed intensity distribution might be said to be expected. Herzberg and Jungen [25] (in their footnote 4) warn, however, that such observed intensity distribution should not be taken as evidence of the validity of the united-atom approximation for the ground-state wavefunction. Nevertheless, the experimental result shows clearly that the ground-state wavefunction has the form of an $s\sigma$ orbital.

Herzberg and Jungen [25] described two methods for treating their data. The first might be described as the matrix approach. Each Rydberg state was represented by a 2×2 block [as in equation (7)] on the diagonal of an infinite matrix which contained all the Rydberg states. Also included in the matrix were elements which connected states with different n but the same v. The exact diagonalization of such a matrix is of course impossible, but by first taking account of the near degeneracies and then treating the effects of the remaining levels by second-order perturbation theory they were able to obtain fairly rapidly convergent solutions. The second method was due to Fano,[26] who made use of the 'periodicity' of the two interacting series. In effect a single exact calculation was made which, since the pattern repeats periodically with n, is valid for all values of n. The success of the approach can be judged by reference to Figure 8 (top), in which the full curves represent the calculated positions of the Rydberg levels as calculated by Fano's theory. Whichever method was used to reduce the data, further refinements were made to the theory in order to account for the rather small perturbations which connect levels with different values of v.

Detailed analyses of the perturbations enabled Herzberg and Jungen [25] to make an experimental determination of the quadrupole moment of the H_2^+ ion in good agreement with a theoretical value obtained by Bates and Poots [27] of 1.5 a.u. Also derived from the analysis were accurate values of

* Actually there is no l-uncoupling interaction for any J in the Π^- levels since they are given by the single element in the 3×3 matrix of equation (7).

[27] D. R. Bates and G. Poots, *Proc. Phys. Soc.* (*A*), 1953, **66**, 784.

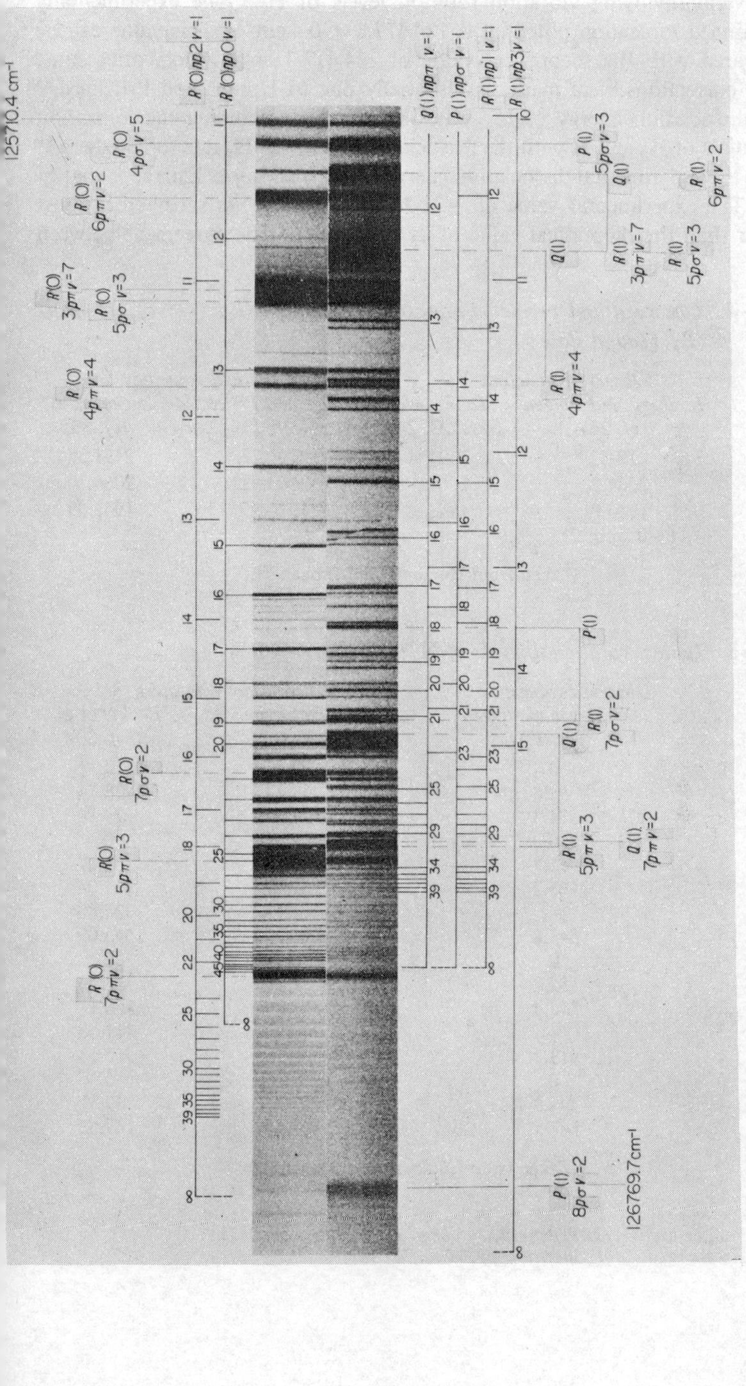

Figure 9 *Spectrogram of Rydberg series in* H_2 *leading to* $v = 1$ *in* H_2^+
(Reproduced by permission from *J. Mol. Spectroscopy*, 1972, **41**, 425)

the ionization potential of H_2, the dissociation energy of H_2^+, and the energies of several low-lying vibration–rotation levels of H_2^+. The experimentally determined ionization potential is 124 417.2 ± 0.4 cm^{-1}. This value can be compared with the theoretical value of 124 417.3 cm^{-1} which, after some small corrections were made, is essentially due to Hunter and Pritchard.[28] The dissociation energy of H_2^+ was obtained by combining the ionization potentials of H_2 and H with the dissociation energy of H_2 due to Herzberg.[29] Since the experimental dissociation energy of H_2 is an upper limit so is that of H_2^+. The experimental value of $\leq$21 379.8 ± 0.4 cm^{-1} is satisfactorily just higher than the theoretical value of 21 379.3 cm^{-1}. The agreement between

Table 4 *Observed and predicted vibrational quanta,* $\Delta G(v + \frac{1}{2})$, *of* H_2^+ *in its* $^2\Sigma_g^+$ *ground state*

	Observed value/cm^{-1}		*Theoretical value*/cm^{-1}	
	Herzberg and Jungen (*ref.* 25)	*Takezawa* (*refs.* 23, 24)	*Hunter and Pritchard* (*ref.* 28)	*Beckel* et al. (*ref.* 30)
$\Delta G(\frac{1}{2})$	2191.2	2191	2191.23	2191.33
$\Delta G(\frac{3}{2})$	2064.2	2065	2063.93	2064.10
$\Delta G(\frac{5}{2})$	1940.8	1941	1940.99	1941.09
	(±0.2 cm^{-1})			

(Taken from Herzberg and Jungen [25])

Table 5 *Lowest rotational levels of* H_2^+

		Observed values/cm^{-1}	*Theoretical values*/cm^{-1}	
v	N	*Herzberg and Jungen* (*ref.* 25)	*Hunter and Pritchard* (*ref.* 28)	*Beckel* et al. (*ref.* 30)
0	0	0	0	0
	1	58.8	58.24	58.24
	2	174.3	174.24	174.24
1	0	0	0	0
	1	55.8	55.17	55.17
	2	165.2	165.07	165.06
	3	—	—	328.80
	4	543.8	—	545.09
2	0	0	0	0
	1	52.2	52.25	52.20
	2	155.4	156.24	156.17
	3	—	—	311.08
	4	513.1	—	515.68
3	0	0	0	0
	1	48.9	49.34	49.31
	2	147.5	147.61	147.53

(Taken from Herzberg and Jungen [25])

[28] G. Hunter and H. O. Pritchard, *J. Chem. Phys.*, 1967, **46**, 2153.
[29] G. Herzberg, *J. Mol. Spectroscopy*, 1970, **33**, 147.

the observed and the theoretically predicted vibration–rotation energy levels of H_2^+ can best be seen in Tables 4 and 5.[23–25, 28, 30] These Tables list observed and theoretical values of the vibrational quanta of the lowest rotational levels of H_2^+, respectively. The agreement is everywhere excellent.

5 Other Singlet *p* Series

CO is isoelectronic with N_2 and might be expected to have Rydberg series similar to the Worley–Jenkins series of N_2. This does not, however, seem to be the case. Part of the reason, no doubt, lies in the fact that the $g : u$ selection rules no longer hold. Nevertheless, the *p* series has been observed and one member, with $n = 4$, has been analysed by Ogawa and Ogawa.[31] They find the Σ and Π components about 220 cm^{-1} apart and obtain a fair fit of the energy levels to the standard *l*-uncoupling theory despite the existence of a strong unidentified absorption band between the two components. An interesting point concerns the branch intensities. Ogawa and Ogawa report that the *R* branch of the Π—Σ component, which lies at higher energies, is stronger than the *P* branch and it seems, from their reproduction of the band, that the *P* branch of the Σ—Σ component is stronger than the *R* branch. Such an intensity distribution might be expected if excitation is from a $p\sigma$ orbital in the ground state, and indeed the united-atom approximation of the outermost orbital of the ground state is expected to be $3p\sigma$ (see Herzberg,[18] Figure 156, page 328).

The 3*p* complex of BH has been observed by Bauer, Herzberg, and Johns,[32] and while the energies of the levels can be accounted for easily, the separation of the two components ($\approx$740 cm^{-1}) is too large to detect the small effect of the *l*-uncoupling on the branch intensities. The intensity pattern should, like CO, be that of an np—$p\sigma$ type transition, as shown in Figure 2b. The transition to the *p* complex in BH is much weaker than to either the 3*s* state or the 3*d* complex, as might be expected from the approximate selection rule $\Delta l = \pm 1$.

What are apparently the 4*p* complexes of AlH and AlD have been observed by Lagerqvist, Lundh, and Neuhaus[33] and by Johns.[34] In this complex the separation between the two components is small ($\approx$28 cm^{-1}); in AlH the Σ state is above the Π state and in AlD the opposite is true. The closeness seems to have been caused by an accidental homogeneous interaction with a non-Rydberg Π state. The interactions have not been quantitatively explained.

[30] C. L. Beckel, B. D. Hansen, and J. M. Peek, *J. Chem. Phys.*, 1970, **53**, 3681.
[31] M. Ogawa and S. Ogawa, *J. Mol. Spectroscopy*, 1972, **41**, 393.
[32] S. H. Bauer, G. Herzberg, and J. W. C. Johns, *J. Mol. Spectroscopy*, 1964, **13**, 256.
[33] A. Lagerqvist, L. E. Lundh, and H. Neuhaus, *Phys. Scripta*, 1970, **1**, 261.
[34] J. W. C. Johns, unpublished results.

6 Rydberg Series of NO

The ground state of the NO^+ ion has neither orbital nor spin angular momentum and Rydberg states of NO are therefore examples of the simplest possible system envisaged in Section 2. At the same time, the ionization potential of NO is relatively low ($\approx$75 000 cm^{-1}, $\equiv$ 1330 Å) so that detailed experimental study is relatively easy. One might therefore have expected that Rydberg series of NO would have been understood as soon as high-resolution spectra became available. As it turns out the spectrum is extremely complicated and it has taken many years of extensive study, mainly by Professor Miescher and his collaborators,[35–43] to understand it.

The main reason for the complexity of the far-u.v. spectrum of NO lies in the fact that the ground state of NO^+ is more tightly bound than the ground state of NO. Consequently the Rydberg series do not consist of strong 0—0 type transitions but rather of short progressions in which the 1—0 bands are probably the strongest. In addition to the Rydberg states, which have internuclear distances closely similar to that of the ion (*i.e.* shorter than the ground state of NO), there are many non-Rydberg states which have internuclear distances larger than the ground state of NO. These states also give rise to progressions of bands in absorption and not only do they overlap the Rydberg absorptions but the two sets of states perturb each other, thus making detailed analysis difficult.

Nevertheless as early as 1963 a relatively clear picture of the Rydberg states of NO emerged, helped considerably by study of the spectra of various isotopic species of NO which enabled unambiguous vibrational quantum numbers to be assigned to the bands. The picture that has emerged is shown schematically in Figure 10. As outlined in Section 2, series of *s*, *p*, *d*, and *f* Rydberg complexes are expected and, as shown in Figure 10, are indeed found. In the following paragraphs we will discuss the *f* series first and then the *s* and *d* series together, and finally the *p* series. The reasons for such a grouping will, it is hoped, become obvious.

The *f* Series.—Despite the fact that $l = 3$ in this series, and that there are consequently Σ, Π, Δ, and Φ components to each complex, these bands are perhaps the best understood in the far-u.v. spectrum of NO. The reason for this is that the *nf* electron is usually very far from the core and, on account of its high *l* value, hardly penetrates the core at all. In other words *l* is almost

[35] A. Lagerqvist and E. Miescher, *Helv. Phys. Acta*, 1958, **31**, 221.
[36] A. Lagerqvist and E. Miescher, *Canad. J. Phys.*, 1962, **40**, 352.
[37] K. P. Huber and E. Miescher, *Helv. Phys. Acta*, 1963, **36**, 257.
[38] A. Lofthus and E. Miescher, *Canad. J. Phys.*, 1964, **42**, 848.
[39] K. Dressler and E. Miescher, *Astrophys. J.*, 1965, **141**, 1266.
[40] A. Lagerqvist and E. Miescher, *Canad. J. Phys.*, 1966, **44**, 1525.
[41] E. Miescher, *J. Mol. Spectroscopy*, 1966, **20**, 130.
[42] Ch. Jungen and E. Miescher, *Canad. J. Phys.*, 1968, **46**, 987.
[43] Ch. Jungen and E. Miescher, *Canad. J. Phys.*, 1969, **47**, 1769.

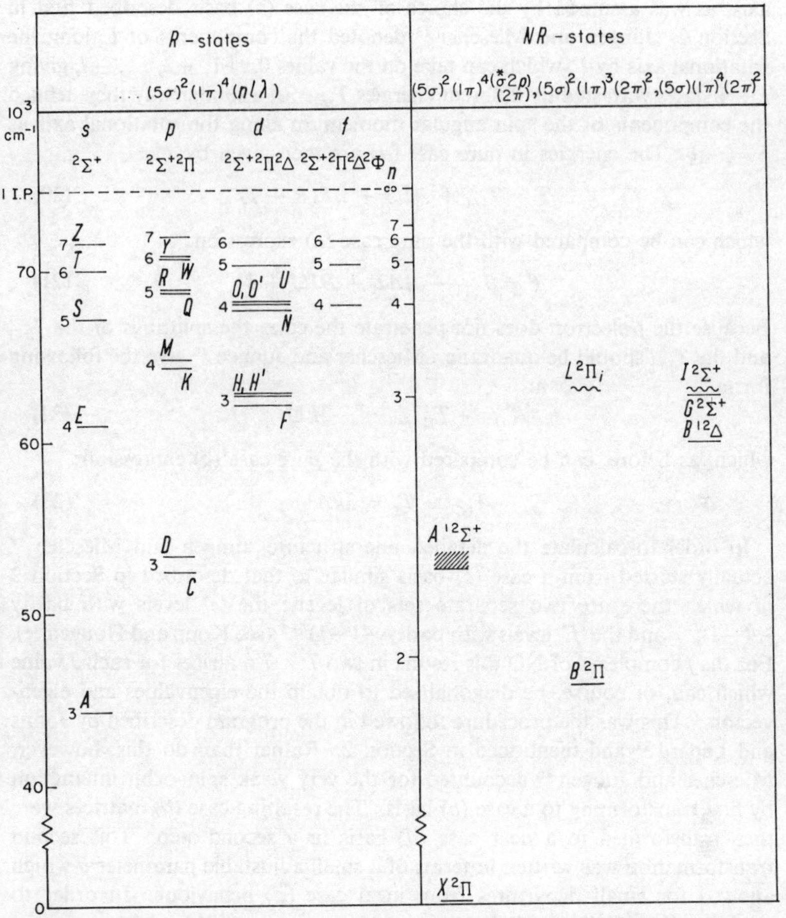

Figure 10 *Energy levels of* NO. *Rydberg states have been grouped to the left; non-Rydberg to the right*
(Reproduced by permission from *J. Chem. Phys.*, 1970, **53,** 4168)

completely uncoupled from the core, and nearly pure case (*d*) results. Jungen and Miescher [43] were thus able to determine relatively simple expressions for the energy levels and for the branch intensities. They proceeded as follows.

In case (*d*), the ion core in its $^1\Sigma^+$ state will rotate with energy

$$F(R) = BR(R+1) \tag{19}$$

and the Rydberg electron will, as it were, feel the effect of the asymmetry of the core only slightly. As a consequence, the orbital angular momentum is quantized along the rotational axis rather than along the internuclear axis, as was assumed by the choice of the case (*a*) basis described first in Section 2. Jungen and Miescher [43] denoted the components of l along the rotational axis by L, which can take on the values 0, ±1, ±2, . . . ±l, giving $l + 1$ states with slightly different energies $T_{|L|}$. In a similar way they defined the components of the spin angular momentum along the rotational axis as $S = \pm\frac{1}{2}$. The energies in pure case (*d*) are then given by:

$$T = T_{|L|} + ALS + BR(R + 1) \tag{20}$$

which can be compared with the pure case (*a*) expression:

$$T = T_{|\Lambda|} + A\Lambda\Sigma + BJ(J + 1) \tag{21}$$

Because the *f*-electron does not penetrate the core, the splittings of the $T_{|L|}$ and the $T_{|\Lambda|}$ should be quadratic. Miescher and Jungen [43] give the following formula:

$$T_{|L|} - T_{|L|=0} = -\tfrac{3}{2}kL^2 \tag{22}$$

which, as before, can be compared with the pure case (*a*) expression:

$$T_{|\Lambda|} - T_{\Sigma} = 3k\Lambda^2 \tag{23}$$

In order to calculate the detailed fine structure, Jungen and Miescher [43] actually started from a case (*a*) basis similar to that described in Section 2 in which there are two separate sets of levels: the '*e*' levels with parity $+(-1)^{J-\frac{1}{2}}$ and the '*f*' levels with parity $-(-1)^{J-\frac{1}{2}}$ (see Kopp and Hougen [44]). For the *f* complexes of NO this results in two 7 × 7 matrices for each *J* value which can, of course, be diagonalized to obtain the eigenvalues and eigenvectors. This was the procedure followed in the program described by Johns and Lepard [11] and mentioned in Section 2. Rather than do this, however, Miescher and Jungen [43] accounted for the very weak spin–orbit interaction by first transforming to a case (*b*) basis. The resulting case (*b*) matrices were then transformed to a near case (*d*) basis in a second step. This second transformation was written in terms of a small adjustable parameter σ which allowed for small departures from ideal case (*d*) behaviour. In order to minimize the off-diagonal elements in this 'near case (*d*)' basis the value of σ was set according to the relation:

$$\sigma = (1 - 3k/2B) \tag{24}$$

where the quantity k has already been defined. Particularly simple relations were obtained for the special case where $\sigma = 0$, *i.e.* when $3k = 2B$, and Miescher and Jungen [43] gave explicit expressions for this situation, which actually turns out to describe the 4*f* complex very well indeed.

The extent to which the formulae reproduce the energy levels is shown in Figure 11. Here the quantity $BR(R + 1)$ has been subtracted from the observed energy levels, which have been plotted as open circles. The energies given by the theoretical formulae are represented by the curves and it can

[44] I. Kopp and J. T. Hougen, *Canad. J. Phys.*, 1967, **45**, 2581.

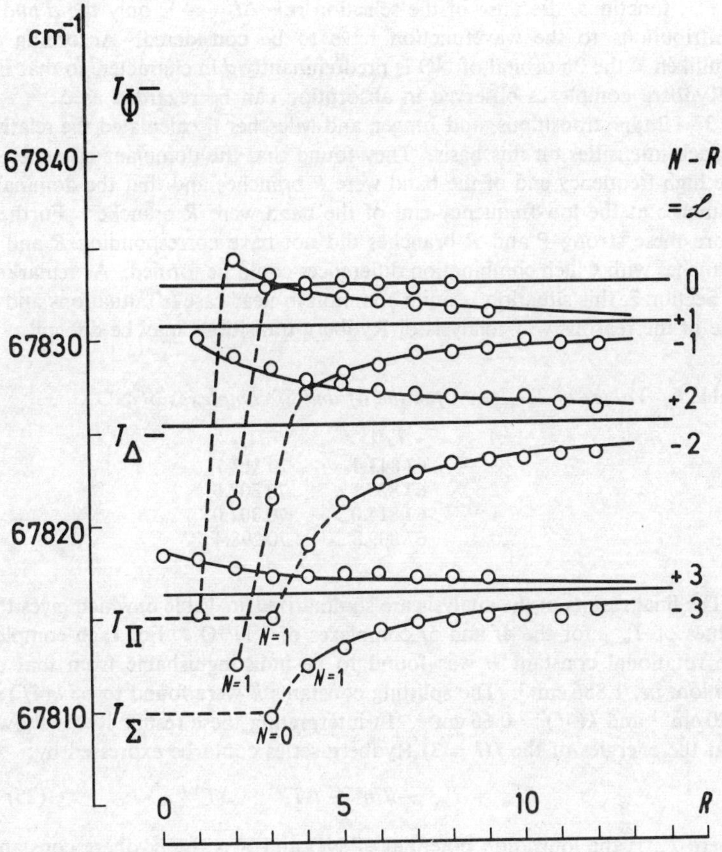

Figure 11 *Energy levels of the 4f complex of* NO *as a function of R after subtraction of BR(R + 1). The circles represent observed energies derived from the rotational analysis and the curves are calculated from the theory*
(Reproduced by permission from *Canad. J. Phys.*, 1969, **47**, 1769)

be seen that the agreement between the two is good. The points in Figure 11 were, of course, obtained from the rotational analysis, which was successful only after the relative branch intensities had been understood. As already indicated in Section 2, the dipole-moment operator, needed for the calculation of the intensities, is a product of a rotational part, a spin part, and an orbital part, and the latter is vital if the relative branch intensities are to be predicted correctly. The excited state has a well-defined value of *l*, namely 3, but it is more difficult to assign a value for *l* to the ground state. In this state the 2π orbital, from which an electron is excited in the transition to the *f* complex, must be considered as being constructed from a linear combination of *p*,

$d, \ldots$ functions. Because of the selection rule $\Delta l = \pm 1$, only the d and g contributions to the wavefunction have to be considered. According to Mulliken [17] the 2π orbital of NO is predominantly d in character, so that the f Rydberg complexes observed in absorption can be regarded as $\Delta l = +1$ or $3 \leftarrow 2$ type transitions, and Jungen and Miescher [43] calculated the relative branch intensities on this basis. They found that the dominant branches at the high-frequency end of the band were P branches and that the dominant branches at the low-frequency end of the band were R branches. Furthermore these strong P and R branches did not have corresponding R and P branches with which combination differences could be formed. As remarked in Section 2, this situation is quite common in near case (d) situations and is one of the reasons why analysis of Rydberg transitions may be difficult.

Table 6 *Values of* $T_{n,\Lambda}/\mathrm{cm}^{-1}$ *for the* 4*f and* 5*f complexes of* NO

Λ	$T_{4,\Lambda}$	$T_{5,\Lambda}$
3	67 843.4	70 317.7
2	67 825.1	70 307.6
1	67 815.0	70 302.0
0	67 809.8	70 298.4

The final results of the analysis are summarized in Table 6, which gives the values of $T_{n,\Lambda}$ for the 4f and 5f complexes of $^{14}N^{18}O$.* For each complex the rotational constant B was found to be indistinguishable from that of the ion, *i.e.* 1.886 cm^{-1}. The splitting constants k were found to be $k(4f) =$ 1.20 cm^{-1} and $k(5f) = 0.66$ cm^{-1}. In interpreting these results it was shown that the energies of the $f(l = 3)$ Rydberg series could be expressed by:

$$T_{n,\Lambda} = T_{\infty} - R/n^2 - \Delta T_n^{(a)} - \Delta T_{n,\Lambda}^{(m)} \qquad (25)$$

where T_{∞} is the ionization potential of NO and R is the Rydberg constant. The quantity $\Delta T_n^{(a)}$ is an atom-like contribution to the energy which results from the spherical polarizability (α) of the core and $\Delta T_{n,\Lambda}^{(m)}$ is a molecular contribution which results from the quadrupole moment (Q_{zz}) and the non-spherical polarizability ($\alpha_{\parallel} - \alpha_{\perp}$) of the core. These quantities are given by:

$$\Delta T_n^{(a)} = R\alpha(n^2 - 4)/(315n^5) \qquad (26a)$$

$$\Delta T_{n,\Lambda}^{(m)} = -R[Q_{zz}/945n^3 + (\alpha_{\parallel} - \alpha_{\perp})(n^2 - 4)/21\,262.5n^5](3\Lambda^2 - 12) \qquad (26b)$$

The molecular contribution vanishes when $\Lambda = 2$ so that the energies of the pure nf states can be identified with the $T_{n,2}$ energies given in Table 6; from equations (25) and (26) one obtains:

$$T_{n,2} = T_{\infty} - R/n^2 - R\alpha(n^2 - 4)/(315n^5) \qquad (27)$$

* This particular isotopic species was analysed because the bands are less overlapped by non-Rydberg bands than are those of the other species.

The two values of $T_{n,\,2}$, one for the 4*f* and the other for the 5*f* state, are thus just sufficient to determine T_{∞} and α for $^{14}N^{18}O$ After making appropriate corrections for zero-point energy differences, Miescher and Jungen [43] found the ionization potential of $^{14}N^{16}O$ to be 74 720 ± 5 cm^{-1}. This value was within earlier error limits but was much more precise. The spherical polarizability of NO^+ was found to be 7.6 a.u., a value which compares favourably with atomic data and with the polarizability of NO in its ground state. The quadrupole moment of NO^+ was determined from:

$$k(nf) = R[Q_{zz}/945n^3 + (\alpha_{\|} - \alpha_{\perp})(n^2 - 4)/21\,262.5n^5] \qquad (28)$$

which can be obtained readily from equations (23) and (26). By assuming that the non-spherical polarizability was one third of the spherical polarizability (as it is observed to be for both N_2 and CO), a value of +0.59 ± 0.04 a.u. was obtained for Q_{zz} of NO^+.

The *s* and *d* Series.—Rotational analyses of the *d*-complexes were attained fairly early by Huber and Miescher [37] and by Jungen,[45] who studied the absorption spectrum, and by Huber,[46] who studied the spectrum in emission. Since all the *d*-complexes have similar properties only the 3*d* complex will be considered here. It was found that the 3*d* complex near 62 000 cm^{-1} did not have the expected properties, and in particular the Σ, Π, and Δ components were not split by the quadratic formula of equations (22) or (5). The Σ and Π (designated $H^2\Sigma^+$ and $H'^2\Pi$, respectively) components were found to be very close together (separated by only 16 cm^{-1} = 8.5*B*) and are 670 cm^{-1} above the Δ component (designated $F^2\Delta$). Three other anomalous features were noted in the detailed analysis. First, the matrix elements which connect the Σ and Π components were found to be too small, suggesting that $l(l + 1)$ was less than 6, which would be required for a *d*-complex with $l = 2$. Second, the branch intensities in the absorption spectrum did not agree with those calculated for a *d*-complex ← $d\pi$ transition and there were similar anomalies in longer-wavelength emission spectra also involving the *d*-complexes. Third, the observed spin–orbit interaction effects in the Σ, Π pair were found to be larger than expected whereas they were small in the Δ component.

Jungen [47] was able to devise a quantitative theory which accounts for all these anomalies in a convincing manner. He first calculated the positions of the various components of the complex according to equation (25), where now

$$\Delta T_n^{(a)} = R\alpha(2/105)(n^2 - 2)/n^5 \qquad (29a)$$

$$\Delta T_{n,\Lambda}^{(m)} = -R[2Q_{zz}/315n^3 + (\alpha_{\|} - \alpha_{\perp})(n^2 - 2)/1653.75n^5](3\Lambda^2 - 6) \qquad (29b)$$

because $l = 2$ rather than 3, as it was in equation (26). The quadrupole moment and the polarizabilities were considered known from the work on the

[45] Ch. Jungen, *Canad. J. Phys.*, 1966, **44**, 3197.
[46] M. Huber, *Helv. Phys. Acta*, 1964, **37**, 329.
[47] Ch. Jungen, *J. Chem. Phys.*, 1970, **53**, 4168.

Table 7 *Observed and calculated positions of the* 3d *and* 4s *Rydberg states of* NO

State	$T_{obs.}$/cm^{-1}	$T_{calc.}$/cm^{-1}	ε/cm^{-1}
$H'\,^2\Pi$ ($3d\pi$)	62 721	62 007	+714
$H\,^2\Sigma^+$($3d\sigma$)	62 705	61 946	+759
$F\,^2\Delta$ ($3d\delta$)	62 043	62 188	−145
$E\,^2\Sigma^+$($4s\sigma$)	60 863	(61 622)	(−759)

f-complexes, thus making it possible to calculate term energies for the *d*-complexes. These are given in the third column of Table 7. Jungen [47] considered that the differences (given in column 4 of Table 7) between the observed and calculated energies were due to the long-range model not taking account of penetration effects. In particular, since the differences were not all of the same sign, it was suggested that not all the penetration effects could have the same origin. According to Jungen [47] the long-range model neglected the following:

'(i) all exchange effects between Rydberg and core electrons, as well as the increased effective nuclear charge felt by the Rydberg electron inside the core;

(ii) the possible non-orthogonality of hydrogenic Rydberg orbitals and core orbitals of the same symmetry;

(iii) the mixing of Rydberg orbitals of the same symmetry due to the non-spherical character of the molecular self-consistent field;

(iv) electron correlation.'

Only the $3d\delta$ orbital can be regarded as orthogonal to the core orbitals, which all have σ or π symmetry. It is thus not surprising that the smallest value of the penetration energy $\varepsilon(\varepsilon = T_{obs} - T_{calc})$ was obtained for the $F^2\Delta$ state. Jungen [47] considered its penetration energy to be due to exchange effects and increased nuclear attraction in the core [see (i) above].

The hydrogenic $3d\pi$ orbital used in the long-range calculation is not orthogonal to the 2π valence orbital [see (ii) above]. As a consequence the $3d\pi$ state is found higher than predicted, *i.e.* ε is $+714$ cm^{-1}. Jungen used this value of ε to estimate the overlap, S, of the hydrogenic $3d\pi$ wavefunction with the valence 2π function and found S to be 0.107. He then showed that the spin–orbit coupling constant, A, in the $3d\pi$ (*i.e.* $H'^2\Pi$) state could be related to the spin–orbit coupling constant A in the 2π (*i.e.* $X^2\Pi$) state by:

$$A(3d\pi, H'^2\Pi) = S^2 A(2\pi, X^2\Pi) \tag{30}$$

from which he obtained $A = 1.4$ cm^{-1} for $H'^2\Pi$, which agrees quite well with the observed value of 0.96 cm^{-1}. Analysis of the fine structure of the $F^2\Delta$ state gave an experimental A value of <0.1 cm^{-1}, which is expected from this model since there is no overlap of the $3d\delta$ orbital with the 2π orbital.

The main reason for the large positive value of $\varepsilon(= +759$ cm$^{-1})$ for $3d\sigma$ appears to arise from mixing with the $4s\sigma$ Rydberg orbital [see (iii) above].

The most obvious consequence of this mixing is to reduce the effective value of l in the $3d\sigma : 3d\pi$ interaction: an effect already noted from the rotational analysis. On the assumption that $3d\sigma$ and $4s\sigma$ form an interacting pair, Jungen [47] calculated the wavefunctions (mixing coefficients) from the observed energies (column 2 of Table 7) and their unperturbed energies, which were assumed to be given by the long-range calculation (column 3 of Table 7). He found that $H\,^2\Sigma^+$ should be represented by $c_1(3d\sigma) + c_2(4s\sigma)$, with $c_1 = 0.768$ and $c_2 = 0.640$. Since only the d character of the $H\,^2\Sigma^+$ state can contribute to the l-uncoupling it was shown that the apparent value of $l(l+1)$ should be given by $c_1{}^2 l^*(l^*+1)$, where l^* has the hydrogenic value of 2. For the $H : H'$ interaction (*i.e.* $3d\sigma : \pi$) the calculated effective value of $l(l+1)$ turns out to be 3.54, which compares well with the experimental value of 3.69.

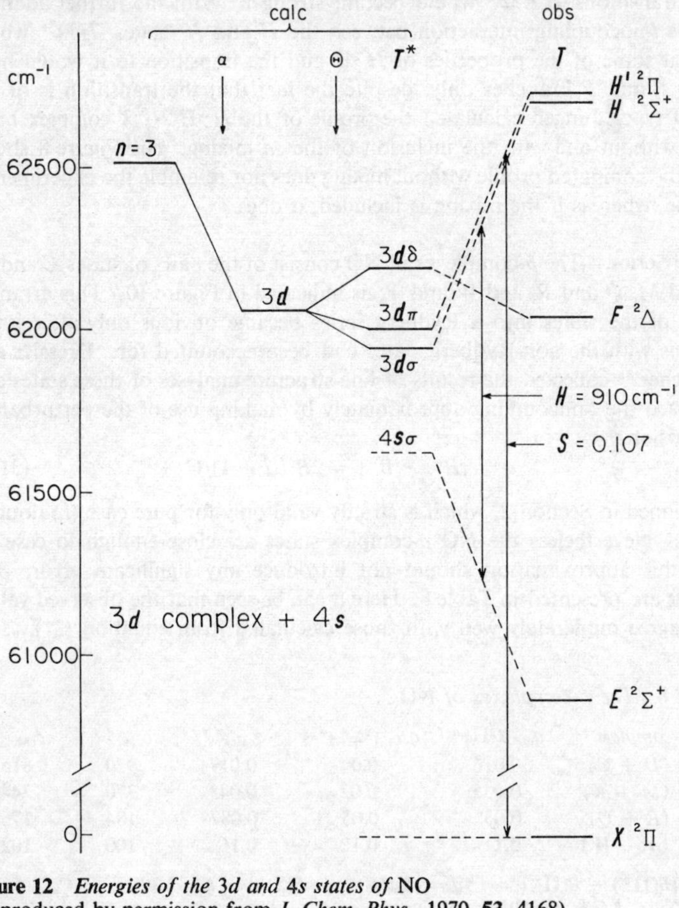

Figure 12 *Energies of the* 3d *and* 4s *states of* NO
(Reproduced by permission from *J. Chem. Phys.*, 1970, **53**, 4168)

The *sd* mixing described above has another important effect on the observed spectrum; it results in some marked intensity anomalies. We shall here consider only the vacuum-u.v. absorption spectrum but there are similar anomalies in the longer-wavelength emission spectrum. Jungen showed that the result of the *sd* mixing was essentially to annihilate the transition moment to the $3d\sigma$ component of the complex while augmenting the transition moment to the $4s\sigma$ band. Before the mixing, he assumed that transitions to the $F^2\Delta$, $H\,^2\Sigma^+$, and $E\,^2\Sigma^+$ states would have about the same intensity and would have weak *P*, *Q*, and *R* branches, and that transitions to $H'^2\Pi$ would have weak *P* and *R* branches only. All branches were expected to be weak because all these transitions violate the approximate selection rule $\Delta l = \pm 1$. After *sd* mixing he showed that the *P*, *Q*, and *R* branches associated with the transition to $H\,^2\Sigma^+$ would disappear and the branches associated with transitions to $E\,^2\Sigma^+$ would become stronger. With the further addition of the *l*-uncoupling interaction between the *H* and *H′* states, $H\,^2\Sigma^+$ would exhibit some of the properties of $H'^2\Pi$ and the transition to it would have weak *P* and *R* branches only, despite the fact that the transition is of the Σ—Π type. Jungen calculated the profile of the $H:H' \leftarrow X$ complex band both without and with the inclusion of the *sd* mixing. His Figure 8 shows that the computed profile without mixing does not resemble the experimental profile, whereas if the mixing is included, it does.

The *p* Series.—The *p*-complexes of NO consist of the pairs of states *C* and *D*, *K* and *M*, *Q* and *R*, and *W* and *Y*, as indicated in Figure 10. This arrangement of the states into a Rydberg series became obvious only after interactions with the non-Rydberg states had been accounted for. Dressler and Miescher [39] collected the results of fine-structure analyses of these states and analysed the *l*-uncoupling approximately by making use of the perturbation expression:

$$q = |B^+ - B^-| = 2B^2 l(l+1)/C \qquad (31)$$

mentioned in Section 2, which is strictly valid only for pure case (*b*) doublet states. Nevertheless the NO *p*-complex states are close enough to case (*b*) that this approximation should not introduce any significant error. The results are presented in Table 8. Here it can be seen that the observed values of *q* agree moderately well with those calculated from equation (31). This

Table 8 *The p-complexes of* NO

n-complex	$q_{obs.}(^2\Pi)^{a,f}$	$q_{obs.}(^2\Sigma^+)^{b,f}$	$q_{calc.}{}^{c,f}$	$C_{obs.}{}^{d,f}$	$C_{calc.}{}^{e,f}$
3 $(D+C)$	0.016	0.0	0.017	920	815
4 $(M+K)$	0.035	0.02	0.043	370	344
5 $(R+Q)$	0.03'	0.05	0.087	184	176
6 $(Y+W)$	0.1	0.12	0.16	100	102

[a] $q = |B(\Pi^+) - B(\Pi^-)|$; [b] $q = |B_0(NO^+) - B(\Sigma)|$; [c] $q = 2B^2 l(l+1)/C$; [d] $C = |\nu_\Pi - \nu_\Sigma|$; [e] $C = 22\,000/n^3$; [f] values in cm^{-1}.

quantity q should also be approximately equal to the difference between the observed B_{eff} of the Σ component of the complex and that of the NO^+ ion. Table 8 shows that this expectation also is obeyed. It is interesting to note that the separation, C, between the Σ and Π components varies very nearly inversely as n^3, just as it does in the p-complexes of N_2. Calculated values of this splitting with $A = 22\,000/n^3$ are included in the Table for comparison.

7 Other Doublet Complexes

NO is the only stable diatomic molecule with a doublet ground state, so that it is hardly surprising that little work has been done on other doublet Rydberg series. Herzberg and Johns [48] have observed a d Rydberg series of CH. Intensity calculations showed that a compact band contour was to be expected and the observed spectrum of the $n = 3$ d-complex confirmed this. No detailed assignments could be made despite the fact that quite a number of sharp lines were resolved. Well-resolved Σ—Π and Π—Π bands were observed near where the $3p$ complex was expected. However, the effective rotational constants of these bands made it clear that they do not constitute a p-complex.

8 The Halogen Hydrides – HCl

The halogen hydrides all have a ground-state electron configuration which can be written

$$\ldots\ \sigma^2\pi^4\ X\,^1\Sigma^+ \tag{32}$$

so that Rydberg states leading to the ground state of the ion have configurations with a π^3, or inverted $^2\Pi$, core just as set down in the last column of Table 1. Up to the present time no extended Rydberg series have been analysed in detail nor indeed has the necessary theory been worked out to account for l-uncoupling effects. Nevertheless, some understanding of the lower Rydberg states has been attained in recent years. In what follows we shall, for convenience, restrict our attention to HCl, although similar spectra are known for HF (DiLonardo and Douglas [49]), HBr (Price,[50] Barrow and Stamper,[51] and Ginter and Tilford [52, 53]), and HI (Tilford, Ginter, and Bass [54]).

Early work on the vacuum-u.v. absorption spectrum of HCl by Price [50] showed that there were many bands, both discrete and diffuse, in the region

[48] G. Herzberg and J. W. C. Johns, *Astrophys. J.*, 1969, **158**, 399.
[49] G. Dilonardo and A. E. Douglas, *Canad. J. Phys.*, 1973, **51**, 434.
[50] W. C. Price, *Proc. Roy. Soc.*, 1938, **A167**, 216.
[51] R. F. Barrow and J. G. Stamper, *Proc. Roy. Soc.*, 1961, **A263**, 259, 277.
[52] M. L. Ginter and S. G. Tilford, *J. Mol. Spectroscopy*, 1970, **34**, 206.
[53] M. L. Ginter and S. G. Tilford, *J. Mol. Spectroscopy*, 1971, **37**, 159.
[54] S. G. Tilford, M. L. Ginter, and A. M. Bass, *J. Mol. Spectroscopy*, 1970, **34**, 327.

below 1350 Å. Tilford, Ginter, and Vanderslice [55] showed that all the bands in the region from 1350 to 1240 Å could be accounted for by absorption to two electronic states, $b^3\Pi_i$ and $C^1\Pi$, together with some associated vibrational structure. Analysis of the bands was completely straightforward except that the splitting of the three components of the triplet was not quite symmetrical on account of interaction between $C^1\Pi$ and the Π_1 component of $b^3\Pi_i$ (this effect is much more important in the spectra of HBr and HI). The observed spectrum thus agrees exactly with what is expected for the lowest Rydberg states in which an electron has been promoted to a σ-orbital (see Table 1). Whether this orbital can be described as $4s\sigma$ or $3d\sigma$ is not clear.

At somewhat higher energies, Tilford and Ginter [56] found another pair of states similar to those discussed above. They were designated $d^3\Pi_i$ and $D^1\Pi$, respectively. The most reasonable assignment for these states is to the configuration

$$\ldots \sigma^2\pi^3\, 4p\sigma \quad d^3\Pi_i,\ D^1\Pi \tag{33}$$

At higher energies still they found states designated $f^3\Delta$ and $g^3\Sigma^-$ which are probably the lowest states of the configuration

$$\ldots \sigma^2\pi^3\, 4p\pi \tag{34}$$

The remaining states expected from this configuration were not observed because Tilford and Ginter [56] used lithium fluoride optics in their experiments, with the result that no spectra were observed beyond about 1200 Å. Complete analysis of this interesting *p*-complex must therefore await the observation of spectra at shorter wavelength.*

The observed bands of the 4*p* complex of HCl were analysed in terms of a set of effective constants without any attempt to take account of spin-uncoupling or *l*-uncoupling. The results are summarized in Table 9. Tilford and Ginter [56] note that the erratic behaviour of the effective values of the distortion constant *D* is evidence of *l*-uncoupling. On the basis of the theory discussed in Section 2, however, it would seem that *l*-uncoupling would first be observed in the effective *B* values, which should be low at the low-energy end of the spectrum and high at the high-energy end of the spectrum, with a mean value close to that of the ion. Examination of Table 9 shows that the *B* values range from 8.6 at the low-energy end to about 10.6 at the high-frequency end. According to Herzberg [18] the *B* value for HCl^+ is 9.79 which, as expected, lies within the observed range above. It is clear, however, that

* The corresponding 5*p* complex of HBr is a little more completely observed on account of the lower ionization potential of this molecule (see Ginter and Tilford [53]).

[55] S. G. Tilford, M. L. Ginter, and J. T. Vanderslice, *J. Mol. Spectroscopy*, 1970, **33**, 505.

[56] S. G. Tilford and M. L. Ginter, *J. Mol. Spectroscopy*, 1971, **40**, 568.

a complete understanding of the Rydberg spectra of the halogen hydrides and similar molecules will only be attained when the theory discussed in Section 2 has been modified to account for a core with angular momentum.

Table 9 *Effective constants of the 4p-complex of* HCl [a]

Configuration	*State*		ν_0/cm^{-1}	B/cm^{-1}	$D \times 10^3$/cm^{-1}
$..\sigma^2\pi^3 4p\sigma$	$d^3\Pi_2$	Π^+	81 601.0	8.59_8	−1.6
		Π^-	81 600.3	8.66_5	−1.2
	$d^3\Pi_1$	Π^+	81 773.1	9.68_8	+0.39
		Π^-	81 773.4	9.84_8	+1.2
	$d^3\Pi_0$	Π^+	82 271.3	9.38_4	−0.34
		Π^-	82 271.8	9.42_4	−0.09
	$D^1\Pi$	Π^+	82 489.6	9.82_5	+2.7
		Π^-	82 489.7	9.76_2	+1.4
$..\sigma^2\pi^3 4p\pi$	$f^3\Delta_3$	Δ^+	81 825.9	9.45	−0.13
		Δ^-	(81 825)	(9.4)	—
	$f^3\Delta_2$	Δ^+	82 015.5	10.83_6	+2.8
		Δ^-	82 015.2	10.86_6	+3.1
	$f^3\Delta_1$	Δ^+	82 523.8	10.27_0	−1.3
		Δ^-	(82 533)	—	—
	$g^3\Sigma^-$	1^+	82 847.5	10.33_0	+1.1
		1^-	82 847.2	10.39	+2.3

[a] Taken from ref. 56.

9 Conclusion

The foregoing examples have been chosen because they illustrated the sort of information which can be gained from the study of the rotational fine structure of Rydberg series. At the same time they also illustrate the extent to which these structures are understood. While the basic theory has been known for a long time it is only recently that sufficiently high resolution has been attained in the vacuum-u.v. so that molecules other than the lightest of hydrides can be studied. As is often the case when it becomes possible to make detailed studies, the early simple theories which seemed adequate at first have had to be refined and extended, and these refinements point the way to understanding new interactions. The *s—d* mixing invoked by Jungen [47] to explain anomalies in the *d*-complexes of NO is a good example. The perturbation of *d* Rydberg complexes by *s* Rydberg states should be a general phenomenon since the quantum defects of *s* and *d* series usually differ by about unity, placing $(n + 1)s$ states close to $(n)d$-complexes. Caton and Douglas [57] were unable to understand completely the 3*d* complex of BF and they attributed the observed anomalies, at least in part, to *s—d* mixing. In

[57] R. B. Caton and A. E. Douglas, *Canad. J. Phys.*, 1970, **48**, 432.

the same way Johns and Lepard [11] were able to explain the non-quadratic dependence of the energies in the *d*-complex of BH, first observed by Bauer, Herzberg, and Johns,[32] as being the result of mixing with the 4*s* state, which lies only about 1000 cm^{-1} above the 3*d* complex.

Every example mentioned so far has been studied by absorption spectroscopy. This does not mean that this is the only way to gain information concerning Rydberg states. Indeed, much of the understanding of the Rydberg states of NO was obtained by Miescher and his co-workers from the study and analysis of long-wavelength emission spectra. All our knowledge of the *s* and *d* Rydberg states of H_2 comes from emission spectra in the visible and infrared regions of the spectrum, which have been studied most extensively by Dieke.[58] The study of the Rydberg states of He_2 can also be carried out only in emission, and Ginter has made many recent contributions (see for example Ginter and Battino [59]).

Almost all the data so far obtained on extended Rydberg series have involved molecules with Σ cores (see Table 1). The more complicated situation of Rydberg states with Π cores (see column 4 of Table 1) has not yet been studied in detail except for the lowest states of the halogen acids. These spectra are particularly interesting, and complicated, for two reasons. First, they are examples of series in which an ion core with angular momentum is involved and for which spectra of high enough quality for detailed analysis seem to be attainable. Second, the absorption spectra of each of these molecules include both singlet and triplet states. This makes an already complicated situation even more difficult from the point of view of analysis. Presumably the reason is that the spin–orbit coupling in the ion core mixes the triplet states with the singlet states so that both appear with comparable intensity. In this regard one might have expected the spectrum of HF to be the simplest. However, according to DiLonardo and Douglas,[49] in spite of the relatively small spin–orbit coupling constant of the core, both singlet and triplet spectra are readily observed at the long-wavelength end of the spectrum. At energies approaching the first ionization potential the spectrum consists of a very large number of sharp lines with no regular structure discernible. Part of the reason may be that the small spin–orbit coupling of the core (about 292 cm^{-1} according to Gewurtz and Lew [60]) must be of the same order of magnitude as the spacing between many of the Rydberg states and is also of the same order of magnitude as the rotational spacings. Another way of stating the problem would be to say that the core shows coupling intermediate between Hund's case (*a*) and case (*b*). In any event the Rydberg spectra of molecules which have ion cores with orbital angular momentum still need much work before they can be understood as well as those without orbital angular momentum.

[58] G. H. Dieke, 'The Hydrogen Molecule Wavelength Tables', Wiley, New York, 1972.
[59] M. L. Ginter and R. Battino, *J. Chem. Phys.*, 1970, **52**, 4469.
[60] S. Gewurtz and H. Lew, to be published.

The few examples given here may be regarded as relatively well behaved and the observed structures are close to those expected from theory. They are probably the exception rather than the rule, and have been analysed simply because perturbations to the structure are not too severe. More often than not, spectra in the region where Rydberg states are expected are so complicated that analysis into series is not readily possible. The progress made recently in understanding the interactions which can occur amongst Rydberg states should eventually make it possible to unravel some of these complex spectra.

9
Molecular Spectra in Stars

BY E. A. MALLIA

1 Introduction

The presence of molecular lines in the spectra of stars has been known since the early days of this century, when such lines were discovered in the sun.[1] The early work was usually directed towards identification, but as knowledge of the molecules themselves improved it was expected that molecular lines would prove an important source of astrophysical information. This expectation has been amply realized, most fully in the past twenty years.

Molecular lines have been observed in the spectra of most types of star which have temperatures below about 6500 K in their outer layers as well as in some stars with considerably higher temperatures. The exact nature of the observable molecular spectra of stars depends on the composition, pressure, and transparency of their atmospheres. The composition of the outer layers of the sun (a star which has not undergone appreciable evolution in its lifetime of 4.5×10^9 years) is generally taken as the standard for comparison purposes (see Table 1). The present composition of stellar atmospheres

Table 1 *Relative solar abundances of some elements*

Element	*Relative solar abundance*
H	1
He	0.12
C	3.5×10^{-4}
N	8.5×10^{-5}
O	5.9×10^{-4}
F	3.6×10^{-8}
Ar	2.5×10^{-6}
Mg	3.5×10^{-5}
Si	3.5×10^{-5}
S	1.6×10^{-5}
C	2.5×10^{-7}
Ca	2.0×10^{-6}
Ti	6.0×10^{-8}
V	6.3×10^{-9}
Fe	3.5×10^{-5}
Ni	2.0×10^{-6}
Cu	1.0×10^{-8}
Zr	1.0×10^{-9}
Nb	2.0×10^{-10}
La	6.3×10^{-11}

[1] G. E. Hale, W. S. Adams, and H. E. Gale, *Astrophys. J.*, 1906, **24**, 185.

depends principally on the age and initial mass of the stars. The older the star the less heavy elements ($A > 6$) it appears to contain with respect to H, while the more massive the star the more rapidly it evolves. Since evolution results from changes in the nuclear reactions taking place in the interior, outward transport of material can alter the composition of the surface layers.

In order of decreasing stellar radius the atmospheric pressures range from 0.1 Torr in supergiant stars to ~1 Torr in giants, 1—5 × 10^2 Torr in normal dwarfs (like the sun), and up to 10^6 Torr in white dwarf stars. Since in the majority of stars the negative hydrogen ion (H^-) is the most important source of opacity, the atmospheric transparency depends on the hydrogen : metal ratio (metals being the chief source of electrons in cool stars) as well as on the temperature and pressure.

2 Observational Limitations

Molecular absorption in the earth's atmosphere blocks out large sections of the spectra at sea level. If one is interested in the cooler stars the ozone cut-off below 3000 Å (300 nm) is perhaps not such a serious nuisance as the absorption by water vapour in the red and the infrared. The disturbance from water vapour can be mitigated by observing from mountain sites, aircraft, or balloons, but extra-terrestrial vehicles will be ultimately required.

The bulk of the observations up to the present have been obtained with standard prism or grating spectrographs and photographic or photoelectric detectors. Photographic plates are very insensitive beyond 9000 Å, though on bright objects and at low resolving power they can be used at wavelengths of up to 11 000 Å. Photoelectron devices are insentitive beyond 12 000 Å and have the further disadvantage of usually being incorporated in scanning spectrographs which are not well adapted to observations of an extended molecular spectrum. Photoelectron imaging devices combining the good points of both types of detector are now coming into general use.[2] For the detection of radiation of wavelength longer than 12 000 Å a wide range of photoconductive materials is available.

Grating spectrographs will give resolving powers of *ca.* 5 × 10^5 at 5000 Å and 10^5 at wavelengths as great as 50 000 Å on the sun. One can hope to attain 10^5 for only the brightest stars at 5000 Å, while in the i.r. resolving powers of 2—3 × 10^4 can be provided only by Michelson interferometers if observing times are to be kept within reasonable bounds. The signal-to-noise ratios attainable at these resolving powers range from better than 500 on the sun to *ca.* 20 on stars. The accuracy of wavelength measurement is directly related to the signal-to-noise level. For the sun it can be as high as ±0.005 Å but for stars it is very seldom as high as ±0.02 Å.

[2] G. Carruthers, *Astrophys. Space Sci.*, 1971, **14**, 332.

3 Computations of Line Strength

In calculating line strengths (atomic or molecular) in a stellar atmosphere a fundamental assumption is usually made, *viz.* that the atmosphere can be considered as being made up of small elements internally in thermodynamic equilibrium, within which excitation, ionization, and dissociation equilibria can all be described by local values of temperature and electron density (local thermodynamic equilibrium or LTE). The fact that such an assumption implies that there is no outward flow of radiation (as the elements do not interact) shows that it cannot hold strictly. It has been widely used, however, partly because of the great simplification it introduces into the computations, partly because it is a good exploratory tool, and partly because the information necessary to a more rigorous approach has not been available. For the approximation to have some validity, the radiative processes which populate the energy levels of interest must have a considerably longer time-scale than collision rates. In general, therefore, the lower the atmospheric densities, the worse the approximation to LTE.

For a diatomic molecule made up of elements α and β the dissociation constant of concentration is defined as

$$K(\alpha\beta) = AT^{3/2} \exp(-D_0/kT) \tag{1}$$

where A contains such parameters as the molecular weight and the atomic and molecular partition functions and D_0 is the dissociation energy. The number of molecules in the lower level of the transition of interest (defined by quantum numbers v'' and J'') is then

$$N(\alpha\beta) = B \exp[-(E + G_{v''})hc/kT] \,.\, \frac{N(\alpha)N(\beta)}{K(\alpha\beta)} \times \exp[-B_{v''}J''(J''+1)hc/kT] \tag{2}$$

where E and G are the electronic and vibrational energies, respectively, and B is a constant. The strength of the absorption line originating from this level, as measured by the amount of energy it abstracts from the continuous spectrum, is then

$$W_{J''} = Cf_{v'v''}S_{J''} \int_0^{\infty} \frac{g(\tau_0)}{K_0} N(\alpha\beta)\, d\tau_0 \tag{3}$$

where $S_{J''}$ is the Hönl–London factor, C is a constant, $f_{v'v''}$ is the band oscillator strength, K_0 the continuous absorption coefficient per gram at some standard wavelength, and $g(\tau_0)$ the weighting function which takes into account the contribution of each element of the atmosphere. The integration is over optical depth τ_0.

Equations (2) and (3) illustrate the molecular parameters in which the astrophysicist is interested: dissociation energies, oscillator strengths or absorption coefficients, Franck–Condon factors, and Hönl–London factors. Franck–Condon factors can be calculated from assumed potential curves and knowledge of the variation of transition moment with internuclear distance; extensive tabulations of formulae for the computation of Hönl–London factors have been given by Kovács.[3]

The dissociation energy enters as an exponent in the dissociation constant. The dependence of the line strength on it is given roughly by the relation

$$\Delta(\log W_{J''}) \approx \Delta D_0 \theta \qquad (4)$$

where $\theta = 5040/T$ and D_0 is in electron volts. In order not to introduce serious inaccuracy D_0 must then be determined to better than 0.1 eV. Though this may not seem a very stringent requirement, there are in fact hardly any molecules of astrophysical importance for which the dissociation energy is known to this accuracy. On the other hand, the various oscillator strengths influence the line strength directly. Accuracies in the range 10—20% are acceptable even for solar applications.

For the astrophysicist, who has no control over his source, accurate wavelengths constitute a powerful discriminant in attempts to identify lines. Laboratory wavelengths are generally of sufficient accuracy for stellar work but this is sometimes not the case for solar work. The same applies to isotope shifts, the more so as these are often computed. It should also not be forgotten that stellar atmospheres are relatively high temperatures sources. Laboratory work with astrophysical implications is rendered much more valuable if, where possible, it is carried out with a high-temperature source.

Comparison between computations and observations can be carried out in various ways. The simplest is based on equation (3). This can be written as

$$\log (W_{J''}/S_{J''}) = \log (W_0/S_0) - 0.625 B_{v''} J''(J''+1)/T \qquad (5)$$

where W_0 is the strength of the (possibly hypothetical) line with $J = 0$. Provided that the lines are not affected by saturation, *i.e.* line strength is a linear function of numbers of absorbers, a plot of $\log (W_{J''}/S_{J''})$ against $J''(J''+1)$ should be a straight line. The gradient provides a rotational temperature and the intercept on the $\log (W_{J''}/S_{J''})$ axis is a measure of the band oscillator strength.

For lines which are affected by saturation it is more usual to utilize a 'curve-of-growth'. This is a plot of the expected line strength against the effective number of absorbers. Its general form is shown in Figure 1. The observed strengths are entered on the plot and the position of the curve is adjusted until it fits the points. Shifts along the horizontal axis reflect differences between the assumed and actual numbers of absorbers and shifts along the vertical axis reflect errors in the assumed total Doppler broadening at the left-hand end and in the damping at the right-hand end. Thus abundances and information about the physical conditions in the atmosphere can be derived from the curve, provided the molecular parameters are known.

[3] I. Kovács, 'Rotational Structure in the Spectra of Diatomic Molecules', Adam Hilger, London, 1969.

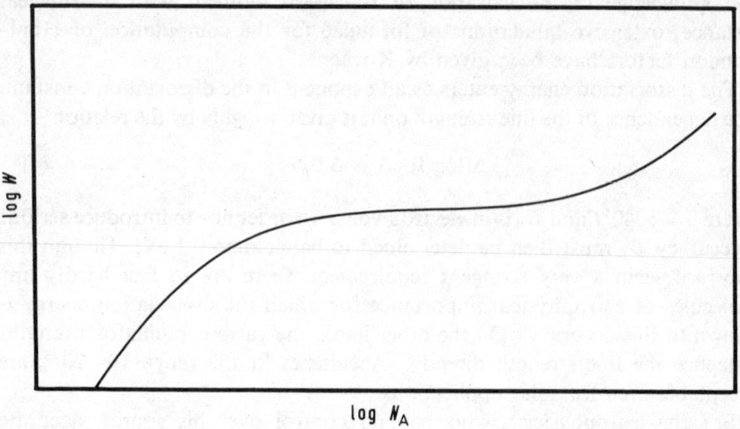

Figure 1 *Curve of growth of line strength (W) with number of effective absorbers (N_A)*

Conversely, if the conditions and element abundances have been otherwise determined such molecular parameters as oscillator strengths and dissociation energies can be found.

For high-quality observations one may consider the detailed profile of lines. Difficulty may be encountered here because of the poor knowledge of the broadening effects of hydrogen on molecular energy levels, which effects determine the shape of the line wings.

The overall accuracy of such computations is difficult to assess. The principal source of uncertainty remains the incomplete knowledge of the physical conditions in the stellar atmospheres, added to which is the doubtful validity of the basic assumption of local thermodynamic equilibrium. Even in the sun the final accuracy of a computed molecular line strength is unlikely to be better than 15%, and for stars it can be anything up to ten times this.

In the most favourable cases abundances in stars can be determined as accurately from molecular lines as from atomic lines.[4] Whereas in the hotter stars like the sun this possibility constitutes a useful alternative source of information, in the cooler stars molecular lines are often the only source of abundances for certain elements. For isotopic abundances atomic lines are generally useless because of the very small shifts for all except the lightest elements.

As far as the derivation of other atmospheric parameters is concerned, it can be pointed out that the physical significance of a rotational temperature is not very great since it is simply a representative temperature that reproduces the intensity distribution among rotational lines. The elevated temperatures

[4] D. L. Lambert, *Monthly Notices Roy. Astron. Soc.*, 1968, **138**, 143.

and low molecular weight of the commonest species also limit the usefulness of molecular lines for determining mass motions in atmospheres. Even so, such lines in solar spectra have been used for this purpose [5, 6] as well as for probing the velocity structure in the photosphere. Furthermore, it can be seen from equation (4) that the strength of lines from molecules with a high dissociation energy will be very sensitive to temperature. The property can be used to determine the detailed temperature distribution in an atmosphere [7] or to obtain mean temperatures by low-resolution measurements of band strengths.[8]

4 Observations

The Sun.—In the sun molecular lines have been identified in three types of spectra:

(i) The absorption spectrum of the normal atmosphere (photosphere), which has rotational temperatures of ~5000 K.

(ii) The emission spectrum of the upper atmosphere (chromosphere), which has rotational temperatures of ~4200 K.

(iii) The absorption spectrum of penumbrae and umbrae of sunspots, which have rotational temperatures of ~4800 and ~3500 K respectively. Photosphere and penumbra have rather similar molecular features in their spectra so discussion will be restricted to the former. The chromospheric spectrum is restricted to a few transitions.

Species and transitions identified in the photospheric spectrum are listed in Table 2. Some of the identifications are not very secure. (When no reference is quoted for a specific result, the source is a review paper entitled 'Molecules in the Sun' in preparation by D. L. Lambert and the Reporter.)

Table 2 *Species and transitions detected in the photospheric spectrum*

Species	*Transitions*	λ/Å
CO	fundamental; 1st overtone vib–rot; $A^1\Pi - X^1\Sigma^+$	46 000; 23 000; 1600
CN	$B^2\Sigma - X^2\Sigma^+$; $A^2\Pi_i - X^2\Sigma^+$	3880; 10 990
C_2	$d^3\Pi_g - a^3\Pi_u$; $A^1\Pi_u - X^1\Sigma_g^+(1,0)$	5160; 10 400
CH	$A^2\Delta - X^2\Pi$; $B^2\Sigma - X^2\Pi$; $C^2\Sigma - X^2\Pi$	4300; 3890; 3150
OH	$A^2\Sigma - X^2\Pi$	3060
NH	$A^3\Pi - X^3\Sigma^-$	3380
MgH	$A^2\Pi - X^2\Sigma$	5205
SiH	$A^2\Delta - X^2\Sigma$	4140
CH^+	$A^1\Pi - X^1\Sigma^+$	4225
SH	$A^2\Sigma - X^2\Pi$	3240; 3280
SiH^+	$A^1\Pi - X^1\Sigma^+(0,0)(0,1)$	3990

[5] E. A. Mallia, *Solar Phys.*, 1968, **5**, 281.
[6] G. L. Withbroe, *Solar Phys.*, 1968, **3**, 146.
[7] D. N. B. Hall and R. W. Noyes, *Bull. Amer. Astron. Soc.*, 1972, **4**, 390.
[8] R. F. Wing, 'Colloquium on Late Type Stars', ed. M. Hack, Oss. Astronomico Trieste, Trieste, 1967, p. 205.

Lines from the $A^2\Sigma$–$X^2\Pi$ transition of SH fall in a very crowded region of the spectrum. There are too few instances of unblended lines to permit a definite identification, the more so as there is no measurement of the oscillator strength. Another doubtful case is that of the Phillips band of C_2. The lines ascribed to the band are very weak and the oscillator strength derived from the solar observations is some three times larger than that found in the laboratory.[9]

The identification of lines from the other listed transitions is well founded. However, for a considerable proportion there are discrepancies between observed and calculated strengths, generally in the sense that the observed lines are weaker than predicted. If R is the ratio of predicted to observed strength, then in the case of CN and OH $R = 3$ and for CH $R = 1.8$. Even more alarming is the fact that lines from the $A^1\Pi$–$X^1\Sigma^+$ transition of CH^+, predicted to appear strongly in the photospheric spectrum, fail to appear at all ($R \approx 60$).[10] A different case is that of MgH, where the lines ascribed to the $A^2\Pi$–$X^2\Sigma^+$ transition are twenty times as strong as expected.[11] This molecule, of considerable astrophysical importance, badly needs an accurate determination of both its dissociation energy and the oscillator strengths of the strongest transitions.

Laboratory information is also required for the higher vibrational bands (4–4) and (5–5) bands of CH $A^2\Delta$–$X^2\Pi$, which are predicted to have observable lines. Another species which may well be present is MgH^+, but the strongest transition falls at a wavelength of <2900 Å and must be looked for in rocket spectra.

As might be expected, a number of species have been observed only in umbral spectra. The umbral spectrum has a very high line density. One cannot rely solely on wavelength coincidences to effect identification. Species thought to be present in umbrae are listed in Table 3. Some of the proposed identifications present difficulties as will be pointed out below.

Under conditions in the umbra the association of C and O into CO is practically complete in all but the deepest visible layers. Therefore carbon-containing molecules are not expected to increase greatly in importance *vis-a-vis* the photosphere. In fact CN and CH are almost unchanged in strength, whereas the Swan band (C_2) probably disappears completely. At wavelengths of up to 10 000 Å the spectrum is dominated by bands of hydrides, such as OH, NH, SiH, MgH, and CaH, and the oxide TiO. Observations of OH and NH bands are as yet inadequate. They fall in a part of the spectrum ($\lambda < 3500$ Å) where the intensity of a sunspot is only 1—2% of that of the photosphere. It is very difficult to obtain a spectrum sufficiently free of contamination by photospheric radiation.

[9] D. L. Lambert and E. A. Mallia, *Bull. Astron. Inst. Czech.*, to be published.
[10] N. Grevesse and A. J. Sauval, *Astron. Astrophys.*, 1971, **14**, 1971.
[11] D. L. Lambert, E. A. Mallia, and A. D. Petford, *Monthly Notices Roy. Astron. Soc.*, 1971, **154**, 265.

Table 3 *Species and transitions thought to be detected in solar umbral spectra*

Species	*Transitions*	λ/Å
TiO	$C^3\Delta$–$X^3\Delta$ $A^3\Phi$–$X^3\Delta$	5167; 7054
	$c^1\Phi$–$a^1\Delta$ $b^1\Pi$–$a^1\Delta$	5597; 8859
MgO	$B^1\Sigma$–$X^1\Sigma^+$ $B^1\Sigma$–$A^1\Pi$	5000; 6060
CaH	$B^2\Sigma$–$X^2\Sigma$ $A^2\Pi$–$X^2\Sigma$	6380; 6946
MgH	higher vibrational bands of $A^2\Pi$–$X^2\Sigma^+$	5500
OH	fundamental + 1st overtone vib–rot	14 000—18 000
H_2O	Ω and X bands	19 000—26 000
HF	fundamental vib–rot	23 000
HCl	fundamental vib–rot	36 000
CuH	$A^1\Sigma$–$X^1\Sigma$	4280
NiH	$A^2\Delta$–$X^2\Delta$ $B^2\Delta$–$X^2\Delta$	6300—6400
FeH	no analysis available	4800—5500
CN	$\Delta v'' = +1$ bands of $A^2\Pi_i$–$X^2\Sigma^+$	16 000

MgH and SiH present interesting problems. If the line strengths in the umbra are computed using laboratory (for SiH) or photospheric (for MgH) molecular parameters one finds that the species are almost an order of magnitude less abundant than expected with respect to the photosphere. As yet it has proved impossible, through a combination of observational and laboratory lacunae, to ascertain if other species suffer from this effect. The case of TiO should prove especially interesting because of its strongly localized region of formation.

Incidentally, the behaviour of CN and CH is in contrast to that of the two metallic hydrides, the change from photosphere to umbra being as predicted. The source of the discrepancies between observations and predictions for both photosphere and unbra is as yet unknown. There is some evidence that the assumption of local thermodynamic equilibrium may be at fault in both cases. Before this can be fully investigated, however, good quality observations of more molecular species are needed.

The log $(W_{N''}/S_{N''})$ *versus* $N''(N''+1)$ diagram for CaH shows unusual structure (Figure 2). For both *A* and *B* bands it is impossible to pass a straight line through the points representing the observed strengths. This fact has been noted by Webber [12] and is apparently present in similar diagrams constructed from laboratory spectra.[13] This anomaly seems not yet to have been explained.

The rather low ionization potential of Ca ensures a supply of ions even in umbrae. It is possible, therefore, that the molecular ion CaH^+ could be present. With plausible assumptions about the dissociation energy and oscillator strength it appears that lines of CaH^+ could well be present in the umbral spectrum. At present the spectrum of the ion has not been observed in the laboratory.

[12] J. C. Webber, *Solar Phys.*, 1971, **16**, 362.
[13] G. Liberale and S. Weniger, *Physica*, 1969, **41**, 47.

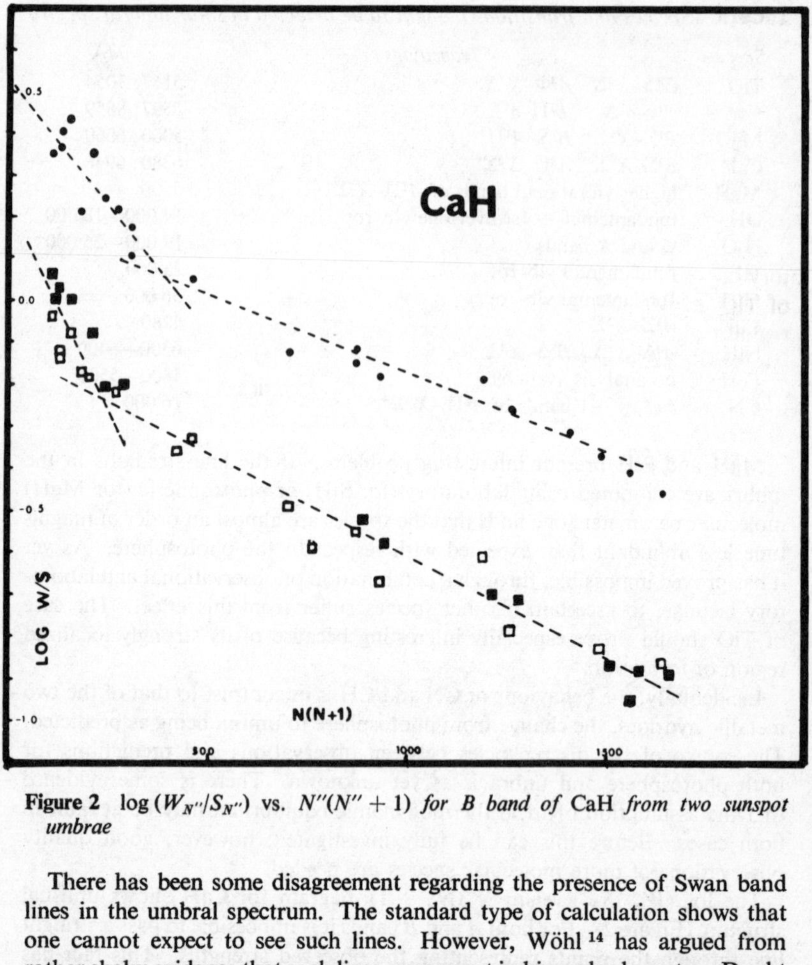

Figure 2 $\log (W_{N''}/S_{N''})$ vs. $N''(N''+1)$ *for B band of* CaH *from two sunspot umbrae*

There has been some disagreement regarding the presence of Swan band lines in the umbral spectrum. The standard type of calculation shows that one cannot expect to see such lines. However, Wöhl [14] has argued from rather shaky evidence that such lines are present in his umbral spectra. Doubt also attaches to the identification of MgO lines. In order to achieve agreement between prediction and observation, Sotirovsky [15] has to assume a dissociation energy of 4.7 eV, which is probably much too high. According to Drowart, Exsteen, and Verhaegen,[16a] $D_0^0 = 3.7 \pm 0.2$ eV, whereas Brewer and Rosenblatt [16b] recommend $D_0^0 = 3.4 \pm 0.3$ eV.

FeH is another doubtful case. Carroll and McCormack [17] have recently identified a complex laboratory spectrum as coming from this molecule.

[14] H. Wöhl, *Solar Phys.*, 1972, **24**, 342.
[15] P. Sotirovsky, *Astron. Astrophys.*, 1971, **14**, 319.
[16] (*a*) J. Drowart, G. Exsteen, and G. Verhaegen, *Trans. Faraday Soc.*, 1964, **60**, 1920; (*b*) L. Brewer and G. M. Rosenblatt, *Adv. High Temp. Chem.*, 1969, **2**, 1.
[17] P. K. Carroll and P. McCormack, *Astrophys. J.*, 1972, **177**, L33.

A comparison of laboratory wavelengths with those of unidentified weak lines in the solar spectrum [18] led them to suggest that lines of FeH were present in both photospheric and umbral spectra. If the lines so identified really belong to FeH, their strength in the umbral spectrum is 60 times less than expected on the basis of the photospheric identifications. Such a large discrepancy must cast some doubt on the reality of the presence of FeH, although it should be kept in mind that the prediction embodies assumptions even about the transition involved, since the spectrum is unanalysed.

Isotopic relative abundances for several species have been determined from umbral spectra. Rather rough values have been obtained [19] for $^{58}Ni/^{60}Ni$, $^{63}Cu/^{65}Cu$,[20] and $^{29,30}Si/^{28}Si$ and more accurate ones for $^{46,47,49,50}Ti/^{48}Ti$ [21] and $^{25,26}Mg/^{24}Mg$.[22] In no case have the ratios been found to be different from the terrestrial ones.

In the i.r. there have been exciting new discoveries, principally through the work of Hall on umbral spectra. As expected, CO is a very important contributor to the spectrum. The great strength of the $^{12}C^{16}O$ lines of the first overtone band has led to the definite detection of $^{13}C^{12}O$ lines.[23] The ratio $^{12}C/^{13}C$ is the same as the terrestrial one. Other important discoveries include the 1.9 and 2.6 μm bands of H_2O and vibration–rotation bands of OH [24] and possibly SiO. The H_2O observations should provide information on the excited states of the molecular since numerous high-excitation lines not observed in the absorption spectrum of the earth's atmosphere have been found.[25]

Two other results, important for abundance studies, have been the detection of the vibration–rotation lines of HF and HCl,[26, 27] F and Cl not having been previously detected on the sun. All of these species should prove of great use in probing the detailed temperature structure of the sunspot atmosphere. Where computed and observed line strengths have been compared for i.r. bands no serious discrepancies have come to light.

Other Stars.—Most of the observations of cool stars have been carried out on bright giants and supergiants, principally because of the comparative ease with which such stars may be observed. Certain types of cool star, however, do not appear to exist in dwarf form. Unless otherwise indicated, the following remarks apply to giants and supergiants.

18 C. E. Moore, M. G. J. Minnaert, and J. Houtgast, 'The Solar Spectrum 2935—8870 Å', N.B.S. Mon. 61, U.S. Dept. of Commerce, Washington, 1966.
19 D. L. Lambert and E. A. Mallia, *Monthly Notices Roy. Astron. Soc.*, 1971, **151,** 437.
20 O. Hauge, *Astron. Astrophys.*, 1970, **10,** 73.
21 D. L. Lambert and E. A. Mallia, *Monthly Notices Roy. Astron. Soc.*, 1972, **156,** 337.
22 R. Boyer, J. C. Henoux, and P. Sotirovsky, *Solar Phys.*, 1971, **19,** 330.
23 D. N. B. Hall, R. W. Noyes, and T. R. Ayres, *Astrophys. J.*, 1972, **171,** 615.
24 D. N. B. Hall, *Kitt Peak Nat. Obs. Contr.*, 1970, no. 556.
25 W. S. Benedict, Paper read at the 3rd Colloquium on High Resolution Molecular Spectroscopy, Tours, 1973.
26 D. N. B. Hall and R. W. Noyes, *Astrophys. Letters*, 1969, **4,** 143.
27 D. N. B. Hall and R. W. Noyes, *Astrophys. J.*, 1972, **175,** L95.

In stars with N(O) > N(C) the evolution of the molecular spectrum with decreasing temperature follows a reasonably well understood path.[28] The increasing importance of CO, H_2, and N_2 (and CO_2 at the lowest temperatures) curtails the increase in numbers of other carbon-, nitrogen-, and hydrogen-containing molecules (see Figure 3). Even so, down to about 3500 K the visible and near-i.r. regions of the spectrum are still dominated by the molecules that one finds in the solar photosphere. This is especially the case in the 1 μm region where the bands of the red CN system reach considerable strength. C_2 is the only species whose lines fade out relatively quickly with decreasing temperature. In the 2 μm region CO bands attain great strength, and at shorter wavelengths OH is an important contributor.

Below 3500 K TiO becomes prominent. Its systems eventually dominate the spectrum from *ca.* 4500 to 11 000 Å. In the coolest stars the 1.06 μm bands of VO appear,[29] as does H_2O absorption [30] at 9280 Å. H_2O, OH, and CO are the principal absorbers out to 3 μm, with CO, CN, and H_2O remaining after that.

The analysis of the TiO spectrum is fairly complete and likely to improve when the Berkeley monograph on the subject becomes available. At present the $\Delta v'' = +1$ bands of the $^1\Pi$–$^1\Sigma$ transition have not been analysed and the three band heads at λλ9208, 9230, and 9248 are unclassified. (In private discussion Dr. B. Lindgren has suggested that they are members of the 2–0 sequence of the $^1\Pi$–$^1\Sigma$ transition.) More serious is the lack of an analysis of the VO i.r. bands. These lines fall in an otherwise clear part of the spectrum and should prove amenable to accurate measurements in stellar spectra.

An important result that has emerged from work on these stars is the small $^{12}C/^{13}C$ ratio in their atmospheres (~6—8 against 89 for the sun).[31, 32] This makes it evident that the products of nuclear reactions in the stellar interior are reaching the surface. Quadrupole lines of the (1–0) band of H_2 have almost certainly been detected in the spectrum of α Herculis,[33] but doubt has been cast on the previous identifications of (2–0) band lines.[34] Recently Pesch [35] has suggested that the band head at 5570 Å, prominent in the spectra of some M dwarfs, is produced by CaOH. This suggestion has not been subjected to detailed examination.

In stars for which N(C) > N(O) (carbon stars), carbon compounds are expectedly prominent in the spectrum. In the case of diatomic molecules laboratory work is extensive enough to enable identifications out to at least 10 000 Å to be fairly complete. Small gaps, such as the higher bands of the $A^2\Delta$–$X^2\Pi$ transition in CH and the $\Delta v = +1$ sequences in the red CN

[28] T. Tsuji, *Ann. Tokio Obs.*, 1964, **9,** 1.
[29] A. McKellar, *J. Roy. Astron. Soc. Canada*, 1955, **49,** 73.
[30] H. Spinrad and R. L. Newburn, *Astrophys. J.*, 1965, **141,** 965.
[31] D. L. Lambert, A. L. Brooke, and T. G. Barnes, *Astrophys. J.*, in the press.
[32] H. Spinrad and R. F. Wing, *Ann. Rev. Astron. Astrophys.*, 1969, **7,** 249.
[33] R. W. Day, D. L. Lambert, and J. Sneden, *Astrophys. J.*, 1973, **185,** 213.
[34] H. Spinrad and M. S. Vardya, *Astrophys. J.*, 1966, **146,** 399.
[35] P. Pesch, *Astrophys. J.*, 1972, **174,** L55.

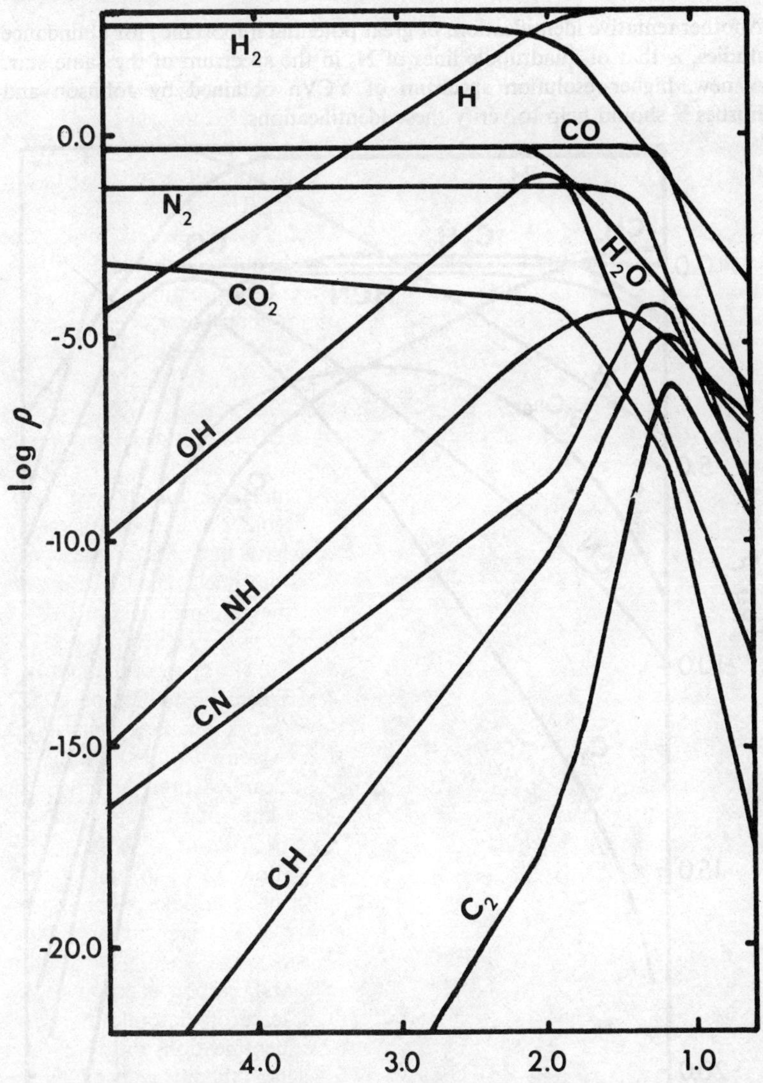

Figure 3 *Molecular equilibria for giant stars with oxygen more abundant than carbon* [N(O) > N(C)]. *p is partial pressure and* $\theta = 5040/T$

system, can be easily filled. In the i.r., laboratory observation of the vibration–rotation bands of CH is lacking although these bands have been tentatively identified in the spectrum of YCVn,[36] using computed wavenumbers.

[36] P. Connes, J. Connes, R. Bouigue, M. Querci, J. Chanville, and F. Querci, *Ann. Ap.*, 1968, **31**, 485.

Another tentative identification, of great potential importance for abundance studies, is that of quadrupole lines of N_2 in the spectrum of the same star. A new, higher-resolution spectrum of YCVn obtained by Johnson and Forbes [37] should help to verify these identifications.

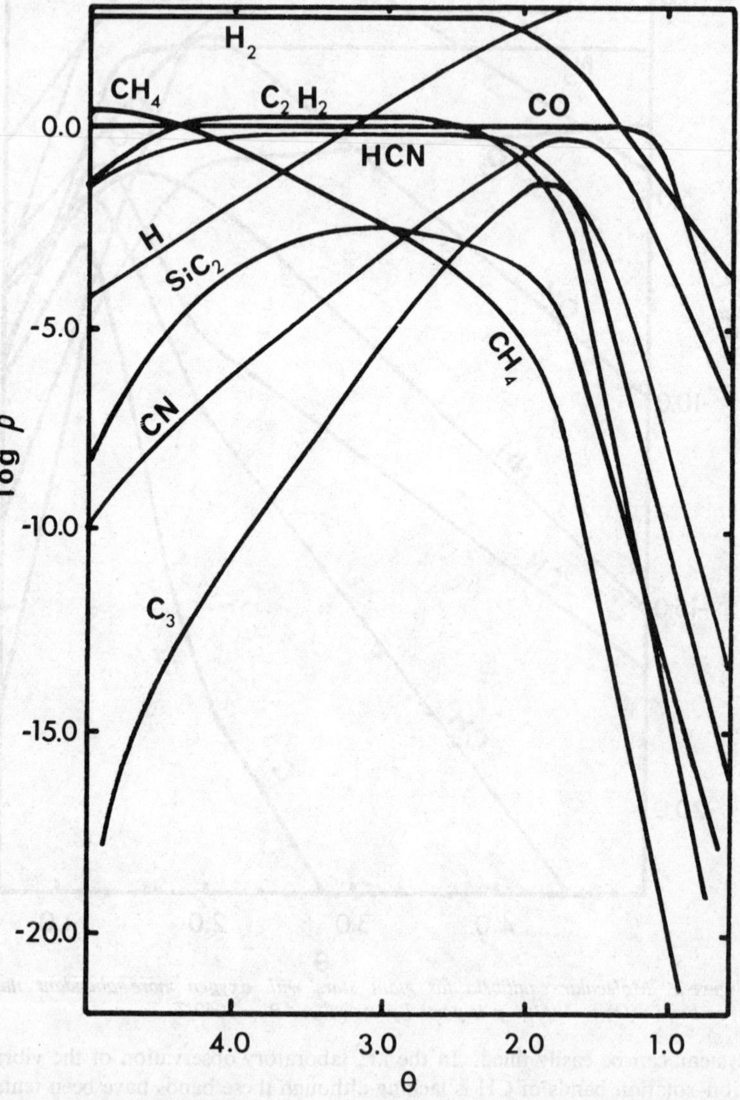

Figure 4 *Molecular equilibria for giant stars with carbon more abundant than oxygen* [N(C) > N(O)]

[37] H. L. Johnson and F. F. Forbes, *Bull. Amer. Astron. Soc.*, 1973, **5**, 336.

Perhaps the single biggest lack of information concerns polyatomic carbon molecules. HCN and C_2H_2 are expected to be abundant in the atmospheres of carbon stars (see Figure 4) but such measurements of wavelength as exist are restricted to low-*J* lines which are usually weak in stellar spectra. Both species have been tentatively identified in the 2 μm region of the YCVn spectrum,[36] but they can more easily be searched for in carbon stars at 1 μm where they also have lines.[38] The only carbon polyatomic molecule seemingly securely identified is SiC_2,[39] which produces the well-known Merrill–Sandford bands.

A prominent feature in the spectra of some carbon stars is the strong continuous absorption that sets in at wavelengths less than 4500 Å. This has been ascribed in the past [28] to C_3, but information from certain hybrid types of star casts doubt on this.[40] Laboratory work on molecular continua is rather sparse.

In S stars the heavy metals are more abundant than the transition metals and (although this is still rather uncertain) $N(C) \approx N(O)$. ZrO bands replace the TiO bands of the 'normal' stars and the prominent H_2O absorptions almost disappear [one piece of evidence that $N(C) \approx N(O)$]. A notable molecular feature in the spectra of S stars is the group of emission lines of AlH which results from inverse predissociation.[41]

The α and γ systems of ZrO have been analysed [42] by Lagerqvist *et al.* (1954) but there are numerous bands beyond 9300 Å which are as yet unanalysed. From observations of the γ band it has been found that the relative abundances of Zr isotopes in S stars are different from the terrestrial ones.[43] LaO bands are prominent and there has been a tentative identification [44] of NbO, but for this species the lack of an analysis has hampered the search. In spite of protracted efforts, several band heads in the spectrum of S stars have resisted identification (Figure 5). These are the Keenan and Wing bands in the near-i.r. From their behaviour in different stars it appears that more than one molecule is responsible.[45]

Recently it has been claimed that vibration–rotation bands of SiO are present in the spectrum of the S star χ Cygni.[46] Wing and Price [47] have argued against such an interpretation, pointing out that it is much more likely that what are observed are bands of CN. In view of the suggestion that $N(C) \approx N(O)$ in S stars, Wing's interpretation would seem to be more likely.

38 T. D. Faÿ, L. W. Frederic, and H. T. Johnson, *Astrophys. J.*, 1968, **152**, 151.
39 B. Kleman, *Astrophys. J.*, 1956, **123**, 162.
40 C. B. Stephenson, *Astrophys. J.*, 1967, **150**, 543.
41 G. H. Herbig, *Pub. Astron. Soc. Pacific*, 1956, **68**, 204.
42 A. Lagerqvist, U. Uhler, and R. F. Barrow, *Arkiv Fysik*, 1954, **28**, 281.
43 A. Schadee and D. N. Davis, *Astrophys. J.*, 1968, **152**, 169.
44 D. N. Davis and P. C. Keenan, *Pub. Astron. Soc. Pacific*, 1969, **81**, 231.
45 R. F. Wing, *Mem. Soc. roy. Sci. Liège*, 1972, 6th Series, **III**, 123.
46 J. Fertel, *Astrophys. J.*, 1970, **159**, L7.
47 R. F. Wing and S. D. Price, *Astrophys. J.*, 1970, **162**, L73.

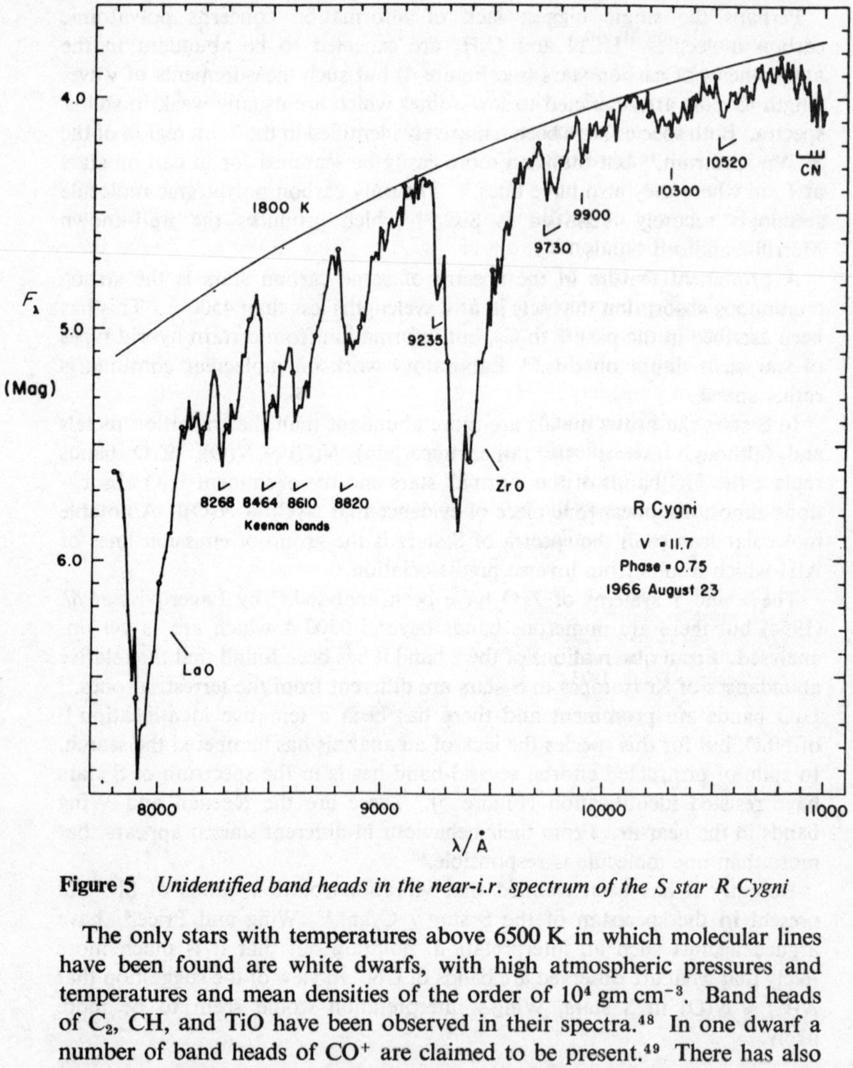

Figure 5 *Unidentified band heads in the near-i.r. spectrum of the S star R Cygni*

The only stars with temperatures above 6500 K in which molecular lines have been found are white dwarfs, with high atmospheric pressures and temperatures and mean densities of the order of 10^4 gm cm^{-3}. Band heads of C_2, CH, and TiO have been observed in their spectra.[48] In one dwarf a number of band heads of CO^+ are claimed to be present.[49] There has also been the suggestion that the preponderance of He in some types of white dwarf and the high atmospheric pressures must lead to the formation of He_2 molecules which would have an appreciable lifetime in a stable excited state. The strong circular polarization observed in the light of He-rich dwarfs could be due to a Zeeman effect in these molecules.[50]

[48] J. L. Greenstein, 'Stars and Stellar Systems', University of Chicago Press, Chicago, 1960, vol. 6, p. 676.
[49] I. Bues and G. W. Wegner, personal communication, 1973.
[50] J. R. P. Angel, *Astrophys. J.*, 1972, **171**, L17.

5 Conclusion

There has been a long and fruitful collaboration between astrophysicists and molecular spectroscopists, which in recent years has been extended into the microwave region of the spectrum. It is true to say, however, that a number of astrophysically interesting molecules have been rather neglected. Theoretical developments have not yet reached the stage where quantities like the dissociation energy or oscillator strengths can be computed with sufficient confidence even for diatomic molecules (H_2 is an exception).[51, 52] As has been pointed out in the case of solar spectra, improvements in the accuracy of certain molecular parameters would enable astrophysicists to draw far-reaching conclusions about the validity of methods currently in use in computations of line strengths. Laboratory observations of spectra of molecular ions are also of considerable interest to astrophysicists. A start has already been made in this direction.[53, 54] For stellar applications there is a considerable amount of classification and analysis that needs to be undertaken. Here, perhaps, the onus is on the astrophysicist to improve his knowledge of stellar atmospheres to the point where he can derive the full benefit from data that are already available.

[51] A. C. H. Chan and E. R. Davidson, *J. Chem. Phys.*, 1970, **54**, 1763.
[52] W. H. Heneker and H. E. Popkie, *J. Chem. Phys.*, 1971, **52**, 4108.
[53] W. J. Balfour, *Canad. J. Phys.*, 1972, **50**, 1082.
[54] G. Herzberg, Paper read at the 3rd Colloquium on High Resolution Molecular Spectroscopy, Tours, 1973.

Author Index